Heat Exchanger Design

Heat Exchanger Design

Second Edition

ARTHUR P. FRAAS

WILEY

A WILEY-INTERSCIENCE PUBLICATION
JOHN WILEY & SONS
New York · Chichester · Brisbane · Toronto · Singapore

Library of Congress Cataloging-in-Publication Data
Fraas, Arthur P.
   Heat exchanger design.

   "A Wiley-Interscience publication."
   Includes bibliographies.
   1. Heat exchangers—Design and construction.
I. Title.
TJ263.F7 1988   621.402′ 5   88-20724
ISBN 0-471-62868-9

Printed in the United States of America

10 9 8 7 6 5 4 3 2

Contents

Handbook

The continuing evolution of new applications and requirements for heat exchangers, not only in the chemical, electric utility, and aerospace industries but also in fields such as food processing and cryogenics, has introduced many problems not treated in the first edition of this text. This, coupled with developments in fluid flow, heat transfer, stress analysis, and computers, together with an increased emphasis on reliability, has made the preparation of a second edition much in order. Although in many cases it is possible to employ conventional commercial units, it is often better, and sometimes essential, to evolve a completely new design. Just to decide on an appropriate course requires a good engineering perspective on and an appreciation for the many subtle problems involved, not to mention sufficient facility in the analytical and experimental techniques required for their solution. This book was designed to help practicing engineers apply and extend their formal theoretical backgrounds to the solution of the practical problems posed by the design, selection, testing, or installation of all sorts of heat exchangers.

The first half of the book is devoted to general problems and basic relations applicable to the design of all types of heat transfer equipment. The second half is concerned with the specialized problems of many different types of heat exchangers. Design and analysis techniques are illustrated by examples that include detailed consideration and calculations for 30 represen-

tative cases. These have been chosen to be sufficiently varied so that a designer should be able to find among them a case in which the techniques used will be applicable to the case at hand. Over 100 charts and tables that have proved useful in design work have been included. These charts are grouped in an appended handbook section except where they are specialized and heavily dependent on the text in a particular chapter, in which case they are placed with that text.

While the third chapter summarizes the basic relations involved in fluid-flow and heat transfer analyses so that the designer will have them immediately available, this is not a text on heat transfer. Rather, the prime objective is to treat heat exchanger design in a comprehensive fashion, looking at the whole complex of technologies involved. In most cases, factors such as vibration, thermal stresses, materials compatibility, corrosion, fabrication techniques, cleaning requirements, and/or reliability will be completely dominant rather than heat transfer or cost. The criteria are too varied and numerous to arrive at an "optimum" design; rather, the objective must be a "well-integrated," well-proportioned design. Thus neither elegance nor rigor in the heat transfer analysis has been a prime objective; the presentation has been made as simply as possible consistent with the development of a sound, effective approach to design work. The reader is presumed to be familiar with a basic textbook such as Ozisik's *Heat Transfer*, and is referred to such books and papers in the literature for more refined treatments of heat transfer phenomena. References have been selected from the tremendous wealth of material that has been published; these provide far more leads to sources of information than could be mentioned in this text.

In preparing this revised edition the author initially planned to include many references to examples of computer programs designed for handling heat exchanger problems. However, so many programs are available, and their number and sophistication are growing so rapidly, that this area has not been emphasized. It soon became evident to the author that the problem for the reader is not primarily how to carry out calculations with a computer program or what programs are available, but rather what one should want to find out from using a computer program. It is vital that the designer be under no delusions with regard to computer-generated designs—the expression "garbage in, garbage out" could hardly be more applicable. After deciding what the key criteria are, one can choose a program, or develop one. The computer then provides a marvelous tool for turning out a whole set of designs from which one can choose that which seems best for the particular case at hand.

It was also difficult to decide how to cope with the gradual transition from English to SI units that has been under way for almost 20 years. In view of the facts that English units are still widely used and English pipe sizes are still standard all over the world, both English and SI units are likely to be in use for some time to come. Thus both sets of units are given throughout this text except in some tables and illustrations that would have become so cluttered as to detract badly from their use. Unfortunately, the use of two sets of units is very confusing, not only to the novice but also to the expert. One of the elements of engineering judgment that comes with experience is a "feel" for the problems, including a good idea of approximately what the right answer to a problem should be. To acquire this "feel" with a repertoire of numbers for heat transfer coefficients, physical properties, mass flow rates, pressure drops, etc., carried in one's head is a major achievement if accomplished with one set of units; few, if any, experts have accomplished this for both sets of units. A major effort has been made to handle this awkward problem in such a way that the reader will be aided in coping with the difficulties inherent in shifting from one set of units to the other.

The author is deeply indebted to the many friends who reviewed portions of the manuscript and contributed invaluable corrections and additional material. Those who made major contributions by reviewing portions of this second edition include M. M. Alsherif, A. E. Bergles, S. M. Cho, E. L. Daman, K. Hannerz, R. S. Holcomb, A. L. London, R. E. MacPherson, J. C. Moyers, M. L. Myers, M. N. Ozisik, G. Samuels, E. E. Stansbury, and R. M. Wilson. In addition, the author is also indebted to those who helped with the first edition, notably M. N. Ozisik, who assisted as coauthor, and G. A. Cristy, J. K. Jones, A. Koestel, M. E. LaVerne, R. L. Maxwell, P. Pasqua, W. M. S. Richards, G. Samuels, T. Sprague, T. Tinker, J. R. Weske, and M. N. Yarosh, who reviewed chapters in their fields and offered many valuable criticisms and suggestions from their extensive professional experience. A special debt is owed M. L. Myers, who supplied valuable material for over half the chapters of this edition.

ARTHUR P. FRAAS

Knoxville, Tennessee
December 1988

Heat Exchanger Types and Construction

Many types of heat exchangers are employed in such varied installations as steam power plants, chemical-processing plants, building heating, air conditioning, and refrigeration systems, and mobile power plants for automotive, marine, and aerospace vehicles. The principal types of equipment employed in these applications are reviewed in this chapter to illustrate the problems with which this book is concerned and to clarify the nomenclature.

FLUID-FLOW ARRANGEMENT

Most heat exchangers may be classified as being in one of several categories on the basis of the configuration of the fluid-flow paths through the heat exchanger. The four most common types of flow path configuration are illustrated diagrammatically in Fig. 1.1. In *cocurrent,* or *parallel-flow,* units the two fluid streams enter together at one end, flow through in the same direction, and leave together at the other end; in *countercurrent,* or *counterflow,* units the two fluid streams move in opposite directions. In *single-pass crossflow* units one fluid moves through the heat transfer matrix at right angles to the flow path of the other fluid. In *multipass crossflow* units one fluid stream shuttles back and forth across the flow path of the other fluid stream, usually giving a crossflow approximation to counterflow.

The most important difference between these four basic types lies in the relative amounts of heat transfer surface area required to produce a given temperature rise

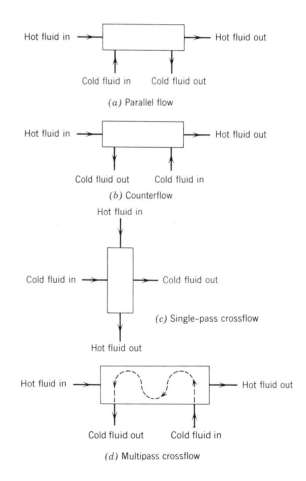

Figure 1.1 Types of flow path configuration through heat exchangers.

for a given temperature difference between the two fluid streams where they enter the heat exchanger. Figure 1.2 shows the relative area required for each type as a function of the change in temperature of the primary fluid for a typical set of conditions. In the region in which the fluid temperature change across the heat exchanger is a small percentage of the difference in temperature between the two entering fluid streams, all the units require roughly the same area. The parallel-flow heat exchanger is of interest primarily for applications in this region. Crossflow units have a somewhat broader range of application, and are peculiarly suited to some types of heat exchanger construction that have special advantages. The counterflow heat exchanger requires the least area throughout the range. Furthermore, it is the only type that can be employed in the region in which the temperature change in one or both of the fluid streams closely approaches the temperature difference between the entering fluid streams. It is interesting to note that nature presents us with one of the best examples of a highly efficient counterflow system in the blood-vessel system in the legs of wading birds such as herons.[1] The warm blood moving outward into the leg from the heart is passed through a system of tiny parallel blood vessels that are interspersed in checkerboard fashion with similar vessels returning from the extremity of the bird's leg, giving one of the world's most effective regenerative heat exchangers. The heat transfer performance of this blood-vessel configuration is so good that the warm blood is cooled almost to the ambient temperature before reaching the region immersed in cold water, and thus the bird loses relatively little heat through the skin of its leg. The extremities of certain other warm-blooded animals, such as penguins and whales, are similarly equipped.

TYPES OF APPLICATION

Heat exchangers are often classified on the basis of the application for which they are intended, and special terms are employed for major types. These terms include *boiler, steam generator, condenser, radiator, evaporator, cooling tower, regenerator, recuperator, heater, and cooler.* The specialized requirements of the various

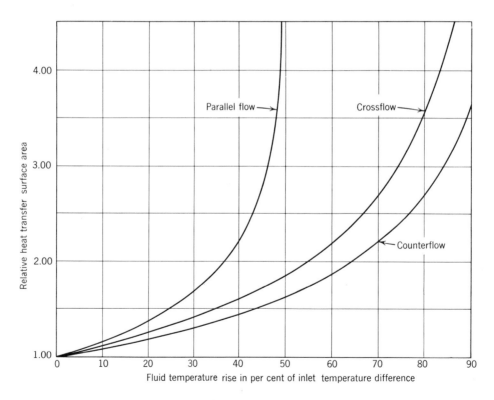

Figure 1.2 The relative heat transfer surface area required as a function of the ratio of the temperature rise (or drop) in the fluid stream having the greater change in temperature to the difference in temperature between the inlet streams.

applications have led to the development of many types of construction, some of which are unique to particular applications. Typical units are described in subsequent sections to illustrate the characteristics and features of the principal types.

Boilers

Steam boilers have been used to produce power for over 200 years, and constitute one of the earliest subjects for the application of engineering principles to heat exchanger design. There is an enormous variety of boilers ranging from many small, relatively simple units for space heating applications to the huge, complex, and expensive boilers in modern central stations; in these, the boiler is thoroughly intergrated with the furnace so that heat losses from the furnace walls are minimized. Layers of tubes are installed in the walls surrounding the combustion zone, and tremendous amounts of surface area in the form of great banks of tubes are exposed to the hot gases. Figure 1.3 shows a typical unit that is about 200 ft high and costs about $50,000,000.

The term *steam generator* is often applied to boilers in which the heat source is a fluid stream other than the hot products of combustion. A steam generator for a pressurized water nuclear reactor power plant is shown in Fig. 1.4. High-temperature, high-pressure water from the reactor is circulated through 0.5-in. diameter U-tubes coaxial with the casing. Steam is generated outside the tubes as the boiling water circulates by thermal convection upward through the tube matrix, where it forms steam at a temperature somewhat lower and a pressure very much lower than in the high-pressure water circuit through the reactor. A completely different type of steam generator unit for a large, low-pressure, gas-cooled reactor is shown in Fig. 1.5. Eight of these multipass, counterflow units are employed with each of the four reactors at the Hinkley Point power plant in England.

Figure 1.3 Partial section through a large modern, coal-fired steam boiler. (Courtesy Combustion Engineering, Inc.)

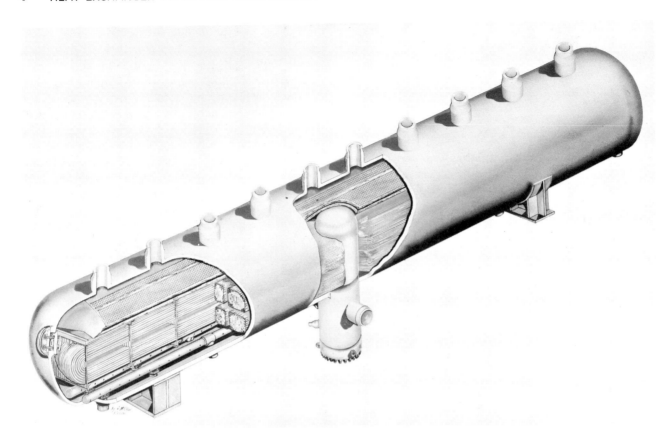

Figure 1.4 Partial section through a 10-ft-diameter, 57-ft-long steam generator for a pressurized water reactor power plant. It has 1,916 stainless steel tubes of 0.625-in. OD to give a total surface area of 16,000 ft². (Courtesy Combustion Engineering, Inc.)

Note that the heat is extracted from the gas in seven stages. This unusual design resulted from limitations on the reactor operating temperature which made it advantageous to employ two steam systems—one at a moderately high pressure and one at a lower pressure—in order to extract as much heat as possible from the gas circulated through the reactor.[2]

Condensers

James Watt more than tripled the thermal efficiency of the steam engines of his day by applying *condensers* so that the steam would be condensed outside rather than inside the engine cylinders. The steam condensers for large modern turbines are built in much the same way as Watt's early units except that they are enormously larger. Figure 1.6 shows a typical modern unit that employs nearly 1,000,000 ft of tubing in 21,850 tubes.

Shell-and-Tube Heat Exchangers

A host of units known as *shell-and-tube* heat exchangers are built of round tubes mounted in cylindrical shells with their axes parallel to that of the shell. These are employed as *heaters* or *coolers* for a variety of applications that include oil coolers in power plants and process heat exchangers in the petroleum-refining and chemical industries. Many variations of this basic type are available; the differences lie mainly in the detailed features of construction and provisions for differential thermal expansion between the tubes and the shell.[3,4]

To point up some of the problems that arise in heat exchanger design, several typical units are presented in Figs. 1.7 through 1.13. The unit shown in Fig. 1.7 is a single-pass tube, baffled single-pass shell, shell-and-tube unit tube, with the flow pattern of Fig. 1.1*d* so that it closely approaches pure counterflow conditions. The unit shown in Fig. 1.8 employs U-tubes to give a more simple

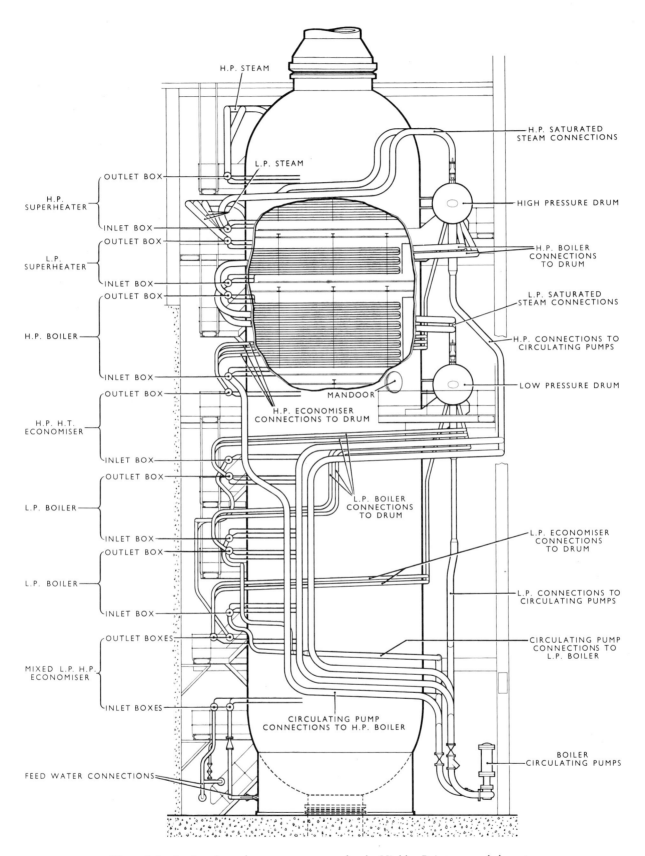

Figure 1.5 Section through a steam generator for the Hinkley Point gas-cooled reactor power plant. The shell diameter is 21.5 ft, and the height 90 ft. CO_2 at 1.24 MPa (180 psi) flows through the shell side at 780 kg/s (1720 lb/s), with a pressure loss of only 11 Pa (1.6 psi), to transfer 164 MW of heat. (Courtesy Babcock and Wilcox, Ltd.)

Figure 1.6   Shop erection of a single-pass condenser for a 300-MW steam turbine. It contains 21,850 admiralty metal tubes 19-mm (3/4-in.) OD and 12.1 m (39.75 ft) long, giving a heat transfer surface area of 15,800 m^2 (170,000 ft^2) in the unit. (Courtesy Allis-Chalmers Manufacturing Co.)

construction, but has a flow pattern less like that of a pure counterflow heat exchanger. The unit of Fig. 1.9 has a flow pattern similar to that of Fig. 1.8, but the construction is more complex to make it convenient to clean the inside of the tubes mechanically, to inspect them, and replace defective tubes. While this configuration does not provide for differential thermal expansion between the tubes and shell, the similar but more complex construction of Fig. 1.10 is adapted to accommodate large differences in temperature between fluids. It is almost as well suited to tolerating large temperature differences between fluid streams as the unit of Fig. 1.8,

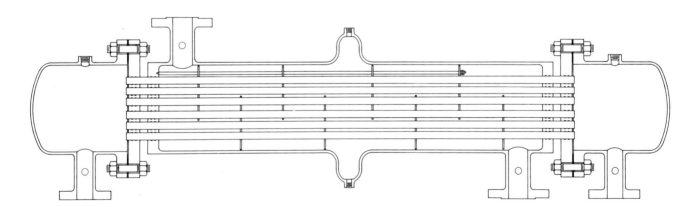

Figure 1.7   A single-pass tube, baffled single-pass shell, shell-and-tube heat exchanger designed to give essentially counterflow conditions. A toroidal expansion joint in the center of the shell accommodates differential thermal expansion between the tubes and the shell, and double header sheets assure that no fluid can leak from one circuit to the other. (Courtesy The Patterson-Kelley Co.)

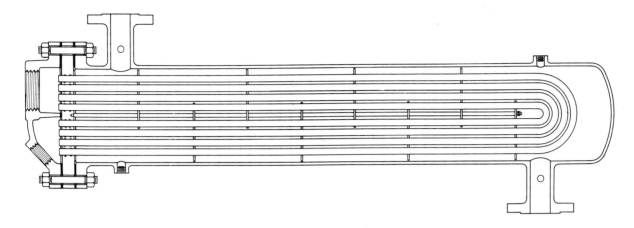

Figure 1.8 A U-tube, baffled single-pass shell, shell-and-tube heat exchanger. (Courtesy The Patterson-Kelley Co.)

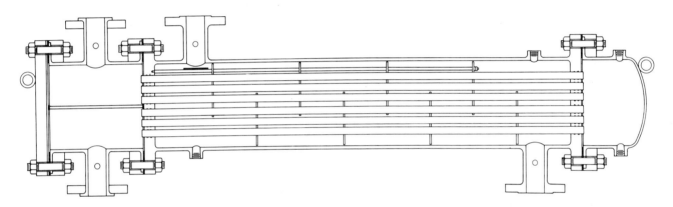

Figure 1.9 A two-pass tube, baffled single-pass shell, shell-and-tube heat exchanger designed for mechanical cleaning of the inside of the tubes. (Courtesy The Patterson-Kelley Co.)

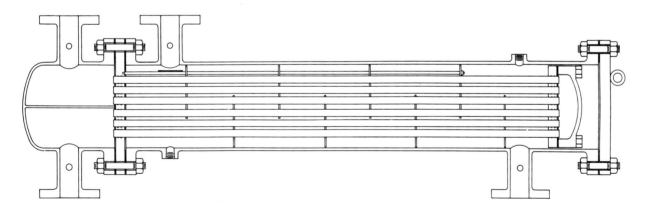

Figure 1.10 A heat exchanger similar to that of Fig. 1.9 except that this one is designed with a floating head to accommodate differential thermal expansion between the tubes and the shell. (Courtesy The Patterson-Kelley Co.)

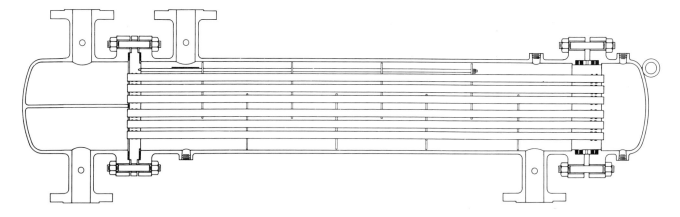

Figure 1.11 A heat exchanger similar to that of Fig. 1.10 except with a different type of floating head. (Courtesy The Patterson-Kelley Co.)

in which the U-bends accommodate differential thermal expansion not only between the tubes and the shell but also between individual tubes if the temperature distribution is not uniform across the tube matrix.

Difficulties sometimes arise with internal leakage through the gasket between the floating head and the header sheet at the right end of units of the type shown in Fig. 1.10. Leakage from one fluid stream into the other can be avoided by the construction shown in Fig. 1.11, in which leakage of either fluid through the packed joints of the floating head goes to the exterior of the shell where it can be detected readily and does not contaminate the other fluid stream. A further variation of this type is shown in Fig. 1.12, which is similar to that of Fig. 1.11 except that a more elaborate packing gland is employed

for the floating head. This type of unit is designed to reduce leakage from the shell-side fluid, and provides a gasketed joint from which any leakage will be to the outside, where it will not contaminate the other fluid stream and can be easily detected.

The shell-and-tube type of construction is well suited to special applications in which the heat exchanger must be made of glass to resist the attack of highly corrosive liquids, to avoid affecting the flavor of food products, or the like. A heat exchanger of this sort, for a process in which a highly corrosive liquid is heated and cooled regeneratively, is shown in Fig. 1.13. In this case the header sheets and baffle plates are made of a high-alumina ceramic. Units are more often made for use with a corrosive liquid on the tube side only and ordinary

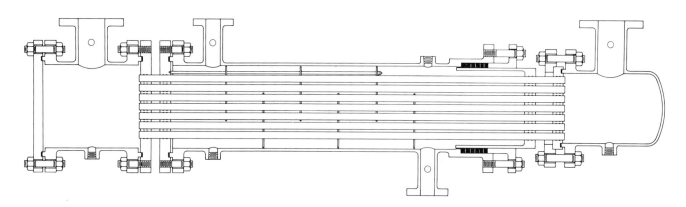

Figure 1.12 Single-pass tube, baffled single-pass shell heat exchanger with a packed joint floating head and double header sheets to assure that no fluid from one circuit leaks into the other (Courtesy The Patterson-Kelley Co.)

Figure 1.13 A five-bank, shell-and-tube heat exchanger made of Pyrex glass for chemical processes. (Courtesy Corning Glass Works.)

steam or water for heating or cooling on the shell side. When this is true, the shell and baffle plates are normally made of steel with provision for differential thermal expansion between the tubes and the shell.

Coolers

The atmosphere is a convenient heat sink for applications where the heat is to be rejected at a temperature 100°F or more above the ambient air temperature or where ample supplies of cooling water are not available. Such applications include petroleum refineries in arid regions, power plants in the arctic (where freezing is a problem), and mobile power plants. For heavy industrial applications the liquid to be cooled is usually circulated through banks of finned tubes similar to the one illustrated in Fig. 1.14. These units may be mounted in air ducts inside the plant or—where large amounts of heat must be rejected—they may be mounted out in the open. In the second case a cooling fan arranged to discharge the air vertically upward, as in Fig. 1.15, gives a low-cost installation with a minimal pumping power requirement. Furthermore, such an installation is insensitive to wind velocity or direction.

Units of the type shown in Fig. 1.14 are commonly used as coolers or heaters in buildings and ships' air-conditioning systems, and for industrial processes where air or gases must be cooled or heated in temperature ranges up to about 260°C (500°F).

Radiators

The term *radiator* is commonly applied to a variety of heat exchangers employed to dissipate heat to the surroundings. Automotive radiators of the type shown in Fig. 1.16 are crossflow units in which the temperature change in either fluid stream is small as compared to the temperature difference. Units of essentially the same construction are employed as condensers in refrigerators or air-conditioning units and, with fans, as heaters for large, open rooms. Aircraft oil coolers have much the same function as automotive radiators, but the premium placed on lightweight construction and compactness has led to the development of the different types of construction shown in Fig. 1.17*a*, *b*, and *c*.

An unusual type of radiator, and one that is truly a thermal radiator, is that shown in Fig. 1.18. This is a panel for the condenser of a potassium vapor power plant

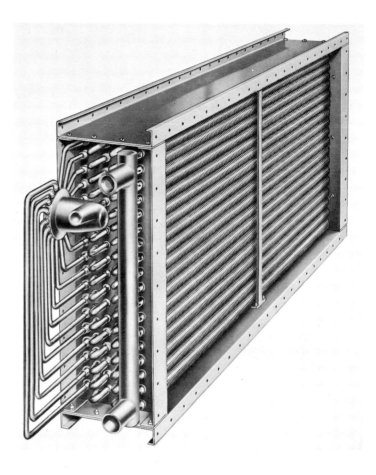

Figure 1.14 Finned tube bank for a large air-conditioning system. (Courtesy Aerofin Corp.)

that was designed for use in space vehicles. For this application heat can be dissipated only by thermal radiation to space, which is at an effective temperature of 4 K, or to the earth, which has a mean temperature of 288 K (520°R). In either case the condenser has been designed to operate red hot in order to dissipate the waste heat of the thermodynamic cycle from a reasonable size and weight of surface.

Cooling Towers

In locations where the supply of water is limited, heat may be rejected to the atmosphere very effectively by means of cooling towers such as that of Fig. 1.19. A fraction of the water sprayed into these towers evaporates, thus cooling the balance. Because of the high heat of vaporization of water, the water consumption is only about 1% as much as would be the case if water were taken from a lake or a stream and heated 10 or 20°F.

Cooling towers may be designed so that the air moves through them by thermal convection, or fans may be employed to provide forced air circulation. To avoid contamination of the process water, shell-and-tube heat exchangers are sometimes employed to transmit heat from the process water to the water recirculated through the cooling tower.

Regenerators and Recuperators

The thermal efficiency of both steam- and gas-turbine power plants can be greatly increased if heat can be extracted from the hot gases that are leaving the steam boiler or the gas turbine and added to the air being supplied to the furnace or the combustion chamber. For a major gain in thermal efficiency it is necessary to employ a very large amount of heat transfer surface area. This is particularly noticeable in gas-turbine plants, where even with counterflow the size of the heat exchanger required for good performance is inclined to be large compared to the sizes of the turbine and

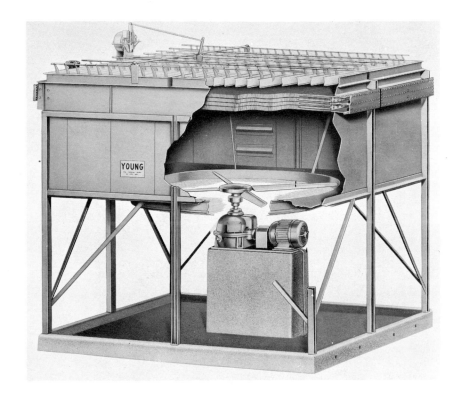

Figure 1.15 Large air-cooled finned tube bank designed for outdoor installation. (Courtesy Young Radiator Co.)

compressor. This characteristic can be observed even in the small, portable gas-turbine plant (about 3 ft in diameter) shown in Fig. 1.20. Note that in this plant the hot combustion gases leave the radial in-flow turbine wheel at the right end of the shaft and enter a set of heat exchanger cores arranged in parallel around the central axis at the right end of the plant. Figure 1.21 shows a close-up view of one of these cores. In each core the hot gases from the turbine flow roughly radially outward through one set of gas passages. Air from the centrifugal compressor wheel at the center of the shaft flows to the right through the space just inside of the outer casing and axially into the other set of gas passages through the core. The air being heated makes two passes, flowing first to the right in the outer portion of the core and then back to the left through the inner portion, thus giving a two-pass crossflow approximation to counterflow. (The flow passages through the combustion chamber are not shown in this view.)

As can be seen in Fig. 1.21, the heat exchanger core is constructed of alternate layers of flat and corrugated sheets. The flat sheets separate the hot and cold fluid streams, while the corrugated sheets act as fins that roughly triple the heat transfer surface area per unit of volume. Note also that the axis of the corrugations is at right angles in alternate layers to provide a crossflow pattern for the two fluid streams.

One of several recuperator units to be mounted in parallel in a much larger gas turbine plant is shown in Fig. 1.22. The hot exhaust gas from the turbine enters vertically at the bottom, flows upward through the heat transfer matrix, and discharges vertically from the top. The air from the compressor enters a large circular port at the top at the right end, flows vertically downward in pure counterflow, and leaves a second circular port at the bottom to flow to the combustion chamber. Figure 1.23 gives a sectional view through the top corner at the air-inlet end to show the internal construction. The hot exhaust gas passages are formed by corrugated sheets sandwiched between flat plates that extend all the way from the bottom to the top of the unit. The air to be heated flows horizontally from the long plenum at the top into the spaces between the walls of the exhaust gas passages. Curved spacer strips guide the air through a 90° bend and then downward between the heated walls. A similar headering arrangement is used at the bottom. Note that both the flowpassage area and the heat transfer surface area for the hot exhaust gas are about three times

Figure 1.16 Radiator for a large truck. (Courtesy Harrison Radiator Division, General Motors Corp.)

as great as the corresponding values for the air being heated. This comes about because the two fluid streams differ in density by a factor of about four.

The air preheaters in steam power plants are usually quite different from the units just described for gas turbines. Rotary regenerators of the type shown in Fig. 1.24 are often used. These consist of a cylindrical drum filled with a heat transfer matrix made of alternately flat and corrugated sheets. The drum is mounted so that a portion of the matrix is heated by the hot gas as it passes from the furnace to the stack. The balance of the matrix gives up its stored heat to the fresh air enroute from the forced draft fans to the furnace. The ducts are arranged so that the two gas streams move through the drum in counterflow fashion while it is rotated, so that the temperature of any given element of the metal matrix fluctuates relatively little as it is cycled from the hot to the cold gas streams.

In the steam- and gas-turbine power plant fields a distinction is sometimes made between air preheaters that involve a conventional heat transfer matrix with continuous flow on both sides of a stationary heat transfer surface and those through which the fluids flow periodically, the hot fluid heating one section of the matrix while the cold fluid is removing heat from another section. Where this distinction is made, the term *regenerator* is applied to the periodic-flow type of heat exchanger, since this term has long been applied to units of this type employed for blast furnaces and steel furnaces, whereas the term *recuperator* is applied to units through which the flow is continuous.

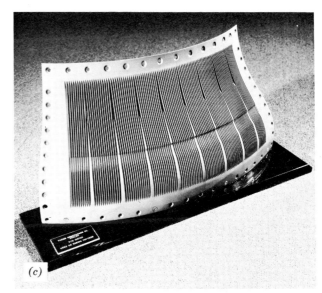

Figure 1.17   Aircraft engine oil coolers fabricated of aluminum for: (a) the DC-7 (tubes expanded into the header sheets); (b) the DC-8 (dip-brazed, plate-fin construction); and (c) the 707 (machined fins on the air side, dip-brazed fins and headers on the oil side). (Courtesy AiResearch Manufacturing Co.)

Figure 1.18   One panel of a radiator for a ground test of a portion of a power plant designed for space vehicle use. The large ends of the copper-finned, tapered, stainless steel tubes are welded into a potassium vapor manifold at the bottom, while the small ends are welded into a series of condensate manifolds at the top. (Courtesy Oak Ridge National Laboratory.)

Figure 1.19 Vertical induced draft cooling tower. (Courtesy Foster Wheeler Corp.)

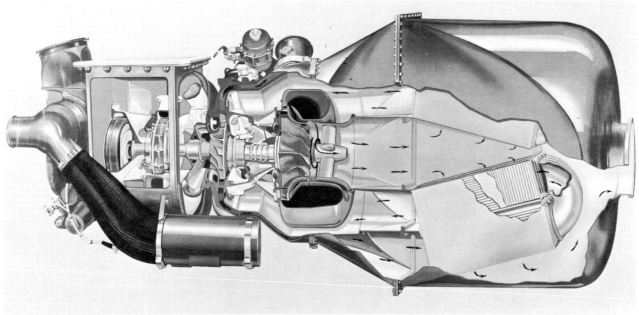

Figure 1.20 A small gas-turbine power plant fitted with a recuperator to improve the fuel economy. (Courtesy AiResearch Manufacturing Co.)

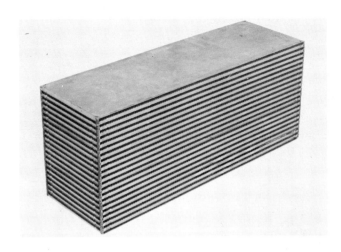

Figure 1.21 A brazed plate-fin recuperator core for the gas turbine of Fig. 1.20. (Courtesy AiResearch Manufacturing Co.)

Plates and Panels

Where it is desirable to build heat transfer surfaces into the walls of a compartment, as in refrigerators, steam chests, or environmental test chambers, one of the simplest and least expensive arrangements is provided by units formed of sheet steel stampings seam-welded to form panels with integral fluid passages. Figure 1.25 shows panels of this sort used to line the interior of a cylindrical compartment.

Immersion Heaters and Coolers

Immersion heaters and coolers provide a convenient means for controlling the temperature of baths or pools. Panel-type heat transfer surfaces arranged in banks of closely spaced parallel panels are often used. Axially finned tubes arranged in banks, as in Fig. 1.26, and mounted with the tube axes vertical constitute another type of heat transfer surface well suited to this purpose. Natural thermal convection usually induces sufficient circulation in the bath to maintain it within the desired temperature limits.

Double-Pipe Heat Exchangers

Two concentric pipes with one fluid in the inner pipe and the other in the annulus between them give a simple heat exchanger construction well suited to some applications. For small laboratory heat exchangers such a unit can be made up of two lengths of copper tubing that fit inside each other with a standard reducing Tee copper tube fitting at either end. A similar construction is often used in large units. If a fluid with a poor heat transfer coefficient, such as oil or air, is to be cooled by water, an axially finned tube can be placed inside of a larger pipe to give the construction shown in Fig. 1.27. Units of this sort can be mounted in both series and parallel to give any desired capacity and heating or cooling effectiveness, so that special requirements can be met by assembling a bank of stock commercial units. This type of construction is particularly advantageous where one or both of the fluids is at high pressure, which would cause the shell

Figure 1.22 A welded steel recuperator for a large gas-turbine power plant. (Courtesy Harrison Radiator Division, General Motors Corp.)

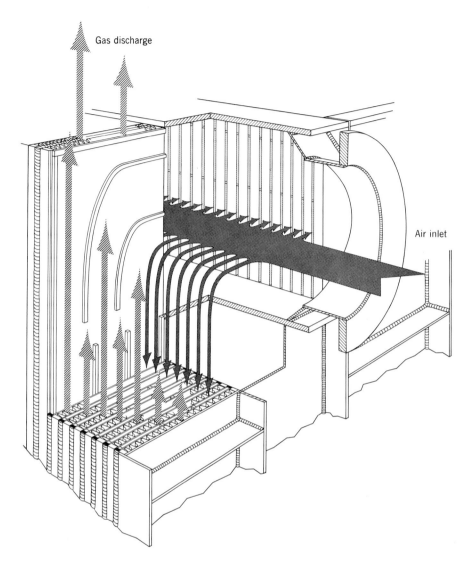

Gas discharge

Air inlet

Figure 1.23 Section through a top corner of the recuperator of Fig. 1.22 showing the construction of the heat transfer matrix and the headering arrangement. (Courtesy Harrison Radiator Division, General Motors Corp.)

wall thickness and cost to be large if a conventional shell-and-tube heat exchanger were employed.

Plate-and-Frame Heat Exchangers

Stringent hygiene requirements in the food and pharmaceutical industries often mandate frequent, thorough cleaning of components such as heat exchangers. To meet this requirement a variation on the Platecoil configuration of Fig. 1.25 has been developed. Commonly called the plate-and-frame heat exchanger, it utilizes stacks of sheet-metal plates that have been pressed to give patterns of ridges and grooves that define the coolant flow passages and provide turbulence-promoting surfaces. Figure 1.28 shows a heat exchanger of this type and the patterns used in two typical plates, while Fig. 1.29 shows the flow paths for the hot and cold fluid streams in the header region. The joints between plates around both the outer edges and the fluid inlet and outlet manifolds are sealed with elastomer gaskets that fit into

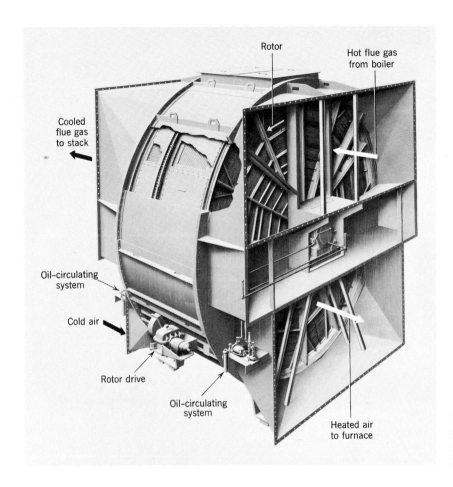

Cooled
flue gas
to stack

Rotor

Hot flue gas
from boiler

Oil-circulating
system

Cold air

Rotor drive

Oil-circulating
system

Heated air
to furnace

Figure 1.24 A rotary regenerator for preheating the air in a large coal-fired steam power plant. (Courtesy Air Preheater Corp.)

Figure 1.25 A Platecoil heat exchanger used to line the inside of a large tank. (Courtesy Platecoil Division, Tranter Manufacturing, Inc.)

Figure 1.26 Heat exchanger designed for immersion in a large vat. (Courtesy Brown Fintube Co.)

grooves pressed into the plate surface. The corrugations in the heat transfer surface region serve not only as turbulators to enhance the heat transfer, but also as spacers between the plates; over 60 different patterns are available from which to select that best suited to the viscosity and Reynolds number for the particular fluids of a given application.[5] To provide a good seal at the gasketed joints and carry the pressure loads the plates are forced together by long tie bolts arranged around the periphery. (In the unit shown in Fig. 1.28 the tie bolts fit in the forks that extend to either side of the front cover plate.)

This type of heat exchanger is sufficiently easy to clean that in some applications the units are disassembled for a thorough cleaning at the end of each day, an operation to which these units are uniquely suited. In recent years the excellent heat transfer performance, compactness, and reasonable cost of this type of heat exchanger have led to its use in many other applications where pressures are relatively low, that is, up to 2.5 MPa (360 psig). The upper temperature limit imposed by the elastomer gasket material may be as high as 230°C (450°F), although for gas-to-gas heat exchangers where leakage is less critical asbestos gaskets have been used at temperatures as high as 285°C (550°F). This type of heat exchanger is particularly attractive for applications requiring plates of expensive alloys such as titanium, tantalum, stainless steels, and high-nickel alloys. Units can be built to give just the right capacity using the proper number of standard stampings to provide surface areas as large as 2,200 m^2 (24,000 ft^2).[5] If more capacity becomes necessary, it is relatively easy to add additional plates.

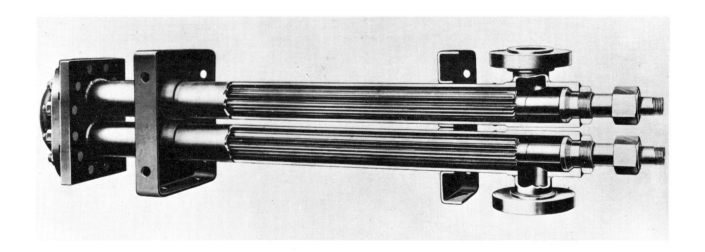

Figure 1.27 A double-pipe heat exchanger with axial fins on the inner pipe. The end fittings are designed so the basic unit shown can be used in series and/or parallel with additional units to give the capacity desired. (Courtesy Industrial Equipment Division, Baldwin-Lima-Hamilton Corp.)

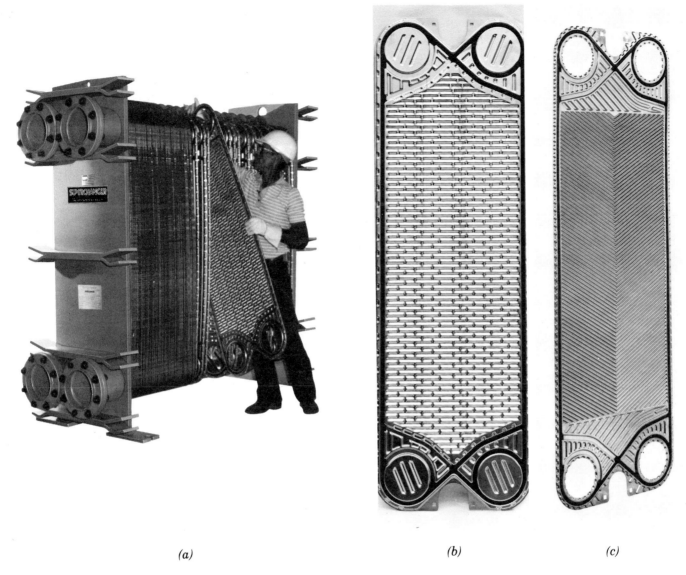

(a)   (b)   (c)

Figure 1.28 Photo showing the installation of a plate in a plate-and-frame heat exchanger together with two other plates having typical patterns of corrugations designed to improve heat transfer. (Courtesy Tranter, Inc.)

Spiral Plate Heat Exchangers

For applications in which the flow rates are relatively low and pure counterflow is required to give a very close approach temperature for the two fluids, it has been found that an effective configuration is obtained with the spiral plate type shown in Fig. 1.30. The plates are similar to those of Fig. 2.1 except that they are very long and the thickness of the passage between the plates must be rather small so that, after the sheets forming the upper and lower surfaces are welded together, the unit can be wrapped into a spiral.

Plastic Heat Exchangers

In applications involving fluids such as acids and concentrated salt solutions that are highly corrosive to any but

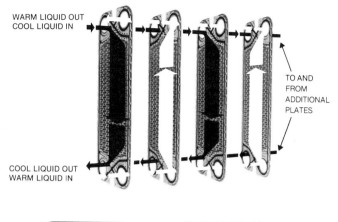

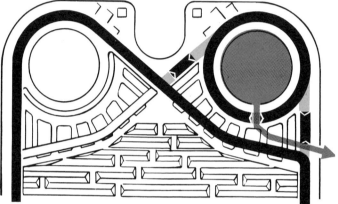

Figure 1.29 Diagrams showing the flow paths for the fluids passing through a plate-and-frame heat exchanger. (Courtesy Tranter, Inc.)

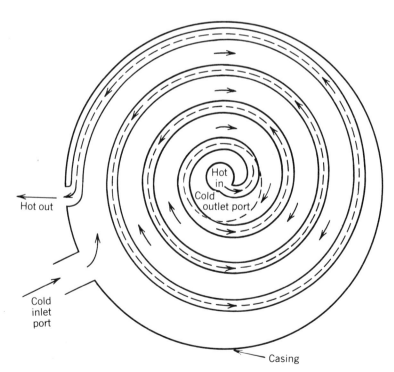

Figure 1.30 Diagram showing the flow paths in a spiral plate heat exchange. (Courtesy A. P. V. Equipment Corp.)

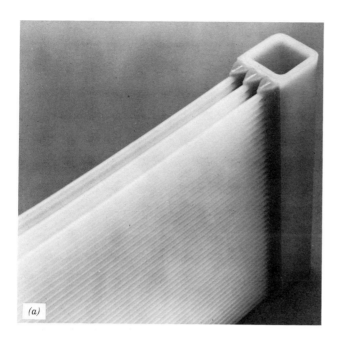

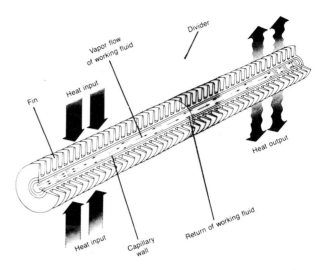

Figure 1.32 Schematic diagram illustrating the way in which a heat pipe operates. (Courtesy Hudson Products Corp.)

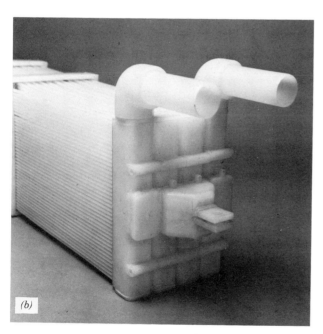

Figure 1.31 Photos showing the construction of some plastic heat exchangers. (Courtesy +GF+Plastic Systems, Inc.)

(roughly 1% that for metals), it is about the same as for glass. Further, in applications involving gases or viscous liquids the heat transfer coefficients may be sufficiently poor that a poor thermal conductivity of the wall material may not detract seriously from the performance. In such cases the lower cost of a plastic unit may make it more attractive than a metal unit. Not only is the cost of the basic material comparatively low, but the fabrication costs can be much lower than for metals. This feature can be envisioned by examining some typical geometries for the header region of a plastic heat exchanger (Fig. 1.31). Plastics used in units commercially available include polypropylene (for salts, alkalies,

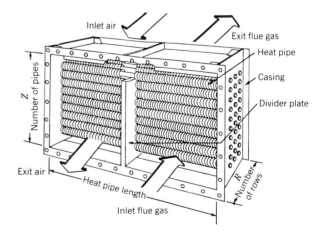

Figure 1.33 Schematic diagram of an air preheater employing heat pipes. (Courtesy Hudson Products Corp.)

the precious metals, plastic heat exchangers have been found to offer a good alternative to glass. Problems with the high rigidity and brittleness of glass are avoided and leaktight joints are much more easily obtained. Although the thermal conductivity of all of the plastics is poor

Figure 1.34 Photo showing shop-fabricated modular sections of a large heat pipe regenerative air preheater being erected for use with the furnace for an ammonia reformer. (Courtesy Hudson Products Corp.)

and non-oxidizing acids) and the more expensive fluorinated polymers for acids and aromatic compounds.

Heat Pipes

The exceptionally high heat transfer coefficients for boiling and condensing coupled with the high latent heat of vaporization have made possible the development of heat pipes such as that shown in Fig. 1.32. The excellent heat transport properties of these units (treated in Chapt. 6) have led to the development of completely new types of heat exchangers in recent years. One such unit for extracting heat from flue gas for preheating air to be supplied to a furnace is shown diagrammatically in Fig. 1.33; Fig. 1.34 shows a large unit of this type being installed for use with the furnace of an ammonia reformer. This type of unit is also used in air-conditioning systems where it not only has the advantage that no moving parts are required to circulate the heat transport fluid between the hot and cold surfaces but there is no motor, blower, or pump that might be an objectionable source of vibration and noise.

REFERENCES

1. P. F. Scholander, "Counter Current Exchange, a Principle of Biology," *Woods Hole Oceanographic Institute Collected Reprints*, Contribution no. 983, 1958.

2. T. B. Webb, "Sizewell Nuclear Power Station: Gas Circuits and Boilers," *Nuclear Power*, September 1961, p. 72.

3. "Standards of the Tubular Exchanger Manufacturers Association," Tubular Exchanger Manufacturers Association, New York, 1959.

4. "Heat Exchangers," The Patterson-Kelley Co., East Stroudsaburg, Pa., 1960.

5. R. K. Shah, "What's New in Heat Exchanger Design," *Mechanical Engineering*, vol. 106(5), May 1984, p. 50.

6. V. Ramamurti, "Design Features of a Spiral Heat Exchanger," ASME Paper no. 82-WA/DE-6, presented at the ASME Winter Annual Meeting, November 14–19, 1982.

2

Heat Exchanger Fabrication

Fabrication techniques are likely to be the determining factor in the selection of a heat transfer matrix geometry. They are a major factor in the initial cost and to a large degree influence the integrity, service life, and ease of maintenance of the finished heat exchanger.

TUBULAR VERSUS FLAT-PLATE CONSTRUCTION

Flat plates have the advantage that sheet stock is less expensive than tubing per unit of surface area, and flat sheets offer many possibilities for constructing heat exchanger flow passages that are aerodynamically clean so that pressure losses can be minimized. Fluid passages can be formed by corrugating or stamping the sheets and soldering, brazing, or welding them together. Unfortunately, units of this construction have the disadvantage that stress concentrations at the seam-welded joints will induce cracks if large pressure differentials exist. Thus flat-plate construction is not suitable for applications involving large pressures. Whereas in many applications it may be possible to balance the pressures in the two fluid circuits so that the pressure differential across the flat plates is small under normal conditions, off-design conditions may still present a problem, since the failure of a pump or a loss of system pressure in one circuit gives a large pressure differential for a short interval. Flat plates also present severe thermal stress problems in many units because of their stiffness in shear, and the internal fluid passages cannot be cleaned mechanically. For these and other reasons tubular heat exchangers are more widely used than flat-plate units.

Panels with Integral Cooling Passages

One of the simplest heat exchanger constructions is that shown in Fig. 2.1. As can be seen in the detail section shown at the bottom of Fig. 2.1, the coolant passages are formed in the sheets by a stamping or pressing operation, and the sheets are then seam-welded or brazed together. The stampings can be designed to give a single, long serpentine passage (as in the panel shown at the top of Fig. 2.1), or any of a variety of arrangements of manifolded passages can be employed, for example, the middle and lower units of Fig. 2.1 or those in Fig. 1.25.

The plates or panels can be formed into many shapes such as circular discs, cylindrical shells, etc., to conform well to the interior or exterior of a tank or chamber. The shape of the fluid passages can be modified in local areas to fit around inlet and outlet openings, projections, supports, and the like. Steam, water, brine, Freon, or other fluids can be circulated through the panels to heat or cool the enclosed space.

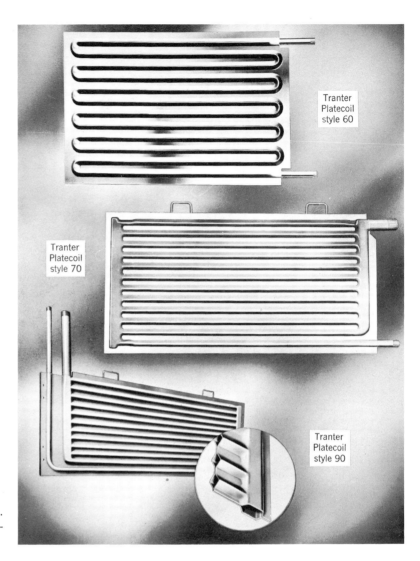

Tranter
Platecoil
style 60

Tranter
Platecoil
style 70

Tranter
Platecoil
style 90

Figure 2.1 Three types of plate-coil panels. (Courtesy Platecoil Division, Tranter Manufacturing, Inc.)

TUBE-TO-HEADER JOINTS

In heat exchangers making use of round tubes to form the heat transfer matrix, hundreds or thousands of tubes must function in parallel. One of the biggest problems in the detailed design of such a heat exchanger is provision for tube headers. This is most often done by bringing the tubes into a flat plate called a *tube-header sheet*. The holes in header sheets are ordinarily drilled; the usual clearances and tolerances are given in Table H7.3. The tubes are sealed in the header sheet by any one of a variety of techniques; those most widely used are discussed in this section.

Rolled Joints

Boilers constructed a century or more ago were usually built by swaging the tube ends into the tube sheets, but since the middle of the last century the bore of the tubes has been expanded by rolling in the manner shown in Fig. 2.2. A rolling tool such as that in Fig. 2.3 is inserted in the open end of the tube and rotated and expanded to roll the tube wall outward and expand its diameter. The tube wall is left with a substantial residual compressive stress to lock it into the header sheet.[1-3]

Tube-rolling tools are ordinarily made with a cluster of tapered rollers grouped around a tapered mandrel with the taper on the mandrel double that on the rollers. As can be seen in Fig. 2.3, this procedure makes it possible to expand the tube diameter uniformly throughout the portion being rolled. The mandrel can be forced inward to expand the rolls by an axial load, or the slots in the cage for the rollers can be canted slightly, relative to the mandrel axis, so that the mandrel tends to be drawn inward as the tool is rotated. This process has an added advantage in that reversing the direction of rotation

Figure 2.2 Photo showing tubes being rolled into the header sheet of a shell-and-tube heat exchanger. (Courtesy The Patterson-Kelley Co.)

quickly frees the tool. The roller cage may be fitted with a collar mounted on a ball bearing to prevent it from being drawn into the tube. Tools of this sort are commercially available for tube internal diameters from 0.15 to 16 in. As many as 15 joints per minute have been rolled by a single operator using a compressed-air driven tool on ¾-in.-diameter condenser tubes. For high-production work, automatic machines have been built with banks of tube-rolling tools. One of these machines has rolled 1200 tubes per hour.

For best results, the tube should be expanded just enough to give a leaktight joint. Overrolling thins down the tube wall and reduces its ductility so that tube failures are more likely. Furthermore, rolling induces a slight axial growth in the tube, and differential growth between the first and last tubes installed may prove troublesome if many tubes are mounted between two header sheets. In extreme cases overrolling may cause an increase in diameter of the header sheet—in some instances as much as 0.5 in. The proper amount of rolling varies somewhat with the physical properties of the tubes and header sheet. Best results are usually given by a reduction in tube-wall thickness of from 4 to 5%. Since there are likely to be some variations in dimensions from hole to hole and from tube to tube, it has been found more convenient to measure the torque on the rolling tool than the reduction in tube-wall thickness. If commercial tooling with torque measuring equipment is employed,

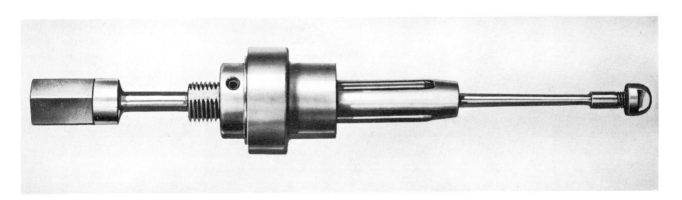

Figure 2.3 Tool for rolling tubes into header sheets. (Courtesy Thomas G. Wilson, Inc.)

a few sample tubes can be rolled into a block drilled to simulate the header sheet, and careful dimensional measurements can be made to establish the relationship between the torque required for the last part of the rolling operation and the reduction in tube-wall thickness. The proper torque to give a 4 to 5% reduction in tube-wall thickness that is determined in this way can then be used for the production operation.

The surface finish of both the tubes and holes influences the leaktightness and holding power of a rolled joint. In general, a smooth finish seems best, although a ductile tube will roll well into a rough hole. The strength of the tube joint under axial loads can be increased by circumferentially grooving the holes in the tube sheet. If the tubes are to be subjected to tensile loads, the joint strength can be increased by flaring the tube end with an additional set of rolls at an appropriate angle. Flaring the tubes has the additional advantage of reducing turbulence losses at the tube-inlet end.

The clearance between the tube and header sheet should be kept to a minimum consistent with reasonable dimensional tolerances and easy insertion of the tubes during assembly. Recommended header-sheet hole sizes are given in Table H7.3.

Hydraulic Expansion

A new technique for expanding tubes into header sheets employs hydraulic pressure instead of rollers. A fitting with O-ring seals at either end in positions in-line with the front and back faces of the header sheet is inserted into the tube. Sufficient hydraulic pressure is applied to impose tensile stresses in the tube wall about 115% of the yield stress and thus expand the tube firmly into the header sheet. This technique has the advantage that it is applicable to any thickness of the header sheet, can be controlled accurately, and deviations in circularity or tube thickness are accommodated. Pressures as high as 483 MPa (70,000 psi) are employed for thick-walled tubes.

Welding

Rolled tube-to-header joints can be made to give a reasonably good degree of leaktightness for most applications involving moderate temperatures. For the higher temperature applications, however, creep tends to relieve the initial stresses set up by rolling, and leads to enough relaxation in the tube wall so that leaks are likely to develop. For such applications, or where a very high degree of leaktightness is required, the tubes can be

attached and sealed at the header sheet by welding.[4] Several types of welded joint are indicated in Fig. 2.4. The first of these represents the most obvious approach, but it has the disadvantage that there may be difficulties in welding the relatively thin tube wall to the much thicker header sheet because of differences in heating and cooling rates and because of the residual thermal stresses that result. These stresses may be sufficiently severe to cause cracking in the weld. This problem can be alleviated to some degree by *trepanning*, as indicated in Fig. 2.4b, which gives a relatively thin section in the header sheet at the weld and reduces the thermal stresses.

Where the differences in header sheet and tube-wall thickness are particularly large, the arrangement of Fig. 2.4c gives substantially better conditions at the weld. The projection from the header sheet may be forged as in Fig. 2.4c or welded in place as in Fig. 2.4d. While expensive, these last two arrangements give the highest strength by far and the greatest integrity in the tube-to-header connection. They are particularly advantageous in applications where forces on the tubes, tube vibration, or thermal stresses may lead to severe stress concentrations at the weld. An indication of the severity of these stress concentrations is given by the greatly enlarged photograph of Fig. 2.5, which shows a section through a weld of the type illustrated in Fig. 2.4a. Imperfections in welds constitute a much less severe, but still important, source of stress concentrations.[4]

Another shortcoming of the simple welds seen in Fig. 2.4a and b is their susceptibility to *crevice corrosion*. This is likely to be serious in steam generators, where evaporation may tend to concentrate dissolved salts in the crevice in back of the weld and thus induce severe corrosion in the crevice. Chloride corrosion in stainless steel heat exchangers has been a particularly nasty problem that has given trouble even with chloride ion concentrations well below 1 ppm.

For some applications, notably condensers for steam turbines, a marked improvement in the leaktightness of rolled joints can be obtained by spraying the header sheet and the inner surface of the outer end of the tube with a rubber latex. This coating is applied after the tube-rolling operation. It may be vulcanized in place to give a tightly bonded coating.

Explosion Welding

Surfaces can be welded together by detonating a carefully chosen charge of explosives to accelerate the lighter element of a joint through a small standoff distance to make it impact at high velocity on the surface to which

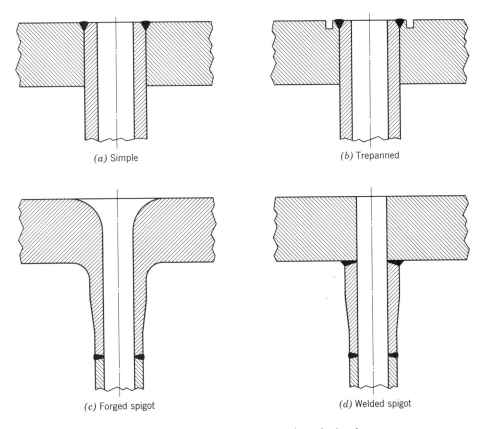

(a) Simple

(b) Trepanned

(c) Forged spigot

(d) Welded spigot

Figure 2.4   Types of welded joint for tube headers.

Figure 2.5   Section through a simple tube-to-header weld showing the crevice under the weld. (Courtesy Oak Ridge National Laboratory)

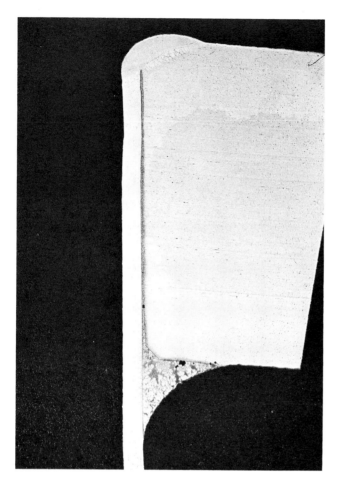

Figure 2.6 Section through a welded Inconel tube-to-header joint that has been back-brazed with a high-temperature brazing alloy. (Courtesy Oak Ridge National Laboratory)

it is to be welded. An excellent bond can be obtained, even between dissimilar metals. The process has been used for repairing tube-to-header joints in inaccessible places, and for joints that must withstand high pressure differentials, that is, over 14 MPa (2000 psi).[5] Explosion welding can also be used for seam welding or for welding a corrosion-resistant alloy liner to a steel vessel.[6] The method is best suited to cases in which a thin sheet is to be welded to a thicker plate.

Brazing

Tubes may be brazed into header sheets by any of a variety of techniques to give tube-to-header joints that are free of the stress concentration and crevice corrosion problems just mentioned.[4] The brazing process is also

well adapted to the fabrication of heat transfer matrices as in the aircraft oil cooler of Fig. 1.17b.

The brazing alloy can be applied to the joint in a variety of fashions. Washers of the alloy can be placed around the tube at the joint, the brazing alloy may be plated on the OD of the tube at the joint, or it may be applied with a brush in a suitable vehicle, such as a low-ash methyl methacrylate cement. A brazing process commonly employed for aluminum heat exchangers entails dipping the assembly in a molten salt bath, which serves both to heat the unit and flux off the oxide films on the surface to be brazed. Steel heat exchangers are often brazed in a furnace with an inert atmosphere, such as nitrogen or argon, or a hydrogen atmosphere may be used to eliminate oxide films by hydrogen reduction. The hydrogen brazing process is particularly well adapted to the brazing of stainless steel and high-nickel alloys. If an especially high degree of integrity is desired, the tubes can be welded into the header sheets, and then the assembly can be back-brazed. An additional advantage of this arrangement is that the tubes are accurately positioned by the welding operation so there is no jigging problem during brazing.

Brazing operations on stainless steels or other similar alloys are ordinarily carried out at temperatures between 2000 and 2200°F using one of several nickel-iron-chrome-silicon-boron brazing alloys in a very dry hydrogen atmosphere.[4] This type of brazing alloy tends to diffuse into the base metal to give a joint that has substantially the same strength as the base metal. As indicated in Fig. 2.6, excellent brazed joints can be obtained, but the brazing alloys are inclined to be brittle. No welding can be carried out close to the brazed joint after brazing, because the thermal stresses induced in the welding operation are likely to crack the braze.

Plain carbon steel can be brazed with copper in a dry hydrogen atmosphere to give an excellent joint from standpoints of both strength and leaktightness. The copper flows nicely to yield good ductile fillets at the joint. Other ductile high-temperature brazing materials include nickel-manganese and gold-nickel alloys.

HEADER-SHEET MOUNTING

Flanged joints are commonly used for mounting header sheets in heat exchanger shells where the pressures and temperatures of the application are not severe. Figure 1.10 shows a typical installation of this sort. An advantage of this arrangement is that the tube matrix and header sheet can be removed as an assembly for replacement or servicing. Flat gaskets or O-rings can be employed to seal the joint. A second approach is to make the header an integral part of the heat exchanger shell—an arrangement which has proved to be essential for many high-pressure and high-temperature operations. Heat exchangers such as that of Fig. 15.9 are built in this manner.

FINNED SURFACES

Such a variety of finned surfaces are manufactured for heat transfer purposes that it is often difficult to choose

(b)

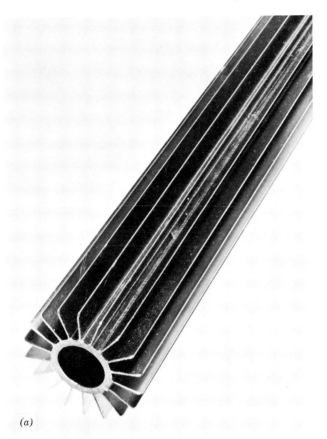

(a)

Figure 2.7 Types of commercial finned tubing: (a) axial aluminum fins rolled into grooves (Courtesy Baldwin-Lima-Hamilton Corp.); (b) and (c) axial steel fins attached by overlapping spot welds (Courtesy Brown Fintube Co.); (d) helical fins upset on a steel tube by a rolling operation (Courtesy Baldwin-Lima-Hamilton Corp.); (e) helical fins formed by winding a copper ribbon on a copper tube and soldering (Courtesy Aerofin Corp.); (f) and (g) helical steel fins formed by winding a flanged ribbon on a steel tube (Courtesy G. M. Jackson and Co.); (h) stamped steel fins pressed onto a steel tube (Courtesy Warren Webster and Co.); (i) stud fins flash-welded to a steel tube. (Courtesy Babcock and Wilcox Co.)

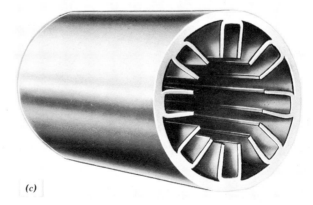

(c)

(d)

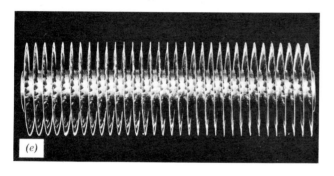

(e)

(g)

(h)

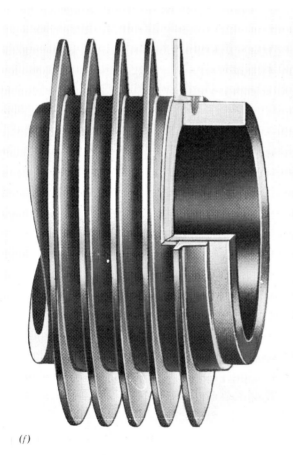

(f)

(i)

Figure 2.7 (*Continued*)

the best type to use for a given application.[7] Some typical examples of finned tubes are shown in Fig. 2.7. Axial fins such as those indicated in Fig. 2.7*a* can be attached by rolling or swaging them into grooves machined in the tubes, by overlapping spot welds at the base of U-shaped strips as in Fig. 2.7*b* and *c*, or they can be made integral with the tube by extrusion. Helically finned tubes with a long pitch can be formed by twisting tubes with straight axial fins. Closely spaced spiral fins can be formed by upsetting the metal in the tube wall with a rolling operation, as in Fig. 2.7d, by machining fins in a thick-walled tube, or by wrapping a ribbon edgewise onto the tube with a special machine to give a finned tube such as that in Fig. 2.7e. A collar at the base of the fin may be used to facilitate attachment to the tube by overlapping spot welds, as in the example shown in Fig. 2.7f.

With helically wrapped fins, the fin height is limited by the amount that the metal at the tip of the fin can be stretched in the wrapping operation. This limitation can be removed by slitting the fins to give the type of surface shown in Fig. 2.7g or by wrinkling the base of the fin as in Fig. 2.7e. The spirally wrapped ribbon may be soldered, brazed, or seam-welded to the tube, or it may be fitted into a groove cut or rolled into the tube for the purpose. The sides of the groove may be closed by a rolling operation that locks the fin rigidly in place. An advantage of the composite construction with a mechanical, soldered, or brazed joint is that the fin can be made of a high-conductivity material, such as copper or aluminum, while the tube material may be a cheaper, stronger, or more corrosion-resistant alloy such as steel or stainless steel. Finned tubes such as that of Fig. 2.7h are made by stamping out round or square fins with a flared skirt around the central hole to space the fins on the tube. These fins may be pressed onto the tube to give a tight mechanical joint or they may be soldered or brazed in place. Pressing the fins onto the tube is cheap and is well suited to low-temperature applications where corrosion is not a problem; the soldering or brazing, while more expensive, may be required for applications where high temperatures or corrosive conditions would loosen a press fit and spoil the thermal bond between the tube and the fins.[8] Stud fins, such as those in Fig. 2.7i, have been used in many boiler applications. They have the advantage of being mechanically rugged and resistant to corrosion and erosion conditions that would soon cause thin plate fins to corrode or burn out.

Plate fin-and-tube heat exchangers have been widely used for automotive and aircraft applications. These are built by stamping fins out of aluminum or copper, stacking the fins in a jig, as in Fig. 2.8, and then pressing the tubes into place. The tubes are usually flattened (Fig. 2.8) to reduce the gas-side flow resistance, and are ordinarily coated with a thin layer of solder or brazing alloy so that the unit can be soldered or brazed in a furnace or salt bath. In automotive radiators, the upper header sheet is normally made an integral part of the expansion tank, as in Figs. 2.9 and 2.10, and the entire unit is soldered or brazed in a single operation. A somewhat similar construction is employed for air conditioning and refrigeration units with flat or corrugated plate fins pressed onto round tubes, as in Fig. 2.11.

Plate-type heat exchangers are fabricated in a variety of forms. Perhaps the simplest of these is made of alternate flat and corrugated sheet with the corrugations in the alternate layers at right angles, as shown in Fig. 2.12, so that a crossflow heat transfer matrix is formed with one fluid passing through one set of passages and the other fluid flowing at right angles through the intermediate passages. As shown in Fig. 2.13, the corrugated sheets constituting the fins may be stamped with all sorts of turbulence promoters, including slits, waves, joggles, and steps.

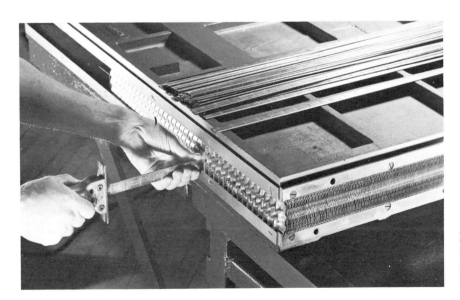

Figure 2.8 Flattened tubes being assembled in a corrugated plate-fin matrix for a truck radiator. (Courtesy Modine Manufacturing Co.)

Figure 2.9 Header sheets assembled on the radiator matrix of Fig. 2.8. (Courtesy Modine Manufacturing Co.)

An essentially similar, but geometrically more complex, type of extended surface is that illustrated in Fig. 2.14, which is one that is used frequently for automotive applications. Copper sheets are stamped in the form shown to provide the water flow passages. Corrugated sheet fins are mounted between these to supply the extended surface on the air side of the radiator, and the assembly is soldered together to give a strong, rigid unit. Another somewhat similar construction consists of flat plates provided with pin fins on the air side. A patented machine is used to form the wire into streamlined fins with flattened feet to provide good attachment to the plates. The assembly is fabricated by brazing.

An interesting construction that has been used for aircraft oil coolers employs copper or aluminum tubes formed by impact extrusion and the ends expanded into somewhat larger hexagons. The tubes are then arranged in a jig so they nest closely together, and the ends of the unit are sealed, usually by dipping them in a bath of solder or brazing alloy. This gives a compact unit with a large percentage of free-flow area on the air side.

Yet another unusual type of extended surface is that shown in Fig. 2.15, which is for recuperator surfaces in

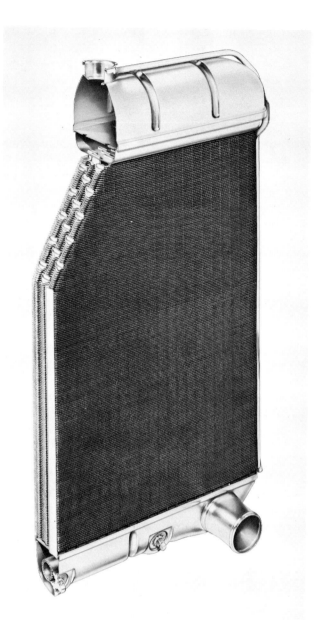

Figure 2.10 Section through a completed radiator similar to those shown in Figs. 2.8 and 2.9. (Courtesy Modine Manufacturing Co.)

the production of liquid oxygen. One stream of fluid is allowed to flow in one direction through the intermediate annulus while the other stream flows in the opposite direction through the inner and outer annuli.

TUBE BENDING AND JOINING

Many of the illustrations in Chapter 1, particularly Figs. 1.3 and 1.5, indicate the complex shapes of tubing often

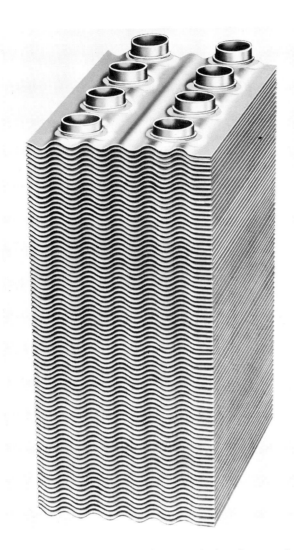

Figure 2.11 Element of a corrugated plate-fin, round-tube Freon evaporator made of aluminum for an air-conditioning system. (Courtesy The Trane Co.)

required in heat exchangers, and imply the importance of tube-bending operations as a factor in heat exchanger fabrication cost. Tubes are usually bent cold, and the metal on the inside of the bend yields in compression, whereas that on the outside of the bend yields in tension. Thus if the plastic deformation in the metal is to be kept to less than 25%, the minimum centerline radius to which the tube can be bent is 2 diameters. The tube material, temper, wall thickness-to-diameter ratio, and tooling all affect the minimum bend radius readily obtainable. Typical values for stainless-steel tubing are given in Fig. H7.4.

Probably the most common method of bending tubes is *roll forming* in the manner indicated in Fig. 2.16. The rolls can be grooved in such a way that the tube wall is supported during the rolling operation, and the tendency

of the tube to be flattened is markedly reduced. If the ratio of the tube-wall thickness to the diameter is rather small, the stress distribution in the tubes is complex and may lead to buckling so that the tubes may have to be filled with sand, plastic, or a low-melting-point lead-tin-bismuth-alloy to provide internal support during the bending operation.[9] For larger radius bends in thick-walled tubes this is not necessary, and conventional rolls can be employed, as in Fig. 2.17. Large radius bends of varying curvature, particularly if in three dimensions, as in the tube bundle of Fig. 2.18, are difficult to fabricate by rolling operations, especially if close dimensional tolerances are required.[10] Such tube shapes can be fabricated readily by stretch forming, and the dimensional stability of such parts is good since there is virtually no spring back; the elastic strain from the stretching operation is essentially axial, and is relieved by axial shrinkage rather than bending. If there is some difficulty with a small amount of spring back from slight differences in the work-hardening between the inner and outer fibers of the bend, the tube can be heated electrically while in the stretching fixture, and the residual strain relieved.

Tubing is ordinarily supplied in straight lengths of 20 to 40 ft. Some tube mills have draw benches that permit fabrication of tubing in lengths up to 65 ft. For some applications, particularly boilers, much longer lengths are often required. Because of shipping problems, the long tubes are prepared by welding shorter tubes end to end in the shop or at the erection site. Several methods employed for joining tubes are shown in Fig. 2.19. A major problem with the joints of Fig. 2.19a and c is the uncertainty of a full penetration weld. If such a weld is not obtained, corrosion or stress cracking may result in some applications. With the flash butt weld indicated in Fig. 2.19b, the metal upset as the pieces are brought axially toward each other in the welding operation tends to obstruct the flow passage, and thus will increase the fluid pressure drop. For tubes 40 ft or more in length, with an internal diameter of about 1.0 in., this increase in pressure drop is a relatively small percentage of the overall pressure drop through the tube, and hence is not objectionable. This obstruction can be avoided with the joint of Figs. 2.19c. Unfortunately, none of the types of weld indicated are well suited to a thorough inspection except by X-ray since the inner surface is inaccessible. The type of joint shown in Fig. 2.19d presents the same objections as some of the tube-to-header sheet joints of Fig. 2.4 in that it is susceptible to crevice corrosion.

Tube Spacing

The spacing of long tubes in large heat exchangers presents many awkward problems. In the higher temper-

Figure 2.12 Brazed stainless steel, corrugated, plate-fin heat exchanger for high-temperature service. (Courtesy Modine Manufacturing Co.)

ature units it is undesirable to tie the tubes together rigidly unless they are bent to permit differential thermal expansion. If lugs are welded to the tube wall, the weld zone represents both a region of stress concentration and one in which the grain structure differs sufficiently from the main part of the tube so that corrosion may be induced under some conditions.

Simple drilled baffle plates similar to those of Fig. 2.2 are widely used in shell-and-tube units.[11] These serve the dual function of spacing the tubes and directing the flow.

This arrangement has been found eminently satisfactory for separating the tubes in shell-and-tube heat exchangers filled with liquids. Tube vibration usually is not a problem since the relatively massive and viscous liquids provide good damping. In units in which relatively high-velocity fluids pass over the tubes in crossflow, however, tube vibration somewhat similar to that of wires humming in the wind may be induced by the Kármán vortices as they break off the trailing side of the tube. This has caused chafing and fretting in some units, and

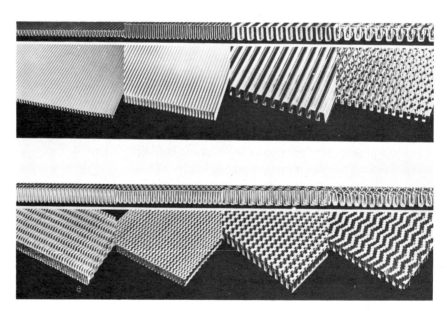

Figure 2.13 Types of corrugated plate fins showing typical ways in which the fins may be undulated or interrupted in the direction of flow to improve the heat transfer coefficient. (Courtesy The Trane Co.)

operations. An indication of the degree of precision possible is given by experience gained in the fabrication of coils for a large, helical-tube, steam generator for a gas-cooled reactor plant. The tubing employed was 2.25Cr1Mo steel having a diameter of 25.4 mm (1.0 in.), a wall thickness of 2.8 to 4.5 mm (0.110 to 0.177 in.), and straight lengths of up to 130 m (425 ft). The tubes were wound into coils having diameters of 3 to 4.2 m (9 to 14 ft) on a conventional 3-roll coiling machine. Measurements after the coils had been formed showed that the deviation from the design radius could be kept to less than 4 mm, or only about 0.2%, an amazingly small deviation.[12] The capability for maintaining such close dimensional tolerances in the tube-bending operation greatly eases the problems of assembling the coils with the support structure and spacers.

Pressure-Vessel Fabrication

The larger heat exchanger shells are ordinarily fabricated by automatic electric arc welding. Since the maximum amount of weld metal ordinarily laid down in one pass is roughly equivalent in cross-sectional area to a ³⁄₈-in.-diameter rod, a large number of passes are required to join heavy plates. The minimum number of passes for a high-quality joint is given by the use of a J-groove such as that of Fig. 2.20. Care must be taken to obtain full penetration in the first pass at the root of the groove, and the surface should be cleaned and inspected after each pass.

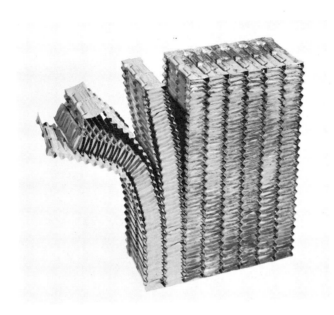

Figure 2.14 Partially assembled core element of an automotive radiator built of sheet metal stampings to give a unit somewhat similar to the flattened tube-and-corrugated plate-fin unit of Fig. 2.10. (Courtesy Harrison Radiator Division, General Motors Corp.)

sometimes has been responsible for wearing holes through the tube wall. This problem is treated in Chapter 9.

Tube spacing problems in the course of assembly are strongly dependent on the accuracy of the bending

Figure 2.15 Spiral copper ribbon-packed concentric annuli for a recuperator in small oxygen plant. (Courtesy Joy Manufacturing Co.)

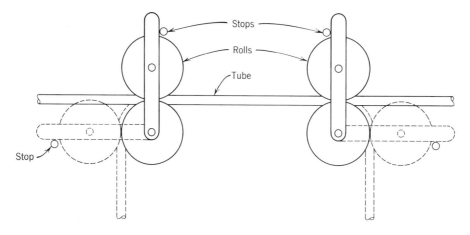

Figure 2.16 Fixture for roll-forming two small-radius bends at the same time. The solid and dotted lines indicate the positions of the tube and rollers before and after bending, respectively. (See also Fig. H7.4.)

Axial welds are made by mounting the two halves of the pressure vessel in a jig, which is carefully aligned to allow for weld shrinkage. A track carrying the welding head is mounted and aligned with the J-groove in which the weld metal is to be deposited, and a number of passes are made. The assembly is then rotated to place the other axial groove at the top, and the process is repeated. Circumferential welds are made by mounting the sections to be joined on rollers, tack-welding, and rotating the assembly under an arc-welding head.

The material in vessels having thicknesses greater than approximately 2.0 in. is ordinarily preheated with gas flames to a temperature of 300 or 400°F to reduce residual thermal stresses and thus avoid thermal cracks in the welds. The welds are inspected visually after each pass, and the entire weld is X-rayed after completion. If defects are found, they are cut out with an oxyacetylene cutting head, and new weld metal deposited after chipping the scale out of the cut. Vessels having walls more than 1 in. thick ordinarily are given a stress-relief heat treatment to about 1200°F, and then are slowly cooled to room temperature to relieve local stresses induced by shrinkage and distortion during the welding operation.

Inspection

A variety of methods for the inspection of tubing, shells, and welded joints are in use. The external surface of tubing can be inspected visually for nicks, flaws, laps, and scratches. The interior of tubing having a diameter greater than ⅜ in. can be inspected visually with a

Figure 2.17 Roll-forming bends in a bank of boiler tubes. (Courtesy Combustion Engineering, Inc.)

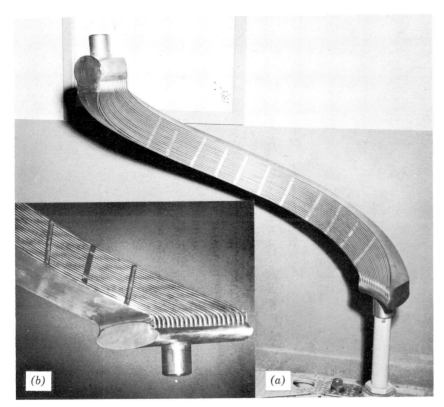

Figure 2.18 One of twelve helical spiral tube bundles for a molten salt-to-NaK heat exchanger designed for an exceptionally compact nuclear power plant in which the heat exchanger formed a spherical shell integral with the reactor shield. The inset (b) shows a close-up view of the header region at the lower end. (Courtesy Oak Ridge National Laboratory)

"boroscope" to depths of 10 ft. A more sensitive technique entails careful cleaning of the tube or other part, dipping it in a red dye penetrant, draining and superficially drying it, and then spraying it with a white, chalky powder. Cracks are revealed by the red dye which bleeds out and discolors the white coating. Tubing can be inspected for internal cracks or flaws by extremely sensitive techniques involving ultrasonic or eddy-current apparatus, which will reveal internal defects such as slag inclusions or cracks only a fraction of an inch long.[13,14]

Tube-to-header joints ordinarily are inspected visually and leak-tested for all applications regardless of the fabrication procedure. If the parts are not too large, a leak test can be made by submerging the unit in water while pressurizing the interior with compressed air. For larger units the surface can be painted or sprayed with a soapy solution. (Preparations developed expressly for this

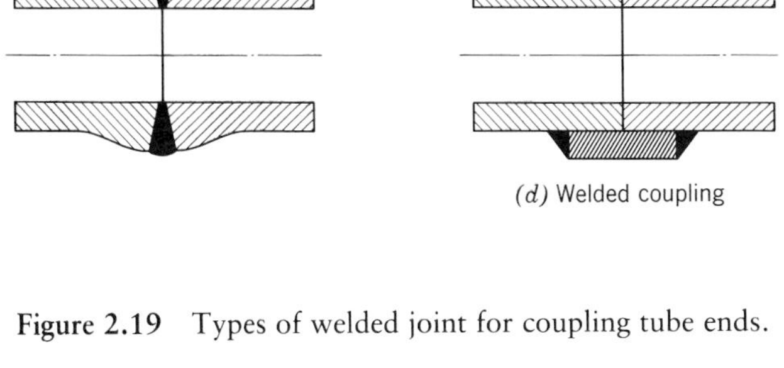

(a) Simple butt

(b) Flash butt

(d) Welded coupling

Figure 2.19 Types of welded joint for coupling tube ends.

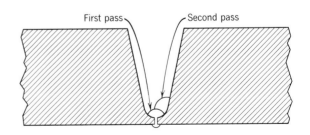

First pass   Second pass

Figure 2.20 J-groove joint for arc-welding heavy steel plate.

purpose are commercially available.) If the tube-to-header joints are welded, X-ray inspection may be required, and the design must be such as to permit it. In inspecting welds visually, a defect known as a *crater* commonly is found at the end of a weld pass. If the welder is properly trained and required to meet tight inspection standards, he can avoid all but a trace of a crater.

REFERENCES

1. E. F. Fisher and G. J. Brown, "Tube Expanding and Related Subjects," *Trans. ASME*, vol. 76, 1954, p. 563.

2. F. E. Dudley, "Electronic-Control Method for Precision Expanding of Tubes," *Trans. ASME*, vol. 76, p. 577.

3. A. Nadai, "Theory of the Expanding of Boiler and Condenser Tube Joints Through Rolling," *Trans. ASME*, vol. 65, 1943, p. 865.

4. G. M. Slaughter, "Welding and Brazing of Nuclear Reactor Components," *AEC Monograph*, Rowman and Littlefield, New York, 1964.

5. R. K. Shah, "What's New in Heat Exchanger Design," *Mechanical Engineering*, vol. 106(5), May 1984, p. 50.

6. L. J. Bement, "Practical Small Scale Welding," *Mechanical Engineering*, vol. 105(9), September 1983, p. 53.

7. J. B. Anderson, "Fuel Elements for Nuclear Reactors," *Trans. ASME*, vol. 79, 1957, p. 29.

8. K. A. Gardner and T. C. Carnavos, "Thermal Contact Resistance in Finned Tubing," *Journal of Heat Transfer, Trans. ASME*, vol. 82-2, 1960, p. 279.

9. L. Beskin, "Bending of Curved Thin Tubes," *Trans. ASME*, vol. 67, 1945, p. A-1.

10. F. J. Gardiner, "The Spring Back of Metals," *Trans. ASME*, vol. 79, 1957, p. 1.

11. A. J. Gram, "Mechanical Design of Heat Exchangers," *Industrial and Engineering Chemistry*, vol. 52, 1960, p. 474.

12. P. Burgsmuller, "Fabrication Experiments for Large Helix Heat Exchangers," *Sulzer Technical Review*, Special number, Nuclex 78, 1978, p. 11.

13. J. A. Tash, "Field Inspection of Boiler Tubes with Ultrasonic Reflectoscope," *Trans. ASME*, vol. 74, 1952, p. 201.

14. D. C. Martin et al., "Evaluation of Weld-Joint Flaws as Reinitiating Points of Brittle Fracture," *The Welding Journal*, vol. 36, 1957, p. 217-S.

Heat Transmission and Fluid Flow

The fields of heat transfer and fluid flow are so broad and complex that they can be reviewed only briefly in this chapter; it must be presumed that the reader is familiar with and has at hand basic books in these fields.[1-3] However, it has been the author's experience that most young engineers have difficulty relating their formal background in fluid mechanics and heat transfer to the practical problems of heat exchanger design. It is hoped that this chapter will help in this transition by reviewing those basic concepts that play major roles in heat exchanger design work and by relating them to typical design problems. This chapter is also intended as an aid in the selection of experimental data from the wealth of papers in the literature. Neither elegance nor rigor have been an object; the presentation is intended to be the simplest possible rationale consistent with an effective approach to the design problems taken up in subsequent chapters.

The nomenclature and symbols used follow common practice in the heat exchanger field. Since individual writers often differ in their choice of symbols for some quantities, Table H1.1 in the appended Handbook presents the nomenclature used here together with the units implied. Symbols are defined in the text only where some ambiguity might otherwise arise. A single set of English units together with the standard SI units is used throughout this text and the Handbook in an effort to help the reader reduce the annoying errors that so often intrude when conversion of units is required.

THERMAL CONDUCTION

The flow of heat through solids to or from fluids is the central problem in most heat exchanger design work. The thermal conductivity of the materials involved is a basic consideration and, for the simplest cases, is related to the rate of heat flow by the basic heat-conduction equation:

$$Q = \frac{kA\,\Delta T}{L} \tag{3.1}$$

Thermal Conductivity

Some notion of the tremendously wide differences in thermal conductivity between materials is given by the curves of Fig. 3.1, which show that the thermal conductivities of the materials of interest differ in value by as much as 100,000. The highest values are given by the metals, the next highest by the dense ceramics,[4] the next by organic solids and liquids, while the lowest values are given by the gases. As might be expected, the thermal conductivity of a porous solid falls between that of the dense solid and that of the gas in the pores.[5] It is interesting to note that the spread in thermal conductivity within any of the above groups of materials shown in Fig. 3.1 falls within a factor of approximately three of the mean value for that group.

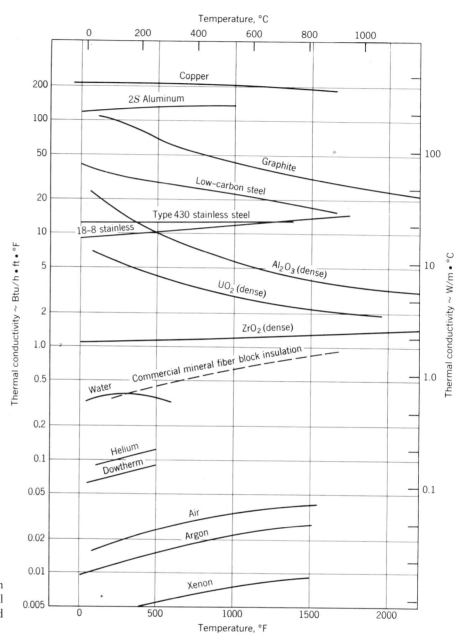

Figure 3.1 Effects of temperature on the thermal conductivity of typical materials. (Replotted from Refs. 1 and 4–6.)

The thermal conductivity of a given material depends on many factors. Small amounts of an impurity usually cause a large loss in the thermal conductivity of a pure metal. Exposure to fast neutron irradiation may reduce the thermal conductivity of metals or ceramics by a factor of two or more. Temperature has large effects, as indicated by Fig. 3.1. Pressure has little effect on the contribution to the thermal conductivity of gas in porous materials unless the gaps between particles are less than the mean free path for the molecules of the gas. As shown in Fig. 3.2, the effect becomes important if the pressure is reduced below about 10 mm Hg.[6] At low temperatures where thermal radiation heat fluxes are small, good thermal insulation can be effected through evacuating the space between two polished surfaces to a pressure of 0.01 mm Hg or less. Even better thermal insulation can be obtained if the evacuated gap is filled with reflective insulation. Exceptionally good thermal insulation for cryogenic equipment can be obtained with thousands of alternate layers less than one-thousandth of an inch thick

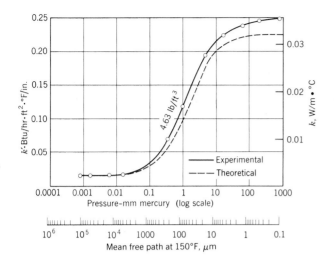

Figure 3.2 Effects of air pressure on the thermal conductivity of a mineral fiber insulation, density = 74.2 kg/m^3 = 4.63 lb/ft^3. (Note that k' is per inch of thickness rather than per foot.) (Versohoor and Greebler, Ref. 6.)

of aluminum foil and plastic film or glass cloth. By evacuation of the space between them, their thermal conductivity at cryogenic temperatures can be made as low as 1.73×10^{-5} W/m · °C (10^{-5} Btu/h · ft · °F).

General Equation

For the general case, the one-dimensional heat-conduction equation is

$$\frac{Q}{A} = -k \frac{dt}{dx} \qquad (3.2)$$

Assuming that the thermal conductivity is independent of temperature, this equation can be integrated to give Eq. 3.1, where A is the mean area normal to the direction of heat flow.

Thermal Resistance

In terms of the *thermal resistance R*, Eq. 3.1 becomes

$$Q = \frac{\Delta t}{R} \qquad (3.3)$$

where

$$R = L/kA$$

Equation 3.3 is similar in form to that for the flow of electric current through a conductor with a potential difference between its ends. By analogy, the heat flow is equivalent to electric current, the temperature difference to potential difference, and the thermal resistance to electrical resistance. Using this electrical analog, a composite slab consisting of a number of layers normal to the direction of heat flow with each layer having a different thermal conductivity can be treated as a system of electrical resistances in series. The heat transfer across such a composite slab with a temperature difference Δt is

$$Q = \frac{\Delta t}{\Sigma R_i} \qquad (3.4)$$

where R_i equals the thermal resistance of the *i*th layer.

Mean Area

Where the heat-flow path area varies in the direction of heat flow, the area used in Eq. 3.3 depends on the geometry. The mean area for the radial flow of heat across the wall of a cylindrical tube is equal to the logarithmic mean of the inner and outer surface areas, that is,

$$A = \frac{A_0 - A_i}{\ln \frac{A_0}{A_i}}$$

The radial flow of heat across a spherical shell is a function of the geometric mean of the inner and outer surfaces, that is, $A = \sqrt{A_i A_0}$. The flow of heat across a slab is a function of the arithmetic mean of the surface areas, that is, $A = (A_i + A_0)/2$.

Fin Efficiency

One important application of the thermal conduction relation is the determination of the efficiency of finned surfaces. The temperature drop between the root and the tip of the fin reduces the fin efficiency, since such a drop reduces the mean effective temperature difference between the heat transfer surfaces and the fluid. The *fin efficiency* has been defined as the ratio of the average temperature difference between the bulk free fluid stream and the surface to the temperature difference between the bulk free stream and the base surface, that is, the root of the fin. This relation can be expressed as follows:

$$\text{Fin eff.} = \frac{\text{bulk fluid temp.} - \text{mean surface temp.}}{\text{bulk fluid temp.} - \text{base surface temp.}}$$

The fin efficiency depends on the temperature distribution in the fin, and hence on the thermal conductivity and the dimensions of the fin as well as on the heat transfer coefficient between the fin surface and the fluid.

Analytical work has shown that the fin efficiency is a function of the parameter $w = \sqrt{h/kb}$, where w is the fin height, h is the heat transfer coefficient, k is the thermal conductivity of the fin material, and b is the fin thickness.[7] Charts for the fin efficiency as a function of $w\sqrt{h/kb}$ for both circular disc fins and rectangular axial fins are presented in Figs. H7.2 and H7.3. Additional charts for tapered circular fins and pin fins are given in Ref. 7 of this chapter. While tapered fins make more efficient use of the fin material than fins of uniform thickness, tapered fins are rarely used except where they can be fabricated by casting.

Contact Conductance

The interface between two solid surfaces not metallurgically bonded together constitutes a resistance to heat flow. The heat flux divided by the temperature drop across the interface is called the *contact conductance* of the joint, and typically runs 567.8 to 5678 W/m² · °C (100 to 1000 Btu/h · ft² · °F). Since solid surfaces are not perfectly smooth, direct contact between the surfaces takes place at a limited number of spots, and the volume enclosed in the voids is usually filled with air or the surrounding fluid. Heat transfer across the interface takes place mainly by conduction through the fluid layer filling the voids and through the surface high spots that are in direct contact. There is no convection since the fluid layer is very thin, and radiation across the gap is negligible for normal temperatures. The contact conductance essentially is composed of two resistances in parallel: the resistance of the fluid layer and that of the spots in direct contact.

To obtain a solution to the problem it is necessary to make a series of assumptions, particularly with respect to the character of the surface roughness, which in practice varies with the fabrication process.[8,9] Using a reasonable set of assumptions, Cetinkale and Fishenden[8] have developed a dimensionless equation for contact conductance. Their predicted values have agreed fairly well with experimental data for contact conductances ranging from 3120 to 71,000 W/m² · °C (550 to 12,500 Btu/h · ft² · °F) for steel, brass, and aluminum surfaces ground to various degrees of roughness, for pressures from 0.131 to 5.51 MPa (19 to 800 psia), with air, spindle oil, or glycol between the surfaces.

An extensive experimental investigation of contact conductance that impressed the authors as being especially useful to the heat exchanger designer has been made by Barzelay, Tong, and Holloway.[10] Aluminum and steel surfaces of various degrees of roughness have been tested at pressures from 34 to 300 kPa (5 to 425 psi) and mean interface temperatures from 93 to 204°C (200 to 400°F). Figures H5.17 and 5.18 were taken from this study to show the effects of interface pressure, surface roughness, mean interface temperature, and insertion of a sandwich material on the contact conductance for aluminum-aluminum and stainless steel-stainless steel joints. It is apparent from these results that the contact conductance increases with the pressure and the mean interface temperature, and decreases with increasing surface roughness. Thin foils of good thermal conductivity inserted between the surfaces improve the contact conductance when the foil is softer than the interface material, but decrease the contact conductance when the foil material is harder. A layer of oxide will, of course, introduce an additional barrier and reduce the contact conductance.

THERMAL RADIATION

There are many applications in which thermal radiation is an important mechanism in heat transmission. Boiler furnace tube arrays, furnaces for metallurgical and ceramic work, heat exchangers for high-temperature chemical-process operations, and radiators for space vehicles are typical examples. For space vehicles, thermal radiation is especially important, since in the vacuum of space there is no material to which heat can be conducted by molecular motion, so that the only mechanism by which heat can be dissipated is by thermal radiation.

The Stefan-Boltzmann law for the emission of thermal radiation from a *perfect black body* is

$$Q = \sigma A T^4 \tag{3.5}$$

where A = surface area, m² (ft²)
  T = absolute temperature, K (°R)
  σ = Stefan-Boltzmann constant
   = 5.67 W/m² · K⁴ × 10⁻⁸
   (= 0.1713 × 10⁻⁸ Btu/h · ft² · °R⁴)

Emissivity, Absorptance, and Reflectivity

The thermal radiation from an actual body is less than that of a perfect black body at the same temperature. In applying the Stefan-Boltzmann law for the emission of thermal radiation, an *emissivity* factor ϵ is employed.

This is defined as *the ratio of the thermal radiation flux emitted by an actual body to that emitted by a perfect black body at the same temperature*. Similarly, a perfect black body absorbs all the thermal radiation incident upon it; whereas an actual body reflects a portion, so that an *absorptance* factor, similar to the emissivity, may be defined. For thermal radiation at any given temperature, the emissivity and absorptance factors are usually equal.

The term *reflectivity* is also used, and indicates the fraction of the incident radiation that is reflected. If the angle at which the reflected ray leaves the surface is equal to that of the incident ray, the reflection is said to be *specular*. The reflection is said to be *diffuse* if the incident ray is reflected uniformly in all directions. The reflectivity is equal to 1 minus the absorptance (or emissivity).

Most polished surfaces have emissivities between 0.05 and 0.2, whereas painted, tarnished, or darkened surfaces have emissivities from 0.3 to 0.9. While the term *perfect black body* is applied to denote a surface having an emissivity of unity, even surfaces that appear black to the eye may have emissivities as low as 0.5. Table H2.5 gives emissivities for many typical surfaces.

The emissivity of a surface depends very heavily on the properties of the outer surface layer.[11,12] A slightly oxidized copper surface has a very much higher emissivity than a polished copper surface, for example.

Form Factor

In carrying out radiant heat transfer calculations for irregular geometries, it is generally convenient to introduce a quantity called the *form factor*. The form factor F_{12} is defined as *the fraction of the total energy emitted from surface 1 that is intercepted by surface 2*.[11] The form factor depends on the geometry and orientation of the surfaces. In deriving form factors, it is assumed that the emitting surface is a diffuse emitter, that is, it emits radiation uniformly in all directions.

The form factor for a flat plate or a convex surface radiating to space is unity, since no portion of the surface "sees" any other portion, and hence none of the energy emitted is reabsorbed by the surface. The form factor for a piece of angle iron is less than unity, since the inner surfaces of the legs emit toward each other. The effective area of these two surfaces is that of the plane bounded by the tips of the two legs. If the surface of a tube were covered with closely spaced disc fins, the form factor would be close to unity if applied to the cylindrical envelope through the tips of the fins. The form factor for simple geometries can be obtained directly from charts.[1,11,13] (A fine set appears in Ref. 13.)

Net Heat Flux

Most engineering problems entail the determination of net radiation heat exchange between two surfaces (i.e., a source and a sink). The net radiation between two surfaces A_1 and A_2 includes the component for heat exchange by direct radiation:

$$Q = A_1 F_{12} \epsilon_1 \epsilon_2 \sigma (T_1{}^4 - T_2{}^4) \tag{3.6}$$

plus the component for multiple reflections and absorptions.

The treatment of multiple reflections and absorptions for complex geometries is a complicated problem. The equation for the net radiation heat exchange between two surfaces (i.e., a sink and a source) in the presence of reradiating surfaces, including the effects of multiple reflections and absorptions, is given by[1,3,11]

$$Q_{12} = A_1 \mathscr{F}_{12} \sigma (T_1{}^4 - T_2{}^4) \tag{3.7}$$

where the factor $\mathscr{F}_{12}$ depends on the emissivity of the surfaces and the form factor F_{12}. The factor $\mathscr{F}_{12}$ can be determined readily for three simple geometries commonly used in engineering problems in which the form factor F_{12} is unity:

1. Where A_1 is small compared with its enclosure A_2, and A_1 does not see itself: $\mathscr{F}_{12} = \epsilon_1$.
   And

$$Q_{12} = \sigma A_1 \epsilon_1 (T_1{}^4 - T_2{}^4) \tag{3.8}$$

2. For two infinite parallel planes (i.e., the distance between the planes is small compared with the plane dimensions):

$$\mathscr{F}_{12} = \frac{1}{\left(\dfrac{1}{\epsilon_1} + \dfrac{1}{\epsilon_2} - 1 \right)}$$

$$Q_{12} = \sigma A_1 \cdot \frac{1}{\dfrac{1}{\epsilon_1} + \dfrac{1}{\epsilon_2} - 1} (T_1{}^4 - T_2{}^4)$$

$$\tag{3.9}$$

3. For two concentric cylinders or spheres, where A_1 is the inner surface,

$$\mathscr{F}_{12} = \frac{1}{\dfrac{1}{\epsilon_1} + \dfrac{A_1}{A_2} \left(\dfrac{1}{\epsilon_2} - 1 \right)} \tag{3.10}$$

Note that the relations given in 1, 2, and 3 assume a vacuum or a nonabsorbing medium between the surfaces. At temperatures above about 2000°F carbon dioxide, carbon monoxide, water vapor, and the hydrocarbons absorb and emit thermal radiation in substantial amounts if the mass of gas in the path of the radiation is equivalent to that in a layer 3 m (10 ft) or more thick at atmospheric pressure.

The value of σT^4 has been evaluated and plotted as a function of temperature in Fig. H5.16.

FLUID FLOW

Fluid-flow problems play an important part in heat exchanger design. Pressure losses, fluid-flow distribution, and mixing are often determining factors in the selection of a basic geometry. Furthermore, in most heat exchangers the principal barriers to heat transfer are the fluid films on the metal surfaces. The structure of these fluid films depends on the fluid flow pattern and the nature of the fluid flow, particularly on the extent and intensity of turbulence.

Laminar Flow

The simplest type of fluid flow is that in which viscous forces are dominant. When this is the case, the fluid particles pass through the flow field in nearly parallel lines with no rotation of the elementary particles. This is called *laminar flow*.

The velocity distribution for laminar flow through a round tube is readily calculated by relating the viscous shear forces to the local velocity and the distance from the wall.[14] A simple parabolic velocity distribution across the channel results—the velocity being zero at the walls and a maximum at the center, as indicated in Fig. 3.3. The mean velocity V through the tube can be related to the pressure drop, the tube length, the viscosity, and the tube ID to give the expression (for SI units):

$$\Delta P = 0.0397 \left(\frac{\mu V L}{d^2} \right)$$

where the units are as defined in Table H1.1. This result, known as the *Hagen-Poiseuille equation* for laminar flow through a tube, can be rearranged to show that the volume flow rate is directly proportional to the pressure gradient and the fourth power of the passage diameter, and inversely proportional to the viscosity.

Potential Flow

In many practical cases viscous forces have relatively little effect compared to dynamic forces. When this is the case, a good approximation to the flow pattern may be obtained by relating the motion of the fluid particles to the pressure and inertial forces, under the assumption that the fluid viscosity is zero. The resulting body of work, known as *potential flow theory*, has provided an excellent insight into many fluid flow problems,[15] notably in aircraft wing theory. The paths of representative fluid particles in potential flow through a passage are called *streamlines*, and can be calculated for many conditions for which the boundaries are smooth curves, free of cusps projecting into the flow. A typical case is shown in Fig. 3.4, for potential flow over a circular cylinder.

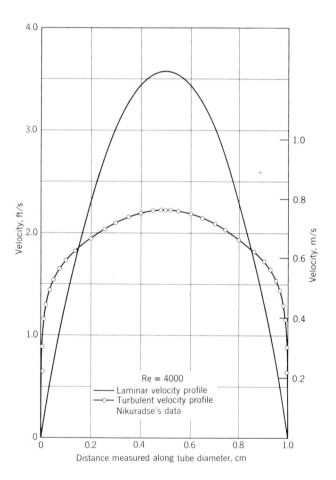

Figure 3.3 Velocity distribution across a circular channel for laminar and turbulent flow conditions at the same average velocity. (J. C. Knudsen and D. L. Katz, *Fluid Dynamics and Heat Transfer*, McGraw-Hill Book Co., New York, 1958.)

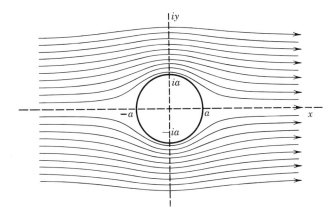

Figure 3.4 Streamlines for potential flow over a cylinder.

Figure 3.5 Flow patterns showing turbulence in the boundary layer along the walls of a long, straight channel: (*a*) camera stationary with respect to the walls; (*b*) camera moving at the velocity of the bulk fluid. (Prandtl and Tietjens, Ref. 14.)

Turbulent Flow

To obtain as much heat transfer capacity as practicable from a unit of a given size, and thus reduce the cost per unit of output, heat transfer fluids are pumped through most heat exchangers at substantial velocities. The resulting flow regime is usually such that the fluid flow patterns are determined more by inertial than by viscous forces, except that viscous forces lead to radial mixing, or eddying, between the neighboring layers. This eddying, or turbulence, is advantageous in that it assists greatly in improving the heat transfer, but it has the disadvantage that the flow patterns are much more complex than those for either laminar or potential flow, and they cannot be completely predicted mathematically even for very simple passage geometries.

A particularly important variety of turbulence may be observed in the region close to the wall of a long tube. Figure 3.5 is a view of the flow pattern in this region. The flow is characterized by a thin laminar flow region immediately adjacent to the wall, with a layer of small vortices between this thin laminar flow region and the bulk free stream. Note that the scale of the turbulence (i.e., the vortex diameter) increases with distance from the wall into the free stream. The velocity distribution across a channel for turbulent flow is shown in Fig. 3.3. In comparing this with the corresponding diagram for laminar flow in Fig. 3.3, note that the velocity is zero at the walls and maximum at the center in both cases, but for the same peak velocity the area under the curve is greater for turbulent flow, and the rate of shear is high near the wall.

While there are different views regarding the terms that ought to be applied and the methods that should be used to handle calculations for turbulent flow, the very

thin laminar flow region immediately adjacent to the wall is usually called the *laminar sublayer*, since viscous forces are dominant in this region. Adjacent to this sublayer is a highly turbulent region called the *transition layer*, in which the mean axial velocity increases rapidly with distance from the wall. The third major region, the *bulk free stream*, differs from the others in that the inertial forces are dominant, and the velocity changes relatively little with the distance from the wall. There is intense fine-grained turbulence in the transitional layer and relatively lower intensity, larger scale turbulence in the bulk free stream. Actually, of course, the bulk of the eddies originate at the wall and move outward into the free stream where they decay. They begin as small, high-velocity vortices and die out as larger, low-velocity vortices. The boundary layer is very thin at the inlet to a passage or at the leading edge of a flat plate, but its thickness increases with distance downstream along the wall as drag forces slow down a progressively greater amount of fluid. This effect is shown in Figs. 3.6 and 3.7.[16,17]

Reynolds Number

About 100 years ago Sir Osborne Reynolds[1] observed that the type of flow pattern was very much dependent on the ratio of inertial to viscous forces, and found that irrespective of the type of fluid, the velocity, or the scale of the system tested, the same flow pattern could be obtained if the ratio of the inertial to viscous forces were kept constant. He expressed this ratio very simply as a dimensionless quantity that has since been called the Reynolds number and is given by the equation,

$$\text{Re} = \frac{\rho D V}{\mu} = \frac{DG}{\mu}$$

In a sense, the Reynolds number might be compared to

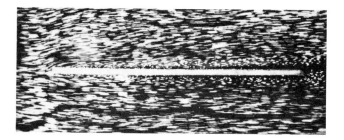

Figure 3.7 Flow pattern showing boundary layer thickening in the course of parallel flow from left to right over a flat plate. (Eck, Ref. 16.)

the relation between the length of a beam and its depth, that is, its stiffness. Under dynamic loads a long, slender beam is inclined to be unstable and vibrate, whereas a short, stiff beam tends to be stable and will not vibrate. Similarly, for high Reynolds numbers the inertial forces are large compared to the viscous forces, so that instabilities and turbulence tend to occur, whereas for low Reynolds numbers the dynamic forces are small compared to the viscous forces, and hence do not tend to induce turbulence.

Small irregularities in the surface of even a nominally straight wall cause distortions in the streamlines. The dynamic forces that result induce vortices, thus initiating turbulent flow.[18] In general, with very smooth passage walls in straight channels free of obstructions, the flow sometimes remains laminar up to a Reynolds number as high as 5000. However, laminar flow is unstable for Reynolds numbers above 2000, and is likely to change abruptly to turbulent flow. If turbulent flow is once initiated at a Reynolds number above 2000, even in a long, straight, smooth passage it will persist and can be eliminated only by reducing the Reynolds number to a value below 2000. If there are bends, irregularities in the passage walls, or obstructions, dynamic forces will disrupt the flow and turbulence will occur at Reynolds numbers far below 2000.

The effect of turbulence on a typical flow pattern is indicated by the set of pictures in Fig. 3.8. These were taken at intervals after the initiation of transverse flow over a cylinder, that is, the initial velocity was zero. Note that immediately after flow starts the flow pattern is essentially the same as that calculated from potential flow theory and presented in Fig. 3.4. A stagnant region then begins to develop in the wake of the cylinder and eddies form and grow, distorting the flow pattern. As shown in Fig. 3.9, the extent to which the flow pattern is affected depends on the Reynolds number. Figure 3.9a shows that

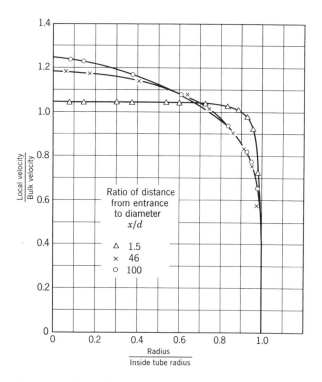

Figure 3.6 Turbulent velocity distributions across a circular channel at several distances from a smoothly rounded inlet. (R. G. Deissler, NACA Report no. TN2138, 1950.)

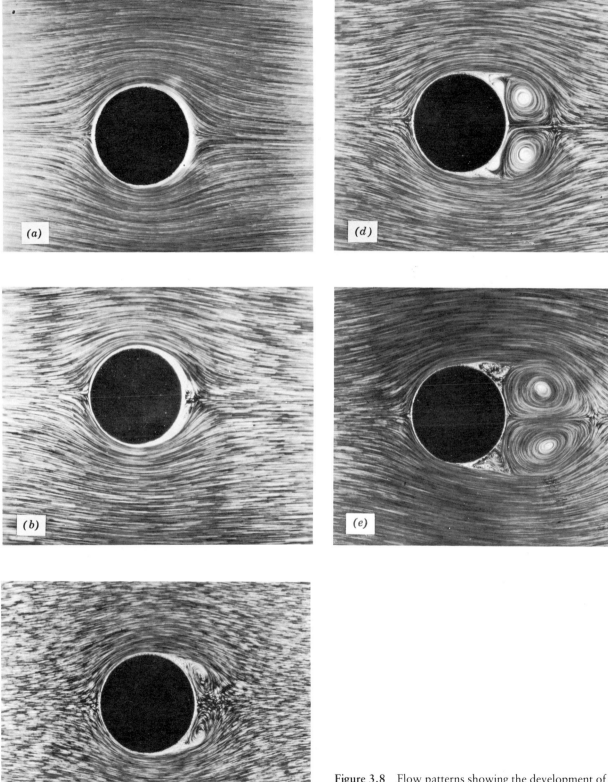

Figure 3.8 Flow patterns showing the development of turbulence at a series of five intervals of time after the initiation of flow across a circular cylinder. The first picture was taken just as the flow began, the last after fully developed turbulent flow was established. (Prandtl and Tietjens, Ref. 14.)

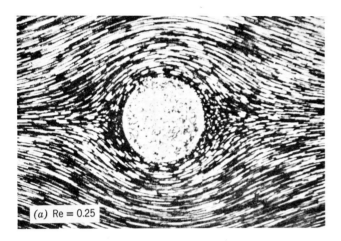

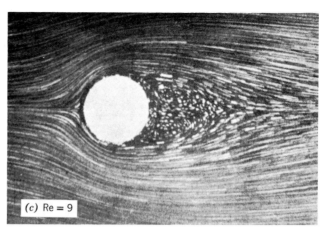

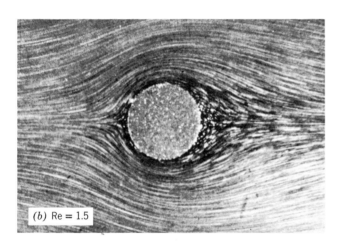

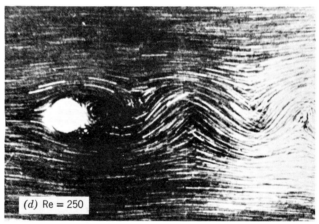

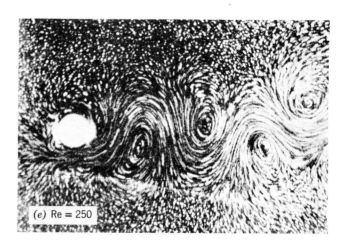

Figure 3.9 Effects of Reynolds number on the flow pattern around a circular cylinder transverse to the flow. The flow pattern in frame (*d*) was obtained using a camera stationary with respect to the cylinder, and that in frame (*e*) was obtained with the camera moving at the velocity of the bulk free stream. (Prandtl and Tietjens, Ref. 14.)

even for a Reynolds number as low as 0.25 there is an appreciable stagnation region in the wake of the cylinder. The size of this stagnation region grows until well-defined eddies become apparent at Reynolds numbers less than 10.0. At still higher Reynolds numbers, these eddies tend to be washed away alternately, first from one side and then the other. The string of vortices forming the wake looks something like a gear train with alternate vortices rotating in opposite directions. This effect can also be seen in Fig. 3.10, which shows the appearance of the flow past an elliptical cylinder as obtained in short time exposures with the camera first at rest with respect to the cylinder, and then with it moving at the same velocity as the bulk fluid. The vorticity is much more obvious in Fig. 3.10*b* than in 3.10*a*.

The train of vortices in the wake of the cylinder of Fig. 3.9*e* is typical of what are known as *Kármán vortex streets*, commonly found in the wakes of flow obstruc-

tions. The low-velocity fluid filaments in the boundary layer cause vortex streets to appear not only in the wake of a flow obstruction such as a cylinder but also in the wake of flat plates parallel to the flow (as in Fig. 3.7) or streamlined bodies, such as airfoils (as in Fig. 3.11). Note the similarity in the vortex streets of Figs. 3.9*e*, 3.10*b*, and 3.11 for three different geometries of flow obstruction.

At Reynolds numbers below 2000 the vortices tend to slow down, and the flow reverts to laminar or viscous flow, the distance required for the reversion increasing with Reynolds number. At Reynolds numbers above 2000 the vortices formed initially tend to break up to form a field of smaller vortices that persists for a long distance downstream. The heat transfer rate between a solid surface and the fluid in such a vortex field may be greatly increased by this turbulence.

In examining Figs. 3.4 and 3.9 it is apparent that the flow pattern for turbulent flow is essentially the same as that predicted by potential flow theory in the region where the fluid is accelerating, but the actual turbulent flow pattern departs from that for potential flow where there is deceleration along the streamlines. This difference is readily understood if it is remembered that the static pressure increases along a streamline where deceleration occurs. Close to a wall, where viscous forces bring the particles practically to rest, inertial effects are small, and the nearly stagnant fluid tends to flow under the influence of the adverse pressure gradient established by the free stream. This pressure gradient drives it in a direction that is inherently opposite to that of the mainstream, thus producing an eddy. The point on the

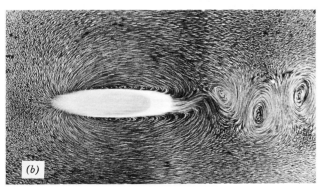

Figure 3.10 Flow patterns around an elliptical cylinder transverse to the flow with (*a*) the camera at rest with respect to the elliptical cylinder, and (*b*) the camera moving at the velocity of the free stream. (Riegels, *Aerodynamische Versuchsanstalt*, Gottingen, personal communication.)

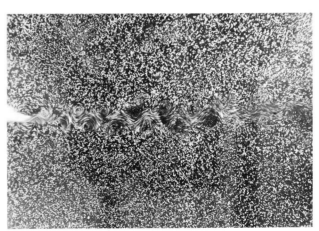

Figure 3.11 Vortex street in the wake of an airfoil. (Riegels, *Aerodynamische Versuchsanstalt, Gottingen*; personal communication.)

wall where the streamline pattern departs from that for potential flow marks the point at which *flow separation* begins and a vortex forms. The vortex tends to be washed off at frequent intervals and a new one develops, so that a series of vortices trail off to form a wake downstream of the point of flow separation. The problems of flow separation are discussed further in Chapter 8.

Friction Factor

It has been found convenient to express the pressure drop under turbulent flow conditions in terms of the *dynamic head*, $q = \rho V^2/2$ (to which the inertial forces in the fluid are proportional), the friction factor, f_d, and the passage length-diameter ratio. Thus, for SI units, the pressure drop is expressed as

$$\Delta P = f_d \left(\frac{\rho V^2}{2} \right) \left(\frac{L}{D} \right) \qquad (3.12)$$

For English units, to obtain the dynamic head, the fluid density must be divided by g, that is,

$$q = \frac{\rho V^2}{2g}, \text{ and } \Delta P = f_d \left(\frac{\rho V^2}{2g} \right) \left(\frac{L}{D} \right)$$

Experimental determinations of the friction factor for flow in pipes show that it varies with the Reynolds number, and, in the turbulent flow region, with the surface roughness. These effects are shown in Fig. H3.4. The friction factor in the laminar flow region is inversely proportional to the Reynolds number (this is consistent with the Hagen-Poiseuille relation of Eq. 3.11), so that, for circular passages, it can be expressed as

$$f_d = \frac{64}{\text{Re}} \qquad (3.13)$$

In the turbulent region, the friction factor for smooth-walled pipes falls off gradually with an increase in Reynolds number, because the intensity of the turbulence is not directly proportional to the velocity but rather to the 0.8 power of the velocity. Thus an approximate equation can be written for the friction factor in the turbulent region for smooth pipes:

$$f_d = 0.2 \text{Re}^{-0.2} \qquad (3.14)$$

Hydraulic Radius

While most of the discussion until now has been concerned with flow through circular passages, many other geometries are important, including annular, square, and triangular passages as well as such irregular passages as those for axial flow between parallel tubes.[19] It has been found that the same turbulence intensity and the same friction factor prevail if the ratio of the flow-passage area to the wetted perimeter is kept constant. This ratio is called the *hydraulic radius* and is given by $R_h = $ flow-passage area/wetted perimeter. For a circular passage this becomes

$$R_h = \frac{\frac{\pi}{4} D^2}{\pi D} = \frac{D}{4}$$

Note that laminar flow pressure-drop data are *not* correlated properly by using the hydraulic radius.

In the course of routine design work it is often convenient to use the term *equivalent diameter* rather than hydraulic radius. The equivalent diameter is used to signify the diameter of a circular passage that would have the same hydraulic radius as the passage geometry in question: $D_e = 4R_h = 4$ (flow-passage area/wetted perimeter).

The friction factor f_d, as used in Eqs. 3.12, 3.13, and 3.14, is defined on the basis of the equivalent passage diameter. Some writers prefer to define it as f_r on the basis of the hydraulic radius. In that case its value is ¼ as great, and Eq. 3.12 must be changed either by replacing D_e (the equivalent diameter) with R_h (the hydraulic radius) or by multiplying Eq. 3.12 by 4, that is, for SI units,

$$\Delta P = f_d \left(\frac{\rho V^2}{2} \right) \left(\frac{L}{D_e} \right) = f_r \left(\frac{\rho V^2}{2} \right) \left(\frac{L}{R_h} \right)$$

$$= 4f_r \left(\frac{\rho V^2}{2} \right) \left(\frac{L}{D_e} \right) = 2f_r \left(\frac{\rho V^2}{2} \right) \left(\frac{L}{D_e} \right)$$

Pressure-Loss Coefficients

The pressure loss in pipes or ducts can be estimated by first calculating the pressure drop for a straight pipe or duct of the same length using the appropriate friction factor from Fig. H3.4, and then adding the pressure losses caused by elbows, valves, tees, changes in section, etc. These losses can be calculated by multiplying the dynamic head by a loss coefficient:

$$\Delta P = C \frac{\rho V^2}{2} \left(\Delta P = C \frac{\rho V^2}{2g} \text{ for English units} \right)$$

$$(3.15)$$

Loss coefficients for typical obstructions in pipes and ducts are given in Fig. H3.6 in terms of the equivalent number of diameters of straight pipe.

Drag Coefficients

Drag coefficients have been widely used to evaluate the drag force on a body in a fluid stream. The drag coefficient C_D is defined in much the same way as the friction factor, that is,

$$\text{Drag force} = C_D A \frac{\rho V^2}{2} \qquad (3.16)$$

The area A for Eq. 3.16 is ordinarily taken as the maximum cross-sectional area of the body in a plane normal to the direction of flow. Figure H3.5 and Table H3.1 give drag coefficients for typical geometries.

A *skin-friction* or *surface-drag coefficient* is often used for flat plates parallel to the flow and for streamlined bodies, for which the only eddy losses are those stemming from skin friction. It is used in the same manner as the drag coefficient of Eq. 3.16 except that it is applied to the total wetted surface area. The surface-drag coefficient is also given in Table H3.1.

While the cases considered until this point have involved relatively simple geometries, more complex geometries such as finned or roughened tubes are also of interest. Flow conditions in the Reynolds number region between 1,000 and 10,000 are particularly sensitive to the effects of relatively small surface irregularities, and widely different friction factors and heat transfer coefficients may be experienced with geometries differing only in the degree of surface roughness. However, for Reynolds numbers well above 10,000, and particularly above 50,000, the effects of surface roughness on the friction factor are usually much less pronounced and do not change markedly with Reynolds number. These effects are discussed in more detail later in this chapter in connection with heat transfer coefficients.

Model Tests

The most important use of Reynolds number is associated with the application to new designs of test data obtained from existing units or from models. In general, if the Reynolds number and the geometry are the same, it is possible to predict with confidence the flow pattern and the pressure drop for seemingly very different conditions. For example, it is possible to apply data obtained with air or water to the design of a heat exchanger for a petroleum oil or even a molten salt. The effects of changes in either the fluid or the velocity have importance only insofar as they affect the Reynolds number. This assists greatly in design and development work, for it permits a design to proceed on the basis of preliminary flow tests, which can be conducted with small wood or plastic models using air or water. The results can be applied to the design of equipment employing fluids difficult to work with, such as liquid oxygen, molten metals, or highly poisonous or corrosive fluids.

Compressibility Effects

One major precaution must be taken in extending results obtained with a liquid to an application involving a gas or vice versa. While liquids are essentially incompressible, the density of gases changes rapidly with pressure. The changes in density can be neglected and gases treated as incompressible if the pressure changes brought about by the motion of the fluid are small compared with the absolute pressure, that is, if the velocity of the gas is not too great.

The compressibility effects in a gas are likely to be important at bulk free stream velocities somewhere between 20 and 50% of the velocity of sound. With most conventional geometries, changes in flow direction in bends or over obstructions usually give small regions in which the local velocities are from two to five times the average, and thus may approach or exceed the velocity of sound if the bulk free stream velocity exceeds 20% of the velocity of sound. If this occurs, compressibility effects in these local regions may induce large distortions in the flow pattern, and large increases in the pressure drop are likely to result. The ratio of the velocity of the gas to that of sound is known as the *Mach number*. Figure 3.12 shows the effects of air velocity on the pressure drop across two of the best of a series of streamlined elbows tested for an aircraft engine induction system. Even after much design and test work with different passage profiles it was not possible to avoid a sharp increase in the pressure drop when the nominal Mach number in the duct reached a value of about 0.35. Note that the onset of compressibility occurred at a higher Mach number for elbow B than for A, even though B gave a higher pressure drop at low Mach numbers. In general, the design of gas-flow passages should be such that bulk free stream velocities do not exceed a Mach number of approximately 0.2 unless careful analysis or testing is planned.

Effects of Density on Pressure Drop

Variations in density are often large enough to have an important effect on the pressure drop. It has been found

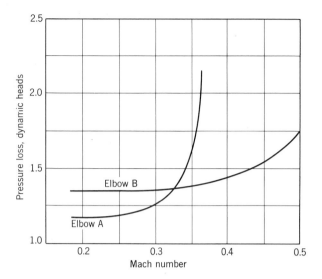

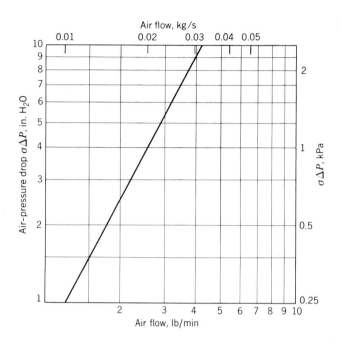

Figure 3.12 Pressure drops across two quite similar, carefully streamlined elbows showing differences in the onset of compressibility effects. (Fraas, SAE paper, unpublished.)

Figure 3.13 Use of $\sigma\Delta P$ as a parameter to relate heat exchanger pressure drop to the gas weight flow for a range of gas densities.

convenient in many heat exchanger design studies to relate the weight flow rate, the pressure drop, and σ—the ratio of the gas density to that of air at standard conditions—by

$$\text{Weight flow rate} \sim \sqrt{\sigma\Delta P}$$

If this is done, the pressure drop for any fluid density can be obtained from a curve such as that in Fig. 3.13. It may also be convenient to simplify calculations for some types of analysis by making use of charts in the form of Fig. 14.10. It should be noted that such a chart is good for any gas if a correction is applied for differences in the Reynolds number resulting from differences in the gas viscosity. Note also that charts for the dynamic head as a function of fluid velocity for a wide range of fluid densities are included in Figs. H3.2 and H3.3 of the Handbook.

HEAT TRANSFER BETWEEN FLUIDS AND SOLID SURFACES

Heat Transfer Coefficient

The rate of heat transfer between a solid surface and a fluid passing over it can be expressed in much the same way as for simple heat conduction through solids:

$$Q = hA\,\Delta T \tag{3.17}$$

where Δt is the temperature drop between the bulk fluid and the wall, A is the heat transfer surface area, and h is the *heat-transfer coefficient*. The last term is equivalent to the term k/L in the simple heat conduction equation.

Equation 3.17 is easy to use except for the evaluation of the heat transfer coefficient h, which depends on the fluid-flow conditions, the thermal properties of the fluid, and the passage size. The greater part of the balance of this chapter is concerned with the evaluation of the heat transfer coefficient.

It is often convenient in heat exchanger work to combine the two surface heat transfer coefficients (i.e., for the hot and cold fluid streams) and the thermal resistance of the tube wall to give a single parameter: the *overall heat transfer coefficient* U, which may be defined as

$$U = \frac{Q}{A\,\Delta T} \tag{3.18}$$

In establishing U for any given case it is convenient to refer first to Eqs. 3.3 and 3.18, from which it follows that

$$\frac{1}{U} = AR \tag{3.19}$$

It is apparent from Eq. 3.19 that, for a given value of R, the area A, on which U is based, should be specified when giving a value for U. Where the areas of the surfaces for the hot and cold fluids are not the same, it does not matter on which surface U is based, because the choice is arbitrary. For example, the overall heat transfer coefficient for the radial flow of heat across a circular tube based on the internal tube surface is

$$\frac{1}{U_i} = \frac{1}{h_i} + \frac{A_i B}{A_{mw} k_w} + \frac{A_i}{A_o h_o} \qquad (3.20)$$

where A_{mw} = mean wall area for heat transfer, m^2 (ft^2)
A_i, A_o = internal and external tube surface areas, m^2 (ft^2)
h_i, h_o = heat transfer coefficients inside and outside the tube, $W/m^2 \cdot °C$ (Btu/h · ft^2 · °F)
k_w = thermal conductivity of the tube wall, $W/m \cdot °C$ (Btu/h · ft · °F)
B = tube-wall thickness, m (ft)

Nusselt and Prandtl Numbers

The heat transfer coefficient was found to be related to two dimensionless parameters important in heat transfer work: the Nusselt and Prandtl numbers. The Nusselt number, Nu, is the term used for the ratio hD/k. This quantity is the ratio of the heat transfer coefficient to the thermal conductance implied by the quantity k/D. It can be seen intuitively, from the simple conduction equation, that the heat-flow rate to a fluid flowing through a passage should be proportional to the thermal conductivity divided by a representative distance in the direction of heat flow, for example, the passage diameter. The Prandtl number, Pr, is the term used for the ratio $c_p \mu / k$. This quantity is the ratio of the molecular diffusivity of momentum (as indicated by the viscosity) to the molecular diffusivity of heat (as indicated by the ratio of the thermal conductivity to the specific heat). As is the case with the Reynolds number, a wealth of experimental and analyical work shows the value of the Nusselt and Prandtl numbers as heat transfer parameters.

Heat Transfer to a Fluid in Laminar Flow

Heat transfer from a surface to a fluid under laminar flow conditions takes place by simple conduction, and thus the heat transfer rate depends on the radial temper-

ature gradient near the heated wall. This temperature gradient depends not only on the velocity distribution and the thermal conductivity of the fluid but also on the extent to which the fluid has been heated in traversing the passage up to the point in question. Analytical expressions have been prepared for such basic geometries as circular and rectangular passages, but usually these cannot be solved explicitly for the heat transfer coefficient. They can be evaluated numerically with a computing machine. The heat transfer coefficients that result depend on the wall temperature distribution assumed. Typical cases include a constant wall temperature, a uniform temperature difference between the wall and the bulk fluid (i.e., uniform heat flux), or a wall temperature varying linearly in the direction of flow.

A classical solution of the problem for laminar flow through circular tubes was given by Graetz assuming a fully developed parabolic velocity distribution and a uniform wall temperature. Graetz' solution was extended to the cases of constant wall heat flux and linearly varying wall temperature.[20] The Nusselt number has been found to depend on the quantity x/D (1/Re · Pr) for circular tubes, and Fig. 3.14 shows this relation for the three sets of boundary conditions. Note that the Nusselt number is constant for values of x/D (1/Re · Pr) greater than about 0.2, so that in this range the heat transfer coefficient depends only on the thermal conductivity and the passage diameter. The values of the limiting Nusselt number are summarized in Table H5.1 for different passage shapes. The heat transfer coefficients can be found easily by using the graph given in Fig. H5.1. These curves were derived from calculations for a fluid whose viscosity can be considered to be independent of temperature.

If the fluid viscosity varies substantially over the temperature range from the wall to the center of the fluid stream, the velocity distribution is altered as indicated in Fig. 3.15. In practice it has been found that this may lead to an increase in the heat transfer coefficient of as much as 40% where a liquid is used to cool a hot surface or where a gas is used to heat a cold surface. Similarly, it has been found that the heat transfer coefficient may be reduced as much as 20 or 30% if a cold surface is heated by a liquid or a hot surface is cooled by a gas. This comes about because the viscosity of a gas increases with temperature, whereas that of a liquid falls off with an increase in temperature.

Analysis of much experimental data indicates that good correlation of experimental data is obtained if the Nusselt number is multiplied by the 0.14 power of the ratio of the viscosity at the wall to that for the bulk free stream.[21] Thus in the region where the temperature

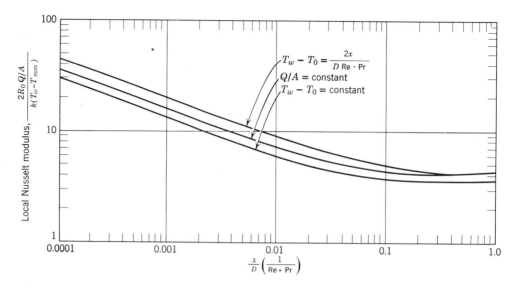

Figure 3.14 Heat transfer relations for laminar flow through a round tube for three different boundary conditions, that is, with the wall temperature varying linearly in the direction of flow, a constant heat flux, and a constant wall temperature. T_{mm} is the local mixed mean temperature. (Sellars, Tribus, and Klein, Ref. 20.)

distribution across the fluid stream does not change with axial position, the Nusselt number can be expressed as

$$\text{Nu} = (\text{Nu})_{\text{id}}\left(\frac{\mu_b}{\mu_w}\right)^{0.14} \qquad (3.21)$$

Note that the values given in Fig. H5.1 and Table H5.1 are for the ideal Nusselt number. To facilitate calculations of the effects of variations in viscosity between the wall and the bulk fluid, the viscosity correction factor $(\mu_b/\mu_w)^{0.14}$ has been plotted against the viscosity ratio in Fig. H5.2.

For short tube lengths or in the entrance region, the velocity distribution differs substantially from a parabola; this difference may lead to an increase in the local heat transfer coefficient over that given by the Graetz solution by as much as 50% for a passage $L/D = 10$.[22]

Example 3.1. Determine the local heat transfer coefficient under constant heat flux conditions 0.915 m (3 ft) from the inlet of a 12.7-mm (0.5-in.) ID circular tube through which water is flowing at the rate of 0.0305 m/s (0.1 ft/s) at 38°C (100°F) while the local tube wall temperature is 15°C (60°F).

Solution. The value of the Reynolds number as determined at the bulk stream temperature is

$$\text{Re} = \frac{\rho VD}{\mu}$$

$$= \frac{983 \times 0.0305 \times 0.0127}{6.86 \times 10^{-4}} = 555$$

or, for English units,

$$\text{Re} = \frac{\rho VD}{\mu} = \frac{61.4 \times (0.1 \times 3600) \times 0.5}{1.66 \times 12} = 555$$

Thus the flow is laminar, and data given in Fig. H5.1 can be used to determine the local heat transfer coefficient. The values of the parameters in this figure are

$$\frac{x}{D}\frac{1}{\rho V c_p} = \frac{0.915}{0.0127}\frac{1}{983 \times 0.0305 \times 4.187} = 0.575$$

$$\frac{k}{D} = \frac{0.619}{0.0127} = 48.8 \text{ W/m}^2 \cdot {}^{\circ}\text{C}$$

or, for English units,

$$\frac{x}{d}\frac{1}{G'c_p} = \frac{3 \times 12}{0.5}\frac{1}{61.4 \times 0.1 \times 1.0} = 11.7$$

$$\frac{k}{d} = \frac{0.358}{0.5} = 0.716 \text{ Btu/h} \cdot \text{ft} \cdot {}^{\circ}\text{F} \cdot \text{in.}$$

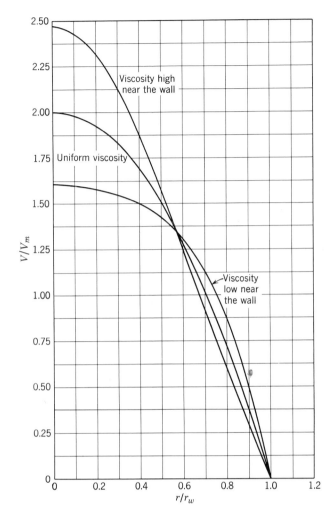

Figure 3.15 Effects of heating and cooling on the velocity distribution under laminar flow conditions. (Jakob, Ref. 3.)

Hence, from Fig. H5.1, the ideal heat transfer coefficient h_{id}, with no allowances for variations in viscosity, is h_{id} = 238 W/m² · °C (42 Btu/h · ft² · °F). The heat transfer coefficient with allowances for variations in viscosity is given by

$$ h = h_{id}\left(\frac{\mu_b}{\mu_w}\right)^{0.14} = 238\left(\frac{6.86}{11.17}\right)^{0.14} $$
$$ = 216 \text{ W/m}^2 \cdot {}^{\circ}\text{C} $$
$$ = 42\left(\frac{1.66}{2.71}\right)^{0.14} $$
$$ = 38.2 \text{ Btu/h} \cdot \text{ft}^2 \cdot {}^{\circ}\text{F} $$

Note that these calculations could have been simplified by using Figs. H3.1 and H5.2.

Heat Transfer to a Fluid in Turbulent Flow

In turbulent flow the bulk of the heat transferred between the center of the fluid stream and the wall is carried by eddies in which the average transverse velocity component is substantially smaller than, but roughly proportional to, the axial velocity. The effectiveness of turbulence in transferring heat through the boundary layer depends on the physical properties of the fluid, including the thermal conductivity, the specific heat, and the viscosity. It has been shown both analytically and experimentally that where liquids are heated under turbulent flow conditions in long, straight, smooth, circular passages, the relationship between the heat transfer coefficient, the fluid properties, and the flow conditions can be approximated quite well as follows:

$$ \text{Nu} = \frac{hD}{k} = 0.023\left(\frac{c_p\mu}{k}\right)^{0.4}\left(\frac{DG}{\mu}\right)^{0.8} \quad (3.22) $$

where the fluid properties are evaluated at the bulk free stream temperature. It is often convenient to employ an explicit expression for the heat transfer coefficient. This expression can be obtained by combining the terms of Eq. 3.22 to give

$$ h = 0.023\left(\frac{c_p^{0.4}k^{0.6}}{\mu^{0.4}}\right)\left(\frac{G^{0.8}}{D^{0.2}}\right) \quad (3.23) $$

Note that the major parameters of Eq. 3.22 are arranged in three groups, each dimensionless (i.e., the Nusselt number hD/k, the Prandtl number $c_p\mu/k$, and the Reynolds number DG/μ). Equation 3.22 indicates that the heat transfer coefficient increases somewhat less than linearly with Reynolds number. This occurs because the transverse mixing velocities associated with turbulence increase somewhat less than linearly with the axial velocity. Since the interchange of heat through the boundary layer depends on the same turbulent mixing process as the interchange of momentum that determines the friction factor, and since the friction factor varies inversely as the 0.2 power of the Reynolds number, it can be deduced that the heat transfer coefficient should increase as the 0.8 power of the Reynolds number.[23]

The effects of Prandtl number are more subtle. While the specific heat and the thermal conductivity of a fluid ordinarily do not change much with temperature, the viscosity does, particularly for liquids. Variations in viscosity through the boundary layer change the velocity distribution, as indicated qualitatively in Fig. 3.15. Since the viscosity of liquids ordinarily drops with increasing temperature, if a liquid is being heated the boundary layer tends to be thinner than would be the case for isothermal flow, and hence the heat transfer coefficient is increased. The reverse is true if the liquid is being cooled. An allowance for these effects is often made by changing the exponent of the Prandtl number in Eq. 3.22 from 0.40 to 0.30 for the cooling of liquids.

The viscosity of a gas ordinarily increases with temperature so that the effects of variations in the bounary-layer thickness are opposite to those observed for liquids. Fortunately, the Prandtl number for gases is close to unity, and for most applications the effects of temperature variations through the boundary layer are small—on the order of a few percent. However, where temperature differences of the order of 555°C (1000°F) or more exist—as in some aircraft, rocket, and nuclear reactor applications—variations in the physical properties through the boundary layer may cause the heat transfer coefficient to differ from that of Eq. 3.22 by 30% or more. Experiments both with air and helium at the NASA Lewis Laboratory have shown that the data are correlated nicely by evaluating the physical properties of the fluid at the arithmetic mean temperature between the wall and the bulk free stream.[24,25] This relation includes not only the thermal conductivity and viscosity in the Prandtl number and the thermal conductivity in the Nusselt number, but also the viscosity and density in the Reynolds number, so that Eq. 3.22 becomes

$$\frac{hD}{k_f} = 0.023 \left(\frac{c_{pf}\mu_f}{k_f} \right)^{0.4} \left(\frac{\rho_f VD}{\mu_f} \right)^{0.8} \quad (3.24)$$

where the subscript f denotes *film*, that is, fluid properties at a temperature midway between the temperatures of the wall and the bulk free stream.

Colburn Modulus and Stanton Number

The Colburn modulus is widely used in heat transfer analysis work for applications in which the Reynolds number may vary from 100 to 10,000, that is, through the laminar, transition, and turbulent regions. Colburn[26] proposed that heat transfer data be correlated by plotting the dimensionless group

$$\left(\frac{h}{Gc_p} \right) \left(\frac{c_p\mu}{k} \right)^{2/3} = J \quad (3.25)$$

(called the *Colburn modulus*) against the Reynolds number. The dimensionless ratio h/Gc_p in the Colburn

modulus is called the *Stanton number* and is also often used in heat transfer work.

Equation 3.22 for turbulent flow heat transfer in circular passages can be expressed in terms of the Colburn modulus by rearranging the terms and using the Prandtl number to the 0.33 power instead of the 0.4 power:

$$J = 0.023 \left(\frac{DG}{\mu} \right)^{-0.2} \quad (3.26)$$

The Colburn modulus has been applied to the correlation of data for heat transfer to and from oils in shell-and-tube heat exchangers where the viscosity has varied by a factor of 5 between the wall and the bulk free stream. Good correlation has been obtained in the Reynolds-number range from 100 to 10,000 by using the viscosity correction term given in Eq. 3.21 for laminar flow. A modified Colburn modulus J' is often used for such cases, that is,

$$J' = \left(\frac{h}{Gc_p} \right) \left(\frac{c_p\mu}{k} \right)^{2/3} \left(\frac{\mu_w}{\mu_b} \right)^{0.14} \quad (3.27)$$

Figure 3.16 shows the modified Colburn modulus plotted as a function of Reynolds number for fluids flowing inside tubes for a series of passage length-diameter ratios.

Heat Transfer Coefficient Charts

To facilitate the routine calculation of heat transfer coefficients under turbulent flow conditions, charts are given in Figs. H5.3 and H5.5 for water and air, respectively, at 93°C (200°F) flowing through long, straight passages. The correction factors in Figs. H5.4 and H5.6 can be applied to these values to give the heat transfer coefficients for other temperatures, and for other liquids and gases, while the correction factor in Fig. H5.9 can be employed to allow for entrance effects as discussed in the following section.

Effects of Passage Shape

The heat transfer correlations given in Eqs. 3.22, 3.23, and 3.24 are for circular passages. Fortunately, these relations also hold quite well for other long, straight passages having rectangular, triangular, or other cross-sectional shapes, if the equivalent diameter of the passage is employed in calculating the Nusselt and Reynolds numbers.[27-29] Where the passage shape may induce flow

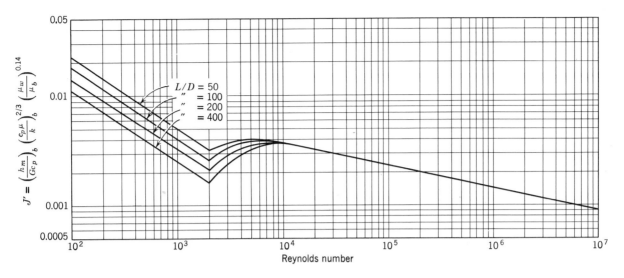

Figure 3.16 Modified Colburn modulus as a function of Reynolds number for a series of passage length-diameter ratios. (J. C. Knudsen and D. L. Katz, *Fluid Dynamics and Heat Transfer*, McGraw-Hill Book Co., 1958.)

separation, as in flow across tube banks or over interrupted or corrugated fins, it has been found that, as with flow patterns and pressure-drop relations, the effects on the heat transfer coefficient of variations in passage geometry cannot be predicted analytically but must be based on experimental data obtained from models or full-scale units.

Entrance Effects

There are pronounced entrance-region effects where the fluid enters a long tube or passes over an obstruction.[30] As might be surmised from the thin boundary layer near the inlet region of Fig. 3.7, the heat transfer coefficient is very high in entrance regions. Thickening of the boundary layer with distance from the entrance leads to a reduction in the heat transfer coefficient. Figure H5.9 shows the effect of this factor on the average heat transfer coefficient plotted as a function of the length-diameter ratio for round tubes.

Effects of Interrupted Surfaces

The high heat transfer coefficient in the entrance region just cited can be exploited by interrupting a surface at intervals in the direction of the fluid flow. It is evident from Fig. H5.9 that the surface should be interrupted at intervals of 5 to 30 passage diameters if a substantial increase in the average heat transfer coefficient is to be obtained. The effects of interrupting the surfaces in the direction of flow depend to some degree on the extent to which the surfaces are aligned or staggered.[31] Figure 3.17 shows some experimental data for a typical series of tests. Note the pronounced increase in the heat transfer coefficient over that for continuous plates. These data indicate the advantages of interrupted fin matrices such as those of Fig. 2.13.

Effects of Surface Roughness

Some indications of the effects of surface irregularities on the heat transfer coefficient are indicated in Fig. 3.18. These data were taken from a series of tests in which spiral wire springs were inserted into long straight tubes and allowed to expand radially so that they rested against the wall. The turbulence induced by these springs increased the heat transfer coefficient at the expense of a greater pressure drop, which in this instance was a higher percent than the increase in the heat transfer coefficient. If pumping power is the controlling consideration in the design of a heat exchanger, the data indicate that it is undesirable to introduce the spiral springs. If, on the other hand, pumping power is a relatively small factor in the overall cost, the amount of heat transfer surface area required can be reduced drastically, and a smaller, lighter, and less expensive unit can be built by incorporating turbulence inducers to improve the heat transfer coefficient.

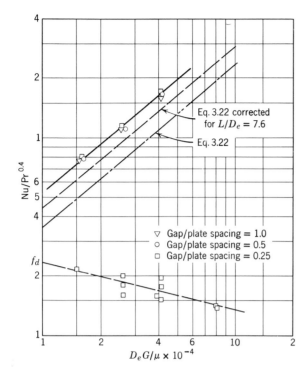

Figure 3.17 Heat transfer performance of stacks of uniformly spaced flat plates for several gaps between stacks. (Sams and Weiland, Ref. 31.)

Local Heat Transfer Coefficients for Single Cylinders in Crossflow

The local heat transfer coefficient varies with angular position around a cylinder in crossflow, as shown in Fig. 3.19. The decrease in the heat transfer coefficient with angular position away from the stagnation point at the front of the tube stems from thickening of the boundary layer and the increase in its temperature, the effect being analogous to the entrance effect discussed in the preceding sections. Beyond the point at which flow separation occurs (see Fig. 3.9), the local heat transfer coefficient increases sharply because of the good mixing produced by the turbulence in the separated flow region.

Flow Across Tube Banks

The most common type of heat exchanger consists of tube banks with one fluid passing through the tubes and the other across the tubes. The more common tube arrays may be classified as either *in-line* or *staggered* with spacing variations as indicated in Fig. H5.10. The in-line tubes tend to give a somewhat lower pressure drop and poorer heat transfer because the flow tends to be channeled into high-velocity regions in the center of the lanes between the tube rows (see Fig. 14.2). The staggered tubes, on the other hand, produce good mixing of the flow over the tube banks, but give a higher pressure drop.

Local Heat Transfer Coefficients in Tube Banks

Figure 3.20 shows the effect of the position of the tube in the tube bank on the local heat transfer coefficient.[32] Note that the turbulence induced by the front bank leads to a higher heat transfer coefficient for the second and subsequent banks than for the first. Thus the pumping work dissipated in turbulence by a tube bank is of value in improving the heat transfer over subsequent banks.

Both the heat transfer coefficient and the friction factor depend on the tube spacing. So many combinations of the tube spacing, both transverse and parallel to the direction of flow, are possible that a large number of curves is required.[33,34] Two such sets of curves are given in Figs. H5.11 and H5.12. In these curves the friction factor, or pressure-loss coefficient, is the fraction of a dynamic head lost per bank of tubes, with the dynamic head based on the nominal fluid velocity through the minimum flow-passage area between the tubes. Note that where the transverse spacing is large and the axial spacing is small the heat transfer coefficient and friction factor approach the values for a smooth passage. Where the transverse spacing is small (i.e., where the gap between the tubes is ¼ or ½ of a tube diameter) the friction factor approaches unity (i.e., the pressure loss per bank approaches that for flow through a flow nozzle with no diffuser to recover the dynamic pressure). The resulting turbulence, of course, greatly increases the heat transfer coefficient and the pumping power required.

The friction factor is sometimes computed, as for other heat transfer matrices, on the basis of either the mean hydraulic radius or the equivalent diameter. When this is done, the hydraulic radius is taken as the flow passage volume divided by the total surface area, and the diameters used in both the Reynolds number and the heat transfer parameter (Nusselt, Stanton, or Colburn) are the equivalent diameter, that is, $4R_h$. The curves for finned tubes given in Fig. H5.13, for example, are based on the equivalent diameter computed in this way.

Performance of Finned Surfaces

Finned surfaces are often employed where the heat transfer coefficient readily attainable with one fluid stream is much higher than that readily attained with the

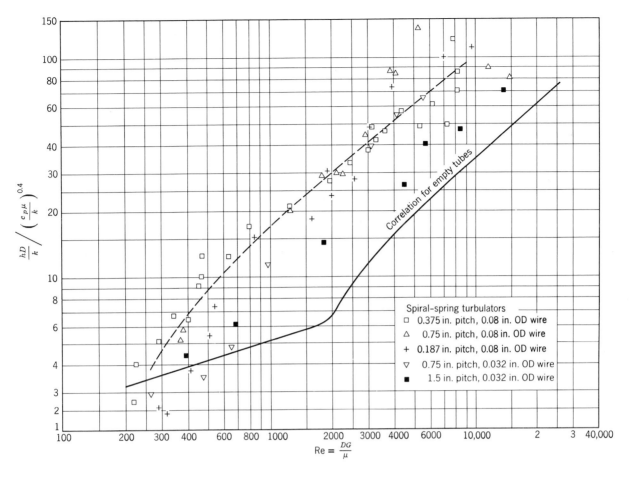

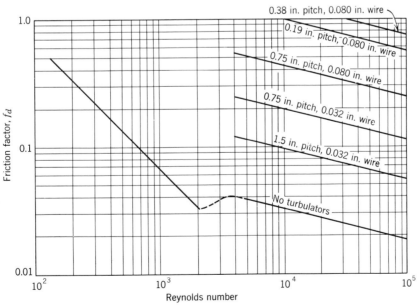

Figure 3.18 Effects of wire spring turbulators expanded inside a 0.562 in. ID tube for several wire diameters and coil pitches. (L. G. Siegel and G. L. Tuve, Case Institute of Technology, September 1, 1946.)

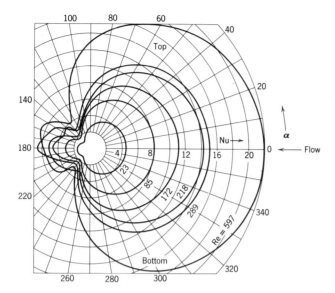

Figure 3.19 Local heat transfer coefficient as a function of angular position around the circumference of a cylinder under crossflow conditions (α = angular position from the stagnation point). (E. R. G. Eckert and E. Soehngen, "Distribution of Heat Transfer Coefficients Around Circular Cylinder," *Trans. ASME*, vol. 74, 1952, p. 346.)

other, for example, in gas-to-liquid heat exchangers. The use of finned surfaces is particularly advantageous when applied to tube banks in crossflow. The principal pressure losses arise from turbulence set up by the rapid changes in flow-passage cross-sectional area in the direction of flow. The presence of circular or helical fins does not have much effect on the turbulence patterns and hence on the pressure loss based on the dynamic head for the minimum flow-passage area between tubes. Thus the wetted area on the outside of the tubes can be increased 5 or 10 times by adding fins; yet the pressure loss will increase by only a factor of 2 or 3, while the average heat transfer coefficient will be reduced by a factor of only about 2. Curves for several typical finned tube-bank configurations are included in Fig. H5.13.

Axially finned tubes with fluid flowing between the fins parallel to the tube axis present special problems. If the tips of the fins on adjacent tubes are separated by a distance greater than the spacing between the fins, the flow tends to "channel," that is, much of it will go through the large hydraulic radius passages outside the fin envelope for each tube. There is relatively little mixing between the fluid stream in the large hydraulic radius channels between the finned tubes and the small hydraulic radius channels between the fins. This causes a loss in heat transfer performance, since the flow

filaments between the fins operate at a temperature considerably above the average stream temperature; hence the effective temperature difference between the finned surface and the immediately adjacent fluid stream is reduced. If the tubes are mounted very closely together so that the hydraulic radii of the various passages are fairly uniform, the velocity distribution also will be reasonably uniform, and the pressure drop and the heat transfer coefficient can be estimated quite well by determining the equivalent passage diameter from the wetted perimeter and the flow-passage area.

Axial flow over round tubes with circumferential circular fins has not ordinarily appeared attractive. Under special circumstances, however, this type of fin configuration may prove advantageous; for example, it was chosen for the fuel elements of the British gas-cooled reactors at Calder Hall. In this instance the designers were trying to extract as much heat as possible from a fuel rod of a given diameter in a channel 20 ft long. Closely spaced axial fins had a disadvantage in that flow channeling, as just discussed, would have limited the heat transfer performance to a lower value than that desired. This problem was avoided by using circumferential fins on the fuel rods to induce the turbulent flow pattern indicated in Fig. 3.21. The high degree of mixing obtained in this manner increased pumping power requirements substantially, but this was justifiable in terms of the overall economics of the system. While the situation is complex, it appears that steep, helically spiraled fins give better performance for such applications than either simple axial fins or circumferential fins. Much work on the heat transfer and pressure-drop characteristics of spiral fins of this character has been conducted.

Heat Transfer to Liquid Metals

The relationships governing the heat transfer to liquid metals are somewhat different from those discussed in the preceding section for the more conventional fluids. This difference comes about because the thermal conductivity of liquid metals is roughly two orders of magnitude higher than that of water or hydrocarbon liquids, and about three orders of magnitude higher than that of gases. Under turbulent flow conditions the rate-controlling mechanism in the transfer of heat from a solid surface to a liquid metal is not almost solely dependent on the lateral transport of fluid particles by turbulence but is primarily one of simple conduction. Since the relative importance of these two heat transfer mechanisms is indicated by the Prandtl number, Fig. 3.22 was prepared to show the temperature distribution across a

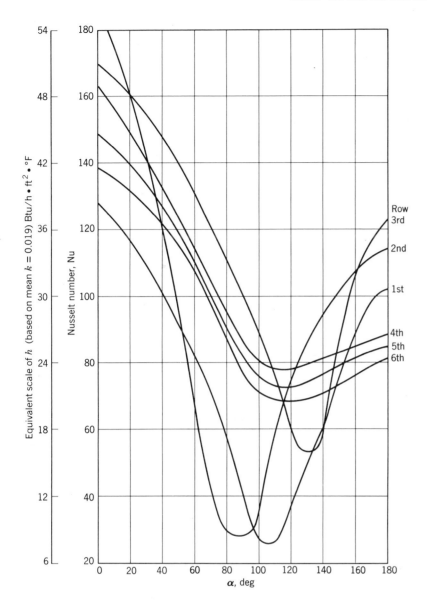

Figure 3.20 Effects of tube position on the local heat transfer coefficient in banks of staggered tubes under crossflow conditions with air at 149°C (300°F) flowing over 19-mm (0.75-in.) OD tubes. The tube pitch is 2 diameters; the Reynolds number is 18,300. (Thompson et al., Ref. 32.)

fluid stream under turbulent flow conditions for a series of Prandtl numbers.[35] Note that the Prandtl number for liquid metals runs from 0.01 to 0.05, whereas it runs from 1 to 10 for water and most commercial liquids, and 0.5 to 1.0 for gases.

For flow through round tubes, the heat transfer coefficient for liquid metals has been correlated with the principle parameters that affect it by the following equation:

$$h = \frac{k}{D}[7 + 0.025(\text{RePr})^{0.8}] \qquad (3.28)$$

for either SI or English units. A more convenient form in some cases is given by

$$h = \frac{k}{D}\left[7 + 0.032\left(\frac{W}{D}\right)^{0.8}\left(\frac{c_p}{k}\right)^{0.8}\right]$$

where W is the flow rate in kilograms per second through a circular passage. For English units, W is the flow rate through the circular passage in pounds per second. Extensive analytical and test work show that these relations hold quite well where $\text{RePr} > 100$ and $L/D > 60$.[36] For convenience, a chart prepared from this equation is presented in Fig. H5.15. This chart will be found to be very helpful when working with sodium, potassium, NaK, cesium, or lithium. Values for the physical properties of these fluids are given in Table H2.1.

Figure 3.21 Turbulent flow pattern for flow parallel to a tube with circumferential fins. (P. Fortescue and W. B. Hall, "Heat Transfer Experiments on Fuel Elements," *Journal of the British Nuclear Energy Conference*, vol. 2, 1957, p. 83.)

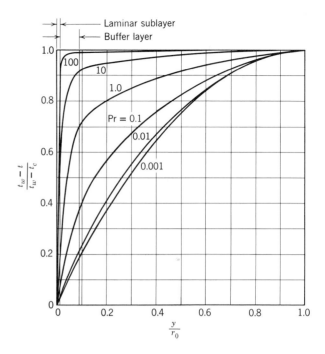

Figure 3.22 Fully developed temperature distribution in a circular tube for heat transfer under turbulent flow conditions for a Reynolds number of 10,000 and a series of Prandtl numbers. (Martinelli, Ref. 35.)

Equation 3.28 holds only if the liquid metal wets the surface so that there is no surface film to act as a thermal barrier. There is ordinarily no problem with the alkali metals since they usually wet structural metals and alloys very strongly, but lead, bismuth, and mercury have given difficulty with failure to "wet" low-alloy and stainless steel surfaces. If this occurs, the heat transfer coefficients may be reduced by as much as a factor of 10. To minimize this effect with mercury, a small amount of magnesium is often added to improve the wetting. Adding too much magnesium may give difficulties with corrosion and mass transfer.

Heat Transfer by Thermal Convection

Thermal convection is an important factor in the design of some types of heat exchangers, for example, the evaporator coils in refrigerators and cold storage vaults, heating and air-conditioning systems, and small steam boilers. It is also important at low loads, under shutdown conditions, and as it affects the heat losses from most heat exchangers at all loads.

A system may be designed deliberately to induce as much thermal convection as possible, or unforeseen conditions may lead to important effects arising from thermal convection in one or both of the fluids involved. A virtually ideal example of such a system is given by the coolant circuit for a gas-cooled reactor. Figure 3.23 shows a schematic diagram for a system of this sort. It is designed to induce as large a natural thermal convection circulation rate as possible by locating the heat source near the bottom of the system, and the heat sink near the top. While blowers are required for high-power operation, in the event of a blower failure the flow rate given by natural thermal convection would be sufficient to assure removal of the fission-product decay heat generated in the reactor, and thus a disastrous meltdown of the reactor would be prevented.

The circulation rate for natural thermal convection can be calculated in much the same way as for forced convection. The driving force is the difference in density between the fluid in the riser and that in the descending leg if a closed loop is employed; or the difference in

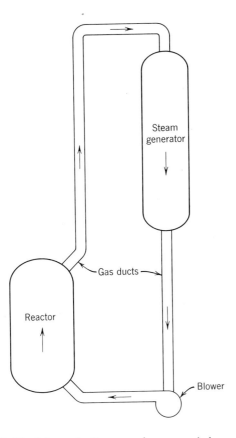

Figure 3.23 Schematic diagram of a gas-cooled reactor and steam generator system designed to give thermal convection for removal of the fission-product decay heat if the blower stops.

density between the fluid in the stack and the surrounding atmosphere is the driving force if an open system with a vertical chimney is employed. It is easy to see that the highest circulation rate is obtained with the heater at the bottom of a hot stack, and the cooler at the top of a descending cold duct. Since the flow may fall in either the laminar or the transition region, care should be exercised in estimating the friction factor and the heat transfer coefficient for each component of the system.

A variety of cases arise in which the fluid flow over a heated or a cooled surface is not constrained by ducts or channel walls to follow any particular flow path. One example of such a condition is given by heat transfer between two horizontal parallel surfaces in which the lower surface is heated. The fluid between them tends to circulate because warm fluid from the lower plate rises, is cooled by the upper plate, and then drops back to the lower plate. The flow pattern tends to take the form of hexagonal cells with upflow in the center of each cell and downflow in the boundary region between cells.

Natural convection heat transfer for partially enclosed spaces between vertical, parallel plates has been investigated by Siegel and Norris,[37] and, for liquids confined between two horizontal plates heated from below, by Globe and Dropkin.[38] A review of natural convection in horizontal fluid layers heated from below is given by Ostrach.[39] This and similar analytical and experimental work has shown that the Grashof number is a significant parameter in relating the heat transfer rate to the dimensions of the system and the fluid properties. This parameter is defined as

$$Gr = \frac{L^3 \rho^2 \Delta t \beta g}{\mu^2} \qquad (3.29)$$

where L is the characteristic dimension and β is the volumetric coefficient of expansion per degree Celsius. For English units, Eq. 3.29 becomes

$$Gr = \frac{L^3 \rho^2 \Delta t \beta g 3600^2}{\mu^2}$$

where L is the characteristic dimension of the system in feet and β the volumetric coefficient of expansion of the fluid per degree Fahrenheit.

Most experimental data for thermal convection heat transfer have been correlated by plotting the Nusselt number as a function of the product of the Grashof and Prandtl numbers, that is, for thermal convection the Grashof number plays much the same role as the Reynolds number plays in forced convection.

For fluids having Prandtl numbers near unity, thermal convection heat transfer data for vertical and horizontal surfaces can be correlated reasonably well by[1,3]

$$Nu = c(GrPr)^n$$
$$\frac{hL}{k} = c\left[\left(\frac{L^3 \rho^2 \beta \Delta t g}{\mu^2}\right)\left(\frac{c_p \mu}{k}\right)\right]^n \qquad (3.30)$$

For English units where μ is in pounds per hour · feet, the factor 3600^2 must be added as in Eq. 3.29. Note that L, a characteristic length, is used in the Nusselt number in place of the passage equivalent diameter. The values of the constant c and the exponent n are given in Table H5.2 for various geometries for both the laminar and turbulent regions. Note that the value of the exponent n is $1/4$ in the laminar region and $1/3$ in the turbulent region. Therefore, the heat transfer coefficient is independent of the characteristic dimension in the turbulent region, that is, $(1/L)(L^3)^{1/3} = 1$, whereas in the laminar region it is inversely proportional to the $1/4$ power of the characteristic dimension. For gases, the Prandtl number is close to unity; hence the Nusselt number becomes a function of the Grashof number only.

Simplified relations for the thermal convection heat transfer coefficient for air for vertical and horizontal surfaces are given in Table H5.3. Note that the characteristic dimension is taken as the height for vertical surfaces, and the mean value of the length of the sides for horizontal rectangular flat plates. The term Δt in Table H5.3 is the temperature difference between the surface and the bulk fluid temperature. The fluid temperatures are evaluated at the arithmetic mean temperature of the surface and the bulk fluid. For a heat flux from one surface to another across an air space, the Δt is the temperature difference between the two surfaces. The coefficient of expansion β for gases is taken as equal to $1/T_g$, where T_g is the absolute temperature of the gas.

Natural thermal convection heat transfer data for liquid metals (Na, Nak, Pb, Pb-Bi, and Hg) have been correlated in the laminar flow range by

$$\text{Nu} = 0.53\left(\frac{\text{Pr}}{0.952 + \text{Pr}}\right)^{1/4} (\text{GrPr})^{1/4} \quad (3.31)$$

where the physical properties are evaluated at a temperature midway between the wall and bulk fluid temperatures.[40]

Example 3.2. Consider a helium-cooled reactor and steam generator system which must be designed so that if there is a failure in the power supply for the helium-circulating blowers, the heat from fission-product decay can be removed from the reactor by thermal convection. As a first approximation consider the system of Fig. 3.23 with a reactor core equivalent to a matrix of 25-mm (1-in.) ID vertical passages, 6.09 m (20 ft) long. Helium at 2.07 MPa (300 psia) flows upward through the core and through a duct to the top of a steam generator. It flows downward through a tube matrix having an equivalent passage diameter on the gas side of 12.7 mm (0.5 in.) with a length of 24.4 m (80 ft), where it gives up the heat removed from the fuel, and then returns to the bottom of the core. At full power the gas velocity is 60.9 m/s (200 ft/s) through the core and 9.1 m/s (30 ft/s) through the steam generator, while the reactor core inlet and outlet gas temperatures are 316 and 649°C (600 and 1200°F), respectively.

Determine the difference in elevation required between the center of the reactor and the center of the steam generator to remove the afterheat (which is roughly equivalent to 2% of the full power output of the reactor) while maintaining the same reactor inlet and outlet gas temperatures. Neglect losses in the ducts, blowers, and phenum chambers.

Solution. To remove 2% of full power with the same gas temperatures, the gas velocity should be reduced to 2%, that is, 1.22 m/s (4 ft/s) in the reactor and 1.83 m/s (0.6 ft/s) in the steam generator. The pressure drops through the core and steam generator can be estimated by using the mean gas temperatures as shown in Table 3.1.

TABLE 3.1   Estimate of the Pressure Drop in a Circuit for a Gas-Cooled Reactor Coupled to a Steam Generator

Condition	Reactor Core	Steam Generator
Gas velocity, m/s (ft/s)	1.22 (4)	0.183 (0.6)
Passage length, m (ft)	6.1 (20)	24.38 (80)
Mean gas temperature, °C (°F)	482 (900)	482 (900)
Gas density, kg/m³ (lb/ft³)	1.33 (0.083)	1.33 (0.083)
Gas viscosity, Pa·s (lb/s·ft)	38×10^{-6} (0.093)	38×10^{-6} (0.093)
Passage equivalent diameter, mm (in.)	25.4 (1)	12.7 (0.5)
$\text{Re} = \dfrac{\rho VD}{\mu}$	1,065 (laminar flow)	80 (laminar flow)
$f = \dfrac{64}{\text{Re}}$	0.06	0.8
Passage $\dfrac{L}{D}$	240	1,920
Dynamic head $\dfrac{\rho V^2}{2}$, Pa $\left(\dfrac{\rho V^2}{2g}, \text{lb/ft}^2\right)$	0.96 (0.0201)	0.0222 (0.000465)
$\Delta P = f_d \left(\dfrac{\rho V^2}{2}\right)\dfrac{L}{D}$, Pa $\left(f_d \dfrac{\rho V^2}{2g}\dfrac{L}{D}, \text{lb/ft}^2\right)$	14.2 (0.296)	34.1 (0.714)

The average difference in density between the hot gas in the riser and the cooler gas in the return passages is $1.698 - 1.083 = 0.615$ kg/m³ ($0.1060 - 0.0676 = 0.0384$ lb/ft³).

The combined pressure drop through the core and the steam generator is $14.2 + 34.1 = 48.3$ Pa ($0.296 + 0.714 = 1.01$ lb/ft²), and this should be equal to the difference in the density multiplied by the difference in elevation between midplanes of the core and the steam generator. Thus the difference in elevation required is

$$\frac{48.3}{0.615} = 7.85 \text{ m, or } \frac{1.01}{0.0384} = 26.3 \text{ ft}$$

Allowances for duct losses would probably make it necessary to increase this by about 50%.

HEAT TRANSFER TO BOILING LIQUIDS

Boiling liquids pose a special set of heat transfer problems. For example, so much fine-grained turbulence is commonly induced close to the hot surface by vapor bubble formation that even in a static pool the heat transfer coefficient is likely to be very high. Both the phenomena involved and the relationships between the principal parameters are so complex that they are covered in a separate chapter (see Chapt. 5).

CONDENSING VAPORS

Condensers constitute an important and widely used type of heat exchanger with unique characteristics. The heat transfer mechanism in a condenser can be visualized by considering the behavior of a molecule of vapor as it strikes a liquid surface that is at a temperature slightly below the boiling point. Such a molecule of vapor loses much of its energy to the molecule of liquid that it strikes, so that it probably does not have sufficient energy to escape from the liquid surface. If the liquid surface can be kept a bit below the boiling point by agitation, extremely high heat transfer rates can be obtained. Steam jets directed into pools of somewhat subcooled water have given fluxes in excess of 3,154 kW/m² (1,000,000 Btu/h · ft²).

Heat Transfer Coefficients

The principal barrier to heat transfer on the vapor side of a condenser is usually the film of liquid covering the heat transfer surface, since the film temperature at the liquid-vapor interface is essentially equal to the condensation temperature at the prevailing pressure. A major problem in condenser design is to keep the liquid flowing off the surface so that the film thickness, and hence the resistance to heat flow, will be minimized. In any given situation the liquid film thickness depends on the surface geometry, the liquid viscosity and density, and the rate at which the condensate flows off the condenser surfaces. The gross flow rate depends on the heat flux and the latent heat of condensation of the vapor. For vertical tubes from which the condensate drains by laminar flow, it is possible to derive an expression for the mean effective heat transfer coefficient from basic heat transfer and fluid-flow relations. This expression with the analytically determined constant for saturated steam for English units is

$$h_m = 0.943 \left(\frac{k^3 \rho^2 g \Delta H_v}{L \mu \Delta t}\right)^{1/4}$$
$$= 1.47 \left(\frac{\pi D_0 k^3 \rho^2 g}{4 W \mu}\right)^{1/3} \quad (3.32)$$

for $4W/\pi D_0 \mu < 2000$. The corresponding relation for the condensation of steam on the outer surfaces of tubes in horizontal banks is

$$h_m = 0.725 \left(\frac{k^3 \rho^2 g \Delta H_v}{N D_o \mu \Delta t}\right)^{1/4} = 0.95 \left(\frac{L k^3 \rho^2 g}{W \mu}\right)^{1/3}$$
$$(3.33)$$

for $2W/L\mu < 2000$, where

D_o = outside diameter of the tube, m (ft)
L = tube length m (ft)
ΔH_v = latent heat of condensation, J/kg (Btu/lb)
N = number of tube rows in the vertical plane
Δt = temperature difference between the tube wall and the saturated vapor, °C (°F)
W = mass flow rate of condensate from the lowest point per vertical bank of tubes, kg/s (lb/h)

The same equations hold for S.I. units if g is omitted. Note also that for English units, $g = 32.2 \times 3600^2$ ft/h². Experimental data are consistent with these equations except that the constants for vertical tubes tend to run about 20% higher in the taller units because waves form on the free liquid surface and reduce the mean effective thickness of the fluid film.

Dropwise Condensation

It has been found in experiments that, if traces of oil are present in steam, highly polished surfaces may give much higher heat transfer coefficents than indicated by the previous equations. The higher rate stems from a change in the character of the condensation process from one giving a continuous liquid film of fairly uniform thickness to one in which the surface is covered with discrete liquid globules, which run off readily with good mixing within the droplet, thus reducing the effective thickness of the liquid on the surface. Figure 3.24 shows an example of the dropwise condensation of steam. Droplets grow, coalesce, and run off the surface, leaving the greater portion of the condensing surface freely exposed to incoming steam. Heat transfer coefficients with dropwise condensation of steam may be as much as 10 to 20 times higher than obtainable for the ideal film condensation case given by Eq. 3.32.

Hampson and Ozisik's data[40] for dropwise condensation of air-free steam on a 3 × 5 in. plane surface show that heat transfer coefficients averaged 137 kW/m² · °C (24,000 Btu/h · ft² · °F) for the vertical surface, 68.5 kW/m² · °C (12,000 Btu/h · ft² · °F) for the horizontal surface with the condensing surface facing downward, and 49.1 kW/m² · °C (9,000 Btu/h · ft² · °F) for nearly horizontal (3° inclined) surfaces facing upward. The reduction in heat transfer coefficient with the horizontal surface should be expected, because the droplets cannot drain off the surface easily, and the amount of metal surface freely exposed to the steam is less than would prevail for a vertical surface. Similarly, with long vertical surfaces a larger portion of the condensing surface is covered with drops, thus reducing the average heat transfer coefficient. The data of Fitzpatrick, Baum, and McAdams[41] for dropwise condensation of steam, with both 3.05- and 1.83-m (10-ft and 6-ft) long vertical evaporator tubes, averaged 68.5 kW/m² · °C (12,000 Btu/h · ft² · °F).

Dropwise condensation can be promoted by the use of any of a number of materials that inhibit wetting of smooth condenser surfaces by the condensate. Promoters such as oleic acid, benzyl mercaptan, and stearic acid are very effective in promoting dropwise condensation of steam if added in small quantities in a suitable solvent. Deposition of a monolayer of material on the condensing surface is sufficient to promote dropwise condensation; an excessively thick layer lowers the heat transfer coefficient. It is difficult to maintain continuous dropwise condensation conditions, because the film that tends to prevent wetting is gradually washed off. Once the surfaces become oxidized or fouled, injection of promotor into the steam is not effective in breaking the condensate film into droplets.[40] Because of this difficulty, it is usually not practicable to design for dropwise condensation, and virtually all condensers are designed for simple filmwise condensation.

Effects of Fluted Surfaces

The average liquid film thickness for filmwise condensation can be reduced substantially by fluting the surfaces vertically or grooving them circumferentially so surface tension concentrates the condensate in the grooves and leaves a thin film of liquid over the sections of surface between the grooves,[42] as indicated in Fig. 3.25. Some

Figure 3.24 Photo showing dropwise condensation of steam under ideal conditions. (Hampson and Ozisik, Ref. 40.)

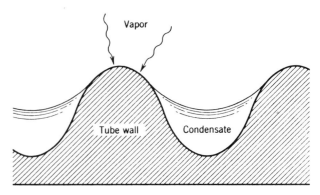

Figure 3.25 Section through a vertical tube with axial flutes showing the effects of surface tension on the configuration of the condensate film. (Lustenader et al., Ref. 42.)

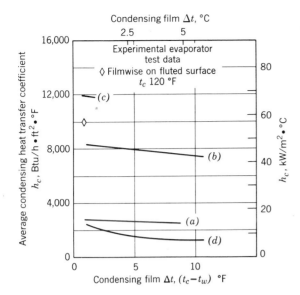

Figure 3.26 Average condensing heat transfer coefficient for steam as a function of the film temperature drop: (a) filmwise on smooth brass; (b) filmwise on a fluted surface at 100°C (212°F); (c) filmwise on a fluted surface at 30°C (86°F); (d) calculated from Nusselt's theoretical derivation for laminar flow in the condensate film for the conditions under which the data for (c) and (d) were obtained. (Lustenader et al., Ref. 42.)

indication of the benefits derived in this way is given by Fig. 3.26, which compares the ideal heat transfer coefficient for laminar film condensation on smooth surfaces with both dropwise condensation on smooth surfaces and film condensation on smooth and fluted surfaces. It should be mentioned that, in the tests from which these data were obtained, an attempt to attain a still higher heat transfer coefficient from fluted surfaces by coating them to produce a dropwise type of condensation actually led to a marked drop in the overall heat transfer coefficient. Extended operation later washed the chemical from the surfaces, and the high heat transfer rate indicated by Fig. 3.26 for the fluted surface was regained.

Effects of Noncondensables

If noncondensable gas is present, the effective temperature of the vapor is its saturation temperature at its partial pressure—not at the total pressure in the condenser. If the heat transfer coefficient is calculated from the saturation temperature corresponding to the total pressure for steam at pressures of around 6.9 K Pa (1.0 psia), the apparent loss in heat transfer coefficient commonly runs about 25% if the vapor contains only 1% by volume of noncondensable gas.

Example 3.3. A 26.7-mm (1.05-in.) OD, vertical tube condenser operates at 15.35 kPa (2.222-psia) pressure and condenses air-free steam at a rate of 0.00253 kg/s (20 lb/h) per tube. Determine the film heat transfer coefficient and the tube length for a temperature drop of 4.45°C (8°F) across the condensate film.

Solution. The saturation temperature of steam for 15.35 Pa (2.222 psia) is 54.4°C (130°F), and the latent heat of vaporization is 2280 kJ/kg (1020 Btu/lb). The physical properties of the condensate are $\mu = 5.32 \times 10^{-4}$ Pa · s (1.29 lb/h · ft); $k = 0.64$ W/m · °C (0.370 Btu/h · ft · °F); $\rho = 986$ kg/m^3 (61.5 lb/ft^3). Checking first for the Reynolds number,

$$\frac{4W}{\pi D_o \mu} = \frac{4 \times 0.00253}{\pi \times 0.0267 \times 5.32 \times 10^{-4}} = 222$$

Or, in English units,

$$\frac{4W}{\pi D_o \mu} = \frac{4 \times 20 \times 12}{\pi \times 1.05 \times 1.29} = 222$$

Hence Eq. 3.32 should be used; that is,

$$h_m = 1.47 \left[\frac{\pi D_o k^3 \rho^2 g}{4 W \mu} \right]^{1/3}$$

$$h_m = 1.47 \left(\frac{0.370^3 \times 61.5^2 \times 4.18 \times 10^8}{222 \times 1.29^2} \right)^{1/3}$$

$$= 1.47(0.218 \times 10^9)^{1/3}$$

$$= 885 \text{ Btu/h} \cdot \text{ft}^2 \cdot °F$$

Since the heat transfer coefficients for vertical tubes are about 20% higher than the theoretical value calculated in the preceding equation,

$$h_m = 1.2 \times 5000 = 6000 \text{ W/m}^2 \cdot °C$$

$$= 1.2 \times 885 = 1060 \text{ Btu/h} \cdot \text{ft}^2 \cdot °F$$

The tube length can be determined by equating the heat given up by the condensing vapor to the heat transfer rate across the condensate film. Hence

$$W \Delta H_v = \pi D_o L h_m' \Delta t$$

$$0.00253 \times 2280 = 3.14 \times 0.0267 \times L \times 6000 \times 4.45$$

$$L = 2.6 \text{ m}$$

Or, for English units,

$$20 \times 1020 = 3.14 \times \frac{1.05}{12} \times L \times 1060 \times 8$$

$$L = \frac{20 \times 1020 \times 12}{3.14 \times 1.05 \times 1060 \times 8} = 8.75 \text{ ft}$$

REFERENCES

1. W. H. McAdams, *Heat Transmission*, 3rd ed., McGraw-Hill Book Co., New York, 1954.

2. M. N. Ozisik, *Heat Transfer*, McGraw-Hill Book Co., New York, 1985.

3. M. Jakob, *Heat Transfer*, vols. 1 and 2, John Wiley & Sons, New York, 1949 and 1957.

4. W. D. Kingery et al., "Development of Ceramic Insulating Materials for High Temperature Use," *Trans. ASME*, vol. 80, 1958, p. 705.

5. R. G. Deissler and C. S. Eian, "Investigation of Effective Thermal Conductivities of Powders," NACA RM E52C05, June 24, 1952.

6. J. D. Versohoor and P. Greebler, "Heat Transfer by Gas Conduction and Radiation in Fibrous Insulation," *Trans. ASME*, vol. 74, 1952, p. 961.

7. K. A. Gardner, "Efficiency of Extended Surfaces," *Trans. ASME*, vol. 67, 1945, p. 621.

8. T. N. Cetinkale and M. Fishenden, *Proceedings of the General Discussion on Heat Transfer*, Institution of Mechanical Engineers, London, 1951, p. 271.

9. N. D. Weills and E. A. Ryder, "Thermal Resistance Measurement of Joints Formed between Stationary Metal Surfaces," *Trans. ASME*, vol. 71, 1949, pp. 259–267.

10. M. E. Barzelay, K. N. Tong, and G. F. Holloway, "Effects of Pressure on Thermal Conductance of Contact Joints," TN-3295, National Advisory Committee for Aeronautics, Washington, D.C., 1955.

11. M. N. Ozisik, *Radiative Transfer and Interactions with Conduction and Convection*, John Wiley & Sons, 1973.

12. J. T. Berans, J. T. Gier, and R. V. Dunkle, "Comparison of Total Emittances with Values Computed from Spectral Measurements," *Trans. ASME*, vol. 80, 1958, p. 1405.

13. C. O. Mackey, L. T. Wright, R. E. Clark, and N. R. Gray, "Radiant Heating and Cooling," Part 1, Cornell University Engineering Experimental Station, Bulletin no. 32, 1943.

14. L. Prandtl and O. G. Tietjens, *Applied Hydro- and Aeromechanics*, McGraw-Hill Book Co., New York, 1934.

15. L. Prandtl and O. G. Tietjens, *Fundamentals of Aero- and Hydromechanics*, McGraw-Hill Book Co., New York, 1934.

16. B. Eck, *Technische Strömungslehre*, 5th ed., Springer-Verlag, Berlin, 1957.

17. H. Schlichting, *Boundary Layer Theory*, Pergamon Press, Elmsford, N.Y., 1955.

18. T. Theodorsen, "Mechanism of Turbulence," Ohio State University Experimental Station, Bulletin no. 149, 1952, p. 1.

19. B. W. LeTourneau et al., "Pressure Drop for Parallel Flow through Rod Bundles," *Trans. ASME*, vol. 79, 1957, p. 1751.

20. J. R. Sellars, M. Tribus, and J. S. Klein, "Heat Transfer to Laminar Flow in a Round Tube or Flat Conduit," *Trans. ASME*, vol. 78, 1956, p. 441.

21. E. N. Sieder and G. E. Tate, "Heat Transfer and Pressure Drop of Liquids in Tubes," *Industrial and Engineering Chemistry*, vol. 28, 1936, pp. 121, 166, 188, 193.

22. W. M. Kays, "Numerical Solutions for Laminar Flow Heat Transfer in Circular Tubes," *Trans. ASME*, vol. 77, 1955, p. 1265.

23. T. Von Karman, "The Analogy between Fluid Friction and Heat Transfer," *Trans. ASME*, vol. 61, 1939, p. 705.

24. B. Pinkel, "A Summary of NACA Research on Heat Transfer and Friction for Air Flowing through a Tube with a Large Temperature Difference," *Trans. ASME*, vol. 76, 1954, p. 305.

25. R. G. Deissler, "Variable Fluid Property Effects," *Trans. ASME*, vol. 82, 1960, p. 160.

26. A. P. Colburn, "A Method of Correlating Forced Convection Heat Transfer Data and a Comparison with Fluid Friction," *Trans. AIChE*, vol. 29, 1933, pp. 174–209.

27. W. M. Kays and A. L. London, "Convective Heat Transfer and Flow-Friction Behavior of Small Cylindrical Tubes—Circular and Rectangular Cross-Sections," *Trans. ASME*, vol. 74, 1952, p. 1179.

28. P. Miller, J. J. Byrnes, and D. M. Benforado, "Heat Transfer to Water Flowing Parallel to a Rod Bundle," *Journal of AIChE*, vol. 2, 1956, p. 226.

29. P. Miller, "Heat Transfer to Water in an Annulus," *Journal of AIChE*, vol. 1, 1955, p. 501.

30. R. G. Deissler, "Turbulent Heat Transfer and Friction in Entrance Regions of Smooth Passages," *Trans. ASME*, vol. 77, 1955, p. 1221.

31. E. W. Sams and W. F. Weiland, Jr., "Experimental Heat Transfer and Friction Coefficients for Air Flowing through Stacks of Parallel Flat Plates," NACA RM E54F11, 1954.

32. A. S. T. Thompson et al., "Variation in Heat Transfer Rates around Tubes in Cross-Flow," *Proceedings of the General Discussion on Heat Transfer*, Institute of Mechanical Engineers and ASME, 1951, Institute of Mechanical Engineers, London, p. 177.

33. O. L. Pierson, "Experimental Investigation of the Influence of Tube Arrangement on Convection Heat Transfer and Flow Resistance in Cross-Flow of Gases over Tube Banks," *Trans. ASME*, vol. 59, 1937, p. 563.

34. C. E. Jones and E. S. Monroe, Jr., "Convection Heat Transfer and Pressure Drop of Air Flowing Across In-Line Tube Banks," *Trans. ASME*, vol. 80, 1958, p. 18.

35. R. C. Martinelli, "Heat Transfer to Molten Metals," *Trans. ASME*, vol. 69, 1947, p. 947.

36. R. N. Lyon, *Liquid Metals Handbook*, United States Government Printing Office, 1952.

37. R. Siegel and R. H. Norris, "Tests of Free Convection in a Partially Enclosed Space between Two Heated Vertical Plates," *Trans. ASME*, vol. 79, 1957, p. 663.

38. S. Globe and D. Droplin, "Natural Convection Heat Transfer in Liquids Confined by Two Horizontal Plates and Heated from Below," *Journal of Heat Transfer, Trans. ASME*, vol. 81-2, 1959, p. 24.

39. S. Ostrach, "Convection Phenomena in Fluids Heated from Below," *Trans. ASME*, vol. 79, 1957, p. 299.

40. H. Hampson and N. Ozisik, "An Investigation into the Condensation of Steam," *Proceedings of the Institute of Mechanical Engineers*, vol. 1B, 1952, p. 282.

41. J. P. Fitzpatrick, S. Baum, and W. H. McAdams, "Dropwise Condensation of Steam on Vertical Tubes," *Trans. AIChE*, vol. 35, 1939, p. 97.

42. E. L. Lustenader, R. Richter, and F. N. Neugebauer, "The Use of Thin Films for Increasing Evaporation and Condensation Rates in Process Equipment," *Journal of Heat Transfer, Trans. ASME*, vol. 81-2, 1959, p. 297.

4

Performance Estimation

A major phase of heat exchanger design work is concerned with the estimation of the size and performance characteristics of heat exchangers for new applications. This chapter is concerned with some of the more important techniques applicable to such performance estimates.

TEMPERATURE DISTRIBUTION AND ITS IMPLICATIONS

Estimating the performance of a heat exchanger directly from the basic heat transfer equation $Q = UA \, \Delta t$ presents difficulties. While the surface area is an obvious function of the basic geometry chosen, and the average local heat transfer coefficient can be determined as indicated in Chapter 3, the evaluation of the effective temperature difference between the two fluid streams presents a special set of problems, because in general it is not the same throughout the heat exchanger. Since the character of the temperature distribution varies widely from one type of heat exchanger to another, the mean effective temperature difference must be estimated with perceptive care.

Typical Temperature Distributions

The mean effective temperature difference between the two fluid streams in a heat exchanger depends on the geometry and the fluid-flow path configuration. The basic relations can be deduced from the curves shown in Fig. 4.1 for a number of idealized cases that give a valuable insight into the basic problems. Note that in each instance the temperature distribution in the heat exchanger is plotted as a function of the distance from the cold fluid inlet end of the heat exchanger. In all instances, the heat transfer surface area per unit of length is assumed constant throughout the heat exchanger and the heat transfer coefficients independent of the axial position, that is, the local fluid temperature.

The simplest case (Fig. 4.1a) is a pure counterflow heat exchanger in which the temperature rise in the cold fluid is equal to the temperature drop in the hot fluid; thus the temperature difference between the two fluids is constant throughout the length of the flow passage. In the rest of the cases the situation is more complex since the temperature difference varies, and hence the heat flux. As a consequence, the slopes of the fluid temperature curves vary with the distance from the inlet. This effect is quite pronounced in the second idealized case, for which the temperature on one side of the heat exchanger is constant irrespective of the distance from the fluid inlet, a condition that would prevail in a condenser. The temperature of the cold fluid rises rapidly at first near the cold fluid inlet end, and progressively less rapidly as the temperature difference between the fluid streams reduces the heat transfer rate per unit of surface area. A similar effect can be noted for the boiler of Fig.

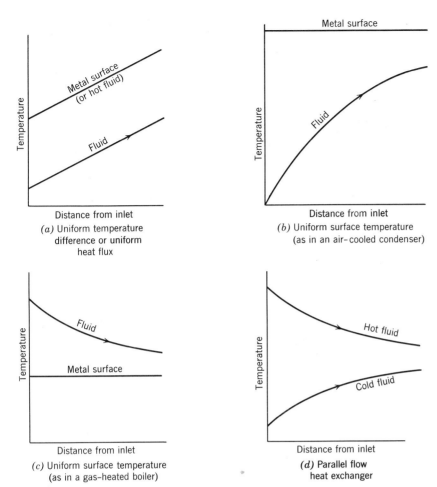

Figure 4.1 Axial temperature distributions in typical heat transfer matrices. (*continued on following page*)

4.1c. In the parallel and counterflow heat exchangers of Figs. 4.1d and e, not only the temperature difference but also the temperatures of both fluid streams vary from one end of the heat exchanger to the other.

A still more complex situation prevails in a once-through boiler in which the water is first heated to the boiling point, then boiled at constant temperature, and finally superheated. Case f presents such a temperature distribution for a steam generator in a gas-cooled reactor plant.

Electrically heated grids and nuclear fission reactors constitute a specialized form of heat transfer matrix having interesting properties. For the simplest case, the amount of power developed per square foot of heat transfer surface is uniform from the fluid inlet to the outlet; hence the temperature difference between the heated surface and the fluid remains approximately constant from one end to the other, as in Fig. 4.1a. In most nuclear reactors the situation is more complex, because the neutron flux tends to be highest at the center

of the reactor; thus the heat flux tends to be highest at the midplane and falls off toward either end. This situation results in a temperature distribution similar to that shown by the solid lines in Fig. 4.1g. If the maximum possible fluid exit temperature is to be obtained for a given maximum allowable fuel element temperature and heat transfer coefficient, the fuel element surface temperature should be constant throughout the length of the reactor. Ideally, the fluid temperature rises exponentially from the inlet to the outlet, while the power per unit of area falls off exponentially from the reactor inlet, and the temperature distribution is similar to that for the condenser of Fig. 4.1b. In practice, fuel element fabrication and nuclear problems make it necessary to compromise on the metal temperature distribution. A two-stage arrangement designed to approach the constant-temperature condition is shown in Fig. 4.1h. In this instance two stages of fuel loading are employed, so there is a higher power loading in the first 60% of the reactor than in the last 40%. In principal, any number

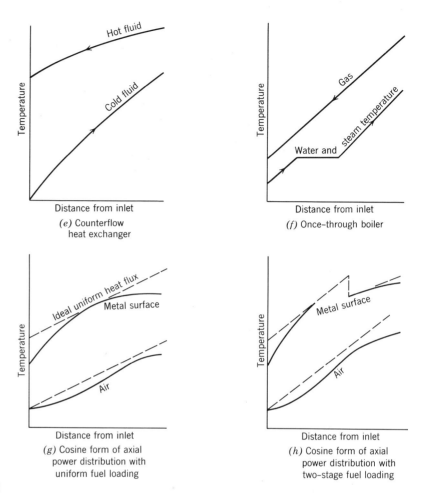

(e) Counterflow heat exchanger

(f) Once-through boiler

(g) Cosine form of axial power distribution with uniform fuel loading

(h) Cosine form of axial power distribution with two-stage fuel loading

Figure 4.1   *(Continued)*

of stages may be used, but more than two or three is ordinarily impractical.

Temperature Distribution in Counterflow Heat Exchangers

In general, the temperature distribution in an idealized parallel or counterflow heat exchanger is as indicated in Fig. 4.1*d* or *e* if there is no change in phase in either fluid. The heat absorbed by the cold fluid equals that given up by the hot fluid:

$$W_1 c_1 \, \delta t_1 = W_2 c_2 \, \delta t_2 \qquad (4.1)$$

If the flow-passage area and heat transfer matrix geometry are independent of length, and if changes in the physical properties with temperature cause negligible variations in the heat transfer coefficients of the two fluid streams, the local heat flux at any given point along the length of the heat exchanger is directly proportional to the local temperature difference Δt between the two fluids.

The local temperatures at any axial position through the heat exchanger can be calculated from the changes in fluid stream temperature resulting from heat transfer. Consider a differential length dx at a distance x from the cold fluid inlet, as shown in Fig. 4.2. Heat added to the cold fluid, as manifested by a temperature rise dt, can be equated to the heat transferred through the increment of surface area in the length dx as $W_1 c_1 dt_1 = (UA/L) \Delta t \, dx$; hence, the differential change in the temperature of the cold fluid becomes

$$dt_1 = \frac{UA}{W_1 c_1 L} \Delta t \, dx \qquad (4.2)$$

Similarly, the differential change in the temperature of the hot fluid becomes

$$dt_2 = \frac{UA}{W_2 c_2 L} \Delta t \, dx \qquad (4.3)$$

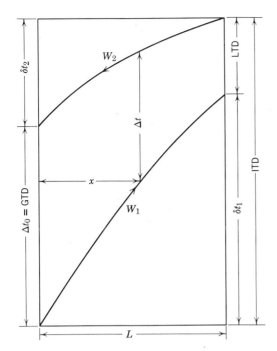

Figure 4.2 Nomenclature for the axial temperature distribution in a simple counterflow heat exchanger.

Subtracting Eq. 4.2 from Eq. 4.3, and noting that $d(t_2 - t_1) = d(\Delta t)$, gives

$$\frac{d\,\Delta t}{\Delta t} = \frac{UA}{L}\left(\frac{1}{W_2 c_2} - \frac{1}{W_1 c_1}\right)dx \qquad (4.4)$$

With the boundary condition that $\Delta t = \Delta t_0$ for $x = 0$, and assuming that U, c_1, and c_2 are independent of x, Eq. 4.4 can be integrated to give

$$\Delta t = \Delta t_0 e^{ax} \qquad (4.5)$$

where

$$a = \frac{UA}{L}\left(\frac{1}{W_2 c_2} - \frac{1}{W_1 c_1}\right) \qquad (4.6)$$

The parameter a can be expressed in a different form by substituting Eq. 4.1 in Eq. 4.6, to eliminate $W_2 c_2$, so that

$$a = \frac{UA}{W_1 c_1 L}\left(\frac{\delta t_2}{\delta t_1} - 1\right) \qquad (4.7)$$

For parallel-flow conditions the parameter a is similar to that for counterflow except that the first term in the parentheses in Eqs. 4.6 and 4.7 has a negative sign.

It should be mentioned that the parameter a is closely related to the "number of heat transfer units," or NTU's, a term introduced by London and Kays and often used by other writers.[1] By definition,

$$\text{NTU}_1 = UA/W_1 c_1 = \text{NTU}_2(W_2 c_2/W_1 c_1),$$

where Wc is the product of the weight flow and the specific heat of a fluid, the subscripts 1 and 2 are used to distinguish between the two fluids, and the area A is that used as the reference area in computing U. Note that if one of the fluids is at a constant temperature, as in a boiler or condenser, the two parameters differ only by the quantity L: $\text{NTU} = \pm aL$, where a is positive if the temperature of the cold fluid is essentially constant, and negative if the temperature of the hot fluid is nearly constant.

Sometimes it is desirable to express the local Δt in terms of the terminal temperature differences. By substituting in Eq. 4.5 the outlet condition $\Delta t = \Delta t_L$, where $x = L$,

$$\Delta t_L = \Delta t_0 e^{aL} \qquad (4.8)$$

Solving for a, we obtain

$$a = \frac{1}{L}\ln\frac{\Delta t_L}{\Delta t_0} \qquad (4.9)$$

Substituting Eq. 4.9 in Eq. 4.5 gives

$$\Delta t = \Delta t_0 \exp\left(\frac{x}{L}\ln\frac{\delta t_L}{\Delta t_0}\right) \qquad (4.10)$$

which reduces to

$$\Delta t = \Delta t_0 \left(\frac{\Delta t_L}{\Delta t_0}\right)^{x/L} \qquad (4.11)$$

Note that this relation applies to either counterflow or parallel-flow conditions since it is independent of the parameter a.

Log Mean Temperature Difference

The detailed temperature distribution through a heat exchanger is usually of academic interest only, but the mean effective temperature difference between the two fluid streams is extremely useful. This mean effective temperature difference involves the logarithm of the ratio of the temperature differences at the two ends of the heat exchanger, and thus has come to be called the *logarithmic mean temperature difference*, or LMTD. This quantity is defined as

$$\text{LMTD} = \frac{1}{L} \int_0^L \Delta t \, dx \qquad (4.12)$$

Substituting Δt from Eq. 4.11 in Eq. 4.12 gives

$$\text{LMTD} = \frac{\Delta t_0}{L} \int_0^L \left(\frac{\Delta t_L}{\Delta t_0}\right)^{x/L} dx$$

$$= \frac{\Delta t_0 - \Delta t_L}{\ln \dfrac{\Delta t_0}{\Delta t_L}} \qquad (4.13)$$

For the temperature distribution of Fig. 4.2, Δt_0 may be referred to as the *greatest temperature difference*, or GTD, and Δt_L as the *least temperature difference*, or LTD. On this basis Eq. 4.13 can be written as

$$\text{LMTD} = \frac{\text{GTD} - \text{LTD}}{\ln \dfrac{\text{GTD}}{\text{LTD}}} \qquad (4.14)$$

This relation applies to either parallel-flow or counter-flow heat exchangers.

Sometimes it is convenient to express Eq. 4.14 in another form. By using the notation of Fig. 4.2, both the GTD and LTD can be expressed in terms of the inlet temperature difference, or ITD, between the two fluid streams entering the heat exchanger:

$$\text{GTD} = \text{ITD} - \delta t_2$$

$$\text{LTD} = \text{ITD} - \delta t_1$$

Substituting these values in 4.14 gives

$$\text{LMTD} = \frac{\delta t_1 - \delta t_2}{\ln \dfrac{1 - \delta t_2/\text{ITD}}{1 - \delta t_1/\text{ITD}}} \qquad (4.15)$$

or

$$\frac{\text{LMTD}}{\text{ITD}} = \frac{\delta t_1/\text{ITD} - \delta t_2/\text{ITD}}{\ln \left(\dfrac{1 - \delta t_2/\text{ITD}}{1 - \delta t_1/\text{ITD}}\right)} \qquad (4.16)$$

Heat Transfer Rate as a Function of the LMTD

Once the LMTD is determined, the heat transfer rate for the heat exchanger as a whole is given by $Q = UA(\text{LMTD})$. It is often tedious to evaluate the expression for the logarithmic mean temperature difference, since it is likely to involve a small difference between large numbers divided by the logarithm of a number close to 1. To obtain three significant figures, it may be necessary to use 8- or 10-place logarithms; hence the detailed calculations are likely to be tedious. A chart such as that of Fig. H4.1 saves much time and annoyance.

Various types of crossflow heat exchangers present more complicated temperature distribution patterns than the simpler cases of Fig. 4.1 (e.g., see Fig. 4.3 for single-pass crossflow heat exchangers). Correction charts have been developed to convert the LMTD for counterflow conditions to the LMTD for typical single- and multi-pass crossflow conditions,[2-4] and a useful set is included in Fig. H4.2.

CALCULATIONAL PROCEDURES

The procedure for carrying out heat exchanger performance and size estimates depends on the design conditions. Usually the inlet and outlet temperatures and flow rates of the two fluid streams are given, and the heat exchanger size is to be estimated. As a rule, restrictions are imposed on the pressure drops of the two fluid streams. Since the pressure drop depends on the fluid velocity, the flow-passage equivalent diameter, and the passage length, the designer is confronted with a set of

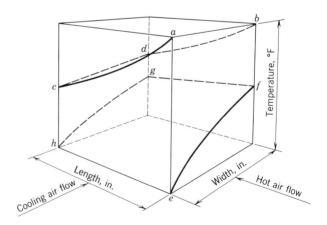

Figure 4.3 Two-dimensional temperature distribution in a typical single-pass crossflow heat exchanger.

relations involving six independent variables. Any given combination of these variables yields a particular set of values for the quantity of heat transmitted and the two fluid pressure drops. Often only one of the infinitude of combinations possible will meet the conditions specified.

Simple Application of Basic Relations

The most commonly used design procedure for coping with this complex situation is simply a cut-and-try approach. A particular heat transfer matrix geometry is first assumed (e.g., the tube diameter and spacing), velocities are assumed for the two fluid streams, and the required heat transfer surface area and the resulting pressure drops are calculated. The results are examined and a second approximation attempted with the geometry modified to give a unit which should come closer to meeting the desired conditions. The procedure is then repeated until a suitable unit is found.

Analytical Solution

The cut-and-try approach is easy to understand but awkward and tedious to apply. It is sometimes possible to make use of a more refined approach. The case at hand must be examined carefully and engineering judgment applied to single out the quantities that can be held constant or related to other quantities. The pressure drops, for example, can be expressed as functions of the tube length and the fluid-flow rates. One fluid-flow rate usually can be expressed as a simple function of the other by using the design fluid inlet and outlet temperatures and equating the heat gained by one fluid to that lost by the other. The LMTD for the unit can be calculated. The tube length can be expressed as a function of the heat to be transmitted, the heat transfer coefficients, and the LMTD. The heat transfer coefficients, in turn, can be expressed as functions of the fluid-flow rates. It is important that the conditions imposed in writing these relations be such that they are necessary and sufficient to define the solution but are not redundant or incompatible. In general, this approach entails a time-consuming development of the various relations and the reduction of these to two simultaneous equations, one from heat transfer and the other from pressure-drop considerations. If the work is done properly, the simultaneous equations can be solved analytically, graphically, or on a digital computer. The analytical approach is so involved, and so dependent on the special conditions that apply to the particular case to be solved, that detailed treatment has been deferred to examples in later chapters (e.g., regenerators in Chapt. 13 and axial flow steam generators in Chapt. 15).

Short Cuts for Performance Estimation

Heat exchanger selection or performance estimation work often begins with a given set of temperature conditions and a basic heat transfer matrix geometry for which experimental data are available. When this is the case, the problem can be reduced to one of sizing the unit to yield the desired temperatures. The ratio of the temperature change in one of the fluids to a major temperature difference has proved to be a potent tool for handling such problems. However, it must be used with discernment because, while there are general similarities, the various types of temperature distribution indicated in Fig. 4.1 have subtle effects on the basic relationships that apply to any particular case.

Uniform Heat Flux Cases

The simplest and most straightforward situation is presented by those cases in which the temperature difference responsible for the heat transfer process is constant over the length of the heat exchanger. This condition holds for a pure counterflow unit in which the temperature rise in one fluid equals the temperature drop in the other. It may also apply to nuclear reactors or to electrically heated surfaces if the heat flux is fairly uniform over the length of the coolant passage. For these conditions the quantity of heat flowing through the heat transfer surface can be equated to the temperature rise in the fluid stream of interest, that is,

$$UA \, \Delta t = Wc_p \, \delta t \qquad (4.17)$$

Furthermore,

$$U = K_1 W^{0.8} \qquad (4.18)$$

and

$$A = K_2 L \qquad (4.19)$$

By substituting,

$$K_1 W^{0.8} K_2 L \, \Delta t = Wc_p \, \delta t$$

or,

$$\frac{\delta t}{\Delta t} = \frac{K_1 K_2 L}{W^{0.2} c_p} = K_3 \frac{L}{W^{0.2}} \qquad (4.20)$$

where K_1, K_2, and K_3 are constants that depend on the heat transfer matrix geometry and the physical properties of the fluid streams. Thus for this case the ratio of the temperature rise or drop to the temperature difference is directly proportional to the passage length and inversely proportional to the flow rate to the 0.2 power.

This relationship can be applied to give a quick, simple, and accurate solution to a variety of problems related to a particular heat transfer matrix geometry, if data are available from calculations or a text even if the data are for only one set of conditions. The known point can be plotted on linear coordinates consisting of $\delta t/\Delta t$ and L. Since $\delta t/\Delta t$ varies as $W^{-0.2}$, additional points for other fluid flow rates can then be calculated and plotted for the base value of L. Straight radial lines can then be drawn through these points, as in Fig. 4.4, to give a performance chart covering a wide range of conditions.

If test data are available for a particular fluid passage length and a range of flow rates (in pounds per square foot of flow-passage area), they can be correlated by plotting $\delta t/\Delta t$ against W on logarithmic coordinates, as in Fig. 4.5. The line through them should have a slope of -0.2, and should establish a better set of values for the chart of $\delta t/\Delta t$ versus L than is given by a single point. It also makes it convenient to prepare a chart for $\delta t/\Delta t$ versus L for the range of values of W that are of interest, since points can be picked directly from the curve of $\delta t/\Delta t$ versus W.

Uniform Wall-Temperature Cases

A more generally applicable situation than the uniform-heat flux case treated in the preceding section is that in which the wall temperature of the heat transfer matrix, rather than the heat flux, is substantially uniform throughout the unit. Cases b and c of Fig. 4.1 for

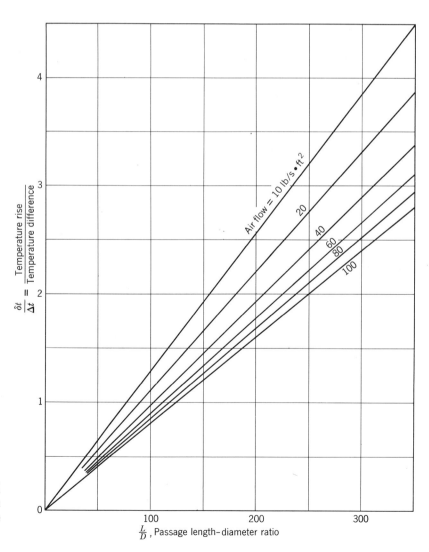

Figure 4.4 Performance chart for a typical series of counterflow heat exchangers in which the temperature difference between the two fluids is uniform throughout the length.

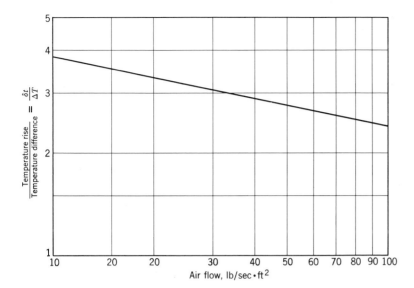

Figure 4.5 Effect of flow rate on the ratio of the temperature rise to the temperature difference for a counterflow heat exchanger with a uniform temperature difference throughout the length and a fluid-passage length-diameter ratio of 250.

condensers and boilers are obvious examples. Since the heat transfer coefficient for boiling or condensing is very high, the tube-wall temperature is essentially constant and almost equal to the temperature of the boiling liquid or condensing vapor.

Effectiveness

One of the most useful parameters in heat exchanger design and performance calculation work is the ratio of the fluid temperature rise (or drop) to the overall temperature difference (i.e., the ITD in Fig. 4.2). This ratio is called the *heating* (or *cooling*) *effectiveness*, and has proved to be most helpful in performance estimation and analysis for all types of heat transfer matrices including reactor cores. By using this parameter it is possible to prepare easily constructed charts that present a comprehensive picture of the performance attainable for a wide range of conditions.

The implications of "heating effectiveness" can be surmised by examining case *b* of Fig. 4.1. For any given matrix geometry in which the surface temperature is constant, from Eq. 4.8 we may write

$$\ln \frac{\Delta t_L}{\Delta t_0} = aL \qquad (4.21)$$

Since $\delta t_2 = 0$ for the case of Fig. 4.1*b*, from Eq. 4.7

$$a = -\frac{UA}{W_1 c_1 L} \qquad (4.22)$$

Since U varies as $W_1^{0.8}$, and the surface area for a given matrix geometry is directly proportional to the flow-passage length L, Eq. 4.22 can be written

$$a = -\frac{K_1}{W_1^{0.2}} \qquad (4.23)$$

where K_1 = a constant. Substituting Eq. 4.23 into 4.21 gives

$$\ln \frac{\Delta t_L}{\Delta t_0} = -K_1 \frac{L}{W_1^{0.2}} \qquad (4.24)$$

It can be seen from Eq. 4.24 that the ratio $(\Delta t_L / \Delta t_0)$ may be plotted against L on semilog coordinates to give a straight line as in Fig. 4.6*a*. If data are available for one heat exchanger passage length, such a line is defined for the matrix because the value of $(\Delta t_L / \Delta t_0)$ at a fluid-passage length of zero is also known. The ratio $(\Delta t_L / \Delta t_0)$ can be related to the heating effectiveness, since Δt_0 is equal to the overall temperature difference for this particular case.

$$\eta = \text{heating effectiveness} = \frac{\text{fluid temp. rise}}{\text{overall temp. diff.}}$$

$$= \frac{(\text{overall temp. diff.}) - (\text{outlet temp. diff.})}{(\text{overall temp. diff.})}$$

$$= 1 - \left(\frac{\text{outlet temp. diff.}}{\text{overall temp. diff.}} \right)$$

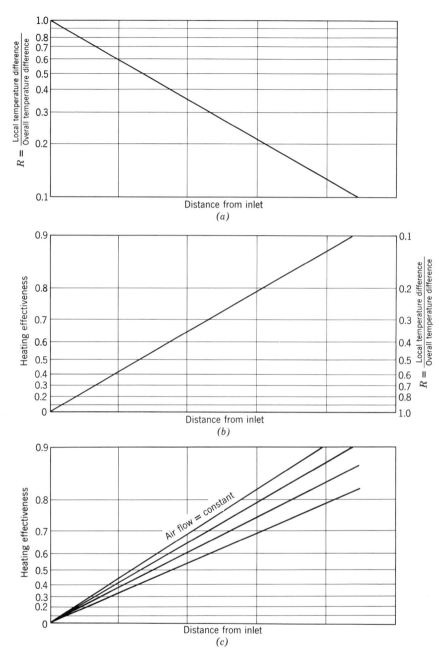

Figure 4.6 Diagrams showing the basis for the construction of a performance chart for a series of air-cooled condensers.

Hence

$$\eta = 1 - \frac{\Delta t_L}{\Delta t_0} \qquad (4.25)$$

Thus a line can be drawn to define the heating effectiveness for a heat transfer matrix by taking the coordinates and line of Fig. 4.6a, inverting the ordinate scale, and labeling it to read in terms of the heating effectiveness, giving Fig. 4.6b.

An explicit relation can be obtained for heating effectiveness in terms of the heat transfer matrix length and the fluid-flow rate if Eqs. 4.25 and 4.24 are solved for $(\Delta t_L / \Delta t_0)$ and equated to give

$$\ln (1 - \eta) = -K_1 \frac{L}{W_1^{0.2}} \qquad (4.26)$$

Solving for η gives

$$\eta = 1 - e^{-K_1}\,\frac{L}{W^{0.2}} \qquad (4.27)$$

It follows from this that, for a given value of η, the flow-passage length is proportional to $W^{0.2}$. This relation can be applied, as it is in Fig. 4.7, to present heating effectiveness as a function of heat transfer matrix length and fluid-flow rate. Thus a few good experimental points for a particular unit make it possible to plot a chart such as

Fig. 4.7 that defines the heating effectiveness of that basic heat transfer matrix geometry for a wide range of flow-passage lengths and a series of fluid-flow rates.

Example 4.1. In a cooler for a closed-cycle gas turbine, cold water flows inside the tubes, and the hot gas from the regenerator flows in the axial direction outside the tubes. The water flow rate is sufficiently high that the tube-metal temperature can be assumed to be constant and equal to the water temperature throughout

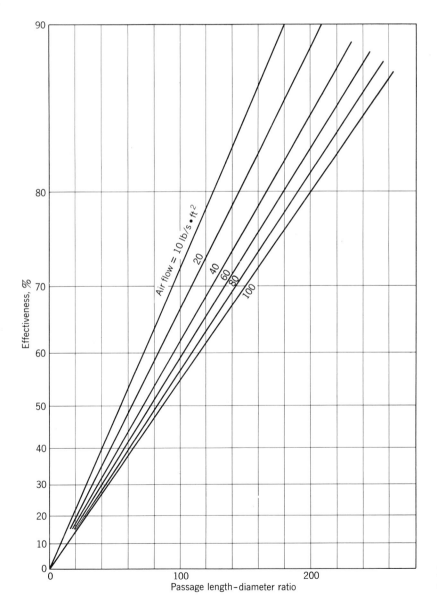

Figure 4.7 Performance for a series of air-cooled condensers.

the tube matrix. (1) If the temperature drop in the hot gas flowing through the matrix is 104°F, and the temperature difference between the gas leaving the cooler and the tube-wall surface is 26°F, what is the cooling effectiveness? (2) To what extent must the gas-flow rate be reduced to increase the effectiveness to 0.85?

Solution. (1) The effectiveness is given as

$$\eta = \frac{\text{fluid temperature drop}}{\text{overall temperature difference}} = \frac{\delta t}{\delta t + \Delta t_L}$$

$$= \frac{104}{104 + 26} = 0.80$$

(2) From Eq. 4.26 the relation between the gas-flow rate and the effectiveness is $\ln (1 - \eta) = -(K_1 L / W^{0.2})$. If the flow rate for 85% effectiveness is W', we may write

$$\left(\frac{W'}{W}\right)^{0.2} = \frac{\ln (1 - 0.80)}{\ln (1 - 0.85)} = 0.849$$

$$\frac{W'}{W} = (0.849)^5 = 0.44$$

Thus the flow rate has to be reduced to 44% of its initial value to increase the effectiveness from 0.80 to 0.85. (Note that this can also be deduced from Fig. 4.7 by finding the passage length-diameter ratio for a flow of 100 lb/s·ft² and an effectiveness of 0.80, and then finding the flow rate to give an effectiveness of 0.85 for the same length-diameter ratio.)

EFFECTIVENESS OF UNITS WITH NONUNIFORM SURFACE TEMPERATURES

Many liquid-to-gas counterflow and crossflow heat exchangers closely approach the uniform wall-temperature condition because they are designed to give little temperature change in the liquid relative to both the gas temperature rise and the inlet temperature difference, and the heat transfer coefficient for the liquid is much higher than for the gas (e.g., in automotive radiators), so that a constant surface temperature condition represents a good approximation. The same can be said for some gas-cooled nuclear reactors (e.g., that of Fig. 4.1*h*) in which the variations in the surface temperature along the crucial latter two-thirds of the heated length amount to only about ±15% of the mean temperature difference. There

is a strong incentive to make such approximations, since they greatly simplify the tasks of performance estimation by making it possible to prepare easily constructed charts that present a comprehensive picture of the performance attainable with a given heat transfer matrix.

The degree to which the nonuniform surface temperature conditions can be approximated by the uniform surface temperature case is shown in Fig. 4.8 for a series of counterflow heat exchangers, for a range of ratios of the hot-fluid temperature drop to the inlet temperature difference between the two fluid streams, or ITD. The upper line for a zero temperature drop in the hot fluid is, of course, straight as indicated by the preceding analysis. The balance of the lines are curved, but straight, dashed lines are drawn through each to show the curvature is small, and that straight lines represent fairly good approximations to the true curves, especially if the temperature drop in the hot fluid is small relative to the inlet temperature difference.

Instead of the representation indicated in Fig. 4.8, the hot-fluid temperature drop may be made a fixed fraction of the cold-fluid temperature rise—as is the case if the mass flow ratio of the two fluid streams is kept constant. Figure 4.9 shows that for this condition the curvature of the lines for the effectiveness increases rapidly with increases in both the effectiveness and the ratio of the hot-fluid temperature drop to the cold-fluid temperature rise. Thus the utility of this type of performance chart is restricted to the range in which the hot-fluid temperature drop is not more than perhaps 30% of the cold-fluid temperature rise, if the simple approximation of a straight line is to be used for interpolation or extrapolation from a few experimental points.

The technique represented by Fig. 4.8 sometimes can be extended to single-pass crossflow units. Figure 4.10 was prepared to investigate the applicability of this approach. Note that straight lines again represent good approximations to the curves representing the actual performance for hot-fluid temperature drops up to about 30% of the inlet temperature difference.

Superposition of Pressure-Drop Characteristics

The pressure-drop characteristics may be presented in separate charts (as in Fig. 3.14) or curves may be superimposed on a chart such as that of Fig. 4.8 to give a chart similar to the one in Fig. 14.10. Such charts give a good insight into the effects of the principal parameters, and thus have proved to be enormously valuable in designing

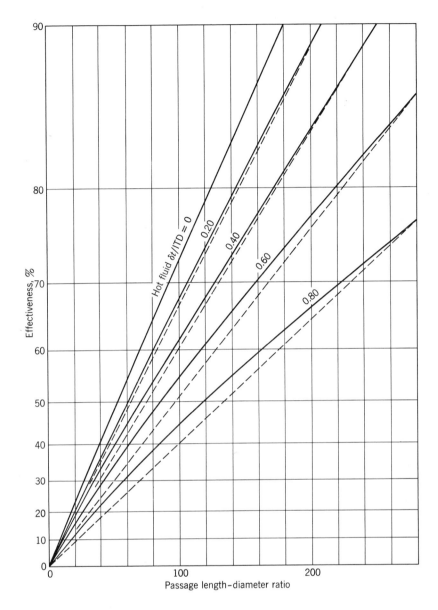

Figure 4.8 Influence of the ratio of the hot-fluid temperature drop to the inlet temperature difference on curves for the effectiveness of a series of counterflow heat exchangers. (The dashed, straight lines are drawn to highlight the amount of curvature.)

heat transfer matrices for applications in which compromises between pressure drop, effectiveness, and matrix size are difficult to make.

Effects of High Gas Velocities

While the problem is too specialized to treat in detail in this book, it should be mentioned that it is sometimes advantageous in mobile power plants to consider design conditions in which the gas velocities are so high that compressibility effects become important. Under these conditions the heat transfer analysis is complicated by

the reduced static temperature of the gas in the high-velocity regions. A clever technique for handling problems of this sort through the use of relatively simple charts has been worked out by A. S. Thompson.[5]

COMPARISON OF HEAT TRANSFER FLUIDS

Sometimes the designer is confronted with a choice between heat transfer fluids. The field is usually narrowed by materials considerations, but there may be

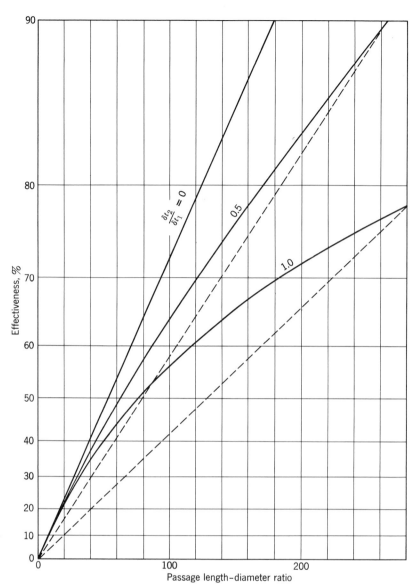

Figure 4.9 Influence of the ratio of the hot-fluid temperature drop to the cold-fluid temperature rise on curves for the effectiveness of a series of counterflow heat exchangers. (The dashed, straight lines are drawn to highlight the amount of curvature.)

the option of using fluids as diverse as helium, water, and sodium. If the problem is looked at purely from the heat transfer and heat transport standpoints, the key questions become the temperature difference between the fluid and the surface, the pressure drop, and the pumping power requirement for a given fluid temperature rise or drop at a given allowable pressure. The associated size of the pipe or duct to convey the fluid from one system component to another, together with the size of the passages in the heat exchanger, is also likely to be important, at least from the cost standpoint.

There are so many parameters that enter into this comparison that one must settle on the most significant or become lost in a morass of numbers. The author has examined many such comparisons.[1,5-8] and from these selected three parameters as the most significant, namely: the pumping power, the heat transfer coefficient, and the flow-passage size.[9] On this basis, a convenient way to compare different fluids graphically is to plot the heat transfer coefficient against the flow-passage size required to remove 100 MW of heat from a heat transfer matrix of a given geometry for a pumping power equal to 1% of the heat removed. Estimates of these parameters were made for a dozen typical working fluids for conditions

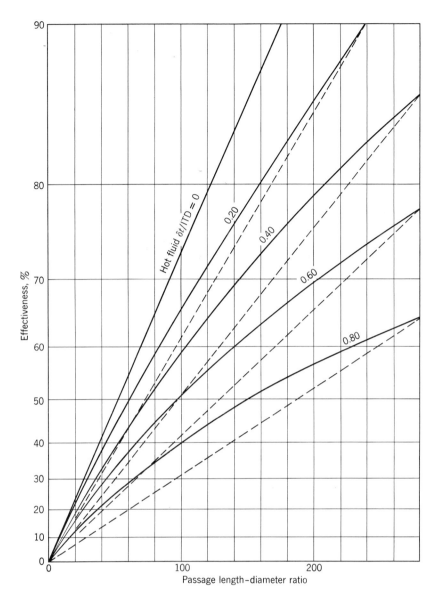

Figure 4.10 Influence of the ratio of the hot-fluid temperature drop to the inlet temperature difference on curves for the effectiveness of a series of single-pass crossflow heat exchangers. (The dashed, straight lines are drawn to highlight the amount of curvature.)

representative of those under which each fluid might be used; the results are presented in Table 4.1 and Fig. 4.11. In examining Fig. 4.11, as one might expect from the physical property data of Table 4.1, air at 1000°F requires the largest flow passage and gives the lowest heat transfer coefficient of the fluids considered, while lithium and water give both the highest heat transfer coefficients and the smallest flow passages, being better than air at atmospheric pressure in either measure by roughly a factor of 1000. Inasmuch as the fluid density and specific

heat have similar effects on both the heat transfer coefficient and the heat transport capacity, it is not surprising that, if one uses these two parameters as coordinates, all of the fluids fall in a scatterband that is not very wide in spite of the enormous spread in physical properties included in this set. The value and validity of this perspective will become much more clear in the course of examination of the many typical cases considered later in this text.

TABLE 4.1 Comparisons of the Characteristics of Typical Heat Transfer Fluids

	Density, lb/ft³ Liquid	Density, lb/ft³ Gas	Specific Heat, Btu/lb·°F	Thermal Conductivity, Btu/h·ft·°F	Viscosity, lb/h·ft	Melting Point, °F	Boiling Point at 1 atm, °F	Heat of Vaporization, Btu/lb	Temperature Rise, °F	Heat Removed, Btu/lb	Volumetric Flow Rate, ft³/s	Dynamic Head, psi	Velocity, ft/s	Flow-Passage Area for 100 MW, ft²	Heat Transfer Coefficient, Btu/h·ft²·°F	Relative Magnitude of Radioactivity from Neutron Activation
Water: no boiling, 200°F	60		1.0	0.393	0.738	32			200	200	7.9	16.2	98	0.08	17,500	Low
boiling, 1 atm, 212°F	60	0.0373	1.0				212	970	30	1,000	2,540	0.0504	220	11.6	2,000	
20 atm, 416°F	53	0.636	1.0					811	30	841	177	0.723	202	0.87	6,500	
200 atm, 692°F	30	11.1	2.8					222	30	306	27.9	4.59	122	0.23	6,500	
Air: 1 atm, 200°F		0.0602	0.241	0.0184	0.0519				200	48.2	25,700	0.00498	54	475	11.2	Low
1 atm, 1,000°F		0.0272	0.263	0.0332	0.0884				200	52.6	66,200	0.00193	50	1,324	6.6	
20 atm, 1,000°F		0.544	0.263	0.0332	0.0884				200	52.6	3,310	0.0387	50	662	77	
200 atm, 1,000°F		5.54	0.263	0.0332	0.0884				200	52.6	331	0.387	50	66	490	
He: 1 atm, 200°F		0.0083	1.25	0.0985	0.0545				200	250	45,600	0.0028	110	415	21.3	Very low
1 atm, 1,000°F		0.00376	1.25	0.176	0.099				200	250	101,000	0.00127	110	917	12.8	
20 atm, 1,000°F		0.0752	1.25	0.176	0.099				200	250	5,050	0.0253	100	46	140	
CO₂: 1 atm, 200°F		0.0915	0.218	0.0127	0.0433				200	43.6	23,800	0.00537	46	518	11.5	Low
1 atm, 1,000°F		0.0415	0.2793	0.0352	0.0827				200	55.9	40,900	0.00313	52	785	10.2	
20 atm, 1,000°F		0.830	0.2793	0.0352	0.0827				200	55.9	2,450	0.0522	52	39	114	
Na: no boiling, 1,540°F	47.3		0.305	30	0.355	208			200	61	32.8	3.9	54	0.61	14,000	High
NaK: no boiling, 1,540°F	45		0.254	16	0.30	63			200	50	42.1	3.04	49	0.86	8,700	High
Li: no boiling, 1,540°F	30		0.9	12	0.65	355			200	180	17.5	7.31	93	0.19	26,000	Low (very low for ⁷Li)[a]
K: no boiling, 1,540°F	39		0.19	17	0.38	144	1425		200	38	64	2.0	43	1.5	6,200	Medium (low for ³⁹K)[a]
boiling, 1,540°F/29 psia	39	0.060	0.19	17	0.38	144		760	30	766	2,060	0.062	192	10.7	6,000	
Cs: boiling, 1,540°F	92	0.41	0.060	10	0.42	83	1274	205	30	207	1,116	0.1147	100	11.2	4,000	Medium
Flinak (LiF–NaF–KF), 1,100°F	132		0.437	2.66	12.6				200	87.4	82.1	1.56	21	3.9	8,000	High
LiF–BeF₂–ZrF₄–UF₄, 1,200°F	141		0.47	0.83	19	813			200	94	71.4	1.79	21	3.4	5,400	High
HTS (NaNO₃, KNO₃, KNO₂), 600°F	115.8		1.85	0.35	7.02				200	370	24.2	5.29	40	0.61	6,400	High
Dowtherm A: no boiling, 600°F	49.3		0.70	0.1037	0.727		495		200	140	137	0.934	26	5.3	2,000	Low
boiling, 600°F/49.3 psia	49.3	0.7237	0.579					114.1	30	132	992	0.129	80	12.4	300	

[a] The high cost of the separated isotopes ⁷Li and ³⁹K can sometimes be justified.

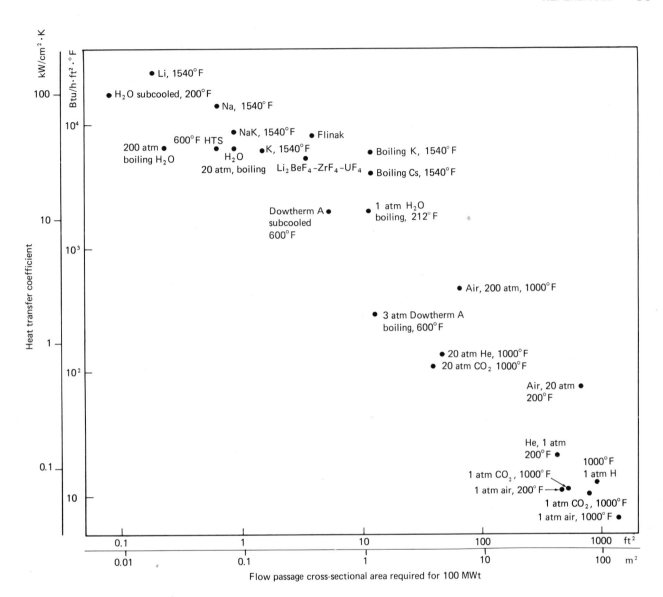

Figure 4.11 Comparison of the heat transfer coefficients obtainable and the flow-passage cross-sectional area requirements for 13 representative fluids used for heat transfer and heat transport purposes. The heat transfer matrix was assumed to consist of 25-mm (1.0 in.) ID passages, the amount of heat removed was taken as 100 MW, and the pumping power was limited to 1% of the heat removed. (Fraas, Ref. 9.)

REFERENCES

1. A. L. London and W. M. Kays, "The Gas Turbine Regenerator—the Use of Compact Heat Transfer Surfaces," *Trans. ASME*, vol. 72, 1950, p. 611.

2. R. A. Bowman, A. C. Mueller, and W. M. Nagle, "Mean Temperature Difference in Design," *Trans. ASME*, vol. 62, 1940, p. 283.

3. K. A. Gardner, "Variable Heat Transfer Rate Correction in Multipass Exchangers, Shell-Side Film Controlling," *Trans. ASME*, vol. 67, 1945, p. 31.

4. R. A. Stevens et al., "Mean Temperature Difference in One, Two, and Three-Pass Crossflow Heat Exchangers," *Trans. ASME*, vol. 79, 1957, p. 287.

5. A. S. Thompson, "Flow of Heated Gases," *Trans. ASME*, vol. 72, 1950, p. 91.

6. W. M. Kays and A. L. London, "Heat Transfer and Flow Friction Characteristics of Some Compact Heat Exchanger Surfaces," *Trans. ASME*, vol. 72, 1950, p. 1075.

7. W. F. Seifert et al., "Organic Fluids for High-Temperature Heat-Transfer Systems," *Chemical Engineering*, October 30, 1972, p. 96.

8. W. M. Kays and A. L. London, *Compact Heat Exchangers*, 3rd ed., McGraw-Hill Book Co., New York, 1984.

9. A. P. Fraas, *Engineering Evaluation of Energy Systems*, McGraw-Hill Book Co., New York, 1982.

5

Boiling Heat Transfer and Flow Stability

Boiling is a complicated business—far more so than is generally realized. In fact, the complexities of the phenomena involved are such that many aspects are not well understood. In spite of the gaps in the information available, the boiling process and related problems are so important that this chapter has been prepared to summarize the state of the art and provide a reasonably good basis for engineering design and development work.

POOL BOILING

The simplest form of boiling is that in which a heated surface is immersed in an open pool. Under boiling conditions the liquid film immediately adjacent to the hot surface ordinarily is heated to a temperature a bit above its boiling point. Once a small bubble forms it grows rapidly as vapor is released from the superheated liquid surrounding the bubble. When the bubble reaches a critical size it breaks away from the surface and moves out into the bulk fluid. Under some conditions the bulk liquid temperature may be sufficiently below that of the hot surface so that the heat of vaporization in the bubble is reabsorbed and the bubble collapses, but boiling nonetheless gives a very high heat transfer coefficient.

HEAT FLUX AND SURFACE TEMPERATURE

The higher the heat flux from the hot surface to the liquid, the greater the amount by which the temperature of the hot surface exceeds the boiling point, that is, the greater the degree of superheat in the boundary layer, and the greater the rate of growth of a bubble. Many measurements have been made of this temperature differential, using many types of surface in many different liquids, under a variety of conditions. A typical set of results is shown in Fig. 5.1 for boiling from a heated wire in an open pool.[1] Heat fluxes of around 300,000 Btu/h · ft^2 are commonly attained with small temperature differences in pool boiling of water. If the surface is heated to too high a temperature in an effort to get a still higher heat flux, the rate of bubble formation becomes so rapid that a vapor blanket tends to form on the surface and separate it from the liquid. Heat transfer then occurs either by conduction and radiation through the vapor blanket or by intermittent contact of the liquid, since instabilities in the movement of the liquid cause waves in the free surface to penetrate the vapor blanket. When this occurs, the local heat transfer rate fluctuates wildly, and severe thermal stresses or vibration may be induced

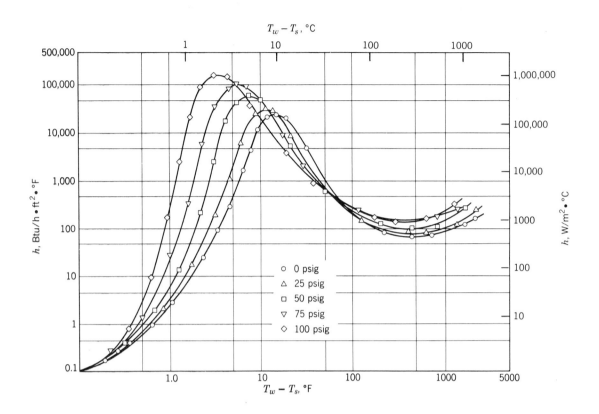

Figure 5.1 Effects of pressure on the relation between heat transfer coefficient and film-temperature drop for water boiling from the surface of a nickel wire. (Farber and Scorah, Ref. 1.)

in the hot surface. To avoid this unstable behavior, boilers are designed so that they do not operate under these conditions; that is, the heat fluxes are kept to values to the left of the peak of the curve in Fig. 5.1.

Nucleate, Transition, and Film Boiling

The stable boiling described in the preceding section, where the stirring action of the bubbles gives rapidly increasing heat fluxes with increasing temperature differences, is referred to as *nucleate boiling*. If an unstable vapor film partially covers the surface and the heat flux tends to decrease with an increase in the surface temperature, the condition is called *transition boiling*. If a vapor film insulates the surface from the liquid so that the heat flux is low even though the surface temperature is much above the boiling point of the liquid, the condition is referred to as *film boiling*.

Burn-Out

Excessive metal surface temperatures may sometimes occur under film-boiling conditions. If the heat flux is essentially independent of temperature, as it is for surfaces heated by thermal radiation in a furnace or by nuclear fission in the fuel elements in a nuclear reactor, the surface temperature may exceed the melting point under unfavorable fluid-flow conditions if the heat flux is too great. The peak of the curve in Fig. 5.1 is often referred to as the *burn-out heat flux*, and also as *departure from nucleate boiling* (DNB).

Relations Between Pool Boiling Regimes

The types of pool boiling just described are indicated in Fig. 5.2. This figure is similar to Fig. 5.1 except that the

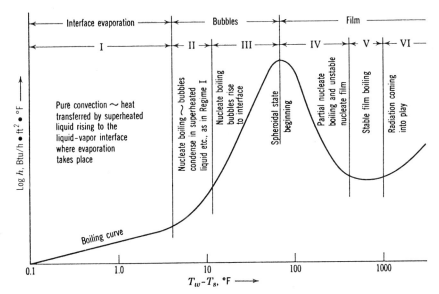

Figure 5.2 Diagram showing the principal pool-boiling regimes and their relative positions on a curve for the heat transfer coefficient plotted as a function of the film temperature drop. (Farber and Scorah, Ref. 1.)

heat transfer coefficient is plotted for a somewhat different set of conditions, thus modifying the apparent shape of the curve so the right end is lowered. Figure 5.2 shows the manner in which the heat transfer mechanism varies with the heat load in a typical system. Note that for low heat loads there are no vapor bubbles in the liquid; thermal convection currents produce sufficient circulation that the heat is removed by evaporation from the free surface. At somewhat higher heat loads, even though bubbles begin to form at the hot metal wall, surface tension forces cause them to contract and disappear when they move upward into liquid having less superheat than that in the zone immediately adjacent to the hot wall. At still higher heat loads, the bubbles rise all the way to the free liquid surface to give conventional pool boiling. If the metal surface temperature is increased still further, an abrupt loss in heat flux occurs, because the surface is no longer wetted by the liquid but becomes increasingly blanketed by a film of vapor. This instability characterizes the transition region between *nucleate* and *film boiling*. The heat flux drops to a minimum with further increases in the surface temperature until no wetted areas remain, and then it rises again as thermal radiation becomes effective in augmenting heat conduction through the vapor film covering the surface.

As the load is increased on a boiler in which the liquid is recirculated, operation ordinarily shifts from Regime I to Regime II during the initial warm-up, and then into Regime III at substantial power. The design of recirculating boilers is normally such that the boiler never operates under conditions to the right of Region III, but

in once-through boilers conditions similar to those of Regime IV may occur.

FORCED-CONVECTION BOILING

The fluid behavior begins to differ from that of pool boiling as the velocity and vapor quality are increased over the low values that prevail under pool-boiling conditions. One of the best ways to visualize these effects is to examine the flow in a single, long tube in which subcooled liquid is first heated, boiled, and then superheated as it progresses from one end of the tube to the other. The characteristics of the flow in each of these regimes is discussed in one of the following sections.

Boiling in Subcooled Liquid

As the liquid temperature rises in passing through the first portion of a heated tube, a region is reached in which the wall temperature appreciably exceeds the boiling point of the liquid, even though the bulk liquid itself has not yet been heated to the boiling point. In this zone bubbles begin to appear on the heated surface, grow, are washed away, and then, as they lose heat to the surrounding liquid, shrink and disappear. If the boundary layer is thick, after leaving the surface the bubbles increase in size as they pass through the superheated liquid in the boundary layer, and then shrink as they move out into the cooler free stream. Figure 5.3a shows flow of this character. The pictures in Fig. 5.3 are

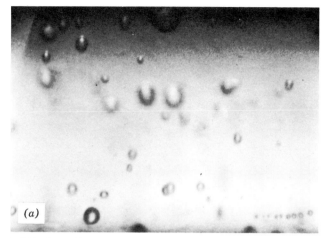

Bubbly flow with subcooled liquid.

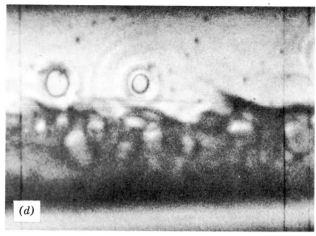

Annular flow with a local vapor quality of 1%.

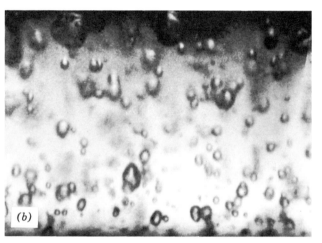

Bubbly flow with saturated liquid.

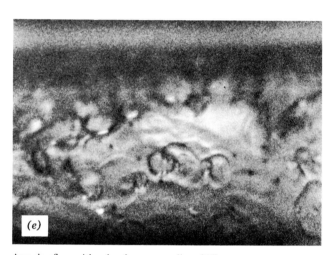

Annular flow with a local vapor quality of 5%.

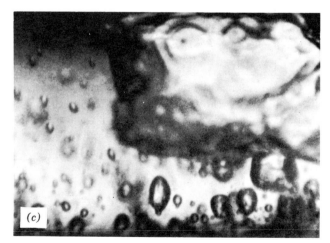

Slug flow with a local vapor quality of 0.1%.

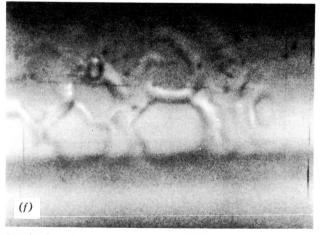

Annular flow with a local vapor quality of 30%.

Figure 5.3   Photos showing the various types of two-phase flow in a horizontal once-through boiler tube using Freon. The flow is from left to right.

frames from a movie taken at 7000 frames/s of boiling Freon flowing through a Pyrex tube. Heat was supplied by high-temperature air flowing through the annulus formed by a concentric surrounding quartz tube.[2] Examination of the frames from high-speed movies of this sort discloses that the bubbles usually form, break free of the surface, collapse, and disappear very rapidly—the entire cycle requiring a period of only about 0.001 s.

Boiling at Very Low Qualities

When the bulk free stream temperature reaches the boiling point, the number of bubbles per cubic inch becomes much greater, since the bubbles do not shrink by losing heat to the surrounding liquid. Instead, the bubbles coalesce in a short distance into larger bubbles that nearly fill the tube, and these move down the tube through an annular region of liquid between slugs of bubbly liquid. Figures 5.3b and c illustrate flow of this type. Note the large bubble in Fig. 5.3c that is moving down the tube between liquid slugs and that, while the quality of the vapor-liquid mixture in frames b and c is about 15% by volume, it is only about 0.1% by weight.

Boiling at Intermediate Qualities

As the volume fraction of vapor in the fluid stream increases to the 50–80% region (depending on operating conditions) the nature of the flow changes markedly, and—if the fluid wets the wall thoroughly—an *annular-flow* regime prevails in which the vapor moves as a continuous stream down through the center of the tube while the liquid adheres to the wall and moves along in an annular film. The flow in this region is shown in Fig. 5.3d. The liquid-vapor interface lies across the middle of this frame. Note that bubbles in the liquid film toward the bottom of the frame give the liquid region a spongy appearance, and that the relatively smooth surface of the liquid film in the upper portion of the frame is marred by the emergence of two bubbles. The bubble near the center of the frame has caused a set of concentric ripples in breaking free of the surface. At the right similar waves can be noted from a bubble that is just outside the field of the camera.

The axial velocity of an annular liquid film is much lower than that of the vapor; that is, the average liquid velocity runs from 3 to 15% of the vapor velocity. Both the vapor and liquid velocities increase with the fraction evaporated as the fluid progresses down the tube, and the liquid film traveling along the wall becomes progressively thinner. Figure 5.3e shows the flow in the region

where the quality is up to about 5% by weight. In this region the vapor-flow rate on a volumetric basis is from 5 to 10 times that of the liquid-flow rate. Figure 5.3f illustrates a similar picture in the region where the quality is up to 30% (i.e., where the volumetric flow rate of the vapor is about 50 times that of the liquid). Depending on the dynamic head and the Reynolds number in the vapor flow, and the Reynolds number and the Froude (or Weber) number in the liquid, waves form on the surface of the annular liquid film, and droplets are torn from the the tops of the waves and are carried off entrained in the vapor.[3] This effect becomes more pronounced as the vapor and liquid velocities increase with increasing vapor quality.

Boiling at High Qualities

The liquid film becomes progressively thinner with further progress along the length of the tube up to the region where the vapor quality runs from 50 to 90%. Then, depending on the surface condition, pressure, flow rate, surface tension, and the wetting properties of the fluid, the flow regime becomes very different in character; dry spots appear on the wall. These grow in number and extent until the rivulets between them dry up (see Fig. 5.3f), and virtually all remaining liquid is in the form of fine droplets suspended in the vapor. The terms *fog* or *mist flow* are ordinarily used to identify this flow regime, although the droplet size is usually of the order of 10 to 100 μm. Because of the turbulent character of the vapor flow and waves in the liquid film, this transition from annular-film flow to dry-wall mist flow moves irregularly back and forth along a limited length of the tube.

The mist present in the vapor appears to originate partly in the transition region between slug flow and annular flow, and partly from droplets torn from the tops of waves in the annular flow region. Surface tension is an important factor in determining the size and quantity of the droplets in the mist.

EFFECTS OF FLOW REGIME ON THE HEAT TRANSFER MECHANISM

Perhaps the most important implications of these various flow regimes are those related to heat transfer. Where the liquid wets the walls, nucleate boiling ordinarily occurs so that the wall temperature seldom exceeds the temperature of the saturated liquid by more than the amount implied by heat transfer data from pool-boiling experiments. In fact, for annular flow, the temperature

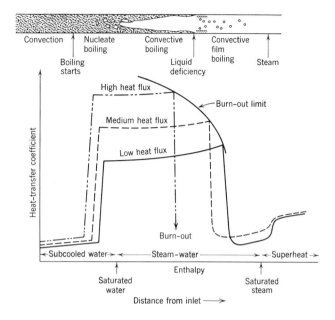

Figure 5.4 Effects of heat flux on the heat transfer coefficient for a once-through boiler tube. (Polomik et al., Ref. 4.)

differential between the wall and the liquid commonly is less than for pool boiling, since direct evaporation from the free liquid surface serves to increase the heat transfer coefficient. This effect increases with increasing vapor quality[4] as the liquid film thickness decreases (Fig. 5.4). In any event, wherever the liquid wets the walls, the heat transfer coefficient is high, irrespective of liquid velocity or whether bubbly flow, slug flow, or annular flow prevail.

In the dry-wall region the heat transfer mechanism becomes drastically different. Usually the heat transfer coefficient between the vapor and the wall is relatively low except at the high mass flow rates obtainable at high pressures (e.g., steam at 2000 psi). At the lower pressures the bulk of the heat transferred is associated with the evaporation of liquid droplets that impinge on the wall. Thus at low pressures the principal factor determining the heat transfer rate may not be heat diffusion through the boundary layer, but may be the rate at which the liquid droplets diffuse from the free stream to the wall. As discussed in a later section, a twisted ribbon, such as that of Fig. 5.5, or other turbulence-promoting device may be very helpful in throwing the liquid droplets against the wall and thus in drying out the mist. This process, of course, increases the heat transfer coefficient by extending the nucleate boiling regime to higher vapor qualities.

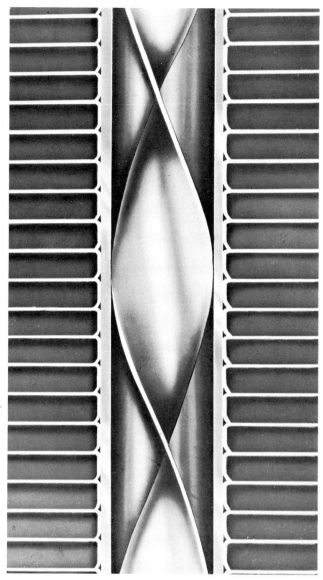

Figure 5.5 Twisted ribbon installed in a finned Freon evaporator tube for an air-conditioning system. (Courtesy The Trane Co.)

LIQUID SUPERHEATING AND THE EFFECTS OF NUCLEATION SITES

The notion that the boiling point of water at standard atmospheric pressure is a fixed and predictable quantity is one of the most sacred of engineering traditions. However, many people are aware that if extremely pure water is placed in a meticulously cleaned glass beaker, it is possible to raise the water temperature to as much as

50°F above its normal boiling point with no sign of boiling. Such a condition is unstable, however, and if boiling once starts, it becomes so violent that it appears to be explosive. This phenomenon of *liquid super-heating* above the boiling point generally has been regarded as a laboratory curiosity. However, in recent years it has been found that "explosive" boiling may occur in engineering equipment when special precautions are taken to maintain the purity of the liquid at a very high level and if the heated surfaces are smooth. While it is not a widespread problem, the phenomenon gives such a good insight into the boiling heat transfer mechanism that it deserves some discussion here.

Bubble Formation

Observation of boiling under many different conditions shows that bubbles invariably start at nucleation sites, usually tiny pits in the hot surface.[4] The bubbles grow, break free, are washed away, and then other bubbles grow from the same nucleation sites. A plume of bubbles streaming to the right of such a nucleation site can be seen in the lower right corner of Fig. 5.3*a*. The other bubbles in that frame originated from nucleation sites upstream, that is, to the left of the field. With highly polished surfaces free of nucleation sites, there is little inclination for a bubble to form, and hence substantial amounts of liquid superheating may occur if the liquid is sufficiently free of impurities so there are no nucleation sites in the form of suspended particles or gas bubbles. Microscopic examination of nucleation sites discloses that they are commonly scratches or pits of such a shape that the release of a bubble does not tend to wash away all the vapor trapped in the nucleation site by inflow of the liquid film, even though it strongly wets the surface. Rather, the geometry is such that the surging liquid surface is stopped at a sharp edge by surface tension forces, and the site is left with a small amount of vapor or noncondensable gas in a pocket. The most effective nucleation sites are not hemispherical or conical depressions in the surface but rather pits having openings smaller in diameter than the void region beneath the surface.[5] This effect can be visualized by examining Fig. 5.6, which shows a typical series of events in the cycle of growth and release of a vapor bubble.

Bubble Size as Affected by Physical Properties

The most critical phase in the growth of a bubble is depicted in Fig. 5.6*b*, where the radius of the free surface

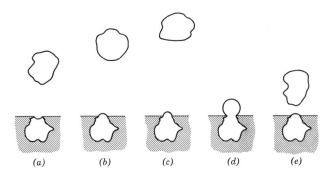

Figure 5.6 Section through a nucleation site showing the principal phases in the formation and release of a vapor bubble under boiling conditions.

surrounding the bubble is at a minimum, and the pressure within the bubble at a maximum. This effect can be seen by relating the surface tension of the liquid and the bubble radius to the difference in pressure between the vapor within the bubble and the pressure of the surrounding liquid. Treating the bubble surface as if it were a spherical shell in tension, we may write

$$P_v - P_l = \frac{2\sigma}{R} \qquad (5.1)$$

where P_v = pressure of the vapor within the bubble, lb/ft^2 (N/m^2)

P_l = pressure of the surrounding liquid, lb/ft^2 (N/m^2)

σ = surface tension of the liquid, lb/ft (N/m)

R = bubble radius, ft (m)

Thus the smaller the radius of the bubble, the higher the pressure that must be generated within the bubble to make it grow beyond the minimum, or critical, radius (which is the one for the condition of Fig. 5.6*b*) and make the nucleation site effective. Similarly, from Eq. 5.1 it can be reasoned that the higher the heat flux (i.e., the greater the temperature difference between the surface and the boiling point of the liquid, and hence the greater the vapor pressure to be generated within the bubble), the smaller the nucleation site that will be effective in generating bubbles.

It has been shown that starting with the relation of Eq. 5.1 it is possible to derive from fundamental thermodynamic considerations the relation between the amount of superheat in the liquid at the wall, that is, the wall temperature minus the saturation temperature, and the principal parameters.[6] One form of this relation, written to apply to all types of liquid, is

$$T_w - T_s \sim \frac{2\sigma T_s (\rho_l - \rho_v)}{\rho_v \rho_l \, \Delta H_v R_i} \qquad (5.2)$$

where ρ_v = density of the vapor, lb/ft³ (kg/m³)
ρ_l = density of the liquid, lb/ft³ (kg/m³)
ΔH_v = heat of vaporization, Btu/lb (kJ/kg)
R_i = the minimum, or initial bubble radius, ft (m)
T_s = saturation temperature of the liquid at the pressure in the liquid, °R (K)
T_w = temperature of the wall, °R (K)

From this it can be deduced that the higher the temperature and the lower the pressure of a boiling liquid, the greater the amount of superheating that may be expected before bubbles are released, and hence the greater the degree to which superheating of the liquid and explosive boiling may become problems. Table 5.1 shows data for a variety of liquids to illustrate the magnitude of the effect. Note that as the pressure in a boiler increases, the expected amount of liquid superheating falls off rapidly; hence the nucleation problem becomes less important.

Table 5.1 shows physical property data for some typical liquids with an indication of their tendency to superheat. From the table it can be seen that boiling alkali metals are inclined to give exceptionally large amounts of superheat, particularly under start-up conditions where the pressures in the boiler are very low. The liquid alkali metals are especially likely to give difficulty; experience shows that in order to avoid corrosion, they must be used in meticulously clean systems, and the liquids must have a very high degree of purity. The amount of superheat in alkali metals may exceed 500°F, and, when this happens, violently explosive boiling occurs.

Boiling Sounds

Experience with water in very clean glass systems shows that the bulk liquid may superheat by about 15°C (27°F), causing intermittent, explosive boiling. Violent surging of the liquid flow through the system occurs, and the initial surge is accompanied by a sound much like a sharp hammer blow. This is followed by a series of lower amplitude "pings" similar to those produced by dropping assorted sizes of small steel balls on a steel plate. The pinging sounds fall off in amplitude and frequency as the liquid loses its superheat by vapor evolution, and cease during the quiescent periods during which the liquid superheats between "explosions." Observation shows that the pinging sounds are associated with the collapse of vapor bubbles, the phenomenon being a sort of implosion. The first loud ping occurs as the liquid surges back into the void left by the first heavy surge induced by the explosion. The pinging sounds are essentially similar in tone quality to those produced in pump cavitation, and are of essentially the same character. Under normal, "smooth" boiling conditions the same sort of sound can

TABLE 5.1   Amount of Superheat Required for Nucleate Boiling at Atmospheric Pressure[a]

Liquid	Normal Boiling Point, °F	Specific Volume Vapor, ft³/lb	Density of Liquid, lb/ft³	Latent Heat of Vaporization, Btu/lb	Thermal Conductivity, Btu/h · ft · °F	Surface Tension, lb/ft	Superheat Required for Typical Nucleation-Site Diameter, °F
Water	212	26.8	59.8	970	0.393	0.00403	30
Mercury	675	4.0	795	126	7.1	0.027	360
Sodium	1618	60.6	46.4	1609	30.1	0.0077	260
Potassium	1400	32.5	41.6	850	18.1	0.0043	125
Rubidium	1270	16.6	82.1	347	11.8	0.003	100
Cesium	1260	10.4	105	214	10.6	0.002	67
Freon-11	112	1.45	89	75	0.052	0.00058	2.4
Benzene (C_6H_6)	176	5.6	54	169	0.08	0.0019	15
Ethyl alcohol (C_2H_6O)	173	9.7	49	367	0.096	0.0015	11

Note: The data for water and the alkali metals were obtained from H. W. Hoffman and A. I. Krakoviak, "Convective Boiling with Liquid Potassium," *Proceedings of 1964 Fluid Mechanics Institute*, Stanford University Press, Stanford, Calif., 1964, p. 19.

[a]As calculated for some representative liquids assuming a heat flux and a set of nucleation sites that yield the same initial bubble diameter in each case as for water boiling with a 30°F superheat. (Calculated from Eq. 5.2.)

be heard, but the amplitude is quite low and the pings come in rapid succession. Since the bubble release is normally random, the sound emerges as a noise, not a tone. However, in some systems the geometry of the boiler may be such that a coupled oscillation occurs, and some element of the boiler vibrates in phase with the bubble-release process to produce a tone. A sound of this sort is called a "boiling song."

Nucleation Devices

In liquid metal systems where explosive boiling often proves to be a problem, it is best to insert nucleation sites deliberately to promote the nucleation of bubbles. If the system is to be kept extremely clean, suspended particles are not acceptable as nucleation sites. Small bubbles of a noncondensable gas, such as air, provide ideal nucleation sites, but this approach cannot be used where difficulties would be posed by corrosion or the presence of noncondensables in the condenser. It is sometimes possible to add a small percentage of a low-boiling-point component dissolved in the main liquid stream to provide the initial small bubbles which can then grow, fed by

vapor from the main component in the liquid stream. The most generally effective arrangement, however, is to make use of pits or crevices in the heated surface. For example, the crevices along the edges of twisted ribbons inserted in tubes may provide excellent nucleation sites.

Investigations show that the size of the active sites decreases with an increase in the heat flux,[6] as should be expected from Eq. 5.2. In this connection it should be pointed out that the surface finish can have a pronounced effect on the boiling heat transfer characteristics. Figure 5.7 shows a set of curves for some typical surface finishes, and illustrates the possible magnitude of the effect.[7] Note that, in this instance, the peak, or burn-out, heat flux is unaffected by the finish. In general, nucleation sites should have effective diameters ranging from 2.5 to 25 μm (0.0001 to 0.001 in.), and should be located near the entrance to the boiling region so that vapor can be evolved from the free liquid surfaces surrounding the bubbles as the flow progresses downstream from the nucleation-site region.

In the annular-film regime, boiling can take place by direct vaporization from the free liquid surface, the heat being conducted from the solid surface into the liquid

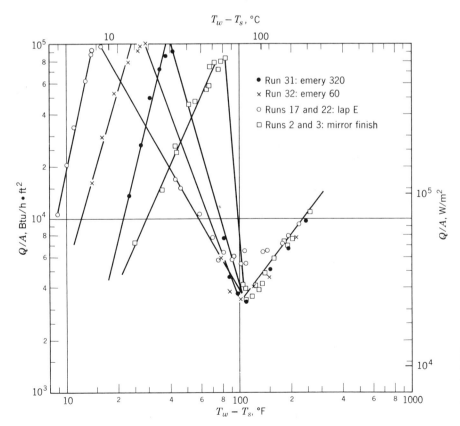

Figure 5.7 Effects of surface finish on the heat transfer coefficient under pool-boiling conditions using pentane boiling from a copper surface. (Berenson, Ref. 7.)

and then through the liquid film to the free surface. This last mode of heat transfer is particularly effective with boiling metals, since the thermal conductivity of the liquid metal is so high that the temperature drop through the annular film is small compared to the amount of superheat required to activate nucleation sites.

Other techniques that can be used to initiate boiling include the use of a hot spot that is deliberately overheated by the amount necessary to initiate bubble formation. However, the disadvantage of such a hot spot is that it may give difficulty with local corrosion. Yet another technique is to employ a throttling orifice near the inlet to the boiling region. The liquid can be super-heated ahead of this orifice so that some flashing occurs in the high-pressure-drop region across the sharp edge of the orifice; thus vapor bubbles are present in the saturated liquid as it flows into the boiler. The difficulty with this arrangement is that vapor bubbles may start to form in the superheated liquid zone upstream of the orifice, in which case extremely violent system-flow fluctuations will occur. These fluctuations result because the "premature" vapor bubbles cannot readily pass through the orifice, and hence the large increase in fluid volume accompanying their formation causes a backward surge of liquid through the liquid-supply system. The resulting flow oscillations are usually much more violent than those that would be experienced with no throttling orifice, and are likely to lead to a burn-out condition.

The flow instabilities associated with liquid super-heating and explosive boiling may be periodic or very irregular, depending on the proportions of the system and the operating conditions. Problems of this sort are usually hard to diagnose, and can be corrected only by installing effective nucleation devices.

HEAT TRANSFER COEFFICIENTS FOR NUCLEATE BOILING WITH FORCED CONVECTION

While boiling heat transfer coefficients depend on a host of factors, values for some typical sets of conditions are very helpful in orienting oneself. As a point of departure, Fig. 5.8 shows the effects of pressure, heat flux, and inlet velocity on the heat transfer coefficient for water for once-through flow in round tubes. It is obvious that the curves of Fig. 5.8 are not directly applicable to other fluids or other conditions, and that it is important to correlate the data available for heat transfer coefficients under forced convection, nucleate-boiling conditions.

The factors affecting the rate at which vapor bubbles remove heat from a hot surface under forced-convection, nucleate-boiling conditions have been investigated analytically in an effort to develop a basis for correlating data for different fluids and operating conditions. Several bases have been developed. Of these, the author prefers one that Levy[8] has developed from the basic equations for bubble-growth rate worked out by Forster and Zuber.[6] It was from these relations that Eq. 5.2 was derived. This correlation is

$$Q/A = \frac{k_l c_{pl} \rho_l^2 (T_w - T_s)^3}{\sigma T_s (\rho_l - \rho_v) B_l} \qquad (5.3)$$

where the notation is as in Eq. 5.2, except for the constants k_l, c_{pl} and B_l. The constants k_l and c_{pl} depend on the fluid and the surface, while B_l is very nearly inversely proportional to $\rho_l \Delta H_v$ so that Eq. 5.3 is closely approximated by

$$Q/A = \frac{k_l c_{pl} \rho_l^3 \Delta H_v (T_w - T_s)^3}{\sigma T_s (\rho_l - \rho_v)} \qquad (5.4)$$

Note the similarity in the quantities involved in Eqs. 5.4 and 5.2.

The validity of this correlation is best indicated by experimental results. Figure 5.9a shows a set of data for a variety of liquids plotted on the conventional Q/A and $(T_w - T_s)$ coordinates together with the same data plotted using the coordinates suggested by Eq. 5.3. Figure 5.9b shows a similar set of curves obtained for water boiling at a wide range of pressures. In both cases note how the correlation of Eq. 5.3 serves to bring the data together to define a single curve. Thus it appears that Eq. 5.3 or 5.4 can be used to predict the relation between the heat flux and the quantity $(T_w - T_s)$ for clean commercial surfaces operating under well-developed nucleate-boiling conditions, irrespective of the liquid used or the operating temperature and pressure.

BURN-OUT LIMITATIONS

The heat transfer coefficient is very high for nucleate-boiling conditions, and is not very sensitive either to the mass flow rate or the vapor quality. In practice, the difficult problem is not so much one of estimating the heat transfer coefficient for nucleate boiling as it is a problem of estimating whether or not there will be a transition from nucleate to film boiling under some unfavorable combination of operating conditions. If this transition

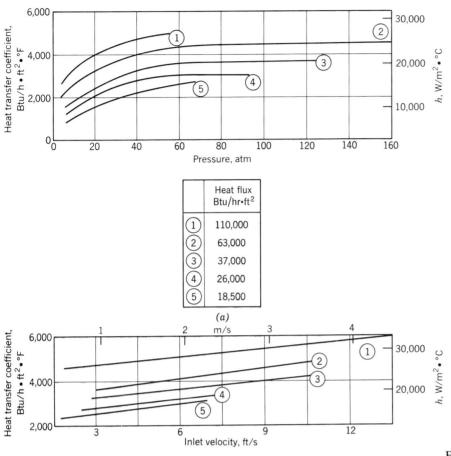

	Heat flux Btu/hr•ft²
①	110,000
②	63,000
③	37,000
④	26,000
⑤	18,500

(a)

	Heat flux Btu/h • ft²	Pressure, atm
①	74,000	120
②	74,000	65
③	70,000	40
④	26,000	62
⑤	30,000	40

(b)

Figure 5.8 Heat transfer coefficients for water flowing through round tubes at low quality as a function of (a) pressure and (b) inlet velocity for a range of heat fluxes. (J. G. Collier, "A Review of Two-Phase Heat Transfer," AERE CE/R 2496, Atomic Energy Research Establishment, Harwell, England. The experimental data are from papers by F. F. Bogdanov, *Izvest. Akad., Nauk., SSSR Otdel. Tekn. Nauk.,* no. 6, 1954, and no. 4, 1955.)

occurs, the heat transfer coefficient drops drastically, because a completely different heat transfer regime prevails.

The transition between nucleate and film boiling seems to depend primarily on the surface heat flux and the local vapor quality, with the mass flow rate, pressure, and surface geometry having important, but less pronounced, effects. Figure 5.10 presents experimental data in this transition region for the burn-out heat flux as a function of vapor quality. Each of the curves shown is for one of a series of tube length-diameter ratios for

axial flow through a round, electrically heated tube.[9] Similar curves illustrated that the burn-out-limited quality for a high heat flux falls off as the mass flow rate is increased at constant pressure.[10] While this at first seems strange, when we consider that fluid friction, turbulence, and wave formation increasingly tend to break up the annular-liquid film and leave dry spots as the mass-flow rate is increased, it is evident that the experimental data are consistent with the behavior that should be expected.

The effects of pressure on the burn-out-limited quality

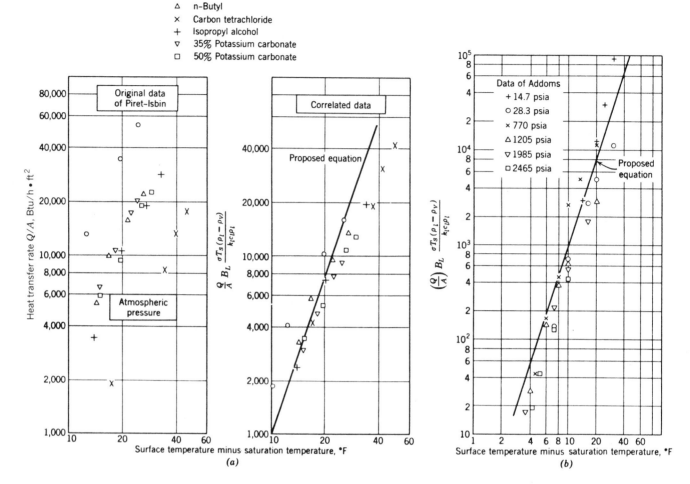

Figure 5.9   Correlation of pool-boiling heat transfer data for (a) a variety of different fluids, and (b) for water at a wide range of pressures. (Levy, Ref. 8.)

at some typical high heat fluxes are shown in Fig. 5.11 for two typical geometries: flow through a round tube and axial flow through a bundle of heated rods.[10] The more complex geometry of the rod bundle apparently leads to a turbulence pattern in the vapor stream that induces dry spots at a much lower vapor quality than for flow through the inside of a simple round tube. As shown in Fig. 5.11, in both cases the burn-out-limited quality increases with pressure.

Under supercritical pressure conditions the density of the vapor is so high that the heat transfer coefficient is very high even with dry vapor, especially in the vicinity of the critical temperature. This effect is indicated by the experimental data in Fig. 5.12.

Effects of Surface Wetting

Difficulties are sometimes experienced with poor wetting even in water systems. In mercury boilers the problem is chronic. The effects are ordinarily small for pool boiling or for bubbly flow, but they drastically reduce the burn-out heat flux obtainable under annular-film flow conditions. Dry spots and burn-out have been observed in rod bundles at heat fluxes as low as 15,000 Btu/h · ft² with water at atmospheric pressure and vapor qualities of around 5%. Figure 5.13 presents some frames from high-speed movies taken at these conditions, and shows dry spots surrounded by the irregular lip of the liquid film. Note that the contact angle both for the liquid film

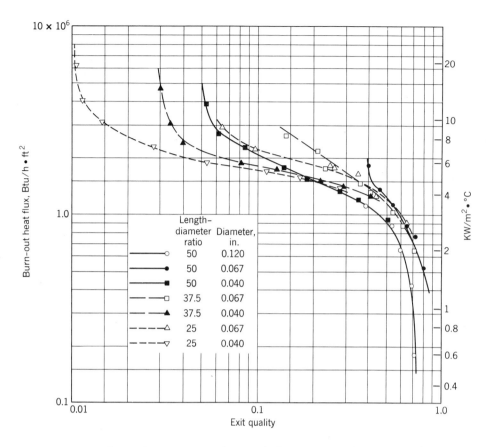

Figure 5.10 Effects of vapor exit quality on the burn-out heat flux for water flowing at 14.7 psia through round tubes having various length-diameter ratios. (Lowdermilk and Weiland, Ref.9.)

around the dry spot and for the droplets skittering across the dry, hot surface clearly shows poor wetting. The condition was corrected by adding a small amount of sodium bicarbonate to the water. Experience indicates that for stainless steel surfaces the water should have a pH of 12.

It is possible to exploit poor wetting to promote nucleation. For example, it has been shown that spraying a steel surface so that it is covered with tiny dots of Teflon increases the heat transfer coefficient to boiling water by a factor of 2 to 5.[12]

PRESSURE DROP UNDER TWO-PHASE FLOW CONDITIONS

Estimation of the pressure drop for two-phase flow is greatly complicated by the variety of possible modes of flow. For bubbly flow, to a first approximation the effect of the bubbles is roughly equivalent to an increase in the viscosity of the liquid. For annular flow, the situation is much more complex since the liquid or the gas flow may be either laminar or turbulent. Thus four possible two-phase, annular-liquid-film flow regimes may prevail (i.e., both components turbulent, both laminar, the gas flow turbulent with the liquid flow laminar, and the liquid flow turbulent with the gas flow laminar). Furthermore, there may be much or little mist entrained in the gas, and this affects the interchange of momentum as droplets enter or leave the gas stream, thus affecting the pressure gradient.

Ratio of Two-Phase and Single-Phase Pressure Drops

One way of correlating two-phase pressure-drop data is to relate it to the pressure drop that would prevail if only the liquid were flowing. This can be done by plotting the

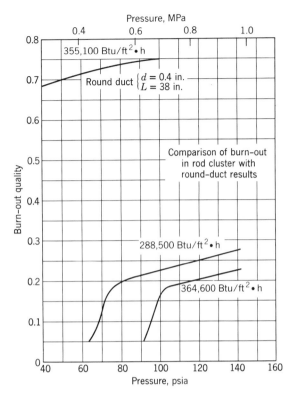

Figure 5.11 Effects of pressure on the vapor quality at burn-out for water flowing axially between 0.50-in.-diameter heated tubes on an equilateral triangular pitch, with an 0.080-in. space between adjacent rods. (Plotted from data given by Becker, Ref. 10.)

ratio of the experimentally determined two-phase pressure drop to the pressure drop for the liquid only against some function of the ratio of the gas and liquid volumetric flow rates, that is,

$$\frac{\Delta P_{TP}}{\Delta P} = f\left(\frac{\rho_l G_g}{\rho_g G_l}\right) \qquad (5.5)$$

The results of such studies indicate that the two-phase pressure drop commonly runs from two to three times the pressure drop for the liquid alone, the factor increasing with the gas-volume flow fraction.[13,14] The situation is complicated in that the pressure drop depends on the flow regime, and this in turn depends on the Reynolds numbers in the two streams. Furthermore, the relative importance of gravitational force varies with the attitude of the flow passage and may have important effects; that is, flow upward in vertical or inclined passages may give values substantially different from those found for downward or horizontal flow.[15]

Bubbly Flow

Under bubbly flow conditions, Wallis[15] found that experimental data for the two-phase pressure drop are correlated well by the relation

$$\frac{\Delta P_{TP}}{\Delta P_l} = 1 + 3\left(\frac{\rho_l G_g}{\rho_g G_l}\right)(G \times 10^{-6})^{0.33} \qquad (5.6)$$

where the subscripts l and g indicate the liquid and gaseous phases, and TP indicates two-phase flow. Figure 5.14 shows data for tube ID's from 3/8 to 7/8 in. and a range of liquid flow rates. The data are plotted on the coordinates suggested by Eq. 5.6. Observations through the transparent walls of the tubes during the tests showed that, where the experimental points fell to the right of the straight line for bubbly flow, the flow regime changed to annular-film flow.

Annular-Film Flow

Under annular-liquid-film flow conditions the liquid moves along the walls at roughly 5% of the gas velocity, and the gas velocity is much more important than the liquid-flow rate in determining the pressure drop. One of the best ways of relating experimental data is to plot the superficial friction factor as a function of the superficial Reynolds number for the gas, calculated as if there were no liquid phase present. Figure 5.15 illustrates curves of this sort for a series of superficial Reynolds numbers for the liquid, calculated as if only the liquid were flowing through the tube. Note that the higher the liquid-flow rate, the higher the superficial friction factor. This is expected, since increasing the liquid-flow rate not only reduces the flow-passage area available for the gas phase, but also increases the wave height in the liquid film and hence the effective surface roughness.

The sharp change in slope of the curves of Fig. 5.15 was found to be associated with the presence or absence of mist in the gas phase. As the gas flow was increased from a low value with essentially no mist in the gas, the wave height in the liquid film increased even though the liquid-flow rate was held constant. As can be seen in Fig. 5.16, this increase in wave height (and the consequent increase in friction factor) continued until droplets began to be torn from the waves, after which the wave height decreased as the increasing shear forces removed more

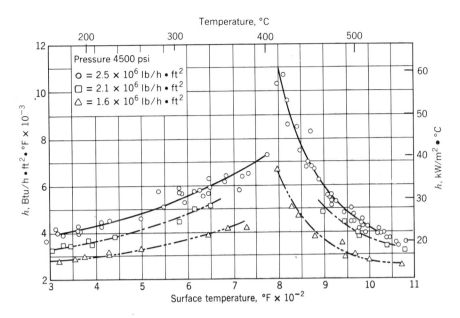

Figure 5.12 Heat transfer coefficients for several mass-flow rates of water in a uniformly heated once-through boiler tube operating at 4500 psia. (Dickinson and Welch, Ref. 11.)

and more liquid droplets from the crests of the waves. The effects of the superficial liquid Reynolds number on the value of the superficial gas Reynolds number at which this transition occurred are shown in Fig. 5.17 for the conditions of Fig. 5.15. The points plotted include both visual observations of the beginning of mist formation and the indications of transition in flow regime indicated by the peaks of the curves in Fig. 5.15.

Comparison of Data for Gas and Vapor

The experimental difficulties associated with both acceleration effects and heat losses in tests with vapor-liquid mixtures have led most research workers to employ air-liquid mixtures rather than the vapor of the liquid with which they were working. However, the data available indicate that this does not introduce any appreciable error.[14]

Total Pressure Drop

The large increase in fluid velocity (and momentum) associated with boiling flow through tubes may introduce a static pressure drop as large as the frictional pressure drop discussed in the preceding section. Gravitational forces are also important, especially in tall boilers. These effects should be included in any general expression for the overall pressure drop. In view of the large changes in flow velocity along the length of a boiler tube, it is best to begin by writing an expression in differential form, that is,

$$-dP = f_{TP} \frac{vG^2}{2g} \frac{dl}{D_e} + \frac{G^2}{g} dv + \frac{\sin \phi \cdot dl}{v} \quad (5.7)$$

where v = specific volume of mixture, ft^3/lb
$\quad f_{TP}$ = two-phase friction factor (dimensionless)
$\quad \phi$ = inclination of the tube axis relative to the horizontal, deg

The first term on the right-hand side of Eq. 5.7 is the frictional component, while the second and third terms are the momentum and elevation components, respectively. The total pressure drop can be obtained by integrating Eq. 5.7, but, in order to perform the integration, a knowledge of the two-phase friction factor and the variation of the quality (or the specific volume of the mixture) with the distance is required. The two-phase friction factor f_{TP} can be determined from the informa-

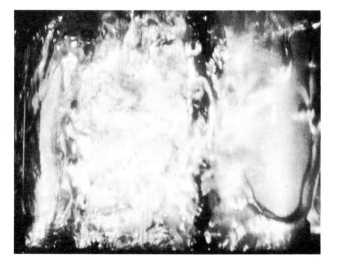

Figure 5.13 Close-up view of two of a cluster of 0.5-in.-diameter electric heaters used to simulate fuel rods for a boiling water reactor. Incipient burn-out conditions caused by

tion given. An average value usually serves, and simplifies the integration. For fluid flowing and evaporating in heated channels, the specific volume of the mixture varies with the distance, because vapor is continuously added to the stream along the heated channel until 100% quality is reached. If the pressure drop along the channel is large enough to affect the saturation pressure, the specific volume changes along the channel because it is pressure-dependent. For cases in which the pressure-dependent properties vary appreciably with the distance along the heated channel, a general expression for evaluating the two-phase pressure drop, as obtained by integrating Eq. 5.7, is[16]

$$\Delta P = \int_{L_1}^{L_2} \frac{fG^2v'}{2gD_e}\left[1 + X\left(\frac{v''}{v'} - 1\right)\right]$$

$$+ \frac{\dfrac{G^2v'}{g}\left(\dfrac{v''}{v'} - 1\right)\dfrac{dx}{dL} + v'\left[1 + X\left(\dfrac{v''}{v'} - 1\right)\right]}{1 + X\dfrac{dv''}{dP}\dfrac{G^2}{g}}\, dL$$

$$\tag{5.8}$$

where
f = single-phase (i.e., liquid) friction factor, dimensionless
L = length in the direction of flow, ft
P = pressure, lb/ft^2
$\Delta P = P_2 - P_1$ = pressure drop, lb/ft^2
G = mass flow rate per unit area, lb/ft$^2 \cdot$ h
v' = specific volume of liquid, ft^3/lb
v'' = specific volume of vapor, ft^3/lb
X = vapor quality at L, lb vapor/lb (vapor + liquid)

A step-by-step integration of Eq. 5.8 is required if the pressure-dependent properties vary appreciably along the channel.

For a linear variation in the quality X, and for X_2 $dV''/dP\, G^2/g < 0.01$, Eq. 5.8 reduces to

$$\Delta P = -\frac{dP}{dL}(L_2 - L_1)\left[1 + A\frac{X_1 + X_2}{2}\right]$$

$$+ AC(X_2 - X_1) + \frac{L_2 - L_1}{X_2 - X_1}\frac{\sin\phi}{Av'}\ln\left(\frac{1 + AX_2}{1 + AX_1}\right)$$

$$\tag{5.9}$$

nonwetting are evident in the form of a dry spot on the fuel element at the right of each frame. The vapor quality was about 5%, and the film speed 3500 frames per second. (Courtesy Oak Ridge National Laboratory.)

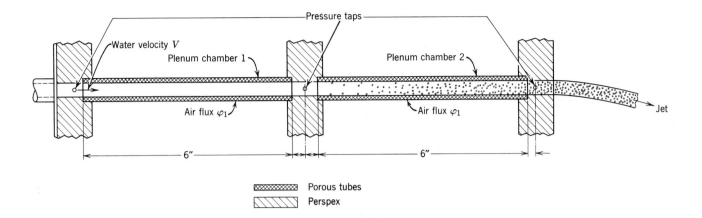

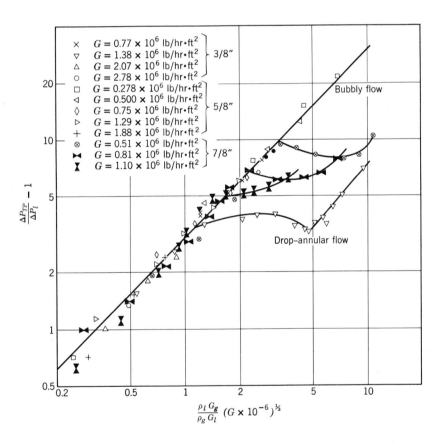

Figure 5.14 Pressure-drop correlation for bubbly flow conditions. The data were obtained isothermally with an air-water mixture in tubes ranging from $\frac{3}{8}$-in. to $\frac{7}{8}$-in. ID. (Wallis, Ref. 15.)

where

$$A = \left(\frac{v''}{v'} - 1\right) \qquad C = \frac{G^2}{g} v' \qquad -\left(\frac{dP}{dL}\right)_{L_1} = \frac{f G^2 v'}{2 g D_e}$$

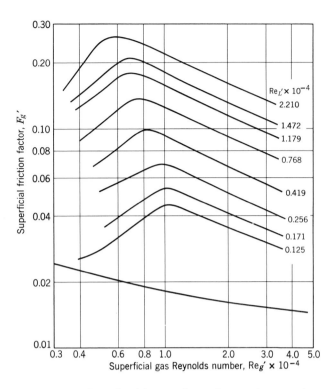

Figure 5.15 Superficial friction factor for annular two-phase flow vertically downward. (Chien and Ibele, Ref. 3.)

A special case of Eq. 5.9 for $\phi = 0$ (i.e., horizontal tubes) and $X_1 = 0$ (i.e., all liquid at the saturation temperature at the inlet to the heated length) gives

$$\Delta P = \frac{fG^2v'}{2gD_e}(L^2 - L_1)\left[1 + \frac{1}{2}X_2\left(\frac{v''}{v'} - 1\right)\right]$$
$$+ X_2\frac{G^2v'}{g}\left(\frac{v''}{v'} - 1\right) \qquad (5.10)$$

For the preheater and superheater sections there is little change in specific volume, and hence the momentum component of the pressure drop is negligible. (In Eq. 5.10 the first term is the friction component and the second term the momentum component.) The ratio of the momentum to friction pressure drop for the boiler section, from Eq. 5.10, is

$$\frac{\Delta P\,(\text{friction})}{\Delta P\,(\text{momentum})} = f\left(\frac{L_2 - L_1}{2D_e}\right)\left[\frac{1}{X_2\left(\frac{v''}{v'} - 1\right)} + \frac{1}{2}\right]$$
$$(5.11)$$

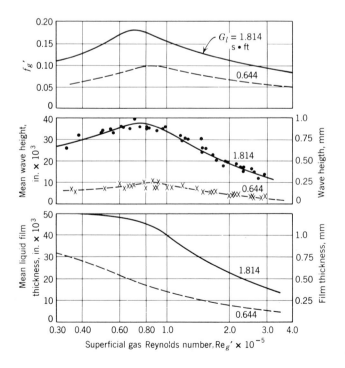

Figure 5.16 Effects of Reynolds number on the superficial friction factor, the mean height of the disturbed liquid layer, and the mean liquid-film thickness for annular two-phase flow. (Chien and Ibele, Ref. 3.)

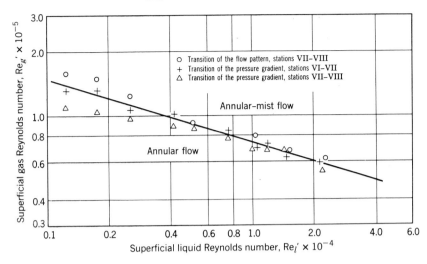

Figure 5.17 Effects of superficial liquid Reynolds numbers on the superficial gas Reynolds number at which droplets begin to be torn from the liquid surface, giving the transition from annular flow to combined annular and mist flow. (Chien and Ibele, Ref. 3.)

where $L_2 - L_1$ = length of the boiler section
X_2 = quality of vapor at the boiler outlet

It is apparent from Eq. 5.11 that for a given vapor quality and pressure, the longer the boiler section (i.e., the lower the heat flux), the less important the momentum component of pressure drop. For a given tube length and vapor quality at the boiler exit, the momentum component of the pressure drop is more important at low pressures (i.e., high values of v''/v') than at high pressures.

HEAT TRANSFER COEFFICIENTS FOR NUCLEATE BOILING IN CRYOGENIC SYSTEMS

Although cryogenics is a highly specialized field, it seems worthwhile to review the heat transfer problems here, partly because the need for work in this area is becoming more common and partly because the experience with cryogenic fluids in recent years provides valuable insights and additional perspective on boiling heat transfer processes.[17] For example, the heats of vaporization and the surface tensions of the cryogenic fluids differ substantially from those of the more conventional fluids, thus providing a wider range of physical properties for appraising correlations. Further, it is inherent in their use that they must extract heat by boiling, for it is impractical to subcool these fluids. This is particularly true for liquid helium which must be used in the 5 K region. Cryogenic systems are also characteristically much cleaner than more conventional systems; there are no other dissolved gases, for example, because they have

all been removed in the preparation process. The metal surfaces are also exceptionally clean because the last step in preparing the system is usually thorough evacuation.

Surface Temperature Difference and Heat Flux

As with conventional fluids, there must be some superheating of the cryogenic fluid to initiate nucleate boiling, the amount of superheat required being proportional to both the surface tension and the boiling point expressed in absolute temperature units, as indicated by Eq. 5.2. Both of these quantities are small, hence boiling is initiated with very little superheat, the amount required increasing with the boiling point of the fluid, as indicated in Fig. 5.18 for the three fluids most used (i.e., helium,

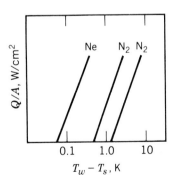

Figure 5.18 Representative nucleate-boiling curves for helium, hydrogen, and nitrogen, showing the reduced wall superheat required for progressively lower boiling temperatures. (Smith, Ref. 17.)

hydrogen, and nitrogen), which have boiling points at atmospheric pressure of 4, 20, and 77 K, respectively. For these fluids the shape of the curve for the heat flux as a function of the difference in temperature between the surface and the local boiling point is also similar to that for conventional fluids, though the peak heat flux is much reduced because it is roughly proportional to the heat of vaporization. This similarity is evident in Fig. 5.19, which was obtained for liquid nitrogen at about 75 K.[18] As with conventional fluids, the shape of the curve and the temperature difference for initiation of boiling depend on the surface finish and treatment.[19] The curves for helium and hydrogen are similar to Fig. 5.19, but difficulties in measuring tiny differences in temperature near absolute zero coupled with the effects of differences in surface finish lead to some spread in the results reported by different investigators. Figure 5.20 presents a comparison of several sets of experiments conducted with helium.[17] Note that the set of curves at the lower left is for the nucleate-boiling region, while the set toward the upper right lies in the film-boiling region. Reference 17 also gives a similar set for hydrogen in which the peak heat flux is higher than for helium by about a factor of 10 and the temperature difference is greater by about 0.3 K, as one would expect from their respective physical properties. It should also be pointed out that with these

cryogenic fluids there is a hysteresis effect of heat load; the difference in temperature between the surface and the boiling point is greater if the condition is approached by increasing the heat flux than if it is approached by reducing the flux.[19] This is another factor that is responsible for the spread of the data in Fig. 5.20.

In summary, experience in the design and testing of cryogenic heat transfer equipment indicates that the same basic heat transfer and fluid flow relations apply as for conventional fluids, but one must make proper allowances for differences in the physical properties.[20]

CHARACTERISTICS OF MIST FLOW

The discussion up to this point has been concerned primarily with nucleate boiling because most boilers are designed to operate in the nucleate-boiling mode. This stems from the fact that the heat transfer coefficient for nucleate boiling is generally 20 to 100 times that for forced convection of the vapor, hence the surface area required is much less for a given heat load. However, in some applications the structure of the steam generator can be simplified by employing a "once-through" design in which feedwater enters one end of a boiler tube and superheated vapor emerges from the other end. The

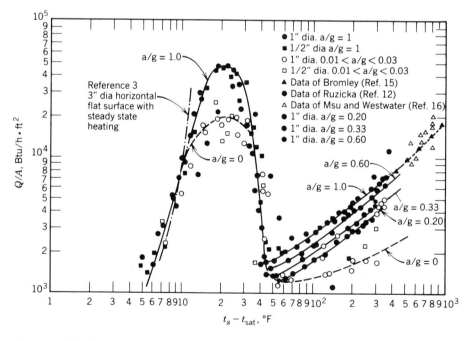

Figure 5.19   Film boiling of nitrogen under standard and fractional gravity conditions, and nucleate boiling under standard gravity and free-fall conditions. (Merte, Ref. 18.)

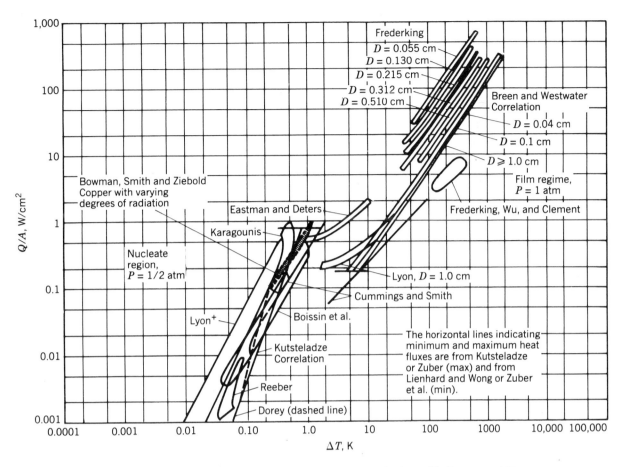

Figure 5.20 Plots of envelopes for data on the nucleate boiling of helium near 4 K as obtained by different investigators. (Smith, Ref. 17.)

reason that this structurally simple approach is not often followed is that, once there is no longer a liquid film on the tube wall, most of the wall area at any given instant is dry, and the heat transfer process is characterized by the low heat transfer coefficient of the vapor. This not only greatly increases the surface area required, but it subjects the tubes to locally severe and fluctuating thermal stresses. There is some enhancement of the heat transfer by droplets striking the wall, but they do so at small angles and are blown off by the vapor generated in the contact zone. While droplets striking the wall increase the heat transfer coefficient by as much as a factor of 10 at low vapor qualities, the effect falls off rapidly as the vapor quality increases. The problem is aggravated by the inhibiting effect of surface tension in the tiny droplets of mist, permitting them to superheat and yet remain in liquid form, so that measurements of the vapor temperature may indicate a superheat of 20 to 50°C when the vapor quality is only 90%. This has been a problem ever

since the first monotube boilers were introduced for motor torpedo boats and steam automobiles nearly 100 years ago. These units, with a blow-torch type of flame directed down the center of a tightly coiled tube, looked wonderfully attractive, but the entrained moisture reduced the effectiveness of the boiler-tube surface and led to problems such as losses in turbine efficiency caused by moisture churning (the interchange of momentum between the rotor and the stator) and turbine blade erosion.[21]

Droplet Size

The droplet size in the mist flow in a boiler tube at the point where the dry-wall condition develops has proved difficult to determine. Fortunately, a very clever experiment has been devised to yield a good picture of the droplet-size distribution.[22] In these tests, feedwater was introduced through thin annular slots at the base of an electrically heated boiler tube to produce the classical annular-liquid-flow regime. The feedwater flow and the heat input were adjusted to bring conditions almost to the dry-wall condition at the tube outlet, and a skimming arrangement was employed to carry off the last bit of liquid on the wall at the tube outlet. A microscope was focused on the mist flow emerging from the tube so that pictures could be taken to determine the drop-size distribution. The annular-film thickness was measured with an electrical resistance probe mounted flush with the inner surface of the tube wall. Figure 5.21 shows the liquid film thickness for operation at various liquid and vapor flow rates leaving the boiler tube at a steam pressure of 2 atm, whereas Fig. 5.22 shows the flow rate in the liquid film as a function of the total liquid flow rate for various steam flow rates. Figure 5.23 shows the droplet-size distribution for a typical set of conditions, while Table 5.2 shows the droplet sizes for a range of conditions. Table 5.3 shows the fraction of the total liquid flow that would be in the size range above the droplet diameter at the peak of a curve of this type for a range of such diameters.

The diameter of droplets shed from the ends of tubes in a recirculating boiler under annular-film flow conditions with a fairly high vapor exit quality would probably run substantially larger than in Fig. 5.23, a matter of

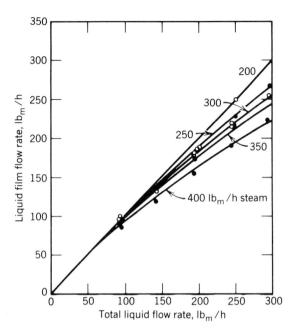

Figure 5.22 Effects of steam- and total liquid-flow rates on the flow rate in the liquid film just before reaching the dry-wall condition. (Pogson, Ref. 22.)

special interest in the design of vapor separators as well as in the design of superheaters. As a first approximation, one would expect the droplet diameter to be roughly equal to the liquid-film thickness at the point

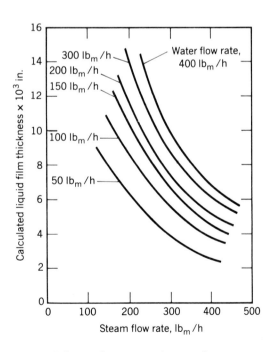

Figure 5.21 Effects of steam- and water-flow rates on the thickness of the water film on the wall of a boiler tube just before reaching the dry-wall condition. (Pogson, Ref. 22.)

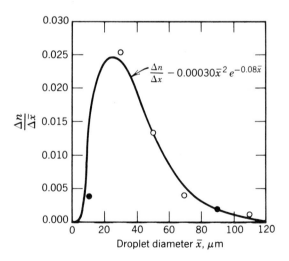

$$\frac{\Delta n}{\Delta \bar{x}} - 0.00030\bar{x}^2 e^{-0.08\bar{x}}$$

Figure 5.23 Test data for the water-droplet-size distribution in the mist flow emerging from the boiler tube of Figs. 5.21 and 5.22 plotted with the empirical curve of Nukiyama-Tanasawa. (Pogson, Ref. 22.)

TABLE 5.2 Droplet-Size Distribution from Photographic Analysis (Pogson, Ref. 22.)

Steam-Flow Rate, $lb_m/\mu m$	Water-Flow Rate, $lb_m/\mu m$	Number of Photos	Total No. of Drops	Drops 0–20 μm	Drops 21–40 μm	Drops 41–60 μm	Drops 61–80 μm	Drops 81–100 μm	Drops 101–120 μm	Mean Dia., μm	Root Mean Cube Dia. μm
300	300	9	70	6	33	19	7	4	1	42	52
300	250	7	97	7	51	27	8	2	2	40	49
300	200	6	43	3	22	11	3	2	2	43	55
300	150	5	33	2	17	7	3	3	1	44	56
300	100	6	27	3	16	7	1	1	0	39	34
Totals		33	270	21	139	71	22	11	6		

where the droplets were shed. The film thickness indicated by Fig. 5.21 is around 0.010 in., or 250 μm. This is much larger than the 40 μm of Table 5.2 for the droplets in the mist flow. Work on the size of droplets shed from the liquid films flowing off the trailing edges of stator blades in turbines has shown that these relatively large droplets are drawn into filaments as they are picked up by the vapor stream, and the filaments are broken up into small droplets by turbulence in the vapor to give droplet diameters in the 100- to 300-μm range.[23] The diameter varies with the passage geometry, the vapor velocity, and the physical properties of the vapor and liquid, and may be correlated by the following expression:

$$D \sim \frac{\mu_l^{0.42}\,\sigma^{0.5}}{\rho_v^{0.67}\,\rho_l^{0.167}\,V_v^{1.33}} \qquad (5.12)$$

where D = particle diameter
  μ_l = liquid viscosity
  σ = surface tension
  ρ_v = vapor density
  V_v = vapor velocity

This correlation was applied to the data of Table 5.2 to yield the estimated droplet sizes of Table 5.4 for some representative conditions of interest for both water-steam and potassium boilers. Air-water spray systems that

TABLE 5.3 Fraction of Liquid Mass Removed If All Droplets Above a Certain Size Are Removed

Most Probable Droplet Size[a], μm	Removal of All Droplets Larger Than				
	2 μm	5 μm	10 μm	50 μm	100 μm
5	96%	94%	89%	31%	0%
15	99%	99%	97%	45%	2%
25	99%	99%	99%	60%	7%
50	100%	100%	100%	99%	60%

95% Mass Point

Most Probable Droplet Size[a], μm	Size of Droplet Above Which 95% of Mass is Found, μm
5	3
15	13
25	23
50	48

[a]The shape of the droplet-size-distribution curve was taken to be that shown in Fig. 5.23. Inasmuch as the droplet size for which this curve peaks will vary with the physical properties of the liquid (i.e., the type of liquid and its temperature) and with both the liquid and vapor flow rates, these tables were prepared for a range of conditions giving droplet diameters at the peak of the curve of 5, 15, 25, and 50 μm.

TABLE 5.4 Mean Droplet Sizes for Typical Cases

| Working Fluid: | Steam | | | Potassium | |
Point in System:	Boiler	Boiler	Turbine Out	Boiler	Turbine Out
Temperature, °F	540	250	100	1540	1040
Pressure, psia	963	30	0.95	30	1.5
Vapor					
Density, lb/ft³	0.2165	0.0727	0.0028	0.0543	0.0038
Viscosity, lb/h · ft	0.053	0.033	0.028	0.0509	0.042
Liquid					
Density, lb/ft³	46.3	58.9	62	42.6	44.05
Viscosity, lb/h · ft	0.22	0.64	1.7	0.288	0.4
Surface tension, lb/ft	0.00158	0.00377	0.0048	0.00412	0.0055
Velocity, ft/s	100	100	1036	100	1036
$\rho_v^{0.67}$	0.3587	0.1727	0.0195	0.1420	0.0239
$\rho_l^{0.167}$	1.90	1.98	1.99	1.87	1.88
$\mu_l^{0.42}$	0.5294	0.8291	1.2497	0.5929	0.6806
$\sigma^{0.5}$	0.0397	0.0614	0.0693	0.0642	0.0742
$V_v^{1.33}$	457	457	10243	457	10243
Droplet-size parameter	0.0000676	0.0003265	0.0002178	0.0003133	0.0001096
Droplet mean dia, μm	17	82	55	79	28

might be used to investigate problems such as those of vapor separators, mist flow in manifolds and headers, and the like yield droplets in the same size range.

HEAT TRANSFER ENHANCEMENT FOR ONCE-THROUGH BOILERS

Considerable work has been done on the problems of once-through boilers because of incentives to use them for various applications such as nuclear power plants employing water-cooled reactors and potassium vapor, Rankine cycle, space power plants. The principal approach has been directed toward keeping the liquid film on the tube wall up to higher vapor qualities by measures such as the provision of grooves in the wall so that capillary forces will be more effective in holding the liquid on the wall. The second approach has been to enhance the heat transfer coefficient in the dry-wall, mist-flow region by means of turbulating devices such as twisted tapes that are intended to centrifuge the liquid droplets to the hot tube wall. Both approaches have yielded some promising geometries at the expense of extra complexity and increased pressure drop.

Ribbed or Fluted Tubes

To provide a safe margin against the development of a dry wall and possible tube burn-out in fossil fuel boilers, the vapor quality at the exit from a boiler tube must be limited to a relatively low level. Typical values recommended by Babcock and Wilcox[24] are shown in Fig. 5.24 for a range of steam pressures and heat fluxes. Providing ribs (or flutes), which may be axial or spiraled much like rifling in a gun barrel, has proved to be an effective way to retain the liquid film on the tube wall to much higher vapor qualities than for a plain tube. Figure 5.25 shows the pronounced increase in vapor quality obtainable with a ribbed tube developed by Babcock and Wilcox; in this instance the amount of liquid in the vapor entering the mist-flow region was reduced by a factor of about 10.[24]

Quite a variety of other enhanced surfaces have been investigated in a series of programs directed at improvements in evaporators for water desalination, "ocean thermal difference" (OTEC) plants, and space power plants.[25–27] This work has disclosed that the performance of some of the devices that looked promising was disappointing; liquid films that deposit on a twisted tape, for example, keep that liquid from reaching the hot tube wall where it would evaporate. Four of the more promising of the many surfaces tested are shown in Fig. 5.26.[27] The most effective of these was found to be the fluted surface of 5.26b whose potential was suggested by work on heat pipes. The effectiveness of these surfaces in delaying dryout of the liquid film on the wall is compared in Fig. 5.27, while the improvement in the heat transfer coefficient for operation with dry steam is shown in Fig. 5.28. (The surface area in Fig. 5.27 was taken as

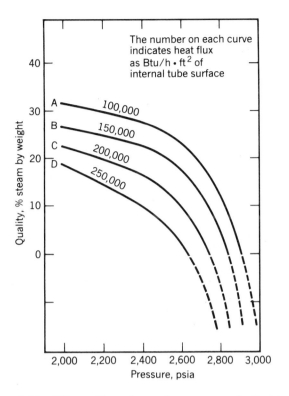

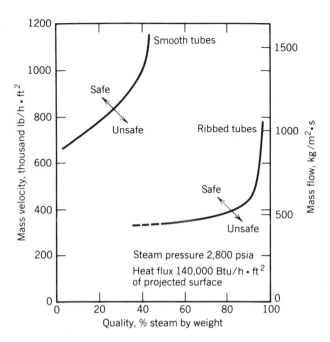

Figure 5.24 Effects of heat flux and pressure on the limiting steam quality for plain boiler tubes. (Courtesy The Babcock and Wilcox Co.)

Figure 5.25 Increase in the limiting steam quality obtainable with ribbed tubes in place of plain tubes in steam boilers. (Courtesy The Babcock and Wilcox Co.)

that for a plain tube of the same internal diameter, while for Fig. 5.28 the total extended area was used.) Note the high vapor qualities obtained with the finely fluted surface either with or without the addition of the wire coil turbulator. Adding the wire coil served to defer wall dryout to somewhat higher vapor qualities and enhanced heat transfer in the subsequent mist flow region, but, as shown in Fig. 5.29, it also increased the pressure drop by a factor of over five. Whether the use of any of these surfaces would be justified in any particular application would depend on balancing the premiums associated with reducing the heat exchanger weight and size against the cost, pumping power, and fabrication complexities of the candidate surfaces as compared to plain tubes.

The test data of Figs. 5.27 to 5.29 were obtained with water; substantially similar results were obtained with potassium with excellent correlation between the performance predicted from the water tests and the actual performance found in the tests with potassium.

Supercritical Pressure Effects

At supercritical pressures the physical properties of a fluid undergo large changes in the vicinity of the critical

temperature. A set of data for water is shown in Fig. 5.30; other fluids such as carbon dioxide and hydrogen have similar characteristics. One consequence of these effects is that the heat transfer coefficient drops off drastically with an increase in the heat flux in the region of the critical temperature, an effect indicated by Fig. 5.31. This has sometimes led to burnout conditions and severe damage. The problems are too complex and specialized to treat here; the reader can probably find a particular problem treated in the extensive literature to which References 28 to 30 provide a good entree.

FLOW STABILITY IN HEAT TRANSFER MATRICES UNDER BOILING CONDITIONS

Several types of flow instability have been experienced in boiling systems. One of these, that associated with liquid superheating when nucleation sites are lacking, was mentioned in a previous section. A second type, referred to as *static instability,* can be analyzed just on the basis of the forces acting on the system under steady-state conditions. The effectiveness of various corrective measures can be deduced readily because the problems are amenable to fairly straightforward solutions. A third type of flow instability in boiling systems, referred to as

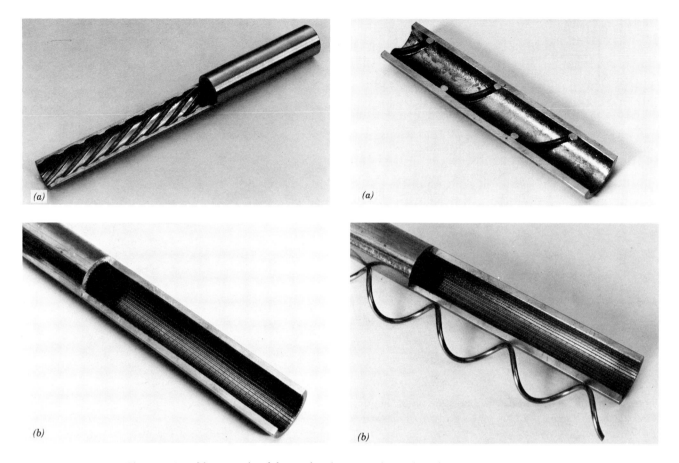

Figure 5.26 Photograph of four tubes having enhanced surfaces that gave the best performance in tests of dryer-superheater tubes. (Courtesy Oak Ridge National Laboratory.)

dynamic instability, stems from dynamic effects that cannot be analyzed without including fluid acceleration and thermal inertia effects. Problems of this type are inherently so complex that no straightforward general analytical solution has been found for them.

Nature of Static Instability

The nature of the static-flow stability problem in boiling systems is indicated by the curves in Fig. 5.32, which show the relative pressure drop across a given tube as a function of the relative flow rate for some representative conditions. The two straight dashed lines that form the upper and lower envelopes are for dry vapor and saturated liquid, respectively, that is, with no boiling. Because of its far higher velocity, the pressure drop for the vapor is, of course, much higher than that for the liquid at any given weight flow rate. The intermediate

curves show the pressure drop across a given tube with a given rate of heat input. While the steam-outlet pressure and temperature are the same throughout, each curve is for a different feedwater inlet temperature. As the flow is increased at any given feedwater inlet temperature, the steam quality at the tube outlet falls off as indicated by the superimposed dotted curves for the exit steam quality. The basic problem can be visualized by examining the upper curve, which is for conditions giving an amount of preheating equal to 10% of the heat required for complete vaporization. Note that at the typical, relative pressure drop indicated by the horizontal dashed line, there are three different possible flow rates. The lowest flow would give superheated vapor at the tube outlet (at point A), while the next highest flow would yield about 3% quality vapor (at point B), and the highest flow would yield water at a temperature below the boiling point (at point C). The intermediate

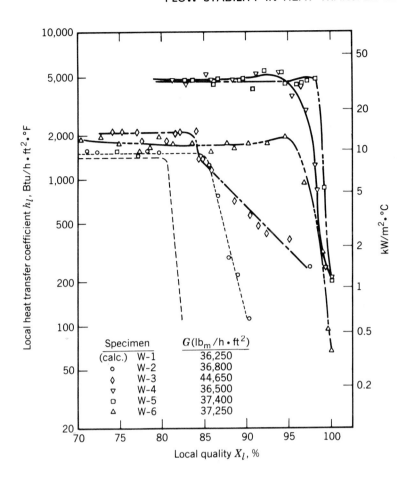

Specimen		$G (lb_m/h \cdot ft^2)$
(calc.)	W-1	36,250
○	W-2	36,800
◇	W-3	44,650
▽	W-4	36,500
□	W-5	37,400
△	W-6	37,250

Figure 5.27 Effects of the local vapor quality on the local heat transfer coefficient for a series of tubes having different internal surfaces, including those of Fig. 5.24. The critical heat fluxes were generally in the range of 3 to 6 W/m² (10,000 to 20,000 Btu/h · ft²). The dashed curve with no points is for Specimen W-1, a plain tube. (Courtesy Oak Ridge National Laboratory.)

point (that giving mostly saturated water with a little steam) is in an unstable region, because any slight disturbance tending to increase the flow leads to a reduction in pressure drop, which gives a further increase in flow. Similarly, the effects of a slight reduction in flow are also cumulative. Flow conditions for the other two points are stable at the given pressure drop, but changes in the pressure drop across the tube under transient conditions during start-up or changes in load could cause abrupt shifts in the flow rate from point A to point C, or vice versa. This could lead to "slugging," "water hammer," or other violent and erratic forms of flow behavior.

An examination of Fig. 5.32 leads to the conclusion that, to obtain stable, predictable flow conditions in a boiler with good flow distribution between parallel tubes, the curves for pressure drop as a function of weight flow should have a substantial positive slope throughout the range of operating conditions that might be encountered in service.

Analytical Relationships for Static-Stability Analyses

The problem is evidently complex, since it must include the effects of such parameters as the specific volumes of the vapor and liquid, the system pressure, and temperature. A discussion by E. F. Leib[31] gives a straightforward approach to the problem. He presents equations for the pressure drop in both the nonboiling and boiling zones, presuming that the rate of heat addition is uniform along the length of the tube and that the friction factor does not vary much with the mass-flow rate. This last assumption seems justified in light of the data presented in the earlier section on two-phase flow.

The first step in establishing the relation between the fluid pressure drop and the flow rate is to consider the simplest cases, namely those in which the heat input is uniform along the length of a heated horizontal passage and the flow is either all liquid or all vapor throughout the heated length. For such conditions, the pressure drop

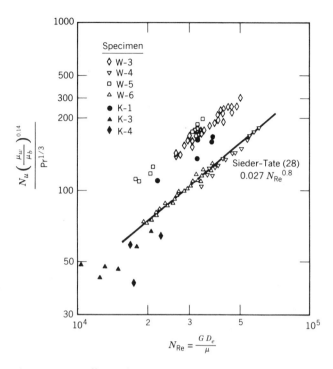

Figure 5.28 Effects of Reynolds number on the conventional heat transfer parameter as determined for dry steam flow through a series of tubes having different internal surfaces, including those of Fig. 5.26. The subscripts w and b refer to the wall and bulk free stream as in Eq. 3.21. (Courtesy Oak Ridge National Laboratory.)

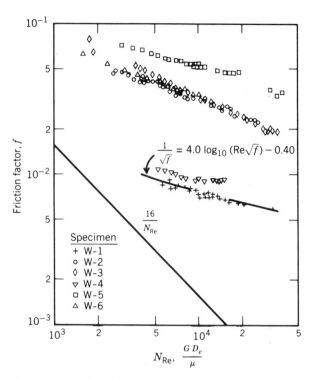

Figure 5.29 Friction factors for single-phase flow through the tubes of Figs. 5.26–5.28. (Courtesy Oak Ridge National Laboratory.)

may be modified to put it in terms of the specific volume v, and, to simplify the mathematics, only the first term of Eq. 5.7 will be used, that is,

$$\Delta P = f \frac{\rho V^2 L}{2gD} = f \frac{\rho}{2g} (vG')^2 \frac{L}{D} = f \frac{vG'^2 L}{2gD} \quad (5.13)$$

This equation shows that, to a first approximation, the pressure drop at a given weight-flow rate is directly proportional to the specific volume.

Leib has extended the relation of Eq. 5.13 to cover conditions where two-phase flow prevails, and he obtains the pressure drop as a function of the vapor quality for cases in which the fluid flows through passages that are continuous from the preheater inlet to the boiler outlet. Since his derivation omits many steps, and since his definitions of symbols differ somewhat from conventional practice, it seems desirable to present a modified form of his derivation. A number of special variables enter the picture and may be defined as follows:

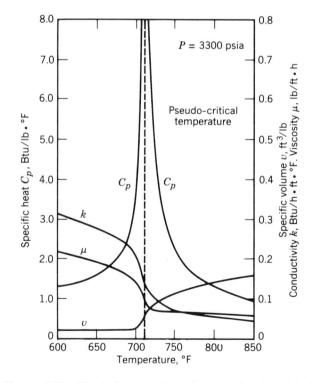

Figure 5.30 Physical properties of water in the critical temperature and pressure region. (Swenson, Ref. 28.)

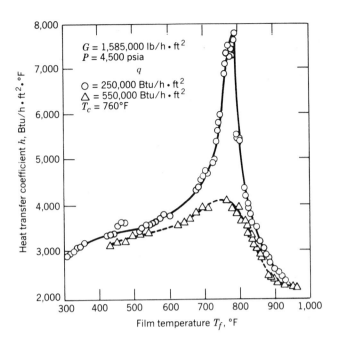

Figure 5.31 Effects of the heat flux on the heat transfer coefficient in the vicinity of the critical temperature at 140% of the supercritical pressure. (Swenson, Ref. 28.)

ΔH_p = enthalpy rise in preheating zone, Btu/lb

H_v = enthalpy of vaporization, Btu/lb vapor

L' = length of tube in which preheating takes place, ft

L = distance from tube inlet, ft

L_t = total tube length, ft

ΔP_p = pressure drop in that portion of the tube in which preheating takes place, psf

ΔP_b = pressure drop in that portion of the tube in which boiling takes place, psf

ΔP_{p+b} = pressure drop across the preheating plus boiling sections, psf

Q = heat supplied to a tube of total length L_t, Btu/s

v' = specific volume of saturated liquid, ft³/lb

v'' = specific volume of saturated vapor, ft³/lb

W = weight flow through tube, lb/s

X = fluid quality at any point in the tube, lb vapor/lb fluid (vapor + liquid)

X_0 = fluid quality at tube outlet, lb vapor/lb fluid (vapor + liquid)

It follows directly that the pressure drop across the preheating zone ΔP_p is given by

$$\Delta P_p = \frac{f v' G'^2 L'}{2gD} \qquad (5.14)$$

The pressure drop across the boiling zone ΔP_b is

$$\Delta P_b = \int_{L'}^{L_t} \frac{f v G'^2}{2gD}\, dL = \frac{f G'^2}{2gD} \int_{L'}^{L_t} v\, dL \qquad (5.15)$$

where

$$v = X v'' + (1 - X)v' = X(v'' - v') + v'$$

$$= \frac{Q}{W H_v}\left(\frac{L - L'}{L_t}\right)(v'' - v') + v' \qquad (5.16)$$

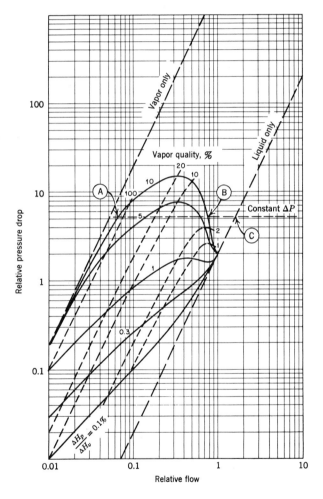

Figure 5.32 The nature of statically determinant boiling instability, as indicated by the relation between the calculated pressure drop and the flow rate for 0.16 MPa (24-psia) water in a once-through boiler tube with a fixed rate of heat input. The solid lines are for 5 different amounts of preheating, that is, 5 values of the ratio of the heat required for preheating to the heat of vaporization $\Delta H_p/\Delta H_v$. (Courtesy Oak Ridge National Laboratory.)

since $X = (Q/WH_v)(L - L')/L_t$. By substituting Eq. 5.16 into Eq. 5.15,

$$\Delta P_b = \frac{fG'^2}{2gD}\left[\frac{Q(v'' - v')}{WH_vL_t}\int_{L'}^{L_t}(L - L')\,dL \right.$$
$$\left. + v'\int_{L'}^{L_t}dL\right] \qquad (5.17)$$

By integrating

$$\int_{L'}^{L_t}dL = L_t - L', \int_{L'}^{L_t}(L - L')\,dL$$
$$= \left[\frac{L^2}{2} - LL'\right]_{L'}^{L_t} = \frac{1}{2}(L_t - L')^2$$

Substituting these values in Eq. 5.17 gives

$$\Delta P_b = \frac{fG'^2}{2gD}\left[\frac{Q(v'' - v')}{WH_vL_t}\frac{1}{2}(L_t - L')^2 + v'(L_t - L')\right] \qquad (5.18)$$

The quality of the vapor at the tube outlet X_0 is

$$X_0 = \frac{Q(L_t - L')}{WH_vL_t} \qquad (5.19)$$

Substituting Eq. 5.19 into Eq. 5.18 gives

$$\Delta P_b = \frac{fG'^2}{2gD}\left[\frac{1}{2}X_0(v'' - v')(L_t - L') + v'(L_t - L')\right] \qquad (5.20)$$

Adding Eqs. 5.14 and 5.20 to obtain the overall pressure drop for the preheating and boiling zones gives

$$\Delta P_{p+b} = \Delta P_b + \Delta P_p$$
$$= \frac{fG'^2}{2gD}\left[\frac{1}{2}X_0(v'' - v')(L_t - L')\right.$$
$$\left. + v'(L_t - L') + v'L\right]$$
$$= \frac{fG^2v'L_t}{4gD}\left[X_0\left(\frac{v''}{v'} - 1\right)\left(\frac{L_t - L'}{L_t}\right) + 2\right] \qquad (5.21)$$

In examining Eq. 5.21 it is apparent that the term

outside the brackets is similar to Eq. 5.14. Within the parentheses, the first term, X_0, is the quality of vapor at the tube outlet. The second term, $(v''/v') - 1$, is the ratio of the specific volumes of the vapor and liquid minus 1, while the last term, $(L_t - L')/L_t$, is the ratio of the length of tube in which boiling takes place to the total heated length.

Sometimes it is convenient to express the vapor quality in Eq. 5.21 in terms of the enthalpy and the tube length so that

$$X_0 = \frac{Q - W\Delta H_p}{WH_v}$$
$$= \frac{Q\Delta H_p}{W\Delta H_pH_v} - \frac{\Delta H_p}{H_v} = \frac{\Delta H_p}{H_v}\left(\frac{L_t}{L'} - 1\right) \qquad (5.22)$$

Substituting this value of X_0 into Eq. 5.21 gives

$$\Delta P_{p+b} = \frac{fG'^2v'L_t}{4gD}\left[\frac{\Delta H_p}{H_v}\left(\frac{L_t}{L'} - 1\right)\right.$$
$$\left. \times \left(\frac{v''}{v'} - 1\right)\left(1 - \frac{L'}{L_t}\right) + 2\right] \qquad (5.23)$$

Calculational Procedure

Using Eq. 5.23 a set of calculations for the pressure drop across a tube in which both preheating and boiling take place is presented in Table 5.5 to illustrate the method used. While the calculations are for water, the results should have general application. In these calculations only the relative pressure drops and relative flow rates are calculated. The flow rate is normalized so that at a relative flow rate of unity the quality of the liquid-vapor mixture leaving the tube is zero; that is, the fluid is heated just to the boiling point. A series of calculations similar to those of Table 5.5 was made for a wide range of conditions, and the results are plotted to give a set of charts. The first of these, Fig. 5.33, is prepared from Eq. 5.13. This figure shows the relative pressure drop as a function of relative flow rate for a range of fluid specific volumes for conditions in which no heat is added to the fluid and no boiling takes place.

Using the same coordinate system as in Fig. 5.33, a second set of curves was prepared and is presented in Fig. 5.34 to show the pressure drop as a function of the flow rate for cases in which the fluid enters the tube as a liquid at its boiling point so that no preheating of the liquid occurs. Equation 5.21 was used for these calculations, and the length of the preheating section was

TABLE 5.5   Calculation of Relative Pressure Drop for Boiling With Preheating (from Eq. 5.23)

$$\text{Condition: } \frac{V''}{V'} = 1001, \frac{\Delta H_p}{H_v} = 0.02$$

(a)	(b)	(c)	(d)	(e)	(f)	(g)
Relative Flow	Portion of Tube Functioning as Preheater, $\dfrac{W\,\Delta H_p}{Q} = \dfrac{L'}{L_t}$	Portion of Tube Functioning as Boiler, $1 - \dfrac{L'}{L_t}$	Quality, $\dfrac{\Delta H_p}{H_v}\left(\dfrac{L_t}{L'} - 1\right)$ $= \dfrac{(c)}{(b)}\dfrac{\Delta H_p}{\Delta H_v}$	$\left(\dfrac{v''}{v'} - 1\right)$(c)(d)	(e) + 2	Relative Pressure Drop, $f \times$ (a)^2
1.0	1.0	0	0	0	2.0	2.0
0.9	0.9	0.1	0.002	0.2	2.2	1.78
0.8	0.8	0.2	0.005	1.0	3.0	1.92
0.7	0.7	0.3	0.009	2.7	4.7	2.3
0.6	0.6	0.4	0.013	5.2	7.2	2.59
0.5	0.5	0.5	0.02	10	12	3.0
0.4	0.4	0.6	0.03	18	20	3.2
0.3	0.3	0.7	0.047	32.9	34.9	3.14
0.2	0.2	0.8	0.08	64	66	2.64
0.05	0.05	0.95	0.38	361	363	0.91
0.02	0.02	0.98	0.98	960	962	0.385

taken as zero ($L' = 0$). The flow rate is normalized so that at a relative flow rate of unity the quality of the vapor at the tube outlet is unity, that is, 100% saturated vapor. In Fig. 5.34, in addition to the flow-rate scale, a second scale is provided across the bottom to show the quality of the mixture leaving the tube. Note that as the flow rate is reduced, the quality increases until 100% quality is reached. Further reductions in flow rate lead to superheating. At low flow rates these curves become asymptotic to the line for vapor only, whereas at high flow rates they become asymptotic to the line for liquid only. In all cases, the slope of the curves is positive, so the flow is stable. However, the slope in some regions is about half of that for flow with no change in phase, so that the static stability is reduced.

A third set of curves has been prepared and is presented in Fig. 5.35 to cover conditions under which some preheating takes place. For this chart the enthalpy rise in the preheating region is taken to be 10% of the enthalpy of vaporization. Again, curves are prepared for a series of vapor-liquid specific volume ratios. For water and water vapor, the values of 1000, 400, 100, 40, and 10 for the ratio $(v'' - v')/v'$ correspond to pressures of 0.16, 0.42, 1.7, 4.0, and 11.3 MPa (24, 62, 247, 576, and 1630 psia), respectively. The significance of the relative flow rate used in this chart can be seen by consid-

ering a typical case. If a water-cooled reactor is operated at a flow which is such that the temperature rises just to the boiling point at the reactor outlet, the relative flow rate in Fig. 5.35 would be unity. The shape of the curves as the flow rate is reduced from this condition depends on the pressure, that is, on the ratio of the specific volume of the vapor to that of the liquid. For this case, with the amount of preheating considered, static-flow instability should not be a problem if the system pressure is above about 600 psia.

Equation 5.23 is employed for making the calculations for flow rates down to those giving saturated steam at the tube outlet. For the lower relative flow rates for which superheating occurs, the calculation of the pressure drop is a bit involved. Basically, the pressure drop through a tube in which preheating, boiling, and superheating occur is simply the sum of the pressure drops in the three regions. It can be estimated directly from the results given in Figs. 5.33 and 5.34 or 5.35 (depending on whether there is preheating). If the ratio of the length of tubing in the superheater section to the length in the boiler and preheater is taken as z, and the relative flow giving saturated steam at the tube outlet in Fig. 5.35 is W_0, the relative flow rate W', with superheating for the same unit tube length and rate of heat input, must be

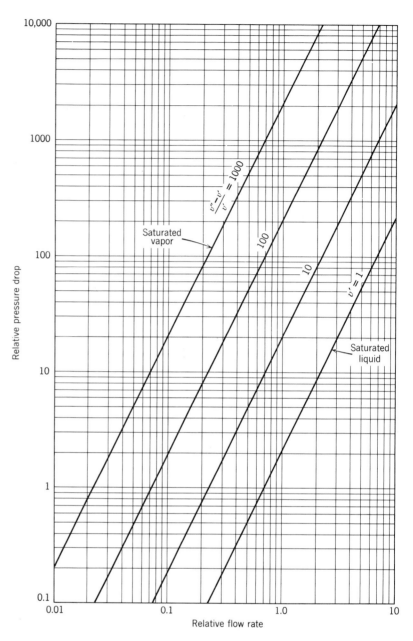

Figure 5.33 Effects of fluid density on the relation between the pressure drop and the flow rate through a tube with no change in specific volume. (Courtesy Oak Ridge National Laboratory.)

$$W' = \frac{W_0}{1 + z} \qquad (5.24)$$

The pressure drop in the preheating and boiling region (ΔP_{p+b}) for flow with superheating thus becomes

$$\Delta P_{p+b} = (\Delta P_{p+b})_0 \left(\frac{W'}{W_0}\right)^2 \frac{1}{1 + z} = \frac{(\Delta P_{p+b})_0}{(1 + z)^3} \qquad (5.25)$$

where $(\Delta P_{p+b})_0$ is the pressure drop for the flow rate giving saturated steam at the tube outlet (i.e., for W_0). The extra factor $(1 + z)$ in the denominator must be applied, because the tube length in the preheating and boiling region is less than the total length of the tube required to produce saturated steam with the flow of W_0 (the total heat input Q to the tube and the tube length are kept constant).

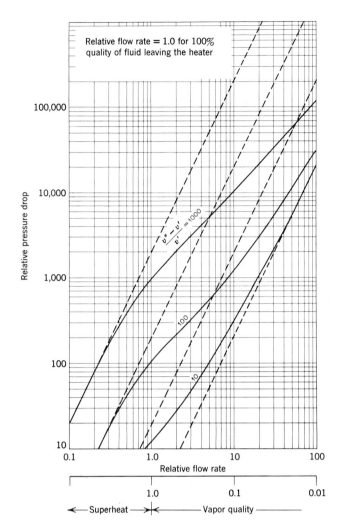

Figure 5.34   Relation between the pressure drop and the flow rate in a once-through boiler tube, with a constant heat input for several specific volumes of the steam for boiling and superheating with no preheating. (Courtesy Oak Ridge National Laboratory.)

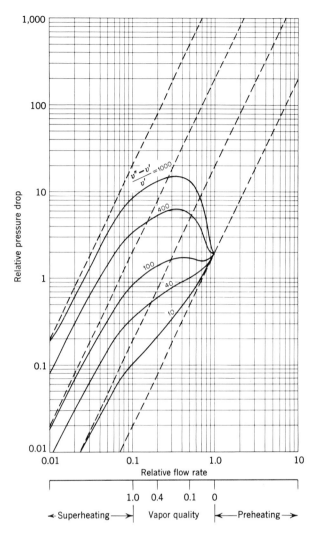

Figure 5.35   Relation between the pressure drop and the flow rate in a once-through boiler tube, with a constant heat input for five specific volumes of the steam for boiling and superheating, and the amount of preheating equal to 10% of the heat of vaporization. (Courtesy Oak Ridge National Laboratory.)

Similarly, the pressure drop in the superheating portion of the tube is given by

$$\Delta P_{sh} = \Delta P_v \left(\frac{W'}{W_0}\right)^2 \frac{z}{1 + z} = \Delta P_v \frac{z}{(1 + z)^3} \quad (5.26)$$

where ΔP_v = pressure drop for vapor flow only at a flow rate of W_0.

Calculations for the relative pressure drop for boiling with both preheating and superheating, using the data of Figs. 5.33 and 5.35, are presented in Table 5.6. In

selecting W_0 it should be remembered that Figs. 5.32, 5.33 and 5.34 are all prepared so that the pressure drop per unit tube length at any given value of relative flow rate, as given on any of the charts, corresponds to that at the same relative flow rate on the other charts. The actual pressure drop can be obtained in each case by multiplying the relative pressure drop by the appropriate value of $(fv' LG^2/4gD)$ and dividing by the square of the relative flow rate.

TABLE 5.6 Calculation of Pressure Drop for Boiling With Preheating and Superheating Using the Data of Figs. 5.33 and 5.35

Condition: $\dfrac{v'' - v'}{v'} = 1001$ and $\dfrac{\Delta H_p}{H_v} = 0.10$

From Fig. 3.35: $(\Delta P_{p+b})_0 = 7.5$ and $W_0 = 0.091$, where the vapor quality $= 1$
From Fig. 5.33: $\Delta P_v = 16$ for $W_0 = 0.091$

z	$1 + z$	$(1 + z)^3$	$\Delta P_{p+b} = \dfrac{(\Delta P_{p+b})_0}{(1 + z)^3}$	$\Delta P_{sh} = \Delta P_v \dfrac{z}{(1 + z)^3}$	Total ΔP
0	1.0	1.0	7.5	0	7.5
0.1	1.1	1.331	5.63	1.2	6.83
0.2	1.2	1.728	4.34	1.85	6.19
0.3	1.3	2.196	3.41	2.2	5.61
0.4	1.4	2.745	2.73	2.34	5.07
0.5	1.5	3.375	2.22	2.38	4.60
0.6	1.6	4.10	1.83	2.34	4.17
1.0	2.0	8.0	0.937	2.0	2.94
2.0	3.0	27	0.278	1.185	1.46
3.0	4.0	64	0.117	0.75	0.867
8.0	9.0	729	0.0103	0.176	0.186

The assumption of constant heat input along the length of the tube is a good approximation for the boiling region of conventional boilers and for reactor-core matrices. Conditions under which large amounts of superheating occur are ordinarily not of interest in reactors, whereas in coal-fired boilers the heat input per foot of length of channel in the superheating region is kept low in an actual design. Thus, for the purposes of this study, the calculation of the pressure drop in the superheating region is made using Fig. 5.33 (which was prepared for saturated steam), because the heat transfer coefficient in the superheating region is sufficiently low that the increase in specific volume in the superheating region does not have a large effect on the shape of the curves.

The set of curves presented in Fig. 5.32 (for $(v'' - v')/v' = 1000$) shows the effects of changing the amount of preheating in a tube in which preheating and boiling take place. Note that the vapor quality is not directly proportional to the flow rate as in Figs. 5.34 and 5.35, because the proportion of the tube functioning as a preheater varies from one curve to another.

In reviewing these curves from the standpoint of stability considerations it is apparent that if the fluid enters a system of heated passages at the boiling point, the flow should be stable, because the curves in Fig. 5.34 are smooth and continuous with a substantial positive slope. However, the slope of the curves is less than for operation with all liquid or all gas; hence the stability is reduced relative to the simpler cases (e.g., Fig. 5.33). Figure 5.35 indicates that severe instabilities could occur in boiling reactors if appreciable amounts of preheating of the liquid are required to raise it to the boiling point, and if the ratio of the specific volume of the vapor to that of the liquid is greater than perhaps 50 (e.g., in reactors such as the Materials Test Reactor (MTR)). Instability occurs at flow rates giving small amounts of vapor in the fluid leaving the reactor. In that flow region, a slight perturbation causing a small reduction in flow rate leads to an increase in pressure drop, which in turn produces a further reduction in flow. For the large vapor-liquid specific volume ratios, it is apparent that, for some pressure drops across the heat transfer matrix, there could be a relatively large liquid flow through one tube while a small flow of vapor and liquid would emerge from another. For vapor-liquid specific volume ratios of approximately 100, there is a broad range in which the flow through any given channel is indeterminant, and at a given pressure drop the relative flow rate could run all the way from 20 to 100% of design flow if the heat transfer matrix were operating so that the average channel would give fluid leaving the reactor at the boiling point. Figure 5.32 shows that, if the preheating amounts to only 1% of the heat of vaporization of the fluid for conditions giving an average vapor quality of 0 to 4%, the flow might be unstable if the specific volume of the vapor is of the order of 1000 times the specific volume of the liquid.

Figures 5.35, and 5.32 do not mean that flow through a heat transfer matrix is unstable if trace amounts of local boiling take place. They do indicate that flow instability is a distinct possibility if the water-temperature rise through the hottest portion of the matrix approaches that required to raise the temperature of the water leaving the matrix to the boiling point at the pressure prevailing at the matrix-outlet face.

Effects of Static Head

In heat transfer matrices in which the direction of flow is either vertically upward or downward, the static head of the liquid column influences the flow-stability characteristics. If the fluid is evaporated to dryness and the heat input per unit length of channel is fixed, the height of the relatively high-density fluid column and the consequent static head imposed at the inlet is directly proportional to the flow rate. The effects of this factor are shown graphically in Fig. 5.36 for a typical case for a low-pressure system in which the flow is upward through

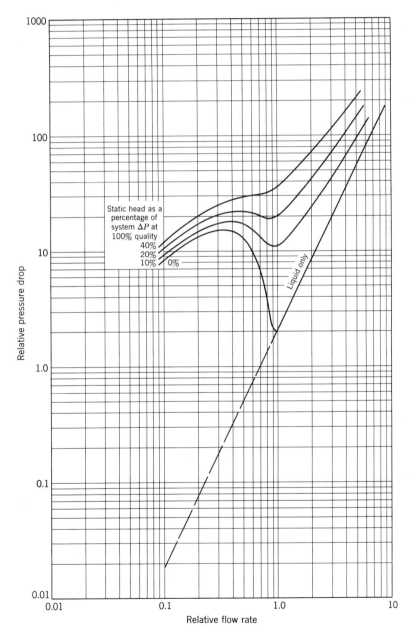

Figure 5.36 Effects of static head on the relation between the pressure drop and the flow rate in a once-through boiler tube with a constant heat input for an amount of preheating equal to 10% of the heat of vaporization. The specific volume of the vapor is that for saturated steam at 0.16 MPa (24 psia). (Courtesy Oak Ridge National Laboratory.)

the channel. The bottom curve is the same as the upper curve of Fig. 5.32; that is, it represents the relative pressure drop-relative flow curve for an amount of preheating equal to 10% of the heat of vaporization, and includes no allowances for a static head. The effects of static head were included by applying perturbations about the 100% vapor quality point for the basic curve. Three additional conditions were considered: static heads at the 100% quality point equal to 10, 20, and 40% of the system relative pressure drop for the basic curve. The relative pressure drop at each of several values of the relative flow was then calculated by taking the relative

pressure drop for the base curve and adding to it the static head, which was taken as directly proportional to the relative flow rate. It is apparent from Fig. 5.36 that the static head can be an important factor in stabilizing upward flow. By the same token, if the flow is downward, the static head has a destabilizing influence.

Effects of Orifices at the Inlet

The effects of orificing the inlet to the economizer are very similiar and can be computed in much the same way. Figure 5.37 shows the results of a set of perturbations to

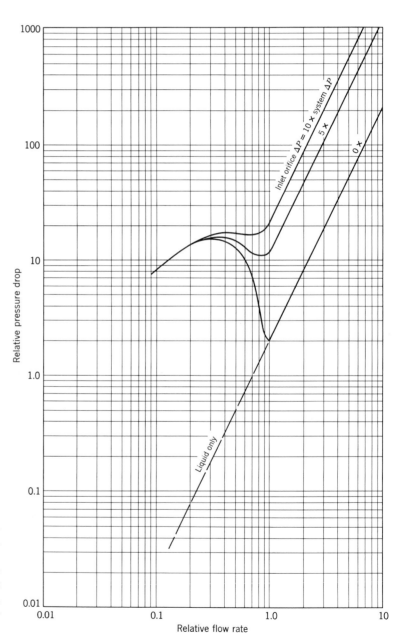

Figure 5.37 Effects of orificing the inlet to a once-through boiler tube on the relation between the pressure drop and the flow rate with a constant heat input and an amount of preheating equal to 10% of the heat of vaporization. The specific volume of the vapor is that for steam at 0.16 MPa (24 psia). (Courtesy Oak Ridge National Laboratory.)

the same base curve as that used for Fig. 5.36. In this instance it was assumed that orifices at the inlet to the economizer are sized to yield pressure drops across the orifice equal to respectively 5 and 10 times the system pressure drop with liquid flow only. The curves are constructed simply by multiplying the relative pressure drop given by the 100% liquid curve at a given relative flow by factors of 5 and 10, respectively, and then taking the orifice ΔP as proportional to the square of the relative flow, and adding the resulting values to the relative pressure drop at the corresponding flow for the base curve. As in the construction of Fig. 5.36, the calculations are so simple that no tabulation is necessary; the results can be plotted directly as they are obtained.

It is evident from Fig. 5.37 that orificing has a pronounced stabilizing influence, and that the effect increases with the ratio of the orifice pressure drop to the pressure drop for the rest of the passage with no boiling in the liquid.

Differences in Heat Addition to Parallel Passages

The previous discussion was concerned primarily with flow in single channels. In applying these relations to heat transfer matrices with a multiplicity of channels in parallel, it is necessary to consider the effects of differences in heat-addition rates between parallel channels coupled with common headers. The effects of such differences in heat-addition rates can be envisioned by examining Fig. 5.38, which illustrates a case representative of conditions in a modern 1600-psia once-through boiler. The base curve used is that for a specific volume ratio of 11; that is, the $(v'' - v')/v' = 10$ curve of Fig. 5.35, for which the preheating is equivalent to 10% of the heat of vaporization. The chart is constructed by taking the base curve from Fig. 5.35 and by sliding the 1.0 relative flow-rate point along the line for 100% liquid so that in each instance the flow rate is changed by a factor equal to the change in the heat-addition rate relative to the base curve. In examining these curves it can be deduced that if there were differences in the heat addition to channels operating in parallel with the same pressure drop, there would be no difficulty with static-flow instability, but some channels would deliver excessively superheated steam, whereas others would deliver a mixture of vapor and water. Although the flow would be stable, overheating of some of the tube walls would result partially from the higher steam temperature and partially from the poorer local heat transfer coefficient. Since the excessively heated vapor would be in the tubes with the highest heat flux, the difference in temperature

between the tube wall and the steam would be greatest in the hot channels. These two effects combined could cause certain tubes to overheat 200 or 300°F.

The principal assumption made in constructing Fig. 5.38 is that the heat-input distribution along the channel is similar for each case, and that the total amount of heat input to each channel differs by the factor noted.

Several possibilities for flow instability in parallel channels exist if the individual channels are to operate in a region that may give difficulty with flow instability. This effect is shown in Fig. 5.39, which was prepared in much the same way as Fig. 5.38, except that the base curve in that figure is for a much lower pressure (i.e., the same base curve as used for Figs. 5.36 and 5.37). Not only may the flow surge erratically in individual channels,

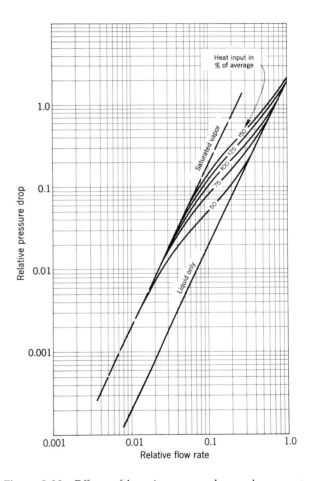

Figure 5.38 Effects of heat input per tube on the pressure drop-flow relationship for once-through boiler tubes for an amount of preheating equal to 10% of the heat of vaporization and a vapor specific volume equal to 11 times that of the liquid (e.g, steam at 11 MPa, or 1600 psia). (Courtesy Oak Ridge National Laboratory.)

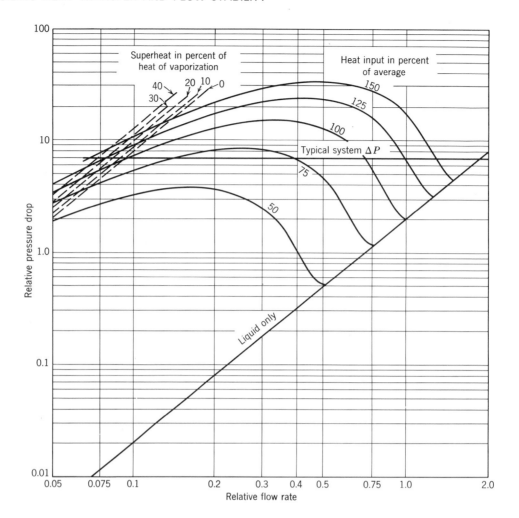

Figure 5.39 Chart for the same conditions as for Fig. 5.38, except for a vapor specific volume 1001 times that of the liquid (e.g., steam at 0.16 MPa, or 24 psia). (Courtesy Oak Ridge National Laboratory.)

but coupled oscillations may develop between parallel channels with the flow becoming large first in one channel and then in another. Very complex oscillations can develop in geometries in which a large number of channels are coupled into common headers because of the large number of degrees of freedom.

DYNAMIC INSTABILITY IN BOILING SYSTEMS

Periodic flow oscillations that occur under certain conditions in boiling systems are difficult to predict and hard to correct if they are encountered. Such oscillations may lead to severe local overheating and damage. Periodic

oscillations in the flow of boiling liquids have been investigated experimentally and analytically.[32-34] A complete analytical treatment of the subject requires that the transient energy and momentum equations for the system be included in the analysis, and this leads to a set of coupled nonlinear partial differential equations—the direct solution of which is not possible except in trivial cases.

For certain simplified models these equations are linearized using a small perturbation technique to obtain operational expressions, and the stability criteria are delineated by applying the standard methods used in servomechanisms. The results of these investigations indicate that the flow stability in boiling systems is a complicated function of system geometry, the amount of subcooling, the heat flux, the system pressure, and the flow conditions. No general rules can be set forth to obtain quantitative criteria for the stability of the flow resulting from various feedback effects. However, qualitatively it can be said that in a natural circulation boiling loop the amplitude of flow oscillations generally increases as the amounts of either subcooling or friction in the hot riser are increased, and the amplitude of these oscillations is reduced by increasing the frictional losses in the cold downcomer.

In analyzing the stability of boiling flow in the more complicated systems, high-speed digital or analog computers provide the only means for solving the resulting set of equations and then usually only for particular, idealized cases. However, if one is astute in combining terms and parameters, it is even possible to prepare a fairly general-purpose program for use of a microcomputer having a 640K RAM.[35]

REFERENCES

1. E. A. Farber and R. L. Scorah, "Heat Transfer to Water Boiling Under Pressure," *Trans. ASME,* vol. 70, 1948, p. 369.

2. P. J. Berenson and R. A. Stone, "A Photographic Study of the Mechanism of Forced Convection Vaporization," presented at October 1963 meeting of the AIChE, San Juan, Puerto Rico.

3. E.-F. Chien and W. Ibele, "Pressure Drop and Liquid Film Thickness of Two-Phase Annular and Annular-Mist Flows," *Journal of Heat Transfer, Trans. ASME,* vol. 86-2, 1964, p. 89.

4. E. E. Polomik, S. Levy, and S. G. Sawochka, "Heat Transfer Coefficients with Annular Flow During 'Once-Through' Boiling of Water to 100% Quality at 800, 1100, and 1400 psi," *Journal of Heat Transfer, Trans. ASME,* vol. 86-2, 1964, p. 81.

5. P. Griffith and J. D. Wallis, "The Role of Surface Conditions in Nucleate Boiling," *Chemical Engineering Progress Symposium Series,* vol. 56, AIChE, 1960.

6. H. K. Forster and N. Zuber, "Growth of a Vapor Bubble in a Superheated Liquid," *Journal of Applied Physics,* vol. 25, 1954, p. 474.

7. P. J. Berenson, "Experiments on Pool-Boiling Heat Transfer," *International Journal of Heat and Mass Transfer,* vol. 5, Pergamon Press, Elmsford, N.Y., 1962, p. 985.

8. S. Levy, "Generalized Correlation of Boiling Heat Transfer," *Journal of Heat Transfer, Trans. ASME,* vol. 81-2, 1959, p. 37.

9. W. H. Lowdermilk and W. F. Weiland, "Some Measurements of Boiling Burn-out," NACA RM-E-54-K-10, 1955.

10. K. M. Becker, "Burn-out Conditions for Flow of Boiling Water in Vertical Rod Clusters," *AIChE Journal,* 1963, p. 216.

11. N. L. Dickinson and C. P. Welch, "Heat Transfer to Supercritical Water," *Trans. ASME,* vol. 80, 1958, p. 746.

12. R. K. Young and R. L. Hummel, "Improved Nucleate Boiling Heat Transfer," *Chemical Engineering Progress,* vol. 60, 1964, p. 53.

13. R. W. Lockhart and R. C. Martinelli, "Proposed Correlation of Data for Isothermal Two-Phase, Two-Component Flow in Pipes," *Chemical Engineering Progress,* vol. 45, 1949, p. 39.

14. S. W. Gouse, Jr., "An Index to the Two-Phase Gas-Liquid Flow Literature," Part 1, DSR-8734-1, MIT Engineering Projects Laboratory report, 1963.

15. G. B. Wallis, "Some Hydrodynamic Aspects of Two-Phase Flow and Boiling," *International Developments in Heat Transfer, ASME,* 1962, p. 319.

16. W. L. Owens, Jr., "Two-Phase Pressure Gradient," *International Developments in Heat Transfer, ASME,* 1962, p. 363.

17. R. V. Smith, "The Influence of Surface Characteristics on the Boiling of Cryogenic Fluids," *Journal of Engineering for Industry, Trans. ASME,* vol. 91(1), 1969, pp. 1217–1221.

18. H. Merte, Jr., and J. A. Clark, "Boiling Heat Transfer with Cryogenic Fluids at Standard, Fractional, and Near-Zero Gravity," *Journal of Heat Transfer, Trans. ASME,* vol. 86(2), 1964, p. 351.

19. K. J. Coeling and H. Merte, Jr., "Incipient and Nucleate Boiling of Liquid Hydrogen," *Journal of Engineering for Industry, Trans. ASME,* vol. 91(1), 1969, p. 513.

20. B. A. Hands, *Cryogenic Engineering,* Academic Press, New York, 1986.

21. A. P. Fraas, *Engineering Evaluation of Energy Systems,* McGraw-Hill Book Co., New York, 1982, p. 347.

22. J. T. Pogson et al., "An Investigation of the Liquid Distribution in Annular-Mist Flow," *Journal of Heat Transfer, Trans. ASME,* vol. 92(2), 1970, p. 651.

23. W. D. Pouchot et al., "Analytical Investigation of Turbine Erosion Phenomena," Interim Technical Report No. 1, vol. 11, Westinghouse Astronuclear Laboratory Report WANL-PR-(DD)-014, November 1966, NASA-CR-81135.

24. *Steam,* 38th ed., Babcock and Wilcox Co., 1972, p. 1–4.

25. A. E. Bergles, Enhanced Single-Phase Heat Transfer for Ocean Thermal Energy Conversion Systems," prepared at Iowa State University for the Oak Ridge National Laboratory, Report no. ORNL/Sub-77/14216/1, April 1977.

26. S. Yilmaz and J. W. Westwater, "Effect of Commercial Enhanced Surfaces on the Boiling Heat Transfer Curve,"

Heat Transfer Symposium, Milwaukee 1981, *AIChE Symposium Series* 208, 1981.

27. J. K. Jones et al., Development of Integrally Finned Dryer-Superheater Tubes for Potassium Rankine Cycle Boilers," Report no. ORNL-TM-3383, Oak Ridge National Laboratory, April 1971.

28. H. S. Swenson et al., "Heat Transfer to Supercritical Water in Smooth Bore Tubes," *Journal of Heat Transfer, Trans. ASME,* vol. 87(2), 1965, p. 477.

29. R. D. Wood and J. M. Smith, "Heat Transfer in the Critical Region—Temperature and Velocity Profiles in Turbulent Flow," *AIChE Journal,* vol. 10(2), 1964, p. 180.

30. B. Shiralkar and P. Griffith, "The Effect of Swirl, Inlet Conditions, Flow Direction, and Tube Diameter on the Heat Transfer to Fluids at Supercritical Pressure," *Journal of Heat Transfer, Trans. ASME,* vol. 92(2), 1970, p. 465.

31. E. F. Leib, Discussion of Van Brunt's "Circulation in High-Pressure Boilers and Water-Cooled Furnaces," *Trans. ASME,* vol. 63, 1941, pp. 344–347.

32. G. B. Wallis and J. H. Heasley, "Oscillations in Two-Phase Flow Systems," *Journal of Heat Transfer, Trans. ASME,* vol. 83-2, 1961, p. 363.

33. E. H. Wissler, H. S. Isbin, and N. R. Amundson, "Oscillatory Behavior of a Two-Phase Natural-Circulation Loop," *Journal of AIChE,* vol. 2, 1956, p. 157.

34. W. H. Lowdermilk, C. D. Lanzo, and B. L. Siegel, "Investigation of Boiling Burnout and Flow Stability for Water Flowing in Tubes," NACA-TN-4382, 1958.

35. A. S. Thompson and B. R. Thompson, "A Model of Reactor Kinetics" *Nuclear Science and Engineering*, vol. 100 (1), 1988.

6

Heat Pipes

A heat pipe conveys heat from one region to another as if it were a solid rod of metal having a thermal conductivity as much as 100,000 times that of copper. The mode of operation is much like that of a vertical refluxing capsule such as that of Fig. 6.1, in which fluid is boiled at the lower end, the vapor rising up the pipe, condensing on the walls at the top, and the condensate draining in rivulets that flow back down the wall to the boiling zone. The high heat of vaporization serves to transport a surprising amount of heat per unit mass of vapor; for water the heat of vaporization yields about 600 times the amount of heat given up by a temperature drop of 1 K, and the surface heat transfer coefficients for boiling and condensing are about 10 times as high as those usual for forced convection.

HEAT PIPE STRUCTURES

In 1962 George Grover at the Los Alamos Scientific Laboratory saw that he could make a device similar to but much more potent than a refluxing capsule by employing capillary forces instead of gravity to return the liquid from the condenser to the boiler. Such a device would have the additional advantage that it could be made to function under 0-g conditions or in any attitude, including upside down. He found that one way to provide the capillary forces was to line the capsule with a wick formed by tightly rolling fine mesh stainless steel screen to a thickness of perhaps 10 layers.[1] Such a system is indicated schematically in Fig. 6.2.

A wide variety of capillary structures such as those of Fig. 6.3 have been employed as wicks. Particles can be sintered to form a tube similar to that formed of screen as described above and shown at the upper left of Fig. 6.3. The tubing for the heat pipe can be formed with internal grooves to act as capillary passages, as in the "open-channels" type in Fig. 6.3. To avoid the retarding effects stemming from the shear forces between the oppositely flowing streams of liquid and vapor, these grooves can be covered with a wick structure, or an "artery" can be employed. This also avoids the tendency for liquid entrainment and thus a loss in capacity stemming from the short-circuiting of a portion of the refluxing stream. The resistance to return flow of the liquid can also be reduced by providing an open annulus between the porous wick and the outer tube or by corrugating the screen. (If either of these approaches is to be employed, the gaps between the outer tube and the porous liner must be sealed at the ends to avoid "breaking the siphon" and thus losing the capillary pumping effect for the annulus.)

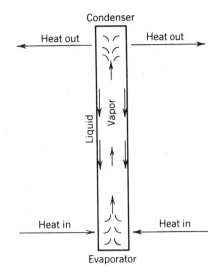

Figure 6.1 Schematic diagram of a refluxing capsule.

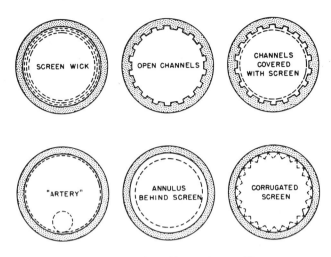

Figure 6.3 Some examples of heat pipe capillary structures. (Courtesy Oak Ridge National Laboratory.)

TYPICAL APPLICATIONS

The outstanding advantages of this clever concept for a wide variety of special applications are indicated by the many papers on heat pipes that soon followed Grover's paper in both the United States and Europe. The ability of heat pipes to function under 0-g conditions coupled with their inherently high reliability made them ideal heat transport systems for spacecraft, both to dissipate waste heat from the power plant and to cool equipment such as motors and electronic components.[2,3] They also provided an ideal means for keeping the permafrost frozen around the footings for the supports of the Alaskan pipeline,[4] and are excellent elements for regenerators and recuperators for power plants, furnaces, and air-conditioning systems,[5] as well as for many other cases.

BASIC RELATIONS

A definitive treatment of the relations between the various parameters affecting the performance of heat

pipes was given by Cotter in Reference 6. In view of the complexities introduced by the many possible geometries, it has seemed best to treat only a simple case here to illustrate the principles of operation and refer the reader to the literature for detailed presentations on specialized geometries such as those mentioned in References 2–17.

Under normal operating conditions a heat pipe functions with the wick structure flooded and some excess liquid spread in a film over the surface in the condensing region. The main driving force for fluid circulation stems from surface tension in the liquid film of the meniscus at the outboard end of each pore in the evaporator region. The situation is the inverse of that treated under nucleate boiling in the previous chapter (see Fig. 5.6 and Eq. 5.1). As indicated in Fig. 6.4, the radius of curvature of the meniscus at the pore outlet becomes progressively smaller as the evaporation rate increases because of the increased suction required to induce flow through the wick as well as to overcome the pressure drop in the vapor flow from the boiler region to the condenser. The pressure differential across the meniscus is given by

$$\Delta P_m = \frac{2\sigma}{R} \qquad (6.1)$$

where ΔP_m = pressure drop across the meniscus, Pa
σ = surface tension, N/m
R = radius of the meniscus, m

The pressure differential generated at the surface of

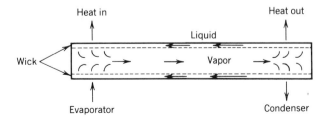

Figure 6.2 Schematic diagram of a heat pipe.

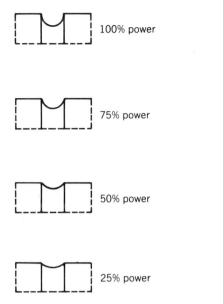

100% power

75% power

50% power

25% power

Figure 6.4 Effects of power on the radius of the liquid meniscus in a heat pipe wick.

the evaporator is equal to the pressure drops through the wick and vapor regions coupled with any effects of gravity that may be operative, that is,

$$\text{Meniscus } \Delta P + \text{gravity } \Delta P$$
$$= \text{liquid pass } \Delta P + \text{vapor pass } \Delta P \quad (6.2)$$

As is evident from inspection of Figs. 6.2 and 6.3, the pressure drop for the condensate return flow through the wick structure is heavily dependent on the wick geometry employed. If a simple wick of sintered powder or densely packed wire screen is used, its pressure-drop characteristics are best determined experimentally because of uncertainties in the pore size and the effective flow-passage area. If an artery or an annulus between the wick and the pipe ID is employed, the flow capacity can be greatly increased because the bulk of the flow will take place in the open passage rather than through the pores of the wick. The pressure drop can be estimated from the Poiseuille relation for laminar flow using the equivalent diameter and cross-sectional area of the open passage, that is,

$$\Delta P_l = \frac{32\,\mu VL}{D^2} \quad (6.3)$$

where ΔP_l = pressure drop in liquid passage, Pa
μ = liquid viscosity, N · s/m^2

V = velocity, m/s
L = passage length, m
D = equivalent diameter, m
$\quad$ = twice annulus thickness, m

Similarly, the pressure drop for the vapor-flow region is given by

$$\Delta P_v = \frac{\rho V^2 fL}{2D} \quad (6.4)$$

where ΔP_v = pressure drop in vapor passage, Pa
ρ = liquid density, kg/m^3
f = friction factor (~ 0.02)

The above relations, combined with the allowance for static-head effects, can be employed to estimate the performance of heat pipes for the region of principal interest. One must hasten to add, however, that this region is limited by several other factors. The most important of these is sonic velocity in the vapor, a factor that becomes dominant in the lower temperature range in which any particular heat pipe working fluid might be employed. As the temperature drops, the pressure of the saturated vapor and hence its density drop very rapidly. Inasmuch as the pressure head available from the capillary forces remains constant, the velocity of the vapor increases more and more rapidly as the pressure is reduced (see Fig. 6.5) until it reaches sonic velocity at the outlet of the evaporator. This constitutes the upper limit for heat transport out of the evaporator. Another factor is a heat flux great enough to cause a sufficiently high radial temperature drop through the wick structure in the evaporator that boiling occurs within the wick. The resulting large pressure drop through the wick caused by the escaping vapor chokes the flow and drastically reduces the output. This limitation can be relieved by increasing the length of the evaporator region, thus reducing the heat flux. The relative importance of these various limitations depends heavily on the choice of working fluid, the operating temperature, and the detailed geometry of the heat pipe.

PERFORMANCE CHARACTERISTICS OF A TYPICAL HEAT PIPE

There are so many different combinations of operating conditions and heat pipe geometries (see Refs. 2–15) that it seems best to simplify the situation by considering the performance characteristics of one representative heat

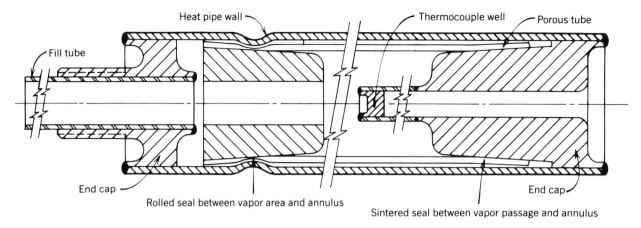

Figure 6.5 Section through a typical heat pipe showing its construction. (Courtesy Oak Ridge National Laboratory.)

pipe geometry. This seems eminently sensible, because in the region of interest the effects of geometry are easy to envision, whereas the effects of the two major variables, the choice of working fluid and the operating temperature, are not. Further, it seems likely that readers will be able to examine the set considered here and find heat pipes that will provide fair approximations to any particular sets of requirements they may have, and then straightforward perturbations can be applied to yield good approximations to their needs.

Representative Heat Pipe

The heat pipe geometry chosen for the base case is one that was employed to absorb heat from an isotope energy source located inside a shield and convey it to a bank of thermoelectric cells to provide electricity for undersea service.[16,17] The unit employed a thin cylindrical wire screen wick surrounded by an annulus, as shown in one example in Fig. 6.2. The ends of the wick were sealed to the outer pipe, as shown in Fig. 6.5. The ID of the wick was 21.5 mm, the wick thickness 0.5 mm, the annulus thickness 0.75 mm, and the pipe length 1.65 m. The wick was fabricated by wrapping 400 mesh stainless steel wire screen on a copper tube mandrel to a thickness of 12 layers, enclosing it with an outer copper tube, and drawing it down through a die to compact the screen. The copper was then dissolved out with nitric acid, and the screen was sintered for 15 min at 927°C (1700°F) in a vacuum. The effective pore size was found to be 30 μm. The heat pipe was assembled and loaded with potassium after thorough degassing.

Effects of Working Fluid and Operating Temperature

The effects on heat pipe performance of the choice of working fluid and the operating temperature were investigated for a set of 10 typical working fluids. The physical properties of the working fluids were selected from References 7, 18, and 19 for the appropriate temperature range in each case and entered in Table 6.1. The maximum pumping head produced by surface tension in the wick was then calculated from Eq. 6.1 and expressed in terms of the height to which the pores in the wick could raise a column of the liquid. Optimization studies[6,8] have shown that it is best to make the pressure drop in the liquid annulus equal to that for the vapor flow through the central passage. For this case the vapor passage length-diameter ratio is a bit less than 100, so that the pressure drop in the vapor passage would be about two dynamic heads. Thus for the full capacity output at any given temperature, the vapor dynamic head would be one-fourth the head developed by the wick. This assumes that the heat pipe would operate either horizontally or under 0-g conditions to avoid the complication of the gravitational head. Thus the vapor velocity was calculated from the wicking height and the vapor density. The heat transport capacity, or limiting heat load, was then calculated from the vapor velocity, the vapor flow-passage area, the vapor density, and the heat of vaporization. The results are presented in Table 6.1 and Figs. 6.7 and 6.8.

As one would expect, all 10 of the fluids considered in Table 6.1 are similarly affected by increasing the

TABLE 6.1 Heat Pipe Working Fluid Physical Properties and Typical Performance Data

Heat pipe OD = 2.54 cm, vapor region ID = 2.14 cm, pore dia. = 0.03 mm
Vapor dynamic head = 0.25 wick head

Fluid	Temp., °C	Latent Heat, kJ/kg	Liquid Density, kg/m³	Vapor Density, kg/m³	Liquid Thermal Conductivity, W/(m · K)	Liquid Viscosity, cP	Vapor Pressure, Bar	Surface Tension, N/m × 100	Wicking Height, m	Vapor Velocity, m/s	Heat Load, kW
Nitrogen	−200	205.0	818	3.81	0.146	0.194	0.74	0.99	0.16	13.0	3.7
	−190	190.5	778	10.39	0.132	0.126	3.31	0.77	0.13	6.9	5.0
	−180	173.7	732	22.05	0.117	0.095	6.69	0.56	0.10	4.1	5.7
	−170	152.7	672	45.55	0.103	0.080	10.07	0.37	0.07	2.3	5.8
	−160	124.2	603	80.90	0.089	0.072	19.37	0.19	0.04	1.2	4.5
Ammonia	−40	1,384.0	690	0.65	0.303	0.290	0.76	3.57	0.69	59.9	19.6
	−20	1,338.0	665	1.62	0.304	0.260	1.93	3.09	0.62	35.3	27.8
	0	1,263.0	639	3.48	0.298	0.250	4.24	2.48	0.52	21.6	34.4
	20	1,187.0	610	6.69	0.286	0.220	8.46	2.13	0.47	14.4	41.6
	40	1,101.0	580	12.00	0.272	0.200	15.34	1.83	0.42	10.0	47.9
Freon-11	0	190.0	1533	2.59	0.108	0.550	0.42	2.18	0.19	23.5	4.2
	20	183.4	1487	5.38	0.100	0.440	0.93	1.92	0.17	15.3	5.5
	40	175.6	1439	10.07	0.097	0.370	1.82	1.66	0.15	10.4	6.7
	60	167.5	1389	16.85	0.094	0.320	3.14	1.40	0.13	7.4	7.5
	80	159.0	1334	30.56	0.089	0.280	5.85	1.14	0.11	4.9	8.7
Methanol	30	1,155.0	782	0.31	0.203	0.521	0.25	2.18	0.37	67.8	8.8
	50	1,125.0	764	0.77	0.202	0.399	0.55	2.01	0.35	41.3	13.0
	70	1,085.0	746	1.47	0.201	0.314	1.31	1.85	0.33	28.7	16.6
	90	1,035.0	724	3.01	0.199	0.259	2.69	1.66	0.31	19.0	21.5
	110	980.0	704	5.64	0.197	0.211	4.98	1.46	0.28	13.0	26.1
Water	60	2,359.0	983	0.13	0.649	0.470	0.20	6.62	0.90	182.4	20.3
	80	2,309.0	972	0.29	0.668	0.360	0.47	6.26	0.86	118.8	28.9
	100	2,258.0	958	0.60	0.680	0.280	1.01	5.89	0.82	80.1	39.4
	120	2,200.0	945	1.12	0.682	0.230	2.02	5.50	0.78	56.6	50.7
	140	2,139.0	928	1.99	0.683	0.200	3.90	5.06	0.73	40.8	63.0
	160	2,074.0	909	3.27	0.679	0.170	6.44	4.66	0.69	30.5	75.1
Dowtherm	200	321.0	905	0.94	0.119	0.390	0.25	2.50	0.37	41.7	4.6
	250	301.0	858	3.60	0.113	0.270	0.88	2.00	0.31	19.1	7.5
	300	278.0	809	8.74	0.106	0.200	2.43	1.50	0.25	10.6	9.3
	350	251.0	755	19.37	0.099	0.150	5.55	1.00	0.18	5.8	10.3
	400	219.0	691	41.89	0.093	0.120	10.90	0.50	0.10	2.8	9.3
Cesium	525	510.2	1710	0.04	19.520	0.200	0.16	5.11	0.40		
	625	495.3	1690	0.13	18.130	0.180	0.57	4.51	0.36	152.9	3.5
	725	485.2	1670	0.26	16.830	0.170	1.52	3.91	0.31	99.1	4.5
	825	470.3	1640	0.55	15.530	0.160	3.41	3.41	0.28	63.7	6.0
Potassium	600	2,000.0	705	0.11	41.810	0.140	0.19	7.86	1.49	215.1	17.3
	650	1,980.0	695	0.19	40.080	0.130	0.35	7.51	1.45	159.5	22.1
	700	1,960.0	685	0.31	38.080	0.120	0.61	7.12	1.39	121.7	27.2
	750	1,938.0	675	0.49	36.310	0.120	0.99	6.72	1.33	95.1	32.5
	800	1,913.0	665	0.72	34.810	0.110	1.55	6.32	1.27	75.9	37.8
Sodium	800	3,977.0	757	0.13	57.810	0.180	0.47	12.30	2.17	244.9	47.4
	900	3,913.0	745	0.31	53.350	0.170	1.25	11.30	2.03	155.3	67.5
	1000	3,827.0	725	0.67	49.080	0.160	2.81	10.40	1.92	100.9	93.5
	1100	3,690.0	691	1.31	45.080	0.160	5.49	9.50	1.84	68.9	120.6
	1200	3,577.0	669	2.30	41.080	0.150	9.59	8.60	1.72	49.4	147.6
Lithium	1130	20,100.0	440	0.01	69.000	0.240	0.17	28.50	8.67		
	1230	20,000.0	430	0.03	70.000	0.230	0.45	27.50	8.56	801.1	162.9
	1330	19,700.0	420	0.06	69.000	0.230	0.96	26.00	8.28	546.0	222.5
	1430	19,200.0	410	0.11	68.000	0.230	1.85	24.00	7.83	381.1	286.8
	1530	18,900.0	405	0.19	65.000	0.230	3.30	22.50	7.43	276.0	365.5

operating temperature; over the temperature range considered in each case the vapor pressure and density increase by roughly a factor of 25, the surface tension and hence the capillary pumping head fall off by 20 to 50%, the vapor velocity produced by the capillary head falls off by a factor of around 6, and the net effect of

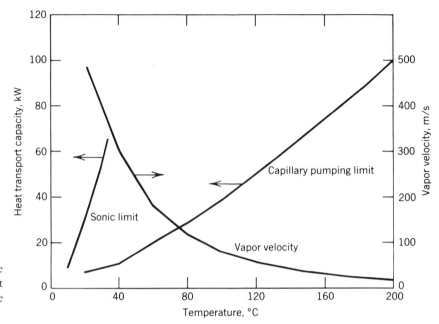

Figure 6.6 Effects of temperature on the limiting vapor velocity and heat transport capacity of the reference design heat pipe operated with water.

these various factors leads to an increase in the heat transport capacity by a factor of 2 to 3. These effects are shown for water in Fig. 6.6. Note that the temperature range for each fluid was chosen so that the vapor pressure at the upper end would be limited to roughly 7 atm because of safety considerations; for example, most building codes impose very severe restrictions on any equipment that operates at pressures above around 7 bar, or ~100 psi. The temperature at the lower end of the range was limited by sonic velocity considerations; the heat transport capability falls off very rapidly if the temperature is reduced much below the lower value used in each case. The sonic velocity was calculated for each fluid from

$$\text{Sonic velocity} = 106 \sqrt{\frac{\text{temperature, K}}{\text{molecular weight}}}, \text{m/s} \quad (6.5)$$

The effects of operating temperature on the capillary head available and the heat transport capacity are shown in Fig. 6.7 for the lower temperature working fluids, while Fig. 6.8 similarly shows these effects for the higher temperature working fluids. In the latter, both water and Dowtherm have been included to facilitate comparison because different scales were used to give a better definition of the values in each of the two regimes. Water and ammonia are clearly superior fluids for the lower temperature region because of their high surface tensions, high heats of vaporization, and low molecular weights,

which give high vapor velocities. The alkali metals give even higher heat transport capabilities in the high-temperature region, with the performance increasing rapidly with reductions in the atomic weight. (The alkali metal vapors are monatomic.)

The Mach number is given at the lower end of each of the curves to indicate the proximity to the sonic limit. For water, for example, reducing the temperature another 20°C would have yielded sonic velocity in the vapor, and, as indicated in Fig. 6.6, any further reduction in temperature would drastically reduce the heat transport capacity. Sonic conditions were avoided in preparing Table 6.1, in part because the approximations used in making the calculations should have included additional terms if vapor velocities were to exceed about Mach 0.5, and partly because the temperature distribution along the length of the heat pipe becomes badly distorted as the vapor velocity approaches the sonic level. (As the vapor pressure at the outlet of the evaporator region drops to half that at the base, there is a corresponding drop in the vapor temperature.) The resulting effect on the axial temperature distribution for a typical case is shown in Fig. 6.9 for a series of heat loads.

Radial Temperature Drop

As indicated in the previous chapter, the heat transfer coefficients for boiling and condensing are very high, and the thermal conductivity of the metal wall of the heat pipe ordinarily introduces little thermal resistance.

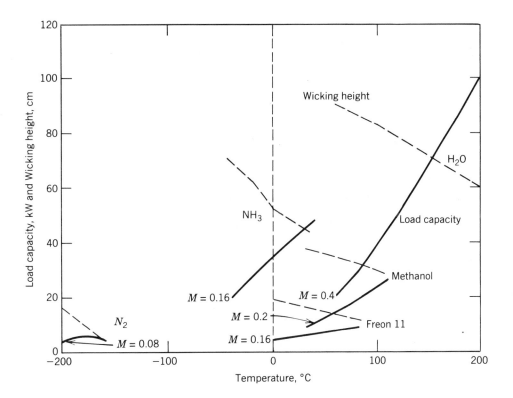

Figure 6.7 Effects of temperature on the heat transport capacity (solid lines) and the wicking height obtainable (dashed lines) in heat pipes employing low-temperature working fluids. The Mach number is given at the lower end of each heat transport capacity curve to indicate the proximity to the sonic limit.

However, depending on the working fluid and the surface heat flux in the evaporator and the condenser, the radial temperature drop through the fluid annulus and the wet wick may be substantial. The magnitude of this factor for a typical set of cases is indicated by Table 6.2, which was calculated for a surface heat flux of 1 W/cm² (3170 Btu/h · ft²). Depending on the fabrication method, the conductivity of the wick may be approximated by considering the metal wick structure as conducting heat in parallel with the liquid that fills the pores, or, at the other extreme, the metal wick structure can be considered as a set of layers in series with layers of liquid. The effective conductivity of the wet wick for these two cases thus becomes[7]

$$k_{(series)} = \frac{1}{\dfrac{1 - \epsilon}{k_s} + \dfrac{\epsilon}{k_l}} \qquad (6.6)$$

$$k_{(parallel)} = (1 - \epsilon)k_s + \epsilon k_l \qquad (6.7)$$

where k = thermal conductivity, W/m · K
  s = subscript for wick
  l = subscript for liquid
  ϵ = void fraction in wick

Inspection of Table 6.2 shows that, for the moderately high heat flux chosen for this example, the radial temperature drop is negligible for the alkali metals, about 11°C for water, 25°C for ammonia, 38°C for methanol, and 80°C for liquid nitrogen, Freon-11, and Dowtherm. Note that while it makes a great deal of difference in the conductivity of the wick whether one uses Eq. 6.6 as opposed to Eq. 6.7, the overall temperature drop across the wick and the annulus is little affected. This is because the conductivity in the metal wick is sufficiently high that it impedes the heat flux very little, whereas the low-conductivity liquids constitute major barriers to heat flow, particularly through the annulus.

The temperature drops listed in Table 6.2 are excessive for all of the fluids except water and the alkali metals

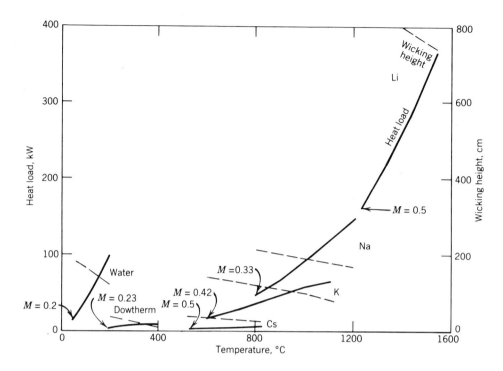

Figure 6.8 Effects of temperature on the heat transport capacity (solid lines) and the wicking height obtainable (dashed lines) in heat pipes employing high-temperature working fluids. The Mach number is given at the lower end of each heat transport capacity curve to indicate the proximity to the sonic limit.

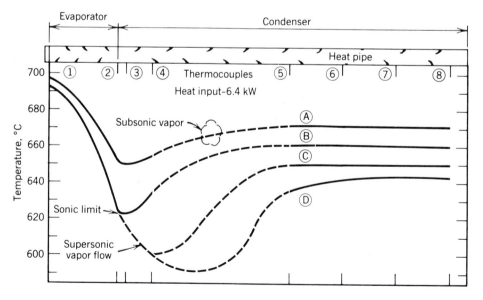

Figure 6.9 Axial temperature distributions in a heat pipe for operation with vapor velocities below, at, and above the sonic velocity. (Dunn and Reay, Ref. 7.)

TABLE 6.2 Temperature Drop Through the Wick and Annulus of the Heat Pipe of Table 6.1[a]

Wick	Liquid	Metal Conductivity, W/m · °C	Liquid Conductivity, W/m · °C	Parallel Conductivity of Wick, W/m · °C	Series Conductivity of Wick, W/m · °C	ΔT Annulus and Wick (Parallel), °C	ΔT Annulus and Wick (Series), °C
Al	Nitrogen	205	0.11	123	0.27	68.2	70.0
	Ammonia	(25°C)	0.28	123	0.70	26.8	27.5
	Freon-11		0.094	123	0.23	79.8	81.9
	Methanol		0.2	123	0.50	37.5	38.5
	Water		0.68	123	1.69	11.1	11.3
	Dowtherm		0.1	123	0.25	75.0	77.0
	Nitrogen	17	0.11	10	0.27	68.7	70.0
Type 304 stainless steel	Ammonia	(25°C)	0.28	10	0.68	27.3	27.5
	Freon-11		0.094	10	0.23	80.3	81.9
	Methanol		0.2	10	0.49	38.0	38.5
	Water		0.68	10	1.60	11.5	11.3
	Dowtherm		0.1	10	0.25	75.5	77.0
	Cs	26	16	22	20.80	0.7	0.5
	K	(800°C)	37	30	29.51	0.4	0.2
	Na		49	35	32.01	0.3	0.2
	Li		69	43	34.63	0.2	0.1

[a]For a heat flux of 1 W/cm², and annulus thickness of 0.75 mm, a wick thickness of 0.5 mm, and a void fraction in the wick of 0.4 is used.

because boiling would tend to occur in the annulus or wick so that the wick would be choked with vapor. This is because the radial temperature drop through the annulus and the wick must be less than the amount of superheat required to initiate boiling from typical nucleation sites (see the right column of Table 5.1 in the previous chapter). Thus the maximum allowable heat flux for heat pipes utilizing these fluids would have to be limited to much less than the 1 W/cm² assumed in calculating Table 6.2.

Freezing

A practical limitation on the choice of fluid for a heat pipe is that of its freezing point; this can force the use of a fluid such as a Freon for some applications where water would otherwise be preferable.

ESTIMATING HEAT PIPE PERFORMANCE FROM TABLE 6.1

The performance of a wide variety of heat pipes can be estimated quickly and easily by selecting a case in Table 6.1 that best approximates the conditions of interest and

applying appropriate perturbations using Eqs. 6.1–6.5. Examples of such perturbations for the cases most likely to be of interest are presented in this section.

Effects of Pore Size

If the size of the pores in the wick is to be changed from that of Table 6.1, Eq. 6.1 shows that the capillary pumping capacity is inversely proportional to the effective pore diameter. If the wick is to be fabricated of wire screen, Fig. 6.10 shows the size of the openings as a function of the mesh in wires per inch.

Effects of Gravity

The effects of gravity can be quite large because the gravitational head generated by the liquid in the annulus may be many times that developed by the meniscus. Estimation of the effect of the gravitational head can be carried out readily by using Eq. 6.2. Figure 6.11 shows the results of such a set of calculations for the heat pipe geometry of Table 6.1 operated with potassium at 575°C. Note the very large effect of the gravitational head.

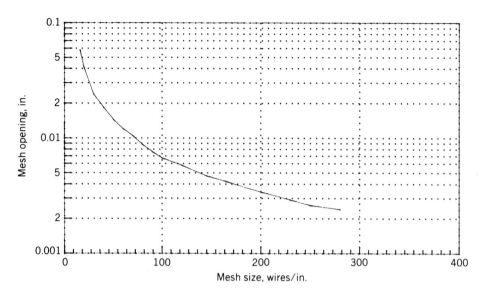

Figure 6.10 Width of opening in wire mesh screen as a function of mesh size.

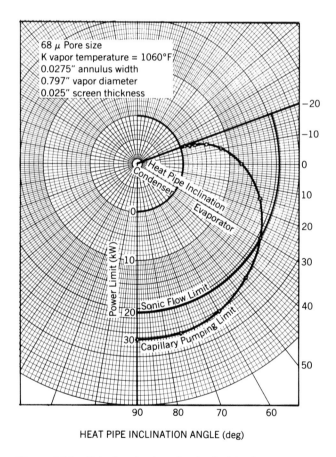

Figure 6.11 Calculated values for the limiting heat transport capacity of a heat pipe as a function of its angle of inclination at sea level. (Courtesy Oak Ridge National Laboratory.)

Effects of Wetting

As pointed out in the discussion of boiling nucleation in the previous chapter, both the surface tension of a liquid and the tenacity with which it adheres to a surface are sensitive functions of trace amounts of impurities, either in the liquid or in microscopic surface films such as an oxide or oil. These effects are subtle and the consequent losses in performance are difficult to avoid. For example, a heat pipe of the geometry for which Table 6.1 and Fig. 6.11 were calculated was built under rigid quality-control conditions and tested in various attitudes with a heat load of 2.7 kW, far below its nominal capacity. It functioned properly until the evaporator end was raised about 6 cm above the condenser end, at which point the condenser temperature began to fall rapidly while the temperature of the evaporator began to rise. Normal operation was restored by dropping the evaporator end below the condenser. This behavior was believed to have been caused by trace amounts of oxygen contaminating the potassium, though it may have stemmed from a pinhole in the joint where the wick was supposed to be sealed to the end cap at the evaporator end. This instance points up the vital importance of extreme cleanliness in the fabrication, charging, and sealing operations, and the need for acceptance testing if the wicking action is to be depended on heavily.

Effects of Diameter and Length

If the proportions of the heat pipe are kept the same, the capacity will be directly proportional to the square of the diameter and inversely proportional to the length. Note that keeping the proportions the same while increasing the diameter will also increase the radial temperature drop, and this may make it necessary to reduce the heat flux in order to avoid boiling inside the pores of the wick or even in the annulus.

MATERIALS COMPATIBILITY

As in every application, the choice of materials for good compatibility and freedom from corrosion is a prime consideration. In the case of heat pipes, good wetting of the wick by the working fluid introduces an additional constraint. A prime example is given by mercury, the vapor pressure of which makes it a tantalizing candidate as the working fluid for applications in the 300 to 600°C range in spite of its toxicity. It was omitted from the set of working fluids in Table 6.1 because it does not wet the iron-chrome-nickel alloys either well or consistently. Extensive experience in work on the mercury vapor topping cycle[20] and nuclear electric space power programs has demonstrated how frustrating this problem can be.[21,22] Much improved wetting can be obtained by adding small amounts of materials such as magnesium to the mercury to remove the oxide film that inhibits wetting, but such additions may lead to increased rates of corrosion, or they may be more than offset by surface contamination from traces of oil that somehow enter the system. The latter problem is greatly increased by the far greater surface-volume ratio in a heat pipe as compared to the corresponding value for a conventional Rankine vapor cycle system.

The indications of materials compatibility given by Table 6.3 provide a good basis for a rough screening of candidate materials from the corrosion standpoint. While good data are available for the surface tensions of most candidate fluids,[7,18] the degree to which they will wet the surfaces of different materials poses problems that are very subtle and difficult to quantify. At first thought, wetting agents might be used, for example, a detergent can be added to water, but this generally reduces the surface tension, defeating the purpose. In general, a good procedure is to clean the heat pipe thoroughly, add a charge of the working fluid by distilling it into the heat pipe to avoid fresh contamination, allow it to stand at the desired operating temperature for perhaps a day, and then drain off the liquid in which the remaining traces of contaminants should have become dissolved. For the alkali metals, this ordinarily entails retorting the heat pipe with the cleaning charge and holding it at a temperature of about 600°C for a few hours to remove traces of oxide from the metal surfaces before draining. It may be necessary to repeat such operations a number of times. In short, the procedures required to yield performance closely approaching the ideal may prove to be involved and expensive for some materials combinations.

Special problems that crop up with some of the otherwise attractive working fluids include the gradual production of an inert gas from the decomposition of the working fluid or reaction of the fluid with material in the wick or pipe; for example, an organic fluid may

TABLE 6.3   Compatibility of Materials for Heat Pipes[a,b]

Working Fluid	Structural Material				
	Aluminum	Copper	Nickel	Type 304 Stainless Steel	Niobium
Nitrogen					
Ammonia	Y	N	Y	Y	
Freon-11	Y				
Methanol	N	Y	Y	Y	
Acetone	D	Y		Y	
Water	N	Y	D	D	
Dowtherm		Y		D	
Cesium		N		Y	Y
Potassium		N		Y	Y
Sodium		N		Y	Y
Lithium		N	N	N	Y

[a]Refs. 7, 15, and 18.

[b]Symbols: Y = yes, well demonstrated; D = depends on surface treatment, temperature region, etc; N = not suitable.

polymerize and give off hydrogen. Because of all of the above factors, it is usually best to choose a set of materials that has been shown by extensive operating experience to make a good combination for service in heat pipes. Table 6.3 provides some information of this sort.

RELATION BETWEEN THE OPERATING TEMPERATURE AND THE HEAT LOAD

Heat pipes may be operated in a wide variety of modes, generally at temperatures above that for the sonic limit and at heat loads below that for the wicking limit. A common mode entails control of the operating temperature by varying the coolant flow over the condenser. Another mode is that prevailing in recuperators where the heat pipe temperature floats at a level intermediate between the local temperatures of the hot and cold fluid streams, with the heat load depending on the heat transfer coefficients in those fluid streams. In most cases, the heat pipe proportions are chosen so that there is an almost negligible temperature drop between the hot and cold ends of the heat pipe.

For some applications it is desirable to obtain a special relation between the heat load and the operating temperature; for example, it may be necessary to maintain the heat source within a narrow temperature range for a wide range of heat rejection rates. One way to accomplish this is to employ a heat pipe in which a small amount of inert gas is introduced in the sealing operation.[15] This gas will collect at the outboard end of the condensing region during operation, making the portion it occupies ineffective. The magnitude of this effect can be varied by providing a reservoir for the inert gas either with a bulb at the condenser end or with a reentrant central tube open only to the outboard end of the condenser. A heat pipe of the latter type was designed for cooling electronic equipment; test results are shown in Fig. 6.12 for operation at a series of power outputs.[15] Note that increasing the heat load by a factor of 6 led to an increase in the source temperature of only about 20°C (36°F).

Some types of heat pipe pose special problems in the start-up process because, for example, the working fluid flowing to the condenser may freeze there so that it cannot return to the evaporator. Such problems are addressed in Reference 11.

SPECIAL EFFECTS

Many special cases present a need for special effects such as those just cited. The degrees of freedom available to the clever designer open up a wide range of possibilities, far too many to include here. However, one of these is of sufficiently general interest to merit mention. For some applications it may be desirable to bend the heat pipe between the heat source and the heat sink, a possibility that seems quite reasonable if the vapor velocity is kept below a Mach number around 0.3 (see Fig. 3.12). This is indeed practicable, as has been demonstrated by operation of heat pipes with bellows in the isothermal section bent to angles as much as 120°.[23] In this instance the wick structure was 400 mesh stainless steel screen

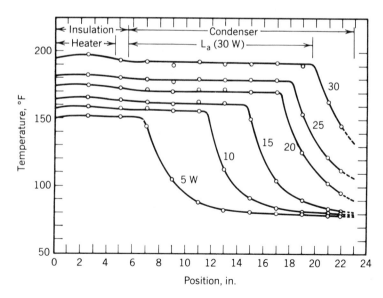

Figure 6.12 Effects of the operating temperature on the heat load for a heat pipe containing a small amount of noncondensable gas. (Marcus, Ref. 15.)

similar to that of the reference heat pipe used for Table 6.1 except that the screen wires were inclined at 45° to the axis of the pipe, that is, they were wound "on the bias" to make the wick more flexible in bending.

REFERENCES

1. G. M. Grover, T. P. Cotter, and G. F. Erickson, "Structures of Very High Thermal Conductivity," *Journal of Applied Physics*, vol. 35, 1964, pp. 1990–1991.

2. J. E. Jones et al., "Multimegawatt Space Nuclear Power Concept," *First Symposium on Space Nuclear Power Systems*, Albuquerque, N.M., January 11–13, 1984.

3. T. Baer, "Space Station Thermal Control," *Mechanical Engineering*, vol. 106(12), December 1984, pp. 22–33.

4. J. W. Galate, "Passive Refrigeration for Arctic Pile Supports," *Journal of Engineering for Industry, Trans. ASME*, vol. 98, May 1976, pp. 695–700.

5. K. T. Feldman and D. C. Lu, "Heat Pipe Heat Exchanger Design Considerations," *Proceedings of the Eleventh Intersociety Energy Conversion Conference*, September 12–17, 1976, pp. 887–892.

6. T. P. Cotter, "Theory of Heat Pipes," Report no. LASL-LA-3246-MS, Los Alamos Scientific Laboratory, March 1965.

7. P. D. Dunn and D. A. Reay, *Heat Pipes*, 3rd ed., Pergamon Press, Elmsford, N.Y., 1982.

8. M. E. LaVerne, "Performance Characteristics of Cylindrical Heat Pipes for Nuclear Electric Space and Undersea Power Plants," Report no. ORNL-TM-2803, Oak Ridge National Laboratory, January 1971.

9. J. E. Kemme, "Ultimate Heat Pipe Performance," *IEEE Transactions on Electron Devices*, vol. ED-16(8), 1969, pp. 717–723.

10. J. E. Kemme, "Heat Pipe Capability Experiments," Report no. LASL-LA-3585-MS, Los Alamos Scientific Laboratory, October 1966.

11. T. P. Cotter, "Heat Pipe Startup Dynamics," *1967 Thermionic Conversion Specialist Conference*, USAEC Report CONF-671045-4, October 1967, pp. 344–348.

12. C. L. Tien and A. R. Rohani, "Theory of Two-component Heat Pipes," *Journal of Heat Transfer, Trans. ASME*, vol. 94, November 1972, pp. 479–484.

13. C. A. Bankston and H. J. Smith, "Vapor Flow in Cylindrical Heat Pipes," *Journal of Heat Transfer, Trans. ASME*, vol. 95, August 1973, pp. 371–376.

14. S. W. Yuan and A. B. Finkelstein, "Laminar Pipe Flow with Injection and Suction Through a Porous Wall," *Trans. ASME*, vol. 78, 1956, pp. 719–724.

15. B. D. Marcus, "Theory and Design of Variable Conductance Heat Pipes," TRW Systems Group, NASA Contractor Report NASA CR-2018, April 1972.

16. A. P. Fraas and M. E. LaVerne, "Reference Design for a Thermoelectric Isotope Power Unit Employing Heat Pipe Modules," Report no. ORNL-TM-2959, Oak Ridge National Laboratory, November 1971.

17. D. B. Lloyd, "Test of a Combined Heat Pipe-Thermoelectric Module," Report no. ORNL-TM-4012, Oak Ridge National Laboratory, April 1973.

18. *Handbook of Thermodynamic and Transport Properties of Alkali Metals*, Chemical Data Series no. 30, International Union of Pure and Applied Chemistry, Blackwell Scientific Publications, Oxford, UK, 1985.

19. *Handbook of Applied Engineering Science*, 2nd ed., The Chemical Rubber Co., 1972.

20. W. L. R. Emmet, "Mercury Vapor for Central Station Power," *Mechanical Engineering*, vol. 63, 1941, p. 351.

21. R. L. Wallerstedt et al., "Final Summary Report-SNAP 2/Mercury Rankine Program Review, vol. 1," NAA-SR-12181, Atomics International Division, Rockwell International, June 15, 1967.

22. P. E. Eggers and A. W. Serkiz, "Development of Cryogenic Heat Pipes," *Journal of Engineering for Power, ASME Trans.*, vol. 93, April 1971, pp. 279–286.

23. D. M. Ernst, Personal communication, January 12, 1984.

Fluidized Beds

7

The ancient practice of panning for gold was probably the first application of a fluidized bed; swirling alluvial sand with enough water to get the particles in suspension produced a fluid of high enough density to float off the sand and silt, leaving the much higher density particles of gold as a residue. In the last century this basic concept of generating high concentrations of solid particles in water by agitation to give fluids of fairly high density was applied in mining operations to separate coal from slate and metallic ore from gangue (rock mixed with the ore). In these floatation processes separation is accomplished with sluices or with agitated pools in which turbulence serves to keep sand suspended in water in sufficiently high concentrations to provide a fluid whose density is greater than that of one set of particles in the mined material but less than that of the other. When used at coal mines, for example, pieces of coal will float to the top where they can be skimmed off, while the rock will sink to the bottom; at copper mines the dense particles of metallic ore will sink to the bottom, while the less-dense particles of rock will float to the top.

Development of fluidized beds of ceramic catalyst (mainly alumina) for the catalytic cracking of petroleum began in 1941. The immediate and outstanding success of this process led to extensive research on the characteristics of fluidized beds and the problems involved in handling dense mixtures of granular solids suspended in turbulent gases.[1-4] The large surface area and the excellent turbulent mixing in gas fluidized beds gives a uniform temperature distribution even for strongly exothermic reactions and high rates of reaction, making them attractive for many applications involving gas-to-solid reactions or gas-to-gas reactions requiring a large amount of catalytic surface. These operations have included not only petroleum cracking and reforming but also roasting of sulfide ores, coal gasification and liquefaction, and the combustion of coal and/or solid wastes.[1-8] The high heat transfer coefficients found in the course of this work have led to the investigation of fluidized solids for use as heat transfer fluids,[9] as slurry fuels for fission reactors,[10] and as heat storage media.[11] Solids fluidized with gases have the major advantage for some applications that they can be operated over a much wider temperature range than any given liquid such as a molten salt or a liquid metal. Thus system design and operation are free of concern for either freezing or boiling, and problems with corrosion can be minimized. On the other hand, although the heat transfer coefficients for gas-fluidized solid particle suspensions are much higher than for gases, they are lower than for molten salts and much lower than for liquid metals. In addition, fluidized solids present some peculiarities that may prove difficult to accommodate; the more important of these are treated briefly in the following discussion.

FLUIDIZATION

Solid particles are easily fluidized with a liquid, and the behavior of the resulting suspension is much like that of the particle-free liquid because the liquid and the solid particles usually differ in density by only a factor of two or three, so that relatively little stirring is required to keep the particles in suspension. Such a fluid can be pumped between plant components or even for long distances as in coal slurry pipelines, though its viscosity and density are much higher than for the fluidizing liquid. However, fluidizing a bed of solid particles with a gas presents a more difficult situation because the densities of the solid and the gas differ by a factor of about 1000. Much higher fluid velocities are required to keep the solid particles in suspension, and the pool is highly turbulent even for fluidizing gas velocities not much above that required to initiate fluidization. Circumstances are such that particle beds fluidized with a gas are the main type of interest for heat exchanger applications, and hence are the principal subject of interest in this chapter.

The behavior of a bed of particles fluidized with a gas can be envisioned by considering the course of events if one begins to force a gas upward through a porous plate into a bed of small particles. At very low velocities the gas percolates upward through the static bed of particles, as in packed, or pebble, beds, which are treated in Chapter 13. As the flow is gradually increased, a point will be reached at which the pressure drop across a typical particle in the bed will begin to exceed its weight, and it will be lifted out of position. As this takes place throughout the bed, the bed expands, the spacing between particles increases, and the bed begins to behave like a fluid having a density equal to that of the mixture of the gas and its entrained solids. The resulting increase in the flow-passage area between particles reduces the local fluid velocities, and hence the pressure drops across the particles so that they tend to settle back into place. The situation for any given particle tends to vacillate between the two limits, the first imposed by a pressure drop across the particle greater than its weight if the particles are too close together, and the second by a pressure drop less than the particle weight when the particles are too far apart. When the gas velocity is increased to perhaps 150% of that for initiation of fluidization, gas bubbles will form with a low particle density in the bubble. These bubbles are subject to a strong hydrostatic force that causes them to rise rapidly to the surface where they burst forth like little volcanic eruptions. As indicated by Fig. 7.1, a photograph of a bubbling bed, the bubble-formation process goes on in

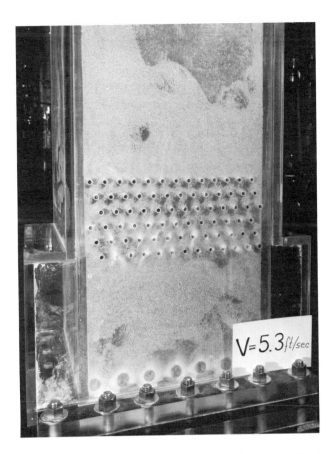

Figure 7.1 High-speed photo of a scale model of a fluidized-bed combustor showing gas bubbles in the bed at a flow rate equivalent to that at full power, that is, just below the point where elutriation would become substantial. (Ref. 12.)

random fashion throughout the bed, with the local variations in bed density agitating the bed and producing turbulent mixing much like that in a pool of violently boiling liquid. The volume fraction in the bubbles, the height of the expanded bed, and the degree of agitation increase with the gas flow rate.

As the gas flow is increased beyond the minimum for fluidization, the bed expands upward. When the flow is increased by a factor of 5 to 50, the bed depth will have increased by about 30% and the upward velocity of the gas leaving the bed will have become sufficient to entrain and carry off some of the particles. In practice, the particle size in the bed is not uniform, hence initially only the smaller particles are carried away, or *elutriated*, by the gas leaving the bed. Inasmuch as the gas velocity in the bed varies from one point to another and with the amount that the bed expands, it is convenient to take the nominal gas velocity up through the exit plenum as representative; this is termed the *superficial velocity*. This may be defined as

$$\text{Superficial velocity} = \frac{\text{volumetric flow rate}}{\text{surface area of quiescent bed}}$$

Figure 7.2 shows the relation between particle size and the superficial velocity for initial fluidization for three typical particle densities in a bed fluidized with air.[12] An additional set of curves also shows the air velocity at which particles will be lifted out of the bed and carried off with the exit gas stream. Examination of these curves indicates that, if all the particles are of the same size and density, the velocity at which elutriation begins is 20 to 90 times the minimum for fluidization, depending on the particle size. However, if the particles differ in size by a factor of, say, 5, elutriation will begin at a velocity only about 4 times that for initiating fluidization.

Estimating the Velocity for Initiating Fluidization

The fluid velocity required to lift a solid particle out of its position in a granular bed depends on its diameter and density and on the density and viscosity of the fluid. One can see intuitively from inspection of the analytically derived dimensionless *Archimedes number*, Ar, that this is a logical parameter. The usual expression for the Archimedes number is

$$\text{Ar} = \frac{g d_p^3 \rho_f (\rho_p - \rho_f)}{\mu_f^2} \qquad (7.1)$$

where Ar = Archimedes number
  g = acceleration of gravity, 9.8 m/s²
  d_p = diameter of particle (equivalent), m
  ρ_p = density of particle, kg/m³
  ρ_f = density of fluid, kg/m³
  μ_f = viscosity of fluid, Pa · s

A number of empirical expressions have been derived from experimental data to relate the Archimedes number to the minimum superficial gas velocity for fluidization.[1-4] One of these that has been evolved in the USSR by Gelperin and Einstein[3] yields the minimum Reynolds number for fluidization in terms of the Archimedes number as follows:

$$\text{Re}_{mf} = \frac{\text{Ar}}{1400 + 5.22\sqrt{\text{Ar}}} \qquad (7.2)$$

where Re_{mf} = minimum Reynolds number for fluidization.

Freeboard Above the Bed

For gas-flow rates appreciably higher than the minimum for fluidization, the top of the bed is not sharply defined but rather is a region in which the particle concentration varies widely as bursting bubbles spurt puffs of particles far up into the plenum over the bed. (See Fig. 7.1.) The irregular eruption of bubbles from the top of the bed causes the gas in the plenum to be very turbulent with local velocities well above the superficial velocity. To provide space in which these irregularities in gas velocity can be dissipated and to give ample opportunity for particles to settle back into the bed, a substantial plenum height is required. To minimize the amount of solids carried off with the exiting gas, the plenum height, or *freeboard* above the bed, is usually made at least 2 m, and to reduce the velocity the diameter of this region is often increased by making this section conical. This substantially increases the size and cost of the casing.

Fast Beds

In some applications it is advantageous to obtain a high degree of particle carry-over in order to circulate some or possibly the greater part of the bed material to another vessel, for example, from a combustion chamber to a heat exchanger, or from a cat cracking reactor to a

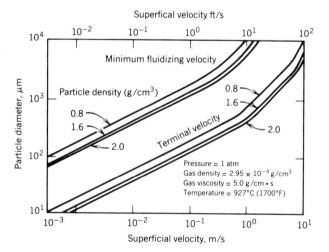

Figure 7.2 Effects of particle size on both the minimum superficial velocity for initial fluidization and the maximum velocity for minor amounts of particle carry-over. Lines are shown for three particle diameters. (Courtesy Oak Ridge National Laboratory.)

combustion chamber for regenerating the catalyst. This is often accomplished by selecting the particle-size range so that the larger particles will remain at the bottom of the primary bed to stabilize it, while the smaller particles are lifted out of the primary bed and carried over to a cyclone separator or allowed to fall out in a region of much reduced velocity to form a secondary bed. The particles in this second bed can be kept fluidized with a much reduced gas-flow rate. For a coal combustion unit, such an arrangement can be used to place the heat transfer surfaces in a cooler zone, or away from the more corrosive zone in which combustion occurs. Systems of this type are often called *fast beds*.

HEAT TRANSFER

The heat transfer characteristics of fluidized beds are quite different from those of conventional fluids in forced convection; they more closely resemble the heat transfer characteristics of pool boiling. The primary heat transfer mechanism stems from rapidly repeated direct contacts between the solid particles and the heat transfer surfaces. Increasing the gas velocity through a fluidized bed increases the frequency with which particles contact the surface, but it also reduces the length of time for the contact. These effects largely offset each other, so that varying the superficial velocity has little effect on the heat transfer coefficient. The net effect is indicated in Fig. 7.3, which shows the heat transfer coefficient increasing

rapidly as the turbulence increases after initial fluidization, reaching a peak, and then falling off slowly as the volume fraction of the bed occupied by gas bubbles becomes larger. The heat flux falls off rapidly at a little higher velocity because the bed density drops sharply. Inasmuch as contacts by solid particles provide the principal heat transfer mechanism, it is not surprising that changing the gas pressure has little effect on the heat transfer coefficient. Changing the particle size does have an effect, the smaller particles giving somewhat higher heat transfer coefficients. Even changing the geometry of the heat transfer matrix has relatively little effect as long as the passages through the matrix are many times the particle size. There can be some effects, however; the effects of horizontal and vertical spacing on the heat transfer coefficient for banks of horizontal tubes are shown in Fig. 7.4. The same study indicated that the heat transfer coefficient is 5 to 15% higher for vertical than for horizontal tube banks.[3] This is probably because the local gas velocity and heat transfer coefficient are higher for vertical than for horizontal surfaces. This effect can be seen in Fig. 7.5, which shows the heat transfer coefficient as a function of the angular position around two tubes, one inclined at an angle of 60° relative to the vertical flow and the other at an angle of 30°. Note the higher coefficients for the top and bottom of the tube inclined at a smaller angle relative to the gas-flow path.

Estimating the Heat Transfer Coefficient

Many investigators have applied conventional heat transfer theory to evolve empirical expressions designed to relate the heat transfer coefficients obtained from many different fluidized beds.[14] These attempt to allow for the effects of particle size and physical properties together with the physical properties and velocity of the gas. The resulting expressions are similar in that they relate the Nusselt number to various powers of the Reynolds, Archimedes, and Prandtl numbers, but they differ in the way these are arranged and in the numerical values assigned to the various coefficients and exponents. They usually avoid differences in the shape of the curve for the heat transfer coefficient as a function of the superficial velocity by taking the peak of that curve as the significant point, hence their expressions are given in terms of the maximum heat transfer coefficient $h_{\max}$. The maximum heat transfer coefficients calculated using some 19 different empirical expressions were compared with experimental data that had been obtained for a variety of particles, including alumina, sand, and ash having

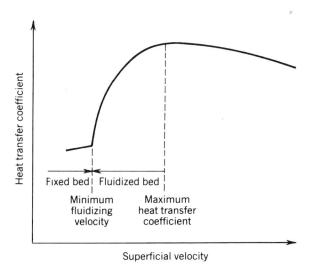

Figure 7.3 Characteristic curve for the heat transfer coefficient in a gas-fluidized bed as a function of the superficial velocity.

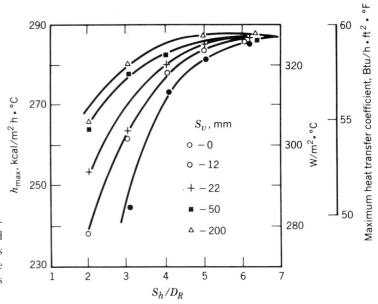

Figure 7.4 Heat transfer coefficients for 20-mm-diameter horizontal tubes in a fluidized bed of sand having a mean particle diameter of 0.35 mm as functions of the ratio of the horizontal spacing to the tube diameter (S_h/D) for a series of vertical spacings S_v in a staggered tube array. (Ref. 3.)

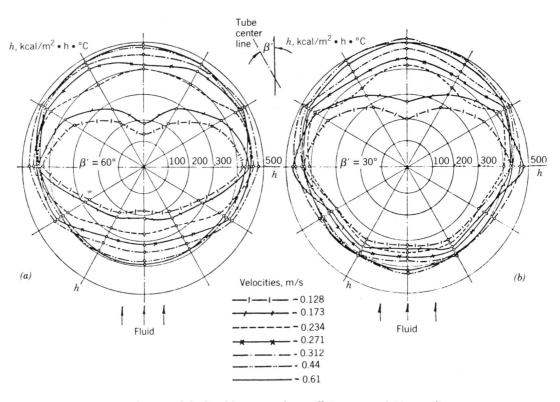

Figure 7.5 Distribution of the local heat transfer coefficient around 20-mm-diameter tubes in banks inclined with respect to the vertical at angles of (a) 60°, and (b) 30°. Data are plotted for a bed of sand fluidized at superficial velocities ranging from 0.128 to 0.61 m/s. (Ref. 3.)

sizes ranging from 462 to 2320 μm. Although the various expressions differ widely in their complexity, examination of the results discloses that the correlation giving nearly the best approximation (generally within less than 20% for the full range of experimental data) is also one of the simplest. This empirical expression, developed by Zabrodsky[15] is

$$h_{max} = 0.88 \frac{k_f}{d_p} Ar^{0.213} \qquad (7.3)$$

where h_{max} = maximum heat transfer coefficient, $W/m^2 \cdot K$

k_f = thermal conductivity of the fluid, $W/m^2 \cdot K$

Note that the superficial velocity at which the maximum heat transfer coefficient will occur is roughly three times the velocity for initial fluidization, hence the velocity for h_{max} can be estimated from Eq. 7.2.

Temperature Effects

The effects of temperature on the physical properties of the gas have a large effect on both the Reynolds number and the heat transfer coefficient; these effects are included in the above correlations. However, radiative heat transfer becomes a factor above 600°C, and commonly adds about 14% to the heat transfer coefficient at 1000°C.[3,16] The amount of this radiative contribution to the total heat transfer coefficient has been estimated in Reference 16 from

$$h_r = \frac{\sigma \alpha_w \left[\left(\frac{T_b}{100} \right)^4 - \left(\frac{T_w}{100} \right)^4 \right]}{T_b - T_w} \qquad (7.4)$$

where h_r = heat transfer coefficient for radiation

σ = Stefan-Boltzmann constant $\times 10^8$

α_w = absorptivity of the wall

T_b = bed temperature

T_w = wall temperature

This assumes that the cavity effect makes the emissivity of the porous bed about 1.0, that is, equal to that of an ideal black body.

Heat Transfer in the Freeboard

The large-scale turbulence in the freeboard together with heat transfer from particles tossed up into this region produces heat transfer coefficients many times higher than what might be estimated from the superficial velocity. The effect is particularly pronounced just above the bed. The resulting heat transfer coefficient is also increased by thermal radiation from both the bed and hot particles in the freeboard and varies with the superficial velocity and other factors, but Fig. 7.6 shows the results from a typical fluidized-bed combustion furnace.[17]

Heat Transfer From Particle Streams Flowing Under Gravitational Force

In some systems it is desirable to recover heat from granular material being dumped from a hot furnace. For example, the ash and spent lime must be dumped either continuously or periodically from fluidized-bed combustors. Cooling this red-hot stream of particles is important from the standpoint of the design and construction of the hopper that is to receive it, and recovering the heat from this material can make an important contribution to the thermal efficiency of the plant. Both theoretical and experimental investigations have shown that the heat transfer coefficients associated with this mode of fluidization are almost as high as for a bubbling bed,[18,19] though they fall off with distance down the channel because the transverse mixing of the particles in the stream is much less than in a bubbling bed.[19] Figures 7.7 and 7.8 show the effects of both distance down the channel and the velocity of the stream of particles for two particle sizes.

PRESSURE DROP

As in any heat exchanger, the pressure drop across the bed is an important design consideration. The pressure drop across the bed itself can be estimated by simply calculating the hydrostatic head imposed by the weight of the bed. To this must be added the pressure drop across the gas distributor plate under the bed. It happens that flow instabilities stemming from the periodic formation of large gas bubbles may cause "chugging" of a bed and disruptive vibrations and/or erosion of the tube matrix, or excessive particle elutriation.[20,21] This is particularly likely if the diameter or width of the bed is less than the height. The problems are too complex to treat here, but they can be avoided as a rule by designing the distributor plate under the bed so that the pressure drop across it will be at least 30% of the total pressure drop across the bed. The situation is analogous to the boiling flow-stability problems treated in Chapter 5.

The practicable bed depth is usually limited by

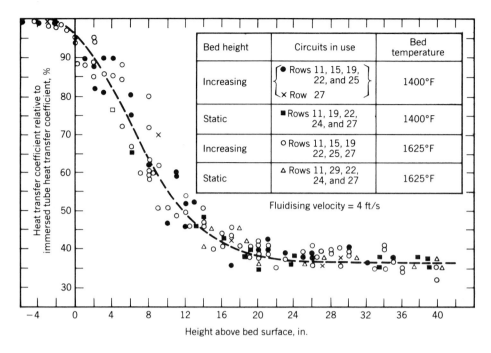

Figure 7.6   Heat transfer coefficients for horizontal tubes in the freeboard above the bed of a fluidized-bed combustor. (Ref. 17.)

pumping power considerations, which in turn depend on the overall requirements of the complete system in which the fluidized bed is being employed. Several representative cases will be examined in later chapters.

SPECIAL PROBLEMS

Fluidized solids differ in behavior from conventional fluids in many ways other than in their heat transfer characteristics.[1,8,22] In fluidized beds, the flow does not eddy above a large obstruction as in a conventional fluid, but rather stagnates so that particles may settle out and deposit on horizontal surfaces such as the tops of large tubes. Other problems arise if the fluidized solids are piped from one element of a system to another, or stored temporarily in tanks, as might be the case if they are to be used as an energy storage medium. Particles dumped into storage bins will settle out and may not flow again

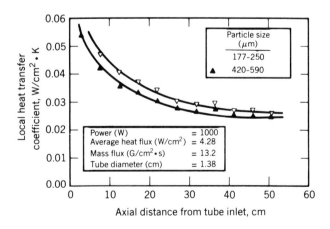

Figure 7.7   Effects of the distance from the tube inlet on the heat transfer coefficient for flow of a stream of particles under the action of gravity. (Ref. 19.)

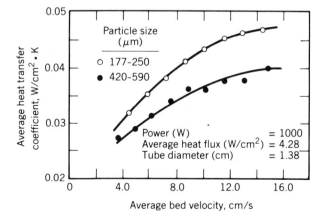

Figure 7.8   Effects of particle size and flow velocity on the average heat transfer coefficient for a stream of particles flowing under the action of gravity. (Ref. 19.)

when a valve at the bottom of the bin is opened. Instead, they may form a bridge over the opening, or only a central core will flow out, a phenomenon known as *ratholing*. Flow through pipes and restrictions will usually be blocked by *bridging* if the opening is less than about eight particle diameters across. Vibrators may be helpful in avoiding these conditions, but their effectiveness depends on the characteristics of the particles. Spherical particles such as microspheres of alumina for chemical processes and the rounded particles of beach sand flow much more readily than angular particles from a crushing operation. Particles of silica and alumina are hard and abrasion-resistant, whereas materials such as coal ash are porous, friable, and decrepitate rapidly into smaller particles that are elutriated from the bed. Some materials such as coal are inclined to be sticky, yielding particles that tend to agglomerate. For coal, the degree to which this occurs varies with both the type of coal and its moisture content. Even hard, dry particles may give trouble because they become charged with static electricity by friction within the bed. Electrostatic forces make them adhere to surfaces, a condition that in one project led to such heavy deposits on the pipe walls that the flow became severely blocked after only a day or two of operation. In that project,[9] the problem proved so severe that even severe vibration and hammering on the pipes served only to delay, not prevent, deposition and flow blockage. Thus while paper analyses indicated that a graphite particle suspension in helium looked highly promising as a reactor coolant, particle agglomeration and sticking to the walls of flow passages proved to be such a formidable obstacle that it was necessary to abandon the concept. In some cases there has been trouble with erosion; for example, the hard angular particles of iron sulfide have caused rather high erosion rates of the boiler tubes employed to remove the heat from the exothermic reaction in fluidized beds for ore roasting operations. In short, the behavior of fluidized solids presents a complex set of problems; this chapter is intended only to indicate some promising possibilities and ways to estimate the performance that might be obtained. If it appears from scouting calculations that a fluidized bed looks promising for a special application, the literature or an expert should be consulted to find out if there may be some peculiar problems that could prove serious.

MECHANICAL DESIGN

One of the first steps in the mechanical design of a fluidized bed is the selection of a geometry for the gas

distributor plate on which the bed rests and through which the gas is introduced to the bottom of the bed. Figure 7.9 shows some typical configurations for the air admission ports for the air distributor plate. A simple plate with drilled or punched holes may be employed provided that the hole size is less than several diameters of the particles in the bed. For very fine particles, and for hoppers or bins used to store fine particles, porous plates of sintered, fine metallic powder or very fine screen such as that used in Micropore filters have given good results in small units. Appreciable spans are usually required, in which case the fine screen can be backed up with coarse screen and/or perforated plate to support the substantial hydrostatic load. However, such provisions tend to be expensive, and they are subject to severe thermal stress and warping if they are used in an application such as a fluidized-bed coal combustor where the bed operates at a much higher temperature than the entering air. In such an application where the distributor plate will normally run at the relatively low temperature of the gas entering the bed, the plate and its support structure will be subject to abrupt and severe heating if the operation is suddenly interrupted and the hot bed material drops down on it. The damaging effects of such transient conditions are commonly avoided in combustors and ore roasting furnaces by using a set of tuyeres that extend upward to inject air 10 to 30 cm above the otherwise solid bed plate. The support structure is then protected from contact with the hot bed material by a layer of firebrick or a quiescent layer of ash. Figure 7.10 shows a distributor plate of this type that employs tuyeres similar to the one shown in the lower right corner of Fig. 7.9.

The forces acting on heat exchanger tubes in a fluidized bed may be quite high because the density of the stream of fluidized particles may be several times that of water, or over 1000 times that of air. The high fluid density together with the severe turbulence in the bed may induce serious tube vibration and surprisingly high stresses.[20,21] In designing the support structure for the tubes to withstand these forces, allowances must also be made for differential expansion between the tubes and the casing, a requirement that often precludes otherwise attractive structures.

Example 7.1. The exothermic heat released in the course of roasting iron sulfide in a fluidized-bed furnace is to be removed by a set of pipes through which boiling water is to be circulated. In view of the corrosive atmosphere together with the abrasive character and high density (5 g/cm^3) of the particles, 4-in. schedule 80 pipes of Inco 800 will be employed in the form of short

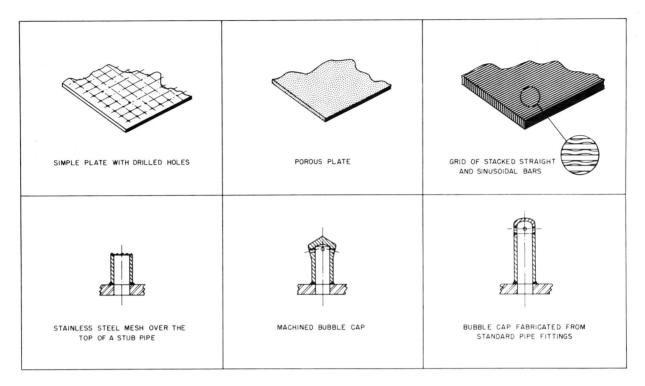

Figure 7.9 Typical designs of air tuyeres that have been used in fluidized beds. (Ref. 12.)

U-bends projecting horizontally into the bed. The overall length of pipe in each will be 4.56 m (15 ft). Estimate the superficial velocity and the heat removed per U-bend for a bed temperature of 700°C, a metal wall surface temperature of 300°C, a furnace pressure of 4 atm, and a mean particle size of 2 mm.

Solution. From Table H7.2 the area of the pipe wall is 1.178 ft²/ft, giving a total area per U-bend of 17.7 ft², or 1.64 m². Using the values for air as given in Table H2.4 as a first approximation, the density of the gas in the bed would be about 1.48 kg/m³ (0.093 lb/ft³), its viscosity about 0.413×10^{-4} Pa · s (0.10 lb · s/ft²), and its thermal conductivity 0.065 W/m · K (0.0378 Btu/h · ft · °F).

From Eq. 7.1 the Archimedes number is

$$\text{Ar} = \frac{9.8 \times 8 \times 10^{-9} \times 1.48 \, (5000 - 1.48)}{(0.413 \times 10^{-4})^2}$$

$$= 380,000$$

From Eq. 7.2 the minimum velocity for fluidization is

$$\text{Re}_{mf} = \frac{380,000}{1400 + 5.22 \; 850,000} = 82.1$$

$$V = \frac{82.1 \times 0.413 \times 10^{-4}}{0.002 \times 1.48} = 1.147 \text{ m/s} \, (3.54 \text{ ft/s})$$

The superficial velocity for normal operation would probably be at least three times this, or about 3.5 m/s (11.5 ft/s).

From Eq. 7.3 the maximum heat transfer coefficient is

$$h_{\max} = \frac{0.88 \times 0.065 \times 380,000^{0.213}}{0.002}$$

$$= 440 \text{ W/m}^2 \cdot \text{K} \, (78.4 \text{ Btu/h} \cdot \text{ft}^2 \cdot °\text{F})$$

Assuming that the emissivity of the tubes in the bed is 0.36, from Eq. 7.4 the contribution from thermal radiation is

Figure 7.10 Photograph of a distributor plate for a 6-ft square fluidized-bed combustor. The tuyeres are of the type shown in the lower right of Fig. 7.9, while the five coal feed ports are in the form of horizontal Xs mounted above the tuyeres. The distributor plate is shown standing on one edge. (Ref. 13.)

$$h_r = \frac{5.67 \times 0.36 \,(8950 - 1080)}{400}$$

$$= 40 \ \text{W/m}^2 \cdot \text{K} \ (7 \ \text{Btu/h} \cdot \text{ft}^2 \cdot {}^\circ\text{F})$$

The heat removed by each U-bend would be

$$Q = (440 + 40)\, 1.64 \times 400 = 315 \ \text{kW}$$

REFERENCES

1. F. A. Zenz and D. F. Othmer, *Fluidization and Fluid Particle Systems*, Reinhold Publishing Corp., New York, 1960.

2. A. M. Squires, "Species of Fluidization," *Chemical Engineering Progress*, vol. 68, April 1962, p. 66.

3. J. F. Davidson and D. Harrison, *Fluidization*, Academic Press, New York, 1971.

4. D. Geldart, "The Fluidized Bed As a Chemical Reactor: A Critical Review of the First 25 Years," *Chemistry and Industry*, September 2, 1967, pp. 1474–1481.

5. J. L. Stollery, "Fundamentals of Fluid Bed Roasting of Sulfides," *Engineering and Mining Journal*, vol. 165(10), October 1964.

6. J. H. Kleinau, "Pulp and Paper Mill Sludge Incineration," Paper presented at the TAPPI First Secondary Fiber Pulping Conference, Dayton, Ohio, October 22–25, 1968.

7. H. R. Hoy and J. E. Stanton, "Fluidized Combustion under Pressure," Paper presented at the Joint Meeting of the Chemical Institute of Canada and the AIChE, Toronto, May 24, 1970.

8. A. M. Squires, *Applications of Fluidized Beds in Coal Technology*, *Alternative Energy Sources*, Academic Press, New York, 1976, Chapt. 4.

9. D. C. Schluderberg et al., "Gaseous Suspensions—A New Reactor Coolant," *Nucleonics*, vol. 19(8), August 1961, p. 67.

10. J. P. McBride and D. G. Thomas, "Technology of Aqueous Suspensions," *Fluid Fuel Reactors*, published by Addison-Wesley Publishing Co., Reading, Mass., for the USAEC, 1958, Chapt. 4.

11. O. W. Durrant and M. J. Braun, "High Temperature Thermal Energy Storage for Power Generation," *Mechanical Engineering*, vol. 106(3), March 1984, p. 71.

12. A. P. Fraas et al., "Design of a Coal-Fired Closed Cycle Gas Turbine System for MIUS Applications," *Proceedings of the Tenth Intersociety Energy Conversion Engineering Conference*, 1975, pp. 260–268.

13. A. P. Fraas and R. S. Holcomb, "Atmospheric Fluidized Bed Combustor Technology Test Program," Paper presented at the Fluidized Bed Combustion Technology Exchange Workshop, Sponsored by the Energy Research and Development Administration and the Electric Power Research Institute, Reston, Va., April 11–13, 1977.

14. J. S. M. Botterill et al., "Temperature Effects on the Heat Transfer Behavior of Gas Fluidized Beds," Heat Transfer Symposium, Milwaukee, Wisc., 1981, *AIChE Symposium Series No. 208*, vol. 77, pp. 330–340.

15. S. S. Zabrodsky et al., *Vesti Akademia Nauk BSSR, Ser. Fiz. Energy Nauk*, no. 4, 1974, p. 103.

16. L. P. Golan et al., "High Temperature Heat Transfer Studies in a Tube Filled Bed," *Proceedings of the Sixth International Conference on Fluidized Bed Combustion*, April 9–11, 1980, Atlanta, Ga., Sponsored by the U.S. Department of Energy, pp. 1173–1185.

17. J. Byam et al., "Heat Transfer to the Cooling Coils in the 'Splash' Zone of a Pressurized Fluidized Bed Combustor," Heat Transfer Symposium, Milwaukee, Wisc., 1981, *AIChE Symposium Series No. 208*, vol. 77, pp. 351–358.

18. A. O. O. Denloye and J. S. M. Botterill, "Heat Transfer in Flowing Packed Beds," *Chemical Engineering Science*, vol. 32, 1977, p. 461.

19. R. E. Nietert and S. I. Abdelk-Khalik, "Thermalhydraulics of Flowing Particle-Bed-Type Fusion Reactor Blankets," *Nuclear Engineering and Design*, vol. 68, 1981, p. 293.

20. J. Baeyens and D. Geldart, "An Investigation into Slugging Fluidized Beds," *Chemical Engineering Science*, vol. 29, 1974, pp. 255–265.

21. A. S. Thompson, "Instabilities in a Coal Burning Fluidized Bed," Report no. ORNL TM-4951, Oak Ridge National Laboratory, February 10, 1975.

22. J. W. Carson, T. A. Royal, and D. S. Dick, "The Handling of Heaps," *Mechanical Engineering*, vol. 108(11), November 1986, pp. 53–59.

8

Flow Distribution Problems

Chapters 3–5 have dealt with fluid-flow and heat transfer problems under conditions giving ideal velocity distributions in the flow approaching the heat transfer surfaces. In practice, large deviations from the ideal, uniform velocity distribution are the rule rather than the exception. These deviations may arise from limitations imposed by fabrication, cost, or space considerations or from poor design. The acute difficulties posed by hot spots in nuclear fission reactors and in coal-, oil-, and gas-fired boilers give particularly striking examples of the consequences of deviations from the ideal velocity and temperature distributions in a heat transfer matrix. This chapter deals first with the flow characteristics of some simple geometries commonly involved in velocity distribution problems, and then presents some typical cases that have given trouble.

TYPICAL VELOCITY DISTRIBUTIONS AND FLOW PATTERNS

A poor velocity distribution across a heat transfer matrix is often induced by poor inlet-flow conditions that induce flow separation. Extensive experience shows that the most effective approach to problems involving flow separation is to employ some sort of a test rig that permits direct observation of the flow pattern.[1–3] It is instructive to examine photos of typical flow patterns obtained in the course of work of this sort.

Sharp-Edged Orifice

Figure 8.1 shows a typical flow pattern for a sharp-edged orifice. Note the smooth flow upstream of the orifice, the flow contraction—or *vena contracta* in the throat of the restriction—and the eddies on either side of the jet downstream of the orifice. It is clear that the mean velocity in the jet is substantially higher than the nominal velocity obtained by dividing the total flow by the area of the full opening in the orifice. It is for this reason that the orifice coefficient for sharp-edged orifices is around 0.6 where the channel diameter is substantially more than the orifice diameter.

Square-Shouldered Inlet

The flow pattern of Fig. 8.2 is quite similar to that of Fig. 8.1 except that the eddies along the side of the jet are limited to a smaller size by the passage wall. Not only do eddies of this sort represent a pressure loss, but in some applications they may cause erosion if abrasive particles, such as sand, are entrained in the fluid. Erosion of the inner surface of condenser tubes just downstream of the inlet is often caused by sand in the cooling water.

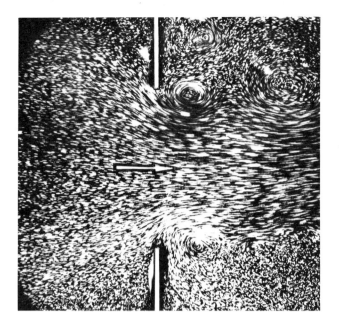

Figure 8.1 Flow pattern for a sharp-edged orifice. (Eck, Ref. 1.)

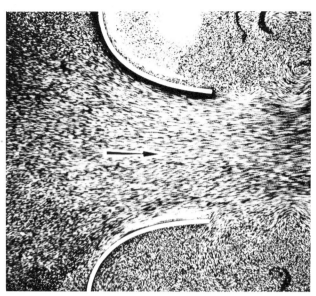

Figure 8.3 Flow pattern for a flow nozzle showing the discharge opening running full. (Eck, Ref. 1.)

Flow Nozzle

Figure 8.3 shows that the smoothly rounded approach provided by a flow nozzle avoids the *vena contracta* of the square-edged inlet, and hence gives a jet velocity that is essentially equal to the average velocity given by the

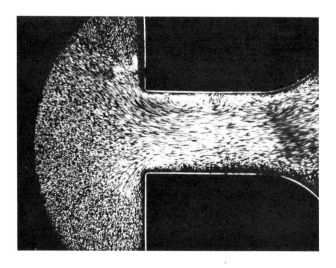

Figure 8.2 Flow pattern for a square-shouldered inlet showing separation just downstream of the entrance. (Eck, Ref. 1.)

aperture size and flow rate. It is for this reason that the orifice coefficient for a flow nozzle is very nearly unity. Note also that if the nozzle outlet were faired smoothly into a straight-walled channel, eddies such as those near the inlet of the channel of Fig. 8.2 would be avoided.

Diffusers

The effects of channel convergence or divergence are shown in Fig. 8.4. Note that in all cases the streamlines at the inlet are smooth, but that at any given station downstream of the inlet the boundary-layer thickness increases with an increase in the amount of divergence in the channel. In Fig. 8.4a, for example, the gradually converging passage walls produce enough acceleration in the flow so that the boundary-layer thickness does not increase beyond a point about one diameter downstream of the inlet. In Fig. 8.4b, the gradually diverging passage produces pronounced thickening of the boundary layer and some small eddies, but no flow separation. In Fig. 8.4c the rapidly diverging passage induced pronounced flow separation with a large eddy along one wall. Note that the boundary-layer thickness along the other wall is about the same as in Fig. 8.4b. The location of a high-velocity jet in a diffuser such as that of Fig. 8.4c tends to shift erratically from one side to the other under the influence of small perturbations in the inlet flow.

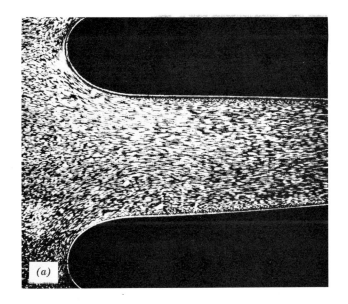

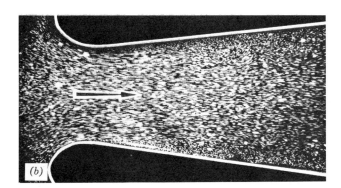

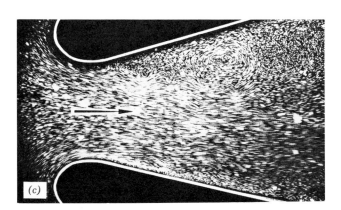

Bends

Flow through two bends in series is shown in Fig. 8.5. Note the flow separation with eddies and backflow on the downstream side of the inner radius of the bend in each case, particularly the first bend for which the complete flow pattern is included in the field of view. A brief consideration of the basic forces acting shows that flow separation should be expected at this point, since centrifugal force gives a substantial gradient in static pressure along a radius in the plane of the bend, with a region of low static pressure along the inside of the bend. Under potential flow conditions the static pressure becomes uniform across the channel downstream of the bend; hence it rises along the wall in the direction of flow. The real fluid tends to give the same pressure distribution, but flow separation occurs and results in some energy dissipation in eddies.

Considerable boundary-layer thickening occurs along the outer wall of the bend, but no flow separation occurs there because the pressure gradient in the direction of flow induces the boundary layer to move in the same direction as the main flow; that is, the pressure gradient is favorable, not adverse.

Plenum Chamber with Baffle

To avoid the jet effect shown in Figs. 8.1 and 8.3, a baffle can be installed concentric with the inlet pipe to improve the flow distribution at the inlet face of a heat transfer matrix in a large duct or a pressure vessel. An

Figure 8.4 Flow patterns for one converging and two diverging channels showing the effects of the degree of divergence on flow separation. (Eck, Ref. 1.)

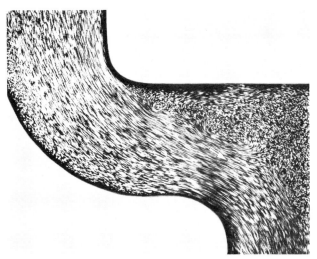

Figure 8.5 Flow pattern for a reverse bend showing regions of separation. The flow enters at the top. (Eck, Ref. 1.)

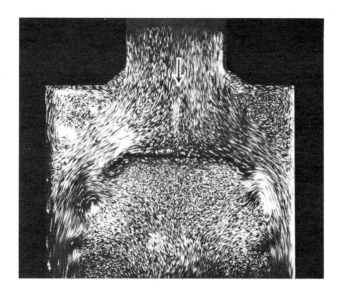

Figure 8.6 Flow pattern for a small plenum chamber with an axial inlet and a central baffle designed to act as a radial-flow, flat-plate diffuser. (Eck, Ref. 1.)

performance. Typical geometries likely to give trouble include abrupt increases in passage size, bends, and plenum chambers. As discussed briefly in Chapter 3, the actual streamline pattern under turbulent flow conditions is essentially similar to that for ideal potential flow except in the boundary layer or in regions in which the flow-passage area is increasing. Where an increase in flow-passage area occurs, the velocity pressure decreases while the static pressure increases in the direction of flow, and backflow tends to occur along the walls from the higher pressure to the lower pressure regions as in Figs. 3.9 and 8.4c. This leads to eddies and large-scale turbulence in which much of the change in velocity energy is dissipated by degradation into heat rather than recovered by conversion into static pressure.

EFFECTS OF CHANNEL DIVERGENCE

Free Jets

Fluid jets emerging into practically infinite reservoirs represent an extreme case of channel divergence having many practical applications. It has been found that the velocity distribution in a typical jet is as indicated in Fig. 8.7 for a jet from a flow nozzle.[4] The virtually flat velocity distribution in the plane of the flow nozzle exit is modified by progressively greater mixing between the portion of the jet around the outer perimeter and the surrounding fluid. The velocity at the center of the jet is unaffected at first, the effect of mixing being simply to reduce the width of the constant velocity region at the center. The outer boundary of the jet spreads with distance from the nozzle in the linear fashion indicated

arrangement of this sort is shown in Fig. 8.6. Note the eddies that have formed in the upper right and left corners; these are similar to those downstream of the first bend in Fig. 8.5. Note also the stagnation region in the center and the eddies behind the two outer edges of the baffle. Whereas this flow pattern leaves much to be desired, it probably reduces the stagnant region to a smaller fraction of the total cross section of the vessel. Some further improvement could probably be made by putting some perforations in the baffle.

The undesirable effects of a poor velocity distribution include increased pressure losses and pumping power requirements as well as impairment of the heat transfer

Figure 8.7 Diagram showing the boundaries of both the constant velocity core and the outer fringe zone for a free jet. Curves for the velocity distribution at two typical stations are superimposed. (Koestel, Ref. 4.)

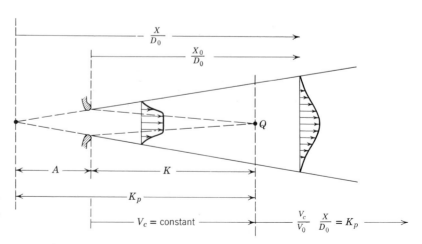

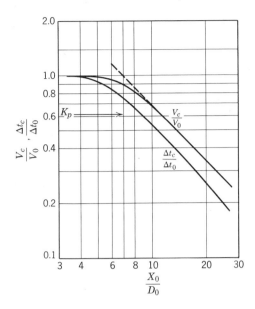

Figure 8.8 Velocity and temperature distributions along the centerline of a heated jet. (Koestel, Ref. 4.)

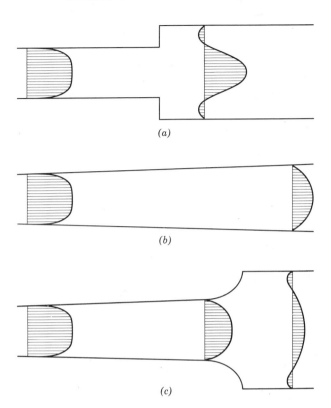

Figure 8.9 Velocity distributions following (a) an abrupt divergence, (b) a long 7° included-angle conical diffuser, and (c) a 7° conical diffuser with a cornet mouth.

by Fig. 8.7. Note the symbols used in Fig. 8.7 for the various quantities involved. At a distance of about eight jet diameters from the nozzle exit the velocity distribution becomes degraded to the point where there is no longer a flat region at the center, and the velocity along the centerline starts to fall off with further travel. This effect is shown in Fig. 8.8, where V_c is the velocity along the jet centerline, V_0 the initial jet velocity, D_0 the nozzle diameter, and X_0 the distance from the plane of the nozzle exit.[4,5]

If the fluid in the jet is at a temperature different from that of the surrounding fluid, the temperature structure in the jet resulting from the mixing process is also of interest. The ratio of Δt_c (the difference in temperature between the fluid at the centerline of the jet and the ambient fluid) to Δt_0 (the initial temperature difference at the nozzle exit) is also shown in Fig. 8.8 as a function of the distance from the nozzle exit. The turbulence governing the mixing process is greatest along the outer boundary of the core of the jet, the size of the eddies increasing with distance from the nozzle.[6]

Abrupt Increases in Channel Cross-Sectional Area

One of the most common types of channel divergence is an abrupt change in cross section, such as that indicated in Fig. 8.9a. Note the changes in the velocity distribution with distance from the abrupt increase in passage size,

and the tendency for backflow to occur along the walls downstream of the divergence.

In estimating the static pressure rise associated with an increase in passage cross-sectional area, it should be remembered that ideally the static pressure rise should be equal to the reduction in velocity pressure, that is,

$$\Delta P_{\text{id}} = \frac{\rho}{2g}(V_1^2 - V_2^2) \qquad (8.1)$$

For an abrupt increase in passage size (e.g., Fig. 8.6), eddy losses make the actual pressure rise much less than the ideal value. The theoretical requirement for conservation of momentum[7] leads to the conclusion that the static pressure rise should be at least that given by the change in momentum, that is,

$$\Delta P_a = \frac{\rho}{g}V_2(V_1 - V_2) \qquad (8.2)$$

Test data indicate that the actual pressure rise is essentially that given by Eq. 8.2; almost all the difference between the two represents the energy lost in eddies.

The efficiency of a flow-passage geometry as a means of recovering the velocity energy in the fluid is called the *diffuser efficiency*, and is defined as

$$\text{Diffuser efficiency} = \frac{\text{actual static pressure rise}}{\text{ideal static pressure rise}}$$

From Eqs. 8.1 and 8.2 it follows that for an abrupt increase in passage size,

$$\text{Diffuser efficiency} = \frac{2V_2}{V_1 + V_2}$$

$$= \frac{2A_1}{A_1 + A_2} \qquad (8.3)$$

Inspection of Eq. 8.3 shows that, where the change in cross-sectional area is small, the diffuser efficiency is good—for example, 80% for an area ratio of 1.5 (which corresponds to a diameter ratio of 1.2). If the ratio of the two diameters is as large as 5, however, the efficiency is extremely low, that is, about 8%.

Straight Diffusers

Much better efficiencies can be obtained with a gradually diverging passage, such as that in Fig. 8.9b.[2] To minimize space requirements and skin-friction losses along the walls, it is desirable to make the divergence as rapidly as possible and yet avoid flow separation and eddy losses. The angle of divergence for maximum diffuser efficiency varies with the diffuser length-diameter ratio, as shown in Fig. 8.10 for both a flat rectangular and a conical diffuser.[2] Curves were run for three different ratios of the diffuser length to the throat size for each case. Note that since the walls of the flat rectangular diffuser diverge in one plane only, the ratio of the length to the throat width is used, whereas for the conical diffuser, in which the passage expands in two dimensions, the ratio of the length to the throat radius is employed. This makes the ideal static pressure gradient along the walls essentially the same for the two cases. As expected, for any given diffuser length, the efficiency at first increases with the divergence angle, reaches a peak, and then falls off as flow separation begins to cause eddy losses. Note also that the greater the ratio of the diffuser length to the throat size, the smaller the angle of divergence for peak efficiency. Furthermore, the conical diffusers seem to be somewhat better than the flat rectangular diffuser, though the difference is not great. Annular

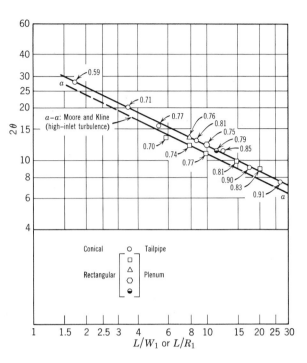

Figure 8.10 Effects of both the length–width ratio for flat diffusers and the length–radius ratio for conical diffusers on the divergence for maximum efficiency. The fraction of the inlet velocity head converted into static pressure is indicated beside each of the experimental points. (Kline, Abbot, and Fox, Ref. 2.)

diffusers in which two concentric cones form the walls have similar characteristics for similar area and flow-passage length-diameter ratios.[2]

For maximum efficiency it has been found best to make the diffuser length as great as the space available permits, using the divergence for maximum efficiency as indicated in Fig. 8.10, and then to employ an abrupt divergence or "cornet mouth" as in Fig. 8.9c. Since the fraction of the velocity energy recovered ideally varies as 1 minus the square of the area ratio, the use of a good diffuser to expand the passage to double its initial area makes it possible to recover most (75%) of the velocity energy ideally available at the inlet to the cone. Doubling the length of such a diffuser makes it possible to recover approximately 70% of the balance, making the ideal recovery around 90% of the total velocity energy at the inlet to the diffuser. These effects together with the corresponding effects for abrupt changes in cross section are shown in Fig. 8.11. It should be noted that there is little point in employing diffuser length-radius ratios greater than about 30, because frictional losses offset the small amount of pressure recovery that might still be effected.

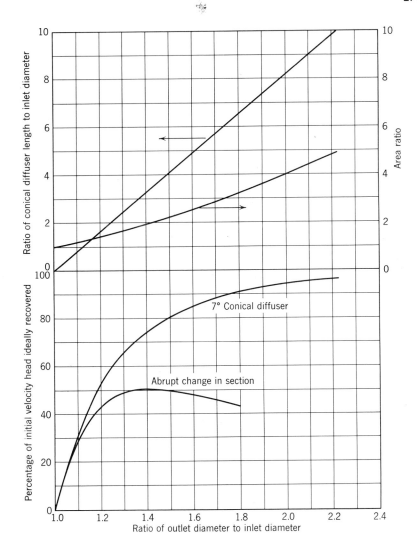

Figure 8.11 Effects of passage–diameter ratio on the area ratio and the ideal pressure recovery for both an abrupt increase in passage size and a 7° conical diffuser.

Flow separation in wide-angle diffusers and its adverse effects on velocity distribution can be avoided through the use of screens located at properly chosen intervals along the length of the divergence.[8] The screens introduce a pressure loss if compared to a long, open diffuser, but in wide-angle diffusers they give a better pressure recovery than if they were not used, and make it possible to maintain a good velocity distribution even for divergence as great as 90° (included angle). The use of screens in straight ducts is discussed later in this chapter.

Radial Diffusers

Where axial space is at a premium it is sometimes convenient to make use of a *flat-plate diffuser* such as that of Fig. 8.12a. This type of unit is ordinarily employed in the form of surfaces of revolution, and good efficiencies are obtainable if the design is made properly. This geometry reduces the effective passage diameter, and thus shortens the passage length required for a given reduction in velocity. The parallel walls of the diffuser region of Fig. 8.12a give a flow-passage area in the diffuser that increases with radius at essentially the same rate as in the 7° conical diffuser of Fig. 8.9b. In the diffuser of Fig. 8.12a the radius of the flare at the inlet is equal to the radius of the inlet duct, and the distance between the parallel plates in the radial-flow region is equal to 31% of the inlet duct radius. The area ratio and ideal pressure recovery for this type of diffuser are given in Fig. 8.13. As in the conical diffuser of Fig. 8.9c, the flat-plate diffuser can be terminated with a cornet mouth at a radius representing a good compromise between space and pressure-recovery considerations.

For a radial diffuser, the diameter required to effect a

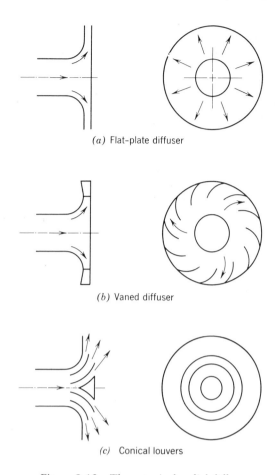

(a) Flat-plate diffuser

(b) Vaned diffuser

(c) Conical louvers

Figure 8.12 Three typical radial diffusers.

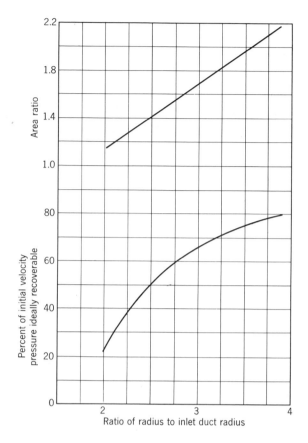

Figure 8.13 The area ratio and ideal pressure recovery for a typical vaneless radial diffuser.

given reduction in velocity can be reduced relative to the simple flat-plate diffuser of Fig. 8.12*a* by employing spiral vanes as in Fig. 8.12*b* and increasing the axial thickness of the diffuser passage with radius. The vanes act to prevent large eddies from forming toward the outer perimeter as a result of irregularities in the circumferential velocity distribution. (These eddies tend to swirl about axes parallel to that of the inlet duct.) In any case, to avoid flow separation, the rate of change in the flow-passage area along a streamline should be about the same as for a 7° cone. Similarly, where the radial space available is limited, there may be some advantage in terminating the diffuser by flaring the vanes to give an effect similar to the cornet mouth of Fig. 8.9*c*.

A basically similar diffuser commonly used for ceiling outlets in ventilating systems makes use of concentric annular vanes, as indicated in Fig. 8.12*c*. Again, the diffuser length is shortened by using vanes to reduce the equivalent flow-passage diameter. While commercial units of this type usually have too small a passage length-diameter ratio to give good pressure recovery, they do improve the flow distribution into a room by breaking up the flow from what would otherwise be a single large jet.

Bends

Immediately downstream of a bend the velocity distribution in a channel is ordinarily poor. As indicated in Fig. 8.14, flow separation occurs on the inside of the bend, secondary flow occurs along the walls, and a high-velocity region persists along the outer wall for some distance downstream.[9] This effect becomes progressively more pronounced as the ratio of the inside radius to the duct radial thickness is reduced. For small radius bends, such as that of Fig. 8.15*a*, the effects can be reduced drastically by employing vanes in the elbow, as indicated in Fig. 8.15*b*. An even better arrangement for rectangular ducts is to increase the flow-passage area by using the same radii for both the inner and outer walls, as in Fig. 8.15*c*, especially if turning vanes are employed. Ideally, the duct walls should be contoured to conform

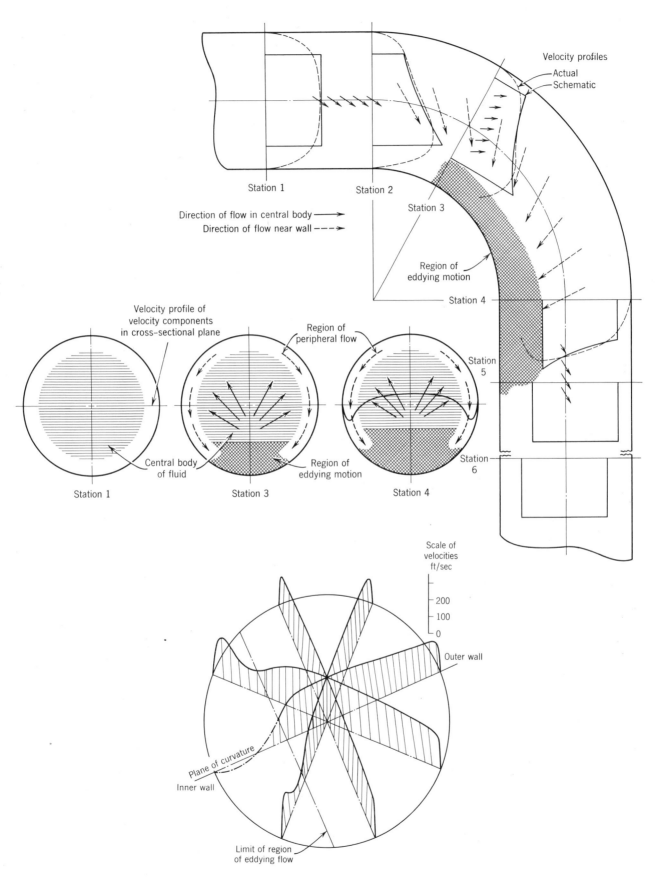

Figure 8.14 Velocity distributions and secondary flow effects in a 6-in. ID, 90° bend for Re = 500,000. (Weske, Ref. 9.)

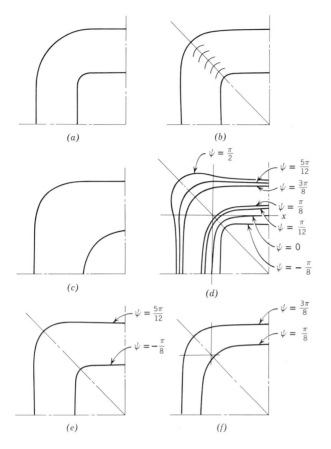

Figure 8.15 Typical bends for rectangular passages: (*a*) conventional bend; (*b*) conventional bend with turning vanes; (*c*) circular arc bend with inside and outside radii equal to the radial thickness of the duct; (*d*) streamlines for potential flow; (*e*) hyperbolic elbow bounded by the ideal streamlines $5\pi/12$ and $-\pi/8$; (*f*) hyperbolic elbow bounded by the ideal streamlines $3\pi/8$ and $\pi/8$.

to the streamlines of Fig. 8.15*d*, which are for ideal potential flow.[10] Elbows of this sort are shown in Fig. 8.15*e* and *f*.

The velocity distributions for some typical elbows are indicated in Fig. 8.16. Note that in the conventional elbow the turning vanes not only greatly improve the velocity distribution, but, as indicated in Fig. 8.16, they also reduce the pressure losses. Well-designed airfoils can reduce pressure losses to as little as 2% of the dynamic head,[11,12] but in most cases elements of circular arcs are entirely adequate. A comparison of Fig. 8.16*c* and *d* shows that an elbow designed to conform to the potential flow streamlines performs well only if the inside radius is large enough to avoid serious flow separation.

If it is desired to make the flow-passage area at the

outlet of an elbow substantially larger than that at the inlet, a vaned elbow such as that of Fig. 8.15*b* can be employed as a diffuser. Simple circular arc vanes with the vane spacing roughly equal to the vane width can be used to give a diffuser efficiency of approximately 70%. Carefully designed cascades of airfoils can be employed to give diffuser efficiencies of as much as 90% for area ratios of as much as 2.7.[11] Thus a cascade can be designed to give not only the shortest but also one of the most efficient of the many types of diffusers. However, a cascade of airfoils can be designed to give an efficient diffuser only if the velocity distribution in the inlet passage is uniform or at least consistent with good cascade design. Acceptable inlet flow conditions are usually difficult to obtain.

Use of Screens

While a poor velocity distribution downstream of an abrupt divergence or a bend can be corrected through the use of a sufficiently long section of straight duct, space considerations ordinarily make this impractical, since a straight length of at least 10 diameters is required to effect a substantial improvement. A much more compact device for flattening a poor velocity distribution is a set of screens, which may be woven wire mesh, a lattice of crossed rods, or perforated plates.

The effectiveness of screens in flattening a poor velocity distribution depends on many factors, such as the flow-passage geometry in the vicinity of the screen, the undisturbed velocity distribution in the duct, and the flow-resistance characteristics of the screen.[13,14] While the problem does not lend itself to an exact analytical solution, Prandtl[7] suggested that the peak velocity V_3 in the jet downstream of a screen can be related to the peak jet velocity V_1 upstream of the screen, and the average velocity V_0 in the duct by

$$\frac{V_3 - V_0}{V_0} = \left(\frac{1}{C_d + 1}\right)\left(\frac{V_1 - V_0}{V_0}\right) \quad (8.4)$$

The drag coefficient C_d for the screen is simply the ratio of the pressure drop across the screen to the dynamic head calculated for the average velocity in the duct.

Equation 8.4 indicates that, to cut the difference between the peak jet velocity V_1 and the average velocity in half, the screen should impose enough resistance so that the pressure drop across the screen is equal to the dynamic head at the center of the incident jet.

Figure 8.17 shows experimental data from tests with three different solidities, that is, ratios of the passage area

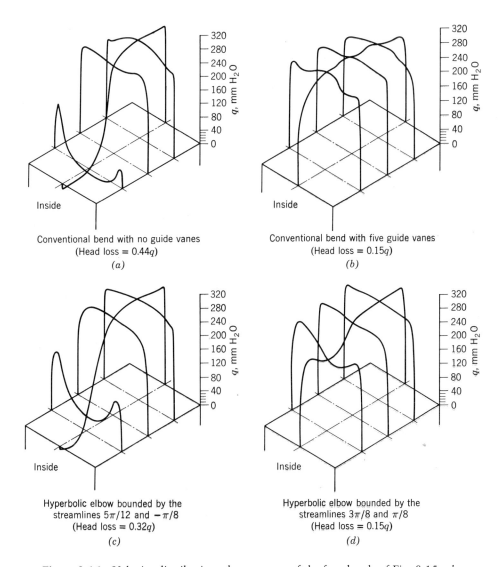

Figure 8.16 Velocity distributions downstream of the four bends of Fig. 8.15*a*, *b*, *e*, and *f*. (Harper, Ref. 10.)

blocked off to the total inlet-face area of the screen. The experimentally determined inlet and outlet velocity distributions are plotted together with an outlet velocity distribution estimated analytically from Eq. 8.4. Note the progressive improvement in the velocity distribution with increasing solidity of the screen, and the relatively good correlation between the experimental data and the analytical estimate based on Eq. 8.4.

While it is not apparent in Fig. 8.17, it is usually best not to use single screens having solidities greater than about 0.5, not only because the ratio of the velocity through the aperture to the nominal average velocity becomes too large but also because small irregularities in

fabrication may adversely affect the velocity distribution. This is likely to be a problem with lattices of bars, although irregularities in hole spacing in perforated plates may also give trouble. Thus if a high screen resistance is required, it is usually best to employ a multiplicity of screens in series with an interval between the screens of 20 or more hole diameters for jet dispersion.

Plenum Chambers

The velocity distribution through a heat transfer matrix may be affected by the manner in which fluid enters the inlet plenum chamber. If a high-velocity fluid stream

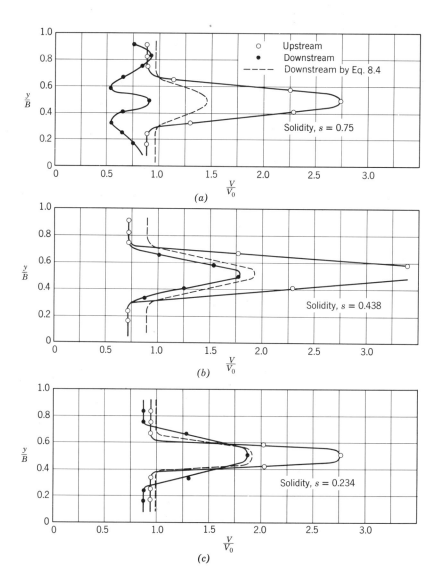

Figure 8.17 Effectiveness of screens having three different solidity ratios in flattening the velocity distribution of flow containing an axially symmetric jet. The duct has the diameter B. (Baines and Peterson, Ref. 13.)

enters a plenum chamber axially, as at the left end of the configuration of Fig. 8.18a, it is apparent that a high-velocity jet will shoot across the plenum chamber and impinge on the center of the heat transfer matrix so that the velocity through the center of the matrix will be higher than through the outer annular zone. If the fluid is brought into a plenum chamber from the side, as at the right end of Fig. 8.18a, a quite different set of effects may prevail. The inlet stream will give a high-velocity jet that creates a low static pressure region across the center of the heat transfer matrix. Impingement of the jet on the wall opposite the inlet pipe will form a zone of high static pressure, and the highest velocity through the heat transfer matrix will occur in line with this zone.

In either of the cases illustrated in Fig. 8.18a the

velocity distribution through the heat exchanger depends on the ratio of the inlet dynamic head to the pressure drop across the heat transfer matrix. In general, the inlet dynamic head should be kept to less than half of the pressure drop across the heat transfer matrix by increasing the duct size, using a diffuser, or by employing screens or baffles in the plenum.

A number of plenum chamber designs have been developed to yield good velocity distributions. That shown in Fig. 8.18b makes use of a torus from which the fluid flows radially inward. The arrangement shown in Fig. 8.18c makes use of a flat-plate diffuser to reduce the dynamic pressure in the inlet fluid stream before discharging into the plenum. This gives a flow pattern similar to that of Fig. 8.6.

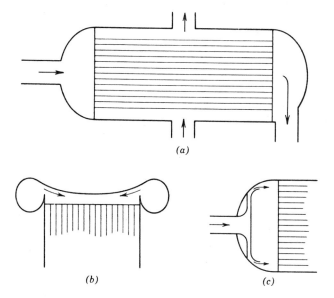

Figure 8.18 Typical inlet duct-plenum chamber configurations for heat transfer matrices.

The probable variation in flow distribution from the mean can be estimated from the inlet dynamic head and the pressure drop across the heat transfer matrix in the same manner as that discussed for screens. For the configuration at the left end of Fig. 8.18a the pressure drop available to force fluid through the central portions is approximately equal to the average pressure drop across the heat transfer matrix plus the inlet dynamic head; hence the ratio of the flow rate through the central portion to the average flow rate is roughly equal to the square root of the two pressure drops. This can be expressed more concisely as

Peak local velocity through a tube
Average velocity through matrix

$$= \sqrt{\frac{\Delta P_{av} + (\rho V^2/2g)_{inlet}}{\Delta P_{av}}}$$

$$= \sqrt{1 + \frac{(\rho V^2/2g)_{inlet}}{\Delta P_{av}}}$$

$$(8.5)$$

The extent to which a poor flow distribution causes a loss in performance can be surmised by considering the probable shell-side flow distribution through the heat exchanger shown in Fig. 8.18a. It is quite obvious that if no baffles were employed and the nominal pressure

drop across the tube bundle was only one or two times the inlet dynamic head, the bulk of the shell-side flow would pass vertically through the middle portion of the tube bundle, completely bypassing the regions at the ends. As a result, these end regions would be ineffective; the heat exchanger might just as well have been made with a tube-bundle length only 20% as great. It is less obvious, but nonetheless true, that the installation of a baffle across the inlet face of the tube bundle would not necessarily ensure a uniform distribution of flow over the tubes. A thorough detailed analysis and/or much testing is commonly required with models or prototype units to obtain a reasonably good flow distribution with configurations of this sort.

Effects of Inclining a Heat Exchanger in a Duct

To reduce pumping power losses a heat transfer matrix may be inclined at an angle in a duct as in Fig. 8.19 so that its inlet-face area is considerably greater than that of the inlet duct. Even if the velocity distribution in the inlet duct is good, unless the passages are properly shaped, much of the flow may tend to bypass the upstream portion of the matrix. This situation comes about because, for the geometry of Fig. 8.19, the velocity is high, and hence the static pressure is low, over the front face at the upstream end, while the velocity falls off and the static pressure builds up toward the downstream end. The reverse is the case for the discharge face; hence the static pressure differential available is greater at the downstream end. The effects are similar to those indicated in Fig. 8.20, and are likely to be especially troublesome if the pressure drop between the two headers is small relative to the dynamic head at the inlet to the header. This is likely to be the case if a thin heat transfer matrix (or a filter) having a low resistance is inclined in a duct in an effort to minimize pressure losses by affording a large flow-passage area. To obtain a good

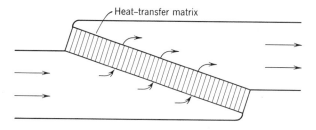

Figure 8.19 Heat exchanger inclined in a duct to increase the flow-passage area in the heat transfer matrix.

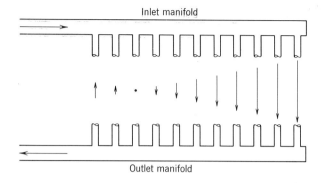

Figure 8.20 Velocity distribution in a bank of tubes with a length–diameter ratio of 50 and the tube and manifold flow areas equal.

velocity distribution across the heat transfer matrix for such a geometry, the installation should be designed so that the pressure drop across the matrix is equal to or greater than twice the dynamic head in the inlet passage. It should be emphasized that the significant factor is the

dynamic head in the inlet passage rather than the nominal dynamic head in the flow passages through the heat transfer matrix.

An excellent set of tests providing a quantitative basis for the design of the inlet and outlet regions for inclined heat exchangers has been reported by A. L. London.[15] The representative geometries examined are shown in Fig. 8.21. The example at the top of Fig. 8.21 shows both a typical rectangular box inlet and a specially shaped inlet. The contour for the latter, shown at the right of Fig. 8.21, was derived from potential flow considerations assuming a rectangular box outlet header. The flow-passage area of the inlet region was restricted to make the local axial velocity in the header high enough to give a uniform static pressure drop across the heat transfer matrix for the full length of the unit. The case at the bottom is for a triangular inlet region that is a fairly good approximation to the theoretically derived contour. The contours for these two cases are shown at the right of Fig. 8.21. The second case from the top at the left of Fig. 8.21 is for an installation in which the outlet stream leaves in the opposite direction from the

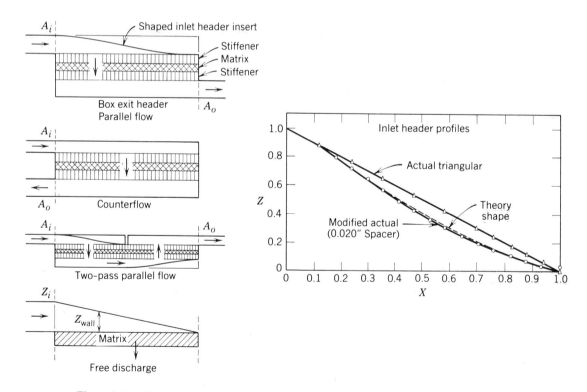

Figure 8.21 Sections through the four header geometries investigated for use with heat exchangers mounted obliquely in rectangular ducts. In the curves at the right the theoretically derived contour for the inlet header sweeps a little below that for the simple triangular approximation also tested. (A. L. London, Ref. 15.)

inlet stream as in Fig. 8.20, whereas the third case from the top is for an installation similar to the first but with a two-pass heat exchanger.

The tests were carried out using 35 layers of wire screen between two layers of honeycomb that were stiff enough to support the screens under the pressure differential across the matrix, which was about double the velocity head at the inlet. The principal results of the tests are summarized in Fig. 8.22, which shows both the velocity distributions over the length of the matrix and the static pressure distributions for both the inlet and outlet regions. Note that for the parallel-flow configurations the theoretically shaped inlet had a peak local velocity through the matrix only about 10% greater than the average, whereas the rectangular box-shaped inlet gave a peak local velocity over 50% greater than the average. The triangular inlet gave a peak velocity intermediate between these two. The two-pass parallel-flow arrangement with the theoretically shaped inlet regions gave about the same velocity distributions as the single-pass case. Perhaps the most interesting result is that the counterflow arrangement gave the best performance of the entire set when the flow area of the rectangular inlet header was made 0.636 as large as the outlet header, a proportion derived from potential flow theory. Even when the inlet and outlet headers had the same flow-passage area, the velocity distribution was much better than for the parallel-flow case with the rectangular box inlet header.

If the flow distribution in the inlet duct is poor, there is likely to be no advantage to inclination of the heat exchanger, since inclination would make the pressure drop even smaller relative to the dynamic head in the high-velocity portion of the inlet stream. The pressure losses associated with screens or other devices that might be used to flatten the velocity distribution would be parasitic, and would probably more than offset any reduction in pressure drop to be gained through a reduction in the velocity through the heat transfer matrix.

Manifolds

A special set of flow distribution problems somewhat similar to those posed by plenum chambers are presented by the design of manifolds for tube banks.[16] These problems are not of much concern where the tube length-diameter ratio is sufficiently large that the bulk of the overall pressure drop occurs in the tube, but they are likely to be important for tube length-diameter ratios under 100, where the pressure drop through the tubes runs two dynamic heads or less while the pressure losses

in the manifolds may run from two to six dynamic heads depending on the tube position.

The pressure drop from the manifold to the tube varies with both the geometry of the junction and the ratio of the velocity in the tube to the velocity in the manifold.[17] Since the design is usually such that the ratio of the velocity in the manifold to the velocity in the tubes is between 0.5 and 1.0, Fig. H3.7 covers this range, and gives approximate values of the pressure loss between the manifold and the tube. Tests show that rounding the corners at the inlet to a tube is about as effective in reducing the pressure loss as using a scoop or a nearly tangential juncture.

The pressure drop in an outlet manifold arising from the turbulence induced by a jet entering from the side may be substantial, especially where there is a long row of tubes in the tube bank and the side-stream velocity is a large fraction of the velocity in the manifold.

In the manifold, frictional losses reduce the total pressure in the direction of flow. Furthermore, where the inlet manifold is of constant cross section the velocity reduction in the direction of flow leads to a conversion of velocity pressure into static pressure, and hence an increase in the static pressure. The reverse occurs in the discharge manifold. Figure 8.23 illustrates, for a typical case, these individual effects together with the overall static pressure difference produced by them between the two manifolds.[16] In this instance the resulting asymmetric pressure distribution tends to give a higher-than-average flow rate through the tubes at the inlet end of the inlet manifold. On the other hand, changing the manifold configuration so that the inlet and exit connections are on opposite sides of the tube bank, as in Fig. 8.20, gives the highest flow through the tubes near the dead end of the inlet manifold. In fact, the flow distribution may be so poor that the flow may actually be reversed in some channels, as in Fig. 8.20.

The calculations for estimating the flow distribution are straightforward but tedious. As a first approximation the manifold can be sized to give an inlet velocity equal to the average velocity through the tubes. If the resulting velocity distribution in the tube bank is not acceptable, the manifold size can be increased. For some installations it may be in order to employ tapered manifolds designed so that the change in velocity along the manifold compensates for the frictional pressure drop.[16]

TYPICAL CASES

Flow distribution problems vary widely, and are peculiar to the particular case under consideration. No general

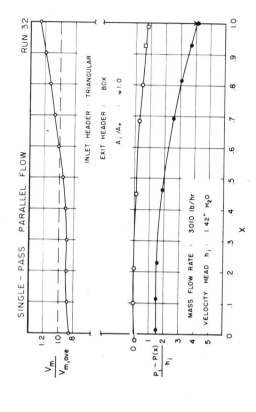

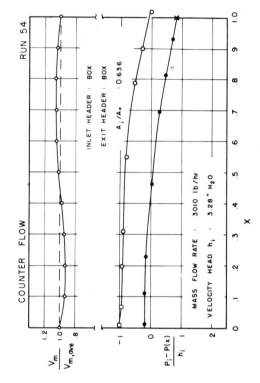

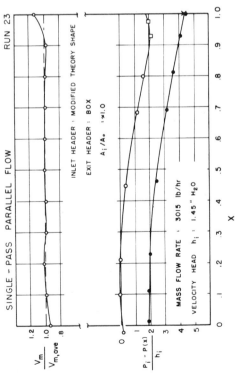

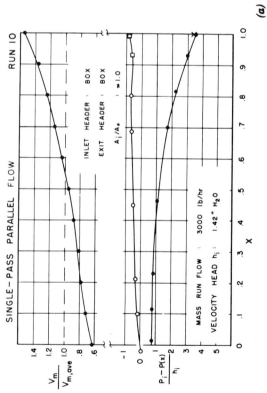

(a)

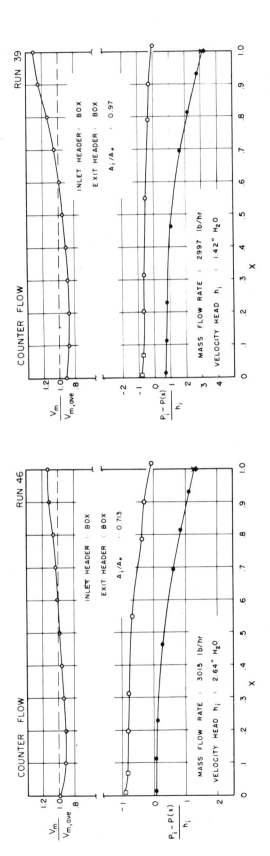

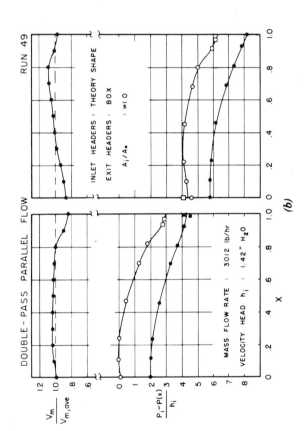

Figure 8.22 Velocity and pressure distributions found in tests of the header configurations shown in Fig. 8.21. (A. L. London, Ref. 15.)

167

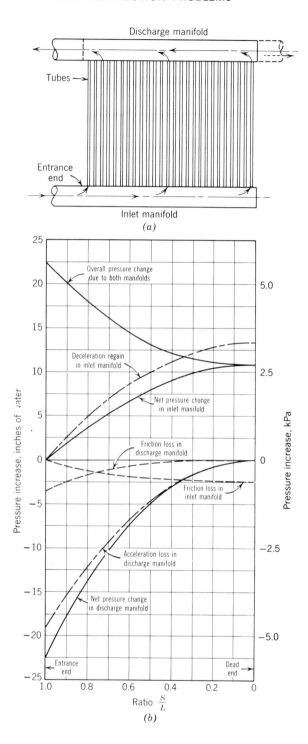

Figure 8.23 Values for the principal components of the pressure variations in the inlet and outlet manifolds of a typical tube bank with manifolds having a uniform cross-sectional area. (Keller, Ref. 16.)

rules can be set forth; each case must be examined carefully to determine how best to handle it. This section presents some typical cases in sufficient detail to indicate the importance of the problem and some possible approaches to evaluating the effects.

Performance Loss in an Aircraft Engine Aftercooler

The Rolls-Royce Merlin engine widely used in British and American fighter aircraft in World War II was equipped with a small, high-performance heat exchanger designed to cool the fuel-air mixture leaving the supercharger in order to increase both the power capacity and the detonation limit. The heat exchanger, known as an *aftercooler*, was mounted as indicated in Fig. 8.24. While this geometry was advantageous in that it was compact and well adapted to the engine structure, the configuration was far from ideal with respect to the fluid-flow distribution through the heat transfer matrix. Not only did the flow-passage area increase by a factor of roughly 6 in the short space between the outlet of the supercharger volute and the inlet face of the heat exchanger core, but the fluid was constrained to turn through a 90° angle. While it had been hoped that the pressure drop across the heat transfer matrix would serve to spread out the flow and give a fairly uniform air-flow distribution, this did not prove to be the case. The velocity distribution was measured experimentally, and is shown in Fig. 8.25. Analysis of this diagram indicated that 60% of the weight flow passed through only about 20% of the inlet-face area of the cooler. It should be noted that the average velocity head at the outlet from the supercharger was 18-in. H_2O, while the pressure drop across the heat transfer matrix, if the air-flow distribution had been uniform, would have been 5.2-in. H_2O. (The heat transfer matrix was a flat tube-and-fin core similar to that of Fig. 14.1. The inlet face was 9 × 12 in., while the depth in the cooling air flow direction was 10 in.)

The effects of the poor velocity distribution on the overall performance of the heat exchanger are indicated in Fig. 8.26, where the upper curve indicates the performance that would have been obtained had the air velocity distribution been uniform across the inlet face; the lower curve shows the actual performance obtained in the engine. The techniques outlined in Chapter 4 were used to compute the upper curve from experimental data obtained with a 3-in. square, 10-in. long core element tested under essentially ideal conditions. The same data were used to estimate the performance of the heat exchanger in the engine for the velocity distribution data of Fig. 8.25, and the results checked closely with the lower curve of Fig. 8.26.

In an attempt to flatten the velocity distribution across the aftercooler core, a splitter vane arrangement was

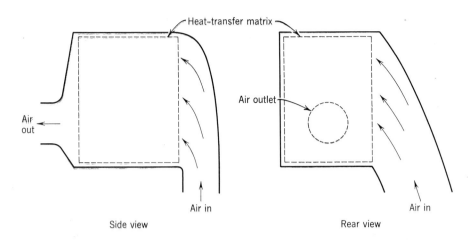

Figure 8.24 Outline of the air passages entering and leaving the aftercooler of the Rolls-Royce Merlin aircraft engine.

worked out experimentally for the inlet plenum. This system proved very effective in distributing the jet with little loss in pressure, and gave such a marked improvement in heat transfer performance, hence in engine

output at high altitudes, that a version suitable for production was built. When this was checked experimentally, however, it was found to give no improvement in performance over the initial version. Careful Pitot traverses disclosed that a strong swirl caused the velocity distribution at the outlet of the supercharger to differ not only between the engines used in the initial and later test work, but also still further in other engines, and shifted from time to time in a random fashion in any given

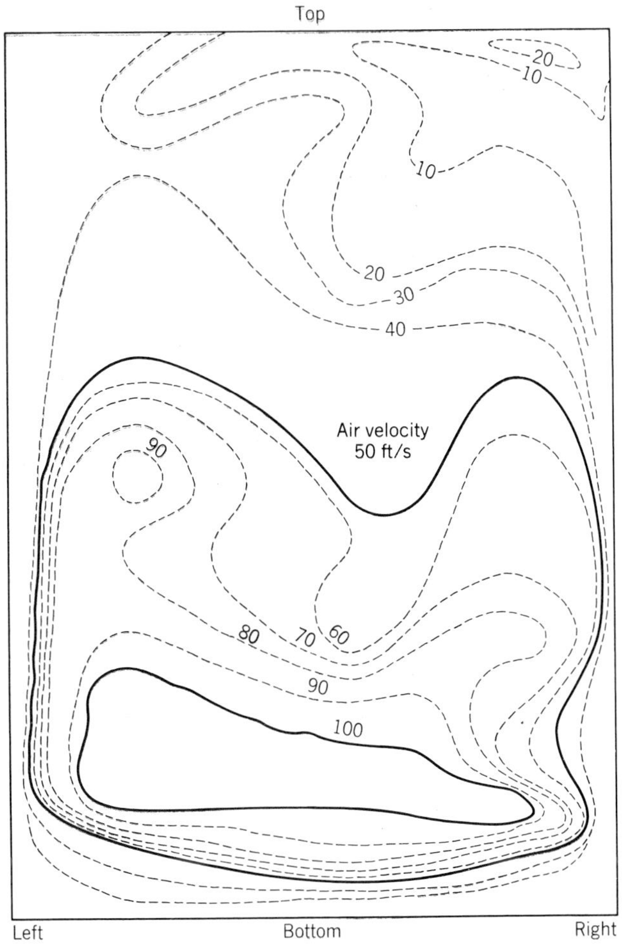

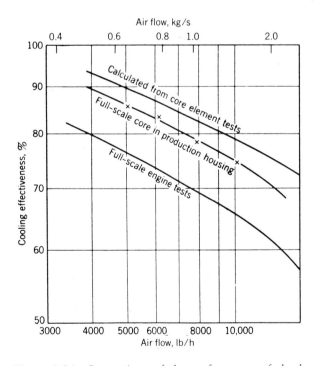

Figure 8.26 Comparison of the performance of the heat transfer matrix of Fig. 8.24 for three inflow conditions: (*a*) as calculated from tests of a core element in a straight duct; (*b*) a full-scale core installed in a housing similar to that of Fig. 8.23, with a blower and duct giving a good velocity distribution in the small inlet duct; and (*c*) as measured in an engine.

Figure 8.25 Air velocity distribution at the outlet face of the heat transfer matrix of Fig. 8.24.

engine. Since the effectiveness of any flow splitter vane is inherently sensitive to its location relative to the incident jet, this erratic shifting in jet position made any baffle or screen arrangement likely to be ineffective much of the time. Thus it was concluded that the poor velocity distribution could not be corrected except by a major redesign of the supercharger diffuser and discharge scroll, and this was not warranted.

Hot Spots in Boiler Furnaces

Hot spots are likely to lead to burned-out tubes in the furnaces of steam boilers. The design of these units is a specialized art in which careful provisions must be made for a good air-flow distribution as well as for charging the fuel. The matter is usually complicated by the requirement that the furnace be designed to operate with two or three different fuels to take advantage of shifts in the relative market prices or to cope with temporary shortages of one type of fuel. Thus the furnace design must be such that good combustion conditions prevail for coal, oil, or gas flames, yet the flow and temperature distributions must be acceptable from the heat transfer standpoint. In all cases, care must be exercised to avoid flame impingement on the tubes and furnace walls and high local velocities in the superheater tube matrix.

An interesting flow distribution problem in boiler design can be seen by examining Fig. 8.27, which shows one way of mounting a bank of superheater tubes. It is

easy to see from potential flow theory that the gas velocity will be much higher close to the bottom of the tube bank than up toward the top, because the velocity distribution will approximate that in a free vortex. Since the gas temperature may run about 2000°F in this region, high local heat transfer coefficients could lead to excessive tube-wall temperatures that would result in tube burn-out.

The condition cited may be aggravated by poor flow distribution on the steam side. If the steam manifolds for the superheater tube bank are not designed properly, the steam flow may be much lower in some tubes than in others, so that poor cooling on the inside coupled with excessive heating on the outside may combine to give tube-wall temperatures much above the acceptable limit. In some cases the flow distribution has actually been as bad as that indicated in Fig. 8.20. Obviously, those tubes in which the steam velocity was nearly zero operated with tube-wall temperatures close to the hot gas temperature.

The foregoing cases constitute just two of many that could be cited. Some of the host of other problems involved in boiler furnace design are discussed in Chapter 15, but a more detailed treatment is beyond the scope of this text.

Hot Spots in Steam Generators for Gas-Cooled Reactors

If a heat exchanger were built in the same manner as the steam generator of Fig. 1.5, but designed to operate with a gas-inlet temperature of perhaps 1350°F and a superheater steam outlet temperature of 1050°F, the tube-wall metal temperature in the jet emerging from the gas-inlet duct would be likely to run much closer to the gas-inlet temperature than to the superheated steam-outlet temperature because of the high local heat transfer coefficients. Since the strength of the tube wall falls off very rapidly from 1050°F to 1350°F, a burst tube would be likely to result.

Several different steps can be taken to relieve the problem. The most drastic is to reduce the gas temperature entering the steam generator to perhaps 50°F above the superheater outlet temperature, thus limiting the maximum possible tube-wall temperature to 1100°F. This solution is disadvantageous in that it greatly increases both the size of the steam generator and the cost of the pumping power required to circulate the reactor coolant. A more attractive approach is to reduce the velocity of the gas entering the plenum ahead of the heat transfer matrix by employing some form of diffuser. The radial diffuser shown in Fig. 8.12a appears well suited to this application. Yet another approach is to

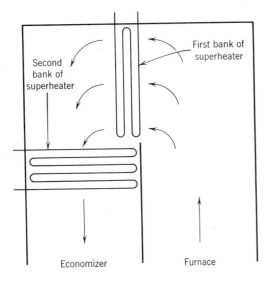

Figure 8.27 A boiler furnace configuration that gave a poor velocity distribution through the banks of tubes in the superheater.

employ screens to distribute the flow uniformly across the inlet to the tube banks. The screens entail an appreciable pumping power loss, and would have to be of a material that is strong at 1350°F. Other effective steps include the use of a parallel-flow rather than a counter-flow configuration for the superheater so that the tube walls in the first tube banks to be struck by the hot gas would be well cooled, and the metal temperature at the hot spots would not exceed 1100°F.

Hot Spots in Nuclear Reactors

Nuclear reactors offer a more convenient example to illustrate the nature of the hot-spot problem than do combustion-fired boilers, because the relationships involved are simpler and the behavior of the system is more predictable. Furthermore, the hot-spot problem is more critical in reactors, since on the one hand the high capital charges for the reactor make it desirable to use the highest possible power density, whereas on the other hand the hazards and costs associated with burn-out at a hot spot are much higher than in conventional boilers.

In all types of reactor it is important to minimize the pumping horsepower, which means that the coolant temperature rise through the reactor core should be as large as possible. This in turn means that regions near the reactor core outlet will be substantially hotter than the average value for the core as a whole. If the axial power distribution through the core is assumed to follow the usual cosine curve, the heat flux and temperatures of both the cooling gas and the fuel element surface may be plotted as functions of distance from the reactor core inlet to give curves such as those of Fig. 8.28. The region yielding the peak fuel element surface temperature is often called the "hot zone," because it is in this zone that excessive fuel element surface temperatures are most likely to occur.

A host of effects lead to variations in the basic temperature pattern indicated by Fig. 8.28.[18] There are both gross and fine radial variations in the neutron flux which lead to variations in the nuclear power distribution and hence in the temperature distribution. Fuel element burn-up and variations in control rod position contribute further aberrations in the local heat flux. These effects may be cumulative in one or a few fuel elements and give local power densities from two to six times the average for the core as a whole.

The gross coolant flow distribution across the reactor core also contributes to the hot-spot problem. In channels in which the coolant flow rate is a few percent below the

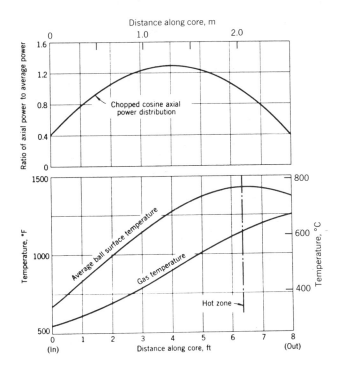

Figure 8.28 Heat flux, average surface temperature, and average gas temperature as a function of the distance from the core inlet of a large gas-cooled pebble bed reactor. (Courtesy Oak Ridge National Laboratory.)

average, the gas temperature rise to the hot zone runs a corresponding few percent above the average. In addition, the lower flow rate leads to lower heat transfer coefficients in that channel, and these cause the difference between the mixed mean temperature of the gas stream and the fuel element surface temperature to run greater than the average. The combined effects of the increases in the enthalpy rise and the film-temperature drop lead to a substantial increase in the hot-zone fuel surface temperature for the low-flow channels.

In reactors making use of moderately complex fuel elements in which there are possibilities of asymmetries, the flow distribution is not likely to be uniform within any given channel. For example, in an electrically heated full-scale model of the fuel element cluster of the EGCR shown in Fig. 8.29, asymmetries gave large variations in the local temperatures of both the gas stream and the fuel element surface. Figure 8.30 shows the velocity distribution prevailing in a typical section.[19] Hot-spot problems in gas-cooled reactors are more acute than in the water-cooled reactors because, in the latter, they are relieved by local boiling. In gas-cooled or organic liquid-cooled reactors no beneficial effects from boiling can be expected.

The fuel element surface temperature variations arising from the irregularities in the velocity distribution of Fig. 8.30 in themselves may not necessarily be objectionable, but they do cause bowing of the fuel elements. It should be noted that the fuel element of Fig. 8.29 was designed with pin-jointed ends for the fuel capsules to avoid structural redundancies in the fuel element assembly and the attendant severe local bending stresses that would otherwise occur in the thin capsule wall. Thus a diametral temperature difference of 30°F across a fuel capsule will not lead to any appreciable thermal stress, but it will cause a 28-in.-long capsule to bow about $\frac{1}{16}$ in. Once such a deflection occurs it tends to restrict the gas flow past the hot side of the fuel capsule, and this leads to a further local increase in both the gas and fuel element surface temperatures, giving a further increase in the lateral deflection of the fuel capsule. Depending on operating conditions, this cumulative process of deflection may reach an equilibrium at some small deflection with a relatively small resultant hot-spot effect, or it may progress and become unstable and lead to touching of adjacent fuel elements and burn-out. Figure 8.29 shows a capsule cluster damaged in this way in an electrically heated test rig. Note the burned spots at various points along the fuel element and the bowed condition of the tubes after the test. They were, of course, quite straight before the test began.

It might be hoped that substantial amounts of transverse mixing would occur where major flow channels

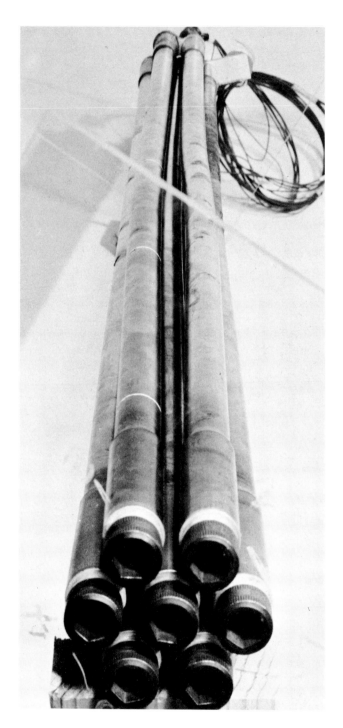

Figure 8.29 Photo of an electrically heated mock-up of a fuel element for the Experimental Gas-Cooled Reactor in which a hot-spot condition caused bowing and burn-out after testing. (Courtesy Oak Ridge National Laboratory.)

through the reactor are subdivided into small parallel channels by the fuel element fine structure. With a fuel rod cluster, for example, it is possible to introduce

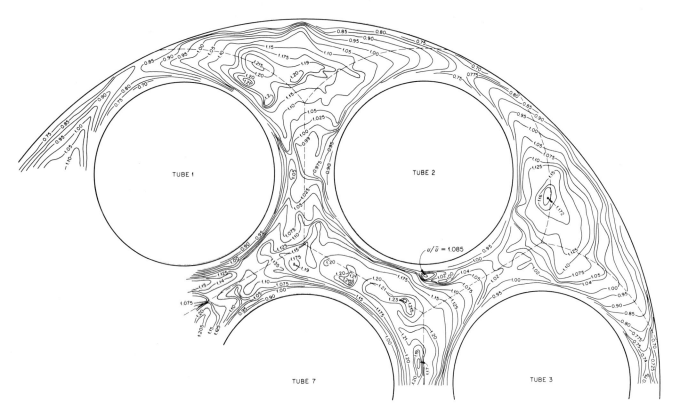

Figure 8.30 Velocity distribution measured at a representative station in an electrically heated full-scale model of a fuel element of the Experimental Gas-Cooled Reactor. (Courtesy Oak Ridge National Laboratory.)

turbulators with the objective of promoting such transverse mixing. Another approach is to interrupt the heat transfer surfaces at intervals in the direction of flow. Tests by a variety of investigators indicate that unless a great deal of pumping power is dissipated in the mixing operations, the transverse mixing does not help greatly in reducing hot-spot effects if the surfaces are in line and the axial gaps between the surfaces are of the order of the channel thickness. Thus it is highly desirable to proportion the fuel elements so that a good coolant-flow distribution is inherent in the design. An important step in this direction is to make use of a relatively coarse type of fuel element with a large spacing between surfaces so that small thermal distortions have relatively little effect on the flow distribution through the parallel channels which make up the fuel element. Unfortunately, for a given reactor core size and power output, this reduces the surface area and increases the heat flux.

Other Causes and Ill Effects of Poor Flow Distribution

Care must be exercised in the design and installation of heat exchangers to avoid asymmetries in the flow and temperature distribution arising from all sorts of irregularities. For example, a shell-and-tube heat exchanger designed to function as a boiler was mounted in an inclined position with water outside the tubes and high-temperature liquid metal inside the tubes. Because of the inclined position, a substantially greater length of some tubes was immersed in the water below the free surface than was the case for others. The average temperature of the liquid metal leaving the more submerged tubes was lower, and this caused bowing of the upper tubes. The resulting bending stresses were cycled with changes in operating conditions, and led to failure of some tubes, since the floating head was not designed to accommodate a substantial temperature variation transversely across the heat transfer matrix.

Friction Factors and Pressure-Loss Coefficients

When estimating the pressure drop across a component or a system it is helpful to make use of sets of data on friction factors and loss coefficients for basic geometries. The component or system at hand can be broken down into elements similar to the basic geometries for which data are available, the pressure drop for each element can be estimated, and the resulting values summed to obtain the total pressure drop. Data for the most common cases are presented in Section 3 of the appended Handbook. A splendid, far more extensive set is available in Reference 20.

Example 8.1. A heater is to be installed in an air duct just downstream of an existing centrifugal blower which is already operating at a discharge pressure close to its capacity. Inspection of the blower confirms the suspicion that the volute discharge height is only half the height of the square flange at the blower discharge, hence the velocity distribution just downstream of the blower would be very poor. Thus while the nominal dynamic head in the duct would be 0.5-in. H_2O (124 Pa), the peak dynamic head would be 2.0-in. H_2O (496 Pa). To obtain a good velocity distribution across the inlet face of the heater would probably require a design pressure drop of 4-in. H_2O (992 Pa) which is two-thirds the nominal capacity of the blower, or 6-in. H_2O (1488 Pa). Determine the length of a diffuser region at the blower outlet to bring the actual discharge velocity down to the nominal velocity in the duct. Inasmuch as a Pitot traverse at the blower discharge shows that the velocity distribution at the volute outlet is within 25% of being flat, use an equivalent cone angle of 5°, which would give a 10° inclination angle for a simple inclined partition in the duct. The tangent of 10° is 0.176, making the length of the diffuser 5.7 times the height of the volute discharge, or 2.85 times the height of the square duct. Examination of Fig. 8.10 indicates that this divergence angle is substantially less than that for maximum efficiency if the inlet velocity is uniform, hence the choice of a 10° angle appears to be appropriate. Figure 8.10 also indicates that almost 70% of the velocity energy at the volute outlet could be recovered with this diffuser, thus making available over 1 in. of H_2O (248 Pa) of additional head for the heat exchanger while at the same time reducing the dynamic head at the inlet to 0.5-in. H_2O (124 Pa). Thus by installing the partition to form a diffuser, it should be not only possible to obtain a good velocity distribution across the heater but also make the heater installation without losing any air-flow capacity in the ventilation system.

REFERENCES

1. B. Eck, *Technische Strömungslehre*, 5th ed., Springer-Verlag, Berlin, 1957.

2. S. J. Kline, D. E. Abbot, and R. W. Fox, "Optimum Design of Straight-Walled Diffusers," *Journal of Basic Engineering, Trans. ASME*, vol. 81, 1959, p. 321.

3. S. J. Kline, "On the Nature of Stall," *Journal of Basic Engineering, Trans. ASME*, vol. 81, 1959, p. 305.

4. A. Koestel, "Computing Temperatures and Velocities in Vertical Jets of Hot or Cold Air," *Heating, Piping, and Air Conditioning*, June 1954.

5. S. Corrsin and M. S. Uberoi, "Further Experiments on the Flow and Heat Transfer in a Heated Turbulent Air Jet," NACA Technical Report no. 998, 1950.

6. J. C. Laurence, "Intensity, Scale, and Spectra of Turbulence in Mixing Region of Free Subsonic Jet," NACA Technical Report no. 1292, 1956.

7. L. Prandtl, *Essentials of Fluid Dynamics*, Hafner Publishing, New York, 1952, p. 80.

8. G. B. Schubauer and W. G. Spangenberg, "Effect of Screens in Wide-Angle Diffusers," NACA Technical Report no. 949, 1949.

9. J. R. Weske, "Experimental Investigation of Velocity Distributions Downstream of Single Duct Bends," NACA Technical Note no. 1471, 1948.

10. J. J. Harper, "Tests on Elbows of a Special Design," *Journal of Institute of Aeronautical Sciences*, vol. 13, 1946, p. 587.

11. J. C. Emery et al., "Systematic Two-Dimensional Cascade Tests of NACA 65-series Compressor Blades at Low Speeds," NACA Technical Report no. 1368, 1958.

12. J. D. Stanitz and L. J. Sheldrake, "Application of a Channel Design Method to High Solidity Cascades and Tests of an Impulse Cascade with 90 deg of Turning," NACA Technical Report no. 1116, 1953, p. 193.

13. W. D. Baines and E. G. Peterson, "An Investigation of Flow Through Screens," *Trans. ASME*, vol. 73, 1951, p. 467.

14. W. G. Cornell, "Losses in Flow Normal to Plane Screens," *Trans. ASME*, vol. 80, 1958, p. 791.

15. A. L. London et al., "Oblique Flow Headers for Heat Exchangers," *Journal of Engineering for Power, Trans. ASME*, vol. 90(1), 1968, p. 271.

16. J. D. Keller, "The Manifold Problem," *Trans. ASME*, vol. 71, 1949, p. A-77.

17. C. Daniels and H. Pelton, "Pressure Losses in Hydraulic Branch-Off Fittings," *Product Engineering*, July 1959, p. 60.

18. B. W. LeTourneau and R. E. Grimble, "Engineering Hot Channel Factors for Nuclear Reactor Design," *Nuclear Science and Engineering*, vol. 1, 1956, p. 359.

19. G. Samuels, "Design and Analysis of the Experimental Gas-Cooled Reactor Fuel Assemblies," *Nuclear Science and Engineering*, vol. 14, 1962, p. 37.

20. I. E. Edelchik, *Handbook of Hydraulic Resistance*, 2nd ed., Hemisphere Publishing, New York, 1986.

9

Stress Analysis

The ability of heat exchanger structures to withstand static weight and pressure loads can be predicted with roughly the same confidence as the heat transfer and pressure-drop characteristics (i.e., with a probable error of 20 to 50%, depending on the complexity of the system), and the problems involved are roughly equivalent in difficulty to those in the fluid-flow and heat transfer fields. The life of a structure under repeated, severe, thermal cycling conditions is much more difficult to predict analytically—so much so that the uncertainty in the life to failure may be as much as a factor of 10. This chapter outlines the more important basic problems, and presents simple design techniques suitable for preliminary design purposes. References are suggested for more refined and thorough treatments for establishing finished designs. Of these, the most widely used is the ASME code for unfired pressure vessels.[1]

PRESSURE STRESSES IN TUBES, HEADER SHEETS, AND PRESSURE VESSELS

Pressure Stresses in Cylinders

For cylinders in which the wall thickness is less than 10% of the diameter, the pressure stress is closely approximated by the simple expression

$$S = \frac{pd_i}{2t} \qquad (9.1)$$

where S = hoop stress in the walls, MPa (psi)
$\quad p$ = pressure differential across the wall, MPa (psi)
$\quad d_i$ = internal diameter, m (in.)
$\quad t$ = wall thickness, m (in.)

This expression can be derived by considering a cylinder split axially on a diameter. The pressure forces acting to force the two halves apart may be equated to the hoop tension forces acting to hold them together. Similarly, the stress in spherical shells can be shown to be half that for a cylinder of the same diameter and wall thickness.

The techniques for handling thick-walled cylinders are much more complex because the stress varies with the radius. Special analytical techniques are necessary to obtain good correlation with experiments.[2,3]

Pressure Vessel Heads

Shells and pressure vessels can be fitted with hemispherical, ellipsoidal, or flat heads. Hemispherical heads are the easiest to stress-analyze. Ideally, the thickness of a hemispherical head should be half that of the cylindrical vessel which it serves to close if the stress in the two is to be the same. If ellipsoidal heads are employed, the degree of oblateness of the ellipsoid is often made such that the stresses in the head do not exceed those in the cylinder for a head thickness equal to that of the cylinder. For this condition the major diameter of the ellipsoid should be approximately twice the minor diameter.

The stresses in a flat, circular disc subjected to a pressure differential across its face depend on the edge restraint conditions (see Fig. 9.1). If the edge is free to rotate, the expressions for the maximum stress and deflection are

$$S_{max} = \frac{0.75r^2p}{t^2} \tag{9.2}$$

$$\delta_{max} = \frac{0.6r^4p}{t^3E} \tag{9.3}$$

where S_{max} = maximum stress, MPa (psi)
r = radius of circular plate, m (in.)
p = pressure differential across head, MPa (psi)
t = thickness, m (in.)
δ_{max} = deflection at center, m (in.)
E = modulus of elasticity, MPa (psi)

If the edges of the disc are fixed rigidly to prevent rotation, the maximum stress and deflection are as follows:

$$S_{max} = \frac{0.5r^2p}{t^2} \tag{9.4}$$

$$\delta_{max} = \frac{0.17r^4p}{t^3E} \tag{9.5}$$

The interaction between the cylindrical wall and the flat head of a welded pressure vessel is an important factor in establishing the stress in the head. Figure 9.1 shows schematically the deflected forms of the structure for three typical cases, namely: (a) where no restraint against rotation is imposed at the edges of the head; (b) where the head is mounted so that the edges of the head are rigidly restrained from rotation (as with a thin head welded into a thick-walled vessel); and (c) where the head is welded into a vessel of about the same thickness so that both would deform under an internal pressure. The last case is the one most nearly representative of the structures usually encountered, but it is too complex to treat in this book other than to say that refined analyses[4] show that the maximum stress in the head will be less than for no edge restraint, while the maximum stress in the shell will be greater.

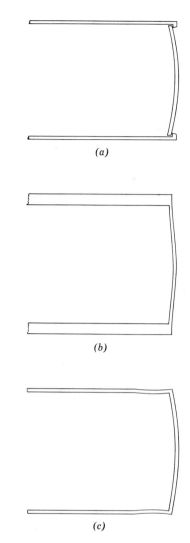

(a)

(b)

(c)

Figure 9.1 Typical edge restraint conditions for flat heads on cylindrical pressure vessels: (a) no rotational restraint at edge of head; (b) edges of head restrained from rotation by rigid attachment to thick cylindrical shell; and (c) edges of head partly restrained from rotation by rigid coupling to cylindrical shell of equal thickness—shell deformed.

The discontinuity stresses between the head and the cylindrical portion of the vessel pose problems not only with flat plates but also with ellipsoidal, hemispherical, or conical heads.[5-7] Consider, for example, a hemispherical head made of the same thickness plate as the cylindrical barrel. When subjected to internal pressure, the diameter of the head tends to increase by an amount equal to only half that of the corresponding diametral growth of the cylindrical barrel. If the head and barrel were not welded rigidly together, but were free to slide

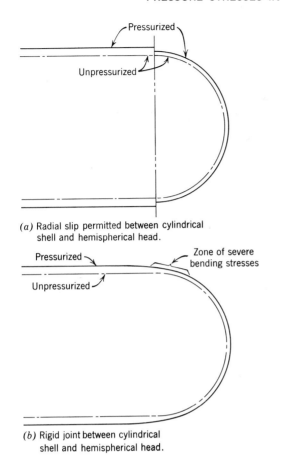

(a) Radial slip permitted between cylindrical shell and hemispherical head.

(b) Rigid joint between cylindrical shell and hemispherical head.

Figure 9.2 Schematic diagrams showing in exaggerated form the relative positions of the inner surface of a cylindrical pressure vessel in the vicinity of its hemispherical head for both unpressurized and pressurized conditions.

in the plane at which they mate, the deflected form of the structure would be as indicated schematically in Fig. 9.2a. However, if the two were welded rigidly together, the deflected form of the structure would be as in Fig.

9.2b, which is characterized by substantial bending stresses which in some zones would add to the pressure stresses. Under certain conditions these bending stresses may result in local stresses that are double the nominal stresses in the wall of the cylindrical vessel.[6] Thus it is evident that the change in section between a thick cylinder and a hemispherical head must be designed with care to minimize the discontinuity stresses associated with the joint. This is specialized work that is best handled in the detail design.

Penetrations

Holes must be provided in pressure vessels not only for the fluid-inlet and -outlet connections but also to provide for drainage, cleaning, and access for inspection and maintenance. If a pressure vessel wall is not thickened around such penetrations, the local stresses induced by an internal pressure are roughly three times the average stress in the vessel even for small penetrations.[8] If the number of penetrations is relatively small and if they are widely dispersed, provision for accommodating these large local stresses is ordinarily made by thickening the wall locally to three times the design value for the unperforated head, by incorporating a forging, or by welding on patches as is indicated in Fig. 9.3a and b.[9,10] If the number of penetrations is large, it is usually less expensive to concentrate the penetrations in a hemispherical or elliptical head, and make its thickness one-and-a-half to two times that of the cylindrical vessel wall. If this is done, the holes should be spaced at least three diameters apart to avoid excessive stresses in the ligaments between the holes. For any of the foregoing arrangements a thick-walled tube welded into a pressure vessel, as in Fig. 9.3c, helps to relieve the stress concentration at the hole, especially if the tube extends for about one tube diameter on either side of the shell. Further extension has little

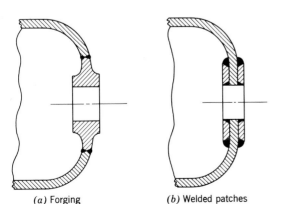
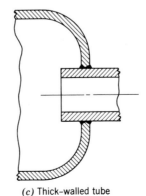

(a) Forging   (b) Welded patches   (c) Thick-walled tube

Figure 9.3 Typical methods of reinforcing pressure vessel penetrations to reduce the local stresses.

effect on the stresses around the penetration because of shear lag effects.

Thermal expansion in pipes connecting two rigidly mounted pieces of equipment may induce severe bending stresses in the casings. In most instances the best procedure is to design the piping to keep the reactions at the casings relatively small, but it is sometimes necessary to strengthen a shell at such a penetration, or it may be necessary to impose limitations on the start-up procedure.[11]

It is sometimes necessary to join two pressure vessel sections having different diameters, or a duct inlet to a pressure vessel may be a substantial fraction of the vessel diameter. The stress analysis techniques are too specialized to be included here, but experience has shown that smoothly contoured transition sections can be designed to minimize the shell stresses in the vicinity of such junctions.[12]

Manholes

The most common form of manhole employed in pressure vessels is elliptical with the closure mounted on the inside of the vessel as indicated in Fig. 9.4. The advantage of this manhole shape is that the closure and its gasket can be removed from the vessel by aligning the short axis of the elliptical cover with the major axis of the hole. Circular closures are sometimes used, in which case they must be installed in the vessel before the vessel head is welded in place, and, if a gasket is used, it must be designed so that it is sufficiently flexible to be sprung through the hole of the vessel when a replacement is required.

The most common size of elliptical manhole is 305 × 406 mm (12 × 16 in.), although 280 × 380 mm (11 × 15 in.) elliptical holes are sometimes used.

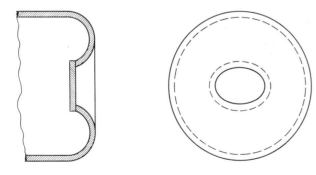

Figure 9.4 An elliptical manhole with its closure flange.

Flanges

It is often desirable to employ a flanged joint between the head of the vessel and the vessel proper so that the entire internals can be removed. Such closures are ordinarily effected with bolted flanges or one of a variety of split-hoop arrangements or clamps, such as that of Fig. 9.5. Flanged joints inevitably set up bending stresses in the shell, and hence require that the vessel be thickened in the vicinity of the joint.[13–15] While this situation is not too serious for relatively thin-walled vessels, it presents major problems for thick-walled vessels, such as those employed in pressurized water reactors, where the basic vessel wall thickness may run up to 200 mm (8.0 in.) In such cases the flanged joint may be responsible for 35% of the total cost of the vessel.

Vessel Support

Shells or pressure vessels may be supported in any of a variety of ways. Relatively small vessels, or those whose wall thickness is a substantial fraction of the diameter, are commonly supported from lugs or feet welded directly to the vessel wall. The most commonly employed arrangement for vertical cylindrical pressure vessels is a skirt attached rigidly to the bottom head as shown in Fig. 9.6a. The thickness of the vessel head should be increased somewhat in the vicinity of the junction with the skirt to accommodate the local bending stresses imposed by the skirt.[16] If the temperature of the base of the mounting skirt differs from the temperature of the vessel, these stresses include both the weight load of the vessel and the interaction of the skirt and vessel. A related problem is discussed later in this chapter in the section called Thermal Sleeves.

Vessels are often supported from brackets attached to a flange for a closure (Fig. 9.6b), since the wall in that vicinity must be thickened anyway to accommodate the bending loads inherent in flanged joints.

Very large, relatively thin-walled vessels require a more refined approach to avoid serious stress concentrations in the shell wall in the vicinity of the supports. Such vessels are sometimes supported by columns equally distributed around the perimeter (Fig. 9.6c), the load being transmitted in shear from the vessel wall to the column.

Header Sheets

The most common method of connecting tubes to a common plenum or passage is the flat header sheet (e.g., that shown in Fig. 1.8). Much effort has been devoted

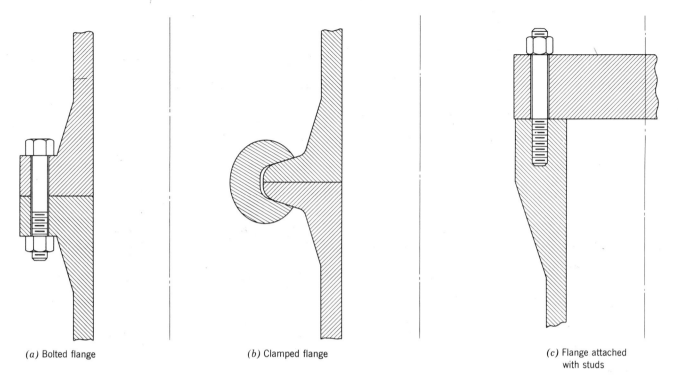

(a) Bolted flange   (b) Clamped flange   (c) Flange attached
with studs

Figure 9.5 Sections through three types of flanged joints.

to the stress analysis of header sheets.[17-22] Fortunately, while the more exact solutions are extremely complex, it has been found that good approximations can be obtained with techniques that are relatively simple to apply. For example, where it is necessary to design header sheets to support a large pressure differential, the

header sheet is likely to be welded into the shell. While the resulting edge restraint strengthens the header sheet, the effect is usually small, because the header sheet thickness is ordinarily much greater than that of the shell. Where this is the case, the stress in the header sheet can be approximated by Eq. 9.2 for a simply supported flat

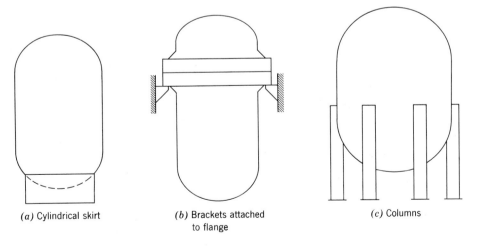

(a) Cylindrical skirt   (b) Brackets attached
to flange   (c) Columns

Figure 9.6 Methods of supporting pressure vessels.

plate with modifications to include allowances for both the stress concentrations around the holes and the reduction in the cross section of the plate caused by removal of material from the holes. In attempting to apply Eq. 9.2, it is evident that the stress in the header sheet is not only directly proportional to both the pressure differential and the square of the header sheet radius-to-thickness ratio, but that it is also a function of the ratio of the hole pitch to the hole diameter. The stress concentration factor for small, widely spaced holes is nearly 3, but it drops off as the ratio of the hole diameter-to-hole spacing is increased. The change in this factor largely offsets the reduction in the effective cross-sectional area as the hole diameter-to-spacing ratio is increased up to about 0.5. Further increases in the hole diameter lead to rapidly increasing stresses. A convenient way to determine the maximum stress is given by Fig. H8.2, which was developed from the ASME boiler code, and includes in a single curve the effects of the stress concentration factor and the loss of material from the holes.

Example 9.1. The use of Fig. H8.2 in designing header sheets can be illustrated by an example. If the allowable stress is 110 MPa (16,000 psi), the pressure differential across the header sheet 6.895 MPa (1,000 psi), and the hole diameter-to-spacing ratio 0.5, the header sheet thickness-to-radius ratio can be determined as follows from Fig. H8.2:

$$\frac{S_{max}}{p} \frac{t^2}{r^2} = 2.56$$

$$\frac{t^2}{r^2} = 2.56 \frac{6.895}{110} = 0.16$$

$$\frac{t}{r} = 0.4$$

Thus a 1000-mm (40-in.) diameter header sheet should be 200 mm (8 in.) thick for the conditions given.

Manifolds

For some types of heat exchangers, particularly boilers, it is desirable to group one or two banks of tubes together and couple them with a manifold. The manifold may be designed as if it were a small header drum, or a special configuration, such as the stack of roughly square cross-section headers of Fig. 9.7, may be used. For cases such as the latter, the complex geometry necessitates a finite element stress analysis, and testing is desirable. If tests are carried out, it is important that the pressure be cycled

Figure 9.7 Stack of manifolds for a tube bank in a coal-fired boiler. Each manifold accommodates two rows of tubes. (Courtesy The Babcock and Wilcox Co.)

in much the same way as it would be in the actual application, because whereas excessive local stresses may be relieved by plastic flow without failure for a few cycles, failure may occur after a large number of cycles. This problem is discussed at greater length in the section on thermal strain cycling, later in this chapter.

Bifurcated Tubes

It is always desirable to minimize the number of field welds, and, especially for field fabrication, it is often preferable to substitute a tube-to-tube joint for a tube-to-header joint. One method of doing this is to employ bifurcated tubes of the type shown in Fig. 9.8. This sort of a connection makes it possible to cut the number of tube-to-header joints in half, and reduces stresses in the header by reducing the number of holes and by increasing the ligament thickness between holes. Although the total number of joints required in the assembly is increased by 50%, the number of joints in the final assembly opera-

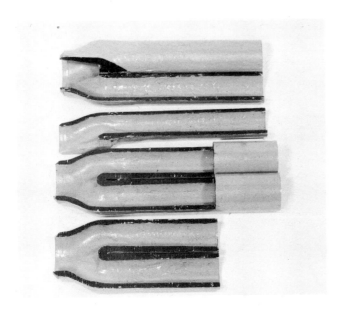

Figure 9.8 Photograph of a bifurcated tube that has been sectioned to show its construction. (Courtesy Combustion Engineering, Inc.)

tion may be cut in half, which may be a major advantage in that it may reduce both mechanical interferences and total assembly time.

Experimental Stress Analysis

The geometric complexities of heat exchanger components such as heads, manifolds and shell penetrations require finite element analysis, perhaps supplemented with model tests during the course of a design. Experience gained from tests of this sort has proved to be enormously helpful in the detail design of components.[23-25]

DIFFERENTIAL THERMAL EXPANSION

Severe thermal stresses may be introduced if temperature differentials in excess of about 60°C appear in structures. This section is intended to present a number of typical cases to indicate the character of the problems.

Differential Expansion Between Tubes and Shell

Differential thermal expansion between the tubes and shell of a heat exchanger of the type shown in Fig. 1.7

presents one of the most obvious examples of a thermal stress problem. For a plain carbon steel heat exchanger in which the coefficient of thermal expansion of the steel is about 11.7×10^{-6} m/m·°C (6.5×10^{-6} in./in.·°F) a temperature difference of 111°C (200°F) between the tubes and the casing tends to give a difference in length of 0.0013 m/m (or in./in.). If, as is usually the case, the cross-sectional area of the casing is much larger than that of the tubes, the deformation is primarily in the tubes. For a modulus of elasticity of 207 GPa (30×10^6 psi) the stress in the tube walls is thus $0.0013 \times 207 = 269$ MPa (39,000 psi), over twice the allowable stress, and more than the yield point. Actually, in the first thermal cycle, plastic strain takes place so that the tubes are under a reversed stress in the isothermal condition. This effect is shown in Fig. 9.9 for an idealized case in which the tubes run at a temperature below that of the casing. Note how the tube stress builds up until it reaches the elastic limit. A further increase in temperature difference causes the tube wall to yield, work-hardening the material. As isothermal conditions are restored, the stress in the tube wall drops linearly through zero to a negative (or compression) stress.

It might be inferred from the idealized diagram of Fig. 9.9 that stresses in subsequent cycles can be kept entirely in the elastic range if the unit is subjected to one abnormally severe cycle entailing a temperature difference of perhaps 300°F at the beginning of its life. While this approach would ease the problem, stress concentrations at the header sheet would increase the local stresses sufficiently so that tube failures would still be likely.

Note that changing the operating temperature of a heat exchanger does not induce thermal stresses as long as it is kept isothermal unless the heat exchanger is built of more than one material, for example, with brass tubes and a carbon-steel shell.

Differential expansion between the tubes and shell can be accommodated in many ways, including the use of a floating header sheet (Fig. 1.12) or an expansion joint in the shell (Fig. 1.7). Since the design and stress analysis of shell expansion joints is quite complex and depends a great deal on conditions peculiar to the case in question,[26,27] no attempt to treat them is included here.

Differential Expansion Between Tubes

Serious thermal stresses may be induced in the tubes and header sheets of a heat exchanger similar to that of Fig. 1.10, especially if the heat transfer coefficient on the tube side is high and the flow distribution through the tubes is not uniform, causing a fairly large variation in the

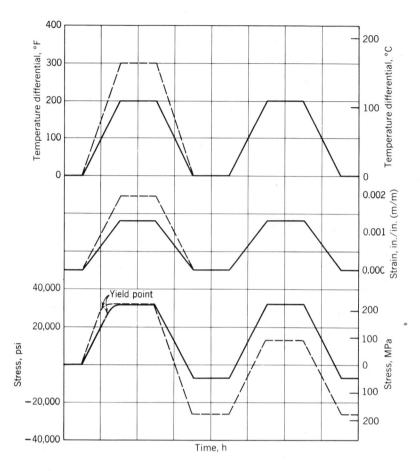

Figure 9.9 Temperature, deformation, and stress as functions of time for an idealized case. The solid lines are for a cycle in which the amplitude of the thermal stress slightly exceeds the elastic limit; the dotted lines are for a case in which the amplitude of the nominal thermal stress exceeds the elastic limit by 50%.

temperature distribution. Even with a uniform velocity and temperature distribution across each half of the heat exchanger, the differential thermal expansion between the two halves may be large if the temperature rise or drop in the tube-side fluid is in excess of 55 to 111°C (100 to 200°F).

The large temperature differences in boiler furnaces make differential thermal expansion a major problem in these units. Experience has shown that the best approach is to design the tube structure to give a substantial amount of flexibility. Even though the high heat transfer coefficient characteristic of boiling tends to give a fairly uniform metal temperature throughout all the tubes in the boiling zone, the tubes are commonly bent to accommodate differential thermal expansion. A more flexible structure in the form of serpentine bends is employed in superheaters where variations in the heat transfer coefficient and flow distribution may lead to large differences in temperature between tubes connected to the same header.

Shear Stresses

An example of what could be a serious thermal stress is presented by a shell-and-tube heat exchanger with double header sheets (Fig. 1.12). If the fluid passing through the tubes is at a much lower temperature than that outside the tubes, and if the heat transfer coefficient inside the tubes is much lower than that outside the tubes—as might be the case if viscous oil were being heated with fairly high temperature steam—the two header sheets may be at substantially different temperatures. The inner header sheet exposed to the condensing steam will run much hotter than the outer header sheet and will tend to expand and shear the tubes. This effect is indicated schematically in Fig. 9.10. Such a configuration might be used satisfactorily for many years in applications involving rather small temperature differences and then, because of the condition indicated in Fig. 9.10, might give trouble if employed in a larger unit or in a higher temperature application giving more severe conditions.

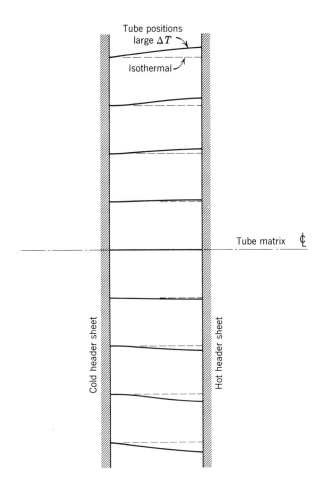

Figure 9.10 Tube bending produced by differential expansion between double header sheets.

In general, no difficulty will be experienced as long as the stresses induced by the differential thermal expansion are in the elastic range. If plastic flow occurs in regions of high stress concentration, however, failure is likely to occur after a limited number of thermal cycles. Normally the most severe stresses in cases such as the one indicated in Fig. 9.10 are not in the thick hot or cold header sheets, but rather are the shear and bending stresses in the tubes between them. The shear stresses can be calculated directly from the relative displacement of the two header sheets, the length of the tube between them, and the modulus of elasticity. The bending stresses can be calculated by considering the tube section between the header sheets as two cantilever beams coupled by a pin joint at the point of inflection midway between the header sheets. (Figure H8.3 presents a chart that facilitates estimates of the bending stresses.) An analysis of just one case, such as that of Fig. 9.10, serves to convince one that much care must be exercised to avoid coupling heavy sections at substantially different temperatures with relatively inflexible members of small cross-sectional area. The criteria for failure in the regions of severe stress concentration are discussed in the section on thermal strain cycling, later in this chapter.

A somewhat similar problem is posed by the heat exchanger of Fig. 1.8. If the heat transfer coefficient between the shell-side fluid and the shell is fairly high, and if the temperature difference between the inlet and outlet fluid streams is substantial, a severe shear stress may be induced in the shell along the temperature discontinuity represented by the baffle. Similarly, if both the heat transfer coefficient and the temperature differential between the inlet and outlet fluid streams on the tube side are high, serious stresses may be induced in the header sheet because of the temperature discontinuity between the two halves.

Thermal Sleeves

A special type of pressure vessel penetration problem is posed in applications in which the temperature of the fluid stream entering or leaving the shell may differ from the temperature of the shell by 55°C (100°F) or more. This problem is likely to be particularly serious if the temperature difference is subject to fluctuations either during on-stream operation or during a start-up or shutdown. The thermal stresses can be greatly reduced by isolating the pipe from the pressure vessel by means of a thermal sleeve, as illustrated in Fig. 9.11. If the temperature difference between the pipe and shell is large, or if the thermal conductivity of the shell-side fluid is high, it may be necessary to provide thermal insulation between the pipe and the shell in the vicinity of the thermal sleeve.

To avoid serious thermal stresses in the thermal sleeve itself, its length-diameter ratio ought to be proportioned according to the temperature difference to be accommodated.[28]

The design of thermal sleeves is concerned mainly with making the axial temperature distribution along the sleeve such that the resulting thermal stresses in the sleeve are acceptable. Several interesting cases may be seen in Fig. 9.12. If heat conduction axially through the thermal sleeve is the principal factor in determining the temperature distribution, the temperature varies linearly with axial position, as in Fig. 9.12a. If the dominant factor is heat transfer across the gap between the thermal sleeve and the pipe within it (as is likely to be the case), the

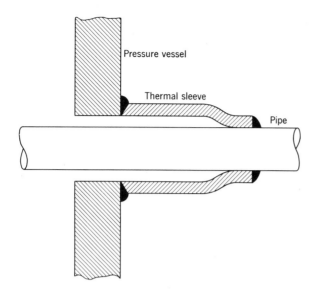

Figure 9.11 Section through a thermal sleeve for a pressure vessel penetration.

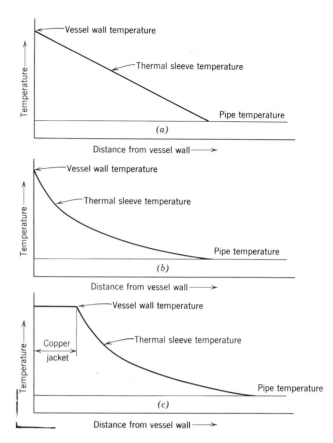

Figure 9.12 Axial temperature distributions in a pipe and thermal sleeve for three typical boundary conditions; (a) heat transfer from the pipe to the vessel by axial conduction in the thermal sleeve; (b) heat transfer mainly by conduction through the fluid between the pipe and the thermal sleeve; (c) base of thermal sleeve maintained at the vessel-wall temperature by a short external copper sleeve, with heat transfer mainly through the fluid between the sleeve and the pipe for the balance of the sleeve length.

sleeve temperature varies exponentially along the length of the sleeve (Fig. 9.12b). If the base of the thermal sleeve is thickened or surrounded with a jacket of a high-conductivity metal, such as copper, a temperature distribution similar to that of Fig. 9.12c is obtained.

The thermal stresses resulting from these temperature distributions can be visualized by considering first the linear temperature distribution of Fig. 9.12a. If the ends of the sleeve were free, the sleeve would deform into a frustum of a cone, and the resulting stresses would be trivial. If the base is welded rigidly to the shell, however, there can be no rotation of axial elements at the base of the sleeve, and both bending and shear stresses are induced. These stresses are most severe at the base of the thermal sleeve, and are directly proportional to the axial temperature gradient. A similar situation prevails for the exponential temperature distribution of Fig. 9.12b. The stresses near the base of the thermal sleeve can be greatly reduced by introducing the nearly uniform temperature section of Fig. 9.12c.

Analytical solutions have been obtained for the stress distributions of Fig. 9.12b and c. While the solutions themselves are too complex to include here, the results can be summarized in tractable form and are presented in Fig. 9.13. This chart gives the parameter $S_{max}/E\alpha T_0$ (where S_{max} is the maximum shear stress in the thermal sleeve) as a function of the length of the constant temperature region and the relaxation length for the exponential temperature distribution region. The symbols used are defined by the diagram at the upper right.

Since the most common case is that in which the temperature of the thermal sleeve varies exponentially with axial distance from the shell, the chart of Fig. H8.6 facilitates preliminary design estimates. In this chart the length-diameter ratio is related to the axial temperature difference for several values of the thickness-diameter ratio of the sleeve and an allowable shear stress of 68.9 MPa (10,000 psi).

If an allowable shear stress other than 68.9 MPa (10,000 psi) is to be employed, the allowable temperature difference given by Fig. H8.6 is directly proportional to the allowable stress. For example, if the allowable shear stress can be increased by 20% to 83 MPa (12,000 psi), the temperature difference can also be

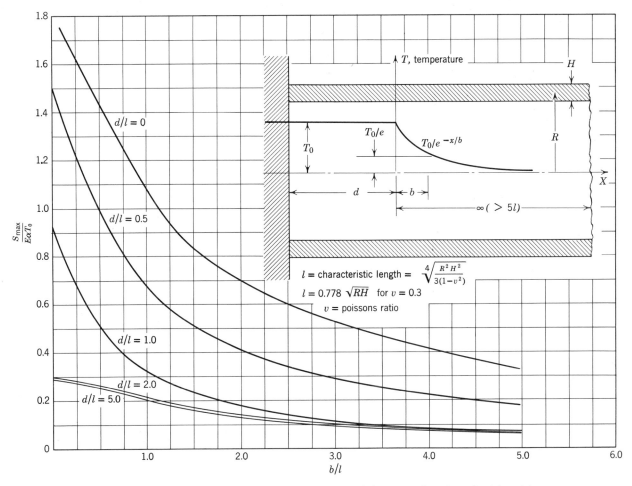

Figure 9.13 Maximum shear stresses in a thermal sleeve as a function of axial position for typical lengths of the constant temperature region at the base. Note the diagram defining the symbols and parameters used. (Cooper, Rocher, and Noble, Ref. 28.)

increased by 20% for a given set of proportions for the thermal sleeve. Similarly, if a material other than plain carbon steel is employed, the maximum shear stress induced is directly proportional to the product of the coefficient of thermal expansion and the modulus of elasticity.

VIBRATION AND NOISE

A variety of difficulties has been experienced with noise and tube vibration in heat exchangers.[29-33] These problems are related to similar difficulties with vibration of pipelines, suspension bridges, and tall, slender smokestacks under crosswind conditions.[34] In each case the exciting force usually is associated with the shedding

of vortices from the downstream side of the cylinder or other obstruction to flow. As a vortex is shed, the flow pattern—and hence the pressure distribution—changes, which leads to oscillations in the magnitude and direction of the fluid pressure forces acting on the tube or other obstruction. If the oscillating force leads to a movement of the obstruction, the motion may affect the vortex shedding process, increase the fluid forces, and lead to a self-excited vibration of large amplitudes. Flutter of aircraft control surfaces and wires humming in the breeze are motions of this type.

Frequency of the Exciting Force

For individual cylinders under crossflow conditions, the frequency with which the vortices are shed is directly proportional to the flow velocity and inversely proportional to the diameter for Reynolds numbers above about 200. A dimensionless parameter known as the *Strouhal number* is used to relate different situations in which vortex shedding is a problem. The Strouhal number is the product of the frequency f at which the vortices are shed and the length or diameter D of the obstruction (measured in the direction of the flow), divided by the flow velocity V, that is, the Strouhal number $= fD/V$. For crossflow over single cylinders, the Strouhal number is commonly about 0.2.[29,31] That is, the vortex shedding frequency is about one-fifth that which might be deduced by dividing the stream velocity by the diameter of the cylinder. This factor implies that the vortex spacing in the wake should be five times the diameter of the cylinder. As can be seen in Fig. 3.9, the spacing is more like three diameters because the wake velocity is roughly half that of the free stream. Note that the Strouhal number is independent of either the fluid density or its viscosity.

Experiments with heat exchanger tube matrices have indicated that the Strouhal number should be modified to employ the tube spacing in place of the tube diameter, since the regular array of tubes forces the vortex shedding phenomenon to conform to a pattern that is typical of the tube array.[29] Figure 9.14 shows the vortex shedding and flow patterns ordinarily encountered with in-line tube banks. The jet stream passing between two closely spaced tubes tends to waver first to one side and then to the other as vortices are shed alternately from the two tubes on either side of a given jet. Several modes of fluid stream motion may occur. The first mode is that in Fig. 9.14a, while the second mode is that of Fig. 9.14b. Note that the overall pressure force acting on a duct wall for the first mode does not change, because the pressure forces generated by alternate banks of tubes are in opposite directions so that the average force on the wall is zero. For the second mode, the forces at the wall all act in the same direction so that their effects are cumulative and have been known to become large enough to buckle duct walls.[31]

If the flow pattern for crossflow over in-line tube banks is similar to that of Fig. 9.14a, the frequency of the oscillation gives a Strouhal number of roughly 0.5 (based on the tube spacing),[31] whereas for the flow pattern of Fig. 9.14b the Strouhal number is around unity (usually from 0.7 to 1.2).[29]

Crossflow over staggered tubes also may induce flow oscillations, although the incidence of difficulty seems to be lower than for in-line tubes. Similarly, while it might be surmised that disc or helical fins would tend to interrupt the vortices so that the vortex shedding process

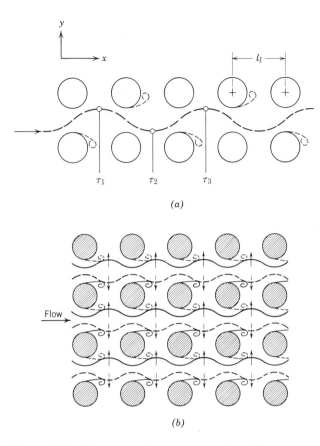

Figure 9.14 Diagrams indicating two modes of flow-pattern oscillation associated with vortex shedding for in-line tube arrays. (Grotz and Arnold, Ref. 29.)

would not be in phase along the entire length of the tube, at least in some cases little difference in the severity of the oscillations has been found between the behavior of bare and finned tubes.

Acoustical Oscillations

A loud noise is likely to be the result if the flow velocity is such that the frequency of either of the two types of oscillation shown in Fig. 9.14 happens to be close to the acoustical frequency of the "closed-end organ pipes," represented by the fluid columns between the tube banks (transverse to the direction of flow). In at least one instance the noise emitted was described as an "unbearable" howl.[29] Noise levels as high as 124 db have been recorded.[31] These lateral acoustical oscillations may also excite large-amplitude lateral vibrations of the tubes or the duct walls if the frequency of the oscillation happens to be close to that of a natural frequency of these members (ordinarily in some bending mode). Pressure fluctuations of 3 psi have been observed in air ducts nominally at atmospheric pressure.[31] A systematic investigation of the effects of this phenomenon using in-line tube arrays in which the transverse and axial tube spacing were varied has shown that the Strouhal number is relatively insensitive to the transverse tube spacing for the spacings commonly used, but it is sensitive to the axial tube spacing.[29,33] If the latter exceeds approximately three tube diameters, the interaction between tube banks becomes relatively small, and it is better to base the Strouhal number on the tube diameter rather than on the axial spacing between the tube banks. The investigation also showed that the fluid velocities giving peaks in the noise amplitude are those for which the vortex shedding frequency is the same as the natural acoustical frequency for some mode of organ pipe oscillation in the air column transverse to both the direction of flow and the axis of the tubes; that is,

$$f = \frac{Cn}{2c} \qquad (9.6)$$

where C is the acoustical velocity in meters per second (or feet per second), c the transverse width of the duct in meters (or feet), and n the mode of the acoustical oscillation (i.e., first, second, third, etc.). Thus the air velocity at which a loud noise of a given frequency is emitted from a given tube matrix geometry varies with the duct width.

Effects of Tube Spacing

A comprehensive series of tests was carried out at Sulzer in Switzerland to determine the effects of tube axial and lateral spacing on the vortex shedding frequency under crossflow conditions in both in-line and staggered arrays of tubes.[33] The effects of these parameters on the Strouhal number are shown in Fig. 9.15. Note that for the close tube spacings of interest for shell-and-tube heat exchangers the Strouhal number falls in the range of 0.4 to 0.5. The larger tube spacings are of interest mainly for air-cooled tube banks.

Tube Vibration Under Crossflow Conditions

Tube vibration may or may not accompany a noisy condition induced by flow across the tube bank.[29-31] By the same token, tube vibration may occur in a tube bank without generating a large amount of noise if there is a great deal of damping associated with motion of the walls of the system. In general, the viscosity of liquids may be

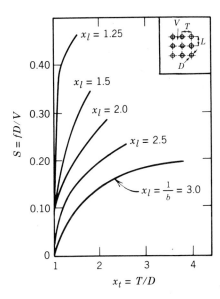

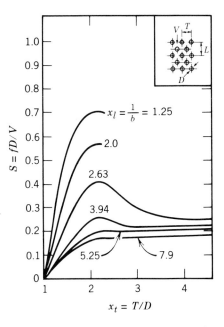

Figure 9.15 Effects of transverse tube spacing on the Strouhal number for a series of longitudinal tube spacings. The set of curves at the top is for inline tube banks, whereas that at the bottom is for staggered tube banks. (Chen, Ref. 33.)

fairly effective in damping tube vibration, whereas the viscosity of gases is insufficient to have any appreciable effect. Similarly, in liquids the compressibility is so low that acoustical vibrations are ordinarily not a problem, but mechanical vibration of the tubes may be serious since the fluid dynamic forces are much larger, and tube deflections are much greater.

An oscillating force on a tube does not cause troublesome vibration unless the forcing frequency is close to the natural frequency of the tube for some mode of vibration. Some typical modes likely to be of interest are shown in Fig. 9.16.

The analysis of vibrating structures poses difficult and specialized problems. Fortunately, general solutions to many of the problems of practical interest have been presented in the form of an excellent set of charts and tables for vibrating beams, plates, and rings.[35] Of these, the most useful to the heat exchanger designer is the one for continuous beams with fixed ends. This table is included as Table H8.3a. While the values presented in Table H8.3a are for steel, corrections for use with other materials are presented in Table H8.3b. The application of these tables to a typical case is shown in a sample problem accompanying Table H8.3b.

Several points should be noted in connection with the use of Table H8.3a. The natural frequency of an empty tube is quite insensitive to wall thickness, because both its weight and stiffness are almost directly proportional to wall thickness. The effects of the weight of a liquid inside of a tube or of fins outside the tube may be substantial, however, and vary with the wall thickness. Note also that we would expect, from Fig. 9.16, that the natural frequency for the third mode of a single span

should be about the same as that for the first mode of a three-span continuous beam of the same overall length. It can be seen from Table H8.3a that this is indeed the case. (It is necessary to multiply the frequency constant for the three-span beam by 9—i.e., the square of the number of spans, to put them on a common basis.)

Since the natural frequency of a tube is principally a function of the tube diameter and the length between supports, Fig. H8.7 shows the natural frequency of bare tubes in first-mode bending as a function of the tube length between supports for tube diameters from 0.5 in. to 1 in. Where tubes act as continuous beams with baffles supporting them at intervals along the span, the natural frequency for a given span between support points may be reduced by as much as a factor of 2 (see Table H8.3a). This situation stems from the absence of rotational restraint at the baffles. The added mass of fins reduces the natural frequency by a factor equal to the square root of the ratio of the mass of the bare tube to the mass of the tube with fins. This factor usually runs about 0.7.

For tube banks in which the tube spacing is less than three diameters, the frequency of transverse oscillations of the type indicated in Fig. 9.14a (the lowest frequency to be expected from the vortex shedding process) is directly proportional to the fluid stream velocity. A second chart for a Strouhal number of 0.5 and several

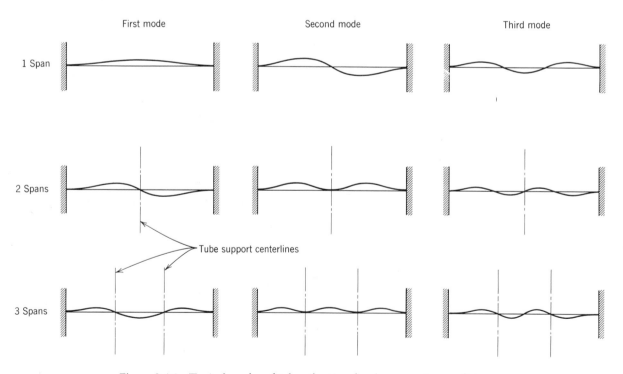

Figure 9.16 Typical modes of tube vibration for three support conditions.

typical tube spacings is presented in Fig. H8.6. It is intended to give a simple, direct indication of the minimum forcing frequency to be expected.

In using the charts of Figs. H8.6 and H8.7, the exciting frequency given by the vortex shedding process can be determined from the tube spacing and the fluid velocity. The tube length between supports can then be chosen to give a higher natural frequency for the tube to avoid tube vibration. Similarly, acoustical vibration can be avoided by choosing a gas-flow passage width, normal to the tube length, sufficiently small so that the frequency for the first mode of acoustical vibration falls above the operating velocity range. This can be done by inserting partitions in the heat exchanger or by inserting a diagonal baffle that breaks up the acoustical waves. A typical example of a problem of this sort is given in Example 14.3.

In applying these charts to shell-and-tube heat exchangers it is apparent that a serious problem may be presented by the jet emerging from the inlet pipe into the shell. A baffle can be used in this region to keep the jet from impinging on the tubes. TEMA standards call for a baffle of this sort if the product of the inlet velocity and the density exceeds 1500 (velocity in feet per second and density in pounds per cubic foot).

Tube Vibration Under Axial Flow Conditions

Some difficulty has also been experienced with axial flow through tube bundles. This again is related to turbulence—probably vortices shed by tube spacers or passage irregularities at the inlet to the tube matrix—but the problem is somewhat different in that sharp resonant peaks are not ordinarily experienced; rather, the amplitude of the vibration tends to increase approximately in proportion to the fluid flow velocity.[36] Analysis indicates that the vibration amplitude should be a function of two parameters: the ratio of the hydrodynamic exciting force to the elastic restoring force and the ratio of the hydrodynamic force to the damping force.[36] These two ratios are

$$\frac{\text{Hydrodynamic force}}{\text{Restoring force}} = \frac{\rho V^2 L^4}{EI}$$

and

$$\frac{\text{Hydrodynamic force}}{\text{Damping force}} = \frac{\rho V^2}{\mu \omega}$$

where ω is the frequency of the vibration in radians per second. Tests with water over a wide range of tube spacings indicate that there is a simple relationship between these two parameters and the ratio of the vibration amplitude to the tube diameter.

$$\frac{\delta}{D} = K_1 \left(\frac{\rho V^2}{\mu \omega} \right) \left(\frac{\rho V^2 L^4}{EI} \right)^{0.5} \qquad (9.7)$$

The effectiveness of this relationship in correlating the test data is indicated by Fig. 9.17.

Tube Failures

Concern for tube vibration is of more than academic interest. Failures of tubes in steam condensers under the action of the high-velocity vapor stream from the turbine have been a problem, for example.[37] The much greater density in liquid streams gives even higher dynamic heads and even more troublesome tube-vibration problems than in steam condensers. Just in the decade following preparation of the previous material for the first edition of this text, tube-vibration problems arose in heat exchangers for six USAEC reactors, namely, Fermi, Hanford N, Hallam, MSRE, ATR, and HFIR. (The steam generator in which tube failures occured at the Hanford N reactor is shown in Fig. 1.4. Note the relatively long span of the tubes between the baffles.) In most of these, tube failures caused by vibration actually occurred, and two cases led to lawsuits involving very large amounts of money because the failures not only led to high costs for replacement but long outages that entailed heavy capital charges. The author was called on to help cope with three cases; in all three the information presented above served to diagnose the problem and point the way to the solution. The key elements of two of these cases do not involve any legal complications, hence they can be summarized here in the following examples.

Example 9.2. After the fuel-to-inert salt heat exchanger for the Molten Salt Reactor Experiment (MSRE) had been fabricated, the project director learned of a serious tube-vibration problem in a somewhat similar heat exchanger and asked the author to appraise the possibility that similar trouble might arise in the new unit. A quick look at the drawings disclosed that the most likely spot for trouble was at the point where the fuel-inlet stream entered the shell and impinged on the U-bends of the tubes carrying the inert salt stream. The geometry was similar to that of Fig. 1.8 except that the fuel entered the end axially rather than from the side, and to minimize the shell-side liquid inventory (the fuel

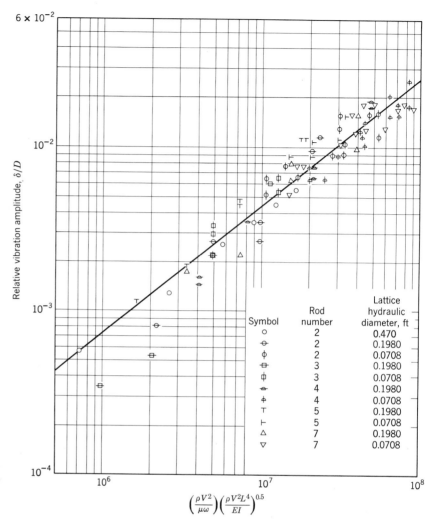

Figure 9.17 Amplitude of tube vibration under axial-flow conditions as a function of a correlation parameter that includes the major design variables. (Burgreen, Ref. 36).

salt on the shell side was both expensive and radioactive), a close-fitting hemispherical head had been installed over the U-bends. Because the head fitted so closely over the U-bends, there was no room for an impingement baffle to spread the stream out radially (like the flat-plate diffusers of Figs. 8.6 and 8.18c) to avoid a high local velocity across the tubes.

The design flow rate for the inert salt corresponded to a velocity of about 1 m/s (3.3 ft/s) in the jet impinging on the tube bundle. The tube diameter and pitch were 12.7 mm (0.5 in.) and 20 mm (0.775 in.), respectively, while the tube-wall thickness was 1.07 mm (0.042 in.). For the first row of tubes in the jet it appeared that the Strouhal number might be around 0.4, hence the vortex shedding frequency at the full flow rate was estimated as follows:

$$\text{Strouhal no.} = fD/V = 0.4 = f \times 0.02/1.0$$

$$f = 20 \text{ Hz}$$

The natural frequency of the unsupported U-bend when empty was estimated to be about 10 Hz, and about 6 Hz if filled with fuel salt, hence vibration was likely to develop at flow rates much less than that for full power. Of course there were uncertainties in the calculations, for example, the Strouhal number chosen, the end restraint conditions assumed for the U-bends, and the degree to which the stream jetting from the inlet pipe would spread out so that the velocity of the salt entering the tube matrix would be less than that used in the estimate.

Cutting the casing open to make changes would be

expensive. In view of the vital importance of ensuring the highest reliability attainable, the author recommended a test with water to determine if changes were required, and if so, to what degree. To obtain the high water flow rate required, a water main was tapped and the cleanliness of the water checked with an extended run through a filter. After this check was satisfactorily completed, the heat exchanger was installed with a flow meter, and a flow test run. The water flow was gradually increased until a pronounced rattling noise developed at about 65% of the rated flow. (If there had been fuel salt in the tubes, the increased mass would have reduced their natural frequency and hence the flow rate at which vibration would have commenced.) The volume of the noise did not drop off with an increase in the flow rate.

The heat exchanger was removed to the shop, a circumferential weld in the casing was cut so that the casing could be removed, and the tubes were inspected. Definite wear had already occurred on two tubes where they had rubbed against a weld bead in the casing in just the short test with water. The casing length was increased about 25 mm (1 in.) by inserting a spacer ring, an impingement baffle was installed, and small lacing bars were inserted between the tubes in the U-bends to stiffen them. The heat exchanger was reassembled and given a final test with water; there was no significant noise from tube vibration up to the maximum water flow obtainable, about 30% above the peak design flow.[38] It was subsequently installed in the reactor system and operated throughout the three-year life of the reactor with no sign of trouble.

Example 9.3. The 100-MWt High-Flux Intensity Reactor (HFIR) at Oak Ridge experienced three tube failures in the set of four heat exchangers employed to transfer the heat from the water in the primary circuit to the water circuit for the cooling towers. Each heat exchanger was 9.2 m (30 ft) high × 1.24 m (4 ft) in diameter and contained 1190 U-tubes 16 mm ($\frac{5}{8}$ in.) in diameter on a 21-mm ($\frac{13}{16}$-in.) equilateral pitch. The configuration was similar to that in Fig. 1.8 except that the vessel axis was vertical instead of horizontal and the shell-side water entered and left the shell at the midplane as in Fig. 8.18a. The central axial baffle split the inlet flow, diverting half the flow upward through the top half of the heat exchanger and the other half of the flow downward. These flows reversed direction at the ends of the shell, returned to the center, and left through the outlet port. Twelve sets of baffles at 650-mm intervals diverted the flow in and out of the tube matrix to give a 12-pass crossflow configuration. The unsupported span

of the tubes was 1300 mm (51 in.). The flow rate through the units was varied to hold the reactor operating temperature constant, which meant that the maximum flow rate was that required during especially hot, humid weather in the summer. The tube failures occurred in the first three years of operation, each of them when the water flow rate was well above the rated flow capacity of 6500 gpm. One failure occurred in each of the three heat exchangers normally used, the fourth exchanger having been held in reserve in case of a failure in one of the others. All three failures occurred in the same position—the tube closest to the shell near the outlet face in the last flow pass going into the outlet port. Because the bundle cross section was roughly hexagonal, this tube position was in the region in which the cross-flow velocity was probably the highest in the tube bundle. (The configuration was similar to that given by Fig. H6.3a if a vertical line were drawn between the third and fourth row of tubes at the left of the layout and the tubes to the left of the line deleted to provide space for axial flow over the baffles.) The first two times a tube failed it was regarded as an isolated case and the leaking tube was simply plugged off so that reactor operation could be continued, but the third failure indicated that the situation was chronic, and the author was called in to look at the problem.

From the circumstances attending the failures it appeared that they had been caused by tube vibration excited by vortex shedding. To check this hypothesis, the average water velocity across the tube row was estimated for the design flow rate and found to be 0.836 m/s (2.74 ft/s). For this velocity and the tube pitch, Fig. H8.6 gives a vortex shedding frequency of 20 Hz. The first mode bending vibration is probably in the form shown in the lower left corner of Fig. 9.16 for a continuous beam on simple supports (no rotational restraint at the baffles) except that it would be for about four spans instead of the three in Fig. 9.16. From Table H8.3a a natural frequency of 31 Hz was obtained for bare, empty stainless steel tubes. Correcting this for the mass of water inside the tubes gave a value of 25 Hz, essentially the same as the 22 Hz estimated for the exciting force from vortex shedding. It should be emphasized that estimates of both the vortex shedding frequency and the natural frequency of the tubes in first mode bending of necessity were approximations, and that, in view of the uncertainties in the flow distribution (see Fig. 12.8), more refined estimates did not appear to be justified.[39]

Tube vibration may induce failures either as a consequence of excessive bending stresses midway between baffles or of chafing and wear on the tubes where they pass through the baffles. In this instance, the baffles were 16 mm (⅝ in.) thick so that there was enough bearing area to make excessive wear unlikely, hence high bending stresses seemed the most likely cause of the failures. Fatigue test data obtained in water for stainless steel with small surface defects indicated that a stress of 90 MPa (13,000 psi) would cause a failure; this corresponds to a deflection, or vibration amplitude, of 6.6 mm (0.26 in.).

A quick and easy way to check this appraisal was to cut a hole about 250 mm (10 in.) in diameter in the shell near the position of the most recent tube failure and remove a portion of the failed tube. Metallurgical examination showed the fracture to be a fatigue failure. A flange with a window was then installed over the opening so that the unit could be operated while observing the tubes with a strobotach. The noise level in the cell was 50 to 60 db because of pump and flow noise in the system, so that conditions were unfavorable for audible detection of a few tubes vibrating in a matrix 4 ft in diameter containing over 1000 tubes. However, by placing one end of a steel rod against the shell and the other against the tragus of one's ear with a thumb tip as a pad over the end of the rod, one could hear a low-intensity but definite rattling noise develop as the water flow was increased between 6500 and 7000 gpm, and the noise intensity continued to increase up to 7500 gpm, the limit imposed by the pump capacity. Observations with the strobotach also showed the tubes starting to vibrate at a flow around 6500 gpm, and gave a tube vibration frequency of 22.5 Hz independent of the flow rate. High-speed motion pictures documented the direct visual observations that some tubes in the field vibrated at one high flow rate and not at another, while the reverse was true for others. As expected, the broken tube vibrated at a much lower frequency—about 2 Hz—because of its much reduced stiffness, and its amplitude was only about 1 mm.

The reactor operating procedure was changed to avoid flows of over 6500 gpm as much as possible. During the next 17 years, the balance of the 20-year design life of the units, there were only a few tube failures per heat exchanger.

Fluid-Elastic Whirling

Tube vibration can also be excited under certain conditions of fluid velocity and tube flexibility as a consequence of a phenomenon called *fluid-elastic whirling*.[40,41] The following brief outline of the problem was drawn largely from Ref. 41. The condition is self-excited, with alternate tubes moving out of phase, first laterally and then axially in the fashion indicated in Fig. 9.18.

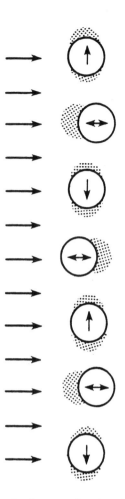

Figure 9.18 Schematic diagram showing the mode of tube motion in fluid-elastic whirling. In the phase shown, the odd tubes are moving transverse to the flow while the even tubes are moving parallel to the flow direction. The two sets of tubes alternate in their direction of motion relative to each other, with their movements being 90° out of phase. Actually, as one can see intuitively from momentum and kinetic energy considerations, the direction of motion of each tube does not change abruptly from the transverse to the parallel direction and vice versa, but rather each tube moves in an elliptical orbit, hence the term *whirling*. (Moretti, Ref. 41.)

The driving force stems from the Coanda effect, which causes fluid streamlines to tend to follow a curved surface. If this occurs asymmetrically in the flow over a tube, the stream flow direction is shifted, and this produces a lateral force on the tubes. Motion of tubes toward and away from each other reinforces the effect, and the elastic restoring forces in the deflected tubes lead to oscillations in the direction in which the flow is deflected. This mode of tube vibration is similar to flutter in aircraft control surfaces in that it requires two degrees of freedom, which in a heat exchanger tube matrix stem from the out-of-phase movement of alternate tubes. It differs from the Kármán vortex-shedding phenomenon in that the frequency of the oscillating force is determined by the natural frequency of the tube, whereas the frequency of vortex shedding and the consequent oscillating force is directly proportional to the fluid velocity, with little feedback from the natural frequency of the vibrating tube. While in both cases the tubes vibrate at their natural frequency, once fluid-elastic whirl starts (at a predictable critical fluid-flow velocity), it does not stop with an increase in flow. Instead, the amplitude builds up rapidly and continuously with further increases in flow rate because of the increasing fluid dynamic forces and kinetic energy input from the fluid stream that overcomes the damping forces. The phenomenon seems less common than the more usual trouble with vortex shedding because it usually appears at a higher velocity.

References 40 and 41 state that the critical velocity at which fluid-elastic whirling is likely to begin can be estimated from an empirical relation that includes the Strouhal number, that is,

$$V/Df = B(m\delta/\rho D^2)^a \qquad (9.8)$$

where V = critical flow velocity
D = tube diameter
f = tube natural frequency
m = tube mass per unit length
a = empirical constant (about 0.5)
B = empirical constant (about 10)
δ = damping parameter
ρ = fluid density

The two empirical constants depend on the geometry of the tube matrix. The range of conditions for which test data are available at the time of writing is not sufficient to define the effects of system parameters on these values. The damping factor depends on the tube support system (e.g., baffle clearances), the viscosity of the shell-side fluid, and the damping characteristics of the tube alloy.

ALLOWABLE STRESSES

Heat exchangers designed for operation with pressures below 1.4 MPa (200 psi) and temperatures below 150°C (300°F) ordinarily present straightforward stress analysis problems. As the operating temperature is increased above 150°C (300°F) to 315°C (600°F), depending on the material, the relations between the allowable stress

and the physical properties of the structural material become progressively more complex, particularly if the pressure is high enough so that thin-shell theory does not give a good approximation. Figure 9.19 shows several measures of the strength of a typical carbon steel as a function of temperature. Note that all five properties—the short-time ultimate tensile strength, the short-time yield strength, the stress for rupture in 10,000 h, the stress for 1% creep in 10,000 h (about 1 yr), and the stress for 1% creep in 100,000 h (about 12 yr)—drop with an increase in temperature above 427°C (800°F). In practice, the creep limitation is ordinarily more important than the rupture limitation; hence the design pressure stresses are ordinarily chosen to give less than 1% creep in 100,000 h. Unfortunately, creep data from 100,000-h tests are meager since they take 12 yr to run. Thus it is usually necessary to work with curves for 1% creep in 10,000 h or assume that the creep rate is insensitive to time, and to use curves for a creep rate of $10^{-5}\%/h$. The stress for rupture in 1000 h gives a good idea of the effects of temperature on strength, and this factor is plotted as a function of temperature in Fig. H8.1 for a variety of structural alloys. Figure H8.1 shows that some aluminum alloys retain their strength to 150°C, or 300°F (ordinary steels are suitable for operation at temperatures up to about 343°C, or 650°F), whereas progressively higher alloy materials are required for operation at the higher temperatures. Note also that the upper temperature limit for highly stressed stainless steels is about 650°C, or 1200°F.

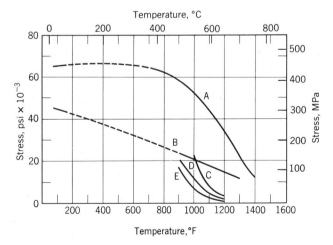

Figure 9.19 Effects of temperature on the short-time tensile and yield strengths (A and B), stress for rupture in 1000 h (C), stresses for 1% creep in 10,000 h (D) and 100,000 h (E) for a 1% chromium, 1/2% molybdenum steel alloy. (Courtesy The United States Steel Corp.)

Selection of Allowable Stresses for Static Loads

The past 100 yr of engineering experience indicate that deviations from ideality are substantial; hence conventional steel structures should be designed so that the nominal, or calculated, stresses are either less than the elastic limit by a factor of two, or less than the ultimate tensile strength by a factor of four. The first of these nominal stresses is the preferred basis, since substantial amounts of plastic deformation seriously affect the usefulness of most structures. Similarly, if creep is a factor, the best basis for an allowable stress is usually the creep strength. In any case, a *factor of safety* (or factor of uncertainty) of at least two should be applied to the applicable property to give the allowable design stress. This factor allows for small variations in the properties of the material, uncertainties in the loading conditions and load distribution, and uncertainties in stress-concentration factors.

Resistance to Thermal Strain Cycling

Thermal stresses differ from pressure stresses in that they are relieved by a small amount of plastic strain. In ductile metals this relief is very effective for a small number of cycles, but if a metal is repeatedly stressed beyond the elastic limit first in one direction and then in the other, the metal eventually cracks and breaks. In a sense this effect is a specie of accelerated-fatigue cracking, where only a few or a few hundred cycles may be required to induce failure instead of several hundred thousand or more. Thermal stresses are especially likely to be troublesome at elevated temperatures, where the elastic limit is much lower than at room temperature.

Good tests to determine the resistance of a material to strain cycling are so expensive and time-consuming that comprehensive data are available for only a few of the alloys for high-temperature applications. For all of those tested it has been found that the significant parameters are the test temperature, the plastic strain per cycle, and the number of cycles.[42–44] One way of expressing the results can be seen in Fig. 9.20, which shows data for type 347 stainless steel. Similar data are available for many alloys, for example, beryllium, Inconel, and Inor-8, a high nickel alloy similar to Hastelloy B.

In designing structures subject to thermal stresses the biggest problem is to identify the operating conditions likely to give trouble and the regions in which plastic strain most frequently occurs.[45,46] Although there is no substitute for a careful analysis, it is always good practice to avoid stress concentrations. As in the design of structures subject to vibration and fatigue, fillets should be as generous as practicable, welds in highly stressed zones should be avoided, and changes in section thickness should be as gradual as possible. For example, tube-to-header joints should be made as in Fig. 2.4c or d rather than as in Fig. 2.4a or b. Welds between materials differing in their coefficients of thermal expansion should be avoided if possible, and, if used, should be in the smallest diameter section practicable.[47,48]

Even carefully prepared designs based on a sound analysis may give trouble if the temperature differentials are large and the geometry complex; hence a test program may be necessary. A good analysis indicates both the operating conditions likely to give trouble and the test program required to investigate the reliability of the structure. It may be possible to carry out significant tests at room temperature by applying known displacements to structural elements in scale models. Electrical

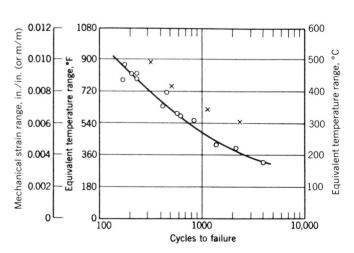

Figure 9.20 Effects of strain range on the number of cycles to failure for type 347 stainless steel cycled to 593°C (1100°F) (Coffin, Ref. 42.)

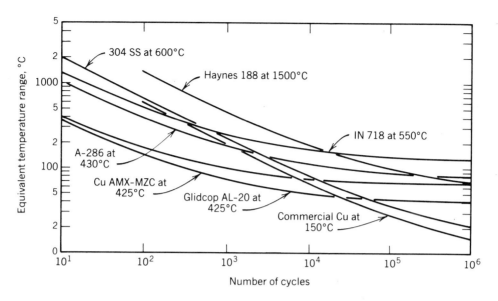

Figure 9.21 Effects of the equivalent temperature range under thermal cycling conditions on the number of cycles to failure for seven typical alloys at temperatures in the range they are likely to be used. (Fraas, Ref. 53.)

heating of some elements of models may be effective, or making the model of materials having different coefficients of expansion may make it possible to induce differential thermal expansion by isothermal heating.

Rapidly fluctuating temperature differences may be induced by unstable flow and heat transfer conditions. For example, under some conditions, the water-side heat transfer coefficient in boilers may fluctuate over a wide range as nucleate- and film-boiling conditions alternately prevail. The high- (or low-) temperature jet from a heat exchanger inlet pipe may weave about through a heat transfer matrix in an erratic fashion and cause local tube- or shell-wall temperatures to fluctuate irregularly. If this occurs, the resulting thermal stresses may induce fatigue failures.

The combined effects of a few rare large temperature excursions, some infrequent moderate temperature variations, and frequent small temperature fluctuations can be approximated by assigning a suitable fraction of the total life to each amplitude of strain cycling.[49]

Estimating the Life of Parts Subject to Thermal Strain Cycling

There are many papers in the literature that give bases for estimating the life of parts subject to thermal strain cycling,[50,51] but they tend to be awkward to use and require physical property data that are hard to find.

Coffin has, in Reference 42, reviewed an extensive set of experimental data on low-cycle fatigue and showed that it could be correlated well through the use of empirical equations relating the number of cycles to failure to the total plastic strain per cycle (the sum of the positive and negative amounts by which the material was deformed plastically each cycle), the elastic strain range per cycle,

Figure 9.22 Ductile failure in a large steel pressure vessel when the pressure was deliberately increased until the vessel burst. (Courtesy The United States Steel Corp.)

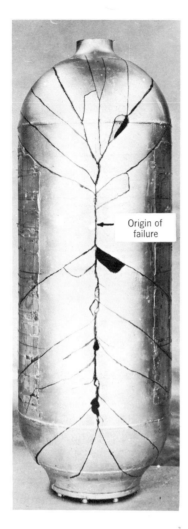

Figure 9.23 Reassembled fragments of a pressure vessel which failed by brittle fracture when the pressure was increased to failure. (Durelli, Dally, and Morse, Ref. 55).

the modulus of elasticity, and the cycle frequency plus six constants that depend on the physical properties of the alloys. A useful approximation for general application where some of the physical test data are not available was worked out by A. S. Thompson[52,53] and may be expressed as

$$N_f = \frac{1}{\left[\dfrac{2}{\epsilon}\left(\dfrac{1 + \mu}{1 - \mu}\right)\alpha\Delta T - \dfrac{2\sigma}{\epsilon E}\right]} \qquad (9.9)$$

where N_f = number of cycles to failure
ΔT = temperature range per cycle
E = modulus of elasticity

μ = Poisson's ratio
ϵ = elongation in a tensile test
σ = yield stress
α = coefficient of thermal expansion

Data for a set of typical alloys for temperatures at which they would be likely to be employed were assembled and the number of cycles to failure in each case was calculated from Eq. 9.9 for a series of temperature ranges. The results were plotted in Fig. 9.21 to provide a reasonable basis for estimating the life of parts under thermal strain cycling conditions.

Types of Failure

Metals may fail in several ways under service conditions.[54] The most common type of fracture under static stresses is a ductile failure, which is characterized by plastic flow and a reduction in area in the plane of the fractured surface. Such a failure occurs in a pressure vessel only when it is severely overstressed, and it ordinarily takes the form of a tear having a limited length, as in Fig. 9.22.

Some steels lose their ductility at low temperatures, and thus may experience brittle failures under moderate loads. Once a crack starts at a stress concentration in a low-ductility material, it is itself a stress concentration so that it is likely to spread rapidly, leading to a large tear. Such tears show little or no reduction in area and

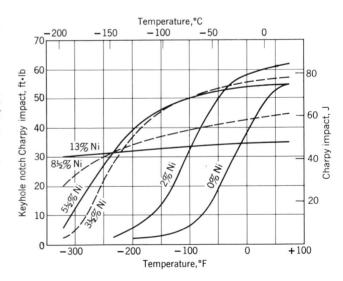

Figure 9.24 Effects of temperature on the Charpy impact strength of normalized low-carbon steels having various nickel contents. (Courtesy The United States Steel Corp.)

Figure 9.25 Magnified section through the wall of an Inconel tube showing a crack induced by severe thermal strain cycling at about 815°C (1500°F). (Courtesy Oak Ridge National Laboratory.)

tend to be much longer than ductile failures in similar structures. The vessel may break into fragments, as in Fig. 9.23,[55] instead of simply splitting open (Fig. 9.22), and the hazards of flying fragments may be a nasty problem. Brittle failures of this sort have occurred principally in large welded structures such as ship hulls, large oil tanks, and bridges. In every instance in which a brittle failure has been investigated thoroughly, the material was characterized by a low ductility at the temperature at which it failed.[56] Work on failed Liberty ships has been especially revealing. By analyzing and testing material from the failed plates and those in which the cracks have stopped, it was found that the Charpy impact strength gives a good basis for correlation.[56,57] The cracks always start in material having a Charpy impact strength of less than 20 J (15 ft·lb) for a standard ¼-in. square notched specimen, and usually stop when they reach material having a Charpy impact strength greater than 20 J (15 ft·lb). Curves such as those in Fig. 9.24 are ordinarily used in preparing specifications. The temperature below which the Charpy impact strength drops rapidly is called the *transition temperature*. Apparently similar heats of steel meeting the same specifications for chemical composition may have widely different transition temperatures; hence a separate specification for the transition temperature must be included if it is likely to be important.

Fatigue cracks induced in heat exchangers by mechanical vibration are infrequent, but those induced by thermal strain cycling are relatively common. Figure 9.25 shows such a crack in a tube wall. Typically, the crack was induced by bending stresses at the stress concentration represented by the header sheet. It progressed very slowly and was noticed only because of a slow leak from one fluid circuit to the other.

REFERENCES

1. *ASME Boiler and Pressure Vessel Code*, ASME, New York.
2. K. P. Singh and A. I. Soler, *Mechanical Design of Heat Exchangers and Pressure Vessel Components*, Arcturus Publishers, Cherry Hill, N.J., 1983.
3. J. H. Faupel, "Yield and Bursting Characteristics of Heavy-Wall Cylinders," *Trans. ASME*, vol. 78, 1956, p. 1031.
4. G. W. Watts and W. R. Burrows, "The Basic Theory of Vessel Heads under Internal Pressure," *Trans. ASME*, vol. 71, 1949, p. A-55.
5. G. W. Watts and H. A. Lang, "The Stresses in a Pressure Vessel with a Flat Head Closure," *Trans. ASME*, vol. 74, 1952, p. 1083.
6. G. W. Watts and H. A. Lang, "The Stresses in a Pressure Vessel with a Hemispherical Head," *Trans. ASME*, vol. 75, 1953, p. 83.
7. G. W. Watts and H. A. Lang, "Stresses in a Pressure Vessel with a Conical Head," *Trans. ASME*, vol. 74, 1952, p. 315.
8. E. Sternberg and M. A. Sadowsky, "Three-Dimensional Solution for Stress Concentration Around a Circular Hole in a Plate of Arbitrary Thickness," *Trans. ASME*, vol. 71, 1949, p. 27.
9. D. E. Hardenbergh, "Stresses of Nozzle Connections of Pressure Vessels," *Proceedings of Society for Experimental Stress Analysis*, vol. 18(1), 1961, p. 152.
10. C. E. Taylor and J. W. Schweiker, "A Three-Dimen-

sional Photoelastic Investigation of the Stresses Near a Reinforced Opening in a Reactor Pressure Vessel," *Proceedings of Society for Experimental Stress Analysis*, vol. 17(1), 1959, p. 25.

11. R. L. Jackson et al., "Importance of Matching Steam Temperatures with Metal Temperatures During Starting of Large Steam Turbines," *Trans. ASME*, vol. 79, 1957, p. 1669.

12. R. A. Struble and B. Schweizer, "Ideal Reducers and Nozzle Flares," *Trans. ASME*, vol. 79, 1957, p. A-137.

13. G. Horvey et al., "Stresses and Deformations of Flanged Shells," *Trans. ASME*, vol. 76, 1954, p. A-109.

14. J. W. Dally and A. J. Durelli, "Stress Analysis of a Reactor Head Closure," *Proceedings of Society for Experimental Stress Analysis*, vol. 17 (2), 1959, p. 71.

15. E. O. Waters and F. S. G. Williams, "Stress Conditions in Flanged Joints for Low-Pressure Service," *Trans. ASME*, vol. 74, 1952, p. 135.

16. N. A. Weil and J. J. Murphy, "Design and Analysis of Welded Vessel Skirt Supports," *Journal of Engineering for Industry, Trans. ASME*, vol. 82, 1960, p. 1.

17. Y. Y. Yu, "Rational Analysis of Heat Exchanger Tube-Sheet Stresses," *Trans. ASME*, vol. 78, 1956, p. A-468.

18. G. Horvay, "The Plane-Stress Problem of Perforated Plates," *Trans. ASME*, vol. 74, 1952, p. A-355.

19. G. Horvay, "Bending of Honeycombs and of Perforated Plates," *Trans. ASME*, vol. 74, 1952, p. A-122.

20. I. Malkin, "Notes on a Theoretical Basis for Design of Tube Sheets of Triangular Layout," *Trans. ASME*, vol. 74, 1952, p. A-387.

21. K. A. Gardner, "Heat Exchanger Tube-Sheet Design," *Trans. ASME*, vol. 70, 1948, p. A-377.

22. K. A. Gardner, "Heat Exchanger Tube Sheet Design— Two, Fixed Tube Sheets," *Trans. ASME*, vol. 74, 1952, p. A-159.

23. G. J. Schoessow and E. A. Brooks, "Analysis of Experimental Data Regarding Certain Design Features of Pressure Vessels," *Trans. ASME*, vol. 72, 1950, p. 567.

24. L. F. Kooistra and R. U. Blaser, "Experimental Technique in Pressure Vessel Testing," *Trans. ASME*, vol. 72, 1950, p. 579.

25. C. W. Lawton, "Strain Gage Test on Model Vessels for Nuclear Power Plant Designs," *Proceedings of Society for Experimental Stress Analysis*, vol. 17(1), 1959, p. 149.

26. G. Murphy, "Analysis of Stresses and Displacement in Heat Exchanger Expansion Joints," *Trans. ASME*, vol. 74, 1952, p. 397.

27. N. C. Dahl, "Toroidal-Shell Expansion Joints," *Trans. ASME*, vol. 75, 1953, p. A-497.

28. W. E. Cooper, M. T. Roche, and J. L. Noble, "Stresses in a Semi-Infinite Thin-Walled Cylinder Caused by an Exponential Temperature Distribution," Report KAPL 973, Knolls Atomic Power Laboratory, March 25, 1955.

29. B. J. Grotz and F. R. Arnold, "Flow Induced Vibrations in Heat Exchangers," AD-104568, Stanford University, 1956.

30. R. L. Solnick and R. H. Bishop, "Noise, Vibration, and Measurement Problems Resulting from Fluid Flow Disturbances," *Trans. ASME*, vol. 79, 1957, p. 1043.

31. A. A. Putnam, "Flow Induced Noise in Heat Exchangers," *Journal of Engineering for Power, Trans. ASME*, vol. 81, 1959, pp. 417–422.

32. Hill and Armstrong, "Aerodynamically Induced Sounds in Tube Banks," *Proceedings of the Physical Society*, vol. 79, part 1, no. 507, 1962, p. 225.

33. Y. N. Chen, "Flow-Induced Vibration and Noise in Tube-Bank Heat Exchangers Due to von Kármán Streets," *Journal of Engineering for Industry, Trans. ASME*, vol. 90(1), 1968, p. 134.

34. R. C. Baird, "Wind-Induced Vibration of a Pipeline Suspension Bridge and Its Cure," *Trans. ASME*, vol. 77, 1955, p. 797.

35. J. N. MacDuff and R. P. Felgar, "Vibration Design Charts," *Trans. ASME*, vol. 79, 1957, p. 1459.

36. D. Burgreen et al., "Vibration of Rods Induced by Water in Parallel Flow," *Trans. ASME*, vol. 80, 1958, p. 991.

37. J. F. Sebald and W. D. Nobles, "Control of Tube Vibration in Steam Surface Condensers," *Proceedings of the American Power Conference*, 1962.

38. R. J. Kedl and C. K. McGlothlan, "Tube Vibration in MSRE Primary Heat Exchanger," Report no. ORNL-TM-2098, Oak Ridge National Laboratory, January 1968.

39. A. P. Fraas, "Tube Vibration in HFIR Heat Exchangers," Report no. ORNL-TM-2467, Oak Ridge National Laboratory, March 1969.

40. H. J. Connors, "Fluid-Elastic Vibration of Tube Arrays Excited by Crossflow," Paper presented at the Winter Annual Meeting of the ASME, 1971.

41. P. M. Moretti, "The Paradox of Flow-Induced Vibrations," *Mechanical Engineering*, vol. 108(12), December 1986, p. 56.

42. L. F. Coffin, Jr., "An Investigation of Thermal Stress Fatigue as Related to High Temperature Piping Flexibility," *Trans. ASME*, vol. 79, 1957, p. 1637.

43. L. F. Coffin, Jr., "Thermal Stress and Thermal Stress Fatigue," *Proceedings of Society for Experimental Stress Analysis*, vol. 15(2), 1958, p. 117.

44. R. U. Blaser, "Thermal Stress Problems in Practice," *Proceedings of Society for Experimental Stress Analysis*, vol. 15(2), 1958, p. 131.

45. R. W. Swindeman and D. A. Douglas, "The Failure of Structural Metals Subjected to Strain-Cycling Conditions," *Trans. ASME*, vol. 81-D, 1959, p. 203.

46. E. L. Robinson, "Steam Piping Design to Minimize Stress Concentrations," *Trans. ASME*, vol. 77, 1955, p. 1147.

47. R. Michel, "Elastic Constants and Coefficients of Thermal Expansion of Piping Materials Proposed for 1954 Code for Pressure Piping," *Trans. ASME*, vol. 77, 1955, p. 151.

48. H. Weisberg, "Cyclic Heating Test of Main Steam Piping Joints between Ferritic and Austenitic Steels," *Trans. ASME*, vol. 71, 1949, p. 643.

49. R. R. Gatts, "Application of a Cumulative Damage Concept to Fatigue," *Journal of Basic Engineering, Trans. ASME*, 1961, vol. 83, p. 529.

50. L. F. Coffin, Jr., "Fatigue at High Temperature," *The Institute of Mechanical Engineers Proceedings*, vol. 188, London, September 1974, p. 109.

51. *1976 ASME-MPC Symposium on Creep-Fatigue Interaction*, ASME G00112, 1976.

52. A. P. Fraas and A. S. Thompson, "ORNL Fusion Power Study: Fluid Flow, Heat Transfer, and Stress Analysis Considerations in the Design of Blankets for Full-Scale Fusion Reactors, Report no. ORNL-TM-5960, Oak Ridge National Laboratory, February 1978.

53. A. P. Fraas, *Engineering Evaluation of Energy Systems*, McGraw-Hill Book Co., New York, 1982, p. 162.

54. D. B. Rossheim et al., "Recent Experience in Examination of High Temperature Catalytic Cracking Pressure Equipment," *Trans. ASME*, vol. 74, 1952, p. 1099.

55. A. J. Durelli, J. W. Dally, and S. Morse, "Experimental Study of Large-Diameter Thin-Wall Pressure Vessels," *Proceedings of Society for Experimental Stress Analysis*, vol. 18(1), 1961, p. 33.

56. P. P. Puzak, E. W. Eschbacher, and W. S. Pellini, "Initiation and Propagation of Brittle Fracture in Structural Steels," *The Welding Journal*, vol. 31, Welding Research Supplement, 1952, p. 561.

57. P. P. Puzak and W. S. Pellini, "Evaluation of the Significance of Charpy Tests for Quenched and Tempered Steels," *The Welding Journal*, vol. 35, Welding Research Supplement, 1956, p. 275-S.

10

Service Life, Reliability, and Maintenance

The value of a high degree of reliability in plant components may be very high indeed if one considers the cost of an outage. Just the unavailability of a 1000-MWe power plant costs around $1,000,000 per day in overhead charges. The costs of an outage in a chemical plant may be very high even for a brief outage if reactions in a stagnant product stream cause fouling of the equipment. Further, the value of high reliability in terms of customer satisfaction and the manufacturer's reputation, while not easily measured, can also be very high. Yet design engineers are usually preoccupied with performance—the design heat load, fluid stream temperatures, and pressure drops. They grudgingly give consideration to weight and fabrication costs, and possibly take part load conditions into account, but reliability doesn't get much attention. One indication of this is given by the subjects of the published papers; thousands appear each year on heat transfer, fluid flow, and stress analysis, but relatively few on reliability, service life, and maintenance. Of the latter set, few contain much good statistical data on reliability, and most of that is in the nuclear power or aerospace fields, where strict governmental requirements make public disclosure necessary. Equipment failures are always a source of embarrassment and possibly legal liability to either the operator or the manufacturer of the equipment, or both, hence in most fields what information is collected is regarded as proprietary. Yet even in the proprietary realm, the amount of data is not great. Surprisingly few companies do a good job of collecting enough data on each failure of their equipment to evolve a comprehensive picture of the incidence of different types of failure and their causes so that they would have good quantitative input into new designs to supplement seat-of-the-pants judgments.

CATASTROPHIC FAILURES

The first failures of industrial equipment to get widespread public attention were the spectacular explosions of steam boilers. Between 1862 and 1879 there were over 10,000 boiler explosions in England, and from 1880 to 1919 there were 14,281 boiler explosions in the United States with the loss of many lives. The ASME was concerned over these failures from the time it was founded in 1880, and initiated an effort to accumulate information on the character and causes of the failures.[1] An indication of the value of good background data on failures is given by Fig. 10.1, which shows that the incidence of boiler failures in the United States stopped increasing by 1890 and began to fall off after 1900. By 1911 sufficient information was available to begin to delineate the ASME Boiler Code, which was formally issued in 1914. In Fig. 10.1 note the rapid drop in the incidence of boiler explosions in subsequent years in spite of the continuing increase in both the number of boilers

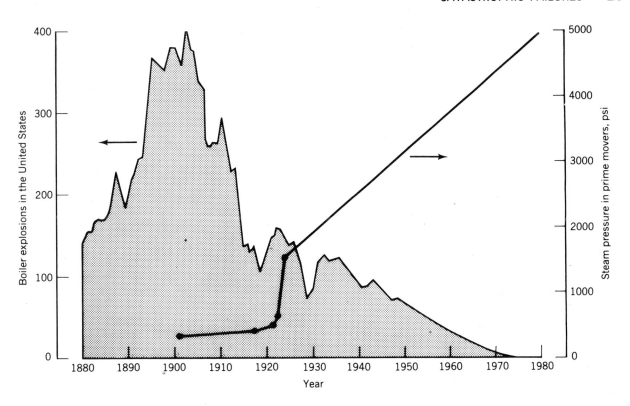

Figure 10.1 Number of boiler failures per year in the United States as a function of year. (Walters, Ref. 1.)

and the steam pressure. The value and soundness of the ASME Boiler Code have made it the basis for boiler design all over the world.

Efforts to quantify reliability information received a major boost under the space program, not because of a hazard to the public but because of the risk of loss of the huge investment required for building a spacecraft and placing it in orbit where it could not be maintained. The enormous number of diodes, resistors, capacitors, junctions, etc., in electronic equipment made the failure probability high just in one component, such as an amplifier. For the complete complex of a unit such as a weather satellite, the failure rate to be expected from the electronic equipment of the 1950s was quite unacceptable. The effort to improve both reliability and the means for predicting it received a further boost in the Apollo program because that mission was completely dependent on the reliability of an enormous complex of electronic equipment. As a result, the reliability of electronic equipment has been improved by orders of magnitude, and a splendid set of reliability data is available for electronic components.[2-5]

Public safety concerns made reliability a major consideration in the nuclear power program from its inception. The accident at the Three Mile Island plant not only aroused an enormous public reaction but demonstrated that a series of small failures, each comparatively innocuous by itself, could lead to an accident that could be financially catastrophic to the plant owner even though it did no harm to the public other than the severe psychological stress generated by activists and sensationalism in the media. One result of this has been the intensification of the effort to collect and organize a fine set of data on the reliability of each of the components used in nuclear plants, including heat exchangers, pumps, valves, motors, electrical controls, and the like.[3-5] These data make it possible to estimate the probability of any hypothesized equipment failure or sequence of failures while in the design stage, and thus make quantitative comparisons of the probability and consequences of an accident for various combinations and arrangements of the equipment in the plant.[6]

Although petroleum refineries are laid out so that they do not represent a hazard to the public, even a spark could ignite a fire that might destroy the plant. Thus there is great financial incentive to get the highest possible reliability in every component, particularly with respect to leaks of flammable fluids. As a result, the American Petroleum Institute (API) has established specifications for refinery equipment that are just as stringent as those for nuclear plants. These require large factors of safety, which make for a high degree of component reliability.

NATURE OF FAILURES AND THEIR EFFECTS

Burst failures such as the boiler explosions that were common a century ago have been virtually eliminated by conformance to the ASME Pressure Vessel Code. Other types of failures, such as leaks, while rarely spectacular, continue to be troublesome and all too often force a plant shutdown. In view of the fact that these troublesome little failures differ in the way and rate at which they develop, a brief survey is in order.

Low-Cycle Fatigue

The phenomenon of low-cycle fatigue discussed in the previous chapter is a major cause of leaks in heat exchangers. Most cases stem from occasional thermal transients in which thermal stresses exceed the elastic limit during rapid power changes, start-up, shutdown, or abrupt forced outages. Usually only certain regions are affected. In a U-bend tube bundle, for example, a large difference between the average temperatures of the two legs will induce bending stresses in the U-bends, and these stresses will be higher in the small-radius bends of the innermost tubes than in the larger radius bends of the outer tubes. As one would expect, leaks stemming from low-cycle fatigue cracks in the U-bends of the inner tubes may develop after perhaps a year of operation; if these inner tubes are plugged, there may be no further trouble with leaks of this type because the thermal stresses in the other tube rows are within the elastic limit. Note that residual stresses from fabricating the bends may contribute to low cycle fatigue failures in bends.

Low-cycle fatigue may also lead to cracks in header sheets, nozzles whose temperature changes more rapidly than that of the shell in the course of a transient, piping, and any region subject to large stresses induced by differential thermal expansion.

Erosion, Fretting, and Wear

Tube walls may be thinned by the abrasive action of particulates in a fairly high-velocity fluid stream, by fretting between tubes and their supports such as baffles as a consequence of flow-induced vibration, and even wear from the insertion of cleaning equipment. Difficulties of this sort are common, for example, in steam plant condenser tubes just downstream from the cooling water inlet.[7] In this region eddies induced by the *vena contracta* at the tube inlet (see Fig. 8.2) cause particles of sand to scrub the tube wall. Erosion is also a problem in fluidized beds, particularly where local gas velocities are well above the average (see Chapt. 15).

Erosion rates depend on the fluid velocity, the angle at which the particle approaches the surface, and the properties of the material. As one would expect, there is usually no erosion for velocities up to some threshold value, but above that there is a rapid increase in the erosion rate with increasing velocity. However, the effects of approach angle are not intuitively obvious, nor are the properties of the material.[8] For some materials the dominant erosion mechanism involves a brittle failure, in which case the erosion rate peaks when the particle trajectory is normal to the surface. For other materials the dominant damage mode is a ductile failure with the incident particle plowing a furrow in the surface, in which case the erosion rate is usually highest for an angle of incidence in the range of 20 to 40°. Figure 10.2 shows these effects for some typical materials. Surprisingly, the dominant failure mode for some brittle materials such as glass seems to be the ductile mode, and for the ductile mode annealed aluminum gave a lower erosion rate than hardened steel!

Fretting damage to a surface may develop where some relative motion occurs between two surfaces that nominally should be static. Galling and/or seizing may occur where small deflections under load cause occasional movement between parts under high contact pressures as in bolt threads or the faces of bolted flanges. Damage is particularly likely if the surfaces are made of aluminum, stainless steel, or nickel alloys. Fretting, galling, and wear rates seem to depend mainly on surface hardness, but other factors have important effects. Making the rubbing surfaces of different materials so that they do not tend to weld together is generally beneficial, particularly if one of the materials is a cobalt-base alloy. The cobalt alloys are exceptionally resistant to wear because of their rapid rate of work hardening and the presence of wear-resistant carbides. The complexities of the effects of the various factors, especially the properties of materials, on erosion, fretting, and wear are too complex to treat further in this text. Controlled tests such as those of References 7 and 8 provide valuable insights, but there is no substitute for extensive service experience.

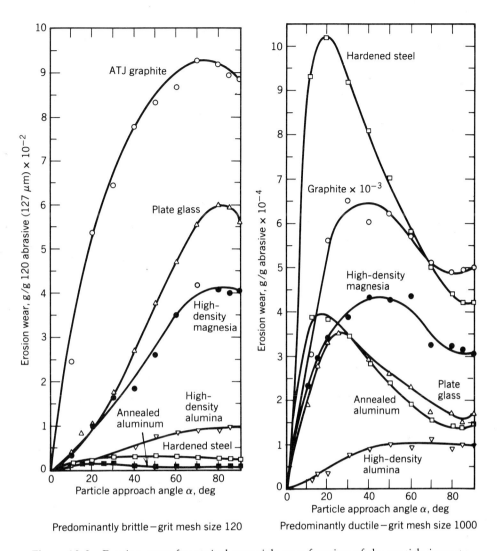

Figure 10.2 Erosion wear for typical materials as a function of the particle impact angle α for silicon-carbide grit projected at a velocity of 152 m/s (500 fps). (Sheldon, Ref. 7.)

Corrosion

Many different corrosion problems in heat exchangers involve water, particularly where it contains both oxygen and salts, acids, or alkalies.[9] Corrosion may take many forms, some obvious, but others frustratingly insidious. The principal types of corrosion are as follows:

Uniform general surface corrosion
Pitting
Catastrophic oxidation
Galvanic corrosion
Crevice corrosion

Intergranular corrosion
Stress-corrosion cracking
Solution corrosion
Erosion corrosion
Cavitation corrosion

Uniform general corrosion is a familiar problem, the effects of which are obvious in even a cursory inspection. In air or water, oxidation attack is commonly general, and the rate falls off with time as a protective oxide coating builds up. The situation is quite different if H_2S is present in reducing atmospheres, particularly at temperatures around 540°C (1000°F); ferritic steels, for example, are rapidly attacked by H_2S, and no protective layer is formed. Sulfur also presents problems when present in stack gases; if temperatures of heat transfer surfaces drop below the dew point (about 180°C, or 356°F) rapid, general attack by sulfuric or sulfurous acid will occur.

Unlike general corrosion, pitting corrosion often is neither obvious nor easy to detect. It occurs when an environment causes local attack on an otherwise protective oxide coating, and may be accelerated by nonuniformities in an alloy. The environment in the pit becomes significantly different from that for the bulk of the surface and is generally conducive to increased local corrosion so that the rate is not likely to fall off with time. It commonly occurs in salt or brackish water if H_2S is present; the metal sulfide corrosion product acts as a catalyst for further corrosion.[9] Stainless steels are particularly sensitive to the presence of chloride ions, the sensitivity decreasing with increasing chromium and molybdenum concentration in the steels. A somewhat similar type of attack known as catastrophic oxidation has been experienced with certain steels at high temperatures where a layer of scale deposit or thermal insulation limits the oxygen access to the surface so that the protective layer of oxide is rather thin. In type 316 stainless steel, in which molybdenum is an important alloying element, molybdenum oxides may form and destroy the protective oxide layer, leading to rapid local attack.

Galvanic corrosion occurs where an electric current is induced to flow between two metals having different electrochemical potentials in contact with the same aqueous environment. The more electrochemically active metal is attacked, usually locally near the line of electrical contact, the distribution and severity of the attack being a function of the conductivity of the water solution. The problem is generally recognized, and corroded areas are usually obvious. Table H7.4 lists the galvanic series of the metals and alloys of interest for operation in seawater.

Crevice corrosion is insidious and hard to detect. It has proved to be a serious problem with stainless steel tubes in boilers if even a small amount of chlorides is present. Water enters the crevice in a tube-to-header joint (see Fig. 2.4a) at low loads and then dries out at higher loads. The process is repeated as the power and temperatures are cycled, leaving an additional trace of salt each cycle. The salt deposit builds up until corrosion occurs at a serious rate. To avoid this type of corrosion, the chloride concentration in the boiler water should be kept below 100 ppb (see Table H7.7), and high-nickel alloys such as Incoloy 600 are used instead of stainless steel

because they are much less sensitive to chloride corrosion.

Intergranular attack may occur along the grain boundaries with little surface evidence of damage. This type of attack is generally observed with stainless steels that have either cooled slowly through or have been slowly reheated through the temperature range from 650 to 870°C (1200 to 1600°F), as can occur during improper heat treatment or during welding. Chromium carbides precipitate in the grain boundaries, lowering the chromium content near the boundary so that it is no longer corrosion-resistant. The problem is greatly reduced by using very low-carbon stainless steels. This type of corrosion is particularly insidious and detectable only through the use of photomicrographs of sectioned material.

Stress corrosion is extremely localized attack under critical conditions of stress and a corrosive environment. In stainless steels and other alloys whose corrosion resistance depends on maintaining a protective oxide film, stress may crack this film, and, if environmental conditions are such that this film cannot immediately reform, corrosion-assisted crack propagation occurs. Stress corrosion may occur intergranularly or transgranularly, the former mode being observed in alloys susceptible to intergranular corrosion where this mode of corrosion initiates the stress-corrosion cracking. With stainless steels, chloride ions are generally the most aggressive initiators of stress cracking. Partly because of this, tight control of the impurities in boiler feedwater includes the requirement that both chloride-ion and oxygen concentrations must be held below 1 ppm, and frequently below 100 ppb to avoid cracking. It can proceed with amazing rapidity. In one striking case a stress-corrosion crack developed in a stress concentration formed by a nick in a strut of the landing gear of a bomber that landed one evening at a seacoast airstrip where salt spray was prevalent; the gear collapsed overnight.

Solution corrosion has been a problem in liquid metal and molten salt systems. It stems from a trace solubility of the structural metal in the circulating liquid. If the solubility is appreciably higher at high temperatures than at low temperatures, metal will be dissolved from surfaces in the high-temperature zones and deposited in the low-temperature regions. This phenomenon is treated in Chapter 17 on heat exchangers for liquid metals and molten salts.

Erosion by particles in the fluid may remove a protective oxide layer and accelerate corrosion. This may be a factor in the wastage of condenser tubes mentioned in the above section on erosion as well as in wastage of the tubes of fluidized-bed combustors, as discussed in

Chapter 15. Cavitation in pump impellers has caused severe erosion, and it may act in conjunction with corrosion to cause severe local damage in high-velocity regions in the vicinity of inlets or return bends in heat exchangers.

Deposits

Deposits on heat transfer surfaces are a common cause of a loss in performance that may be so severe as to cause a forced outage. They may be taken into account by including a fouling factor in the design (see Table H5.4), and they may be kept to tolerable levels by cleaning procedures. This may not be too difficult in some cases, but in others, such as ash and slag deposits in steam boilers, they may prove exceedingly difficult to remove.[10] Figure 10.3 shows such as case. Sludge deposits in low-velocity regions are often a problem in water systems and may lead to concentrations of impurities that in turn lead to serious corrosion.

Electrical Components

The most common type of failure in electrical equipment is a short circuit caused by deterioration in the electrical insulation as a consequence of age, moisture, or excessive temperatures, usually from overloading. Moisture and excessive temperature are also major factors in the failure of solid-state electronic equipment. In some components the life may be definitely limited, as it is in light bulbs and vacuum tubes, where vaporization of their high-temperature filaments continues with no apparent ill effects until the filaments become too thin, their resistance increases, their temperature approaches the melting point, and they abruptly burn out. This occurs after about the same amount of time for all units of a similar model.

INCIDENCE OF FAILURES IN TYPICAL SYSTEMS

The incidence of failures in typical systems gives some perspective on the relative importance of different types

(a)

(b)

Figure 10.3 Ash fouling of a finned tube in the economizer of a furnace fired with No. 2 fuel oil. (Courtesy Escoa Corp.)

of failure. At this point it seems best to define the various terms in common use in the field. First, definitions of what constitutes a failure vary. While ruptured pressure vessels or broken shafts are obviously failures, the term is also commonly used for any malfunction that causes a forced outage. Thus a fault in an electric motor that makes it stop abruptly is termed a failure even though there is no sign of mechanical damage. Similarly, a motor-operated valve that fails to function when the proper button is pushed is said to have failed even though it may still be operated by hand. The probability of such failures is commonly referred to as the *mean time between failures*, or MTBF. For the system as a whole, the mean time between forced outages, or MTBFO, is the comparable figure of merit. An even more significant parameter for a complete system is its availability, that is, the fraction of the year in which the system is *available* for use even though it may be on standby or operating at a low load. The *capacity factor* is the ratio of the actual total yearly output divided by the ideal output if it operated at its full design load throughout the year.

Combustion Gas Turbines

A large amount of combustion gas-turbine experience in utility service has been accumulated by EPRI.[11-13] The operating conditions are much less favorable than for gas turbines used for cogeneration in chemical plants because the utility units are used mainly for peaking service with many starts and stops. Operating conditions vary widely from one utility to another both in the fuel quality (and hence hot corrosion) and in the quality of the maintenance. Keeping these qualifications in mind, it is interesting to review Table 10.1 to appraise the relative importance of the principal causes of forced outages for a set of 12 gas-turbine units, about the best of those surveyed. The data show that the electrical control system was by far the the principal cause of outages, with 69% of the outages caused by instrumentation, control, or electrical equipment. Interestingly, the reliability of the complete gas-turbine units increased with operating time, as shown by Fig. 10.4, indicating that service experience led to correction of some of the design or manufacturing defects and that there was no adverse effect of age in the period covered by the data.

Water Reactors

A set of forced outage data for water reactors similar to that for gas turbines is presented in Table 10.2.[14] Interestingly, as was the case for the gas turbines of Table 10.1, 52% of the forced outages were caused by failures in the instrumentation, control, and electrical equipment. Note also that only 3.4% of the forced outages were caused by troubles with the steam generators. However, defects in the steam generator are likely to take much longer to diagnose and repair than a faulty element in an amplifier, hence the outage time may not be so different as the percentage of forced outages might imply.

Steam Generators for Pressurized Water Reactors

In view of the far more benign conditions, one might expect that the steam generators for nuclear reactors would have shown a lower incidence of failures than fossil fuel boilers. However, this has not been the case in the early years of steam generators for either the gas-cooled reactors in England of the type shown in Fig. 1.5, or the pressurized water reactors in the United States. While the hot water is far cleaner than the combustion gas from coal and there are none of the local hot spots that have proved so troublesome in open combustion furnaces, some peculiar new, quite unforeseen, problems have arisen in these new applications. Not surprisingly, the difficulties with the steam generators such as that of Fig. 1.5 for gas-cooled reactors have been quite different from those in the PWR steam generators, but both sets of troubles have stemmed from conditions different from those in fossil fuel plants. The PWR case provides a particularly good illustration of the way in which a new application can lead to unforeseen service troubles. A review of these failures is a humbling experience—most stemmed from engineering errors or oversights that most of us might easily have made.

A highly instructive detailed survey of tube failure experience in steam generators for water-cooled reactors is presented in Reference 15. The period covered was from 1971 to 1981, with the number of reactors in the set studied increasing from 24 in 1979 to 110 by 1981. The set investigated included several different designs and several different manufacturers. The results are summarized in Fig. 10.5. All of the failures involved tube leaks or incipient tube leaks that were corrected by plugging the defective tubes. Initially, the main cause of failure was stress-corrosion cracking (SCC) of the tubes in the small-radius U-bends in the center rows in the tube bundles. This type of failure faded dramatically after this failure mode was recognized and these tubes were plugged, or the cause of these defects was eliminated in the later designs as they came into service.

By 1973 the major cause of failures became tube wastage caused by corrosion in phosphate sludge

TABLE 10.1   Combustion Turbine Component Performance Summary for HSF Unit Turbines, by CT Code[a]

Component	U80-G[b]	T80-G	T78-G	S80-G	R80-G	079-G	I80-G	U79-G	S78-G	S79-G	P80-G	Q79-G
CT unit, general	0	1	0	0	0	0	1	0	0	1	0	0
		137					142			156		
		1.0					22.3			72.0		
Combustion section (other)	0	0	0	0	0	0	1	0	1	0	2	0
							142		209		279	
							22.9		1146.7		72.2	
Fuel oil flow dividers	0	0	0	0	0	0	0	0	0	0	0	0
Fuel oil pumps	0	0	1	0	0	0	0	0	0	0	0	0
			148									
			156.0									
Fuel oil filters	0	0	0	0	0	0	0	0	0	0	0	0
Fuel oil systems (other)	0	0	0	0	0	0	0	0	0	0	0	0
Starting system	0	0	0	0	0	0	0	0	0	0	0	0
Lube oil system	0	0	0	0	0	0	0	0	0	0	0	0
Air system	0	0	0	0	0	0	1	0	0	0	0	1
							142					165
							40.9					2.0
Electrical controls (other)	4	3	2	4	4	5	4	5	4	3	3	6
	502	411	296	402	687	871	566	635	835	467	419	989
	13.0	4.3	7.5	14.1	3.6	39.8	8.3	10.6	8.2	1.0	8.2	4.3
Compressor	0	0	0	0	0	0	0	0	0	0	0	0
Inlet guide vanes	1	1	0	0	0	0	0	0	0	0	0	0
	137	125										
	5.4	5.7										
Turbine exhaust section	0	0	0	0	0	0	0	0	0	0	0	0
Turning gear system	0	0	0	0	0	0	0	0	0	0	0	0
Fuel nozzles	0	0	1	1	0	0	0	0	0	2	0	0
			148	201						311		
			21.0	16.0						5.0		
Fuel gas system	0	0	1	0	0	0	0	0	0	0	1	0
			78								140	
			4.0								7.0	
Liners	0	0	0	0	0	0	0	3	0	1	0	0
								381		156		
								5.5		96.0		
Cross fire tubes	0	0	0	0	0	0	0	0	0	0	0	0
Transition ducts	0	0	0	0	0	0	0	0	0	0	0	0
Fire detection system	0	0	0	0	0	0	0	0		0	2	0
											279	
											25.0	
Service factor	0.91	0.83	0.77	0.77	0.90	0.66	0.80	0.90	0.55	0.73	0.82	0.69
Fired hours per start	625	370	286	357	238	333	286	303	116	128	345	81

[a] The first entry in the columns is the number of failure events; the second is the number of failures per million hours of operation (λ); the third entry is mean downtime in hours.

[b] Key: first letter = turbine identifier; number = year; G = gas-fired.

deposits on the top of the tube header sheet.[16] This problem was corrected by using high-pressure water jets to remove the sludge deposits, and by shifting from a phosphate cleaning solution to cleaning solutions with a volatile cleaning agent. By 1975 a new problem reared its head—tube denting. This stemmed from galvanic corrosion of the plain carbon steel of the tube support plates in the crevices between the plates and the alloy tubes. The increase in volume associated with the formation of the iron oxide in these close clearances acted to

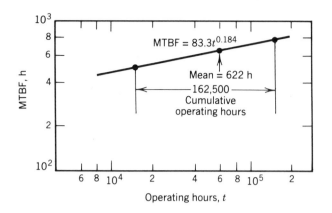

$$MTBF = 83.3t^{0.184}$$

Mean = 622 h

162,500 Cumulative operating hours

Figure 10.4 Gradual increase in the reliability of 13 natural gas-fired gas turbines during their first few years of electric utility service. (Brown and Gardner, Ref. 11.)

pinch the tubes and actually dent them inward so that they could no longer slip axially in the plates and thus accommodate differential expansion. Stress-corrosion cracking followed. The problem was corrected by better control of the feedwater chemistry, better support plate design, and better materials.

By 1981 the bulk of the failures were from stress-corrosion cracking, largely as an indirect consequence of earlier phosphate wastage and tube denting. Table 10.3 shows the location of the tube defects found in 1981, the last year of the survey. Note that 80% of the tube defects were in three regions: the U-bends, just above the tube sheet, and at the tube support plates. These are the regions where bending stresses from differential thermal expansion are greatest. About half of the remaining defects were in the crevice at the tube sheet.

The above data are for the entire group of 110 steam generators containing 1,500,000 tubes. While the numerous failures indicate the importance of the problem, one must add that some of the plants operated with few or no defects in the steam generator, thus indicating that extremely high reliability is achievable once one has the operating experience needed to bring out subtle problems that are easily overlooked or not fully appreciated in an initial design.

FAILURE RATES

To make quantitative estimates of failure rates one must first accumulate, organize, and correlate failure data. Correlation entails the development of mathematical models for the principal types of failure. After the data

TABLE 10.2 Distribution of Equipment Failures and Defects in Soviet VVER-440 PWR Power Plants (a standard approach for collecting data was adopted in 1977)[a]

Equipment Group	No. of Faults/ Unit/Year	Faults as % of Total
Primary Circuit Equipment		
Reactor	5	4.3
Steam generator	4	3.4
Main circulation pumps	2	1.7
Main shutoff valves	1	0.9
Pipelines	1	0.9
Total for group	13	11.3
Turbo-Unit Equipment		
Turbines	1	0.9
Condensers	3	2.6
Separators and steam superheaters	3	2.6
Regenerative heaters	6	5.2
Total for group	13	11.3
Pump equipment of all types	9	7.8
Fittings (excluding main shutoff valves)	10	8.7
Ventilation assemblies	7	6.1
Compressor assemblies	4	3.5
Electrical Equipment		
Turbogenerators	1	0.9
Electrical drives of pumps	3	2.6
Electrical drives of regulating equipment	2	1.8
Switches and circuit breakers	9	7.8
Total for group	15	13.1
Control and Measuring Instruments and Automatic Equipment		
Primary instruments	9	7.8
Secondary instruments	23	20.0
Communication lines	12	10.4
Total for group	44	38.2
Overall total	115	100.0

[a]Ref. 14. (Courtesy *Nuclear Engineering International*.)

are organized, analyzed, and correlated, one has a sound basis for estimating the reliability that one might reasonably hope to achieve in a component or a system. However, the steam generator experience cited above must be kept in mind, for in any new design or appli-

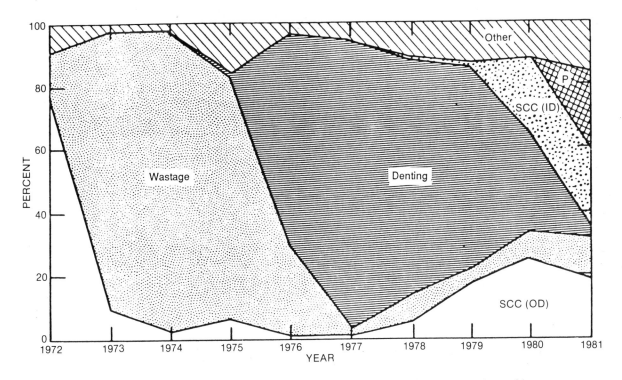

Figure 10.5 Changes with time in the factors responsible for plant outages chargeable to the steam generators in water-cooled nuclear reactor plants. (Tatone and Pathania, Ref. 15.)

TABLE 10.3   Location of 1981 Tube Defects[a]

Location of Defect	No. of Reactors Affected	Tube Defects	
		No.	Percentage
Tube-sheet crevice	8	428	9.1
Above tube-sheet	20	1496	31.9
Tube supports	18	682	14.5
U-bend	14	1551	33.1
Other	7	525	11.2
Undetermined	6	10	0.2

[a]Ref. 15. (Courtesy *Nuclear Safety.*)

cation there may be subtle factors and failure modes that are hard to foresee; extensive testing should be planned to bring out these problems.

Frequency Distributions

The incidence of failures as a function of time varies with the type of failure.[17] Where a component such as a light bulb has a limited life, the failure probability is similar to the death rate in human beings, and may be expressed as

$$P = \frac{1}{\sqrt{2\pi}\,\sigma} \exp\left[-\frac{1}{2}\left(\frac{t-\mu}{\sigma}\right)^2\right] \quad (10.1)$$

where P = probability of death (or failure) in any given year

t = age at death (or time to failure), years

σ = standard deviation, years

μ = average age at death for the population (or mean time between failures), years

This gives the classical Gaussian distribution of Fig. 10.6a, often called the *normal distribution*. The time interval can be years as above, hours, or any other as long as the units are consistent. The standard deviation determines whether the curve of Fig. 10.6a is sharply peaked around the average age at death, as is the case for light bulbs, or more spread out, as it is for humans. Once t, σ, and μ are determined from statistical data, Eq. 10.1 can be applied in analysis work.

If there is no deterioration with age, the failures may

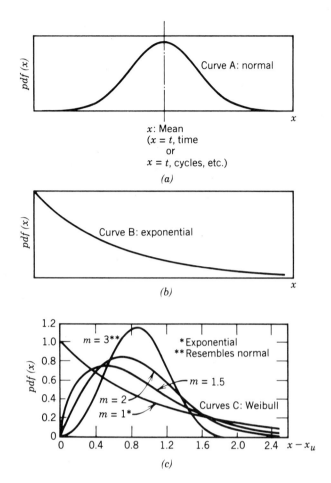

Figure 10.6 Major types of failure frequency distribution. (Rubel, Ref. 17.)

operating time, whereas smaller defects will take longer to progress to the point where they will cause a failure. After an extensive examination of data on the failure rates in different materials,[18] Weibull suggested that the Gaussian expression for the failure probability be modified to give the form

$$P = \frac{m(t - t_u)^{m-1}}{t_0} \exp\left[\frac{-(t - t_u)^m}{t_0}\right] \quad (10.3)$$

where P = probability of failure per unit time
m = the shape parameter
t = time at failure
t_u = the location parameter (commonly zero)
t_0 = the scale parameter

Typical curves for failure probability distributions derived from Eq. 10.3 are shown in Fig. 10.6c. Note that if $m = 1$ and t_u and t_0 are fixed, the resulting failure probability is exponential, while as m is increased the distribution approaches the normal Gaussian distribution.

Aging of Systems

The quite different relations for the probability of a failure as a function of operating time for different types of components have some important implications when considering aging effects in systems. A given system may include a wide variety of parts and components, some of which are subject to purely random failures, whereas others have a definitely limited life. As a consequence, the forced outage rate for a system commonly takes the form shown in Fig. 10.7, with many "infant-mortality"-type failures whose frequency declines rapidly as the system is shaken down. The time required for shakedown may run a day or two for a microcomputer, a month or so for an auto, or one to five years for an electric utility plant. The shakedown is followed by the normal operating portion of the system life, a period in which random failures occur and are repaired. As the system ages, a point may be reached at which components begin to wear out, corrode through, or otherwise deteriorate so that the incidence of forced outages begins to increase with time. At some point economics may dictate that the system be given a major overhaul with extensive replacement of worn-out components, or the entire system may be junked and replaced. Good statistical data from the service records is vitally important in making the associated decisions. In some types of systems there may be a gradual upgrading process through good

occur in a completely random fashion, in which case the exponential distribution of Fig. 10.6b applies. Poisson suggested the mathematical expression for this type of incident, that is,

$$P = \lambda e^{-\lambda t} \quad (10.2)$$

where λ is commonly called the decay constant. Note that this expression is the same as that for the decay of a radioactive isotope, and is also the same as that for the drop in temperature per unit length as a function of the inlet temperature difference between a hot fluid and the constant temperature wall of a fluid-flow passage. (See Eq. 4.5.)

In some cases the character of the defects causing the failures may be such that the normal Gaussian distribution of Fig. 10.6a is skewed. In the case of ceramic turbine blades, for example, the largest defects to escape detection will cause blade failures after relatively little

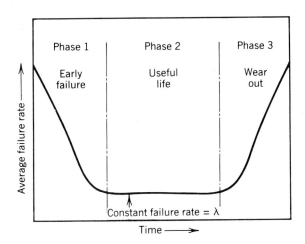

Figure 10.7 Typical curve for the failure rate of mechanical equipment as a function of operating time. (Rubel, Ref. 17.)

maintenance, replacement of failure-prone parts with parts of improved design, improved operating practice, etc., in which case the mean time between forced outages may actually increase with time, as in the gas turbines of Fig. 10.4. Studies indicate that at the time of writing the quality-control efforts of manufacturers have reached the point where about 80% of the systems in use in the United States fall in this category.[19,20] For such systems the lowest-cost approach is to replace components only when there is definite evidence of deterioration rather than replacing parts whether they need it or not in a major overhaul.

Generic Data for the Reliability of Components

Quite a number of studies in recent years have reviewed a wide variety of operating experiences from which generic data on the reliability of components could be collected.[3,21–24] As one would expect, the more comprehensive the study, the greater the spread in the data. In the most comprehensive report, a 1424-page bound volume compiled by the IEEE, three suggested failure rates are given for each type of component, with the "low" and "high" values often differing from the recommended mean by a factor of 10 to 100.[3] Part of this huge disparity stems from differences in the products of different vendors, but the greater part stems from differences in the operating conditions, such as the ambient air temperature and humidity, vibration, the ratio of the normal loads to the design capacity, dirt and dust, transient overloads, etc. A statistical study of this problem for valves has examined the effects on reliability

of the type of valve (gate, globe, ball), type of operator (hand, motor, solenoid, pneumatic), size, etc., and found that each of these parameters led to differences in reliability by factors of as much as 13.[24]

In reviewing this mass of generic data it is striking to find that the upper limit for high-quality components operated under benign conditions is a mean time to failure of 10^6 h irrespective of whether the component is an electric motor or transformer, a pump or valve, a turbine or generator. Of course the mean time to failure of individual elements in a component is much higher; for the exceptionally high-quality welds for the end caps of fuel elements in water reactors it is about 10^9 h. A similarly high value has been demonstrated for the tube-to-header sheet welds in some of the steam generators for water reactors. On this basis, in a 1000-MWe PWR steam generator having 25,000 tubes and 50,000 tube-to-header joints, the mean time to a tube-to-header weld failure could be expected to run 20,000 h, or about three calendar years, if the system is on-stream 80% of the time. Other data for heat exchangers vary widely, but one significant set is available for a particularly difficult application, regenerators in small mobile gas turbines. These not only must be light and compact but they are subject to vibration as well as severe thermal transients when there are rapid changes in power. The AiResearch Manufacturing Co. had trouble with leaks induced by thermal stresses in plate-and-fin units of the type shown in Fig. 1.21; in each there are thousands of brazed joints. The leakage problems in these units were corrected by analyses, changes in the detailed design, and tests. This effort was so successful that, at the time of this writing, over 600,000 h of operation had been obtained in the field with units of this type without a leak.[25] Examples such as this and the PWR steam generators are cited here to show that very high reliability can be obtained in heat exchangers through a good design and development program.

Estimating the Reliability of Systems from Generic Data

In attempting to estimate the reliability of a system from generic data it is essential to emphasize several caveats. First, no one has ever built a system even as simple as a household refrigerator and succeeded in getting a high degree of reliability the first time around; there are simply too many things that can (and do) go wrong. Thus extensive shakedown tests are required to bring out defects in details of the design or fabrication. This is true even for a new but conventional plant; for example, it took TVA three years to get the coal-fired Bull Run plant through its shakedown tests, yet not many years later that plant had the highest availability of any comparable plant in the world. Second, a meaningful estimate aimed at achieving a high reliability in equipment of a new design must assume that the design, fabrication, installation, and operation are all carried out by thoroughly competent people with an adequate budget and a realistic time schedule. Third, even after the first demonstration system has been shaken down and operated successfully for a number of years, peculiar circumstances may lead to unforeseen difficulties in subsequent units. An example of the latter is the tube corrosion problem caused by sludge from phosphate cleaning operations in steam generators for PWR plants, as cited above and shown in Fig. 10.5.

In making an estimate for a specific system, the first step is to list each of the components in the system likely to cause a forced outage. This must include items such as the vital sensing elements in the instrumentation and control system, electrical switch gear, and elements in the supply lines for utilities such as electricity, water, and gas. The failure probability for each item on the list should then be selected from a source of generic data and tabulated. The failure probabilities for the full set of components can then be added to give the probability of a forced outage for the system. Table 10.4 shows a typical set of probabilities as estimated in this way for a small nuclear power plant designed for a remote installation.

At some added cost, the availability of a system can be increased by incorporating a number of components in parallel so that the system can continue to operate after one of the components fails. This provision for redundancy has been the key to the success of completely unattended spacecraft which have so many electronic components that many random failures must be expected. For a ground-based system, the availability also depends on the time required for repair or replacement of a component. Thus the overall availability of a repairable system also depends on the time for repair as well as the mean time between failures of its components and provisions for redundancy. For a simple case in which the system reliability is dominated by one component and n is the number of redundant units of that component, the effects on the system availability of these three factors can be estimated from

$$\text{Availability} = \left(\frac{\text{repair time}}{\text{MTBF} + \text{repair time}} \right)^{n} \quad (10.4)$$

The relative effects of each of the three key factors on the availability of a typical system is illustrated by Table 10.5. This was prepared for a set of diesel generators arranged to operate in parallel to provide a highly reliable power supply for a computer installation that had averaged eight outages per year because of the effects of lightning on the utility power lines. Note that a repair time of 10 h with an MTBF of 1000 h serves to give an availability of 0.9999 with only two diesel generator units, whereas four would be required if the time for repair were 100 h.

TABLE 10.4 Estimated Potential Reliability of a Well-Developed Pebble-Bed Reactor Coupled to a Closed-Cycle Gas Turbine

Time to Service	Component	No. of Critical Units	Failure Rate/10^6 h (Per Unit)	Failure Rate/ 10^6 h (For Category)
100 h	Turbine generator	1	10	10
	Recuperator	1	3	3
	Cooler	1	2	2
	Valves in heater system	7	4	28
10 h	Water pump	1	1	1
	Water pump controls	1	2	2
	Water system valves	4	1	4
	Reactor controls	20	10	200
	Turbine generator controls	20	10	200
	Helium control valves	6	2	12
	Total system			462

TABLE 10.5   Effects of MTBF, Time for Repair, and Redundancy on Availability

Time for repair		100 h			10 h	
MTBF, h	500	1000	2000	500	1000	2000
Availability with						
1 unit	0.8333	0.9091	0.9524	0.9804	0.9901	0.9950
2 units	0.9721	0.9917	0.9977	0.9996	0.9999	
3 units	0.9953	0.9992	0.9999			
4 units	0.9992	0.9999				

MAINTENANCE

Both the direct charges for man hours and materials for maintenance and for system downtime are expensive. The procedures yielding the lowest overall costs depend on the application and are not necessarily obvious. Periodic shutdowns for a major overhaul with widespread replacement of possibly faulty components are widely practiced and appear to be the mark of superlative management, but this practice may not give the lowest costs.[19,20] The first step in devising an appropriate maintenance program is to obtain good records of the principal problems requiring maintenance. Once these are identified, steps can usually be devised to detect impending trouble before it becomes serious. It can then be corrected either during part load operation without shutting down, or in a scheduled shutdown when the inconvenience and costs will be minimal.

Adopting operating procedures that minimize stress on the system, deposits, and degradation by corrosion and deposits can cut costs and improve availability. Draining and flushing a system, and possibly "mothballing" it for extended shutdowns by filling it with a preservative fluid or blanketing it with an inert gas may make a world of difference in component life. During operation, it may be important to exercise tight control over conditions such as ambient air temperatures or water chemistry. The vital effects of the latter in reducing tube failures in PWR steam generators were cited above, and they are important in many other applications, particularly where the formation of $CaCO_3$ scale may be problem.[26] Table H7.7 indicates the high degree of water purity that can be achieved with ion-exchange resins.[27]

Cleaning

Fouling of heat transfer surfaces is a widespread source of difficulty that can be readily detected by tracking the heat transfer performance of a unit. Often fouling can be reduced by appropriate design and/or operating procedures, for example, filters are more easily cleaned than heat exchangers. Biofouling by cooling water can be reduced through the use of screens to keep out marine life such as jellyfish (which have sometimes caused forced outages[28]), or by chlorination (with strict controls to meet EPA standards) to reduce the growth of algae. Mechanical cleaning procedures include scrubbing with long slender brushes, or special tooling using little cutters similar to that of Fig. 10.8 may be required for hard deposits. For soft deposits such as algae, pumping plastic balls of the right size through the tubes may serve. Deposits outside the tubes can often be removed by using high-pressure jets of air, steam, or water from lances maneuvered through openings provided for that purpose. Soluble deposits may be removed by circulating solvents,

Figure 10.8   Tool for mechanically cleaning heat exchanger tubes. (Courtesy Thomas G. Wilson, Inc.)

acids, or other cleaning fluids through the system during shutdowns. The cleaning procedure should be chosen with care, for severe chloride corrosion has been experienced after using cleaning solutions containing chlorides in stainless steel systems, and the ill effects of phosphate cleaning solutions were cited above as a major source of trouble in PWR steam generators.

Inspection

Visual inspection is the oldest, most informative, and generally applicable inspection procedure, but it has obvious limitations. It has been supplemented with dye and magnetic particle inspection for small surface cracks together with X-ray photographs for subsurface cracks and inclusions since the 1930s. Other nondestructive inspection and testing techniques have been developing rapidly with the introduction of a variety of clever new electronic sensors and equipment.[29] The amount of work in this field is so extensive that the American Society of Nondestructive Testing has been organized. While much of the nondestructive examination and testing (NDE) is directed toward quality control in manufacturing, some of it is concerned with field inspection during both operation and shutdowns. Eddy-current electromagnetic probes are well suited to the inspection of thin metal walls and have proved very valuable in detecting wall thinning and cracks in the tubes of condensers and PWR steam generators.[16,29] The inspection operation lends itself to automation so that it can be carried out rapidly with remote handling equipment to inspect large numbers of tubes expeditiously. Ultrasonic techniques provide a good means for detecting cracks normal to the direction of sound beams in either thin or thick walls and lend themselves to automation and rapid scanning for defects in tubes and pressure vessels. Fiber optics equipment permits visual examination of inaccessible defects detected by eddy-current or ultrasonic probes. In short, methods are available for detecting deterioration and incipient failure, so that repairs can be made near the end of the useful life of a part but before a failure that might cause an expensive forced outage.

Repairs

Many different measures can be taken to extend the life of heat exchangers.[30] Tube plugging at the header sheet is a simple, fast procedure that is applicable where only a small percentage of the tubes is affected, as in the above-cited case of the PWR steam generator inner-row tubes having small-radius U-bends. Where local tube-thinning, crevice corrosion, or low-cycle fatigue cracking occurs near the header sheet, the tube life can be greatly extended by inserting a short sleeve and expanding, brazing, or welding the inner and outer ends of the sleeve to the tube.[16,28,30] For heat exchangers similar to that of Fig. 1.9, tubes may be removed and replaced *in situ*, possibly using a more corrosion-resistant material.

The entire heat exchanger can be replaced, a procedure that is a formidable undertaking for large units such as a 375 ton PWR steam generator similar to that of Fig. 15.9. The first such replacement operation was carried out at the Surry Unit 2 of the Virginia Electric and Power Co. in 1979; it required about nine months, including several months for the final check-out.[31] The upper portion containing the vapor separators did not require replacement, so the shell was cut through in the conical region just above the tube bundle, and the upper section, 15 ft in diameter by 25 ft high and weighing 125 tons, was lifted off and set to one side. The lower portion, 14 ft in diameter, 41 ft tall, and weighing 250 tons, was then removed. These operations were rendered especially difficult because a substantial amount of radioactive structural material had been removed by mass transfer corrosion from the hot reactor core and deposited on the lower temperature surfaces of the steam generator. Much of this radioactivity could not be removed by cleaning processes, so personnel were strictly limited in the time they could work close to the vessel. Remarkably, the planning for the replacement work was so well thought out and the work itself so well executed that the total dose to personnel was within 3% of the original estimate of 2070 man-rem.

Design for Maintenance

The prime reason for including this chapter in a text on design is to call attention to the importance of maintenance problems and to indicate their character. Hopefully, this will help the designer to minimize their occurrence, provide good access for cleaning and inspection, and make the design such that repairs can be carried out as readily as possible. Probably the most important point to remember in design work is that, if something might go wrong, it probably will! Valuable insights into problems in the field and corrective measures can be obtained by participating in user's groups that meet periodically to pool their experience.

REFERENCES

1. S. Walters, "The Beginnings," *Mechanical Engineering*, vol. 106(4), April 1984, p. 38.

2. C. G. Messenger, "A Review of the Reliability Design Handbook," *1982 Proceedings, Annual Reliability and Maintainability Symposium*, ASME, January 26–28, 1982, p. 297.

3. "IEEE Guide to the Collection and Presentation of Electrical, Electronic, Sensing Component, and Mechanical Equipment Reliability Data for Nuclear Power Generating Stations," IEEE Std 500-1984.

4. "Reliability Design Handbook," RDH-376, Reliability Analysis Center, Griffiss Air Force Base, N.Y., 1976

5. "Reliability/Design Thermal Applications," MIL-HDBK-251, 1978.

6. D. Okrent, "Risk-Benefit Analysis for Large Technological Systems," *Nuclear Safety*, vol. 20(2), March–April 1979, p. 148.

7. G. L. Sheldon, "Similarities and Differences in the Erosion Behavior of Materials," *Journal of Basic Engineering, Trans. ASME*, vol. 92(2), 1970, p. 619.

8. "Assessment of Condenser Leakage Problems," Report no. EPRI-NP-1467, Electric Power Research Institute, August 1980.

9. G. Pini and J. Weber, "Materials for Pumping Sea Water and Media with High Chloride Content," *Sulzer Technical Review*, February 1979, p. 69.

10. W. J. Marner, "Gas-Side Fouling," *Mechanical Engineering*, vol 108(3), March 1986.

11. H. W. Brown and N. J. Gardner, "Reliability and Availability Assessments of Selected Domestic Combined-Cycle Power Generating Plants," Report no. EPRI AP-2536, Electric Power Research Institute, August 1982.

12. "High Reliability Gas Turbine Combined-Cycle Development Program: Phase II," Westinghouse Electric Corp., Final Report, Report no. EPRI-AP-2321, Electric Power Research Institute, March 1982.

13. "High Reliability Gas Turbine Combined-Cycle Development Program: Phase II," United Technologies Corp., Pratt and Whitney Aircraft, Final Report, Report no. EPRI-AP-2226, Electric Power Research Institute, January 1982.

14. "Dealing with Equipment Failure in Soviet PWRs", *Nuclear Engineering International*, vol. 27(332), September 1982, p. 49.

15. O. S. Tatone and R. S. Pathania, "Steam Generator Tube Performance: Experience with Water-Cooled Nuclear Power Reactors During 1981, *Nuclear Safety*, vol. 25(3), May–June 1984, p. 373.

16. W. D. Fletcher and D. D. Malinowski, "Operating Experience with Westinghouse Steam Generators," Paper prepared at Nuclear Energy Systems, Westinghouse Electric Corp. and presented at the International Conference on Materials for Nuclear Steam Generators, September 9–13, 1975.

17. P. Rubel, "Reliability Engineering Methods in Reactor Safety Technology," Report no. ORNL CF 72-3-39, Oak Ridge National Laboratory, June 1972.

18. W. Weibull, "Statistical Theory of the Strength of Materials," *Ingenious Vetenskaps Akademieu Handlinger*, no. 151, 1939 (in English).

19. T. D. Matteson, "Overhauling Our Ideas about Maintenance," *Mechanical Engineering*, vol. 108(5), May 1986, p. 86.

20. K. L. Wong, "Unified Field (Failure) Theory—Demise of the Bathtub Curve," *1981 Proceedings Annual Reliability and Maintenance Symposium*, ASME, ISSN 0149–144X, p. 402.

21. W. K. Kahl and R. J. Borkowski, "The In-Plant Reliability Data Base for Nuclear Plant Components: Diesel Generators, Batteries, Chargers, and Inverters," Report no. ORNL-TM-9216, NUREG/CR-3831, Oak Ridge National Laboratory, January 1985.

22. R. J. Borkowski et al., "The In-Plant Reliability Data Base for Nuclear Plant Components: Interim Report—The Valve Component," Report no. ORNL-TM-8647, NUREG/CR-3154, Oak Ridge National Laboratory, December 1983.

23. J. P. Drago et al., "The In-Plant Reliability Data Base for Nuclear Plant Components: the Pump Component," Report no. ORNL-TM-8465, NUREG/CR-2886, Oak Ridge National Laboratory, December 1982.

24. R. J. Beckman and H. F. Martz, "A Statistical Analysis of Nuclear Power Plant Valve Failure-Rate Variability—Some Preliminary Results, Report no. LA-10396-MSNUREG/CR-4217, Los Alamos National Laboratory, May 1965.

25. A. E. Hause, Garrett Corp., Personal communication to A. P. Fraas, June 16, 1986.

26. A. P. Watkinson and O. Martinez, "Scaling of Heat Exchanger Tubes by Calcium Carbonate," *Journal of Heat Transfer, Trans. ASME*, vol. 97(2), 1975, p. 504.

27. Preventing Tube Degradation in Japan," *Nuclear Engineering International*, vol. 30(365), February 1985, p. 43.

28. "How Aquatic Life Can Shut Down a Plant (and How to Prevent It)," *Nuclear Engineering International*, vol. 30(375), 1985, p. 20.

29. E. R. Reinhart, "Fossil Plant Availability with NDE," *Mechanical Engineering*, vol. 106(2), February 1984.

30. S. Yokell, "Extending the Life of Tubular Heat Exchangers," *Chemical Engineering*, July 20, 1987.

31. A. L. Parrish III, "Replacing the Steam Generators at Surry Unit 2," *Nuclear Engineering International*, vol. 25(299), May 1980, p. 21.

11

General Design Considerations and Approaches

The selection of most heat exchangers follows a routine pattern established by many years of heat exchanger design and operating experience, but changes in operating conditions or completely new types of application may justify or require a fresh approach. In some seemingly conventional applications, the amount of money involved or the unusual requirements may call for an extensive engineering effort to work out and evaluate new designs. This book is especially concerned with such cases, and with the selection of criteria and their use in the development of new designs.

DELINEATION OF REQUIREMENTS

The first step in preparing the design for a new type of heat exchanger is a clear statement of the requirements. Where possible, the relative importance of each of the various factors and requirements should be established, and the areas in which a wide latitude of choice exists should be indicated. Attention should be directed toward those areas in which special care must be taken to effect good compromises, and the incentives for improving particular performance characteristics should be established as well as possible.

Heat Transfer Performance

The specification of the inlet and outlet temperatures for each of the two fluid streams is the first step in establishing the requirements. Where a range of temperatures is under consideration, the incentives to reach the more desirable end of each range should be indicated. Once the inlet and outlet temperatures are defined, the heat exchanger effectiveness can be estimated. This is important since it will give a good indication of the flow-passage length-diameter ratios required and the feasibility of using parallel or crossflow units as opposed to counterflow units.

Fluid-flow rates must then be established for each of the two fluid streams. Since liquid velocities are ordinarily kept between 2 and 20 ft/s, and gas velocities between 10 and 100 ft/s (with each usually near the middle of the range given), the flow rates give a good indication of the flow-passage cross-sectional area required for each of the two fluid streams. It is sometimes necessary to restrict the fluid velocity to avoid difficulties with such problems as erosion, tube vibration, flow stability (as in boiler tubes), or noise (as in air-conditioning units).

Sludge or other deposits form on the heat transfer

surfaces in some types of application. The extent and thickness of these deposits should be estimated together with their effects on the heat transfer coefficient and fluid friction factor. Such allowances may significantly affect the size of the heat exchanger required because, if heavy deposits are anticipated, they may require that rather low unit heat fluxes be employed to avoid excessive temperature drops. The tube diameter is also influenced since it is ordinarily impracticable to use small tubes if heavy deposits are anticipated. This last consideration will also determine in substantial measure the geometry of the heat exchanger. If periodic mechanical cleaning of the tubes will be required, provisions for such cleaning must be made. If chemical solvents or special cleaning compounds are to be employed to remove deposits, the need for these may affect the choice of the materials of construction. For example, it may be necessary to use stainless steel to withstand an acid cleaning solution even though plain carbon steel would serve for the process fluids.

Size Restrictions

It is often important to limit the length, height, width, volume, or weight of a heat exchanger because of requirements peculiar to a particular application. These limitations may apply not only to the heat exchanger itself but also to provisions for maintenance. For example, it may be essential that the heat exchanger casing be installed in such a way that individual tubes or the entire tube bundle be removable simply by opening a flange at one end of the heat exchanger. The space available is often such that the length of the tube bundle that can be handled is limited. The fluid inventory is likely to be an important consideration for expensive, toxic, or combustible fluids. It may also be necessary to impose special requirements for drainage, vertical removal of the tubes or the tube bundle, or the like.

Cost Factors

Cost factors are often dominant in the choice of a heat exchanger. These include not only the initial capital cost of the heat exchanger but also the costs of plant operation and maintenance. These effects may be complex as, for example, in a chemical-processing plant where the value of one or more products may be affected. The situation in a steam power plant is somewhat simpler, because the capital cost of the equipment can be related directly to the efficiency of the power plant and hence to the costs of producing electrical power.[1-3]

An interesting example showing one way of coping with the problems involved is given by a study of steam generators for a gas-cooled reactor power plant in which materials considerations limited the gas outlet temperature from the reactor. Figure 11.1 shows the effects of the steam-outlet temperature for which the steam generator is designed on the parameters associated with the heat exchanger size and hence the relative cost. Note that, as the heat exchanger size is increased to increase the steam-outlet temperature, the cost of the steam generator per kw of plant output drops a little at first because of increasing plant efficiency, and then levels out and begins to increase rapidly as the steam-outlet temperature approaches the temperature of the hot gas coming from the reactor.

To achieve minimum overall costs it is necessary to balance operating costs against capital charges. The problem can be illustrated by considering a particular case. Once a basic heat transfer matrix geometry is selected on the basis of fabrication, performance, and maintenance considerations, the amount of heat that will be transferred from one fluid to the other for a given set of fluid temperatures in a given size of heat exchanger will depend largely on the flow rates of the two fluid streams. If the flow rates are doubled, the capital charges will be cut almost in half, but the pumping power requirements will be increased by approximately a factor of eight. The principal parameters are shown graphically for a typical gas-turbine regenerator in Fig. 11.2. Similar curves can be prepared for any application. Note that not only the costs of power for pumping ought to be included but also the capital charges dependent on the pumping equipment, since these charges may be a substantial fraction of the cost of the power required. Although individual cases differ widely, it has been found that, for a pumping power requirement between 0.5 and 1.0% of the heat transmitted through the heat exchanger, the overall cost is usually fairly close to the minimum obtainable. Note that this is the case not only for the regenerator of Fig. 11.2, but, as shown by Fig. 11.3, it is also true of the steam generator of Fig. 11.1.

Stress Considerations

Stress considerations are usually relatively unimportant in the design of heat exchangers unless system pressures above 200 psi or metal temperatures above 300°F are employed. For pressures over 1000 psi or temperatures above 1000°F, stress considerations probably will be dominant. In these regions the tube headers are particularly likely to be the controlling factor in the selection of the heat exchanger geometry, and stress considera-

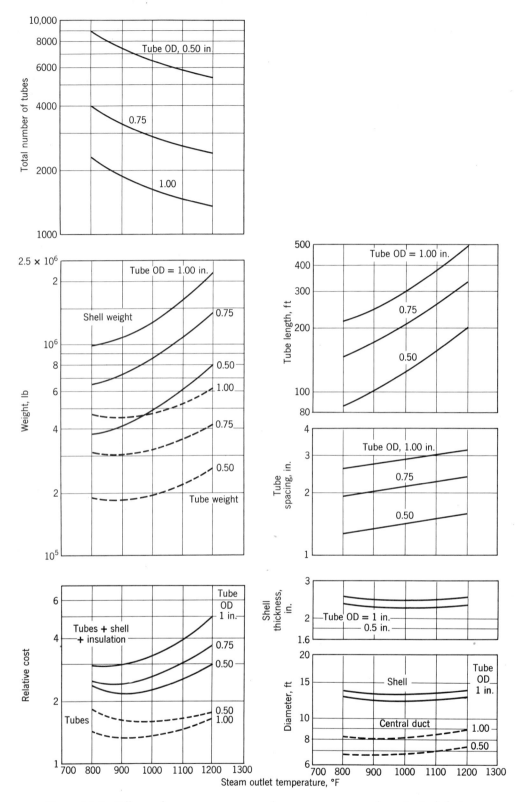

Figure 11.1 Effects of steam generator outlet temperature on the principal design parameters and the capital charges for the steam generator for one type of gas-cooled reactor power plant. (Courtesy Oak Ridge National Laboratory.)

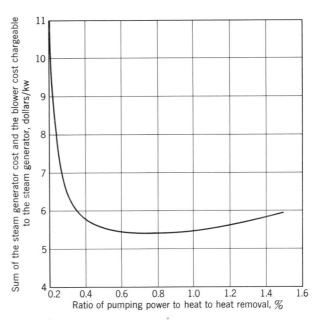

Figure 11.3 Effects of pumping power-to-heat removal ratio on the sum of the steam generator cost and that portion of the blower cost chargeable to the steam generator.

Net electrical output: 500 MW
Plant thermal efficiency: 0.40
Helium pressure: 500 psia
Steam pressure: 2500 psia
Helium-inlet temperature: 1250°F
Helium-outlet temperature: 650°F
Feedwater-inlet temperature: 520°F
Steam-outlet temperature: 1050°F
Tube outside diameter: 0.50 in.
Mass flow rate of steam: 222.9 lb/ft².

(Courtesy Oak Ridge National Laboratory.)

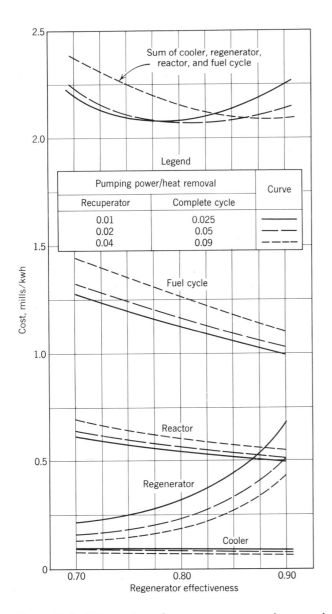

Figure 11.2 Effects of gas-flow rate on pressure losses and costs for a regenerator in a gas-turbine power plant. (Courtesy Oak Ridge National Laboratory.)

tions are a major factor in determining the choice of the material to be employed. (The effects of temperature on the stress for rupture in 1000 h are shown for typical materials in Fig. H8.1.)

Differential thermal expansion is likely to pose important limitations on the design of the heat exchanger if temperature differences of 100°F or more are to be expected between the tubes and the shell; hence it is important to note this item in outlining the requirements to be met. (The mean coefficient of thermal expansion for typical materials is given in Table H2.1.)

Material Requirements and Fabrication Techniques

Corrosion problems are almost always important in the selection of a material for a heat exchanger application. If rather corrosive fluids are involved, it may be necessary to balance the cost of more corrosion-resistant materials against the higher maintenance and replacement costs of less expensive materials. The subject is far too complex to treat in a general fashion in the limited space available in this book. The special problems of widely used types of heat exchanger, such as steam boilers, are treated in later chapters. Tables H7.4 and 7.5 have been included to aid in selecting suitable materials for most fluids likely to be of interest.

Fabrication problems may also be an important consideration in the choice of a material. If the unit is to be soldered, brazed, or welded, the materials chosen must be well suited to the fabrication operation, and a price premium to minimize fabrication difficulties is often justified. If tubes are to be rolled into header sheets, the tube material should be carefully selected and specified to ensure good ductility and suitable work-hardening characteristics.

Leaktightness

The leaktightness required for any given application should be specified both for leakage from one fluid stream to the other and for leakage from either fluid stream to the surroundings. Methods for measurement of leaktightness are outlined in the chapter on heat exchanger testing (Chapt. 20). Heat exchanger specifications ordinarily should include a statement of the method to be employed in checking the leaktightness.

The degree of leaktightness required varies widely with the application, but it is helpful to review some of the typical cases included in Table 11.1. The first case listed is a conventional coal-fired steam plant located on a river having a relatively low level of dissolved solids, so that there is no strong incentive to prevent in-leakage through the condenser or to conserve water. The second

and third are nuclear plants in which tighter control on the water chemistry is required to minimize both corrosion and the transport of radioactivity out of the reactor core. Further, inasmuch as the water leaking out might contain some radioactivity, most of the leakage is caught and returned to the system. The fourth system is a helium-cooled reactor for which there is a strong incentive to minimize the leakage, in part because of the cost of the helium and in part because of the possibility that it might contain a serious amount of radioactivity if there were defective fuel elements. The fifth example is a large reactor containment shell designed to retain whatever radioactive material might be released from the reactor in the event of any conceivable accident. The sixth example is a heavy water reactor for which the controlling consideration in minimizing leakage is the very high cost of the heavy water together with the need to avoid contaminating it with ordinary water. The last two examples are systems making use of a molten salt and a liquid metal; in both cases in-leakage of air would lead to serious internal corrosion, and leakage of fluid out of the system could lead to serious external corrosion. Furthermore, the molten salt in this case is highly radioactive, and the liquid metal presents a fire hazard. These examples are cited to show both common leakage rates that are relatively high where there is no special premium on reducing them, and rates lower by a factor of 10,000 where there is a high premium to keep leakage exceedingly low.

Turning from leakage from a complete system to leakage in a common type of heat exchanger, it is interesting to consider steam condensers for utility power plants, a matter that has been receiving close attention in recent years. In some cases a tube leak of 1 gpm (83 g/s, or 8 lb/min) has been tolerated in condensers cooled with fresh water from rivers or lakes. This amounts to about 5000 kg/day, or roughly 1% of the total inventory of water in the steam system of a 1000-MWe steam plant. A tube leak of 0.1 gpm has been the maximum tolerated in condensers cooled with seawater because of the much greater corrosion associated with such leaks. The strong incentive to minimize corrosion in PWR steam generators has led to leak detection techniques that serve to locate tubes leaking only 0.01 gpm (0.8 g/s, or 0.08 lb/min). To achieve and maintain that high degree of leaktightness requires that the tubes be welded into the header sheets rather than rolled, and makes it economically worthwhile to employ titanium tubes with header sheets of titanium-clad steel. (See Ref. 16 of Chapt. 10.)

The degree of leaktightness achievable depends not

TABLE 11.1   Observed Working Fluid Leakage Rate from Typical Systems

Plant	Leakage Rate, %/day
Coal-fired steam plant (3500 psi, 850 MWe, TVA—Bull Run)	0.5
Pressurized water reactor (160 MWe, Yankee)	0.2[a]
Boiling water reactor (52 MWe, Humboldt Bay)	0.2[a]
Helium-cooled reactor (50 MWe, Peach Bottom)	0.2
Reactor containment shell (normal test requirement)	0.1
Boiling heavy water reactor (200 MWe, Marviken)	0.00055
Molten-salt reactor (8 MWe, MSRE)	<0.000024
Boiling potassium power plant (0.4 MWt, ORNL-LPS)	<0.000024

[a]Most of this leakage is caught and recycled.

only on the fabrication process but also on the techniques used for leak detection. Typical methods are described at the end of Chapter 20.

Safety and Environmental Requirements

The growing body of governmental regulations and legal liability cases has introduced an exceptionally difficult set of demands on equipment designers. The health and safety of employees in the manufacturer's plant and in plants in which the product may be installed as well as the general public have become dominant considerations. In effect, companies are now supposed to foresee any possible adverse condition that might arise not only from a design fault, but also from any conceivable mistake or mishandling by a system operator or someone engaged in maintenance. In fact, in some cases the manufacturer may be held liable even in the case of deliberate sabotage. A review of litigation stemming from product failures shows that accidents are caused less often by mechanical failure than by the designer's failure to consider how the equipment might be used.[4]

The environmental effects of leakage to the atmosphere or hydrosphere of process fluids, blowdown, and wastes from cleaning introduce complex problems. An enormous number of articles, papers, and reports have been published in this field, but little of it provides a good perspective on these enormously formidable problems. Even a minimal treatment of the subject is beyond the scope of this text, but Reference 5 gives a good summary of the environmental effects of heat rejection and process wastes, while Chapters 9 and 22 of Reference 6 give a comprehensive perspective of environmental problems.

Governmental regulations generally impose limits that are as tight as it is technically possible to achieve, and juries are inclined to impose heavy penalties even if there are substantial uncertainties in the causes or magnitudes of adverse effects. Thus the designer must try to envision every unpleasant contingency, an exercise that requires much effort and a fertile imagination stimulated by a review of product liability cases and OSHA regulations in the areas involved. From the environmental standpoint, items that must be given particular attention include the handling and disposal of wastes from blowdown, sludge, and cleaning fluids.

Servicing, Repair, and Maintenance

Maintenance requirements differ widely from one application to another and often impose important limita-

tions on the design. Some of these limitations are discussed in the section on shell-and-tube heat exchangers in Chapter 1, especially the influence of maintenance requirements on the design of the shell and the header-sheet installation, while others are treated in the section on maintenance in Chapter 10. Types of fouling and methods for cleaning are treated in detail in References 7–9.

System Operating and Control Requirements

The response characteristics of a heat exchanger to changes in load often have important effects on the performance of a plant. The rate at which a plant may be started up or shut down, or the output level changed, may depend in large measure on the characteristics of the heat transfer equipment.[10,11] All too often, in new types of plants, problems of this character are not recognized until after the plant is built and in operation. Wherever possible it is desirable to investigate the response characteristics desired, not only at the design point but throughout the load range for which good control is required. This is particularly true where there may be difficulty in obtaining stable operation of a system. When this is the case, the performance characteristics of the major components and of the instrumentation and automatic control equipment must be considered.

Where system stability and control may be a problem, analysis of the dynamic characteristics of the system should be carried out so that the characteristics of the heat exchanger may be specified. Such an analysis—usually most conveniently made with computer—may lead to major changes in the choice of the operating conditions for the plant as a whole and the specification of unusual characteristics for the heat exchanger. For example, to obtain sufficiently rapid rates of response to changes in temperature conditions, it may be necessary to design the heat exchanger to give fairly high fluid velocities at low outputs and to accept a higher pumping power penalty at full-power conditions than would appear proper from a simple cost study that ignored control problems.

DESIGN APPROACHES TO THE SELECTION OF A HEAT TRANSFER MATRIX GEOMETRY

Once the design requirements are established by preparing a rough design specification as just outlined, the first step in the selection of a heat exchanger geometry is to examine commercially available units. If none is available in the size or operating temperature range desired, it is often possible to prepare charts as described in Chapter 4 to indicate the performance of special units scaled up or down from those commercially available.

Perturbations on Existing Designs

When the procedure just described does not yield a unit that meets the desired specifications, or when cost and environmental considerations justify an attempt to design an improved type of heat exchanger, it may be in order to explore the possibilities of solving the problem with an unconventional or unusual design. Of necessity the approach must be intuitive, but the point of departure will usually be some type of heat exchanger that has proved to be satisfactory for other applications and shows special promise of meeting the desired specifications.

The graphic outline of Fig. 11.4 provides an excellent, comprehensive perspective on the relationship of the key elements in the methodology of heat exchanger design. Developed by A. L. London,[1] this shows both the sequence of the various steps and the feedback paths for iterations. The two steps at the top on the left are particularly important in the design of compact heat exchangers for which enhanced surfaces are likely to be attractive. These include plate-and-frame heat exchangers, automotive and aircraft radiators, and elements of air-conditioning systems. The great diversity in the surfaces for which good data are available in these areas is likely to lead to many iterations, particularly in the selection of the matrix for a high-production item, where a small advantage can add up to a big difference in costs. In fact, as pointed out in Reference 1, even for a single large unit, a small advantage in terms of reduced pumping power can in a few years justify even a large difference in capital cost.

At the top on the right, the lower box emphasizes that, if there are several candidate fluids, one of the vital steps is the evaluation of the effects of differences in their physical properties. Or, in searching for a heat exchanger that is in production, one of the essential steps is to compare the physical properties of the fluids employed in previous designs with the properties of the fluids to be used. It is usually more important that the density, viscosity, and thermal conductivity of the fluids be similar than that the heat exchanger be for a similar application. The physical properties of the two fluids in large measure establish the desirability of using finned surfaces, large- or small-diameter tubes, and the like. After thus

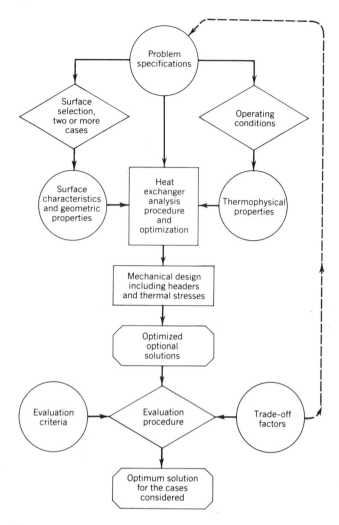

Figure 11.4 Methodology of heat exchanger design and optimization. (London, Ref. 1.)

narrowing the field of possible basic heat transfer matrix geometries, a tentative selection of a matrix geometry may be made on the basis of fabrication and maintenance considerations, particularly the extent to which heat transfer surface fouling is to be expected and the requirements for its removal.

Cut-and-Try Design

Once a basic heat transfer matrix geometry has been selected, it is desirable to investigate the effects of changing such parameters as the tube diameter and spacing. This can be done in several ways. It is possible to carry out a cut-and-try procedure by varying the major parameters in the region of interest, in each instance arriving by a series of successive approximations at a unit that meets the specifications. This procedure is clumsy and is ordinarily not very effective except in the hands of an engineer experienced in the design of somewhat similar equipment.

Parametric Studies

Because of the difficulty in proportioning the various dimensions of a heat exchanger to give the desired heat transfer performance together with the desired fluid system pressure drops, it is sometimes convenient to carry out a parametric study by varying the factors of greatest interest and evaluating the balance to meet the specifications. It is usually best to perform such calculations in a fashion that minimizes the number of calculational operations, for example, by using the techniques presented in Chapter 4 to prepare charts. The characteristics of the units meeting the desired specifications can then be plotted as functions of such factors as the tube diameter or tube length to yield curves showing the effects of greatest interest.

Optimization

The large capital investment in some types of heat exchangers often justifies a systematic selection of the key factors affecting the heat exchanger performance and an attempt to obtain an explicit solution for the design requirements at hand. It is usually possible to reduce most of the many factors just cited to a single common figure of merit such as cost or weight. A computer program can be set up to define the heat exchanger proportions for which the cost, weight, or other parameter have the optimum value. It is usually not possible to obtain a heat exchanger that is at once both the least expensive *and* the lightest, for example; but the calculations can be set up to optimize first for one and then the other, and the results can then be compared.

In setting up a computer program for optimizing a heat exchanger, great care must be exercised to hold just enough parameters constant to give a solution without fixing so many conditions that a solution is impossible.[11] Any approach is heavily dependent on the particular requirements for the application at hand, and hence no general procedure can be suggested. Some specific examples of such a treatment are given in Chapter 15 on steam generators, where techniques are presented for obtaining explicit solutions to the tube length required to meet a given set of conditions.

While optimization studies can be carried out by hand, it is usually best to do them with a computer. Many programs for carrying out calculations of this sort are available, but these should be used with caution, since the optimum set of proportions for one application is seldom the optimum for another. It is usually easier to use a PC spreadsheet program such as Lotus 1-2-3 or Supercalc and tailor the calculations to the case at hand.

EVALUATION OF PROPOSALS

An important set of problems are posed when making an engineering evaluation of several proposals involving widely different types of heat exchanger. Often such proposals may have been prepared on the basis of inadequate specifications. The first step in evaluating proposals of this type is an estimate of the heat transfer performance to be expected. It is unfortunate but true that, under the pressure of competitive bidding conditions, proposals are often submitted offering equipment that is not adequate to meet the specifications because the equipment designed was deliberately undersized in an effort to cut costs. Thus the lower initial costs of such units must be compared with possible increases in operating costs or losses in revenue that would be entailed by reductions in the heat exchanger capacity relative to the specifications. The proposals must also be evaluated for relative ease in servicing, probable relative leaktightness characteristics, corrosion problems, service life, ease of drainage, and stability and control.

Although not immediately obvious, flow distribution may pose service problems in some types of heat exchangers and lead to serious losses in heat exchanger performance because of special problems such as hot spots, tube vibration, or flow instabilities. These problems are very difficult to evaluate analytically and ordinarily require testing. The probable extent of such problems should be considered in appraising competing proposals.

COST ESTIMATION

It is often important to make a rough estimate of the cost of a heat exchanger when preparing a preliminary design for a plant. This section aids in formulating such preliminary rough estimates. The problems of making reliable estimates are complicated by the fact that not only do companies generally consider their basic cost data proprietary—an understandable point of view—but cost estimators as individuals usually guard their "secrets" jealously. Some data have been published, of which the most useful have appeared in a series of articles in *Chemical Engineering*. Information from these articles, through 1984, has been summarized and is available in book form.[12] These data, together with information from the author's files, have served as the basis for this section on costs. It must be emphasized that costs are very sensitive to special requirements; hence the data presented here should be used only for very rough approximations. Seemingly small factors such as rigorous quality control can easily double the cost of a unit.

The values presented in Table 11.2 are for standard units of moderate size in moderate volume production. It has been found that for virtually all types of material and equipment, the unit cost is roughly inversely proportional to the square root of the annual production rate; that is, the cost per pound of equipment falls off by a factor of 10 for every factor of 100 increase in the annual production rate. This relation holds whether the increase is in the size of each item or in the average number of items produced per year. Even though the annual production rate is beyond the designer's control, he can sometimes choose a volume production unit in preference to a special design.

Methods for Estimating Costs

Costs may be estimated in many different ways. This is fortunate, because costs are sensitive to many factors and subject to numerous uncertainties; hence it is usually desirable to obtain estimates via two or more independent methods. If a standard production item can be employed, one or more manufacturers can be contacted for list prices. If there are special requirements, it may be possible to obtain rough price estimates from vendors, but they are often reluctant to give such numbers for fear that they may be misconstrued; they prefer to make firm bids on the basis of firm specifications. If preliminary cost estimates are to be made to compare a variety of approaches to a system design, it may be easier to obtain

data from a handbook on equipment costs.[12,13] Table 11.2 shows a set of data derived in this way. Note that not only do the unit costs vary with the type of heat exchanger, but they are given in terms of different measures of capacity. For general-purpose heat exchangers, the usual measure is the heat transfer surface area, but for boilers it is more convenient to employ the weight of steam produced per hour, while for air-conditioning installations tons of refrigeration is the most convenient measure of capacity.

The usual weight per unit of capacity is also given in Table 11.2, partly because this may be an important consideration in locating a unit in a structure, and partly because it can also serve as an indication of the cost. Many cost estimators have found that they can apply a unit cost in dollars per pound, or per kilogram, with good results. Standard high-production mechanical equipment such as pickup trucks and refrigerators have similar costs per unit weight; at the time of writing the unit cost is about $3/lb. The cost of the material is a major factor; if the equipment is made of stainless steel, the price would be about four times as great as for carbon steel. This factor of four is applicable if the production rate were substantial, but would be higher if it were a special order. Another example is electronic equipment; a TV set costs about $10/lb, while a personal computer costs about $30/lb. Bulk commodities cost much less; sheet steel roofing for a house costs about $0.50/ft^2, or $0.50/lb, whereas sheet aluminum roofing costs about $1.00/ft^2, or $3.30/lb. The spread in these prices per unit weight for different commodities is really not great when one considers the vast differences in the complexity of the manufacturing processes.

Requirements are sufficiently well defined for electric utility plants so that costs for the equipment such as heat exchangers can be estimated from the plant design power output. There is a wealth of information on various

TABLE 11.2 Rough Estimating Values for Unit Costs in 1987 Dollars and Weights of Typical Heat Exchangers (for Plain Carbon Steel Units and Capacities of About 10^7 Btu/h)[a]

Type of Heat Exchanger	Cost	Weight
Shell-and-tube (liquid-to-liquid)	$25.00/ft^2 tube surface	8 lb/ft^2 total surface
Double pipe, finned tube	$12.50/ft^2 total surface	5 lb/ft^2 total surface
Banks of finned tubes (liquid-to-air, Al fins)	$3.50/ft^2 total surface	1 lb/ft^2 total surface
Plate coils (liquid-to-air or liquid-to-liquid)	$10.00/ft^2 total surface	2.6 lb/ft^2 total surface
Steam condensers (Admiralty metal tubes)	$30.00/ft^2 tube surface	6 lb/ft^2 tube surface
Steam boilers (small-gas-fired)	$10.00/lb steam per hour	
Cast-iron cascade coolers	$20.00/ft^2 total surface	16 lb/ft^2 total surface
Cooling towers	$32.00/ton of refrigeration	

[a]Data for mid-1987.

methods for making cost estimates together with basic cost data for these applications; typical reports are given in References 14–16.

For completely new types of heat exchangers it is desirable to make a detailed estimate for each of the major elements of the cost of fabricating a unit. The first step in such a procedure is to obtain costs for the material—tubing, fins, headers, shell, etc. The costs of the shop operations, such as drilling the header sheet, tube bending, making the tube-to-header joints, assembly, and inspection, can then be estimated and added to the materials cost to get the basic fabrication cost. Engineering and general overhead costs may then be estimated if desired; these are almost always much higher than one would expect and quite beyond the scope of this text. However, some basic data on the costs of tubing, pipe, and fabrication processes are included in Section 9 at the end of the appended Handbook to provide a point of departure for such estimates.

Costs of Commercially Available Heat Exchangers

It is not necessary to design a heat exchanger of a conventional type if one wishes to estimate its cost; one need only estimate the surface area required. For a given application, the surface area can be estimated by choosing a standard tube size and estimating the fluid velocity obtainable with the available pressure drop or pumping power assuming an overall pressure drop of three or four dynamic heads. The shell-side flow rate and velocity can be estimated from heat balance considerations, and the heat transfer coefficients can then be estimated. These together with the log mean tempera-

ture difference should give a fair estimate of the area required. If desired, the result can be checked to see if the pressure drops and temperature changes in the two fluid streams are close to the values desired, and, if they are not, a second iteration can be made.

Once the area is defined for a shell-and-tube heat exchanger, its cost, if made of carbon steel or stainless steel, can be obtained directly from Fig. H9.2.

Figure H9.2 was prepared for intermediate size, low-pressure units with simple header sheets. If the area required is outside the size range of Fig. H9.2, or if relatively high pressures, more complex headers, and/or other materials are required, a better estimate of the cost that takes these factors into account can be obtained from a set of empirical relations. These were derived from a critical analysis of a large collection of cost data by Corripio et al.[17] The basic relation they evolved is

$$\text{Cost per unit area} = C_B F_D F_P F_M \qquad (11.1)$$

Two expressions for the base coat C_B in dollars per unit of heat transfer surface area for low-pressure, carbon-steel heat exchangers are given in Table 11.3. In addition, expressions for the heat exchanger design, or type, factor F_D, and the design-pressure factor F_P are also presented. Table 11.4 presents a similar set of expressions for the material cost factor F_M for three stainless steels, four nickel alloys, titanium, and Hastelloy. While Reference 17 does not give cost factors for other common alloys, an older but similar set of data for cost estimates[18] gives factors of 1.1 for plain aluminum, 1.4 for copper, 1.6 for cupro-nickel, 2.0 for Monel, and 2.1 for nickel to be applied to carbon-steel, shell-and-tube units.

A similar and less complex empirical expression for

TABLE 11.3 Correlations for Costs of Heat Exchangers[a]

English units	SI Units
Base cost for carbon-steel, floating-head, 100-psig exchanger: $C_B = \exp[8.551 - 0.30863 (\ln A) + 0.06811 (\ln A)^2]$	Base cost for carbon-steel, floating-head, 700-kN/m² exchanger: $C_B = \exp[8.202 + 0.01506 (\ln A) + 0.06811 (\ln A)^2]$
Exchanger-type cost factor: Fixed-head: $F_D = \exp[-1.1156 + 0.0906 (\ln A)]$ Kettle reboiler: $F_D = 1.35$ U-tube: $F_D = \exp[-0.9816 + 0.0830 (\ln A)]$	Exchanger-type cost factor: Fixed-head: $F_D = \exp[-0.9003 + 0.0906 (\ln A)]$ Kettle reboiler: $F_D = 1.35$ U-tube: $F_D = \exp[-0.07844 + 0.0830 (\ln A)]$
Design-pressure cost factor: 100–300 psig: $F_P = 0.7771 + 0.04981 (\ln A)$ 300–600 psig: $F_P = 1.0305 + 0.07140 (\ln A)$ 600–900 psig: $F_P = 1.1400 + 0.12088 (\ln A)$ A in ft²; lower limit—150 ft², upper limit—12,000 ft²	Design-pressure cost factor: 700–2100 kN/m²: $F_P = 0.8955 + 0.04981 (\ln A)$ 2100–4200 kN/m²: $F_P = 1.2002 + 0.07140 (\ln A)$ 4200–6200 kN/m²: $F_P = 1.4272 + 0.12088 (\ln A)$ A in m²; lower limit—14 m², upper limit—1,100 m²

[a]Ref. 17. (Courtesy *Chemical Engineering*.)

TABLE 11.4 Material-of-Construction Cost Factors for Heat Exchangers[a]

Material	English Units A in ft^2 $F_M = g_1 + g_2 (\ln A)$		SI units, A in m^2 $F_M = g_1 + g_2 (\ln A)$	
	g_1	g_2	g_1	g_2
Stainless steel 316	0.8608	0.23296	1.4144	0.23296
Stainless steel 304	0.8193	0.15984	1.1991	0.15984
Stainless steel 347	0.6116	0.22186	1.1388	0.22186
Nickel 200	1.5092	0.60859	2.9553	0.60859
Monel 400	1.2989	0.43377	2.3296	0.43377
Inconel 600	1.2040	0.50764	2.4103	0.50764
Incoloy 825	1.1854	0.49706	2.3665	0.49706
Titanium	1.5420	0.42913	2.5617	0.42913
Hastelloy	0.1549	1.51774	3.7614	1.51774

[a]Ref. 17. (Courtesy *Chemical Engineering*.)

estimating the cost of shell-and-tube, plate-and-frame, and spiral heat exchangers made of type 304 stainless steel has been presented by Kumana:[19]

$$\text{Cost} = k(\text{area})^n \qquad (11.2)$$

where the cost is in dollars, the area is in square feet, and the constants k and n are given in Table 11.5. While Reference 19 does not give cost factors for other materials, the material cost factors of Table 11.4 should provide a good approximation. To apply these, the cost obtained from Eq. 11.2 should be multiplied by the ratio of the cost factor for the chosen material to that for type 304 stainless steel as given in Table 11.4.

In reviewing the recent literature the author was unable to locate comparable relations for the costs of

TABLE 11.5 Coefficients for Cost-Estimation Correlation[a]

Exchanger Type	Materials of Construction	Applicable Size Range, ft^2	Coefficients	
			k	n
Shell-and-tube	All type 304 stainless steel	400–9,000	235	0.665
Plate-and-frame	Frame: carbon-steel; plates: 304 stainless steel	100–5,000	100	0.778
Spiral-plate	All type 304 stainless steel	100–1,500	660	0.590

[a]Ref. 19. (Courtesy *Chemical Engineering*.)

finned-tube heat exchangers for applications such as air heaters or coolers. However, Reference 20 indicates that the costs of these units with aluminum fins on carbon-steel tubes run about 25% the cost of conventional carbon-steel shell-and-tube heat exchangers. This factor of 25% is based on the gas-side surface area as the measure of the heat exchanger size in place of the shell-side surface area.

Escalation

Prices increase with time. Adam Smith showed that the inflation process had increased the costs of basic commodities by roughly a factor of three per century from the twelfth to the eighteenth century.[21] This rate of increase continued until World War I, but after 1914 U.S. prices have been increasing at double the earlier rate, an effect shown in Fig. H9.1. The reader can update that curve using the cost index data published regularly in the *Engineering News Record*, and thus provide a basis for correcting the cost data presented here to include the effects of inflation. Other valuable cost indexes are published in *Chemical Engineering* and by the Department of Commerce, which puts out a monthly report[22] with detailed indexes for items such as "Bare tube heat exchangers," "Finned tube heat exchangers," "Finned coils" (for air conditioners), etc.

Effects of Special Requirements

Costs increase rapidly with requirements for special materials, inspection, or other quality-control measures. For example, expensive high-temperature materials and exacting quality-control specifications made the cost of a plate-fin and round tube radiator for a nuclear power plant (see Fig. 17.12) ran about 10 times as much per square foot as the value indicated by Fig. H9.3 for aluminum fins on plain carbon-steel tubes.

REFERENCES

1. A. L. London and R. K. Shah, "Costs of Irreversibility in Heat Exchanger Design," *Heat Transfer Engineering*, vol. 4(2), April–June 1983.

2. H. A. Kuljian and W. J. Fadden, Jr., "A New Way to Simplify the Steam Power Plant," *Trans. ASME*, vol. 79, 1957, p. 1115.

3. D. H. Fax and R. R. Mills, Jr., "General Optimal Heat Exchanger Design," *Trans. ASME*, vol. 79, 1957, p. 653.

4. S. Gibson-Harris, "Looking for Trouble," *Mechanical Engineering*, vol. 109(6), June 1987, p. 36.

5. A. M. Kanury, "Environmental Aspects," *Mechanical Engineering*, vol. 105(6), June 1983, p. 28.

6. A. P. Fraas, *Engineering Evaluation of Energy Systems*, McGraw-Hill Book Co., New York, 1982.

7. O. P. Bergelin et al., "The Fouling and Cleaning of Surfaces in Unfired Heat Exchangers—Panel Discussion," *Trans. ASME*, vol. 71, 1949, p. 871.

8. H. E. Bethon, "Fouling of Marine-Type Heat Exchangers," *Trans. ASME*, vol. 71, 1949, p. 855.

9. R. C. Butler and W. N. McCurdy, Jr., "Fouling Rates and Cleaning Methods in Refinery Heat Exchanger," *Trans. ASME*, vol. 67, 1945, p. A-1.

10. J. W. Rizika, "Thermal Lags in Flowing Systems Containing Heat Capacitors," *Trans. ASME*, vol. 76, 1954, p. 411.

11. J. R. Flower and B. Linnhoff, "A Thermodynamic-Combinatorial Approach to the Design of Optimal Heat Exchanger Networks," *AIChE Journal*, January 1980, p. 1.

12. "Modern Cost Engineering: Methods and Data," Compiled and edited by *Chemical Engineering*, 1984.

13. "Building Construction Cost Data 1986," *Means*, 1987.

14. S. Baron, "Costing Thermal Electric Power Plants," *Mechanical Engineering*, vol. 104(10), October 1982, p. 41.

15. J. R. Ball et al., "A Handbook for Cost Estimating," NUREG/CR-3971, ANL/EES-TM-265, U.S. Nuclear Regulatory Commission, October 1984.

16. R. W. Foster-Pegg, "Capital Cost of Gas-Turbine Heat Recovery Boilers," *Chemical Engineering*, July 21, 1986.

17. A. B. Corripio et al., "Estimate Costs of Heat Exchangers and Storage Tanks via Correlations," *Chemical Engineering*, January 25, 1982.

18. C. H. Chilton, *Cost Engineering in the Process Industries*, McGraw-Hill Book Co., New York, 1960.

19. J. D. Kumana, "Cost Update on Specialty Heat Exchangers," *Chemical Engineering*, June 25, 1984.

20. K. M. Guthrie, *Process Plant Estimating, Evaluation and Control*, Craftsman Book Company of America, Carlsbad, Ca., 1974.

21. A. Smith, *Wealth of Nations*, Modern Library, New York, 1937.

22. Producer Price Indexes, Department of Commerce, Washington, D.C.

12

Liquid-to-Liquid Heat Exchangers

The majority of liquid-to-liquid heat exchangers are of the shell-and-tube type described in Chapter 1 and illustrated in Figs. 1.7 through 1.12. This type of heat exchanger is especially well suited to applications in which the heat transfer matrix is large relative to the pipes carrying the inlet and exit fluid streams, and where the heat transfer coefficients readily obtainable with the two fluids are within a factor of two or three of each other, so that there is little incentive to employ extended surfaces. Liquid-to-liquid heat exchangers commonly fall in this region, and thus are usually of a shell-and-tube construction. This chapter presents the special problems and complex relationships involved in such exchangers, together with techniques and charts for estimating their performance.

HEAT TRANSFER PERFORMANCE

Flow Configuration

A variety of flow configurations are employed in shell-and-tube heat exchangers; those more commonly used are illustrated in Fig. 12.1. The choice of flow configuration is related to the choice of means for supporting and spacing the tubes in the heat transfer matrix. Some means of spacing the tubes should be provided at intervals of 30 to 40 diameters along their lengths to ensure a uniform spacing between the tubes, so that the flow and heat transfer distribution through the tube bundle are reasonably uniform. One of the most convenient arrangements is shown in Fig. 12.1a, in which the tubes are spaced with baffles having a hole pattern identical with that used in the header sheets. This arrangement can be used to provide crossflow with either a multiplicity of entrance and exit nozzles (as in Fig. 12.1d), or the tube bundle must be carefully baffled (as in Fig. 12.1c) to give a reasonably uniform flow distribution across the heat transfer matrix. As indicated in Fig. 12.1a, where baffles are used, they may extend all the way across the tube bundle or they may extend only a portion of the way (as in Fig. 12.1b), so that the shell-side flow is partially axial and partially crossflow. The baffles also may be provided with orifices to permit axial flow on the shell side (as in Fig. 12.1e), but this arrangement is not often used because it gives a less favorable relationship between heat transfer and pressure drop.

If axial flow is desired on the shell side, the baffle arrangements just described inevitably lead to rather large pressure losses, which can be avoided by the use of other types of spacer. The incentive to use hydrodynamically more refined spacer arrangements has not been sufficient to lead to their application except in a few specialized types of heat exchangers, such as those for gas-turbine recuperators where the plant performance is very sensitive to the pressure drop. Two such special

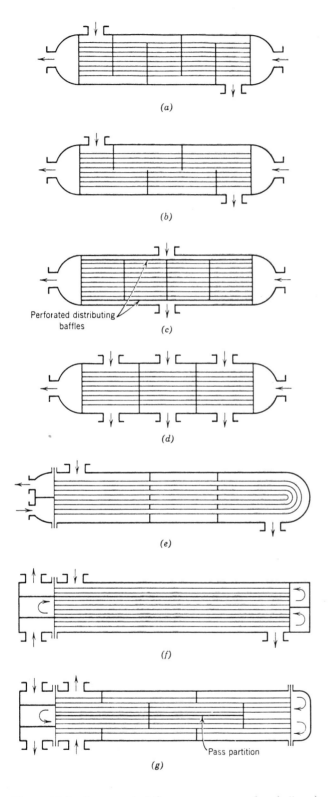

applications and the tube spacer arrangements employed are discussed in later chapters, one in connection with gas-turbine regenerators in Chapter 13, and one in connection with the liquid metal and molten salt heat exchangers discussed in Chapter 17.

The tube-side flow path may be a single-pass arrangement (as in Figs. 12.1a through 12.1d), a two-pass arrangement (as in Fig. 12.1e), or a multipass arrangement (as in Fig. 12.1f). The allowable pressure drop for the tube-side fluid stream is a major factor in determining the number of flow passes chosen.

The flow path configuration chosen for any particular application ordinarily represents a compromise between that giving the maximum LMTD (as discussed in Chapt. 4) and that best suited to manufacturing techniques or best meeting other requirements. For example, while a pure counterflow configuration may give a greater effective temperature difference than the single-pass shell, U-tube configuration of Fig. 12.1e, provision for differential expansion may be a vital consideration.

Header-Sheet Hole Patterns

Figure 12.2 shows the three principal tube arrays employed in shell-and-tube heat exchangers, namely, equilateral triangular, square, and staggered square. The equilateral triangular arrangement gives the strongest header sheet for a given shell-side flow-passage area, whereas the square arrangements simplify some fabrication and some maintenance operations.

The geometry in square arrays is so simple that there is little need for charts to aid in the selection of hole patterns, but the equilateral triangular arrangements are sufficiently complex that Figs. H6.3 and H6.4, together with Table H6.1, are included in the Handbook section to aid in preliminary design work.

Heat Transfer and Pressure Loss

The heat transfer and pressure loss for the fluid flowing inside the tubes can be estimated in a straightforward fashion (e.g., from Figs. H5.3, H5.4, and H3.4). The

Figure 12.1 Seven typical flow arrangements for shell-and-tube heat exchangers: (a) single-pass tube side; baffled, single-pass shell side; counterflow, full baffles; (b) single-pass tube side; baffled, single-pass shell side; counterflow, segmental baffles; (c) single-pass tube side; single crossflow pass on the shell side, full baffles; flow-distributing baffles across the inlet and outlet faces of the tube matrix; (d) single-pass tube side; single crossflow pass on the shell side, full baffles; multiple inlet and outlet connections on the shell side; (e) U-tube with single axial pass on the shell side, annular orifice baffles; (f) four-pass tube side; single axial pass on the shell side, no baffles; (g) four-pass tube side, baffled two-pass shell.

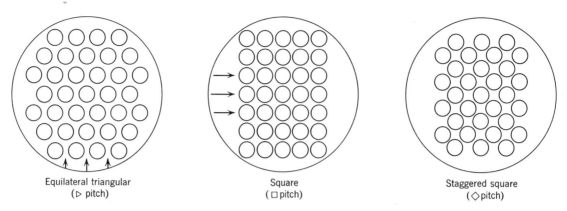

Equilateral triangular
($\triangleright$ pitch)

Square
($\square$ pitch)

Staggered square
($\diamond$ pitch)

Figure 12.2 Principal types of hole patterns used in tube header sheets.

situation for the crossflow region on the shell side is much more complex, especially if it is necessary to work with a viscous fluid (such as a lubricating oil for which the Reynolds number runs well below 2000). Fortunately, under crossflow conditions, the turbulent mixing induced by the irregular geometry is still substantial even at Reynolds numbers of 100 or less. Furthermore, the flow path length over the surface of a tube in crossflow is relatively small as compared to the effective hydraulic radius for the shell-side fluid. Thus there is no sharp break in the curve for the crossflow heat transfer coefficient in the Reynolds number range of from 10 to 10,000.[1-6] Figures H5.11 and 5.12 give correlations for the heat transfer performance for crossflow over bare tubes. (Figure H5.11 gives correlations for the tube rows in line with the direction of flow, and H5.12 with the tube rows staggered.)

In comparing the curves of H5.11 and 5.12 with those for flow through the inside of round tubes given in Fig. H5.3, it is apparent that it is advantageous to place the more viscous fluid, tending to give the lower Reynolds number, on the shell side and the less viscous fluid, giving the higher Reynolds number, on the tubeside. In this way advantage can be taken of the higher heat transfer coefficients at low Reynolds numbers given by crossflow conditions, and the two heat transfer coefficients can be made more nearly the same to give a well-proportioned heat exchanger. If the difference in heat transfer coefficient between the two fluids is still quite large, it may be advantageous to employ finned tubes. If the fin height is small, the ends of the tube can be expanded to a diameter a little larger than the fins so that the tube can be passed through the header sheet (see Fig. 12.3).[7,8]

For most heat exchanger applications, the pressure drop in the two fluid streams is limited to between 35

and 350 kPa (5 and 50 psi) to avoid both excessive pumping power losses and excessive pressures in the shells and plumbing systems. Figure H3.8 shows the approximate Reynolds number and flow rate obtainable as a function of pressure drop for flow through long, straight tubes having a length-diameter ratio of 200 and an internal diameter of 12.7 mm (0.5 in.)

ANALYTICAL APPROACH

There are so many parameters involved in the design of shell-and-tube heat exchangers that at first glance it appears hopeless to attempt to obtain an analytical solution that will give the desired combination of characteristics. Because of this problem, most designers employ empirical relations with a cut-and-try approach that depends on their judgment and experience for convergence on a new design by extrapolation from tested units. For an experienced designer, this is indeed an effective

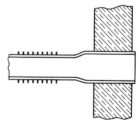

Figure 12.3 Section through a finned tube showing the way in which the end is expanded so that the hole in the header sheet is large enough to pass the fins.

approach if the unit on which he is working employs fluids similar to those with which he has had experience. However, an analytical approach is easier to follow for the less experienced designer, since it shows the basic relationships. Furthermore, an analytical approach can prove to be extremely worthwhile even for the experienced heat exchanger designer if the fluids have unusual properties or if the application calls for performance characteristics substantially different from those of run-of-the-mill heat exchangers.

The nature of the problem is best indicated by summarizing the parameters that are ordinarily given for a particular case, the quantities that are directly determined by them, and the quantities that must be determined from heat transfer and fluid-flow relationships. Ordinarily, the heat exchanger is designed to heat or cool a primary stream for which the flow rate, an acceptable pressure drop, and the inlet and outlet temperatures are defined. The inlet temperature of the secondary stream is ordinarily also defined. While its outlet temperature is likely to be a variable, experience usually indicates that an appropriate value lies in a relatively narrow range. If this should be the case, the LMTD will depend only on the flow configuration, and the flow rate in the secondary stream will be defined by heat balance considerations.

Practical considerations such as fabrication and maintenance ordinarily determine the tube diameter and geometric arrangement, that is, square or equilateral triangular tube pitch and tube centerline spacing. Once the geometric pattern for the tubes is fixed, the relations between the fluid mass flow rate outside of the tubes and both the pressure drop and the heat transfer coefficient are then defined.[9]

Estimate of Number of Tubes Required

The total number of tubes n is dependent on tube-side flow conditions. It is related to the tube length and diameter together with the allowable pressure drop and the total tube-side flow rate, that is,

$$\Delta P_1 = \frac{f_1 G_1'^2 L N_1}{2\rho_1 g D_1} = \frac{f_1}{2\rho_1 g}\left(\frac{N_1 W_1'}{n D_1^2 0.786}\right)^2 \frac{L N_1}{D_1} \quad (12.1)$$

with the nomenclature as defined in Table 12.1; for example, the subscript 1 denotes tube-side conditions. Rearranging to give an explicit relation for n, the total number of tubes, leads to

$$n = \left(\frac{f_1 W_1'^2 L N_1^3}{2\rho_1 g D_1^5 0.786^2 \Delta P_1}\right)^{1/2} \quad (12.2)$$

$$= \frac{W_1'}{C_u}\sqrt{\frac{f_1 L N_1^3}{\rho_1 D_1^5 \Delta P_1}} \quad (12.3)$$

For English units, $C_u = 6.31$; for SI units, with ΔP in Pascals, use $C_u = 1.111$. Since W_1', ρ_1, D_1, N_1, ΔP_1, and f_1 ordinarily can be fixed, n varies as the square root of L. Inasmuch as practical considerations usually make it possible to estimate the tube length within plus or minus 50%, the square root relationship means that n can thus be estimated from Eq. 12.3 within about 25% which is good enough for a first approximation.

Relation from Heat Balance Considerations

If all of the preceding parameters can be defined, only three remain to be determined. These are the shell-side fluid mass flow rate across the tube matrix, the tube length and N_2, the number of passes for the crossflow fluid. Three independent equations are required to solve for these three unknowns. One such relation can be obtained by expressing the total amount of heat added to or taken from the shell-side fluid in terms of the product of the crossflow passage area per pass, the shell-side fluid mass flow rate per unit of flow passage area, and the temperature rise or drop in the shell side fluid, that is,

$$Q = C_m n^{0.5} l\left(\frac{s - d_o}{12}\right) G_2\, \delta t_2 c_{p2} \quad (12.4)$$

(For SI units, delete the number 12.) Note that l, the tube length per shell-side pass, and s, the tube spacing, are defined in Fig. 12.4, and that the use of $n^{0.5}$ with the coefficient C_m is based on the assumption that the tube bundle is rectangular in cross section. The coefficient C_m is the factor by which the square root of the total number of tubes must be multiplied to obtain the number of tubes in a row transverse to the crossflow direction. The coefficient C_m usually has a value of about 1.1, since tube arrays are made quasirectangular (as in Fig. 12.2) in order to give an adequate flow-passage area between flow passes.

Relation from Heat Transfer Considerations

A second basic relationship expresses this same amount of heat as the amount transferred from the tubes to the shell-side stream, that is,

TABLE 12.1 Special Nomenclature for Chapter 12[a]

Part 1

A_L = flow-passage area for leakage around the tube bundle, (shell side)

A_M = flow-passage area for flow through the tube bundle, (shell side)

a, b = exponents in Eq. 12.8

b_w = tube-wall thickness, in.

C_a, C_b = coefficients in Eq. 12.8

$$C_m = \left(\frac{\text{average number of tubes per transverse row}}{\text{number of tube rows}}\right)^{1/2}$$

d_b = baffle OD, in.

d_h = tube hole diameter in baffles, in.

d_m = diameter of circle circumscribed around tube matrix, in.

d_o = tube OD, in.

D_1 = tube ID, ft (m)

D_o = tube OD, ft (m)

d_s = shell diameter, in.

D_s = shell diameter, ft (m)

f_2 = crossflow friction factor (shell side)

F_h = fraction of the shell-side flow passing through the tube matrix for determination of the heat transfer coefficient

F = fraction of the shell-side flow passing through the tube matrix for determination of the pressure drop

H = baffle window height, in, (m)

G_1 = tube-side mass flow rate (total flow rate divided by the tube-side flow-passage area per pass), lb/ft² · h (kg/m² · h)

G_2 = nominal shell-side mass flow rate through the tube bundle (obtained by dividing the total shell-side weight flow by the transverse flow-passage area between the tubes in the tube matrix), lb/ft² · h (kg/m² · h)

l = tube length between baffles, ft (m)

L = tube bundle length, ft ($L = l\,N_2$) (m)

M = ratio of the effective flow-passage area for crossflow through the tube matrix to the total flow-passage area

n = total number of tubes

N_1 = number of tube-side passes

N_2 = number of passes across the tube bundle made by the shell-side fluid

N_p, N_h = factors shown in the tables in Figs. 12.9, 12.10, and 12.11

ΔP = pressure drop, lb/ft² (Pa)

$\text{Re}_h = \dfrac{G_2 D_o}{\mu}\dfrac{F_n}{M}$ = Reynolds number for heat transfer coefficient calculations

$\text{Re}_p = \dfrac{G_2 D_o}{\mu} F_p$ = Reynolds number for friction factor calculations

W = total flow rate, lb/h (kg/h)

W' = total flow rate, lb/s (kg/s)

Y = a factor which when multiplied by s/d_s gives the ratio of the baffle window pressure drop to the tube matrix pressure drop for the shell-side flow.

Subscripts

1—refers to tube side
2—refers to shell side
L—with leakage
NL—with no leakage

Part 2

The factors given in the tables in Figs. 12.9, 12.10, and 12.11 are based on the following clearance ratios:

$$\frac{d_s}{d_m} = \text{shell-to-tube matrix diameter ratio} = 1.075$$

$$\frac{d_h - d_o}{d_o} = \text{baffle hole-to-tube clearance ratio} = 0.0045$$

$$\frac{d_s - d_b}{d_s} = \text{baffle-to-shell clearance ratio} = 0.008$$

[a]This is the nomenclature used by Tinker, Ref. 12, with the author's additions of SI units in parentheses.

$$Q = h_2 A_2 \Delta T_2 = h_2 n\pi \frac{D_0}{12} l N_2 \Delta T_2 \quad (12.5)$$

where N_2 is the number of passes across the tube bundle made by the shell-side fluid, that is, $N_2 l = L$, the tube length.

Relation from Pressure-Drop Considerations

A third basic relationship is that between the pressure drop and the fluid flow across the tubes. If the baffles extend all the way across the tube bundle, and the pressure drops across the baffle windows are neglected,

$$\Delta P_2 = \frac{f_2 G_2{}^2 n^{0.5} N_2}{2 C_m g 3600^2 \rho^2} \quad (12.6)$$

or

$$N_2 = \frac{2 C_m g\, 3600^2 \rho_2 \Delta P_2}{f_2 G_2{}^2 n^{0.5}} \quad (12.7)$$

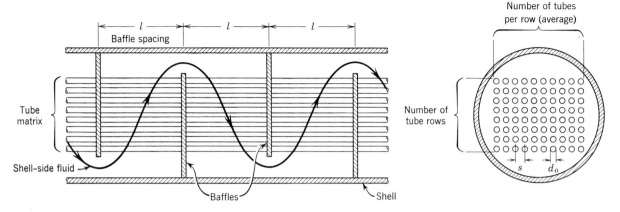

Figure 12.4 Sections through shell-and-tube heat exchanger showing the nomenclature commonly employed.

Solution for l, N_2, and G_2

Equations 12.4, 12.5, and 12.7 constitute a set of three independent equations that can be solved for the three unknowns l, N_2, and G_2. The solution of these equations is complicated because the quantities f_2 and h_2 are functions of G_2, and Δt_2 is a function of G_2 and G_1. Explicit relations in terms of G_2 are needed for each of these three quantities.

If h_2 and f_2 are expressed in the form

$$h_2 = C_a G_2{}^a \qquad (12.8a)$$

$$f_2 = C_b G_2{}^b \qquad (12.8b)$$

The coefficients C_a and C_b and the exponents a and b may be expressed for limited flow ranges by linear approximations determined from a chart (such as that in Fig. 12.9), and Δt_2 is given by

$$\Delta t_2 = \frac{U}{h_2} \Delta t_m \qquad (12.9)$$

where

$$U = \left(\frac{1}{h_1} + \frac{b_w}{12k_w} + \frac{1}{h_2} \right)^{-1}$$

The tube-wall thickness b_w and the thermal conductivity k_w are usually implicit in the specifications. The heat transfer coefficient h_1 for the tube-side fluid is a function of the tube-side flow rate G_1. Since G_1 is known, with a given set of values for n, N_1, the tube-side fluid inlet and outlet temperatures, and the shell-side inlet temperature, the temperature difference Δt_2 becomes a function solely

of G_2. Once either Δt_2 or G_2 is fixed—by making a reasonable assumption if nothing else—Δt_m can be established.

Equating Eqs. 12.4 and 12.5, and substituting the value of N_2 from Eq. 12.7 gives

$$G_2{}^3 = 2\pi g 3600^2 \frac{h_2}{f_2} \left(\frac{\rho_2 \Delta P_2}{c_{p2}} \right) \left(\frac{d_o}{s - d_o} \right) \left(\frac{\Delta t_2}{\delta t_2} \right) \qquad (12.10)$$

Substituting h_2, f_3, and Δt_2 (from Eqs. 12.8 and 12.9) in Eq. 12.10, we may obtain an explicit relation for G_2, that is,

$$G_2{}^{3+b-a} = 2\pi g 3600^2 \left(\frac{C_a \rho_2 \Delta P_2}{C_b c_{p2}} \right) \left(\frac{d_o}{s - d_o} \right)$$
$$\cdot \left(\frac{U}{h_2} \right) \left(\frac{\Delta t_m}{\delta t_2} \right) \qquad (12.11)$$

Correction for Variable Viscosity

If the viscosity of the shell-side fluid varies substantially over the temperature range in the heat exchanger—as is the case for most organic liquids—the correction term $(\mu/\mu_w)^{0.14}$ should be applied to allow for the change in viscosity between the wall and the bulk free stream. (See Chapter 3.) On this basis, Eq. 12.11 becomes

$$G^{3+b-a} = 2\pi g 3600^2 \left(\frac{C_a \rho_2 \Delta P_2}{C_b c_{p2}} \right) \left(\frac{d_o}{s - d_o} \right)$$
$$\cdot \left(\frac{U}{h_2} \right) \left(\frac{\Delta t_m}{\delta t_2} \right) \left(\frac{\mu}{\mu_w} \right)^{0.14} \qquad (12.12)$$

Application of Solution to the General Case

The first step in solving Eq. 12.11 is to estimate the number of tubes n from Eq. 12.3. This value of n defines G_1 and hence h_1. The wall resistance is known, and h_2 may be estimated. Using the resulting value for U, Eq. 12.12 can be solved to give a first approximation for G_2, because all the other quantities are known. The major uncertainty in making this approximation is the effect of G_2 on the ratio U/h_2. Since this ratio is rather insensitive to the value of G_2, a reasonable value for the ratio can be assumed for a first approximation, and convergence is rapid in carrying out a second or third approximation. Substitution of the calculated value of G_2 in Eqs. 12.4 and 12.7 gives the values of l and N_2, so that the tube length L may be calculated, since it is the product of l and N_2.

The calculated value of L and the assumed value of n should be substituted in Eq. 12.1 (or 12.2) to see whether the resulting pressure drop for the tube side is within acceptable limits. If not, the calculations can be repeated with a revised value of n.

Of the various quantities that were fixed in order to make it possible to obtain a solution, the tube spacing is most likely to be subject to variation. The effects of variation in the spacing (or in any of the other arbitrarily fixed quantities) can be investigated by substituting the appropriate values in the equations and by solving in a manner similar to that just outlined.

Approximate Solution Where the Shell-Side Δt is Dominant

Perhaps the most common type of shell-and-tube heat exchanger is that in which an organic liquid having relatively poor heat transfer properties is heated or cooled by water which has much better heat transfer properties. In heat exchangers of this sort the water usually flows through the tubes, and the organic fluid is on the shell side. The temperature drop through the tube wall is ordinarily quite small. The overall heat transfer coefficient is insensitive to the water-side heat transfer coefficient, and depends primarily on the heat transfer coefficient of the shell-side fluid. This makes possible a simplifying approximation: namely, that the overall heat transfer coefficient is roughly equal to the heat transfer coefficient on the shell side. Thus the term U/h_2 in Eqs. 12.11 or 12.12 may be taken as a little less than unity, and an explicit solution can be obtained. (To simplify the arithmetic, $U/h_2 = 1.0$ can be used as a first approximation.)

Effects of Baffles

The preceding analysis included no allowance for the pressure drop across the baffle from one flow pass to the next. This loss includes two components: first, a loss associated with turning the fluid 90° as it emerges from the exit face of the tube bank, and second, the orifice loss associated with the flow through the opening between the baffle and the shell. The magnitude of this loss can be reduced by increasing the flow-passage area by extending the baffle only part way across the tube bundle so that there is axial flow between a portion of the tubes. It is generally good practice to make the net flow-passage area past the baffle approximately equal to that for crossflow through the bundle so the dynamic head is about the same in both restricted regions. If the baffle must be cut back into the tube matrix to provide sufficient flow-passage area, the actual pressure drop may run somewhat higher than the rough value obtained by assuming that the pressure loss is equal to the sum of the dynamic head of the fluid stream leaving the flow passage between the tubes at the exit face of the tube matrix and the dynamic head calculated from the average velocity through the flow-passage area of the window.

Axial leakage through the clearance between the baffle and the shell and between the baffle and the tubes may permit a substantial fraction of the flow to bypass the tube matrix. This is particularly likely if there is a substantial space between the shell and a relatively small bundle of closely spaced tubes. This effect has been investigated, and it has been found that a good estimate of the bypass flow can be obtained through a straight-forward comparison of the flow through the equivalent parallel orifices represented by these clearances and the flow through the matrix for the same pressure drop.[10,11] The clearances both between the baffle and the shell and between the tubes and the baffle are ordinarily not uniform annularly, but have the crescent shape of the gap between a cylinder placed within a hole so that the cylinder is tangent at one side. The orifice coefficient for this geometry is presented in Fig. 12.5. This coefficient differs as much as 10% from that for a concentric annulus.[10]

The extent to which bypass flow of this character can reduce both the effective heat transfer coefficient and the pressure drop through the tube matrix is indicated by Fig. 12.6, which shows a loss coefficient plotted as a function of the ratio of the leakage flow-passage area around the baffle to the transverse flow-passage area through the tube matrix. This curve provides a simple means for obtaining a rough estimate of the effects of clearance variations on the performance of shell-and-tube heat exchangers.

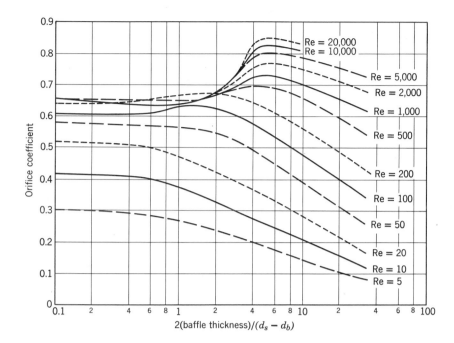

Figure 12.5 Orifice coefficients for the gap between a circular disc baffle and the bore of a shell if the baffle is tangent to the shell at one side. (Bergelin, Bell, and Leighton, Ref. 10.)

If the envelope for the tube bundle is appreciably smaller in diameter than the ID of the shell, a substantial amount of flow may bypass the tube matrix by flowing through the gap between the tube bundle and the shell. The resulting loss in performance may be substantial, particularly for tube bundles for which the tubes are closely spaced. Bypass flow of this sort may be reduced through the use of radial baffles arranged parallel to the axis of the shell, as in Fig. 12.7.

Effects of Deviations from Ideality

The accumulative effects of deviations from ideality are ordinarily quite large in shell-and-tube heat exchangers. Factors that may lead to large variations in heat exchanger performance include nonuniformities in the velocity distribution across the tube matrix (as indicated in Fig. 12.8), flow bypassing (as just discussed), uncertainties in pressure loss and heat transfer coefficients, and

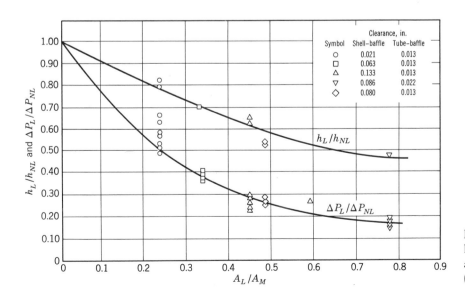

Figure 12.6 Relative pressure drop and heat transfer as functions of the leakage area ratio A_L/A_M for a 5.25-in. ID shell. (Bergelin et al., Ref. 10.)

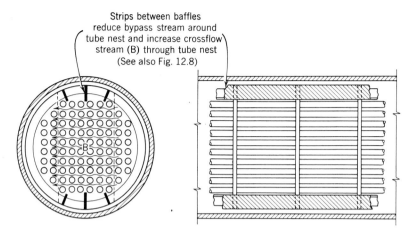

Figure 12.7 Radial baffles designed to reduce the amount of bypass flow through the gap between the sides of the tube matrix and the shell. (Tinker, Ref. 12.)

geometric variations within the limits of manufacturing tolerances. These effects are so complex that there is no simple way to include them. Even empirical methods for handling these effects are complex and difficult to apply.

CALCULATIONAL TECHNIQUE INCLUDING ALLOWANCES FOR DEVIATIONS FROM IDEALITY

A rather complex but effective technique for including the effects of deviations from ideality has been evolved and reported by Tinker.[12] Although his analysis is too complex to include in detail in this book (it involves over 70 different symbols), it represents a much better approximation than the idealized analysis presented in Eqs. 12.1 to 12.12, and provides such a good insight into the effects of the major geometric parameters causing deviations from ideality that a simplified form of this analysis is presented here. Since many of Tinker's symbols differ from those used elsewhere in this book, and because some have been modified to simplify the analysis presented here, the special symbols required are defined in Table 12.1. The first part of the table lists the basic quantities, whereas the second lists the parameters that play prominent roles in the application of this technique to design problems.

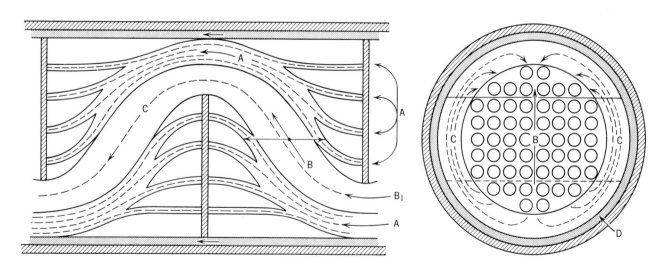

Figure 12.8 Diagram indicating leakage paths for flow bypassing the tube matrix, both through the baffle clearances and between the tube matrix and the shell. (Tinker, Ref. 12.)

Parameters for Bypass Flow Area Ratios

While Tinker's basic derivations are reasonably general, they are cumbersome unless the range of possible combinations of the geometric parameters is greatly reduced. He does this by adopting representative values for each of the three geometric ratios in the second part of Table 12.1 (i.e., the ratio of the clearance between the tubes and holes in the baffles to the tube OD, the ratio of the clearance between the baffle and the shell to the shell ID, and the ratio of the shell ID to the OD of the tube matrix). These quantities are closely related to the various bypass flow rates, and generally are kept as small as possible, consistent with manufacturing tolerances and the clearances required for assembly.

Once the foregoing ratios are fixed, Tinker shows that the various bypass flow rates become functions of the ratio of the shell diameter to the tube length between baffles, the ratio of the tube pitch to the tube OD, the ratio of the baffle window height to the shell diameter, and the ratio of the number of tube rows traversed per pass to the shell ID divided by the tube pitch (i.e., the number of tubes that ideally might be installed along a diameter of the shell). For a well-proportioned unit the last two ratios depend on the first two; hence Tinker has prepared tables for values of the parameters N_h and N_p that he uses to relate the fraction of the flow that passes through the tube matrix (i.e., the total flow minus the bypass flow) to the geometric ratios just discussed. Simplified versions of these tables are presented in Figs. 12.9, 12.10, and 12.11.

Effects of Bypass Flow on Heat Transfer

The analysis presented previously for the idealized case with no bypass flow can be modified to include the effects of bypass flow by using the foregoing factors. The relations given in Eqs. 12.2 and 12.3 for the tube-side flow, of course, require no correction. The relation for the heat added to or taken from the shell-side fluid given in Eq. 12.4 remains the same. However, in determining h_2 for Eq. 12.5, G_2 should be multiplied by the factor F_h, the fraction of the total flow that passes through the tube matrix. This correction factor depends on the various clearance ratios cited, and is given by

$$F_h = \frac{1}{1 + N_h \sqrt{(d_s/s)}} \qquad (12.13)$$

where representative values for N_h are given in the tables

in Figs. 12.9, 12.10, and 12.11. Thus the shell-side heat transfer rate given in Eq. 12.5 is applicable if the heat transfer coefficient is obtained through the use of a Reynolds number modified to allow for bypass flow effects. This may be defined as

$$\text{Re}_h = \frac{G_2 D_o}{\mu_2} \frac{F_h}{M} \qquad (12.14)$$

where the coefficient M is a correction factor close to unity that allows for variations in the effective flow passage area with changes in geometry. Values for M and N_h are given as functions of D_s/l, H/d_s, and s/d_o in tables at the top of the charts of Figs. 12.9, 12.10, and 12.11. Thus the values of F_h and Re_h can be calculated, and h_1 can be determined from the appropriate chart. Note that Figs. 12.9, 12.10, and 12.11 are for tube matrices in which the tube pitch is square, staggered square, and equilateral triangular, respectively. Note, too, that in the staggered-square array Tinker considers the tube pitch to be that measured along the side of the square, not that transverse to the flow (i.e., across the diagonal).

Effects of Bypass Flow on Pressure Drop

The relation for the pressure drop given in Eq. 12.6 for the idealized zero-bypass case can be modified in a manner similar to the one described in the preceding paragraph, except that an allowance should be made for the pressure drop across the baffle windows and the effects of a portion of the flow moving axially through the tube matrix where partial baffles are employed. With these corrections, together with the addition of the term $(\mu_w/\mu)^{0.14}$ to allow for differences in viscosity between the fluid at the wall and the bulk free stream, Eq. 12.6 becomes

$$\Delta P_2 = \frac{f_2(F_p G_2)^2 n^{0.5} N_2}{2g 3600^2 \rho_2 C_m} \left[1.075 \left(1 - \frac{H}{d_s} \right) \right]$$
$$\cdot \left(1 + Y \frac{s}{d_s} \right) \left(\frac{\mu_w}{\mu} \right)^{0.14} \qquad (12.15)$$

The term $1.075(1 - H/d_s)$ in Eq. 12.15 allows for the reduction in pressure drop that occurs if the baffle window extends down into the tube matrix so that a portion of the flow bypasses some of the tube matrix. The term $1 + Y(s/d_s)$ allows for the pressure drop across the baffle windows; that is, the quantity $Y(s/d_s)$ is the

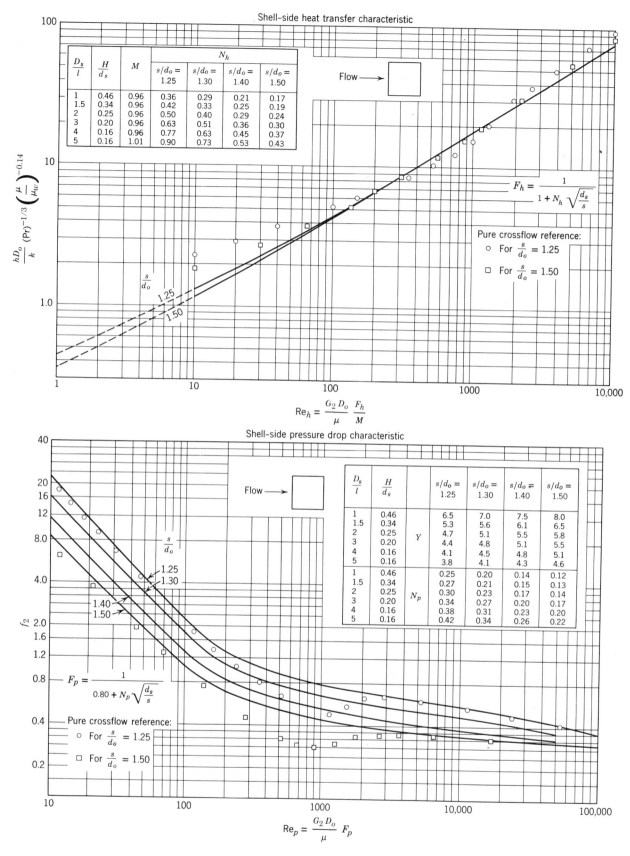

Figure 12.9 Shell-side heat transfer and pressure-drop characteristics for multipass, crossflow, shell-and-tube heat exchangers with the tubes arranged on a square pitch. (Tinker, Ref. 12.)

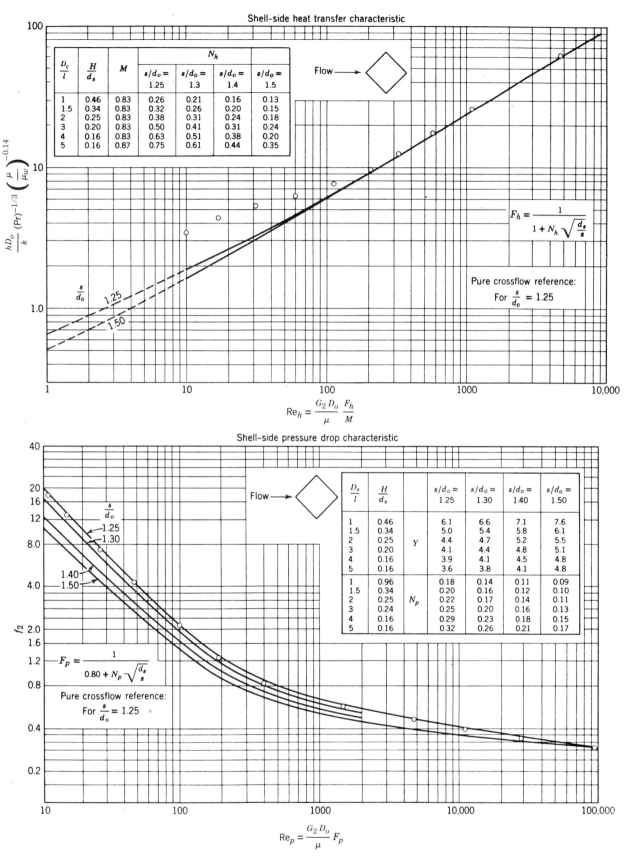

Figure 12.10 Shell-side heat transfer and pressure-drop characteristics for multipass, crossflow, shell-and-tube heat exchangers with the tubes arranged on a staggered square pitch. (Tinker, Ref. 12.)

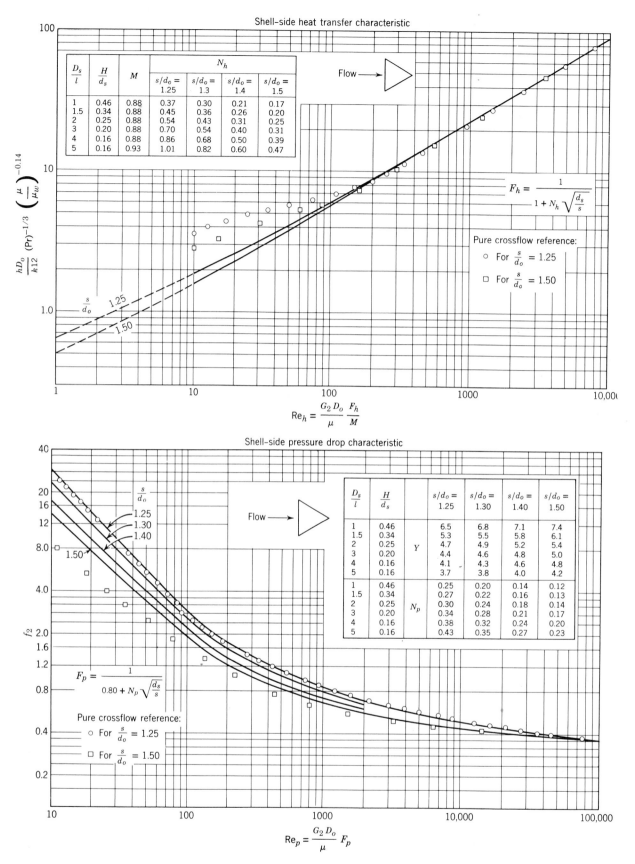

Figure 12.11 Shell-side heat transfer and pressure-drop characteristics for multipass, crossflow, shell-and-tube heat exchangers with the tubes arranged on an equilateral triangular pitch. (Tinker, Ref. 12.)

ratio of the baffle window pressure drop to the tube matrix pressure drop.

Curves for the crossflow friction factor f_2 in Eq. 12.15 are presented in Figs. 12.9, 12.10, and 12.11. Note that the flow fraction F_p used in determining the crossflow friction factor differs somewhat from the flow fraction F_h used in determining the heat transfer coefficient. This comes about because of several effects. For example, the flow through the clearance between the tube bundle and the shell contributes to the pressure drop across the baffle window, though it does not contribute to the heat transfer. The Reynolds number is similarly affected; hence the flow fraction and Reynolds number used to obtain f_2 from Figs. 12.9, 12.10, and 12.11 are defined as

$$F_p = \frac{1}{0.80 + N_p \sqrt{(d_s/s)}} \qquad (12.16)$$

$$\mathrm{Re}_p = \frac{F_p G_2 d_o}{12\mu_2} \qquad (12.17)$$

Values for N_p and Y are given in small tables in Figs. 12.9, 12.10, and 12.11 to help evaluate Eqs. 12.15 and 12.16.

Solution for G_2 with Allowances for Bypass Flow

In Eq. 12.12 an explicit relation for G_2 was derived for the idealized case with no bypass flow. This relation can be modified to include the effects of bypass flow by correcting the pressure drop in the right side of the equation. The appropriate correction factor is the ratio of the pressure drops for the "no bypass" and "with bypass" cases, that is, Eq. 12.6 multiplied by the factor $(\mu_w/\mu)^{0.14}$ and divided by Eq. 12.15. This modification yields

$$G_2^{3-a+b} = \frac{1.86 C_a \pi g 3600^2 d_o \Delta P_2 \rho_2}{C_b (s - d_o) F_p^2 (1 - H/d_s)[1 + Y(s/d_o)] c_{p2}}$$
$$\times \frac{\Delta t_m}{\delta t_2} \frac{U}{h_2} \left(\frac{\mu}{\mu_w}\right)^{0.14} \qquad (12.18)$$

Since all the parameters in this equation except G_2 are either known or can be estimated fairly well, it provides a convenient means for obtaining a good approximation for G_2. The constants a, b, C_a, C_b, F_p, and Y can be estimated for the appropriate geometry from Figs. 12.9, 12.10, or 12.11. The value for ΔP_1 is usually specified by the requirement. By estimating the ratios $\Delta t_m/\delta t_2$ and

U/h_2 rather than values for each of the individual quantities in these ratios, more rapid convergence can be obtained in making subsequent approximations, since these ratios are insensitive to the value of G_2.

Tinker, in applying his calculational technique to typical heat exchangers, found that the accumulation of uncertainties may easily lead to deviations of as much as 30 or 40% between calculated and experimentally determined values.[12] Thus the technique presented here—which omits many small refinements that Tinker included—may give values that will be found to deviate by as much as 50% from experimental test results. Such disparities must be expected where a system involves so many complex relationships.

Example 12.1. *Design of a Marine Oil Cooler.* A lubricating oil cooler for a large marine diesel engine presents a typical set of problems in the design of a shell-and-tube heat exchanger. The oil temperature should be kept below 65°C (150°F) to minimize oil oxidation and sludge formation in the engine. On the other hand, the oil temperature ought not be allowed to drop below 55°C (130°F), or the increased viscosity will result in greater friction and there will be much more corrosion from moisture accumulating in the oil.

The temperature of the cooling water available is that of the surrounding sea, which would rarely exceed 32°C (90°F). Since plenty of water is available, the temperature rise in the water can readily be kept to 5 to 11°C (10 to 20°F). While the water-flow rate can be increased further, this results in only a few degrees increase in the mean temperature difference between the oil and the water—probably not enough to justify the increase in the pipe size and the pumping requirements.

The flow configuration found most suitable for this type of application is one or two passes on the water side to give the desired water pressure drop-flow rate combination, and four or more passes on the shell side to give a sufficiently high velocity in the oil to obtain good heat transfer coefficients. The tube diameter should be sufficient to facilitate cleaning the tube side; that is, it is desirable to use a tube diameter of about ¾ in., although smaller diameter tubes may be used to reduce the size and the weight. For good resistance to corrosion, the tubes should be of Admiralty metal or Monel, the former being the less expensive. The tube spacing is usually about 1.25 times the tube OD.

. If the total pressure drop on the water side is to be held to about 103 kPa (15 psi), and if of this two-thirds is allocated to the plumbing and heat exchanger header inlet and outlet losses, about 35 kPa (5 psi) will be available for the pressure drop through the tubes. The pressure drop available for oil flow through the crossflow matrix for the shell side is ordinarily about 138 kPa (20 psi).

The design requirements and related parameters fixed by these considerations and the design heat load of 211 kW (720,000 Btu/h) are given in Table 12.2.

Solution. The first step in sizing the heat exchanger is to estimate n, the number of tubes. Equation 12.3 can be used for this purpose, since W_1, ρ, D_1, and the maximum allowable value for the tube-side pressure drop ΔP_1 are known, and, after a little experience, f_1 and L can be estimated readily. On this basis a value of 38 appears appropriate for the number of tubes per tube-side pass. Table 12.3 shows the steps in making this and the remaining calculations. Note that in the first trial the value for the shell ID-to-baffle spacing D_s/l is chosen arbitrarily as 1.5 to calculate the various bypass flow

factors. Using the data in the table in Fig. 12.11 the factors F_b and F_p were then calculated. In evaluating G_2 from Eq. 12.18 it is assumed that the shell-side film Δt is dominant; hence the factor (U/h_2) is taken as unity. The exponents a and b and the coefficients C_a and C_b were calculated from Fig. 12.11 assuming that the shell-side Reynolds number lies between 100 and 1200.

The calculated value for the baffle spacing l is 0.758 ft, the number of shell-side passes N is 17, and the tube matrix length is about 12.9 ft. The resulting tube-side pressure drop does not exceed the allowable value of 35 kPa or 5.0 psi. The calculated value of D_s/l ratio (step 31 of Table 12.3) is less than the value assumed in the first approximation (i.e., the shell diameter was not large enough); hence a new value of the ratio D_s/l is selected from Fig. 12.11, and the calculations repeated. Note that

TABLE 12.2   Parameters Fixed for a Marine Oil Cooler

1. Heat load	211 kW (720,000 Btu/h)				
2. Flow configuration	Two-pass tube-side, multiplicity of passes shell side				
3. Matrix geometry	Equilateral triangular pitch				
4. Tube size	19-mm (0.75-in.) OD, 1.25 mm (0.049 in.) thick, 16.6-mm (0.652-in.) ID				
5. Tube spacing	1.25 × tube OD = 23.8 mm (0.938 in.)				
6. Tube material	Admiralty (conductance = 112.7 W/m^2 · °C or 65 Btu/h · ft · °F				

	Shell side		Tube side		
7. Fluid	SAE-30 oil		Seawater		
	°C	°F	°C	°F	
8. Temperature in	65.6	150	32.2	90	
9. Temperature out	60.0	140	37.8	100	
10. Temperature rise or drop	5.6	10	5.6	10	
11. Temperature (mean)	62.8	145	35.0	95	
	Pa	*psi*	*Pa*	*psi*	
12. Pressure (mean)	275,792	40	48,264	7	
13. Allowable pressure drop	137,896	20	34,474	5	
	kg/m^3	*lb/ft^3*	*kg/m^3*	*lb/ft^3*	
14. Density (mean)	849	53	993	62	
	kJ/kg · °C	*Btu/lb · °F*	*kJ/kg · °C*	*Btu/lb · °F*	
15. Specific heat (mean)	2.1	0.5	4.2	1	
	Pa · s	*lb/ft · h*	*Pa · s*	*lb/ft · h*	
16. Viscosity (bulk)	0.031	75	0.00075	1.8	
	W/m · °C	*Btu/h · ft · °F*	*W/m · °C*	*Btu/h · ft · °F*	
17. Thermal conductivity	0.156	0.09	0.614	0.355	
18. Prandtl number	$\dfrac{c_p\mu}{k}$ 418.0		$\dfrac{c_p\mu}{k}$ 5.1		
	mm	*in.*	*mm*	*in.*	
19. Equivalent passage diameter	14	0.54^a	17	0.652	
	kg/s	*lb/s*	*kg/s*	*lb/s*	
20. Total flow rate	18.2	40	9.1	20	

^{a}See items 3, 4, and 5, together with Fig. H.6.1 (for axial flow).

TABLE 12.3 Design Calculations for a Marine Oil Cooler Quantity

Quantity	Source	Units	First Trial	Second Trial	Second Trial (SI Units)
1. N_1, number of tube-side passes	Table 12.2, item 2		2	2	2
2. n, total number of tubes	Estimated from Eq. 12.2		38	38	38
3. d_m, matrix diameter	Table H6.1	in.	7.2	7.2	183 mm
4. d_s, shell diameter	$1.075 \times d_m$	in.	7.7	7.7	196 mm
5. C_m	Estimated		1.05	1.05	1.05
6. D_s/l	Chosen arbitrarily from Fig. 12.11	ft/ft	1.5	1.0	1.0
7. H/d_s, ratio of baffle window height to shell ID	Fig. 12.11	in./in.	0.34	0.46	0.46
8. s/d_o, ratio of tube pitch to tube OD	Table 12.2	in./in.	1.25	1.25	1.25
9. N_h	Fig. 12.11		0.45	0.37	0.37
10. N_p	Fig. 12.11		0.27	0.25	0.25
11. F_h	$(1 + N_h \sqrt{d_s/s})^{-1}$		0.437	0.485	0.485
12. F_p	$(0.8 + N_p \sqrt{d_s/s})^{-1}$		0.632	0.660	0.660
13. M	Fig. 12.11		0.88	0.88	0.88
14. Y	Fig. 12.11		5.3	6.5	6.5
15. $(\mu/\mu_w)^{0.14}$	Fig. H5.2 (μ @ 140°F, μ_w @ 100°F)		0.908	0.908	0.908
16. $(Pr)^{1/3}$ (shell side)	Table 12.2, item 18		7.45	7.45	7.45
17. a, slope of curve for h	Fig. 12.11		0.56	0.56	0.56
18. b, slope of curve for f_2	Fig. 12.11 $100 < Re_p < 1200$		−0.53	−0.53	−0.53
19. C_a, coefficient in $h = C_a G_2^a$	Fig. 12.11		0.0555	0.0555	0.0555
20. C_b, coefficient in $f = C_b G^b$	Fig. 12.11		1,820	1,820	1,820
21. Δt_m, log mean temperature difference	Fig. H4.1	°F	50	50	28°C
22. LMTD correction factor	Fig. H4.2		1.0	1.0	1.0
23. G_2, shell-side flow rate	Eq. 12.18	lb/ft$^2 \cdot$ h	1.86×10^6	1.78×10^6	2,420 kg/m$^2 \cdot$ s
24. Re_p—Reynolds number for friction factor calculations	$(G_2 D_0/\mu_2)F_p$		980	975	975
25. f_2, shell-side friction factor	Fig. 12.11		0.85	0.85	0.85
26. Re_h, Reynolds number for heat transfer coefficient calculations	$(G_2 D_0/\mu)(F_h/M)$		800	825	825
27. h_2, shell-side heat transfer coefficient	$C_a G_2^a$	Btu/h $\cdot$ ft$^2 \cdot$ °F	180	175	995 W/m$^2 \cdot$ °C
28. l, baffle spacing	Eq. 12.4	ft	0.758	0.792	248 mm
29. N_2, number of shell-side fluid passes	Eq. 12.15		17	19	19
30. L, tube length	lN_2	ft	12.9	15.0	4.56 m
[a]31. D_s/l	Based on calculated value of l	ft/ft	0.85	0.91	0.91
32. G_1', tube-side flow rate	$W_1' \times 144 N_1/n \frac{\pi}{4} d_i^2$	lb/ft$^2 \cdot$ s	455	455	2,220 kg/m$^2 \cdot$ s
33. h_1, tube-side heat transfer coefficient	Figs. H5.3 and 5.4	Btu/h $\cdot$ ft$^2 \cdot$ °F	1,470	1,470	8,360 W/m$^2 \cdot$ °C
34. Re_1, tube-side Reynolds number			49,400	49,400	49,400
35. f_1, tube-side friction factor	Fig. H3.4		0.023	0.023	0.023
36. ΔP_1, tube-side pressure drop	Eq. 12.1	psi	3.90	4.55	31.4 kPa
37. Tube conductance	$(k_w 12/b_w) = (65 \times 12/0.049)$	Btu/h $\cdot$ ft$^2 \cdot$ °F	15,900	15,900	90,300 W/m$^2 \cdot$ °C
38. U	$1/h_2 + 1/h_1 + (b_w/12k_w)$	Btu/h $\cdot$ ft$^2 \cdot$ °F	159	155	880 W/m$^2 \cdot$ °C
39. U/h_2			0.885	0.885	0.885

[a]This is the step at which to check the value of D_s/l that it was necessary to assume for item 6 to proceed with the calculation. It provides the basis for the second iteration.

in evaluating G_2 in the second trial the U/h ratio was chosen as 0.9 (step 23), and this value is very close to the calculated value (step 39).

If the tube matrix length calculated is considered unsuitable, it can be shortened or lengthened by increasing or decreasing the number of tubes.

Refinements in Performance Estimates

As pointed out above, the complexities in the geometries of shell-and-tube heat exchangers and consequent devia-

tions from ideality often lead to uncomfortably large differences between the estimated performance and test results. The availability of computers has made it practicable to take into account many more details in the geometry than were included in the above design procedure, and there is extensive literature on these procedures and programs. Special surfaces for heat transfer enhancement, including the use of low-height fins as indicated in the latter part of Chapter 3 and referenced there, also are sometimes attractive. Space does not permit inclusion of even excerpts from this wealth of material, but good sets of information for the above type of unit may be found in References 13 and 14.

Ring-and-Disc Baffles

While the most widely used type of baffle is the plate-type treated above, ring-and-disc baffles are also employed because they make it possible to fill the entire cross section of the shell with the heat transfer matrix and thus reduce the size and weight of the unit. In these units, the outer half of the shell-side cross-sectional area of the casing is blocked by a ring baffle having an OD just enough smaller than the ID of the shell to permit assembly. Intermediate between the ring baffles are disc baffles that block off the central half of the cross-sectional area so that the shell-side fluid flows radially in and out between the ring and disc baffles. Methods for estimating the performance of this type of unit are presented in References 15 and 16 together with test data for some typical cases.

Rod Baffles

If a square pitch is employed, the tubes can be kept at the proper spacing by grids of rods so that the shell-side fluid flows axially through the spaces between the tubes rather than zigzagging across the tubes in the usual multipass crossflow pattern. Axial flow gives a more uniform flow distribution and reduces the pumping power requirement. It also reduces the turbulence, but there is still enough small-scale turbulence generated by the rod spacers to shift the transition between laminar and turbulent flow to a Reynolds number well below 2000; axial flow makes the transition gradual, so that, as in crossflow, the curves for both the heat transfer coefficient and the pressure drop are smooth, with no sharp discontinuity in slope through the transition region.

One way to avoid excessive blockage of the flow, and thus reduce the pressure drop across the grids of rod spacers, is to use sets of parallel rods that extend through the tube bundle vertically in one baffle plane and horizontally in the next, with the baffle spacing half that for conventional full-plate baffles. Flow-passage blockage can be reduced further by passing the rods through only every other one of the spaces between the tubes in any given grid; rods in a second grid extend through the intermediate spaces. If this is done, the grids must be spaced half as far apart as for grids with a rod in every gap between tube rows. Methods for estimating the performance of this type of heat exchanger are presented in References 17 and 18 together with test data on some typical units.

If the ratio of the length to the equivalent diameter on the shell side is to be high, the tubes must be close together and the rod diameter small. This tends to give relatively high contact pressures between the spacers and the tubes, raising the possibility of damage to the tubes caused by fretting between the tubes and the spacers. The condition can be alleviated by using flattened wires for the spacers to provide a line rather than a point contact, an approach treated in Chapter 17 on Heat Exchangers for Liquid Metals and Molten Salts. Also included in that chapter are test data on the use of staggered arrays of flattened wire spacers designed to reduce the pressure drop across a grid of spacers sufficiently so that spacers can be placed between every row of tubes rather than between alternate rows, thus cutting the number of grids in half. Note that extensive endurance testing with the flattened wire spacers has yielded no sign of fretting between the tubes and the spacers.

PLATE-AND-FRAME HEAT EXCHANGERs

Plate-and-frame heat exchangers of the type shown in Fig. 1.28 have become competitive with shell-and-tube heat exchangers for low-pressure applications, particularly for rather viscous fluids. Over 60 different patterns of turbulence promoters pressed into the plates provide a wide range of choices from which to choose that best suited to the fluid viscosity and Reynolds number that best fits any given set of requirements.[19,20] Inasmuch as the heat transfer and pressure-drop characteristics of any given plate form that is commercially available are peculiar to that plate, information of that sort is too specialized to include here; it should be obtained from the manufacturer.

REFERENCES

1. H. S. Gardner and I. Siller, "Shell-Side Coefficients of Heat Transfer in a Baffled Heat Exchanger," *Trans. ASME,* vol. 69, 1947, p. 687.

2. O. P. Bergelin et al., "Heat Transfer and Fluid Friction During Viscous Flow Across Banks of Tubes," *Trans. ASME,* vol. 72, 1950, p. 881.

3. O. P. Bergelin et al., "Heat Transfer and Fluid Friction During Flow Across Banks of Tubes—IV, A Study of the Transition Zone Between Viscous and Turbulent Flow," *Trans. ASME,* vol. 74. 1952, p. 953.

4. O. P. Bergelin, G. A. Brown, and A. P. Colburn, "Heat Transfer and Fluid Friction During Flow Across Banks of Tubes—V," *Trans. ASME,* vol. 76, 1954, p. 841.

5. F. L. Test, "A Study of Heat Transfer and Pressure Drop

Under Conditions of Laminar Flow in the Shell-Side of Cross-Baffled Heat Exchangers," *Trans. ASME*, vol. 80, 1958, p. 593.

6. B. E. Short, "A Review of Heat Transfer Coefficients and Friction Factors for Tubular Heat Exchangers," *Trans. ASME,* vol. 64, 1942, p. 779.

7. R. M. Armstrong, "Heat Transfer and Pressure Loss in Small, Commercial Shell-and-Finned-Tube Heat Exchangers," *Trans. ASME*, vol. 67, 1945, p. 675.

8. R. B. Williams and D. L. Katz, "Performance of Finned Tubes in Shell-and-Tube Heat Exchangers," *Trans. ASME*, vol. 74, 1952, p. 1307.

9. D. S. Morton, "Thermal Design of Heat Exchangers," *Industrial and Engineering Chemistry*, vol. 52, 1960, p. 474.

10. O. P. Bergelin, K. J. Bell, and M. D. Leighton, "Heat Transfer and Fluid Friction During Flow Across Banks of Tubes—VI, The Effect of Internal Leakages Within Segmentally Baffled Exchangers," *Trans. ASME*, vol. 80, 1958, p. 53.

11. F. L. Test, "The Influence of Bypass Channels on the Laminar Flow Heat-Transfer and Fluid Friction Characteristics of Shell-and-Tube Heat Exchangers," *Journal of Heat Transfer, Trans. ASME*, vol. 83-2, 1961, p. 39.

12. T. Tinker, "Shell-Side Characteristics of Shell-and-Tube Heat Exchangers," *Trans. ASME*, vol. 80, 1958, p. 36.

13. E. U. Schlunder, ed., "Heat Exchanger Design Handbook," *Chemical Engineering*, 1982 (2080 pages).

14. W. J. Marner, ed., "A Reappraisal of Shellside Flow in Heat Exchangers," ASME Bk. no. G00252 HTD-vol. 36, 1984.

15. B. Slipcevic, "Designing Heat Exchangers with Disk and Ring Baffles," *Sulzer Technical Review*, vol. 58(3), 1976, p. 114.

16. B. Slipcevic, "Shell-Side Pressure Drop in Shell-and-Tube Heat Exchangers with Disk and Ring Baffles," *Sulzer Technical Review*, vol. 60(1), 1978, p. 28.

17. C. C. Gentry et al., "RODbaffle Heat Exchanger Thermal-Hydraulic Predictive Methods for Bare and Low-Finned Tubes," Heat Transfer—Niagara Falls, 1984, *AIChE Symposium Series No. 236*, vol. 80, 1984, p. 1041.

18. C. C. Gentry, "RODbaffle Exchanger Thermal-Hydraulic Predictive Models over Expanded Baffle-Spacing and Reynolds Number Ranges," Heat Transfer—Denver, 1985, *AIChE Symposium Series No. 245*, vol. 81, 1985, p. 103.

19. J. D. Usher, "Evaluating Plate Heat Exchangers," *Chemical Engineering*, February 23, 1970, p. 90.

20. L. Caciula and T. M. Rudy, "Prediction of Plate Heat Exchanger Performance," Heat Transfer, Seattle, Wash., 1983, *AIChE Symposium Series No. 225*, vol. 79, 1983.

Gas-to-Gas Heat Exchangers

13

The requirements and problems of gas-to-gas heat exchangers differ in many respects from those of liquid-to-liquid heat exchangers. Although the heat transfer coefficients on opposite sides of the heat transfer surface are usually within a factor of 3 or 4 of each other, the absolute values usually are lower than corresponding values for liquid-to-liquid heat exchangers by a factor of 10 to 100; thus a much larger volume of heat transfer matrix is required to transmit a given amount of heat. On the other hand, in most gas-to-gas heat exchangers, up to 4% leakage from one fluid side to the other is often not objectionable, and a lighter and less rugged construction may be employed.

Types of Units Used and Their Applications

To reduce costs it is desirable to keep down the weight of metal in the heat transfer surfaces, which implies that the metal should be in base surface or in high-efficiency fins. Thus the principal types of heat transfer matrices are bundles of plain tubes, round tubes with both internal and external fins, or stacks of alternate flat and corrugated plates with the two fluid streams passing between alternate layers of the flat sheets. In this last group, the corrugated plates serve both as spacers and fins; the effective fin height is sufficiently short that the fin efficiency is high.

Good illustrations can be found in air preheaters for steam power plants and gas-fired heaters for industrial processes. Tubular air heaters are often employed for air preheating in steam plants, where the hot exhaust gases from the furnace are directed over the tubes en route to the stack, while the fresh air from forced draft blowers is directed through the tubes en route to the furnace.[1] A special type of heat exchanger also widely used for this application is the Ljungström regenerative air heater.[2] These units are commonly made of alternate layers of flat and corrugated sheet stacked around a central spindle to form a cylindrical matrix, as shown in Fig. 1.24. This is mounted in such a way that one fluid stream passes axially in one direction through one side of the cylinder while the other fluid stream passes axially in the other direction through the other side. The cylinder is rotated so that the heat deposited in the matrix by the hot gas on one side is given up to the cool gas on the other. There is, of course, a certain amount of leakage between the two fluids because of imperfect sealing of the joint between the rotating cylinder and the inlet and outlet gas ducts, and gas is carried from one fluid stream to the other in the passages rotating under the splitter between two sides of the matrix. This type of unit is feasible for gas-to-gas heat exchangers because the heat capacity of the heat transfer matrix is hundreds of times greater than the heat capacity of the low-density gas contained in the flow passages. Such a unit is not suitable for liquid-to-liquid heat exchangers, where the heat capacity of the heat transfer matrix is much less than that of the contained fluid.

Gas-turbine plants afford a major application for gas-to-gas heat exchangers, since the overall thermal efficiency can be nearly doubled through the use of a regenerator.[3] Both tubular and flat plate units are commonly employed for this application.[4,5] While Ljungström-type units are beginning to be used, the gas leakage loss from the high-pressure to the low-pressure side is a major problem because of the large pressure differential between the two fluid streams.[6] This leakage is not much of a problem in steam power plant applications because the pressure differential between the flue gas and the combustion air supply is relatively small. Other applications for gas-to-gas heat exchangers in gas-turbine plants include their use as intercoolers between stages of the compressor and as coolers in the main gas streams in closed-cycle applications. Gas-to-gas heat exchangers are also employed in a host of other applications, including oxygen production and coal hydrogenation plants.[7-10]

The type of surface used for gas-turbine applications is likely to differ widely between mobile and stationary power plant installations. There is a strong incentive to minimize both weight and volume for mobile plant installations, whereas low cost and rugged dependability are the principal considerations for the stationary types. Because of these pronounced differences in requirements and the great incentive to obtain the minimum weight and volume of heat transfer matrix for mobile gas-turbine plants, extensive tests have been carried out, and a wealth of data is available for suitable heat transfer matrices.[11]

Temperature Limitations Imposed by Corrosion

The minimum gas temperature to which the products of combustion are cooled is an important consideration in the design of air preheaters for combustion gases. Ordinarily, it is considered good practice to design so that the temperature of the combustion gas leaving the preheater does not drop below 300°F, because if this is not done, corrosion may occur as the result of condensation of sulfurous or sulfuric acid.[12] These substances are formed from the sulfur present in most fuels and the water vapor from combustion.

Oxidation of carbon steel becomes a serious problem for gas temperatures in excess of about 1000°F. As a result, the air preheaters used for steel and blast furnaces are ordinarily made of ceramic checkerwork. A multi-

plicity of units are employed, the hot gas and cold air being valved alternately to one heat transfer matrix or another, so that hot gas leaving the furnace is heating some units while others are giving up their heat to the incoming air.

TUBULAR REGENERATORS FOR GAS-TURBINE CENTRAL STATIONS

In designing tubular regenerators for gas-turbine installations the high-pressure gas logically belongs inside the tubes, because this relieves the shell of the high pressure and gives a larger tube centerline spacing. Even with this larger tube spacing, however, tube headering presents difficult problems, especially in providing adequate gas flow-passage area into the tube matrix on both the shell and the tube sides. This is especially difficult at the hot end of the heat exchanger, where the combination of high pressure and temperature on the one hand, and low allowable stresses on the other, severely restrict the choice of geometries. One clever approach to the problem is to group the tubes into bundles with a large number of small headers (Fig. 13.1), an arrangement worked out by Escher-Wyss, Ltd. The number of tube bundles required for the arrangement of Fig. 13.1 is still so large that a means for reducing it was sought in an effort to reduce the number of header welds and the consequent expense. Since the passage equivalent diameter should be kept small to yield a large amount of heat transfer surface area per unit of volume, it was logical to employ finned surfaces. The tube bundle design shown in Figs. 13.2 and 13.3 was evolved to give a large amount of surface area per tube-to-header joint. Since the heat transfer coefficient for the low-pressure gas outside the tubes is substantially lower than that inside, the larger surface area outside the tubes gives a desirable set of proportions. Note in Fig. 13.3 that triangular filler blocks prevent flow bypass through the interstices between the finned tubes, and an inner filler tube plays a similar role inside the tube. An axial cross section through the complete heat exchanger showing the tube bundles installed in the shell is given in Fig. 13.4. Note also the thick header sheet required to support the high pressure in the gas stream flowing into the header pipes leading to the individual tube bundles. The end view of this region is presented in Fig. 13.5 to show how little the shell-side flow is obstructed as it enters and leaves the heat transfer matrix.

Example 13.1. A simple bare tube axial flow regenerator is to be used in a gas-cooled reactor gas-

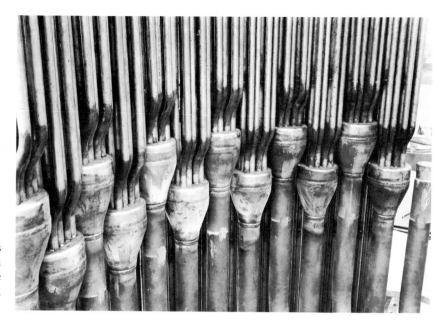

Figure 13.1 Tube bundles and headers being assembled for installation in a regenerator for a stationary gas-turbine power plant. (Courtesy Escher-Wyss, Ltd.)

turbine plant. High-pressure helium from the compressor enters the tubes at 68.9 bar/127°C (1000 psia/260°F) while low-pressure helium from the turbine enters the shell side at 11.7 bar/524°C (170 psia/975°F). The allowable pressure drops are 1.5% of the total pressure for the high-pressure side and 4.5% for the low-pressure side. The 0.4-in. ID, 0.5-in. OD tubes are arranged on an equilateral triangular pitch. Determine the tube spacing, tube length, and the total number of tubes for a heating effectiveness of 80% and a total heat load of 50 MW.

Solution. Since the specific heat of helium is constant, the temperature rise of the high-pressure gas passing through the regenerator is equal to the temperature drop of the low-pressure gas. Using English units, $\delta t = \eta \cdot \overline{\text{ITD}} = 0.80(975 - 260) = 572°F$, and the temperature difference between the hot and cold gas is $\Delta t = (975 - 260) - 572 = 143°F$.

The tube spacing can be related to the allowable pressure drops. The pressure drop for flow inside the tube can be obtained by combining Eqs. 3.12 and 3.14 and evaluating the Reynolds number to give

$$\Delta P_1 = \frac{0.2}{2g3600^2} G_1^{1.8} \frac{\mu_1^{0.2}}{\rho_1} \frac{L}{D_1^{1.2}}$$

where subscript 1 refers to the conditions inside the tube. Assuming a perfect gas, $\rho_1 = P_1 M / 1544 T_1$. Substituting

Figure 13.2 Tube bundle similar to that of Fig. 13.1 except that the tubes are larger and fitted with fins. (Courtesy Escher-Wyss, Ltd.)

Figure 13.3 Section through the tube bundle of Fig. 13.2 showing both the internal and external fins together with the triangular filler blocks that prevent bypass flow through the interstices between fin envelopes. (Courtesy Escher-Wyss, Ltd.)

and arranging the result in terms of the fractional pressure drop gives

$$\frac{\Delta P_1}{P_1} = \frac{0.2 \times 1544\mu^{0.2}}{2g3600^2 M} \frac{T_1 G_1^{1.8}}{P_1^2} \frac{L}{D_1^{1.2}}$$

A similar relation for the shell side is

$$\frac{\Delta P_2}{P_2} = \frac{0.2 \times 1544\mu^{0.2}}{2g3600^2 M} \frac{T_2 G_2^{1.8}}{P_2^2} \frac{L}{D_{e2}^{1.2}}$$

Dividing these two equations we obtain

$$\frac{(\Delta P_1/P_1)}{(\Delta P_2/P_2)} = \frac{T_1}{T_2} \left(\frac{P_2}{P_1}\right)^2 \left(\frac{G_1}{G_2}\right)^{1.8} \left(\frac{D_{e2}}{D_1}\right)^{1.2}$$

The flow rates inside and outside the tubes are related by

$$\frac{G_1}{G_2} = \frac{A_2}{A_1} = \frac{(\pi/4)D_0 D_{e2}}{(\pi/4)D_1^2} = \frac{D_0 D_{e2}}{D_1 D_1}$$

Substituting gives

$$\frac{(\Delta P_1/P_1)}{(\Delta P_2/P_2)} = \frac{T_1}{T_2} \left(\frac{P_2}{P_1}\right)^2 \left(\frac{D_o}{D_1}\right)^{1.8} \left(\frac{D_{e2}}{D_1}\right)^3$$

Substituting the numerical values, we obtain

$$\frac{0.015}{0.045} = \frac{260 + 460 + 286}{975 + 460 - 286} \left(\frac{170}{1000}\right)^2 \left(\frac{0.5}{0.4}\right)^{1.8} \left(\frac{D_{e2}}{D_1}\right)^3$$

$$\frac{1}{3} = \frac{1006}{1149} \times 0.0289 \times 1.493 \times \left(\frac{D_{e2}}{D_1}\right)^3$$

$$\frac{D_{e2}}{D_1} = \sqrt[3]{8.85} = 2.07$$

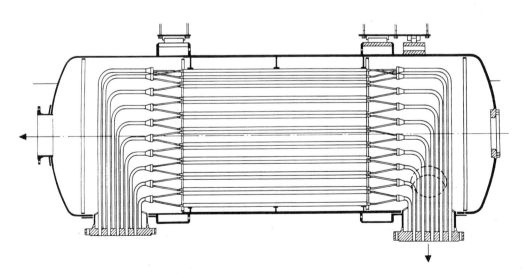

Figure 13.4 Section through a complete regenerator employing tube bundles similar to that of Fig. 13.2. (Courtesy Escher-Wyss, Ltd.)

Figure 13.5 View of the tube header region for the regenerator of Fig. 13.4. (Courtesy Escher-Wyss, Ltd.)

or

$$\frac{D_{e2}}{D_o} = \frac{D_{e2}}{D_1}\frac{D_1}{D_o} = \frac{2.07}{1.25} = 1.66$$

From Fig. H6.1 we obtain $S/d_o = 1.55$ for $d_{e2}/d_o = 1.66$. Hence the tube pitch is $S = 0.5 \times 1.55 = 0.775$ in.

The tube length should satisfy both the heat transfer and the pressure-drop requirements. These relations should provide two simultaneous equations in which the tube length and the flow rate are the two unknowns. The first equation can be obtained by equating the heat transferred across the tube wall to the heat added or subtracted from the gas, that is,

$$\pi D_1 U_1 L \Delta t = \frac{\pi}{4} D_1^{\,2} G_1 c_p \,\delta t$$

$$L = \frac{c_p D_1}{4} \frac{\delta t}{\Delta t} \frac{G_1}{U_1}$$

If we assume that the tube wall resistance to heat flow is small as compared to the gas side resistances, the overall heat transfer coefficient U_1 should be approximated quite well by

$$\frac{1}{U_1} = \frac{1}{h_1} + \frac{D_1}{D_0 h_2} = \frac{1}{h_1}\left(1 + \frac{h_1}{1.25 h_2}\right)$$

Since the same gas is flowing inside and outside the tubes, and the flow outside the tubes is axial, the mean temperatures are not significantly different. Thus the ratio h_1/h_2 can be expressed as

$$\frac{h_1}{h_2} = \left(\frac{G_1}{G_2}\right)^{0.8}\left(\frac{D_{e2}}{D_1}\right)^{0.2}$$

It has already been determined that

$$\frac{D_{e2}}{D_1} = 2.07 \quad \text{and} \quad \frac{G_1}{G_2} = \frac{D_o D_{e2}}{D_1 D_1}$$

$$= 1.25 \times 2.07 = 2.59$$

Hence

$$\frac{h_1}{h_2} = (2.59)^{0.80}\,(2.07)^{0.2} = 2.47$$

Substituting this result in the equation for $1/U_1$ gives

$$\frac{1}{U_1} = \frac{1}{h_1}\left(1 + \frac{2.47}{1.25}\right) = \frac{2.97}{h_1}$$

The heat transfer coefficient h_1 can be obtained from Eq. 3.22 to give

$$h_1 = 0.023 k_1 \frac{\mathrm{Pr}_1^{0.4}\,G_1^{0.80}}{\mu_1^{0.8} D_1^{0.2}}$$

$$h_1 = 0.023 \times 0.132 \times \frac{0.867}{0.124} \times \frac{G_1^{0.8}}{0.506} = 0.0418 G_1^{0.8}$$

Hence

$$\frac{1}{U_1} = \frac{2.97}{0.0418 G_1^{0.8}} = \frac{71}{G_1^{0.8}}$$

Substituting this result in the equation for the tube length gives

$$L = \frac{c_p D_1 \, \delta t}{4 \, \Delta t} \times 71 \times G_1^{0.2}$$

$$L = \frac{1.25 \times 0.4 \times 572}{4 \times 12 \times 143} \times 71 \times (3600 \, G_1')^{0.2}$$

$$L = 15.2(G_1')^{0.2}$$

where $G' = $ flow rate inside the tube, lb/ft$^2 \cdot$s.

The second equation for the tube length can be obtained from the relation for the pressure drop inside the tubes, that is,

$$\frac{\Delta P_1}{P_1} = \frac{0.2 \times 1544}{2g3600^2} \frac{T_1}{P_1^2 M} G_1^{1.8} \, \mu_1^{0.2} \frac{L}{D_1^{1.2}}$$

Substituting the numerical values gives

$$0.015 = \frac{0.2 \times 1544}{2 \times 32.2 \times 3600^2} \times \frac{260 + 460 + 282}{(1000 \times 144)^2 \times 4}$$
$$\times (3600 G_1')^{1.8} \times 0.593 \times \frac{L}{0.017}$$

Hence the second equation for the tube length is $L = (3.82 \times 10^4)/(G_1')^{1.8}$. Eliminating L between the two simultaneous equations gives

$$(G_1')^2 = 251$$
$$G_1' = 15.85 \text{ lb/s} \cdot \text{ft}^2 \ (77.3 \text{ kg/s} \cdot \text{m}^2)$$

Thus the tube length is $L = 15.2(15.85)^{0.2} = 26.4$ ft. The heat transfer rate per tube is

$$Q = \frac{\pi}{4} D_1^2 G_1 c_p \, \delta t$$

$$Q = 0.786 \times \left(\frac{0.4}{12}\right)^2 (3600 \times 15.85)(1.25)(572)$$

$$Q = 3.56 \times 10^4 \text{ Btu/h} \cdot \text{tube} \ (10.42 \text{ kW/tube})$$

The total number of tubes n is

$$n = \frac{\text{total heat transfer rate}}{Q \text{ per tube}}$$
$$= \frac{50 \times 10^3 \times 3413}{3.56 \times 10^4} = 4800 \text{ tubes}$$

REGENERATORS FOR MOBILE GAS-TURBINE PLANTS

Experience has shown that to meet the stringent requirements for low specific weight and volume in mobile gas-turbine plants, it is necessary to employ a heat transfer matrix having a very large surface area per unit of volume. Of course, a disadvantage of such a unit is that the small flow passages are subject to fouling and are difficult to clean. But the use of a regenerator is attractive only if it can be made very light and compact; hence there is no alternative to small passages. (One suitable matrix is shown in Fig. 1.21.) The passages in such a matrix are quite small—the equivalent diameter being in the range from 0.05 to 0.20 in., so that the Reynolds number is likely to be in the range from 100 to 1000. However, the heat transfer coefficient is still good because of the small equivalent passage diameter, and the fin efficiency is high even with thin stainless steel fins because the fin height is so small.

The design of units of this type presents such a different set of problems that an illustrative example seems in order. In practice, an extensive optimization study is required because of the complex interrelationships between the heat exchanger characteristics and the overall power plant. Thus the following example represents only a small portion of the work required for a comprehensive design study:

Example 13.2. A small gas turbine has an air flow of 1.0 lb/s, which leaves the compressor at 450°F and 80 psia. A regenerator is to be designed to give a heating effectiveness of 75% using exhaust gas that leaves the turbine at 1250°F and 16 psia. Fabrication studies have shown that 0.005-in. thick stainless steel sheet can be brazed to give a heat transfer matrix similar to that of Fig. 1.21. The air passages on the high-pressure side will have an equilateral triangular cross section 0.0866 in. on a side, while those on the low-pressure side will have the same passage width but double the passage height. Assuming a two-pass crossflow approximation to counterflow (as in Fig. 1.20) with the two passes on the high-pressure side, determine the length, width, and breadth of the heat transfer matrix. Also determine its weight and volume.

Solution. As a first approximation, take the physical properties of air at 850°F as representative and neglect the added weight of the products of combustion. Assume smooth, straight passages with no turbulators and, hence, laminar flow. (The Reynolds number should be checked after the design is completed to make sure that the last assumption is valid.) For good cycle efficiency the pressure drop must be kept to about 1.0 psi. Neglecting entrance and exit losses, the pressure drop on the high-pressure side may be chosen as 1.0 psi and that on the low pressure side as 0.5 psi.

From the Hagen-Poiseuille relation (Eq. 3.11) the pressure drop in pounds per square foot is given by $\Delta P = 0.0397 \, (\mu V L / d^2)$, where d is in inches, the mean velocity V is in feet per second, the passage length is in feet, and the viscosity is in pounds per foot-hour. The equivalent diameter of the high-pressure triangular passages is $d_1 = (4s^2 \sqrt{3}/4 \times 3s) = s/\sqrt{3} = 0.050$ in. Thus we may write

$$\Delta P_1 = 1.0 \times 144 = \frac{0.0397 \times 0.0825 V_1 L_1}{(0.05)^2}$$

Similarly, for the low-pressure side

$$\Delta P_2 = 0.5 \times 144 \frac{0.0397 \times 0.0825 V_2 L_2}{(0.10)^2}$$

For pure counterflow, from Eq. 4.25 the effectiveness is given by

$$\eta = \frac{UA/Wc_p}{1 + UA/Wc_p}$$

The effectiveness of the assumed crossflow configuration is reduced somewhat relative to that for pure counterflow; hence a correction factor should be obtained from Fig. H4.7a for a value of $W_1 c_1 / W_2 c_2 = 1.0$. While the value of UA/Wc_p has yet to be determined, it is clear from the foregoing equation that if the effectiveness is to be 75% and allowance is made for the crossflow correction factor, UA/Wc_p will be over 4.0; hence the correction factor indicated by Fig. H4.7d is probably about 0.93.

The heat transfer coefficient can be determined from Fig. H5.1 and Table H5.1. The parameter k/d for the high-pressure side is $k/d = 0.0308/0.050 = 0.616$. For the high heating effectiveness desired, the passage length-diameter ratio will be large, and the operating regime throughout the bulk of the tube will fall near the right edge of the chart in Fig. H5.1 where the heat transfer coefficient is independent of both the length-diameter ratio and the flow rate. Thus for $k/d = 0.616$, $h = 32$ Btu/h $\cdot$ ft$^2 \cdot$ °F for circular passages. Correcting this by the ratio of 3.0 to 4.3 (from Table H5.1) to allow for the difference between circular and triangular passages

gives a coefficient of 22.5 Btu/h $\cdot$ ft$^2 \cdot$ °F for the triangular passages on the high-pressure side. The equivalent flow-passage diameter on the low-pressure side will be twice as great, thus cutting the heat transfer coefficient there to 11.3 Btu/h $\cdot$ ft$^2 \cdot$ °F. However, the surface area will be nearly doubled; hence the product hA will be roughly the same for both the high- and low-pressure gas streams. Thus, neglecting the wall resistance and assuming a fin efficiency of 100%, the overall heat transfer coefficient U is about 11.3 Btu/h $\cdot$ ft$^2 \cdot$ F.

Using the values in the preceding paragraph, the expression for the crossflow effectiveness becomes

$$\eta = 0.93 \frac{11.3A/Wc_p}{1 + 11.3A/Wc_p} = 0.75$$

Solving for $11.3A/Wc_p$ gives

$$\frac{11.3A}{Wc_p} = \frac{0.75/0.93}{1 - 0.75/0.93} = 4.2$$

Note that this gives a crossflow correction factor of 0.935, which is close to the value of 0.93 used as a first estimate.

The surface area required can now be determined; that is,

$$11.3A/Wc_p = 4.2; \quad A = 4.2Wc_p/11.3$$

$$= (4.2 \times 3600 \times 0.25)/11.3;$$

$$A = 335 \text{ ft}^2$$

The volume of the heat transfer matrix on the high-pressure side can be determined from the surface area. If half of the 0.005-in. thick wall is charged to each surface, the surface-volume ratio for a single channel 1.0 ft long is given by

$$\frac{\text{Surface, ft}^2}{\text{Volume, ft}^3} = \frac{3 \times 0.0866/12}{0.25 \sqrt{3} (0.0871)^2/144} = 945 \text{ ft}^2/\text{ft}^3$$

Thus the matrix volume required for the high-pressure passages becomes $335/945 = 0.354$ ft^3, of which approximately 90% is flow-passage area (i.e., $0.0866^2/0.0871^2$), and 10% is metal. Similarly, the matrix volume for the low-pressure passages becomes 0.708 ft^3, so that the total matrix volume is 1.062 ft^3.

The average density on the high-pressure side is 0.1655 lb/ft^3, whereas that on the low-pressure side is 0.0332 lb/ft^3. Thus for $W_1 = W_2 = 1.0$ lb/s, the volume flow rates are 6.04 ft^3/s and 30 ft^3/s, respectively.

Three quantities remain to be determined, and three relations are available from which to determine them. From the pressure-drop relations the product of the high-pressure air velocity and the passage length is $V_1 L_1 = 110$. Letting A_1 be the high-pressure flow-passage area, the high-pressure air flow rate is $V_1 A_1 = 6.04$ ft^3/s. The flow-passage volume for the high-pressure air (after allowing for the fact that 10% of the matrix volume is metal) becomes $L_1 A_1 = 0.90 \times 0.354 = 0.319$. Solving the first two of the preceding equations for V_1, and equating, gives

$$\frac{110}{L_1} = \frac{6.04}{A_1}; \quad L_1 = \frac{110 A_1}{6.04}$$

Substituting this value for L_1 in the third equation and solving for A_1 gives

$$\frac{110 A_1^2}{6.04} = 0.319$$

$$A_1 = \sqrt{0.0176} = 0.133 \text{ ft}^2$$

Hence $L_1 = 2.4$ ft; $V_1 = 45.5$ ft/s.

The corresponding quantities on the low-pressure gas side can be determined in a similar fashion, that is,

$$V_2 L_2 = 220$$

$$V_2 A_2 = 30 \text{ ft}^3/\text{s}$$

$$L_2 A_2 = 0.90 \times 0.708 = 0.637 \text{ ft}^3$$

$$A_2 = \left(\frac{0.637 \times 30}{220} \right)^{1/2} = 0.295 \text{ ft}^2$$

$$L_2 = 2.16 \text{ ft}$$

$$V_2 = 98.3 \text{ ft/s}$$

The matrix dimensions now can be established to satisfy the preceding conditions. By referring to the initial conditions, the length in the low-pressure gas flow direction is 2.16 ft, and that for each pass of the high-pressure air is $2.4/2 = 1.2$ ft. These values coupled with the total matrix volume just determined give a matrix width transverse to both flow directions of $1.062/(2.16 \times 1.2) = 0.41$ ft.

Several assumptions were made in starting the problem, and these should be checked. By referring to Fig. H3.1, the Reynolds number for the high-pressure air is 1100, while that for the low-pressure gas is 1000, thus confirming the expectation that the flow would fall in the laminar regime for both streams. The values for the parameter x/dG' are 84.5 and 85 for the high- and low-pressure streams, thus confirming the basis for estimating the heat transfer coefficient. The fin efficiencies as determined from Fig. H7.3 are about 93% and 89% for the high- and low-pressure gas streams, respectively. Thus the initial simplifying assumption of 100% introduced an error of 10%; a second iteration is in order.

The matrix weight can be estimated from the matrix volume of 1.062 ft^3 and from the fact that the metal surfaces constitute 10% of the total volume. Since steel weighs 500 lb/ft^3, the matrix weight is $1.062 \times 500 \times 0.10 = 53$ lb. The casing and headers would add substantially to this, depending on the installation.

Performance Characteristics of Rotary Regenerators

The heat transfer techniques applicable to the rotary regenerator are essentially similar to those applicable to the more conventional heat transfer matrices discussed previously, except that the periodic flow introduces several new variables. For conventional heat exchangers it is sufficient to define the inlet and outlet temperatures, the flow rates, the heat transfer coefficients, and the surface areas for the two sides of the heat exchanger. For the rotary regenerator it is also necessary to relate the heat capacity of the rotor to that of the fluid streams for the fluid-flow rates and rotative speed in question. The solution of the heat transfer relations is complicated by the introduction of a new variable to account for the heat capacity of the rotor. Furthermore, the coupling between the heat transfer coefficients and flow rates in conventional heat exchangers is such that these can be reduced to two variables instead of four variables, whereas for rotary regenerators all four variables must be handled separately. Generalized differential equations relating these parameters can be set up, but no general solution to these equations has been worked out. Particular solutions for many cases of interest have been obtained by graphical and numerical methods, and analytical solutions have been obtained for the simpler particular cases.[2,13,14] The most convenient of these solutions depends on the simplifying assumptions that the heat capacity of the rotor is infinite and the heat transfer surface areas, the mass flow rates, and the specific heats of the two fluid streams are equal. For this case, the relations reduce to that for a conventional counterflow heat exchanger, as given in Eq. 4.25, that is,

$$\eta_1 = \frac{UA/W_1 c_1}{1 + UA/W_1 c_1}$$

Since the thermal resistance of the wall is negligible, this becomes

$$\eta_1 = \frac{hA/2W_1c_1}{1 + hA/2W_1c_1} \qquad (13.1)$$

Equation 13.1 gives the heating effectiveness for the cold-air side of the regenerator in terms of the heat transfer coefficient, the surface area exposed to the gas flow, the gas-flow rate, and the gas heat capacity. When used in air, rotary regenerators can be designed to approach this ideal performance rather closely.

Effects of Combustion Products

Most rotary regenerators are applied to the recovery of heat from combustion gases by transferring that heat to the fresh, cold air, thus increasing the furnace temperature and improving combustion. Addition of fuel to the air in the combustion process leads to an increase in the mass flow rate on the hot gas side on the order of 7% relative to the mass flow of the cold air. At the same time, the heat capacities of the products of combustion (i.e., CO_2 and H_2O) are higher than that of the cold air. Depending on the fuel-air ratio, these factors may make the heat capacity for the combustion products as much as 11% greater than that of the cold air, that is, $W_1c_1/W_2c_2 = 0.90$.

Effects of Rotor Mass and Speed

The temperature of any given element of the metal surface fluctuates as the rotor revolves, as indicated in Fig. 13.6. The amount by which the temperature varies is inversely proportional to both the rotor mass and its speed.

In making a quantitative evaluation of the effects of rotor mass and rpm on the overall performance, the material in the rotor can be considered as if it were a third fluid stream transporting heat from the hot gas to

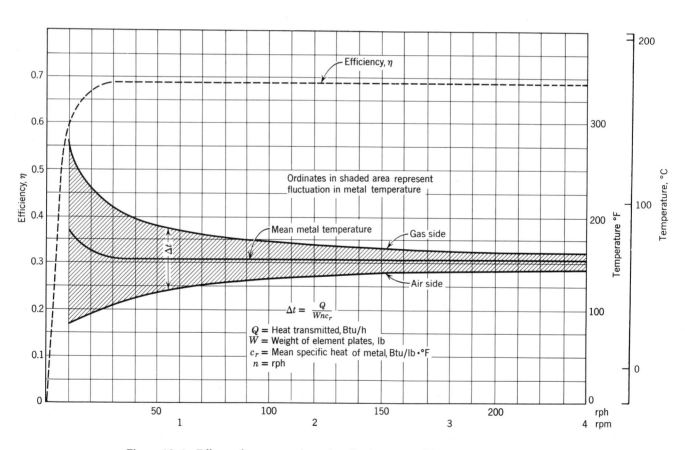

Figure 13.6   Effects of rotor speed on the effectiveness and local metal temperatures at the cold-air inlet end of a large rotary regenerator for a coal-fired steam power plant. (Karlsson and Holm, Ref. 2.)

the cold air.[14] In this sense it can be seen that the "flow rate" of the rotor material should be such that its change in temperature will be small; that is, the product of the rotor mass and specific heat should be several times that of the corresponding quantities for the gas flows. In analyses it has been found convenient to use the parameter $W_r c_r / W_1 c_1 \theta$, where θ is the period of revolution of the rotor in hours, W_r the rotor mass in pounds, and c_r its specific heat.[13]

The effects of this parameter on the performance of rotary regenerators have been evaluated for $W_1 c_1 / W_2 c_2 = 1.0$ and 0.9;[13] the results are presented in Figs. 13.7 and 13.8. Note that the upper curve in each of these two figures corresponds to the case covered by Eq. 13.1. These curves show that for good performance the parameter $W_r c_r / W_1 c_1 \theta$ should be 5 or more. With air or other gases this is ordinarily an easily satisfied condition. Note also that there is little difference between the curves for $W_1 c_1 / W_2 c_2 = 1.0$ and those for $W_1 c_1 / W_2 c_2 = 0.9$. Thus it is evident that for values of $W_r c_r / W_1 c_1 \theta$ greater than 5, Eq. 13.1 is a good approximation, even if the effects of combustion products on the ratio $W_1 c_1 / W_2 c_2$ are neglected and it is taken as unity.

Effects of Carry-Through

The discussion until this point has been concerned with idealized cases which include no allowances for the

transport of cold gas into the hot-gas region or hot gas into the cold-gas region. The loss resulting from transport of this nature is directly proportional to the ratio of the rotor speed to the gas-flow rate. This factor is directly proportional to the ratio of the heat capacity of the gas within the heat transfer matrix to the heat capacity of the heat transfer matrix itself, a ratio that is ordinarily less than 0.001 for regenerators in gas systems. Since there is no incentive to run the rotor at a speed such that the ratio $W_r c_r / W_1 c_1 \theta$ is greater than 10, the losses associated with transport of gas from the cold to hot sides of the regenerator, and vice versa, are ordinarily less than 1%. However, this does indicate another reason for not using rotary regenerators for liquids; the heat capacity of the liquid contained in the rotor would certainly be equal to or greater than the heat capacity of the rotor itself; hence the fluid carry-through would lead to large losses in performance.

Effects of Lateral Leakage

Most rotary regenerators have been employed in conjunction with combustion furnaces where the pressure difference between the cold-air duct and the hot flue gas has been a matter of a fraction of a pound per square inch. The leakage problem becomes much more serious if a rotary regenerator is applied to gas-turbine installations in which the pressure ratio across the

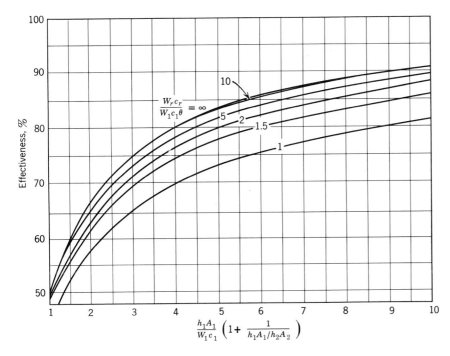

Figure 13.7 Effects of heat transfer design parameters on the effectiveness of rotary regenerators for $W_1 c_1 / W_2 c_2 = 1.0$ for a series of values of $W_r c_r / W_1 c_1 \theta$. (Coppage and London, Ref. 13.)

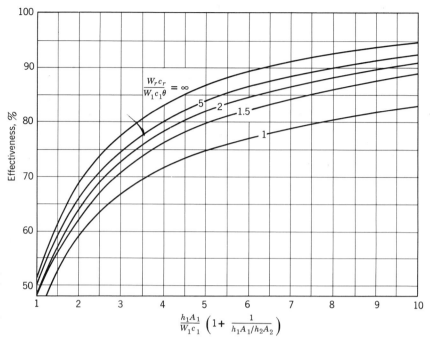

Figure 13.8 Effects of heat transfer design parameters on the effectiveness of rotary regenerators for $W_1 c_1 / W_2 c_2 = 0.9$ for a series of values of $W_r c_r / W_1 c_1 \theta$. (Coppage and London, Ref. 13.)

compressor and turbine is on the order of two to six. For such applications the entire performance, in fact the feasibility of a rotary regenerator, hinges on minimizing the leakage of gas from the cold to the hot stream through the clearances between the rotor and the casing. Various devices have been patented and applied to minimize this leakage. In all cases the leakage flow rates are heavily dependent on the details of the design, manufacturing tolerances, wear rates in operation, and related factors, so that no generally applicable treatment can be attempted here. It should be pointed out, though, that the seal design problem is made more difficult by axial differential thermal expansion between the cold and hot sides of the rotor and between the rotor and the casing.

Effects of Matrix Thermal Conductivity

The length of the heat flow path within the heat transfer surface is so short that thermal conductivity has a negligible effect on the performance. By the same token, from the heat transfer standpoint, the effect of the use of ceramic rather than metal plates and of deposits of soot or coke on the surfaces of a rotary regenerator are quite small. In fact, in some units the mass of a deposit may add significantly to the heat capacity of the rotor, and thus actually improve its heat transfer performance. Such

deposits do, however, restrict the air flow and hence increase the pressure drop through the unit, so that provisions for cleaning them off should be made.

Axial conduction from the hot to the cold end may introduce appreciable losses in performance by reducing the mean temperature. If steel is used for the matrix, the thermal conductivity in the axial direction tends to reduce the performance of a unit slightly—perhaps 0.6% for a large air preheater in a steam plant. If copper is used, the effect becomes appreciable—perhaps as much as 6%, even in a large unit.

Lateral conduction from the hot to the cold zone is advantageous in that its effect is similar to that of increasing the rotor speed without increasing the carry-through losses.

Effects of Matrix Geometry

A close-up view showing the construction of a rotor for a large steam plant is presented in Fig. 13.9. As in most types of gas heat exchanger, pumping power considerations tend to give fluid-flow conditions in or near the laminar flow regime. Thus it is often worthwhile to introduce some form of turbulation in the heat transfer matrix rather than to use the simple configuration of Fig. 13.10. Tests have shown that employing undulated plates in place of flat plates to give the configuration of Figs. 13.11 and 13.12 makes possible improvements in

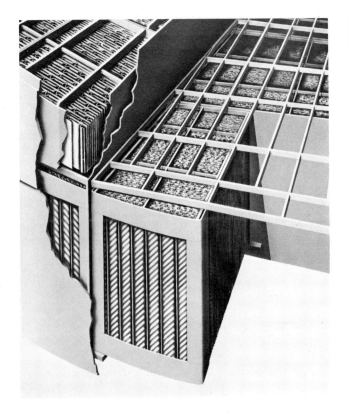

Figure 13.9 Cutaway view showing the construction of the rotor for a large rotary regenerator such as that shown in Fig. 1.24. (Courtesy The Air Preheater Co.)

performance. The pressure losses associated with this geometry are sufficiently large, however, that less severe turbulating devices are sometimes preferred.

Effects of Matrix Material

Where weight is an important factor, ceramic heat transfer matrices have an advantage in that their specific heat is much higher than that of steel. Furthermore, it

has been found possible to fabricate such matrices with passages having a very small equivalent diameter using an aluminum oxide base ceramic having a good thermal conductivity and a very low coefficient of thermal expansion. A radial-flow regenerator drum of this construction for a small gas turbine is shown in Fig. 13.13, and a close-up end view of the heat transfer matrix for an axial flow disc-type unit is illustrated in Fig. 13.14. It is even possible to fabricate these units with the passages undulated in the direction of the gas flow, as in Fig. 13.15.

Tests of a set of three units similar to those of Figs. 13.13–13.15 having the proportions shown in Table 13.1 were reported in Reference 15. Figure 13.16 summarizes the test results for one of these units both in the new, clean condition and after 17 and 31 h of operation in a gas turbine. These test results are very important, for they show that one can indeed achieve the high heat transfer coefficients and low-pressure drops ideally obtainable with very fine passages operated in the laminar flow range at low Reynolds numbers in the region from 50 to 800. That is, the data show that the fabrication technique gave such a high degree of uniformity and such smooth surfaces in these tiny passages that the actual performance approached the ideal within about 7%. Although the possibility of this excellent performance has been recognized for some time, it has not been widely utilized because even a relatively thin layer of dirt or crud deposited on the surface would roughen the surface and obstruct an appreciable fraction of the flow passage, causing a large increase in the pressure drop. Fortunately, for this application in a gas turbine the gas flow reverses direction twice or more during each revolution of the heat transfer matrix, thus providing a self-cleaning action that should minimize dirt deposition. Then too, in some applications the fluid may be sufficiently clean that serious fouling will not occur, or the only fouling might be carbonaceous and the temperatures high enough to burn it off with little buildup of carbon or ash.

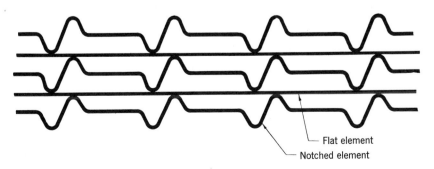

Flat element
Notched element

Figure 13.10 Detail showing the form of the axial corrugations used for spacing the plates in the heat transfer matrix at the cold-air inlet end of the regenerator of Fig. 13.9 (Courtesy of The Air Preheater Co.)

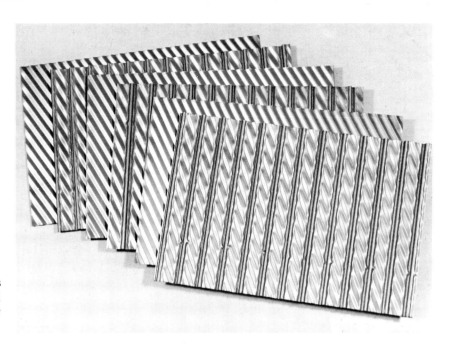

Figure 13.11   Heat transfer matrix plates with turbulence-promoting diagonal corrugations. (Courtesy The Air Preheater Co.)

Selection of Rotary Regenerator Proportions

As with other types of heat exchanger, the optimum proportions of a unit are heavily dependent on the overall economics of the plant in which it is to be installed. Factors influencing the design include fuel costs, capital charges, space availability, pumping power requirements, the extent to which soot and ash may be deposited from the fuel to be employed, and corrosion of the matrix by contaminants in the fuel to be used.

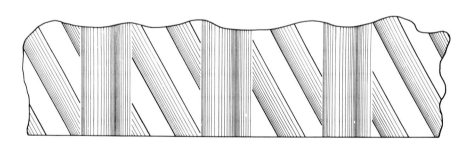

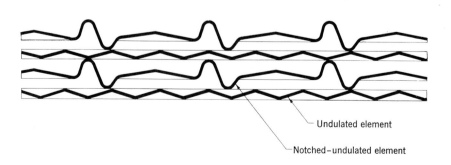

Undulated element

Notched–undulated element

Figure 13.12   Side and end views of the air passages formed by the plates of Fig. 13.11. (Courtesy The Air Preheater Co.)

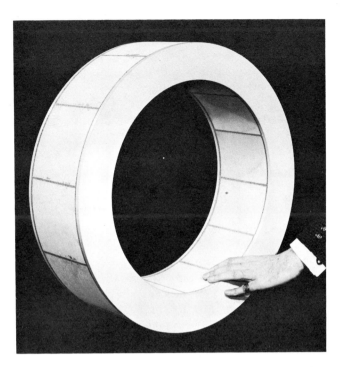

Figure 13.13 Radial-flow, heat transfer matrix for a small gas-turbine regenerator. The rim, matrix, and bonding materials are glass-ceramic compositions with coefficients of expansion very close to zero. (Courtesy Corning Glass Works.)

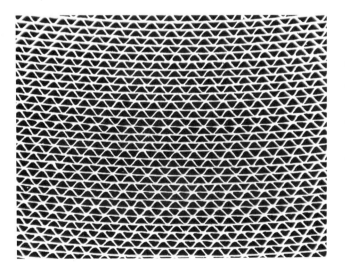

Figure 13.14 End view of an alternate flat and corrugated plate glass-ceramic matrix for an axial-flow rotary regenerator. The equivalent diameter of the air passages is only about 0.050 in. (Courtesy Corning Glass Works.)

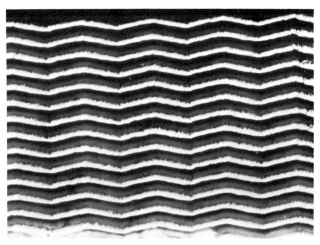

Figure 13.15 Axial section through an undulated passage for a glass-ceramic heat transfer matrix somewhat similar to that of Fig. 13.14. (Courtesy Corning Glass Works.)

PERIODIC-FLOW STATIC HEAT TRANSFER MATRICES

There are many applications in the metallurgical and ceramics industries in which coal-, oil-, or gas-fired furnaces must operate at temperatures of 1370°C (2500°F) or higher, that is, temperatures above those obtainable if the combustion air is not heated before it enters the burners. Preheaters may also be used to reduce fuel costs for processes requiring heat at intermediate temperatures where the combustion products would

TABLE 13.1 Data for the Geometry of the Heat Transfer Matrix in Glass-Ceramic Rotary Regenerator Units Similar to That of Fig. 13.13 That Were Tested at Stanford University[a]

	505A	503A	504A
Core no. (Stanford):			
Corning designation:	L768	L689	L612
Cell count, N, cells/ in.^{2}	526	1008	2215
Porosity, p	0.794	0.708	0.644
Hydraulic dia., $4r_h$, (10^{-3} ft)	2.47	1.675	1.074
Area density, α, ft^2/ ft^3	1285	1692	2397
Cell height/width, d^*	0.731	0.709	0.708
$L/4r_h$	101	149.5	233

[a]Ref. 15. (Courtesy of Stanford University.)

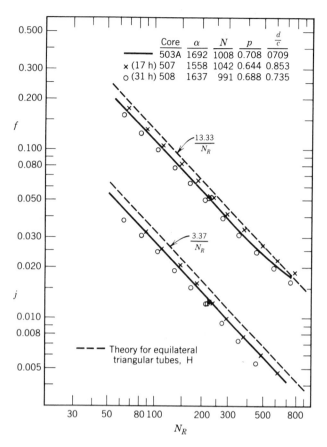

Core	α	N	p	$\frac{d}{c}$
——— 503A	1692	1008	0.708	0709
x (17 h) 507	1558	1042	0.644	0.853
o (31 h) 508	1637	991	0.688	0.735

Figure 13.16 Results of tests of Core 503A of Table 13.1 for operation when new and clean as well as after 17 and 31 h of operation in a gas turbine. (London, Ref. 15.)

otherwise be exhausted at temperatures of 538°C (1000°F) or more. Where the heat transfer matrix of a preheater must operate at temperatures exceeding about 316°C (600°F), the matrix is commonly made of ceramic material. The basic relationships on which the design of such a unit depends are those presented in the preceding section where effects of matrix materials were discussed. Although the size of the unit required is inversely proportional to the heat transfer surface area per unit of volume so there is an incentive to make use of thin sections, fabrication and maintenance problems are likely to favor a coarse matrix. Brick checkerwork, for example, has been widely used. It gives a bulky unit, but the construction is simple, inexpensive, and requires little maintenance. Crossflow over screens or layers of spaced rods has been used for some applications.[16,17] Bundles of rods or tubes with the flow parallel to their axes have also been employed. Although the use of either rods or tubes makes possible a large surface-to-volume ratio, the

flow-passage area obtainable with tightly packed rods in an equilateral triangular array is only about 9% of the total matrix cross section, whereas this parameter can be increased to about 50% with tubes, thus increasing the flow capacity per unit of volume.

The heat transfer performance of all the foregoing types of regenerators can be estimated by applying the basic techniques outlined in the preceding section.

Packed Beds

An interesting type of heat transfer matrix, and one especially attractive for some applications, is provided by a bed of spheres, cylindrical slugs, pebbles, or granular particles randomly packed by pouring them into a cylindrical shell. Beds of this sort can be used statically in periodic-flow regenerator systems, or the beds can be allowed to flow so that the solid particles act as the heat transport medium. A system of the second type is shown in Fig. 13.17. Spheres or rounded pebbles are ordinarily used for such applications to improve the flow characteristics of the bed.

Packed beds are simple in construction and give large heat transfer surfaces for a given matrix volume, but the pressure drop through them tends to be high relative to that of units with smooth straight passages. Because of the many variables, a general comparison is difficult to make; for example, the void fraction in a bed of spheres is fixed, whereas it can be varied over a wide range in a matrix of prismatic blocks. A good comparison can be made for a particular case, however, once the boundary conditions are established. An interesting study of a ceramic nuclear reactor application has been made in which the heat transfer performance of beds of spherical fuel elements was compared to that of a matrix of hexagonal prismatic fuel elements having a central cylindrical hole.[18] In this comparison the void fraction, the film temperature drop, and the matrix height were determined for a bed of 2.5-in. diameter spheres at each of a series of pumping power-to-heat removal ratios. (The importance of the pumping power-to-heat removal ratio as a parameter was discussed briefly in Chap. 11.) The diameter of a matrix of prismatic blocks to give the same pumping power-to-heat removal ratio was then determined. To satisfy the boundary conditions it was necessary to allow the hole diameter in the prismatic blocks to vary. The results are shown in Fig. 13.18. The internal temperature drop in the ceramic fuel elements together with the resulting thermal stresses is also presented. Note that in every respect the hexagonal blocks give more favorable characteristics. The smaller diameter of the prismatic matrix results in a heat exchanger volume

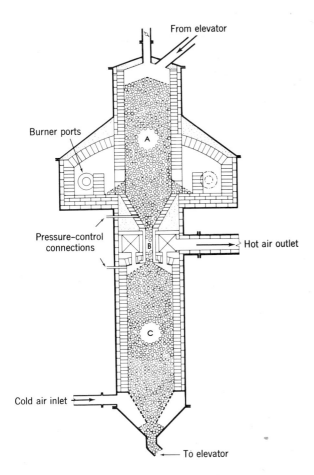

Figure 13.17 A flowing pebble-bed regenerator for preheating the air supplied to a furnace. (Courtesy The Babcock and Wilcox Co.)

about one-third that of the bed of spheres. Thus the cost of the larger vessel, and greater amount of material for a packed bed of spheres, must be weighed against the reduced cost of installation and ease of replacement of the spheres in choosing between the matrices. For some installations the ease with which the heat transfer matrix can be installed and removed is the controlling consideration, in which case the packed bed is probably the best choice.

The pressure-drop and heat transfer performance of packed beds depend on the size and geometry of the packed particles, which in turn determine both the heat transfer surface area and the flow-passage area. The void space in percent of the total volume is more easily measured (or calculated) than the surface or flow-passage areas; hence the voidage is more commonly used in making pressure-drop and heat transfer calculations. For beds of spheres (having the diameter D_s) the surface area

per unit of volume is given by $A = 6(1 - \epsilon)/D_s$. The voidage ϵ in ordered arrangements of spheres varies from 0.26 for rhombohedral arrays to 0.48 for simple cubic arrays.[19] If spheres are poured into cylindrical containers to give randomly packed beds, the average voidage is about 0.38 if the ratio of container diameter to sphere diameter is greater than about 20.[19,20] The corresponding value for randomly packed cylindrical slugs is 0.25.[20]

The extent of local variations in randomly packed beds has been investigated extensively.[19,20] Figure 13.19 shows the radial variation of voidage with distance from the wall for spheres, cylinders, and toroids, randomly packed by simply pouring them into cylindrical vessels.[20] Note that for spheres and cylindrical particles the voidage near the wall is greater than that near the center, and that the cyclic variation in the voidage extends into the matrix from the wall for more than two particle diameters, the amplitude decreasing with distance from the wall.

The radial variation in voidage close to the wall of a randomly packed bed leads to variations in the velocity distribution across the diameter in the manner shown in Fig. 13.20.[21]

The local heat transfer coefficient has been found to vary from one point to another around a ball in a bed of packed spheres, the smallest values occurring at the points where the spheres touch each other. However, the minimum value is not zero at the contact point, but is about 40% of the average[19]—much more than one would expect. The average heat transfer and pressure-drop characteristics for flow through randomly packed beds of spheres have been determined by several investigators,[22,23] and Fig. H5.14 shows some typical results. Note that in the correlation used in Fig. H5.14 the mass flow rate of the coolant is based on the total inlet face area of the bed, and the data are valid for the particular void fraction of the experiment, that is, 0.37.

If an "upflow" configuration is used, "floating" of the bed may prove a problem at high flow rates. Tests indicate that for beds of spheres, floating becomes a problem when the pressure drop across the bed runs from 83 to 87% of the weight of the bed.

Example 13.3. Helium at 98.5 bar (1000 psia) and an average temperature of 482°C (900°F) is to be heated by passing it through a randomly packed bed of 38.1-mm (1.5-in.) diameter spheres 2.438 m (8 ft) high. The mass flow rate through the bed is to be 30.1 kg/s · m² (22,200 lb/ft² · h). Determine the surface area per unit of volume, the Reynolds number, the pressure drop, and the average heat transfer coefficient.

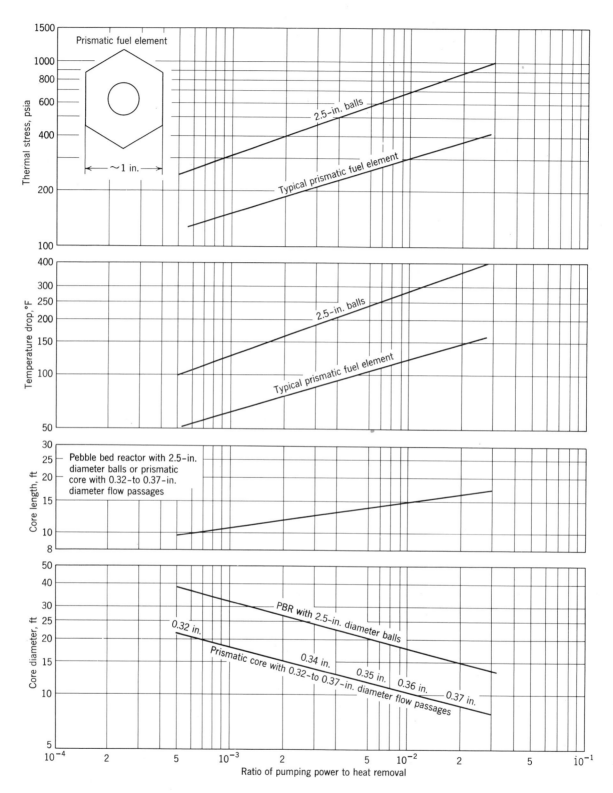

Figure 13.18 Comparison of the heat transfer performance of aligned prismatic block and random pebble-bed reactor (PBR) heat transfer matrices. (Courtesy Oak Ridge National Laboratory.)

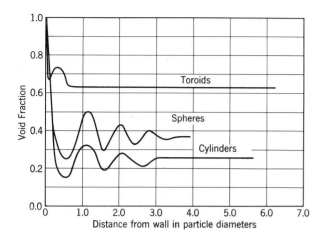

Figure 13.19 Effects of the radial distance from the wall on the local void fraction for spheres, cylinders, and toroids poured randomly into cylindrical containers. (Roblee, Baird, and Tierney, Ref. 20.)

Solution. Using SI units and assuming a void fraction $\epsilon = 0.4$, the surface area per unit of volume is

$$A = \frac{6(1 - \epsilon)}{D_s} = \frac{6 \times 0.6}{0.0381} = 94.5 \ \text{m}^2/\text{m}^3 \ (28.8 \ \text{ft}^2/\text{ft}^3)$$

The Reynolds number is

$$\text{Re} = \frac{G_s D_s}{\mu} = \frac{30.1 \times 0.0381}{0.386 \times 10^{-4}} = 29,600$$

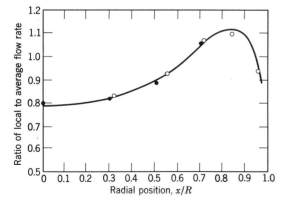

Figure 13.20 Ratio of the local flow rate to the average flow rate through a 4-in. diameter bed of randomly packed cylinders having a diameter and length of 0.25 in. (Schwartz and Smith, Ref. 21.)

The Stanton number and friction factor given by Fig. H5.14 are 0.034 and 2, respectively. The pressure drop thus becomes

$$\Delta P = \frac{fG^2 L}{2\rho g D_s} = \frac{2 \times (30.1)^2 \times 2.438}{2 \times 4.39 \times 0.0381}$$

$$= 13.2 \ \text{kPa} \ (275 \ \text{psf})$$

The average heat transfer coefficient as determined from the Stanton number is

$$h = 0.034 \ G c_p = 0.034 \times 30.1 \times 5.234$$

$$h = 53.5 \ \text{kW/m}^2/ \cdot \ ^\circ\text{C} \ (945 \ \text{Btu/h} \cdot \text{ft}^2 \cdot \ ^\circ\text{F})$$

Temperature Limitations

At first thought it might appear that a packed bed of ceramic pebbles or perforated blocks could be operated at temperatures up to close to the sintering temperature (about 75% of the melting point of the ceramic in absolute temperature units). This does not seem to be the case. The most common application of this type of unit is for blast furnace stoves, which ordinarily heat the air fed to the furnace to 980°C (1800°F), far below the sintering temperature of the firebrick with which they are packed. Higher temperatures would be advantageous for at least some applications, but the highest temperature that the author has found in the literature was 1300°C (2370°F) in experiments in Germany. This was achieved only with clean hot gas obtained with a natural gas burner; gas from burning coal caused rapid deterioration of the ceramic packing. Another singularly significant effort has been carried out in the magneto-hydrodynamic generator program, in which there is an extremely strong incentive to obtain the highest possible temperature in the air fed to the combustor. This work has shown that both MgO and Al_2O_3 deteriorate rapidly at temperatures above 1300°C (2370°F) when operated on the products of combustion of coal, although spinels of these materials show somewhat better resistance to attack.[24] In principle, it appears possible to use alumina and magnesia in an inert gas up to their nominal sintering temperatures, about 1450°C (2640°F) for Al_2O_3 and 1800°C (3270°F) for MgO.

RECUPERATORS UTILIZING HEAT PIPES

Heat pipes provide an excellent basis for the design of recuperators for many applications, particularly if there must be no leakage from one fluid stream to the other. The pressure drops are low, the ductwork and headering are simple and straightforward, the units are compact, cleaning is not difficult, and they have become competitive in cost.[25] Thus it seems in order to consider the design of a unit similar to that of Fig. 1.34.

Example 13.4. A recuperator is required for heating air for a dryer in a chemical plant by extracting heat from the flue gas emerging from a gas-fired low-pressure boiler. The furnace is fired with sulfur-free natural gas, so a stack gas temperature as low as 93°C (200°F) should not lead to difficulties with corrosion. The flue-gas temperature entering the unit will be 177°C (350°F), the air entering the unit will be 38°C (100°F), and the discharge air temperature desired is 121°C (250°F). The weight flow rates for the two gas streams will be the same.

Solution. Choose as the first possibility a set of finned heat pipes having the same geometry as the 0.774-in. OD finned tubes of the matrix of Fig. H5.13. The key dimensional data using the B spacing are

Tube OD = 19.65 mm (0.774 in.)

Fin OD = 37.2 mm (1.463 in.)

Fin thickness = 0.305 mm (0.012 in.)

Fin spacing = 2.81 mm (0.1105 in.)

Fin area/total area = 0.835

Free flow area/inlet face area = 0.572

Heat transfer area/total volume = 279 m^2/m^3 (85.1 ft^2/ft^3)

Equivalent passage diameter = 8.21 mm (0.0269 ft)

Tube spacing = 50.4 × 44.5 mm (1.982 × 1.75 in.)

For a first trial use a Reynolds number of 3000 for which the chart gives a value of 0.00715 for the heat transfer parameter. The physical properties of the fluid streams for an average heat pipe metal temperature of 93°C (200°F) can be obtained from Table H2.4, that is, the specific heat is 1.011 kJ/kg · °C (0.241 Btu/lb · °F), the viscosity 0.2146 × 10^{-4} Pa · s (0.0519 lb/h · ft), and the Prandtl number 0.685. Using these values we obtain

$$G = 3000 \times 0.2146 \times 10^{-4}/0.00821$$
$$= 7.85 \text{ kg/s} \cdot \text{m}^2 \text{ (5780 lb/h} \cdot \text{ft}^2)$$
$$h = 0.00715 \times 7.85 \times 1011/0.777$$
$$= 73 \text{ W/m}^2 \cdot \text{°C (12.8 Btu/h} \cdot \text{ft}^2 \cdot \text{°F)}$$

Estimating the fin efficiency from Fig. H7.2 yields a value of 0.94, which can be combined with the ratio of fin area to total area of 0.835 to give an effective heat transfer coefficient 95% of the above value. The temperature drop within the heat pipe can be estimated quickly from Table 6.2 which gives a radial temperature drop of 11.1°C for a water heat pipe at a heat flux of 10 kW/m^2. In this case, the heat flux is approximately 27.7 × 0.073 = 2 kW/m^2, so that the radial temperature drop would be about 2°C, reducing the effective temperature difference from 27.7 to 25.7°C (46°F).

For the pure crossflow configuration of Fig. 1.33, and with the same weight and mass flow rates in the two streams, the temperature difference between the metal surface of the heat pipes and each gas stream will be uniform throughout the matrix, running 25.7°C (46°F) if the small temperature drop within the heat pipe is included. The weight flow per unit of inlet face area is 57.2% of the mass flow rate through the passages between the tubes, hence the amount of heat transferred in traversing a unit volume of the heat transfer matrix is

$$Q = hA \ T = 73 \times 279 \times 25.7$$
$$= 523 \text{ kW/m}^3 \text{ (50,500 Btu/h} \cdot \text{ft}^3)$$

The gas temperature rise per unit length would be

$$\Delta T = Q/Wc_p = 523/7.85 \times 0.572 \times 1011$$
$$= 115 \text{°C/m (63°F/ft)}$$

The length of the matrix would be

$$L = 83.3/115 = 0.725 \text{ m (2.4 ft)}$$

If the length of heat pipe in each gas stream is chosen to be 3 m (9.85 ft), the heat load on each heat pipe can be estimated by first calculating the number of heat pipes per meter of height and then the heat load per pipe:

Number of heat pipes/m of height
$$= 0.725/(0.0504 \times 0.0445) = 323$$
Heat load/pipe = 523 × 3 × 0.75/323
$$= 3.63 \text{ kW}$$

Referring to Fig. 6.7 to appraise the capacity of a water heat pipe, it can be seen that the capacity of a 25.4-mm diameter heat pipe at the lowest temperature in the matrix for this application, namely, 65°C, would be over 20 kW. The capacity would vary as the square of the diameter, hence the capacity of the 17.65-mm tubes contemplated here would be about 60% that of the units of Fig. 6.7, or about 12 kW for this case. Thus the size chosen for this example has more than sufficient capacity.

REFERENCES

1. E. F. Rothmich and G. Parmakian, "Tubular Air-Heater Problems," *Trans. ASME*, vol. 75, 1953, p. 723.

2. H. Karlsson and S. Holm, "Heat Transfer and Fluid Resistances in Ljungstrom Regenerative-Type Air Preheaters," *Trans. ASME*, vol. 65, 1943, p. 64.

3. D. B. Harper and W. M. Rohsenow, "Effect of Rotary Regenerator Performance on Gas-Turbine-Plant Performance," *Trans. ASME*, vol. 75, 1953, p. 759.

4. C. Keller, "Operating Experience and Design Features of Closed-Cycle Gas-Turbine Power Plants," *Trans. ASME*, vol. 79, 1957, p. 627.

5. W. Hyrnisak, *Heat Exchangers*, Academic Press, New York, 1958.

6. D. B. Harper, "Seal Leakage in the Rotary Regenerator and Its Effect on Rotary Regenerator Design for Gas Turbines," *Trans. ASME*, vol. 79, 1957, p. 233.

7. C. Simpelaar and D. Aronson, "Gas-to-Gas Heat Exchangers as Applied to an Oxygen Plant," *Trans. ASME*, vol. 72, 1950, p. 955.

8. P. W. Laughrey, et al., "Design of Preheaters and Heat Exchangers for Coal-Hydrogenation Plants," *Trans. ASME*, vol. 72, 1950, p. 385.

9. S. C. Collins, "Reversing Exchangers Purify Air for Oxygen Manufacturing," *Chemical Engineering*, vol. 53, 1946, p. 106.

10. P. R. Trumpler and B. F. Dodge, "The Design of Ribbon-Packed Exchangers for Low-Temperature Air Separation Plants," *Chemical Engineering Progress*, vol. 43, 1947, p. 75.

11. W. M. Kays and A. L. London, *Compact Heat Exchangers*, 3rd ed., McGraw-Hill Book Co., New York, 1984.

12. H. Karlsson and W. E. Hammond, "Air-Preheater Design as Affected by Fuel Characteristics," *Trans. ASME*, vol. 75, 1953, p. 711.

13. J. E. Coppage and A. L. London, "The Periodic Flow Regenerator—A Summary of Design Theory," *Trans. ASME*, vol. 75, 1953, p. 779.

14. T. J. Lambertson, "Performance Factors of a Periodic Flow Heat Exchanger," *Trans. ASME*, vol. 80, 1958, p. 586.

15. A. L. London et al., "Glass-Ceramic Surfaces, Straight Triangular Passages—Heat Transfer and Flow Friction Characteristics," Technical Report no. 70, Dept. of Mechanical Engineering, Stanford University, Prepared for the Office of Naval Research, Contract No. 225(91), (NR-090-342), September 1968.

16. L. S. Tong and A. L. London, "Heat Transfer and Flow Friction Characteristics of Woven Screen and Crossed Rod Matrices," *Trans. ASME*, vol. 79, 1957, p. 1558.

17. A. L. London, J. W. Mitchell, and W. A. Sutherland, "Heat Transfer and Flow Friction Characteristics of Crossed Rod Matrices," *Journal of Heat Transfer, Trans. ASME*, vol. 82, 1960, p. 199.

18. A. P. Fraas et al., "Design Study of a Pebble Bed Reactor Power Plant," Report no. ORNL-CF-60-12-5, May 1961.

19. J. Wadsworth, "Experimental Examination of Local Processes in Packed Beds of Homogeneous Spheres," NRC-5895, National Research Council of Canada, February 1960.

20. L. H. S. Roblee, R. M. Baird, and J. W. Tierney, "Radial Porosity Variation in Packed Beds," *Journal of AIChE*, vol. 4, 1958, pp. 460–464.

21. C. E. Schwartz and J. M. Smith, "Flow Distribution in Packed Beds," *Industrial Engineering Chemistry*, vol. 45, p. 1209, 1953.

22. W. H. Denton, "The Heat Transfer and Flow Resistance for Fluid Flow Through Randomly Packed Spheres," *Proceedings of the IME-ASME General Discussions on Heat Transfer*, September 1951, p. 370.

23. J. J. Martin, W. L. McCabe, and C. C. Monrad, "Pressure Drop Through Stacked Spheres," *Chemical Engineering Progress*, vol. 47, 1951, pp. 91–94.

24. R. R. Smyth et al., "Progress in Testing of Refractories for MHD Heater Applications," *Proceedings of the 16th Symposium on the Engineering Aspects of MHD*, Pittsburgh, Pa., May 16–18, 1977, p. IV.7.40.

25. F. D. Rees, W. J. Spengel, and A. I. Shah, "Heat Pipe Air Heater Replacement at Baltimore Gas & Electric Company's Charles P. Crane Station," Paper presented at the 1988 American Power Conference, April 18–20, Chicago, Ill.

14

Liquid-to-Gas Heat Exchangers

One of the most important markets for heat exchangers is for units in which heat is exchanged between a liquid and a gas, usually water and air. Typical units include automotive radiators, aircraft oil coolers, refrigeration and air-conditioning equipment, space heaters for homes and industrial buildings, intercoolers and aftercoolers for compressors, and gas-turbine inlet air coolers and intercoolers. In most instances the heat transfer coefficients on the gas side are much lower than those on the liquid side, and hence finned gas-side surfaces are advantageous.

COMPARISON OF FIN GEOMETRIES

Features of Typical Matrices

Some of the wide variety of finned surfaces commercially available are shown in Figs. 2.7–2.15. Several points should be noted. The flow-passage area on the gas side usually must be many times that on the liquid side. One widely used means of accomplishing this is to employ banks of round tubes with circular disc fins, as in Fig. 1.14. The heat transfer and pressure-drop characteristics of this type of heat transfer matrix are discussed briefly in Chapter 3, and performance curves are presented in Fig. H5.13. If space is not a problem, an advantage of this type of heat transfer matrix is that the tube spacing

can be varied over a wide range to give the desired gas-side pressure drop. However, if both space and pumping power are at a premium, the flattened tubes of Figs. 2.8 and 14.1 are preferable to the round tubes of Fig. 2.11, because they give a greater gas flow-passage area per unit of inlet face area. Furthermore, as can be seen in Fig. 14.2, the flattened tubes give an aerodynamically cleaner matrix, so that the eddy losses are lower. In addition, for in-line tube arrays, the fin material is used to better advantage with the flattened tubes. This comes about because the flow tends to channel down the lanes between tubes, as can be seen in Fig. 14.2 and leaves relatively low-velocity regions in the wakes behind the tubes.[1] This represents a substantial fraction of the matrix surface for in-line, round, finned tubes. Staggered round tubes can be used to reduce this effect greatly, as is evident in Fig. 14.2, but the pumping power losses are increased by the greater turbulence.[2–4] On the other hand, the flattened tubes cannot withstand as high an internal pressure (although the reinforcing effects of the fins help enormously) and are not as well suited to some fabrication processes. For example, round tube units can be fabricated inexpensively by installing the fins with a press fit on the tubes, whereas the flattened tubes must be brazed or soldered to the fins to obtain a good thermal bond. Although round tubes with circular disc fins may give too flimsy a heat transfer matrix structure, this

Figure 14.1 Flattened tube-and-plate-fin core for a water-to-air aftercooler in a turbocharged engine for heavy agricultural equipment. (Courtesy Modine Manufacturing Co.)

condition can be corrected by employing continuous plate fins as in Fig. 2.11.

Heat Transfer and Pressure-Drop Performance

Many efforts to compare the heat transfer performance of different geometries have been made, and these are helpful in showing relative heat transfer and pressure-drop performance. Unfortunately, such comparisons tend to be confusing and are rarely conclusive. In general, increasing the amount of turbulence by wrinkling, louvering, or interrupting the fins (as was done with the fins in Figs. 2.7–2.9, 2.11, and 2.13) increases the heat transfer coefficient at a given gas-flow rate, but it also increases the pumping power required. In fact, more often than not, the increase in the pumping power is greater than the increase in heat transfer because only a part of the increased turbulence is effective in promoting heat transfer; the balance is wasted in ineffective eddies. Thus if the pumping power available is fixed and the matrix volume is to be kept to a minimum, the aerodynamically clean flattened tube and the flat plate-fin

matrix (shown in Fig. 14.1 and at the right of Fig. 14.3) is likely to give close to the best performance obtainable. On the other hand, if the heat transfer matrix cost or weight is the prime consideration, it is usually best to employ some mild (but not drastic) turbulating device to increase the heat transfer coefficient on the gas side and thus reduce the amount of surface area required.

A nice comparison of some typical surface geometries was made by Kays and London, who used air to cool steam-heated test cores.[5] The 13 different core geometries tested are shown in Fig. 14.3. One set of 8 cores (similar to that of Fig. 1.21) was constructed of stacks of alternately flat and corrugated sheet, while the other set of 5 cores (similar to that of Fig. 14.1) was constructed of flat plates stacked on flattened tubes.

For the plain-fin cores, at the left of Fig. 14.3, the number of fins per inch was used to designate each surface tested. A second set of digits has been added to the designations for the louvered and interrupted surfaces to indicate the axial distance between interruptions. For the flattened tube-plate fin surfaces the number designating the surface includes first the number of fins per inch and second the width of the flattened tubes measured in the air flow direction. In addition, the letter S has been added where the tubes are staggered and the letter R where the fins are ruffled.

The test results are summarized in Figs. 14.4–14.7. Inspection of Fig. 14.4 shows that the heat transfer coefficients and friction factors for the plain-fin surfaces are consistent with the corresponding quantities determined for flow through smooth circular tubes. (Note the dashed curves for the latter.) Figures 14.5 and 14.6 show that, as we would expect, interrupting or louvering these surfaces to take advantage of entrance effects (see Figs. 3.17 and H5.9) leads to increases in both the heat transfer coefficient and the friction factor. Similarly, in the flattened tube-plate fin geometries, staggering the tubes or ruffling or wrinkling the fins increases both heat transfer coefficient and the friction factor. Note, too, that these factors for the in-line tube, plain-fin surface number 9.68-0.870 of Fig. 14.7 are substantially the same as for the plain-fin surface number 19.86 of Fig. 14.4, as shown by the dashed curves in Fig. 14.7. A more extensive set of data for finned heat transfer matrices is given in Ref. 11 of Chapter 13.

Pumping Power as a Performance Parameter

While it is not possible to compare all of the features of different surfaces with a single set of parameters, a particularly worthwhile comparison can be obtained by

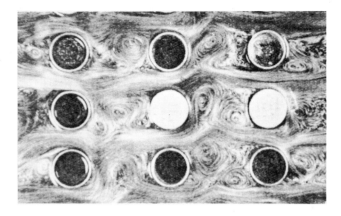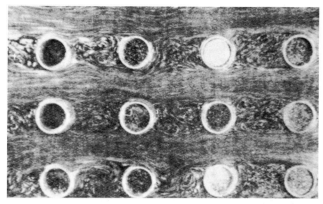

(a)

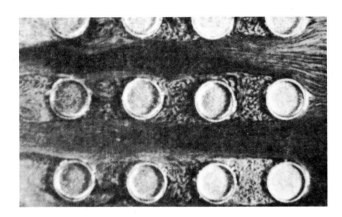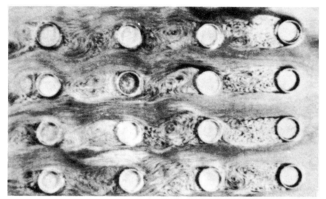

Figure 14.2 Flow patterns for crossflow over typical tube arrays with in-line, staggered square, and equilateral triangular pitches with various tube spacings. The flow is from right to left. (R. P. Wallis, Ref. 1.)

plotting the amount of heat removed per cubic foot of air-flow passage through the heat transfer matrix as a function of the pumping power per cubic foot of air-flow passage through the matrix. Figure 14.8 shows such a comparison of the plain and louvered fin surfaces of Figs. 14.4 and 14.5. The parameter β is the air-side surface area per cubic foot of air-side matrix volume, and the parameter E_{STD} is proportional to the pumping power required on the air side at standard temperature and pressure conditions. Note that over the range investigated, louvering the fins has increased the heat transfer performance somewhat over that obtained with plain fins, and that for any given number of fins per inch (i.e., any given core weight per unit of volume) the louvered surfaces are somewhat better that the plain-fin surfaces. (A later series of tests bears this out.[6]) The slopes of the curves indicate, however, that this is probably not the

case for close fin spacings at Reynolds numbers below the range covered in Fig. 14.8.

Figure 14.9 shows a similar comparison for the flattened tube-and-fin matrices. In this instance α, the ratio of the heat transfer surface area on the air side to the total volume of the matrix, was used instead of the parameter β employed in Fig. 14.8. As in the previous case, it is evident that the surfaces have been well proportioned so that relatively small increases in turbulence yield an increase in heat transfer somewhat greater than the increase in pressure drop. Although not included in this series of tests, the data of Figs. 14.8 and 14.9 suggest that louvering the fins of a flattened tube-plate fin core would give even better performance than any of the five geometries of Fig. 14.9.

The data in Figs. 14.3–14.9 are very useful in themselves, but, for the reader with a substantial interest

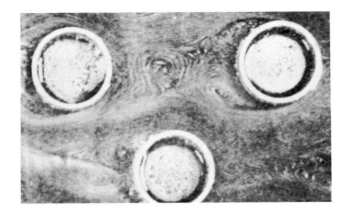

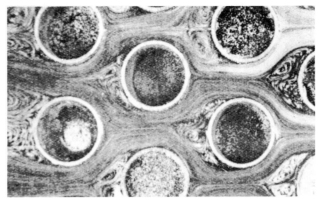

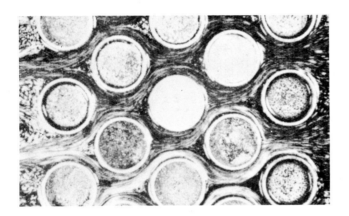

(b)

in the field of compact heat exchangers, they are just an introductory sample of the exhaustive set of information developed at Stanford University between 1947 and 1971 in a program for the Office of Naval Research. This program was concerned primarily with compact heat exchangers for marine and aircraft applications, but also included much related work and has important implications in many other areas. A total of 92 reports and 47 publications were produced,[7] for which Reference 11 of the previous chapter provides a comprehensive summary. It contains the most useful set of information available for applications in this area.

Additional Data for Special Surfaces

In addition to the data presented above for compact heat transfer matrices, there has been a tremendous amount of work on special surfaces designed to enhance heat transfer, and there is a wealth of such data in the literature. Ingenious investigators have devised and tested a remarkable variety of turbulence-promoting geometries, including a wide variety of corrugations, perforated and louvered fins, pin fins, short fins formed by gouging and upsetting metal from base surfaces thick enough to permit this, and wires and screens mounted on or close to the surface.[8-13] Over 2500 reports and papers on enhanced surfaces were issued between 1950 and 1982.[10] This tremendous complex of information presents the designer with an overwhelming array of possibilities. Further, when one attempts to assess any set of information, one finds that there are not only 12 independent geometric variables, but numerous performance criteria such as pressure drop, pumping power, volume, weight, cost, and sensitivity to fouling in service. The situation is further confused because the different investigators use a variety of dimensionless parameters that are useful to

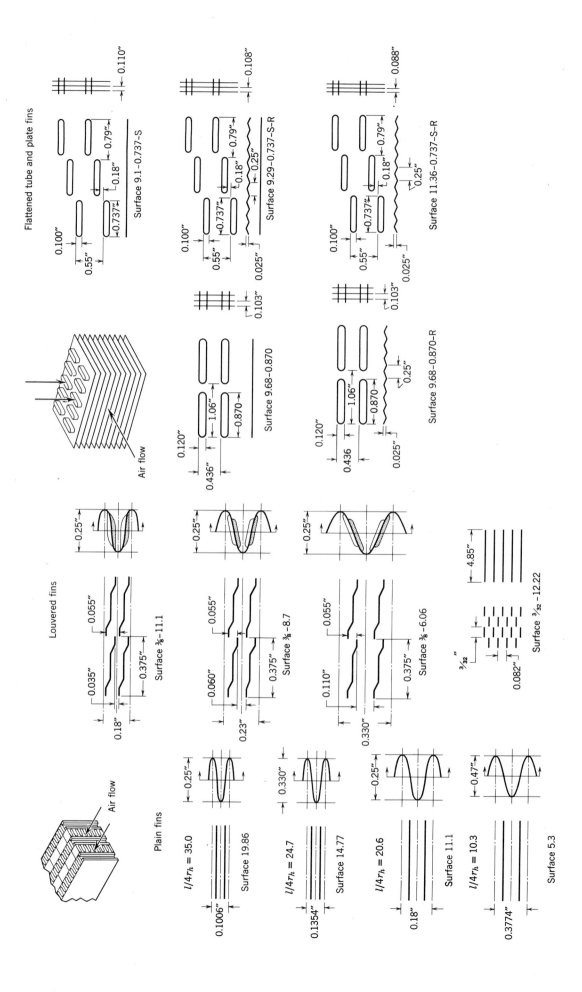

Plain fins

Air flow

$l/4r_h = 35.0$

Surface 19.86

0.1006"

$l/4r_h = 24.7$

0.330"

Surface 14.77

0.1354"

$l/4r_h = 20.6$

0.25"

Surface 11.1

0.18"

$l/4r_h = 10.3$

0.47"

Surface 5.3

0.3774"

Louvered fins

0.055"

0.035"

0.375"

0.18"

Surface 3/8-11.1

0.25"

0.055"

0.060"

0.375"

0.23"

Surface 3/8-8.7

0.25"

0.055"

0.110"

0.375"

0.330"

Surface 3/8-6.06

0.25"

4.85"

3/32

0.082"

Surface 3/32 -12.22

Air flow

0.120"

1.06"

0.870"

0.436"

Surface 9.68-0.870

0.120

1.06"

0.870"

0.25"

0.436

0.025"

Surface 9.68-0.870-R

0.103"

0.103"

Flattened tube and plate fins

0.110"

0.100"

0.79"

0.18"

0.737"

0.55"

Surface 9.1-0.737-S

0.108"

0.100"

0.79"

0.18"

0.737"

0.25"

0.55"

0.025"

Surface 9.29-0.737-S-R

0.088"

0.100"

0.79"

0.18"

0.737"

0.25"

0.55"

0.025"

Surface 11.36-0.737-S-R

270

Geometry of the Different Surfaces

Surface Designation	Fins per in.	Hydraulic Radius, r_h, ft	Plate Spacing, b, in.	Tube or Fin Thickness, in.	Extended / Total Area	Area (Volume between Plates) β, ft²/ft³	Area Core volume α, ft²/ft³	Free flow Frontal Area σ
Plate-fin type:								
Plain fins								
5.3	5.3	0.00504	0.470	0.006	0.719	156		
11.1	11.1	0.00253	0.250	0.006	0.730	334		
14.77	14.77	0.00212	0.330	0.006	0.831	369		
19.86	19.86	0.001495	0.250	0.006	0.833	455		
Louvered fins								
³/₈ –6.06	6.06	0.00365	0.250	0.006	0.623	239		
³/₈ –8.7	8.7	0.00299	0.250	0.006	0.687	288		
³/₈–11.1	11.1	0.00253	0.250	0.006	0.730	289		
³/₃₂ –12.22	12.22	0.002941	0.485	0.004	0.862	302		
Fin-flat-tube type:								
9.68–0.870	9.68	0.00295		0.004	0.795		229	0.697
9.68–0.870-R	9.68	0.00295		0.004	0.795		229	0.697
9.1–0.737-S	9.10	0.00345		0.004	0.813		224	0.788
9.29–0.737-S-R	9.29	0.00338		0.004	0.814		228	0.788
11.32–0.737-S-R	11.32	0.00288		0.004	0.845		270	0.780

Figure 14.3 Heat transfer matrix geometries for which test data are presented in Figs. 14.4–14.7. (Kays and London, Ref. 5.)

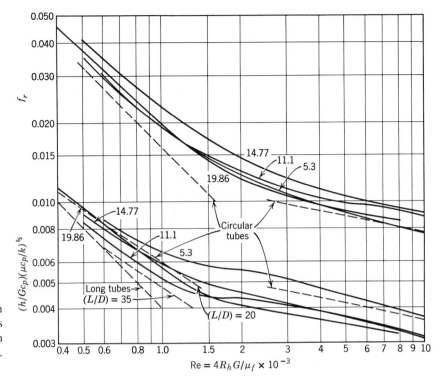

Figure 14.4 Curves for the Colburn modulus and friction factor as functions of Reynolds number for the four plain plate-fin, heat transfer matrices of Fig. 14.3. (Kays and London, Ref. 5.)

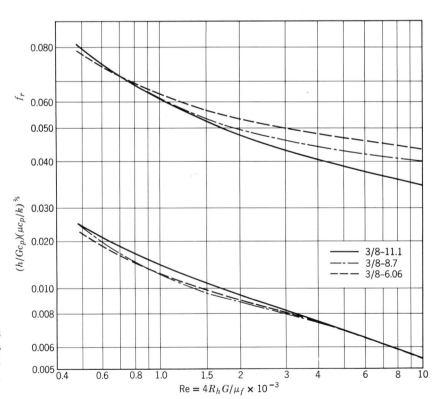

Figure 14.5 Curves for the Colburn modulus and friction factor as functions of Reynolds number for the three louvered plate-fin, heat transfer matrices of Fig. 14.3. (Kays and London, Ref. 5.)

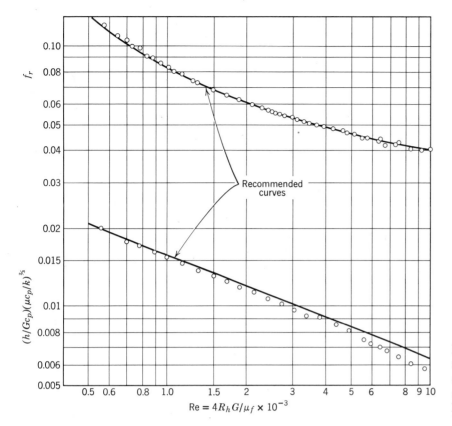

Figure 14.6 Curves for the Colburn modulus and friction factor as functions of Reynolds number for the short interrupted surface of Fig. 14.3. (Kays and London, Ref. 5.)

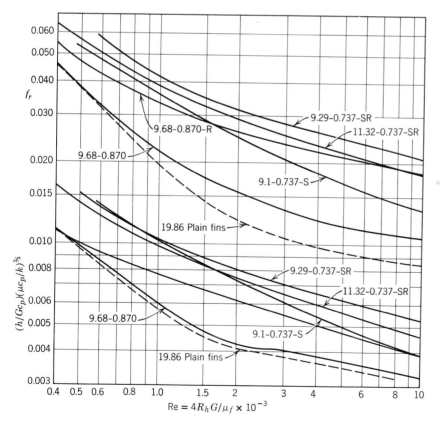

Figure 14.7 Curves for the Colburn modulus and friction factor as functions of the Reynolds number for the flattened tube-plate fin surfaces of Fig. 14.3. (Kays and London, Ref. 5.)

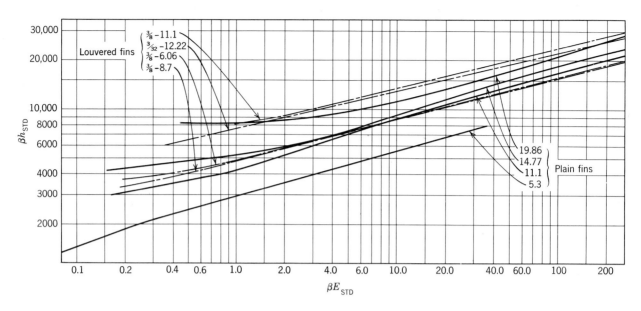

Figure 14.8 Amount of heat transferred per cubic foot of heat transfer matrix as a function of the pumping power required for the plate-fin matrices of Fig. 14.3. The symbol h_{STD} is the heat transfer coefficient for an arbitrary set of conditions taken as standard for the test. (Kays and London, Ref. 5.)

experts but hopelessly confusing to those not intimately familiar with their application. The material included in this text has been selected to meet the needs of most

readers; for those with a more detailed interest in the special field of heat transfer enhancement, the references at the end of this chapter should provide a good intro-

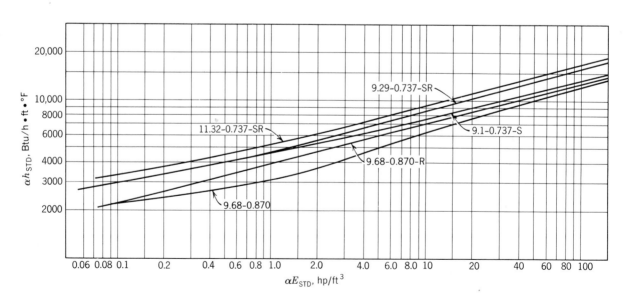

Figure 14.9 Amount of heat transferred per cubic foot of heat transfer matrix as a function of the pumping power required for the flattened tube-and-plate fin matrices of Fig. 14.3. (Kays and London, Ref. 5.)

duction to the extensive literature. Note also the references on enhanced surfaces for other special applications given in Chapters 2, 3, 5, 13, 16, and 17.

Problems in the Use of Turbulence Promoters

It should be emphasized that the improvements in performance obtained by turbulating devices in the foregoing set of tests represent exceptionally good results; more often than not, the increase in pressure drop is found to be greater than the increase in heat transfer. This effect usually stems from excessive turbulence promotion. In any event, it should not be assumed that all turbulating devices give better heat transfer for the same expenditure of pumping power; tests must be run to establish the effects.

In the design of compact heat exchangers for use with air at atmospheric pressure, the small hydraulic radius of the air passages coupled with pumping power limitations commonly lead to operating Reynolds numbers in the 1000 to 5000 range, that is, in the transition region from laminar to turbulent flow. For operation in this region it is usually best to select a heat transfer matrix geometry that induces some turbulence at low Reynolds numbers. The curves of Fig. 14.7 indicate that there are enough irregularities in the geometry of the flattened tube and ruffled-fin heat transfer matrix number 9.68–0.870 to induce sufficient turbulence in the air flow to improve the heat transfer coefficient down to a Reynolds number of 500, where the heat transfer coefficients of the plain and ruffled fins become the same (although the friction factor is still a bit higher for the ruffled fin). Note, too, that the slope of the friction factor curves in Fig. 14.7 become steeper at Reynolds numbers below about 2000, indicating that although the flow is predominantly turbulent, the laminar portion of the boundary layer is thickening at a higher rate than for fully turbulent flow.

On the whole, the relative importance of criteria such as pumping power, cost, weight, inlet face area, matrix depth in the gas flow direction, etc., vary so much from one installation to another that it is not possible to generalize. While it is in a sense an oversimplification, it can be said that the most commonly used geometries are the flat tube and plate-fin surface of Fig. 14.1 and the finned round tubes of Fig. 1.14. The first type is especially well suited to mobile applications in which weight, size, and pumping power are at a premium; the second type is better suited to heavy stationary industrial applications where very long life, reliability, and ease of cleaning and maintenance are the principal considerations. Typical cases involving an application for each of these two types

of heat transfer matrix are given later in this chapter to illustrate the problems involved, useful calculational techniques, and the proportions that result.

DESIGN OF FINNED MATRICES

Selection of Fin Proportions and Material

The use of extended surfaces interjects so many additional variables into the heat transfer and fluid-flow relations that it is not possible to obtain general solutions of the type employed in Examples 12.1 and 13.1. Briefly, it is necessary to converge on a fin matrix geometry before attempting to proportion a heat exchanger to meet a given set of design conditions. Experience is an invaluable guide, but—failing this, or supplementing it—a number of basic considerations are generally applicable. First, the most efficient use of the heat transfer matrix volume is usually obtained when the ratio of the surface areas for the gas and liquid films is roughly inversely proportional to the ratio of their respective heat transfer coefficients. This rough rule of thumb should be modified to allow for the drop in fin efficiency with increasing fin area. Because of this, the fin surface area is ordinarily made roughly half as great as the value implied by the preceding approximation. For example, in water-to-air heat transfer matrices, the heat transfer coefficient on the water side usually runs 2.84 to 5.68 kW/m^2 · °C (500 to 1000 Btu/hr · ft^2 · °F), while that on the air side usually runs 56 to 113 W/m^2 · °C (10 to 20 Btu/h · ft^2 · °F), and the air-side surface area is generally made from 10 to 30 times the water-side surface area.

The fin spacing is usually determined by manufacturing tolerances, the value ordinarily running from 6 to 15 fins per inch depending on the incentive to minimize the volume of the heat transfer matrix. The fin spacing is also related to the fin height in that the gas flow-passage area should be in proper proportion to the liquid flow-passage area. For example, in a double-pipe heat exchanger (such as that of Fig. 1.27) the inner and outer pipe diameters should be proportioned on the basis of the flow-passage area required. This fixes the fin height so the fin spacing is then dependent on the surface area requirement. Similar considerations apply for other types of finned surface.

As discussed briefly in Chapter 3, the fin efficiency depends on the fin shape, height, material, and surface heat transfer coefficient. Charts showing the effects of these parameters for a wide variety of fins have been prepared[13] and two for simple round and rectangular fins are given in Figs. H7.2 and H7.3. Tapering the fins so that they are thicker at the root than at the tip yields a reduction in fin weight and an increase in gas-flow-passage area,[13] but the increase in manufacturing cost is so great that this approach is rarely used except for integral fins formed by casting, rolling, or machining. For applications in which the heat transfer coefficient for the fin side is low, the thermal conductivity of steel is good enough to allow an adequate fin efficiency to be obtained with a reasonable fin thickness. For higher fin-side heat transfer coefficients and large fin heights, the thickness of steel fins becomes excessive, and it is usually worthwhile to employ copper or aluminum fins—the choice of material depending on corrosion problems, manufacturing costs, and the importance of weight. The effectiveness of a fin material is given directly by the ratio of its thermal conductivity to its density. Table H2.2 presents data for typical fin materials.

The fin efficiency may be adversely affected if the fin is not either integral or metallurgically bonded to the tube, that is, soldered, brazed, or welded in place. This is usually not a problem if the heat loading is not too high, and the fin is firmly pressed or swaged in place.[14]

Where the cost of the matrix is the controlling consideration, the coarseness of the tube and fin structure employed depends in part on the headering problem. The larger the tubes in a round tube and fin matrix, for example, the smaller the number of tube-to-header joints. Similarly, the greater the ratio of fin-side to tube-side surface area, the fewer the header connections, and the less the cost of making these joints. Because of the difficulties in estimating manufacturing costs, the best proportions can be determined only by manufacturing experience, and the development of new tooling or equipment may lead to large changes in optimum proportions.

Example 14.1. A 25.4-mm (1.0-in.) OD tube is to be fitted with 75-mm (3-in.) OD fins spaced 394/m (10/ in.) and operated at 94°C (200°F). Determine the fin thickness for both plain carbon-steel and copper fins to give a fin efficiency of 80% for a fin-side heat transfer coefficient of (*a*) 34.1 W/m^2 · °C (6 Btu/h · ft^2 · C°F), and (*b*) 171 W/m^2 · °C (30 Btu/h · ft^2 · °F).

From Fig. H7.2, a fin efficiency of 80% and a fin outer-to-inner radius ratio of 3.0 defines the value of the parameter $w\sqrt{2h/kb}$ as 0.65. Solving for the fin thickness b yields:

$$b = \left(\frac{0.0254}{0.65}\right)^2 \frac{2h}{k} = 0.00305\left(\frac{h}{k}\right)$$

Substituting numerical values gives:

Fin Material	h W/m^2·°C	h Btu h·ft^2·°F	Fin Thickness mm	Fin Thickness in.
Steel				
k = 58.8 W/m·°C	34.1	6	1.77	0.069
= 34 Btu/h·ft·°F	171	30	8.87	0.347
Copper				
k = 372 W/m·°C	34.1	6	0.28	0.011
= 215 Btu/h·ft·°F	171	30	1.4	0.055

In reviewing the results it is evident that, for the lower heat transfer coefficient, the thickness of the steel fin is quite reasonable; however, for the higher coefficient the fin thickness is so great that it would obstruct the air flow severely. If steel fins are to be used with this high heat transfer coefficient, it is clear that the fin height should be cut to half or less to reduce the fin thickness to perhaps 2 mm (0.080 in.) or less. By the same token, the copper fins for the low heat transfer coefficient condition are quite thin and would not withstand rough handling, but are well proportioned for the higher coefficient.

DESIGN OF AUTOMOTIVE RADIATORS

Automotive radiators are so widely used and so familiar to everyone that they afford one of the best illustrations for the application of design techniques well suited to work with compact liquid-to-gas heat exchangers. A variety of compact heat transfer matrices suitable for automotive applications have been built and tested. Although each of these has its own peculiar advantages and disadvantages, much experience indicates that the flat tube-and-plate fin surface (Fig. 2.8) gives close to the optimum heat transfer performance obtainable, is light and compact yet adequately strong, and is among the least expensive to fabricate.

A major question in the choice of detailed dimensions for a heat transfer matrix of this type is the appropriate fin spacing. As the number of fins per inch is increased, the unit can be made progressively more compact. At the same time, however, it becomes more sensitive to clogging by dirt, insects, and the like, and more sensitive to small irregularities in the plate spacing. In practice it has been found that for most automotive applications, 10 to 12 fins per inch represents a good compromise.

Fourteen fins per inch can be fabricated with little increase in cost per unit of surface area, and this closer fin spacing may be justified for some installations, particularly if space is at a premium and fouling of the surfaces is not a problem.

Design Requirements

It is important to control the temperature to a uniform value throughout a reciprocating internal combustion engine to avoid thermal distortion and its effects on bearing alignment, cylinder circularity, etc. Furthermore, the temperature should be sufficiently high that the water vapor in the gases blowing past the piston rings into the crankcase does not condense but is vented through the breather. At the same time, the temperature should not be so high that the lubricating oil deteriorates from oxidation or cracking. To minimize the radiator size it is desirable to operate the coolant system at as high a temperature as possible to give the maximum practicable temperature differential between the liquid coolant and the cooling air. On the other hand, to minimize coolant evaporation losses, it is desirable to keep the system temperature below the boiling point of the coolant. This implies that the system should be pressurized within the limits imposed by simple rubber hose connections. While the situation is obviously complex, experience indicates that a coolant system operating temperature between 82 and 93°C (180 and 200°F) represents a good compromise between these two requirements.

The air should pass through the radiator before reaching the engine rather than vice versa to avoid oil fouling of the radiator surfaces and loss in heat transfer performance. The most difficult design condition is presented by a hot summer day in which the air temperature entering the radiator may easily reach 38°C (100°F). For gasoline engines excessive air temperature around the fuel pump and carburetor induces boiling of the gasoline and vapor locking. To avoid this effect, and to maintain the LMTD as large as possible, the air temperature rise through the radiator should be limited to 8 and 12°C (15 or 20°F).

Both fan power requirements and noise considerations make it desirable to limit the pressure drop across the radiator to about 900 Pa (20 psf). While a substantial additional ram pressure head can be made available at high speeds, this is not available for climbing hills at low speeds and high powers, which represents the critical design condition.

Example 14.2. The design requirements derived from the preceding considerations as applicable to a typical truck radiator design problem are summarized in Table 14.1 together with the dimensional data for the heat transfer matrix to be employed.

Preliminary Design Calculation. The performance of the heat transfer matrix can be estimated from the basic data given in Fig. 14.7 for a flattened tube-plain plate fin core. One good approach to the problem is to estimate the performance for one particular set of conditions in the region of interest and then, using the techniques of Chapter 4, extrapolate from this single design calculation to obtain a performance chart that can be used to select the proportions of the unit desired. A set of calculations for the performance at an arbitrarily chosen point in the range of interest is summarized in Table 14.2.

The first step in making the calculations of Table 14.2 was to assume a reasonable value for one of the air-flow parameters, that is, the air velocity, the weight flow rate, or the Reynolds number. To facilitate reading the curve of Fig. 14.7, as the first step a value was chosen arbitrarily for the Reynolds number. For the reasons given in Chapter 3, the physical properties of the air were taken at the mean surface temperature rather than at the bulk free stream air temperature. From step 6 in Table 14.2 it is apparent that the fin efficiency is so high in this instance that its effects can be neglected when extrapolating to higher air-flow rates in constructing a performance chart. In fact, for this application, it is evident that the transverse tube spacing could be increased to reduce the number of tubes and thus give a less expensive unit. A radiator depth, or air-flow-passage length, of 3 in. was chosen as likely to be close to a suitable value, since automotive radiators commonly have about this depth. Note that this is a simple crossflow heat exchanger with no large-scale mixing of either fluid as it traverses the matrix. From the upper right diagram of Fig. H4.2 the correction factor to be applied to the LMTD for pure counterflow is found to be about 0.99, hence it can be neglected. The LMTD was estimated from the specifications for a first approximation. Had the resulting air temperature rise of step 12 differed appreciably from that of the specifications of Table 14.1, steps 10–12 would have been recalculated and a new value for the LMTD estimated. Fortunately, the arbitrary choices for both the air flow and the radiator depth yield both an air temperature rise and an air-pressure drop close to the design targets.

Performance Chart Construction. A chart presenting heating effectiveness as a function of air-flow-passage

TABLE 14.1   Design Specifications for a Truck Radiator

	SI Units	English Units
Heat load	105.3 kWt	360,000 Btu/h
Water temperature in	82°C	180°F
Water temperature out	74°C	165°F
Air temperature in	38°C	100°F
Air temperature out	46°C	115°F
Air temperature rise	8.3°C	15°F
Inlet temperature difference	44.5°C	80°F
Air pressure	101,300 Pa	14.7 psi
Air-pressure drop	239 Pa	5 psf
Air density	1.14 kg/m³	0.071 lb/ft³
Core matrix	Staggered flat tube plain plate fin	
	(surface number 9.1–0.737-S of Fig. 14.3)	
Fin spacing	3.58 fins/cm	9.1 fins/in.
Tube size	2.54 × 18.43 mm	0.10 × 0.737 in.
Fin material	Copper	
Fin thickness	0.101 mm	0.004 in.
Surface area (air side)	735 m²/m³	224 ft²/ft³
Fin area/total area (air side)		0.813
Free flow area/frontal area		0.788
Air-flow-passage hydraulic radius	0.00105 m	0.00345 ft
Air-flow-passage equivalent diameter	4.21 mm	0.166 in.

TABLE 14.2   Summary of Design Calculations for a Truck Radiator

Line Number	Item	Source[a]	Value
1	Reynolds number	Assumed	4,000
2	G, lb/h · ft²	$\mathrm{Re}\,\dfrac{\mu}{4r_h} = \dfrac{4000 \times 0.050}{4 \times 0.00345}$	14,500 (19.7 kg/s)
3	$(h/Gc_p)(\mu c_p/k)^{2/3}$	Fig. 11.7	0.0054
4	$\mu c_p/k$	$0.050 \times 0.24/0.0175$	0.685
5	h (air side), Btu/h · ft² · °F	③ × ② × 0.24 × ④$^{0.67}$	14.6 (82.8 W/m² · °C)
6	Fin efficiency	Fig. H7.3	0.99
7	Air-flow-passage length, in.	Assumed	3.0 (76.1 mm)
8	Surface area, ft²/ft³	Table 14.1	224 (735 m²/m³)
9	Surface area, ft²/ft² inlet-face area	⑧/4	56
10	LMTD, °F	Fig. H4.1	64 (35.6°C)
11	Heat transfer, Btu/h · ft² inlet-face area	$hA\,\Delta T =$ ⑤ × ⑨ × ⑩	52,400 (165 kW/m²)
12	Air temperature rise given by heat transfer rate, °F	⑪/[② × 0.24]	15 (8.3°C)
13	Inlet-face area required, ft²	360,000/⑪	6.9 (0.64 m²)
14	f_r (air side)	Fig. 14.7	0.0185
15	Air-pressure drop, psf	$\dfrac{0.0185(14{,}500/2600)^2 0.25}{2 \times 32.2 \times 0.071 \times 0.00345}$	4.75 (213 Pa)
16	Air-passage length for 15,000 lb/h, in.	$3(15{,}000/14{,}500)^{0.44}$	3.042 (77.1 mm)
17	Air-passage length for 20,000 lb/h, in.	$3(20{,}000/14{,}500)^{0.44}$	3.46 (87.9 mm)
18	Air-passage length for 10,000 lb/h, in.	$3(10{,}000/14{,}500)^{0.44}$	2.55 (64.7 mm)
19	Air-passage length for 5,000 lb/h, in.	$3(5{,}000/14{,}500)^{0.44}$	1.88 (47.7 mm)

[a]Circled numbers are symbols to indicate the line from which to obtain the quantity to be used in the operation.

length for a series of air-flow rates seems well suited to this design problem. Although this is a single-pass cross-flow unit, the temperature changes in both the hot and cold fluids are less than 20% of the overall temperature difference, and Fig. 4.8 indicates that this places the operating point in a region in which the performance can be represented by a straight line on these coordinates with little error. Thus the performance estimated by the calculations of Table 14.2 can be plotted in Fig. 14.10, and a straight line can be drawn through this point to the origin to give the effectiveness as a function of air-passage length for a hot-fluid temperature drop 18.7% of the inlet temperature difference. For conditions in which the ratio of the air-flow rate to the water flow rate is kept constant, the effect of the air flow rate on the heating effectiveness can be determined using the approach developed in Eq. 4.21 of Chapter 4, which showed that, if the effectiveness is to be kept constant, the air-flow-passage length must be inversely proportional to the air-flow rate to an appropriate power (see Eq. 4.24). This relation was developed for flow in the fully developed turbulent region, and the exponent must

be modified to suit the slope of the curve for the heat transfer modulus of Fig. 14.7. This slope is about -0.44 for the transition flow region of interest, instead of -0.2 as it would be for fully developed turbulent flow. Thus, following the procedure used in Chapter 4, it follows from Fig. 14.7 that

$$\left(\frac{h}{Gc_p}\right)\mathrm{Pr}^{2/3} = C_1\mathrm{Re}^{-0.44} = C_1\left(\frac{4R_hG}{\mu}\right)^{-0.44} \quad (14.1)$$

For the conditions of interest, c_p, Pr, R_h, and μ are constant; hence

$$\frac{h}{G} = C_2G^{-0.44}$$

$$h = C_2G^{0.56} \quad (14.2)$$

The amount of heat transferred can be equated to that added to the gas to give

$$hA_s\Delta T = Q = GA_fc_p\,\delta t \quad (14.3)$$

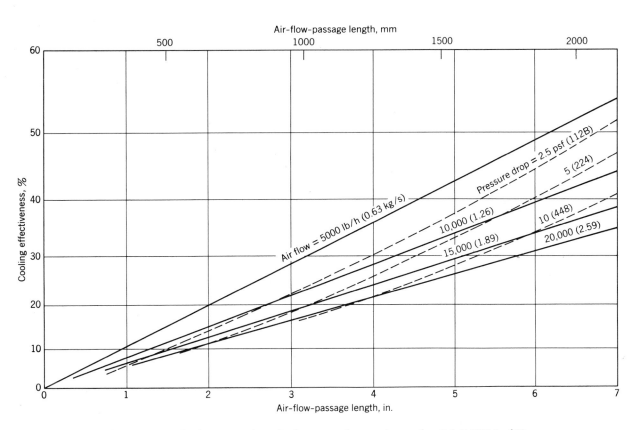

Figure 14.10 Performance chart for heat transfer matrix number 9.1-0.737-S of Fig. 14.3 as designed for the conditions of Tables 14.1–14.3.

Substituting from Eq. 14.2 yields

$$C_2 G^{0.56} A_s \Delta T = G A_f c_p \, \delta t$$

$$C_2 \frac{A_s \Delta T}{A_f \delta t} = G^{0.44} c_p \qquad (14.4)$$

If the temperature difference and the temperature rise are kept constant, and the heat transfer surface area per unit of gas-flow-passage area is directly proportional to the gas-flow-passage length,

$$L = C_3 G^{0.44} \qquad (14.5)$$

The air-flow-passage lengths for other air-flow rates were calculated from Eq. 14.5 for an effectiveness of 0.191, and are recorded at the end of Table 14.2. The points were plotted in Fig. 14.10, and constant air-flow lines were drawn through them to the origin. The air-pressure drop through the unit varies as the product of the friction factor and the square of the air velocity. Since a measurement of the slope of the curve in Fig. 14.7 shows that the friction factor falls off roughly as the -0.44 power of the Reynolds number in the region of interest, the air-pressure drop should vary as the 1.56 power of the velocity. The entrance and exit losses may be taken as roughly equal to the difference in velocity heads between the restricted region in the matrix and that leaving the outlet face. While they will vary as the square of the velocity, they will be small. On the basis of these relations, but neglecting the small change in the relative importance of the entrance and exit losses, Table 14.3 is calculated to permit cross-plotting lines of constant pressure loss on Fig. 14.10. Note that the first step is to find air-passage lengths for several air-flow rates and a constant pressure drop of 5.0 psf (224 Pa). The corresponding air-passage lengths for pressure drops of 2.5 and 10.0 psf are taken, respectively, as twice and one-half that for the 5.0-psf pressure drop at any given air-flow rate.

Selection of Radiator Dimensions. In examining the design requirements of Table 14.1 and the performance chart of Fig. 14.10, it is apparent that the desired proportions are given by the intersection of the 5-psf pressure-drop line with the 0.187 heating effectiveness line. This gives an air-flow rate of 15,000 lb/h (1.89 kg/s) and a heat transfer matrix depth of 3.0 in. (76 mm). Values for any other set of conditions can be obtained directly from Fig. 14.10 by a similar procedure. The inlet-face area can then be calculated directly from the total air flow required. The width and height of the radiator matrix depend on the space available for the installation, but these measurements probably would be made roughly equal to give a well-proportioned and sturdy unit.

The water-flow-passage area must be checked and the water-side pressure drop estimated to be sure that it will be acceptable. The steps in making these calculations are straightforward; they entail simply a determination of

TABLE 14.3 Summary of Calculations for the Constant Pressure-Drop Lines of Fig. 14.10

Air-Passage Length, in.	Air Flow, lb/h · ft²	$\dfrac{G}{G_0}$	$\left(\dfrac{G}{G_0}\right)^{1.56}$	Air-Pressure Drop[a], psf (for Air Density = 0.071 lb/ft³)
3.0	14,500	1.00	1.0	4.75
1.91	20,000	1.38	1.653	5.0
2.99	15,000	1.035	1.055	5.0
5.63	10,000	0.690	0.56	5.0
16.6	5,000	0.345	0.19	5.0
3.82	20,000	1.38	1.653	10.0
5.98	15,000	1.035	1.055	10.0
11.26	10,000	0.690	0.56	10.0
33.2	5,000	0.345	0.19	10.0
0.96	20,000	1.38	1.653	2.5
1.5	15,000	1.035	1.055	2.5
2.82	10,000	0.690	0.56	2.5
8.3	5,000	0.345	0.19	2.5

[a]Note: $\Delta P \sim f \cdot G^2 \cdot L$
$f \sim G^{-0.44}$
Therefore, $\Delta P \sim G^{1.56} \cdot L$.

the water-flow rate from heat balance considerations, the water velocity in the tubes from the water-flow-passage area, and the calculation of the pressure drop from these data, the passage equivalent diameter, and the passage length.

In addition, a correction should be made for the simplifying assumption in line 11 of Table 14.2 that the film-temperature drop on the water side would be negligible. Although it couldn't have been calculated at that point without making some assumptions that would have appeared equally rough, it can be calculated once the size of the radiator has been established so that the number of water passages can be defined. If the radiator width is made 910 mm (30 in.), for the 80-mm matrix thickness in the air-flow direction the water-flow-passage area becomes 0.00929 m² (14.4 in.²). The resulting water velocity is 0.326 m/s (1.07 ft/s), and the heat transfer coefficient is 3620 W/m² · °C (637 Btu/h · ft² · °F). After allowing for the fact that the air-side surface area is 5.15 times the water-side surface area, one finds that the film-temperature drop on the water side is 9.9% of the total of the two film temperature drops.

The temperature drop through the water-passage wall was also considered negligible in line 11. The validity of that assumption can now be shown; the conductance of the 0.0102-mm (0.004-in.) thick copper wall would be 3,400 kW/m² · °C (600,000 Btu/h · ft² · °F), or about 1,000 times the heat transfer coefficient on the water side.

DESIGN OF A BUILDING AIR HEATER

A case study of a building air heater seems in order, since so many heat exchangers are used in this field and it serves to illustrate the design problems associated with applications in which ruggedness and reliability are more important than compactness or light weight.

Example 14.3. For the case at hand, hot process water is available at 160°F for use in heating air for a large dryer. The hot water supplied is to be cooled from 160 to 80°F, while the air passing through the heat exchanger is to be heated from 60 to 140°F. Specifications call for banks of 5/8-in. OD, 0.035-in. wall, helically finned copper tubes similar to that of Fig. 2.7*f* to handle a heat load of 2,000,000 Btu/h. Some air blowers are available, and these can produce a pressure drop of 2.0 in. of water (498 Pa) across the heat transfer matrix with the existing pulleys in the belt drive. The basic conditions given are summarized in the first 13 lines of Table 14.4.

Design Calculations. A preliminary survey shows that tube matrix A of Fig. H5.13*b* with 1.21-in. OD fins spaced 8.7 to the inch is well suited to this application. Note that the collar at the root of the fins increases the apparent tube OD to 0.645 in. Data from Fig. H5.13*b* can be used to fill in lines 17, 28, and 31 of Table 14.4.

The high heating effectiveness required indicates that something very close to a counterflow configuration should be employed. The multipass crossflow configuration of Fig. 1.14 seems to meet this requirement. An examination of Fig. 4.4 indicates that, for the constant temperature difference of this case and a temperature rise equal to four times the temperature difference, the length-diameter ratio for a continuous circular passage on the air side ought to be roughly 300. The higher heat transfer coefficient for crossflow over finned tubes should cut this to about one-half. Using an $l/d = 150$, the air velocity can be estimated roughly from the allowable pressure drop, since the friction factor (based on the equivalent passage diameter) as given by Fig. H5.13*b* is around 0.13. Thus

$$\Delta P = f \frac{\rho V^2}{2g} \frac{l}{d} = 0.00212 \, V^2 = 10.4 \text{ psf (498 Pa)}$$
$$V = \sqrt{491} = 22.2 \text{ ft/s (6.76 m/s)}$$

This value is a little high from the standpoint of noise—for building heating systems the air velocity should be kept below 20 ft/s (6.1 m/s)—but for this application noise is not objectionable. A first approximation can be calculated by entering this value in line 14 of Table 14.4 together with a value of 5 ft/s for the water velocity through the tubes. The second value was chosen to give a reasonable pressure drop for the water side. The flow rate for line 15 is simply the product of lines 13 and 14 for the water and air, respectively, and the Reynolds number can be obtained from Fig. H3.1. The friction factor and the heat transfer modulus can be obtained from Fig. H5.13*b* for the air side, and from Figs. H3.4, H5.3, and H5.4 for the water side. The overall heat transfer coefficient can be calculated as indicated in lines 21–27. Note that the water-side heat transfer coefficient was multiplied by the ratio of the water side to the air-side surface areas in obtaining the value in line 24.

The total flow rate can be calculated from the heat load, the fluid temperature rise (or drop), and the specific heat for the water and air, and the results entered in line 29. The number of parallel passages required on the water side can be calculated by dividing the total water-flow rate by the flow per tube. While this yields the value 13.2, a round number must be used; hence a value of 13

TABLE 14.4 Design Calculations for an Air Heater Using Hot Water

Flow configuration: multipass tube side, single-pass air side

Design heat load: 2,000,000 Btu/h (586 kW)

Heat transfer matrix: helically finned Cu tubes 0.625-in. OD, 0.035-in. wall, Cu fins 1.121-in. OD, staggered pitch, 1.35-in. axial and 1.23-in. transverse spacing (Configuration A of Fig. H5.13b)

Line Number	Quantity	Source[a]	Hot Fluid	Cold Fluid
1.	Fluid		Water	Air
2.	Mean specific heat, Btu/lb · °F		1.0 (4.187 kJ/°C)	0.24 (1.0 kJ/kg · °C)
3.	Internal flow area per tube, in.^{2}	Table H7.1	0.2419 (156 mm^2)	
4.	Tube surface area, ft^2/ft of length	Table H7.1 and Fig. H7.1	0.1453 (0.044 m^2/m)	1,187 (0.571 m^2/m)
5.	Temperature in, °F		160.0 (71.1°C)	60.0 (15.5°C)
6.	Temperature out, °F		80.0 (26.7°C)	140.0 (60°C)
7.	Temperature rise (or drop), °F		80.0 (44.4°C)	80.0 (44.4°C)
8.	Inlet temperature difference, °F		20.0 (11.1°C)	20.0 (11.1°C)
9.	Greatest temperature difference, °F		20.0 (11.1°C)	20.0 (11.1°C)
10.	Least temperature difference, °F		20.0 (11.1°C)	20.0 (11.1°C)
11.	LMTD, °F		20.0 (11.1°C)	
12.	Pressure (mean), psia		30.0 (2.07 bar)	14.7 (1,013 bar)
13.	Density (mean), lb/ft^3		62.3 (998 kg/m^3)	0.071 (1,138 kg/m^3)
14.	Flow velocity through minimum flow area, ft/s		5.0 (1.52 m/s)	22.2 (6.76 m/s)
15.	Flow rate, lb/ft^2 · s	⑬ × ⑭	312.0 (1520 kg/s · m^2)	1.58 (7.71 kg/s · m^2)
16.	Flow rate per passage, lb/s		0.525 (0.238 kg/s)	28.9 (13.15 kg/s)
17.	Equivalent diameter of flow passage, in.	Fig. H5.13b	0.555 (14.1 mm)	0.2155 (5.47 mm)
18.	Re	Fig. H3.1	32,000	2,050
19.	Friction factor f_d	Figs. H3.4 and H5.13b	0.025	0.14
20.	$(h/Gc_p)Pr^{2/3}$			0.0096
21.	$Pr^{2/3}$	Table H2.4		0.777
22.	Heat transfer coefficient, Btu/h · ft^2 · °F	Figs. H5.3 and H5.13b	1120 (63.6 W/m^2)	16.8 (95.5 W/m^2 · °C)
23.	Dynamic head, psi	Figs. H3.2 and H3.3	0.165 (1140 Pa)	0.0036 (24.8 Pa)
24.	1/h (based on air-side surface area)	1/ ㉒	0.0073	0.0595
25.	Fouling factor	Table H5.4	0.001	
26.	1/U	㉔ + ㉔ + ㉕	0.068	(0.0197)
27.	U, Btu/h · ft^2 · °F		14.7	(83.5)kW/m^2 · °C)
28.	Matrix surface per unit of volume, ft^2/ft^3	Fig. H5.13b		98.7 (324 m^2/m^3)
29.	Total flow rate, lb/h	(Heat load)/ ② × ⑦	25,000 (33.9 kg/s)	104,000 (141 kg/s)
30.	Number of passages required	㉙ / ⑯ × 3600	13.0	1.0
31.	Free flow area fraction	Fig. H5.13b		0.443
32.	Matrix inlet-face area, ft^2	⑯ / ㉛ × ⑮		41.6 (3.87 m^2)
33.	Total surface area required, ft^2	(Heat load)/ ㉗ × ⑪		6800 (632 m^2)
34.	Tube matrix length, ft	㉝ / ㉜ × ㉘		1.67 (0.509 m)
35.	Number of tube banks	㉞ × 12/1.35		15.0
36.	Pressure drop, psi	㉓ × ⑲ × ㉞ × 12/ ⑰		0.0467 (322 Pa)
37.	Pressure drop, in. H$_2$O	㊱ × 27.8		1.3

[a]Circled numbers are symbols to indicate the line from which to obtain the quantity to be used in the operation.

was entered in the table. The matrix inlet-face area on the air side is given by the total air flow divided by the flow rate per unit area (line 15) and the fraction of the total cross section available as flow passage. The total heat transfer surface area required is simply the total heat load divided by the overall heat transfer coefficient and the LMTD. The matrix length, number of tube banks, and the air-pressure drop follow directly.

In reviewing the calculations, it is apparent that the air and water velocities chosen yielded reasonable proportions for the heat transfer matrix. While the air-pressure drop is a little higher than the value specified, if desired this amount can be easily reduced by calculating a second iteration using a lower air velocity; 20 ft/s probably would be a good choice.

Limitations Imposed by Tube Vibration. Several series-parallel tube arrangements on the water side would satisfy the conditions of Table 14.4. However, it may be necessary to limit the tube length to avoid tube vibration. The frequency of the aerodynamic forces tending to excite tube vibration can be estimated roughly from Fig. H8.6, and this gives 100 cps. From Fig. H8.7 the length of bare, empty, 5/8 in.-OD steel tubes for a natural frequency of 100 cps in the first bending mode is 38 in. The frequency of the finned tube would be lower since the weight of the bare tube is about 0.221 lb/ft (see Table H7.1), and the weight of the fins would be about 0.3 lb/ft. (The fin weight can be estimated by taking half of the 1.4 ft²/ft total fin surface area—both sides of the fin—given by Fig. H7.1 and multiplying it by the fin weight per square foot. For 0.010-in. thick fins this value is about 0.45 lb/ft².) The weight of water in the tube would be about 0.1 lb/ft. Thus the total weight of the finned tube would be about 0.6 lb/ft, or about three times the weight of the bare tube. This factor would reduce the natural frequency by the square root of three. In addition, the higher density and lower modulus of elasticity of copper would further reduce the natural frequency by a factor of 1.48 (see Table H8.3b). Thus to place the natural frequency above the exciting frequency, the equivalent frequency of a bare tube should be greater by the factor $1.48 \times 1.73 = 2.56$. From Fig. H8.7, a length of 24 in. would give this frequency, that is, 256 cps. To avoid the possibility of tube vibration, the length should be reduced sufficiently to raise the tube natural frequency at least 20% above the forcing frequency. This gives a tube length of 22 in.

Limitations Imposed by Acoustical Resonance. Severe noise might also be a problem. From Eq. 9.6, the duct width for the first mode acoustical resonance is given by $c = Cn/2f = (1160 \times 1/2 \times 100) = 5.8$ ft. Thus the width of the air passages transverse to the tubes should be kept to about 20% less than this amount, or about 58 in., to avoid acoustical resonance and possible difficulties with duct vibration.

While damping in the system might be sufficient—particularly with staggered finned tubes—so that neither noise nor vibration would be a problem, it is usually best not to take a chance. Therefore the required inlet face area could be obtained by using four individual units 22 × 57 in. Since the use of four units would increase both the cost and the space required for the return bends, it might be better to use one or two units with the air passage divided by partitions that would also serve as tube supports. The tubes should be held firmly by the partitions so that chafing would not occur. Note that the use of partitions gives a continuous beam, and this reduces the natural frequency somewhat over that for a single span (see Table H8.3a). Reducing the tube span between partitions to 15 in. should take care of this difficulty.

This vibration problem points up the advantage of plate fins of the type shown in Fig. 2.11. These tie the tubes together to give a stiff matrix, thus raising the natural frequency of the tubes far above the vortex-shedding frequency.

REFERENCES

1. R. P. Wallis, "Photographic Study of Fluid Flow between Banks of Tubes," *Engineering*, vol. 148, 1939, p. 423.

2. S. L. Jameson, "Tube Spacing in Finned Tube Banks," *Trans. ASME*, vol. 67, 1945, p. 633.

3. A. Y. Gunter and W. A. Shaw, "A General Correlation of Friction Factors for Various Types of Surfaces in Crossflow," *Trans. ASME*, vol. 67, 1945, p. 643.

4. E. A. Schryber, "Heat Transfer Coefficients and Other Data on Individual Serrated-Finned Surfaces," *Trans. ASME*, vol. 67, 1945, p. 683.

5. W. M. Kays and A. L. London, "Heat Transfer and Flow Friction Characteristics of Some Compact Heat Exchanger Surfaces," *Trans. ASME*, vol. 72, 1950, p. 1075.

6. W. M. Kays, "The Basic Heat Transfer and Flow Friction Characteristics of Six Compact High-Performance Heat Transfer Surfaces," *Journal of Engineering for Power, Trans. ASME*, vol. 82, 1960, p. 27.

7. A. L. London, "Final Report—Stanford University Office of Naval Research Project on Compact Heat Exchangers and Thermodynamic Investigations," Technical Report no. 76, Department of Mechanical Engineering, Stanford University. November 1971.

8. J. R. Mondt and D. C. Siegla, "Performance of Perforated Heat Exchanger Surfaces," *Journal of Engineering for Power, Trans. ASME*, vol. 96A(1), 1974, p. 81.

9. A. R. Wieting, "Empirical Correlations for Heat Transfer and Flow Friction Characteristics of Rectangular Offset-Fin Plate-Fin Heat Exchangers," *Journal of Heat Transfer, Trans. ASME*, vol. 97C(2), 1975, p. 488.

10. R. L. Webb and A. E. Bergles, "Heat Transfer Enhancement: Second Generation Technology," *Mechanical Engineering*, vol. 105, no. 6, June 1983, p. 60.

11. B. A. Brigham and G. J. VanFossen, "Length to Diameter Ratio and Row Number Effects in Short Pin Fin Heat Transfer," Paper no. 83-GT-54, *Journal of Engineering for Power*, 1983.

12. D. G. Thomas, "Forced Convection Mass Transfer: Part I. Effect of Turbulence Level on Mass Transfer through Boundary Layers with a Small Favorable Pressure Gradient," *AIChE Journal*, May 1965, p. 520.

13. J. T. Davies, *Turbulence Phenomena*, Academic Press, New York, 1972.

15

Steam Generators

Scotch marine and locomotive boilers are just two of the many types of boiler that have represented major milestones in the development of the art of engineering.[1] Since the construction of the older units is discussed extensively in other books, and since their design no longer presents particularly challenging problems from the standpoint of heat transfer, fluid flow, or stress analysis, the design of these units is not treated in this chapter; the space available is devoted primarily to the problems of modern, high-performance units. If properly designed and operated, a modern drum-and-tube unit, such as that illustrated in Fig. 15.1, will surpass the splendid dependability records of even the best of the bulky, old, reliable, low-temperature, low-pressure boilers.[2,3]

DESIGN FEATURES OF A TYPICAL MODERN BOILER FOR A LARGE COAL- OR OIL-FIRED STEAM POWER PLANT

The more important problems involved in the design of a coal-fired furnace and boiler for a large modern central station are illustrated by the unit shown in Fig. 15.1. (See also Fig. 1.3). An open furnace provides a large volume for completion of the combustion reactions, a process that is facilitated by preheating the combustion

air to from 400 to 600°F (204 to 316°C). Heat losses are reduced by lining furnace walls with boiler tubes, and, to avoid local hot spots, the burners are placed so the flames do not impinge on the walls.[1] The bulk of the heat that goes into preheating and boiling the water is transmitted by thermal radiation from the oil or powdered coal flames to the banks of tubes that form the furnace walls (see Fig. 15.2).[4-6] After losing something like half of their heat to the furnace walls, the products of combustion enter banks of boiler tubes at the top of the furnace where the high heat transfer coefficient characteristic of boiling is effective in further reducing the temperature of the hot gas (which may be very hot in some regions) without danger of overheating the tube walls. The gases then flow downward at a reduced and more uniform temperature through the *superheater*, the *reheater*, the *economizer*, and the *air heater* to the base of the stack. The steam drum and the various manifolds and joints serving to connect the tubes are kept away from the flames and are not exposed to the very high-temperature gases. There is little or no high-temperature brickwork through which heat can leak to the surrounding atmosphere. An induced draft fan is often used at the base of the stack so that the furnace is slightly below atmospheric pressure. Heat losses from the superheater and economizer are also minimized through the use of water-cooled walls. The heat load on the economizer is kept low by employing regenerative

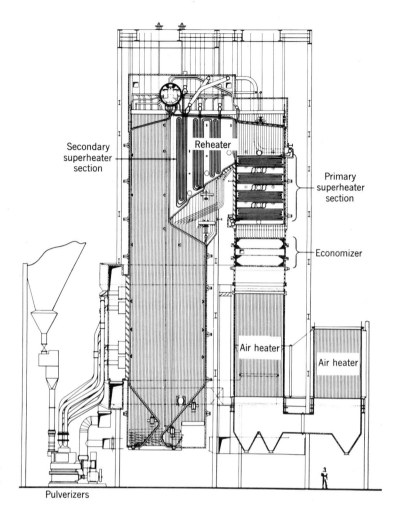

Figure 15.1   Section through a steam boiler for a large central station. (Courtesy The Babcock and Wilcox Co.)

feedwater heating with steam bled from the lower stages of the turbine. Excessive temperatures in the superheater tube walls are avoided by placing them in a region in which the gas temperatures are only moderately higher than the steam.[1]

The temperature of the steam leaving the superheater may be controlled by regulating the amount of combustion gas allowed to bypass the superheater through the economizer; or it may be regulated through the use of a *desuperheater*, or *atemperator*, which is located between the two halves of the superheater.[1] This location for a desuperheater is best, since the steam is at a sufficiently high temperature for good control, yet overheating of the tube walls at the superheater outlet can be prevented, because the steam temperature at the intermediate station does not exceed the maximum rated temperature. Many types of desuperheater are employed, including use of direct injection of a water spray, bundles of tubes in the steam or water drum, and valving arrangements that allow steam to bypass the first stage of the superheater.

The air heater is a counterflow heat exchanger that reduces the amount of heat lost to the stack gases to a low value, and, at the same time, improves combustion by increasing the flame temperature in the furnace. This effect also increases the amount of radiant heat transmission to the furnace walls, thus reducing the heat transfer surface area required and, hence, the cost.

The entire furnace is supported from steel framework extending over the top. The boiler drum (see Fig. 15.3) is hung from heavy U-bolts. The tube banks and tube panels forming the walls are supported on a system of beams and hanger rods carefully designed to minimize stresses from differential thermal expansion between the various elements and yet give a sturdy structure to withstand the large pressure forces produced by the induced and/or forced draft fans.[1]

This type of furnace provides adequate space for good combustion at a moderate gas velocity, and takes advantage of thermal radiation to minimize the pumping power required to force air over the heat transfer surfaces. The tall vertical tubes in the furnace walls can be designed to give good flow stability under boiling conditions with either natural or forced convection circulation of the water-steam mixture (see Chapt. 5), and the average heat flux to the tube surfaces can be kept fairly high to minimize the weight and cost of the heat transfer surfaces.

MAJOR PROBLEM AREAS IN BOILER DESIGN

A host of factors must be considered in the design of steam boilers. The choice of steam conditions is heavily dependent on the type and size of power plant, the performance requirements, the varieties of fuels to be used, the design of the burner and furnace, and the feedwater purity that can be obtained. The choice of materials also presents many complex problems.[7-11] To avoid increasing the scope of this book further, the balance of this chapter is limited to the problems of designing boilers as heat exchangers, and does not include the related problems of furnace design and steam plant equipment.

Figure 15.2 Tube panel being raised into position for installation in the furnace wall of a large boiler similar to that of Fig. 15.1. (Courtesy Combustion Engineering, Inc.)

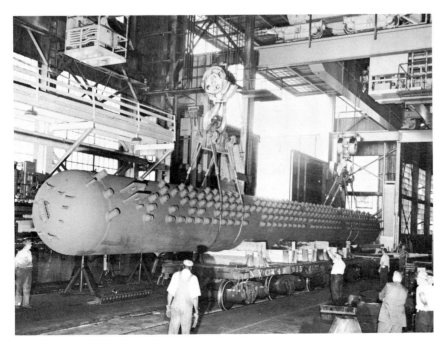

Figure 15.3 Boiler drum 60-in. ID by 85 ft long for a large steam power plant. (Courtesy Combustion Engineering, Inc.)

Natural Convection Circulation

Boilers can be designed (Fig. 15.4) so that the difference in density between the steam-water mixture in the risers and the bubble-free water in the downcomers produces circulation.[12–14] To avoid local hot spots and tube burnout, natural circulation units are ordinarily designed for no more than 50 to 70% by volume vapor at the tube outlet—depending on the system pressure. The resulting density difference may give water velocities in the tube bank of as much as 5 to 10 ft/s (1.5 to 3 m/s). The average heat flux for a good design commonly runs 50,000 Btu/ft² · h (16 W/cm²), while peak local heat fluxes may be as much as 250,000 Btu/h · ft² (80 W/cm²).

Forced Circulation

Where the vertical height available and/or the reduced difference in density at high pressures do not give a circulation rate as high as that desired, circulating water pumps can be employed to increase the capacity obtainable with a given volume. The circulation rate is usually designed to give water velocities at the tube outlets of from 10 to 20 ft/s (3 to 6 m/s) with 40 to 70% steam by volume at the outlet, and average heat fluxes of around 50,000 Btu/h · ft² (16 W/cm²). Because of the extra cost and complication that they entail, forced circulation systems are employed only where the reduction in the capital investment in heat transfer surface and boiler drums, or, in ships, the reduction in size, more than justifies the extra initial and maintenance costs of the pumps.

The recirculating pumps must be located at a level far enough below the steam drum so that the static head acting on the water at the inlet to the pump is sufficient to avoid difficulty with cavitation. This requirement constitutes a major problem and may add substantially to the cost.

Economizers and Superheaters

The larger boilers are designed so that the feedwater is heated to the boiling point in a separate section known as an *economizer*, while the saturated steam is superheated in a separate unit known as the *superheater*. This approach minimizes difficulties both with water-flow instability (see Chapt. 5) and with deposits of solid impurities entering with the feedwater. The concentration of these impurities is kept sufficiently low to avoid difficulty with the formation of deposits in the economizer section where the feedwater is heated to the boiling point. Since the steam-side heat transfer coefficient in the superheater is generally much lower than in the boiler,[15] all but the first portion of the superheater must be located in a zone where the gas temperature is not too high. If this is not done, local overheating is likely to occur.[16]

Blowdown

The boiling process in recirculating boilers tends to concentrate the impurities in the recirculating water. The installation is ordinarily designed so that a small stream is bled from the zone of maximum concentration. This continuous blowdown is usually controlled so that it runs

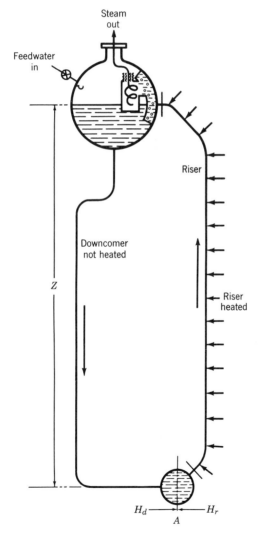

Figure 15.4 Schematic diagram showing the principal elements in a natural circulation boiler. (Courtesy The Babcock and Wilcox Co.)

about 0.5% of the feedwater-flow rate into the boiler. To avoid an appreciable heat loss from the system, the blowdown stream is ordinarily used for feedwater heating.

Vapor Separators

The top drum of a recirculating boiler is designed to separate the steam from the steam-water mixture partly to provide bubble-free water for recirculation to give good natural convection, but mainly to ensure that the steam supply to the superheater is free of water droplets. This last consideration is important both to avoid the formation of solid deposits in the superheater and thermal stresses that would result from the intermittently high local heat transfer rates that occur if slugs of water are carried into the superheater. While many vapor-separating devices are employed,[17] the one shown in Fig. 15.5 indicates the basic approach ordinarily used. A baffle that lines the lower portion of the steam drum directs the steam-water mixture coming up from the boiler tubes into the vanes of cyclone vapor separators where gravity, centrifugal force, and baffles act to direct the steam into the upper part of the drum and the water into the lower portion. The drum normally operates

about half full of water. The lower portion of each cyclone separator usually operates submerged so that the water draining from the walls will flow into the drum without splashing and entraining steam bubbles.

A *demister* is often employed in the steam drum with a secondary vapor separator to remove dust particles and tiny water droplets that might contain substantial concentrations of dissolved solids. Such a unit depends on low flow velocities with repeated changes in flow direction in corrugated baffles to obtain good steam separation.

The steam flow rate through a steam drum may be as high as 6,000 lb/h · ft (9,000 kg/h·m) of drum length in 200 or 300 psi (1.4 to 2.1 MPa) boilers with good vapor separation. Since the steam flow rate for a given dynamic head increases with pressure, flow rates as high as 30,000 lb/h · ft (45,000 kg/h · m) of drum length have been used successfully at pressures of 2400 psi (16.5 MPa). There is not much point in going to still higher flow rates, since a substantial supply of saturated water must be maintained in the steam drum to help accommodate abrupt changes in load. It is also essential to have a substantial water supply in the drum so that even in power transients it is adequate to ensure flooding of all of the boiler tubes.

Once-Through Subcritical Boilers

Developments in feedwater purification techniques, particularly ion-exchange resin treatments, have made possible extremely high-purity feedwater.[3] The total concentration of solids is commonly held to less than 1.0 ppm, and sometimes to less than 0.1 ppm. Exceptionally high feedwater purities are readily achievable if the condensers are cooled with fresh water. With salt or brackish condenser cooling water it is necessary to minimize the amount of in-leakage by using all-welded condensers, or employ full-flow polishing demineralizers to remove the minerals from traces of in-leakage.

It has been possible to take advantage of the very pure feedwater conditions obtainable in recent years to eliminate the steam drum and many of the tube-to-header joints by going to a *once-through* boiler.[18] These units entail much more difficult design problems because they must be carefully proportioned both to give a good water-flow distribution and to avoid trouble with flow instability. Furthermore, the heat flux in the economizer, boiler, and superheater zones must be sufficiently high so that the tube-wall surface is employed to good advantage and yet not so high as to give difficulty with tube burn-out. An examination of the chart of Fig. 5.35 indicates that the flow stability problem is likely to be

Figure 15.5 Section through a boiler drum showing provisions for vapor separation. (Courtesy The Babcock and Wilcox Co.)

much less difficult in the higher pressure units. In practice, it has been found uneconomical to design once-through boilers for pressures below 1500 psi (10 MPa); systems designed for pressures of the order of 2400 psi (16.5 MPa) and higher have given excellent performance. This situation stems in part from the much smaller difference in density between saturated liquid and saturated vapor for the higher pressure units, and also partially from the excellent heat transfer coefficient obtainable with saturated vapor at high pressures. For example, as can be seen in Fig. 5.12, a heat flux of 300,000 Btu/h · ft^2 (95 W/cm^2) can be obtained at 2400 psi (16.5 MPa) with a temperature difference of about 60°F (33°C) between the metal and the vapor near the saturated region.

To reduce the possibility of flow instabilities and hot spots, the economizer and boiler portions of the tubes are commonly made nearly vertical to take advantage of the stabilizing effects of gravity forces. Much care must be taken in the design to assure a uniform heat flux to all of the tubes functioning in parallel. Since this is difficult to achieve, orificing of the tubes at the inlet is helpful (and normally essential) to match the water-flow rate to the heat input to each tube and thus avoid flow instabilities. This and associated problems are discussed in Chapter 5.

Experience has indicated that it is best to design for vapor velocities of 30 to 60 ft/s (10 to 20 m/s) in the tubes at the outlet of the boiler region for full-power conditions. Designs based on lower full-power velocities are likely to give difficulty with flow instability at low-power outputs during start-up or part load conditions. Higher velocities give excessive pressure drops.

Many well-designed once-through boilers with good feedwater purity control have operated for periods of as much as four years without internal cleaning, and yet the maximum thickness of deposits found in the tubes has been only a few thousandths of an inch. Apparently the solids tend to concentrate in the last tiny droplets of water to evaporate, and form a floculent dust that is more likely to give difficulty with deposits on the turbine buckets than deposits in the boiler or superheater. The deposits in the turbine are likely to be water soluble, and, if so, can be washed off by flushing the turbine with very wet steam when idling it prior to a shutdown for maintenance purposes.

Supercritical Boilers

Once-through boilers have been designed and built to operate at pressures above the critical pressure of water, that is, 3206 psia (22.13 MPa).[19–21] In these units there

is no sharp line of demarcation between the liquid and vapor phases as the fluid flows through the boiler. There is no sharp change in the physical properties of the fluid (see Figs. H2.4–2.6), and no sharp change in the heat transfer coefficient. Figure 5.12 shows heat transfer coefficients for typical conditions.

Stress problems are particularly acute in supercritical boilers because both the pressure and the temperature are so high that it is not easy to avoid excessive stresses in even the strongest of the stainless steels. Some of the new refractory alloys show great promise, particularly high-nickel chrome-iron-molybdenum alloys that have good strength at high temperatures and are relatively insensitive to chloride corrosion.

Special Problems

There are many special problems related to boilers for steam power plants that deserve at least to be mentioned; two such are start-up and shutdown.[22,23] In practice, the principal factor limiting the rate at which a boiler can be started up, and the load built up, is thermal distortion in the turbine, which commonly makes it necessary to limit load changes to a rate of not more than 2% per minute. However, even for this seemingly slow rate care must be exercised to proportion the boiler and its control system so that the pressures, temperatures, and water inventory distribution through the system are well behaved throughout the start (which takes many hours).

FLUIDIZED-BED COMBUSTORS

The energy available from U.S. resources of coal is about 10 times that from resources of petroleum, but the bulk of the coal contains substantial percentages of sulfur. The ill effects of acid rain have been well publicized. Even though they have been exaggerated sometimes, for example, activists ignore the fact that there are more lakes in the Adirondacks too alkaline for fish than there are lakes too acid for fish, it has become clear that the sulfur in coal presents serious problems that must be faced. Thus worldwide interest was aroused following a 1970 report of H. R. Hoy of the British Coal Utilization Research Laboratory in Leatherhead, England. In this he presented test data demonstrating that emissions of sulfur can be dramatically reduced if the coal is burned in a fluidized bed of calcined limestone or dolomite so that CaO captures and retains the sulfur as calcium sulfate.[24] The small effort spent on fluidized-bed combustion up to that time had been directed toward the utilization of low-grade fuels and wastes. Hoy's paper led to a greatly

TABLE 15.1   Summary of Data on the 129 Fluidized Bed Combustion Plants Installed or On Order in the United States between 1981 and 1986[a]

Fuel	Product	Size (MWt)	No. of Bubbling Beds	No. of Fast Beds
Coal	Steam	1–10	11	1
		10–100	17	7
		>100	1	2
	Power	10–100	9	5
		>100	3	12
Petroleum coke	Power	10–100	2	4
Lignite	Steam	10–100		1
Tails	Power	10–100	1	12
Tires	Power	10–100	1	
Waste	Power	0–100	5	1
Wood	Steam	1–10	26	
		10–100	8	
	Power	10–100	3	1

[a]Data summarized from Reference 25.

expanded effort, with the emphasis on developing a better method for reducing sulfur emissions than wet stack gas scrubbing. The enormous volume of literature on fluidized-bed combustion (FBC) that has proliferated in the subsequent 27 years is ample testimony to the importance that governments, research labs, and corporations have placed on this field. Many thousands of reports and papers have been produced; space permits referencing only a few of the more significant.

The extensive literature and the slow rate of commercialization of fluidized-bed coal combustion are good indications that the problems involved in the design of steam generators utilizing fluidized-bed coal combustion are multidisciplinary to a far greater degree than for any other heat exchanger application. The complexities and subtleties are even greater than for the conventional coal-fired boilers of the previous section. These have had the benefit of over a century of operating experience that has served as a sort of Darwinian selection process in the evolution of designs for our present coal-fired boilers. Inasmuch as the potential U.S. market in coming decades for fluidized-bed units has been estimated to run in the trillions of dollars, the whole complex of problems deserves attention in this chapter. Thus, before treating the heat transfer aspects, it seems essential to provide some perspective by briefly outlining the more important of the difficulties involving chemistry, thermodynamic cycle analysis, turbomachinery, fuel metering, corrosion and erosion, and system operation and control that dominate, and in large measure determine, the design of

heat exchangers for fluidized-bed combustors. (Note that the basic heat transfer and fluid-flow characteristics of fluidized beds are treated in Chapt. 7.) A good indication of the rate and extent of the progress in developing fluidized-bed combustion systems is given by Table 15.1. This summarizes data on the 87 new fluidized bed combustion units put in service in the United States, that were included in a comprehensive set for the United States, Europe, and Japan during the period between 1981 and 1986.[25]

Sulfur Retention and Combustion Efficiency

If a fluidized-bed combustor is used as just another way to burn fuel, it may be operated over a wide range of temperatures and the operating temperature may be chosen to suit any of several requirements. However, for good sulfur retention the temperature must be around 850°C (1562°F) if limestone is used as the sorbent, and around 900°C (1652°F) if dolomite is used, apparently because MgO plays a role in the reactions. As will be treated later, the bed-temperature requirement for good sulfur removal imposes a major restraint on the design of the heat transfer surfaces.

The fluidized-bed combustor can be viewed as a chemical reactor in which the reaction rate is limited by diffusion processes in the solid particles of coal and lime (the limestone or dolomite is calcined and the $CaCO_3$ converted to CaO as it is heated to bed temperature). To assure a strong excess of calcium in the bed to absorb the sulfur almost as fast as it is released, the calcium feed rate is usually about double the sulfur feed rate.[24,26] The time required for the diffusion and reaction processes in the solid particles is such that roughly 80% of the sulfur released by combustion will be captured by CaO in the fraction of a second required for the gas to transit a bubbling bed 1 m deep. If the bed is 2 m deep, about 80% of the remaining sulfur will be captured as the gas rises through the second meter, giving an overall capture fraction of 0.8 + 0.8 × 0.2 = 0.96, or 96% of the sulfur in the coal. Similarly, making the bed depth 3 m will serve to capture over 99% of the sulfur. At the same time, the combustion efficiency improves because there is more time for the carbon to be burned out of the pores of ash particles before they are elutriated. In 1-m deep beds, for example, combustion efficiencies as low as 85% have been observed at high superficial velocities. By recycling the elutriated material removed from the exhaust gas with cyclone separators, the combustion efficiency can be increased to about 95%.

Corrosion and Erosion

Corrosion and/or erosion of tubes in fluidized beds has often proved to be a serious problem, sometimes causing tube failures in less than 100 h of operation. It is significant that local failures may develop rapidly in one region of a bed while no appreciable damage has occurred in other regions. An explanation for this is that under steady-state conditions oxidation of chromium steels progresses very slowly in either strongly oxidizing or strongly reducing conditions, but progresses very rapidly if these conditions alternate, or if the thin layer of protective oxide is removed by erosion. Inasmuch as the fuel-air mixture must be fuel-rich in the vicinity of the coal injection ports, it follows that, with the bed as a whole supplied with excess air, there must be a zone above each coal feed port in which there is a transition from reducing to oxidizing conditions around the particle plume above the port. Fluctuations in the coal feed rate, or the sloshing movements of a strongly bubbling bed such as those evident in Fig. 7.1, will move the boundaries of this plume, and severe corrosion of any tubes in this region should not be surprising.

The large bubbles that ascend rapidly through a violently bubbling bed induce particle velocities in their wakes about double the superficial gas velocity, thus presenting a much more severe erosion potential than might at first be expected, and providing an explanation

for the instances of severe erosion in some beds and not in others. Even larger particle velocities result if slugging occurs, that is, periodic oscillations of the bed, usually acoustical in nature, as a consequence of coupling between disturbances in the bed and the acoustical characteristics of the air plenum and supply duct system.[27,28] The pressure drop across the air distributor plate must be made large enough to snub such oscillations.

The combustion rate for char is relatively slow, running on the order of 100 s for particles about 1 mm in diameter. The strong turbulent mixing in the bed disperses particles widely in a few seconds so that coal particles will not tend to give small local reducing regions once the volatiles are vaporized. However, the volatiles burn as rapidly as they are released, hence the size of the fuel-rich plume that they will produce can be estimated from the rate at which volatiles are evolved from a rapidly heated particle of coal. Flash pyrolysis tests indicate that the bulk of the volatiles are released from 70-μm coal particles heated to 1000°C (1832°F) in 0.1 s.[29] For larger particles, heat transfer considerations indicate that it would take about 3 s to bring a 1-mm particle up to bed temperature if the cooling effect of the volatilization is included. Tests indicate that particles injected into a fluidized bed are dispersed vertically at a rate about 10% and laterally at about 2% of the superficial gas velocity. From these factors it appears that the fines in the injected coal will release their volatiles within a few centimeters of the injection point, while 1-mm particles will release most of their volatiles within about 60 cm of the feed port. This implies that tube banks should be kept at least 60 cm above the coal feed ports, which would leave only about 40% of the bed volume available for heat transfer surface in a 1-m deep bed with the coal feed ports at the bottom. However, if there is a large gap between the plane of the air tuyeres and the bottom of the tube matrix, depending on the fluidizing velocity and particle size, large bubbles may form and throw particles against the tubes at velocities several times the peak particle velocity within the tube bank, and thus cause severe local erosion of the tubes at the bottom. Analysis aids in understanding problems such as these, but their solution is heavily dependent on extensive test experience.

If the coal feed ports are too far apart there will not be enough oxygen available to burn the volatiles as rapidly as they are distilled out of the coal particles, and the size of the fuel-rich plume can be substantially greater than the region in which the volatiles are released. Although the author has not been able to find any systematic tests designed to determine the proper spacing of coal feed ports, operating experience seems to indicate that they ought not be more than about a meter apart.

Table 15.2 gives one of the best summaries of corrosion/erosion experience world-wide with different alloys in various FBC furnaces that is available at the time of writing.[30] Note that the corrosion-erosion rates are high, running about 1 mm in 10,000 h for the best cases. This is excessive, and must be reduced by better design and improved materials.

Coal Metering and Feed

Liquid fuels can be pumped easily and the flow rate can be regulated accurately by controlling the pressure drop across an orifice of the proper size. Pumping and metering streams of solid particles is far more difficult. Solid particles can be fluidized and conveyed through pipes, but the particle density in a gas suspension is hard to control. Further, for a pressurized fluidized bed pumping the solid particles from atmospheric to furnace pressure is so difficult that it usually has been accomplished with lock hoppers, that is, the gas-fluidized solids at low pressure are directed into a bin with its discharge valve closed, the bin is filled to the proper level, and the inlet valve is closed, after which the bin is pressurized to provide a source of gas-fluidized particles at a high pressure. At least two such bins are required so that one serves to supply the furnace while the other is being filled, and troubles with sticking or leaking valves are common. Screw feeeders are also used to move streams of solid particles from a low- to a high-pressure zone, but how well they function largely depends on the characteristics of the particles, such as their tendency to cake; the highly variable characteristics of coal present problems.

Dividing and metering streams of solid particles has been accomplished in many ways, including use of vibrating tables and belts; suffice it to say that even with exceptionally cleverly designed systems it has proved to be exceedingly difficult to avoid irregular variations in the solids-flow rate in at least some of the divided streams. These problems are of little consequence in feeding the stream of limestone or dolomite because of its long residence time in the bed, but they are vitally important in metering the coal because of possible effects on corrosion. Note that the slow combustion rate for char permits much greater irregularities in the fuel-flow rate to any given feed port in the bed.

To avoid the above difficulties, some coal-fired fluidized beds have been operated with the coal injected over the top of the bed. This approach requires relatively large coal particles from which most of the fines have been removed by screening. Otherwise, small particles would be elutriated before there was sufficient residence time for good sulfur absorption or carbon combustion. Another approach is to prepare a water slurry that can be pumped and metered much as if it were a liquid. This greatly eases pumping and metering, and eliminates the need for the coal-drying operation required before pneumatic handling can be carried out properly. The principal disadvantage of coal-water mixtures is that the weight of the water injected into the bed is about two-thirds the weight of the coal, and the heat required to vaporize the water leads to a loss in thermal efficiency by about 5%, or about 2 points out of the plant thermal efficiency of perhaps 36%. (This is a bit less than the losses associated with limestone scrubbers.) If the system is used in a combined cycle, the addition of water vapor to the combustion products passing through the gas turbine increases its output by a few percent, thus reducing the net loss.

Fast Fluidized Beds

Both the fuel metering and the corrosion problems can be eased through the use of a circulating, or fast, fluidized bed.[31] As indicated in Chapter 7 and shown here in Fig. 15.6[32] which shows the key relations between the Archimedes number, the particle Reynolds number, and the Fronde number over wide ranges of these parameters, the bubbling fluidized bed is only one of a range of possible modes of operation. As indicated in Fig. 15.6 a system can be designed to operate with gas velocities of 6 to 9 m/s, as compared to the 1 to 2 m/s for the bubbling mode of fluidization. Fast fluidized-bed combustors may take any of several forms. Often the bed is stabilized by a layer of larger particles (usually silica gravel 10 to 20 mm in diameter) perhaps 300 to 600 mm thick, while the bulk of the coal particles are smaller so that they are entrained and carried up through a tall furnace to a cyclone separator. They may be recirculated directly to the dense bed, or they may drop into a second bed containing heat transfer surfaces and fluidized with a gas stream at a low velocity. The combustion chamber temperature is held to the desired level by thermal radiation to the water-cooled furnace walls and by controlling the recirculation rate of the material in the entrained bed. The recirculated material may be cooled by the walls of the return circuit and/or by a heat exchanger in a bubbling bed. In the latter case,

TABLE 15.2   Summary of the Data on World-wide Corrosion-Erosion Experience in Fluidized Bed Combustion Units Available up to February 1987[a]

Unit, Location	Operating Time h	Fluidizing Velocity m/s	Wear Rate nm/h
Georgetown Univ. Washington, DC		2.5	
Waterwalls	1000		51–127
In-bed tubes	5850		755
Peoples Republic of China	1200	2.7	833
	5000	2.74	460 max
In-bed tubes	—	3.0	1300
Side-walls	—		100
Stork/TNO, Holland		1–3	
Tube Bundle I	1800		2278 max
	600		4000
	2400		1292
Tube Bundle II	532		850
Nitrided tubes			320 max
Tube Bundle III	342		690 max
Babcock & Wilcox 6′ × 6′ AFBC Alliance, Ohio		2–2.5	
	2414		420
	>300		100,000
Northern States Power Minnesota		4	
In-bed evaporator	>2000		254
Cebu, Phillipines		2.8	
	2500		1800
Allonnes, France		2.2	
Inclined tubes	1000		4000
Membrane wall	1700		600
Wakematsu, Japan			
Evaporator	3200		3000
Superheater	3200		100
Volklingen, Germany		1.3	
	3000		330
Studsvik, Sweden		2.2	
	4000		1000
Gibson Wells, U.K. Unit 1			
Plain tubes	5100	2.3	790 max
Finned and ball-studded	11900	3.2	240 max
Fins alone	5000		170 max
Unit 2			
Plain tubes	3700	2.5	700
Ball studded	5500	3.7	950 max
Finned tubes	3000	3.1	1145 max
Unit 3			
Ball studded	7200	2.2	0
TVA/EPRI 20 MW Padukah, Ky.		2.5	
B&W Tube Bundle	5558		
Water wall			112
In-bed evaporator			250
In-bed superheater			96

TABLE 15.2 (*Continued*)

Unit, Location	Operating Time h	Fluidizing Velocity m/s	Wear Rate nm/h
Corrosion rack specimen			630
C–E tube bundle evaporator	2823		1500
	3900		1150
Central Soya			
Marion, Ohio		2.4–3.7	
In-bed tubes	11,000		454
Water wall	14,000		366
Great Lakes		2.1	
In-bed evaporator	5300		127
In-bed evaporator	4722		356 max
			742
Battelle Columbus, Ohio		2.5	
Water-cooled tubes	1100		5600
Chalmers University, Sweden		2.5	
Evaporator	1210		1000
Superheater			600
Vertical tubes			1200
General Electric Long-Term Materials Test Facility, PFBC Malta, N.Y. Water-cooled in-bed helix		0.7–0.9	
First helix	1050		1974
Second helix	1199		1195
Third helix	104		1270
Nova Scotia Power Co.			
Point Tupper, Canada	10,000	2.4	300 max
Grimethorpe 2m × 2m PFBC, U.K.			
Tube Bank A	864	2.5	1700 max
Tube Bank C	470	1.5	>3000
	250	1.5	>2800
	547		~3000
Tube Bank CZ	1519	1.5	500 max

[a](Data extracted from Ref. 30.)

Note 1. 1 mil/1000 h = 25.4 μm/1000 h = 25.4 nm/h. Typically a maximum corrosion wastage for acceptable tube life is 35 nm/h. Minimum acceptable tube life could be as little as 50,000 h, but not much less. For a 6-mm tube wall, with a maximum loss of 4 mm, the erosion rate would be 80 nm/h.

Note 2. These data have been collected from a variety of sources, and do not always refer to the same thing: Some refer to average wastage, some to maximum wastage averaged over a tube, some to maximum local wastage. While in most cases incidents produced by a local effect, such as a jet, have been excluded, in some cases the report does not make it clear what the source of the problem was. At times, two reports from the same unit give slightly different numbers for metal loss: The differences are not usually great, but somewhat different values may appear in other sources.

roughly half the heat released by combustion is transmitted by thermal radiation to the water-cooled walls of the furnace, while the balance is removed by a tube bank in the bubbling bed where a secondary air stream completes combustion of the remaining char.

Fast fluidized beds have operated satisfactorily with a wide range of fuels using a single feed point for fuel energy inputs of as much as 100 MWt as compared to around 100 feed points for a bubbling bed of the same output. A high combustion efficiency is obtained even with low-grade fuels because of the high recycle ratio—as much as 20:1. Excellent sulfur retention is obtained by using fine particles of limestone. The useful load range is around 5:1, substantially better than for bubbling

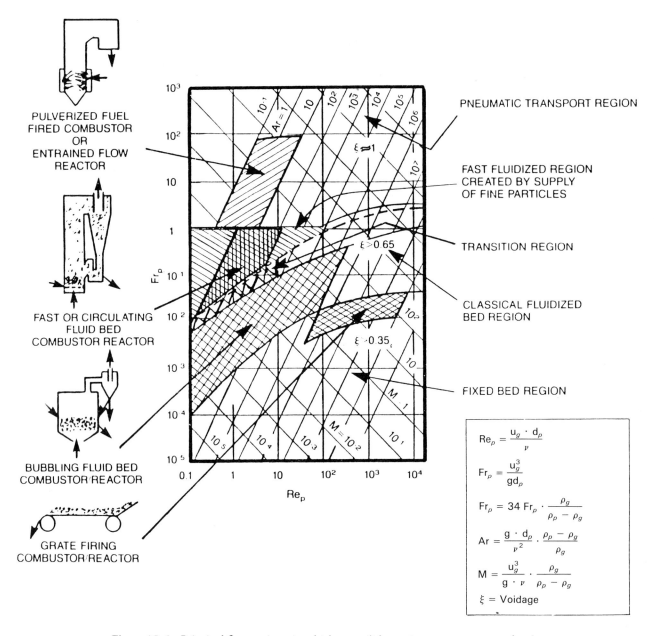

Figure 15.6 Principal flow regimes in which gas-solid reaction systems operate for the full range of conditions from a traveling-grate coal furnace to a pulverized coal-fired furnace. The dimensionless numbers Re_p and Fr_p are based on the particle size. (Courtesy Daman, Ref. 32.)

fluidized beds. The principal disadvantages of the circulating bed combustor are the large height, the large cyclone separators, higher capital costs, and scaling problems that make the system less suitable for steaming rates below 50,000 or above 1,500,000 lb/h (6.3 or above 19 kg/s).

Atmospheric Versus Pressurized Beds

The pressure drop through bubbling fluidized beds is directly proportional to bed depth, so that the resulting pumping power requirement becomes unacceptably high for atmospheric-pressure fluidized-bed combustors (AFBCs) if the depth is increased beyond about 1 m. Hoy recognized this, and in Reference 24 planned to employ the fluidized bed as the combustion chamber for a gas turbine in a compound cycle which would not only serve to pressurize the bed but would also give a higher efficiency system in which about 30% of the power would be produced by the gas turbine and the balance by a conventional steam turbine. He was aware that gas turbines are very subject to erosion and deposits caused by even trace amounts of small particles in the hot gas stream, but felt that the ash particles elutriated from the pressurized fluidized-bed combustor (PFBC) would be friable "snowflakes" that would be much less abrasive than the tiny glassy cinders from a pulverized coal burner. However, microscopic examination of the ash elutriated from FBCs shows a substantial incidence of abrasive glassy cinders, and friable ash seems as likely to form deposits on turbine blades as glassy cinders.

On the surface, it appears reasonable to remove small ash particles with some sort of cyclone separator and filter system, and this is true for the larger ash particles. The catch is that it becomes progressively more difficult to remove particles as their size is reduced, particularly for the range below 10 μm, and even 1-μm particles are erosive. Further, even much smaller particles can give trouble with deposits. An excellent series of tests was carried out in the United States during the 1940s and 1950s by the Locomotive Development Committee (LDC) in an effort to solve this problem—with discouraging results.[26] In all cases serious trouble was encountered with excessive deposits on the turbine blades or excessive erosion, or both. At first glance it may seem that it should be possible to get just the right balance between the erosion and deposition processes, but in practice the composition and hence the tendency of the ash to be sticky at temperatures above about 550°C (1022°F) varies widely from one coal seam to another. Further, a thorough study by R. W. Foster-Pegg of the effects of dust carried into oil- and gas-fired conventional gas turbines has shown that the dust level must be kept below about 1 mg/m^3 to avoid serious trouble with erosion or deposits.[33] This very low dust concentration has proved to be difficult to achieve even with what seems to be reasonably dust-free inlet air. As a consequence, the bulk of the work on fluidized-bed combustion in the 1970s and 1980s has been directed toward atmospheric-pressure beds, but a strong interest in the use of pressurized beds in a combined cycle has continued.

In the course of a detailed review of the original data from the LDC tests, the author noted that there had been no serious difficulty with either erosion or deposits in the lower stages of the turbine where temperatures were below 550°C (1022°F). This proved to be consistent with experience with the gas turbines employed to pressurize fluidized-bed catalytic cracking units which have operated successfully for many years. In these units it has been necessary to keep the turbine inlet temperature below 600°C (1110°F), the temperature level below which the hardness of the high-alloy turbine blades is sufficient to keep erosion down to an acceptable rate. (The aluminum oxide particles used as the catalyst are hard and abrasive, but the temperature at which they might become sticky is above 1500°C (2732°F) so they do not form deposits on turbine blades). This review led the writer to propose that fluidized-bed combustors be pressurized with a gas turbine similar to that employed in diesel engines.[26] Turbines of this type operate with turbine inlet temperatures in the 400 to 540°C (752 to 1002°F) range, well below the level at which serious erosion or deposits are likely in operation with a fluidized-bed combustor. Pressurizing the system in this way not only improves the sulfur retention and combustion efficiency, but it also increases the heat transfer coefficient for heat recovery from the hot gas leaving the bed and makes it possible to get good control of the output from zero to full power. Further, the capital costs of the system appear to be substantially less than for an atmospheric bed because the bed cross-sectional area is inversely proportional to the furnace pressure. This reduces the size and weight of the furnace sufficiently that it can be shop-fabricated, the number of coal feed points and hence the cost of the coal metering system can be greatly reduced, and the size of the building required is reduced. It appears that the resulting savings will reduce the cost of this type of fluidized-bed coal combustion system sufficiently so that it will be less than that for a conventional pulverized coal-fired steam generator equipped with stack gas scrubbers.[34]

Some small PFBC pilot plants have been built and operated for periods of a few thousand hours, and two PFBC demonstration plants of around 100 MWe are under construction in the United States at the time of this writing, one intended for combined cycle operation and the other for the more feasible goal of operation with a turbocharger.

Typical Plant Design

There are numerous variations on the various fluidized-bed combustion systems outlined above, each with

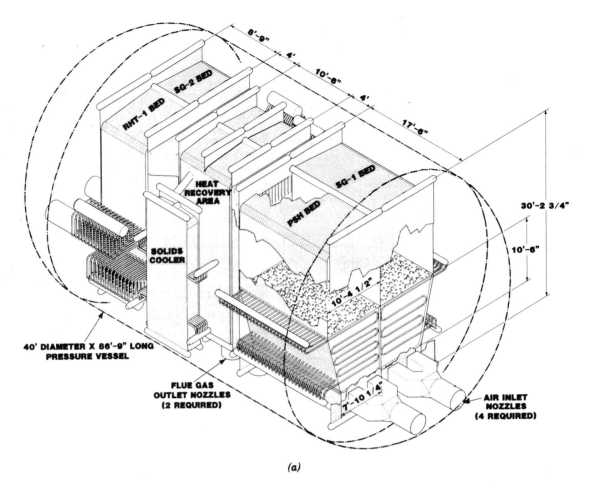

Figure 15.7 Cutaway views of a Foster Wheeler turbocharged fluidized-bed combustion boiler showing the arrangement of the major components in a unit designed to supply an 80-MWe plant with 575,000 lb/h of steam at 950°F and 1475 psi with reheat of 494,000 lb/h to 950°F and 356 psi. The bulk of the boiling occurs in the SG-1 and SG-2 beds shown in the overall view (a). The exploded view of (b) shows some details of the convection passes for the finishing reheater, finishing superheater, and the economizer together with the ash cooler. (Courtesy Foster Wheeler Corp.)

certain advantages and disadvantages.[32] The space available will permit a fairly detailed look at only one of these concepts, a recent plant design that shows how the many different requirements outlined above can be met in an integrated system. The general layout is shown in Fig. 15.7a and 15.7b.[35] The furnaces are pressurized with gas turbines driving only compressors, so that the turbine inlet temperature can be kept down around 400°C (752°F). The fluidized bed is split into four parts, one pair consisting of the primary boiler, or steam generator (SG-1 bed), having the same length as that for the primary superheater (PSH bed), and the other pair

consisting of the balance of the boiler (SG-2 bed) and the primary reheater (RHT-1 bed). The hot gas from these furnaces flows to the central region, where it gives up much of its heat to the finishing stages of the reheater and superheater, and then to the economizer where it is cooled to the turbine inlet temperature. A solids cooler is mounted next to these units to recover heat from the ash and spent dolomite leaving the furnace system. The coal is pumped to the furnaces in the form of a 70:30 (by weight) coal-water mixture. Each of the two large beds has eight feed points, while each of two smaller beds has four. The jet of coal-water mixture is broken into

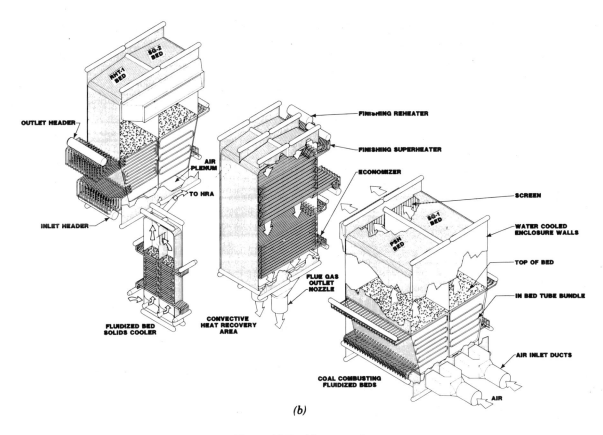

(b)

Figure 15.7 (*Continued*)

small particles by a concentric air stream as it is introduced into the hot bed. The complete boiler assembly is installed in a large cylindrical vessel. Housing the fluidized beds in this large vessel with few penetrations and gasketed joints minimizes problems with leaks of dirty gas. (Plant operators have been plagued by leaks of hot dirty combustion gas from the myriad joints in the walls of pulverized coal-fired furnaces designed to operate with forced draft fans at pressures only a little above atmospheric.)

Example 15.1. An example that provides some excellent insights into the limitations with which one has to work in designing fluidized-bed combustors is instructive. This case, taken from a conceptual design study made by the author, included an estimate of the surface area and footage of tubing required for a steam generator employing a fluidized-bed coal combustor pressurized to 3 atm with a supercharger operating with a turbine inlet temperature of 1000°F. The plant output was 200 MWe, with the same steam conditions as employed in the 500-MWe TVA Widows Creek plant.

The bed temperature chosen was 1650°F, the bed air-inlet temperature was 675°F, and the net heat release to the combustion air from the coal combustion was 1072 Btu/lb after allowing for 10% excess air and the heat require to bring the coal and limestone feed up to the temperature of the bed. The bed depth was chosen to be 15 ft. The heat transfer matrix was to be made of 2-in.-diameter vertical tubes of Croloy having a wall thickness of 0.1 in. in the economizer, evaporator, and reheater, and a wall thickness of 0.2 in. in the superheater. To keep them well above the reducing zones just above the coal feed ports, the tubes in the bed were extended only 12 ft down into the bed. An additional 8 ft of tubing was provided to extend up through the freeboard above the bed to the header drums, thus giving an overall tube length of 20 ft. The heat transfer contribution in the latter region was neglected. The superficial velocity leaving the bubbling bed was taken as 6 ft/s.

The first step was to establish the ratio of the heat transferred in the bed to that given up by the hot gas between the bed and the turbine. (The study from which this case was taken showed that the best way to utilize the heat given up between the turbine exhaust and the stack was for preheating the combustion air and to supplemental feedwater heating.) Table 15.3 shows the system pressures and temperatures together with a heat balance for the steam conditions of the Widows Creek plant. Note that the specific heat of the combustion gas was taken as 0.276 Btu/lb · °F, and that the steam flow to the reheater was 71.5% of the feedwater flow because of steam bled from the turbine for feed heating. (In view of the fact that most of the other studies in this field employ English units, English units are employed here rather than SI units to facilitate comparisons. In this case, including both would have made the table awkward.)

Inspection of the column giving the fractions of the total heat to the steam required for the different sets of heat transfer surfaces discloses that, if the surfaces for the economizer and evaporator were placed in one bed with the reheater in the hot gas stream flowing from that bed to the turbine, the heat loads for these components would be properly balanced. Similarly, the first 80% of the heat input to the superheater could be transferred to tubes in a second bed and the remaining 20% could be handled with a tube array in the hot gas flowing from that bed to the turbine. This arrangement places the two regions having both the highest metal temperatures and the lowest heat transfer coefficients on the steam side into regions with the lowest hot-gas temperatures and lowest heat transfer coefficients, thus reducing the possibility of

overheating the tubes. Thus this arrangement was chosen.

The heat transfer coefficients in the fluidized bed were estimated from Eqs. 7.3 and 7.4, whereas that for boiling was chosen from Fig. 5.8. To simplify the calculations, that value was also used for the economizer region by taking the effective water temperature there as being at the boiling point. The heat transfer coefficients for steam were estimated from Figs. H5.7–9, and that for the hot gas from Fig. H5.11 after estimating appropriate mass flow rates for reasonable pressure drops. A second iteration produced the values presented in Table 15.4. Although the design study of which these tables were a part was made 10 years earlier than the design of Figs. 15.6 and 15.7, and the design conditions are somewhat different, the relative proportions of the various sections of the heat transfer matrices are similar. This is no coincidence because the various boundary conditions force one to use similar divisions of the furnace and similar surface areas and proportions.

STEAM GENERATORS FOR NUCLEAR POWER PLANTS

This chapter has given some notion of the enormously complex problems involved in the design of modern coal- or oil-fired boilers. While hardly mentioned, the design

TABLE 15.3 Cycle Conditions and Heat Balance for a 200-MWe Steam Plant Based on a Fluidized-Bed Combustor Pressurized with a Turbosupercharger[a]

	Pressure, psia	Temp., °F	Enthalpy, Btu/lb	h, Btu/lb	Flow, lb/h × 10^{-6}	Heat to Steam, %[a]
Feedwater in		531	523		3.8	
				240		21.6
Evaporator in		682	763		3.8	
				312		28.2
Evaporator out	2750	680	1075		3.8	
				419		37.7
Superheater out	2450	1050	1494		3.8	
Reheater in	476	644	1327		2.72	
				194		12.5
Reheater out	454	1000	1521		2.72	
Combustion air in	52	675		1072	4.05	
				780		
Bed	41	1650			4.33	
				195		
Turbine in	40	1000				

[a]Ratio of heat transferred in bed to that from hot gas = 780/195 = 4.0; ratio of combustion air flow to steam flow = 1.137 lb/lb; ratio of combustion gas flow to steam flow = 1.137 × 1.09 = 1.24 lb/lb.

TABLE 15.4 Estimate of the Heat Transfer Surface Areas for the Economizer-Evaporator, First and Second Stage Superheaters, and First and Second Stage Reheaters for the Conditions of Table 15.3

	Evaporator	Reheater	Superheater
Fluidized-Bed Heat Transfer Matrix			
Bed-side heat transfer coefficient, Btu/h · ft² · °F	70		70
Steam-side heat transfer coefficient, Btu/h · ft² · °F	6,000		400
Tube-wall conductance, Btu/h · ft² · °F	2,000		1,000
Overall heat transfer coefficient, Btu/h · ft² · °F	66.6		56.1
Heat load, Btu/h × 10⁻⁶	835		505
Steam-side inlet temperature, °F	531		680
Steam-side outlet temperature, °F	680		920
LMTD, °F	965		850
Surface area required, ft²	13,000		10,600
Number of tubes	2,070		845
Total footage of tubing (including freeboard), ft	40,000		17,000
Number of header joints	4,140		1,690
Hot-Gas Heat Transfer Matrix			
Gas-side mass flow rate, lb/s · ft²		3.31	3.31
Gas-side heat transfer coefficient, Btu/h · ft²		24.2	24.2
Gas-side pressure-drop coefficient, ΔP/q		0.0135	0.0135
Tube spacing, in. (transverse × axial)		2 × 1.5	2 × 1.5
Steam mass flow rate, lb/s · ft²		38	55.6
Tube-wall conductance, Btu/h · ft² · °F		2,000	1,000
Overall heat transfer coefficient, Btu/h · ft² · °F		20.9	21.8
LMTD, °F		475	447
Heat load, Btu/h × 10⁻⁶		210	127
Surface area required, ft²		21,200	3,000
Total footage of tubing, ft		40,500	4,800
Number of tubes		420	100
Number of header joints		840	200
Tube matrix inlet face, ft		10 × 20	6 × 20
Tube matrix height, ft		26	25.25

of such units is greatly complicated by the problems associated with burner and furnace design, ash removal, and the like. In many respects, the design of steam generators for some of the nuclear power plants is much simpler. Since furnace problems are not involved, the design of these units can follow procedures more like those used in the design of other types of heat exchangers. Furthermore, there are many other types of boiler that present essentially similar problems, for example, boilers in petroleum refineries and chemical-processing plants, evaporators for refrigeration and air-conditioning systems,[42] and steam generators in binary vapor cycle power plants.[43] For these reasons, the two design cases considered in the following sections are from the nuclear power plant field.

STEAM GENERATORS FOR PRESSURIZED WATER REACTORS

Steam generators for pressurized water reactors are not only vital components and major capital cost items in PWR plants but they present a fascinating set of design problems with nuances that are both particularly challenging and highly instructive.[1,38-43] While they operate under far more benign conditions than the steam generators in coal-fired plants, extensive experience has shown that they are subject to annoying failures that, although seemingly minor, have proved difficult to handle because small deposits of highly radioactive materials produce radiation fields of 5 to 40 rem/h at the steam generator after shutdown.[41] This severely restricts maintenance operations and greatly increases outage time—which costs roughly $1,000,000/day in plant overhead. Further, if it becomes necessary to replace a steam generator, the time required is on the order of six months.[42] Thus an exceptionally high degree of reliability is a prime design requirement. At the same time, the large size and high cost of these components provide enormous incentives to evolve a highly refined design. This section is concerned primarily with the heat transfer problems, because they, more than any other factor, determine the size and cost of the units.

Design Requirements and Restraints

A whole complex of design restraints limits the outlet temperature of PWRs to about 330°C (626°F),[26] a temperature much lower than the steam temperatures commonly used in fossil fuel plants. Carnot cycle efficiency considerations show that, with the peak cycle temperature in this region and a condenser temperature of around 38°C (100°F), an increase in steam temperature of 10°C represents roughly 4% of the available temperature drop in the cycle, and hence a 4% improvement in the thermal efficiency. This provides a tremendous incentive from both the capital and operating-cost standpoints to design for a high steam temperature. Thus, from the heat transfer standpoint, the prime design requirement is to produce steam at as high a temperature as possible consistent with cost considerations. In attempting to do this, one finds that the key limitation is

not the reactor outlet temperature but rather the temperature difference at what is called the *pinch point*, the point in the heat exchanger at which the temperature of the primary system water and that of the steam most closely approach each other. This condition is best visualized by considering the temperature distribution shown in Fig. 15.8. Not only is it impossible to transfer heat from the hot primary water to the cooler steam if one were to try to make the steam temperature higher than the water temperature at this point, but the surface area required in this region is inversely proportional to the local temperature difference, and increases at an unacceptable rate as the temperature difference at the pinch point is reduced below about 10°C. As a consequence of these various restraints, the various steam generators for PWRs are quite similar in performance, a point evident in the data of Table 15.5 for three typical units.

Typical Configurations

Quite a number of configurations have been employed in PWR steam generators, but that evolved by Westinghouse,[38-42] shown in Fig. 15.9, is one of the most widely used.[38-40] The high-pressure primary system water enters at one side at the bottom of the unit, flows upward through U-shaped tubes, and back down to the outlet on the other side. Feedwater enters at the bottom, where the primary water leaves the unit, and flows in a multipass crossflow-counterflow pattern upward between the tubes. Steam is released from the free liquid surface above

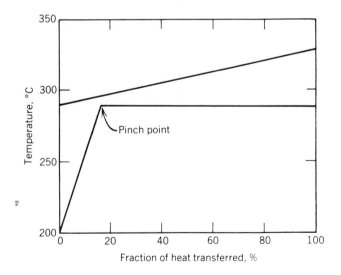

Figure 15.8 Curves showing the temperature distribution in a steam generator for a PWR and the pinch point.

TABLE 15.5   Summary of Design Data for Some Typical PWR Steam Generators

Type of PWR	Once-through	Recirculating	Recirculating	Recirculating
Designer	Babcock and Wilcox	French	Westinghouse	
Designation	Oconee 1[1]	N4[38]	Series 51[37]	Ex. 15.4
Tube shape	Straight	U	U	U
Fluid in tube	Primary	Primary	Primary	Primary
Type of tube	Plain	Plain	Plain	Plain
Capacity, MWt	1,295	1,068	850	500
Steam flow, kg/s	706	601.3	466	204
Temperatures, °C				
Primary in	343	329.5	319	330
Primary out	289	292	283	292
Feedwater	235	229.5	222	230
Steam out	299	289	271	289
Superheat	20	0	0	0
Pinch ΔT	10	10		10
Tube OD, mm	16	19	22.3	19
Tube ID			19.7	16
Tube length, m	15	21.8	20.2	18.7
No. of tubes	15,530	5,600	3,383	3,310
Surface area, m^2	11,600	7,300	4,780	3,410
Surface, m^2/kW	8.9	6.85	5.6	6.8

the top of the U-tubes and passes upward through two stages of vapor separators to the outlet at the top. The water draining from the vapor separators flows downward through the annulus between the tube bundle and the outer shell to the bottom of the unit for recirculation through the tube matrix so that the vapor quality of the steam-water mixture flowing up out of the top of the U-tubes is far below that for the burn-out heat flux. In the design of Fig. 15.9, two stages of vapor separators are employed to reduce the moisture content of the steam delivered to the turbine to much less than 1%. Note that the volume required inside the pressure vessel for the vapor separators is almost as great as that for the boiler tubes.

Once-Through Steam Generators

Babcock and Wilcox have evolved a steam generator design for PWRs that replaces the vapor separators of Fig. 15.9 with heat transfer surface that dries and superheats the wet steam emerging from the boiler region.[1,43] As can be seen in Fig. 15.10, the hot primary water from the reactor enters a plenum at the top of the vessel, flows downward through the inside of the straight tubes, and leaves from a plenum at the bottom. The feedwater enters the vessel a little above the midplane and flows downward through an annulus between the vessel inner wall and the lower baffle (or shroud) surrounding the lower portion of the tube bundle. The bulk of the liquid in the steam-water mixture that rises to this level under thermal convection forces spills over the top of the lower baffle to flow downward through the outer annulus with the feedwater so that the region in the tube matrix where water travels radially to the outer perimeter operates as a crude vapor separator. The steam flowing up from this region into the upper portion of the tube bundle enters with a vapor quality of about 90%. By the time it leaves the top of the tube bundle it has been dried and superheated by about 20°C (36°F). Essentially pure counterflow conditions obtain, making possible the small amount of superheat which eases moisture problems in the turbine. Note that while the primary water follows a once-through path downward through the steam generator, the greater part of the water in the boiling region leaves the tube bundle just above the midplane to return to the bottom so that this section operates as a recirculating boiler. One effect of this is that dissolved solids tend to concentrate in this region, just as in the boiler of Fig. 15.9, and these must be removed periodically with a blowdown or bypass cleanup system.

The layout has been cleverly arranged so that the mean temperature of the tubes is nearly the same as that of the shell, hence no special mechanical provisions such as the U-tube arrangement of Fig. 15.9 are required to accommodate differential thermal expansion. However, the relatively small water inventory in the boiler region

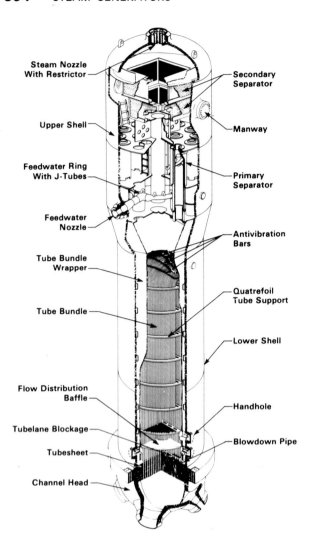

Figure 15.9 Section through a Westinghouse Model F Steam Generator for a PWR showing the arrangement of the principal components. Design data are included in Table 15.5. (Courtesy Westinghouse Electric Corp.)

has led to some difficulties with thermal stresses in the system when dry-out occurred in the boiler region in the course of unusual transients involving an interruption of the feedwater flow. This is mentioned here only to remind the reader that a host of very subtle considerations associated with system operation under odd transients may come to light in the course of operation and prove to be vitally important from the operational standpoint.

Figure 15.10 Section through a Babcock and Wilcox once-through steam generator for a PWR showing the arrangement of the principal components. Design data are included in Table 15.5. (Courtesy The Babcock and Wilcox Co.)

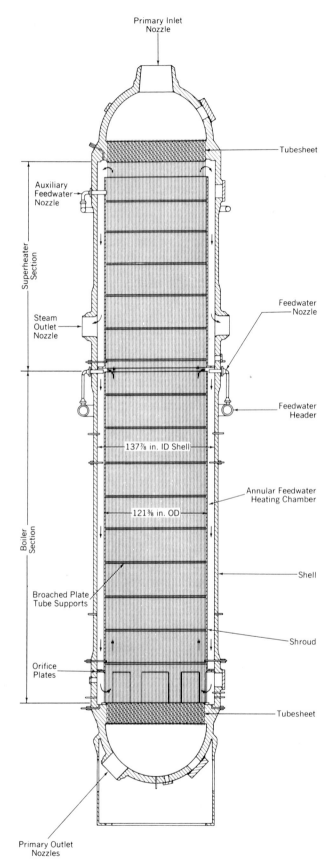

Example 15.2. Estimate of the Tube Length for a Conventional PWR Steam Generator. Steam generators of the type shown in Fig. 15.9 pose some interesting and instructive problems. From the heat transfer standpoint, the key parameters are those that determine the length of the U-tubes. Thus the design of this type of steam generator can be exemplified by considering a single tube. Typical design conditions as derived from References 38 to 40 are listed at the top of Table 15.6. The primary circuit water velocity through the tubes is limited by pumping power and pressure-drop considerations to around 5 m/s; a velocity of 4.81 m/s, or a flow of 0.67 kg/s per tube, was chosen as representative of several designs examined. A heat balance using this flow and the enthalpy changes in both the primary circuit water and the feedwater-steam yielded a feedwater flow of 0.056 kg/s with a heat load per tube of 151 kW. The heat transfer coefficient for the feedwater in the preheating region was first estimated assuming that the flow would be axial between the tubes, but this gave such a low water velocity and heat transfer coefficient that a multipass crossflow arrangement (similar to that of Fig. 15.9) was assumed with a water velocity sufficient to give the same heat transfer coefficient for the feedwater as obtained for

TABLE 15.6   Estimate of the Tube Length Required for a PWR Steam Generator

Temperature
Primary water in   = 330°C (626°F)
Primary water out = 292°C (558°F)
Feedwater in   = 230°C (446°F)
Steam out   = 289°C (552°F)
Heat load = 151 kW/tube
Feedwater flow rate = 0.056 kg/s per tube (0.123 lb/s)
                              = 278 kg/s · m² (56.8 lb/s · ft²)
Primary circuit flow rate = 0.67 kg/s per tube (1.472 lb/s)
                                    = 3,320 kg/s · m² (679 lb/s · ft²)

Tubing
Stainless steel
Tube OD = 19 mm (0.48 in.)
Tube ID = 16 mm (0.406 in.)
Mean wall area = 0.055 m²/m (0.180 ft²/ft)

Heat transfer coefficients
    Primary circuit water h = 30,000 W/m² · °C (5,270 Btu/h · ft² · °F)
    Tube-wall conductance = 12,660 W/m² · °C (2,230 Btu/h · ft² · °F)
    Feedwater preheating h = 30,000 W/m² · °C (5,270 Btu/h · ft² · °F)
    Nucleate boiling h = 28,400 W/m² · °C (5,000 Btu/h · ft² · °F)
    Preheating overall u = 6,890 W/m² · °C (1,210 Btu/h · ft² · °F)
    Boiling overall u = 6,780 W/m² · °C (1,192 Btu/h · ft² · °F)
Length increment per 2°C increment in primary circuit temperature = 144,500/(Q/A)

T_1, °C	T_2, °C	Q/A, W/m² · °C = $u(T_1 - T_2)$	ΔL, m	L, m	Q, %
293	240	365,170	0.40	0.40	5.263
295	261	234,260	0.62	1.02	10.526
297	282	103,350	1.40	2.41	15.789
299	289	67,800	2.13	4.55	21.052
301	289	81,360	1.78	6.32	26.315
303	289	94,920	1.52	7.84	31.578
305	289	108,480	1.33	9.18	36.841
307	289	122,040	1.18	10.36	42.104
309	289	135,600	1.07	11.43	47.367
311	289	149,160	0.97	12.40	52.630
313	289	162,720	0.89	13.28	57.893
315	289	176,280	0.82	14.10	63.156
317	289	189,840	0.76	14.86	68.419
319	289	203,400	0.71	15.57	73.682
321	289	216,960	0.67	16.24	78.945
323	289	230,520	0.63	16.87	84.208
325	289	244,080	0.59	17.46	89.471
327	289	257,640	0.65	18.11	94.734
329	289	271,200	0.62	18.73	100.000

the primary water flow. The tube-wall conductance was calculated using the conductivity of a typical Fe-Cr-Ni alloy such as a type 304 stainless steel or Inconel 600; this gave a much greater resistance to heat flow than the water films, but no alloy is available that would have a higher conductivity and yet would have suitable strength and corrosion resistance. The heat transfer coefficient for nucleate boiling was taken from Fig. 5.8. The overall heat transfer coefficients for the preheating and boiling regions were then calculated. Figure 15.11 was then plotted to help visualize the temperature distribution. While the LMTDs for the preheating and boiling regions could have been used to estimate the tube length required, it was decided to employ a more general approach that would lend itself to the solution of additional cases with more complex temperature distributions and would also give some insights into the greatly varying conditions along the length of the tube. For example, the possibility of mixing the recirculating flow from the vapor separator with the feedwater where it enters the tube matrix was considered as a means of increasing the water velocity and heat transfer coefficient in the preheating region, but was quickly dropped because this would reduce the temperature difference and

increase the area required in that region. Thus the velocity of the feedwater through the bottom of the tube bundle on the primary water outlet side was increased by using a closer baffle spacing, and the recirculating flow was assumed to enter the tube matrix on the side opposite the preheating region.

With the temperature drop in the primary circuit running 38°C, it was convenient to divide the unit into 19 increments with an even 2°C temperature drop in each increment. The calculations of Table 15.6 were then organized as shown using a standard spreadsheet program (Supercalc 3). The heat flux was calculated as simply the product of the local heat transfer coefficient and the local temperature difference. The increment in length was then the heat load per tube divided by the number of increments and the heat flux. The increments in both length and the fraction of heat added from the feedwater inlet end were then summed for each position so that the principal parameters could be plotted as a function of either the position along the tube or the amount of heat added. Figures 15.11 and 15.12 were then plotted using a routine in the spreadsheet program. The tube length compares well with the value given in Reference 37, i.e., 18.7 versus 20.2 m, with much of the

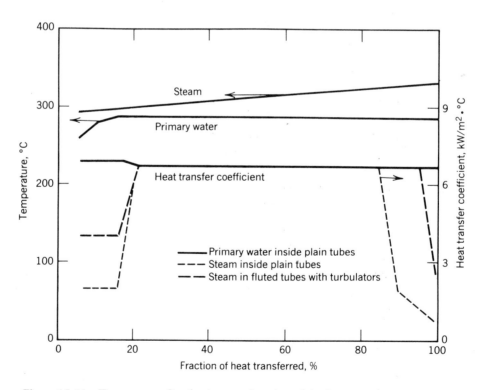

Figure 15.11 Temperature distribution as a function of the fraction of the heat transferred for the PWR steam generators of Examples 15.2 and 15.3.

difference stemming from the 11.6% lower heat load used in Table 15.5.

Example 15.3. *Estimate of the Tube Length for a Once-Through Unit with the Primary Water Outside the Tubes.* It has been suggested that the large volume required for the vapor separators of Fig. 15.9 might be avoided in a somewhat different way than in the "once-through" design of Fig. 15.10 if the primary circuit water were circulated outside the tubes and the feedwater directed through the tubes and evaporated to dryness as in what many people think of as once-through boilers. (This would eliminate the recirculation in the boiler portion of Fig. 15.10.) It is instructive to pursue this once-through flow possibility by modifying the input to the basic table of Table 15.6. The flow-passage area outside the tubes was taken to be the same as that inside the tubes so that the heat transfer coefficient for the primary water stream would be the same as in Table 15.6. The conductance of the tube wall would also be the same, as would the boiling heat transfer coefficient (at least to a first approximation). The very low flow rate for the feedwater, however, leads to a much reduced heat transfer coefficient in the preheating region (see Fig. H5.3).

A difficult problem arises in trying to estimate the heat transfer coefficient in the higher quality region beyond the point where nucleate boiling would break down and a mist prevail. As discussed in Chapter 5, the point at which the annular-flow regime with nucleate boiling breaks down to give a dry-wall condition with drastically reduced heat transfer rates depends on the heat flux, the mass flow rate, and the character of the surface. The data of Fig. 5.29 indicate that, for the conditions of interest here, the dry-wall condition would develop in plain wall tubes at a vapor quality of about 80%. However, the same series of tests showed that the fluted tubing of Fig. 5.27*b* with a bore of 16 mm having 40 internal flutes 0.5 mm wide and 0.5 mm deep retained the wet-wall condition up to a vapor quality of 95% because of surface tension effects similar to those in heat pipes. The same tests also showed that inserting a helically coiled wire with a coil pitch of about 1 diameter created sufficient turbulence to increase the heat transfer coefficient in the dry wall and preheating regions by a factor of about 3. (This is consistent with the data given in Fig. 3.18.) Thus two cases were considered, the first with a plain wall tube and the second with a fluted tube similar to that of Fig. 5.27*b* having helical coil turbulators in both the preheating and mist-flow regions. Heat transfer coefficients for the preheating and mist-flow regions were taken from Figs. 5.28 and 5.29; the results

are presented in Tables 15.7 and 15.8, and are plotted in Figs. 15.11 and 15.12. Note that the tube length for the plain tube once-through case is almost double that for the recirculating boiler, whereas for the fluted-tube once-through case the tube length is just a little greater. Note, too, how easily these calculations can be set up using a standard microcomputer spreadsheet, and how quickly perturbations can be made once the basic format has been set up.

STEAM GENERATORS FOR GAS-COOLED REACTORS

In a gas-cooled reactor power plant the steam generator is one of the most important components from the standpoint of plant layout and overall costs. The general layout employed depends on the specific application. Typical configurations include axial-flow, crossflow, and U-bend arrangements with either finned or bare tubes. (A typical crossflow finned-tube unit for a low-pressure application is shown in Fig. 1.5.)

Example 15.4. *A Steam Generator for a High-Pressure Gas-Cooled Reactor.* To illustrate the approach to the design of a steam generator for a gas-cooled reactor a simple once-through, axial-flow unit is considered here. This basic approach has also been employed for crossflow configurations such as that of Fig. 1.5, but the analysis is more complex.[44] The hot gas from the reactor flows axially over the tubes and heats the water flowing in the opposite direction inside the tubes to generate the superheated steam supplied to the turbines.

The design parameters fixed for this case are the type of gas, the gas-side and steam-side system pressures and temperatures, the tube diameter, the pumping power-to-heat removal ratio for the gas side, and the design power output. Table 15.9 gives the specifications for the case at hand, and Fig. 15.13 shows the temperature distribution for the gas and water as a function of the percentage of the heat transferred from one fluid to the other for the temperatures specified. In Table 15.9 the value chosen for the pumping power-to-heat removal ratio for the gas side represents a near-optimum compromise between the capital cost of the steam generator and the operating cost of the gas-circulating blowers. The pumping power chargeable to the water side is very small as compared with that for the gas side for reasonable water flow rates; hence this factor does not ordinarily represent a limitation on the design.

The principal dependent variables are the tube

TABLE 15.7 Estimate of the Tube Length Required for a PWR Steam Generator With the Primary Water Flowing Outside the Tubes and Once-Through Flow of the Feedwater-Steam Inside the Tubes with Plain Tubes

Temperature
Primary water in = 330°C (626°F)
Primary water out = 292°C (558°F)
Feedwater in   = 230°C (446°F)
Steam out   = 289°C (552°F)
Heat load = 151 kW/tube

Tubing
Stainless steel
Tube OD = 19 mm (0.48 in.)
Tube ID = 16 mm (0.406 in.)
Mean wall area = 0.055 m²/m (0.18 ft²/ft)

Feedwater flow rate = 0.056 kg/s per tube (0.123 lb/s)
$\qquad$ = 278 kg/s · m² (56.8 lb/s · ft²)
Primary circuit flow rate = 0.67 kg/s per tube (1.472 lb/s)
$\qquad$ = 3,320 kg/s · m² (679 lb/s · ft²)
Heat transfer coefficients
$\quad$ Primary circuit water h = 30,000 W/m² · °C (5,270 Btu/h · ft² · °F)
$\quad$ Tube-wall conductance = 12,660 W/m² · °C (2,230 Btu/h · ft² · °F)
$\quad$ Feedwater preheating h = 2,555 W/m² · °C (450 Btu/h · ft² · °F)
$\quad$ Nucleate boiling h = 28,400 W/m² · °C (5,000 Btu/h · ft² · °F)
$\quad$ Preheating overall u = 1,987 W/m² · °C (339 Btu/h · ft² · °F)
$\quad$ Boiling overall u = 6,780 W/m² · °C (1,192 Btu/h · ft² · °F)
Length increment per 2°C increment in primary circuit temperature = 144,500/(Q/A)

T_1, °C	T_2, °C	Q/A, W/m² · °C $= u(T_1 - T_2)$	ΔL, m	L, m	Q, %	u, W/m² · °C
293	240	105,311	1.37	1.37	5.263	1,987
295	261	67,558	2.14	3.51	10.526	1,987
297	282	29,805	4.85	8.36	15.789	1,987
299	289	67,800	2.13	10.49	21.052	6,780
301	289	81,360	1.78	12.26	26.315	6,780
303	289	94,920	1.52	13.79	31.578	6,780
305	289	108,480	1.33	15.12	36.841	6,780
307	289	122,040	1.18	16.30	42.104	6,780
309	289	135,600	1.07	17.37	47.367	6,780
311	289	149,160	0.97	18.34	52.630	6,780
313	289	162,720	0.89	19.23	57.893	6,780
315	289	176,280	0.82	20.04	63.156	6,780
317	289	189,840	0.76	20.81	68.419	6,780
319	289	203,400	0.71	21.52	73.682	6,780
321	289	216,960	0.67	22.18	78.945	6,780
323	289	230,520	0.63	22.81	84.208	6,780
325	289	71,748	2.01	24.82	89.471	1,993
327	289	49,400	3.39	28.21	94.734	1,300
329	289	28,000	5.98	34.20	100.000	700

spacing, the tube length, and the number of tubes. The first step is to choose a reasonable flow rate for the water. Once this is done the power output per tube and the total number of tubes can be calculated. Boiler design experience indicates that for good flow stability at light loads, the steam velocity at the exit of the boiler section of the tube should be 30 to 60 ft/s (10 to 20 m/s) for the full-power condition. In this instance it is desirable to keep the tube length as short as possible, hence the lower value should be used. This gives a water-side mass flow rate G_i of 223 lb/ft²·s (101 kg/s). From this it follows that

Power output per tube

$$= \frac{\pi}{4} D_i^2 G_i (\Delta H_p + \Delta H_v + \Delta H_s)$$

$$= \frac{\pi}{4}\left(\frac{0.4}{12}\right)^2 (223 \times 3600)(218.9 + 365.7 + 398.7)$$

$$= 6.89 \times 10^5 \text{ Btu/h·tube}$$

where the subscripts p, v, and s refer to preheat, vaporization, and superheat, respectively).

Number of tubes

$$= \frac{\text{total power}}{\text{power per tube}} = \frac{300 \times 1000 \times 3413}{6.89 \times 10^5} = 1485$$

The problem is now reduced to finding the tube spacing and the tube length that satisfies the heat transfer

TABLE 15.8 Estimate of the Tube Length Required for a PWR Steam Generator With the Primary Water Flowing Outside the Tubes and Once-Through Flow of the Feedwater-Steam Inside the Tubes with Fluted Tubes Having Helical Coil Inserts in the Preheating and Mist-Flow Regions

Temperatures
Primary water in   = 330°C (626°F)
Primary water out = 292°C (558°F)
Feedwater in   = 230°C (446°F)
Steam out   = 289°C (552°F)
Heat load = 151 kW/tube
Feedwater flow rate = 0.056 kg/s per tube (0.123 lb/s)
                    = 278 kg/s · m² (56.8 lb/s · ft²)
Primary circuit flow rate = 0.67 kg/s per tube (1.472 lb/s)
                    = 3,320 kg/s · m² (679 lb/s · ft²)

Tubing
Stainless steel
Tube OD = 19 mm (0.48 in.)
Tube ID = 16 mm (0.406 in.)
Mean wall area = 0.055 m²/m (0.18 ft²/ft)

Heat transfer coefficients
   Primary circuit water h = 30,000 W/m² · °C (5,270 Btu/h · ft² · °F)
   Tube-wall conductance = 12,660 W/m² · °C (2,230 Btu/h · ft² · °F)
   Feedwater preheating h = 7,670 W/m² · °C (1,353 Btu/h · ft² · °F)
   Nucleate boiling h = 28,400 W/m² · °C (5,000 Btu/h · ft² · °F)
   Preheating overall u = 4,110 W/m² · °C (726 Btu/h · ft² · °F)
   Boiling overall u = 6,780 W/m² · °C (1,192 Btu/h · ft² · °F)
   Mist-flow overall u = 1,693 W/m² · °C (299 Btu/h · ft² · °F)
Length increment per 2°C increment in primary circuit temperature = 144,500/(Q/A)

T_1, °C	T_2, °C	Q/A, W/m² · °C $= u(T_1 - T_2)$	ΔL, m	L, m	Q, %	u, W/m² · °C
293	240	217,830	0.66	1.37	5.263	4,110
295	261	139,740	1.03	2.40	10.526	4,110
297	282	61,650	2.34	4.75	15.789	4,110
299	289	67,800	2.13	6.88	21.052	6,780
301	289	81,360	1.78	8.66	26.315	6,780
303	289	94,920	1.52	10.18	31.578	6,780
305	289	108,480	1.33	11.51	36.841	6,780
307	289	122,040	1.18	12.69	42.104	6,780
309	289	135,600	1.07	13.76	47.367	6,780
311	289	149,160	0.97	14.73	52.630	6,780
313	289	162,720	0.89	15.62	57.893	6,780
315	289	176,280	0.82	16.44	63.156	6,780
317	289	189,840	0.76	17.20	68.419	6,780
319	289	203,400	0.71	17.91	73.682	6,780
321	289	216,960	0.67	18.57	78.945	6,780
323	289	230,520	0.63	19.20	84.208	6,780
325	289	244,080	0.59	19.79	89.471	6,780
327	289	257,640	0.65	20.44	94.734	6,780
329	289	67,720	2.47	22.92	100.000	1,693

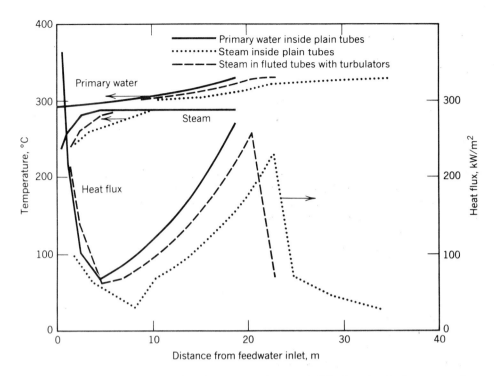

Figure 15.12 Temperature distribution as a function of the length of tube traversed by the steam for the PWR steam generators of Examples 15.2 and 15.3.

requirement and the gas-side pumping power-to-heat removal ratio specified in Table 15.9. A convenient way of solving this problem is to assume an equivalent passage diameter for the gas side (i.e., tube spacing) and to calculate the corresponding tube length and the pumping power-to-heat removal ratio; calculations can then be repeated with one or two other equivalent passage diameters. If the tube length and the equivalent passage diameter are then plotted against the pumping power-to-heat removal ratio on a logarithmic scale, almost a straight-line relation is obtained. From this chart the values of the tube length and the equivalent passage diameter for the specified value of the pumping power-to-heat removal ratio can be readily obtained. Table 15.10 shows the procedure followed in performing these calculations for equivalent passage diameters of 1 in. and 2 in., corresponding to 0.83-in. and 1.07-in. tube spacings, respectively.

In these calculations the water-side heat transfer coefficient in the preheater section was calculated from

$$h_{ip} = 0.023 \left(\frac{c_p^{0.4} k^{0.6}}{\mu^{0.4}} \right) \frac{G_i^{0.8}}{D_i^{0.2}} \quad (15.1)$$

and that for steam in the superheater from

$$h_{is} = 0.0266 \left(c_p \mu^{0.2} \right) \frac{G_i^{0.8}}{D_i^{0.2}} \quad (15.2)$$

where the subscripts p and s refer to the preheater and superheater sections, respectively, and i refers to the inner, or water side. The water-side heat transfer coefficient for the boiler section is high compared with the gas-side heat transfer coefficient, and is relatively insensitive to vapor quality up to about 80%. At the 2500-psi (17.2-MPa) pressure of this example, the heat transfer coefficient for saturated steam does not vary much with vapor quality; hence a constant value of 5000 Btu/h · ft² · °F (2.84 W/m² · °C) was used for the boiling region. The heat transfer coefficient for the gas side h_o is

$$h_o = 0.023 \left(\frac{c_p^{0.4} k^{0.6}}{\mu^{0.4}} \right) \left(\frac{G_o^{0.8}}{D_{eo}^{0.2}} \right) \quad (15.3)$$

where the subscript o refers to the outer, or gas side, of the tube. In this relation the only unknown is G_o, and

TABLE 15.9 Design Specification for an Axial-Flow Steam Generator for a Gas-Cooled Reactor

Total power output = 300 MW thermal
Pumping power-to-heat removal ratio for the gas side = 0.5%

	Preheater	Boiler	Superheater
Fluid inside tubes	Water	Saturated water and steam	Super-heated steam
Gas outside tubes	CO_2	CO_2	CO_2
Pressure inside tubes, psia	2500	2490	2450
Pressure outside tubes, psia	1000	1000	1000
Inside fluid temperature, °F			
Inlet	520	667	667
Outlet	667	667	1050
Gas temperature, °F			
Inlet	650	813	1079
Outlet	813	1079	1350
Log mean temperature differential Δt_m, °F	138	256	353
ΔH_i, Btu/lb	218.9	365.7	398.7
ΔH_o, Btu/lb	43.2	72.6	78.7
ρ_o, lb/ft³	3.49	2.96	2.49
c_{po}, Btu/lb · °F	0.2642	0.2765	0.2903
k_o, Btu/h · ft² · °F/ft	0.0258	0.0305	0.0361
μ_o, lb/ft · h	0.0711	0.0804	0.0914
c_{pi}, Btu/lb · °F ($\Delta H_i/\delta t_i$)	1.49		
k_i, Btu/h · ft² · °F/ft	0.293		
μ_i, lb/ft · h	0.2085		
$c_{pi}\mu_i^{0.2}$ (mean)			1.0
$c_{po}^{0.4} k_o^{0.6}/\mu_o^{0.4}$	0.185	0.204	0.219
$c_{pi}^{0.4} k_i^{0.6}/\mu_i^{0.4}$	1.045		
k_{wall}, Btu/h · ft² · °F/ft	12	12	12
Tube outside diameter, in.	0.5	0.5	0.5
Tube inside diameter, in.	0.4	0.4	0.4

this value can be evaluated from a heat balance for the two fluid streams, that is,

$$A_o G_o \Delta H_o = A_i G_i \Delta H_i \qquad (15.4)$$

$$\frac{A_i}{A_o} = \frac{\frac{\pi}{4} D_i^2}{\frac{\pi}{4} D_o D_{eo}} = \frac{D_i^2}{D_o D_{eo}} \qquad (15.5)$$

Hence,

$$G_o = G_i \frac{\Delta H_i}{\Delta H_o} \frac{D_i^2}{D_o D_{eo}} \qquad (15.6)$$

Substituting Eq. 15.6 in Eq. 15.3 gives

$$h_o = \frac{0.023}{D_{eo}} \left(\frac{c_p^{0.4} k_o^{0.6}}{\mu^{0.4}} \right) \left(G_i \frac{\Delta H_i}{\Delta H_o} \frac{D_i^2}{D_o} \right)^{0.8} \qquad (15.7)$$

This relation was used in calculating the gas-side heat transfer coefficient.

The overall coefficient of heat transfer for the preheater, boiler, and superheater sections was calculated from the individual heat transfer coefficients. The tube length was calculated from the relation obtained by equating the heat transferred through the tube wall to

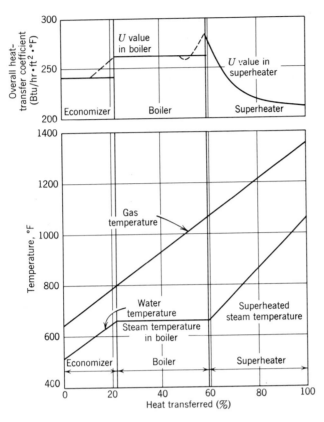

Figure 15.13 Axial temperature distribution in a once-through steam generator for a gas-cooled reactor:

CO₂ flowing between 0.4-in. ID, 0.5-in. OD tubes
Pumping power-to-heat removal ratio = 0.005
Gas at 1000 psia   Steam at 2500 psia
Gas in at 1350°F   Gas out at 650°F
Water in at 520°F   Steam out at 1050°F
Mass flow rate of steam: 229.9 lb/ft² · s

the heat added to the water side, that is,

$$(\pi D_i L)U_i \Delta t_m = \left(\frac{\pi}{4} D_i^2\right)G_i \Delta H_i$$

$$\text{or } L = \frac{G_i \Delta H_i D_i}{4\Delta t_m U_i} \tag{15.8}$$

where U_i is the overall heat transfer coefficient based on the internal tube surface.

The tube lengths thus calculated for the preheater, boiler, and superheater sections are shown in Table 15.10 for the two different values of the equivalent passage diameters chosen. The total length of the steam generator was obtained by summing the lengths of these three sections.

The pumping power-to-heat removal ratio F_o was calculated from

$$F_o = \frac{\text{pumping power for gas flow}}{\text{heat transferred}}$$

$$= \frac{(A_o G_o \Delta P_o / 778\rho_o)}{A_o G_o \Sigma \Delta H_o}$$

$$= \frac{\Delta P_o}{778\rho_o \Sigma \Delta H_o}$$

The pressure drop for the gas side ΔP_o can be calculated from

$$\Delta P_o = \frac{0.2}{2g3600^2} G_o^{1.8} \frac{\mu_o^{0.2}}{\rho_o} \frac{L}{D_{eo}^{1.2}} \tag{15.9}$$

and G_o can be obtained from Eq. 15.6.

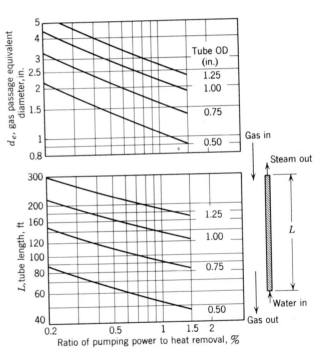

Figure 15.14 Design chart for a series of axial-flow, bare-tube steam generators for a gas-cooled reactor application:

Gas: CO₂   Steam out at 1050°F (566°C)
Gas at 1000 psia (6.89   Water in at 520°F (271°C)
  MPa)
Steam at 2500 psia (17.2   Steam flow rate: 222.9 lb/ft² · s
  MPa)   (1088 Kg/m² · s)
Gas in at 1350°F (732°C)   Tube OD/tube ID = 1.25
Gas out at 650°F (343°C)

(Courtesy Oak Ridge National Laboratory.)

TABLE 15.10 Summary of Preliminary Design Calculations for an Axial-Flow Steam Generator for a Gas-Cooled Reactor

	Economizer	Boiler	Superheater	Total
Water flow rate, $lb/ft^2 \cdot s$ (assumed)	223	223	223	
h_o, $Btu/h \cdot ft^2 \cdot °F$				
For d_{eo} = 1 in.	543	600	642	
= 2 in.	272	300	321	
h_i, $Btu/h \cdot ft^2 \cdot °F$	2570	5000	2790	
Tube-wall conductance, $Btu/h \cdot ft^2 \cdot °F$	2880	2880	2880	
U_i, $Btu/h \cdot ft^2 \cdot °F$ of internal tube surface				
For d_{eo} = 1 in.	452	531	511	
= 2 in.	272	311	312	
L, length, ft				
For d_{eo} = 1 in.	23.4	18.0	14.8	56.2
= 2 in.	38.9	30.7	24.2	93.8
F_0, pumping power-to-heat removal ratio				
For d_{eo} = 1 in.	0.00383	0.00406	0.00492	0.01281
= 2 in.	0.000794	0.000865	0.001007	0.00267

Adaptation of the Procedure to a Computer. The preceding analysis was developed in part because there was a need for a parametric survey to investigate the effects of the many variables on the size and cost of steam generators for use with gas-cooled reactors.[44] Thus a program was developed from the foregoing approach so that the calculations could be carried out with a machine. Data obtained from computer calculations are presented in Fig. 15.14 for 0.5-, 0.75-, 1.00-, and 1.25-in. diameter tubes. Note that the tube length from the computer calculations is less than that obtained from the hand calculations of Table 15.10. This is because the computer calculations were refined by dividing the tube into many small sections and using the local temperatures to evaluate the local heat transfer coefficients for each small section. In the hand calculations of Table 15.10 the tube was divided into only three sections: the preheater, boiler, and superheater. In the superheater section, particularly near the inlet, the physical properties of the steam vary rapidly with the temperature. The resulting high heat transfer coefficient near the superheater inlet gave a shorter tube length in the machine calculations. Note also that the curves in Fig. 15.14 are almost straight lines, and that those for different tube diameters are nearly parallel to each other. Therefore, the calculations in Table 15.10 can be extended easily to investigate the effects of tube diameter, because only one or two points are needed to define each curve.

The data in Table 15.10 indicate that for the design conditions specified in Table 15.9, a tube length of 70 ft and an equivalent passage diameter of 1.5 in. (i.e., 0.95-in. tube spacing) satisfy the requirement for a pumping power-to-heat removal ratio of 0.5%. (The curves given by the computer yield a somewhat shorter tube length and a slightly closer spacing.)

REFERENCES

1. *Steam*, 38th ed., The Babcock and Wilcox Co., 1972.
2. C. D. Shields, *Boilers: Types, Characteristics, and Functions*, F. W. Dodge, 1961.
3. V. J. Calise and C. Dallman, "Status of Condensate and Feedwater Purification in Today's Utility Power Plants," *ASME Technical Reprint* T-195, 1963.
4. J. T. Bevans and R. V. Durkle, "Radiant Interchange within an Inclosure," *Journal of Heat Transfer, Trans. ASME*, vol. 82-2, 1960, p. 1.
5. A. T. Mumford and R. C. Corey, "Variation in Heat Absorption in a Natural-Gas-Fired Water-Cooled Steam Boiler Furnace," *Trans. ASME*, vol. 74, 1952, p. 1191.
6. R. A. Sherman, "Heat Transfer by Radiation from Flames," *Trans. ASME*, vol. 79, 1957, p. 1727.
7. L. Cohen and W. A. Fritz, Jr., "Heat Transfer Studies of Naval Boilers," *Trans. ASME*, vol. 80, 1958, p. 683.
8. H. J. Kerr and F. Eberle, "Graphitization of Low-Carbon

and Low-Carbon-Molybdenum Steels," *Trans. ASME,* vol. 67, 1945 (special appended section), p. 1.

9. H. A. Grabowski, "Corrosion of Steel in Boilers—Attack by Dissolved Oxygen," *Trans. ASME,* vol. 77, 1955, p. 433.

10. R. M. Curran and A. W. Rankin, "Austenitic Steels in High-Temperature Steam Piping," *Trans. ASME,* vol. 79, 1957, p. 1398.

11. C. L. Clark et al., "Metallurgical Evaluations of Superheater Tube Alloys after Six Months Exposure at Temperatures of 1100 to 1500°F," *Journal of Engineering for Power, Trans. ASME,* vol. 82-1, 1960, p. 35.

12. A. A. Markson, T. Ravese, and C. Humphreys, "A Method of Estimating the Circulation in Steam Boiler Furnace Circuits," *Trans. ASME,* vol. 64, 1942, p. 275.

13. O. J. Mendler et al., "Natural Circulation Tests with Water at 800 to 2000 psia under Nonboiling, Local Boiling, and Bulk Boiling Conditions," *Journal of Heat Transfer, Trans. ASME,* vol. 83-2, 1961, p. 261.

14. J. P. Holman and J. H. Boggs, "Heat Transfer to Freon 12 Near the Critical State in a Natural Circulation Loop," *Journal of Heat Transfer, Trans. ASME,* vol. 82-2, 1960, p. 221.

15. W. H. McAdams, W. E. Kennel, and J. N. Addoms, "Heat Transfer to Superheated Steam at High Pressures," *Trans. ASME,* vol. 72, 1950, p. 421.

16. J. H. Hoke and F. Eberle, "Experimental Superheater for Steam at 2000 psi and 1250°F," *Trans. ASME,* vol. 79, 1957, p. 307.

17. C. S. Schlea and J. P. Walsh, "Deentrainment in Evaporators," *Industrial and Engineering Chemistry,* vol. 53, 1961, p. 695.

18. J. Gastpar, "European Practice with Sulzer Monotube Steam Generators," *Trans. ASME,* vol. 75, 1953, p. 1345.

19. W. H. Rowand and A. M. Frendberg, "First Commercial Supercritical-Pressure Steam Generator for Philo Plant," *Trans. ASME,* vol. 79, 1957, p. 409.

20. J. H. Harlow, "Engineering the Eddystone Plant for 5000 psi, 1200°F Steam," *Trans. ASME,* vol. 79, 1957, p. 1410.

21. C. A. Dauber, "Avon No. 8—A Supercritical-Pressure Plant," *Trans. ASME,* vol. 79, 1957, p. 927.

22. J. C. Falkner et al., "Latest Technique for Quick Starts on Large Turbines and Boilers," *Trans. ASME,* vol. 72, 1950, p. 1111.

23. R. L. Jackson et al., "Importance of Matching Steam Temperatures with Metal Temperatures during Starting of Large Steam Turbines," *Trans. ASME,* vol. 79, 1957, p. 1669.

24. H. R. Hoy and J. E. Stantan, "Fluidized Bed Combustion under Pressure," Paper presented at the American Chemical Society Meeting, Toronto, Canada, May 24, 1970.

25. J. F. Thomas et al., "Atmospheric Fluidized Bed Boilers for Industry," Report no. ICTIS/TR35, IEA Coal Research, London, November 1986.

26. A. P. Fraas, *Engineering Evaluation of Energy Systems,* McGraw-Hill Book Co., New York, 1982.

27. W. J. G. Little, "Pulsation Phenomena in Fluidized Bed Boilers," *ASME Proceedings of the 1987 International Conference on Fluidized Bed Combustion,* May 3–7, 1987, p. 561.

28. R. O. Vincent et al., "Dynamic Forces on Tubes Immersed in Bubbling AFBC," *ASME, Proceedings of the 1987 International Conference on Fluidized Bed Combustion,* May 3–7, 1987, p. 567.

29. E. M. Suuberg et al., "Product Compositions and Formation Kinetics in Rapid Pyrolysis of Pulverized Coal—Implications for Combustion," *Proceedings of the XVII Symposium on Combustion,* The Combustion Institute, 1979.

30. J. Stringer, "Current Information on Metal Wastage in Fluidized Bed Combustors," *ASME Proceedings of the 1987 International Conference on Fluidized Bed Combustion,* May 3–7, 1987, p. 685.

31. B. N. Gaglia and A. Hall, "Comparison of Bubbling and Circulating Fluidized Bed Industrial Steam Generation," *ASME Proceedings of the 1987 International Conference on Fluidized Bed Combustion,* May 3–7, 1987, p. 18.

32. E. L. Daman, "USA Overview of Fluidized Bed Combustion Developments," *VDI Berichte* Nr. 601, 1986, p. 357.

33. R. W. Foster-Pegg, Personal communication, June 1978.

34. A. P. Fraas, "Comparison of Various Fluidized Bed Combustor-Gas Turbine Systems," ASME Paper no. 85-IGT-46, Presented at the 1985 Beijing International Gas Turbine Symposium, Beijing, China, September 1–7, 1985.

35. "Developing the Turbocharged Pressurized Fluidized Bed Combustion Boiler," *Heat Engineering,* Foster Wheeler Corp., September–December 1986, pp. 100–107.

36. M. Altman, R. H. Norris, and F. W. Staub, "Local and Average Heat Transfer and Pressure Drop for Refrigerants Evaporating in Horizontal Tubes," *Journal of Heat Transfer, Trans. ASME,* vol. 82-2, 1960, p. 189.

37. A. R. Smith and E. S. Thompson, "The Mercury Vapor Process," *Trans. ASME,* vol. 64, 1942, p. 625.

38. R. M. Wilson and J. D. Roarty, "Westinghouse Aims to Improve Reliability with Its Model F Design," *Nuclear Engineering International,* vol. 30(375), October 1985, p. 28.

39. "Thermal-Hydraulic Characteristics of a Westinghouse Model F Steam Generator," EPRI Report no. EPRI NP-1719, Electric Power Research Institute, March 1981.

40. J. Bellet et al., "The N4 Plant: Culmination of French PWR Experience," *Nuclear Engineering International,* vol. 30(365), 1985, p. 26.

41. C. W. Vernon and J. D. Cohen, "Steam Generator Dose

Rates on Westinghouse Pressurized Water Reactors," EPRI Report no. NP-2453, Electric Power Research Institute, 1982.

42. H. S. McKay, "Steam Generator Replacement at Surry Power Station," *Nuclear Safety*, vol. 23(1), 1982, p. 72.

43. "Thermal-Hydraulic Analysis of Once-through Steam Generators," EPRI Report no. NP-1431, Electric Power Research Institute, June 1980.

44. A. P. Fraas and M. N. Ozisik, "Steam Generators for High Temperature Gas-Cooled Reactors," Report no. ORNL-3208, Oak Ridge National Laboratory, April 1963.

16

Condensers

The design of condensers for such varied applications as steam power plants, chemical-processing plants, and nuclear electric plants for space vehicles involves a variety of heat transfer and fluid-flow problems associated with the condensation of vapors. Some typical analytical relationships and experimental data showing the effects of the more important parameters are presented in the first portion of this chapter. Several preliminary designs for typical applications are outlined in the later sections.

HEAT TRANSFER FROM CONDENSING VAPORS

As indicated in Chapter 3, the principal barrier to heat transfer from a condensing vapor to a cool solid surface is the film of liquid that forms on the solid surface. This film ordinarily increases in thickness until the action of gravity or fluid skin friction induces it to flow along the surface. The equilibrium thickness of the film, and hence its thermal resistance, depend on the rate of condensation, the forces acting on the film, its resistance to flow along the surface, the nature of that flow (i.e., whether it is laminar or turbulent), and the amount of surface upstream of the flowing film. Thus in the design of a condenser, the most important step in evaluating the heat transfer coefficient on the vapor side is the determination of the character and average thickness of the liquid film

of condensate. The complexity of these relations is such that the designer should be even more than usually careful not to seize on and use a formula or a curve indiscriminantly. It is essential that the conditions anticipated be examined carefully and compared with the cases for which there is information available, so that the most applicable data can be employed in the problem at hand. In doing this the designer should attempt to appraise the uncertainties involved and include appropriate allowances.

Heat Transfer Coefficients for Filmwise Condensation

Some fundamental information on heat transfer for condensing vapors is given in Chapter 3. This information includes expressions for the heat transfer coefficient for vapor condensing outside horizontal and vertical tubes. This section represents an extension of that discussion to point up the problems that face an engineer confronted with condenser design work.

Condensation inside of long tubes involves relations that show clearly the importance of dynamic forces in the vapor on condensing heat transfer coefficients.[1-4] A study of both theoretical relationships and experimental data has led Colburn[5] to recommend that the local condensing heat transfer coefficient h_x for filmwise condensation inside tubes be estimated from the relation

$$h_x = a \left(\frac{c_p \mu}{k}\right)^{1/2} \frac{k}{\mu} (F\rho)^{1/2} \qquad (16.1)$$

All the symbols used are usually as defined in this book except that in English units F, the force acting on the liquid film in pounds per square foot, is multiplied by g in feet per hour squared. The total forces acting include the friction of the vapor passing over the surface of the film, the gravitational force on the film, and the momentum input to the film by molecules of condensing vapor, that is,

$$F = F_v + F_w + F_m \qquad (16.2)$$

These three components of the force acting on the film parallel to a vertical heat transfer surface are as follows:

$$\text{Frictional } F_v = \frac{f_r G_v^2}{2\rho_v} \qquad (16.3)$$

$$\text{Gravitational } F_w = \frac{g\Gamma}{10\sqrt{F/\rho_w}} \qquad (16.4)$$

$$\text{Momentum } F_m = \frac{G_v}{\rho_v}\left(\frac{d\Gamma}{dx}\right) \qquad (16.5)$$

where the suscripts w and v refer to the liquid and vapor, respectively. The quantity Γ is the liquid flow rate in kilograms per second per meter of wetted perimeter, or, in English units, in pounds per hour per foot of wetted perimeter.

As discussed in Chapter 5, the friction factor f for vapor flowing through a pipe in which the walls are covered with a liquid may be as much as 10 times the friction factor for a gas flow through a dry pipe. This is partly because of the increase in the mean gas density associated with the entrained liquid droplets, and partly because the liquid film tends to form waves that induce turbulence in the gas flow. Interchange of droplets between the gas stream and the liquid film on the wall also contributes to the loss of substantial amounts of kinetic energy from the gas. Friction factors for gas flow inside pipes with wetted walls are given in Fig. 5.15. These factors hold for two-phase flow irrespective of whether it entails boiling, condensing, or simply two-phase flow with no boiling or condensing (e.g., a mixture of air and water).

The validity of Eq. 16.1 is indicated by the data plotted in Fig. 16.1 for condensation of five different fluids. The line through the experimental data is defined

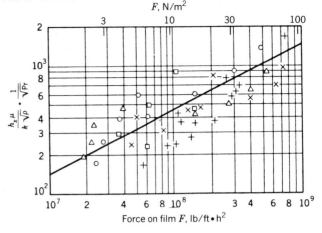

Figure 16.1 Condensing heat transfer performance of a variety of fluids as a function of the force acting to sweep the liquid film off the surface. (Colburn, Ref. 5.)

by Eq. 16.1 if the coefficient $a = 0.045$. It should be noted that F_m and F_v are often much larger than F_w, so that gravitational effects are often minor except in the lower parts of the tube. Experience has shown that this expression is likely to be more appropriate for most condensers than Eqs. 3.31 and 3.32. It also provides a basis for the design of condensers for space vehicle power plants in which gravitational forces will not act to induce drainage of the condensate from the condenser surfaces.

Equation 16.1 and Fig. 16.1 apply to local heat transfer rates. Since the effects of geometry make evaluation of an average heat transfer coefficient tedious, and because these effects entail a variation of only about 30% for most of the geometry range of interest, Colburn has recommended that the heat transfer coefficient for vapor condensing inside tubes be approximated by

$$h = 0.065 \left(\frac{c_p \rho k f_r}{2\mu\rho_v}\right)^{1/2} G_m \qquad (16.6)$$

This equation was developed using the friction factor that would prevail if there were no condensate on the walls, and the average vapor-flow rate G_m is given by

$$G_m = \left(\frac{G_1^2 + G_1 G_2 + G_2^2}{3}\right)^{1/2} \qquad (16.7)$$

where G_1 and G_2 are the vapor mass flow rates for the tube inlet and outlet, respectively. Note that c_p, ρ, k, and μ apply to the condensate, while f_r, ρ_v, and G_m apply to the vapor, and if condensation is complete at the tube outlet, $G_m = 0.58G_1$.

Equation 16.6 is a particularly convenient form since

most organic fluids have properties that are so nearly the same for fluids of similar viscosity that, irrespective of the fluid, nearly the same coefficient can be applied to the mass flow to obtain the heat transfer coefficient. Figure 16.2 shows data for water and a variety of organic fluids plotted on the same coordinates. The dashed lines represent the values indicated by Eq. 16.6, while two solid curves obtained by other investigators (who did not use the correlation of Eq. 16.1 in their own work) are superimposed on the scatter band for organic liquids. Note that the condensing coefficient for water is about six times that for the organic liquids.

Effects of Condenser Tube Configuration

Condensers are ordinarily built so the vapor flows transversely into banks of horizontal tubes with a large inlet flow-passage area for the vapor. The vapor velocity varies widely from the outer to the inner portion of the tube bank, whereas the water flow through the tubes is maintained at a uniform moderate velocity. This arrangement is advantageous in that the heat transfer coefficient on the cooling water side is sensitive to the water velocity, and is normally much lower than that on the vapor side in steam condensers; thus it is likely to be the determining factor in establishing the surface area requirements.

The vapor-side heat transfer coefficient for horizontal tube arrangements depends on the thickness of the liquid film covering the tube surfaces. This in turn depends on the vapor velocities across the tubes tending to sweep off the condensate, turbulence in the condensate film induced by turbulent flow or by dripping from one tube to another, the extra surface area provided by subcooled droplets splashing downward from one tube to the next, and related factors. Where the vapor velocities are low, Eq. 3.32 represents a good approximation.

The vast amount of experimental work carried out since Colburn evolved Eq. 16.6 has provided a greatly expanded data base. Efforts to correlate these data have led to the formulation of better, though more complex, empirical expressions. These generally include the terms of Eqs. 16.6 and 16.7 with somewhat different exponents and some additional terms. The complexity of the expressions can be reduced by using a different relation for each fluid. Of these expressions, two that appear to give particularly good correlation of the available data have been developed by Bergles,[1] the first expressly for steam (Eq. 16.8), and the second for refrigerants (Eq. 16.9). These include the average vapor quality $\bar{x}$ as a variable, and are as follows:

For steam

$$\bar{h}_i = 0.0265 \frac{k_l}{D_h} \left(\frac{G_e D_h}{\mu_l} \right)^{0.8} \Pr_l^{0.33} \left[160 \left(\frac{H^2}{lD_i} \right)^{1.91} + 1 \right]$$

(16.8)

where

$$G_e = G \left[(1 - \bar{x}) + \bar{x} \sqrt{\frac{\rho_l}{\rho_g}} \right]$$

For refrigerants

$$\bar{h} = 0.024 \frac{k_l}{D_k} \left(\frac{GD_h}{\mu_l} \right)^{0.8} \Pr_l^{0.43} \left(\frac{H^2}{lD_i} \right)^{-0.22}$$

$$\times \frac{1}{2} \left[\left(\frac{\rho}{\rho_m} \right)_{in}^{0.5} + \left(\frac{\rho}{\rho_m} \right)^{0.5} \right]$$

(16.9)

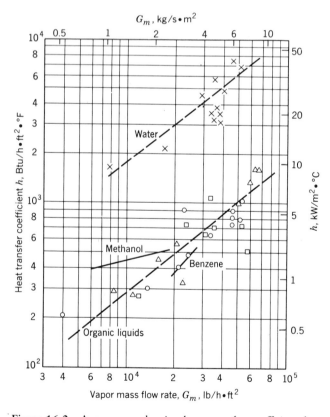

Figure 16.2 Average condensing heat transfer coefficient for flow of both water and a number of organic liquids inside long straight tubes. (Colburn, Ref. 5.)

where

$$\frac{\rho}{\rho_m} = 1 + \frac{\rho_l - \rho_g}{\rho_g} \bar{x}$$

Enhanced Surfaces

Efforts to correlate heat transfer data for condensing vapors are complicated by the large effects of enhanced surfaces which overwhelm smaller effects such as variations in thermal conductivity and viscosity with temperature for a given fluid. A wide variety of devices has been tested. Of these, as is the case for boiling liquids (see Chapt. 5), the most effective have been grooved or fluted surfaces. Figure 16.3 presents a nice set of data for steam at 8-bar (108-psia) condensing inside tubes with six different enhancement geometries which are compared with data for a plain, smooth tube. Geometric data for these tubes is given in Table 16.1. Note that the heat transfer and pressure-loss coefficients are based on the nominal wall surface area, and that the best of the grooved walls more than doubled the nominal heat transfer coefficient with only a 40% increase in the wall surface area introduced by the low-height "fins."

CONDENSERS FOR STEAM POWER PLANTS

Perhaps the most important application of condensers is for large steam power plants. This section includes a description of a typical unit, a discussion of the principal design considerations, and a typical design problem.

Construction

Figure 1.6 shows a condenser for a large steam turbine, and Fig. 16.4 shows sections through a unit of this type. Since the steam pressure at the turbine exit is only (3373 to 6746 Pa) (1.0 to 2.0 in. Hg abs), the steam density is very low, and the volume flow rate is extremely large. To minimize pressure losses, the condenser is normally mounted beneath and attached directly to the turbine to give a short connecting passage having a large flow-passage area. The turbine casing is relieved of most of the weight of the condenser by a spring mounting. In the unit of Fig. 16.4 the steam flows vertically downward from the large central opening at the top, and passes transversely over the tubes, which extend horizontally between header sheets at either end. As can be seen in

Section A-A, numerous steam lanes are provided in the tube bank to distribute the steam uniformly throughout the tube matrix with a minimal pressure loss. As can be seen in the axial section at the left in Fig. 16.4, water goes horizontally through the upper set of tube banks, flows downward through the water box at the left, and returns through the lower tube banks to the outlet at the bottom of the water box at the right. This arrangement serves to reduce the volume of the entering steam as rapidly as possible by exposing it to the coolest water. At the same time, the subcooled droplets of condensate dripping from the upper tubes tend to increase the effective surface area for absorption of steam. Since it is desirable to have the condensate as close to the entering steam temperature as possible to reduce both heat losses and oxygen absorption in the water, the advantage of this arrangement is that the warmest cooling water is just above the hot well. Note that baffles have been installed around vertical slab-shaped groups of tubes at the center to aspirate cool air from the regions just above the center of the hot well. This is important not only to reduce the back pressure on the turbine but also to increase the effectiveness of the condenser, since the presence of noncondensable gases in a condensing vapor reduces the effective temperature difference.

Design Considerations

In practice, the proportions of a particular steam condenser depend on the circumstances peculiar to the power plant in which it is to be installed. The temperature and availability of cooling water, the cooling water pumping requirements, the cost of fuel for the power plant, the purity of the cooling water, and a host of other factors influence the choice of design parameters. However, if conventional design practice is followed, the resulting design usually is not far from the optimum, and from 8 to 10 lb/h of steam can be condensed per square foot of condenser surface area.

Temperature Range

The difference in temperature between the steam and cooling water streams entering the condenser is ordinarily referred to as the *temperature range*. The temperature range is ordinarily kept to about 20°F (11°C). The temperature rise of the cooling water depends on water availability, pumping power requirements, and related considerations, but is usually made about 5°F (3°C) less than the temperature range.

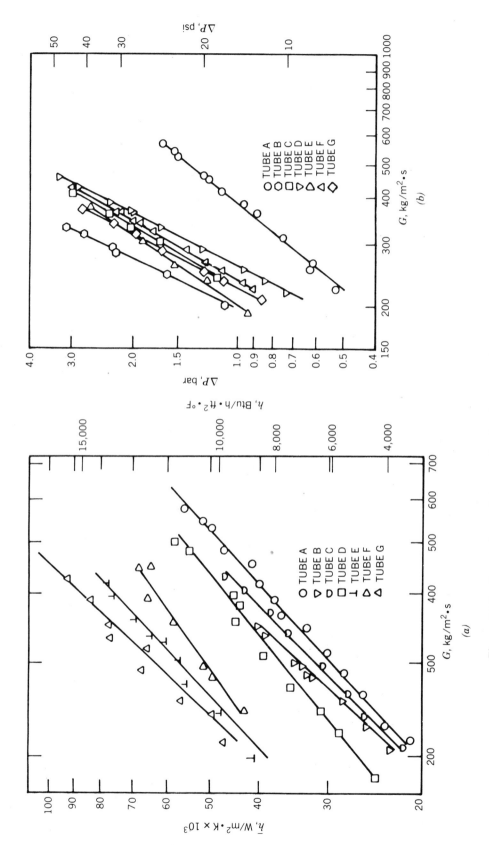

Figure 16.3 Comparison of the condensing heat transfer and total pressure-drop characteristics of the 3.66-m (12-ft) long condenser tubes of Table 16.1 for operation on 8-bar (108 psia) steam. The lowest curve in each case is for Tube A, the plain tube. (Courtesy Bergles, Ref. 1.)

Tube Diameter, Length, Thickness, and Material

While there is an incentive to make use of small-diameter tubes to obtain the maximum heat transfer surface area per unit of volume, the huge quantities of water required make it impractical to do anything other than take relatively dirty water from a river, lake, or harbor and use it with little treatment. Because of this problem, difficulties with tube plugging and fouling generally make it desirable to use tube diameters of at least 5/8 in. and preferably of 3/4 in. In the larger sizes of condenser the tube diameter is usually increased to 7/8 in. or 1.0 in. so that the tube length can be increased without causing excessive pressure losses. Table 16.2 shows the tube diameter and length commonly employed over a large range of condenser sizes. The shorter lengths in the ranges indicated are employed for two-pass units, whereas the greater lengths are employed for single-pass units. Note that the tube lengths are ordinarily chosen to be an even multiple of 2 ft.

The tube-wall thickness is commonly 18 Bwg (0.049 in., or 1.22 mm). For freshwater installations tubes are ordinarily made of a copper alloy such as admiralty metal (70% copper, 29% zinc, 1% tin). For saltwater installations the expense of titanium or a nickel alloy such as Monel may be justified. In some cases the choice of material may be determined by a need to keep contamination of the condensate to a very low level.[6-8]

Cooling-Water Velocity

Since the principal barrier to heat transfer is the heat transfer coefficient on the water side, there is a strong incentive to increase the cooling-water velocity through the tubes. For most installations a good compromise between heat transfer and pumping power requirements is obtained with a water velocity of from 1.8 to 2.4 m/s (6 to 8 ft/s). Somewhat lower velocities are often used for installations in which erosion by sediment is a problem, and for saltwater where erosion is likely to accelerate corrosion. In other installations, higher velocities may be in order to reduce fouling by fine silt.

Overall Heat Transfer Coefficient

In banks of smooth horizontal round tubes in the configurations and with the high vapor velocities normally

TABLE 16.1 Selected Geometric Parameters of the Experimental Tubes of Fig. 16.3[a,b,c]

Tube number	A	B	C	D	E	F	G
Type	Smooth	Twisted Tape #1	Twisted Tape #2	Fin #1	Fin #2	Fin #3	Fin #4
Material	Cu	Cu, SS	Cu, SS	Cu	Cu	Cu	Cu
Outside diameter	1.5875	1.5875	1.5875	1.5875	1.2776	1.2776	1.5900
Inside diameter	1.3843	1.3843	1.3843	1.4707	1.1811	1.1532	1.3970
Equivalent diameter	1.3843	0.8252	0.8252	0.8260	0.7602	0.7564	0.6767
Wall thickness	0.1016	0.1016	0.1016	0.0584	0.0490	0.0622	0.0965
Total wetted perimeter	4.3487	7.0419	7.0419	7.8750	5.3310	5.1891	8.1979
Nominal wetted perimeter	4.3487	4.3487	4.3487	4.6203	3.7104	3.6228	4.7437
Total perimeter/ nominal perimeter	1.0	1.6193	1.6193	1.7045	1.4368	1.4324	1.7279
Cross-sectional area	1.5052	1.4529	1.4529	1.6264	1.0129	0.9648	1.3864
Tape thickness	n.a.	0.0378	0.0378	n.a.	n.a.	n.a.	n.a.
Fin height	n.a.	n.a.	n.a.	0.0599	0.1735	0.1631	0.1448
Fin base width	n.a.	n.a.	n.a.	0.0475	0.1669	0.1394	0.1245
Fin tip width	n.a.	n.a.	n.a.	0.0277	0.0478	0.0434	0.0462
Number of fins	n.a.	n.a.	n.a.	32	6	6	16
Pitch, cm/180°	n.a.	4.57	9.65	30.48	17.15	Straight	27.94

[a]Ref. 23. (Courtesy Bergles, Ref. 1.)
[b]All dimensions in cm or cm², as appropriate.
[c]n.a. = not applicable.

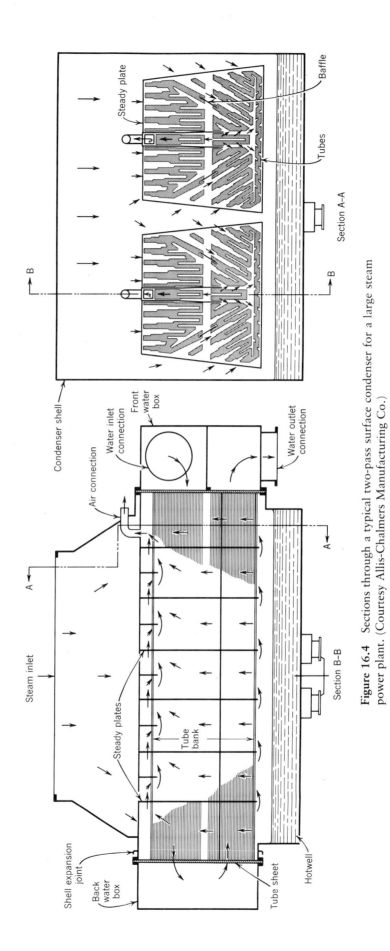

Figure 16.4 Sections through a typical two-pass surface condenser for a large steam power plant. (Courtesy Allis-Chalmers Manufacturing Co.)

Steam inlet

Air connection

Shell expansion joint

Back water box

Steady plates

Tube bank

Tube sheet

Hotwell

Section B–B

Water inlet connection

Front water box

Water outlet connection

A

A

B

B

Condenser shell

Steady plate

Tubes

Baffle

Section A–A

322

TABLE 16.2 Ranges of Tube Length Commonly Employed in Steam Condensers[a,b]

Condenser Heat Transfer Surface Area, ft²	Tube Diameter, in.			
	5/8	3/4	7/8	1
100	6 to 10	6 to 10		
1,000	8 to 14	8 to 14		
3,000		12 to 18	12 to 18	
5,000		14 to 20	14 to 20	14 to 20
14,000		16 to 22	16 to 22	16 to 22
30,000			22 to 28	22 to 28
100,000			24 to 30	24 to 30

[a]Reprinted by permission of Allis-Chalmers Mfg. Co.
[b]Tube length is given in feet.

employed in steam condensers, the overall heat transfer coefficient is primarily a function of the cooling-water velocity through the tubes. Figure 16.5 facilitates calculations for this important set of conditions. It gives the overall heat transfer coefficient as a function of water velocity for clean, bright, new, horizontal tubes with no contamination on either the steam or cooling water sides. If the tubes are of a material other than admiralty metal, or have a wall thickness other than 18 Bwg, the correction factors indicated in the small table in Fig. 16.5 should be employed. Note that Fig. 16.5 includes the small effects of changes in the tube diameter on the water-side heat transfer coefficient. Figure 16.5a shows a curve for a temperature correction factor similar to that given in Fig. H5.4. Note that the heat transfer coefficient changes substantially with temperature and tends to offset the benefits of a reduction in cooling-water inlet temperature.

A fouling factor of 0.85 is commonly applied to the heat transfer coefficient curves of Fig. 16.5, which are for clean tubes, to allow for the accumulation of films of sludge, corrosion products, and other material on the heat transfer surfaces. A set of fouling factors for typical plant sites is given in Table H5.4. It should be noted that in freshwater plants in warmer areas the cooling water is commonly chlorinated before it reaches the condenser to reduce the fouling of condenser surfaces by algae. Growth of most algae and other marine organisms can be prevented by running the surfaces at temperatures above 60°C (140°F), but this is usually uneconomical.

Air Removal

Noncondensable gas is normally present in the condensing vapor, and tends to accumulate at the condensing surface where it reduces the effective temperature difference. At the inlet, the partial pressure of the noncondensable gas is small compared with the total pressure; hence the concentration of gas near the condensing surface is small, but it increases progressively in the direction of the vapor flow. Therefore the inlet to the air suction pumps should be located near the end of the vapor-flow path where the greatest gas density occurs. Note the manner in which this has been done in the condenser of Fig. 16.4. The sweeping effect of steam flowing through the condenser is an important factor in collecting the noncondensable gas at the bottom of the condenser; stagnant zones where the gas may collect should be avoided by maintaining the steam velocity above 30 m/s (100 ft/s), except in the last portion of the steam-flow path.

The amount of air entering the steam system of a typical coal-fired steam plant in good condition is commonly about 25 g/min (1 cfm) at STP per 100 MWe. To reduce the size of the air pumping equipment, provision should be made for cooling the suction air. This can be done by shrouding some of the tubes as in the condenser of Fig. 16.3, and drawing the air from the bottom of the condenser up over the tubes. An additional advantage of this procedure is that it serves to remove most of the moisture remaining in the air.

Steam Velocity

A steam inlet velocity of 30 to 60 m/s (100 to 200 ft/s) through the first row of tubes is good practice at low vacuum (i.e., 26 in. Hg), while a velocity of 60 to 120 m/s (200 to 400 ft/s) may be employed at a higher vacuum (i.e., 29 in. Hg). A network of lanes through the tubes (as in Fig. 16.4) should be provided to distribute the steam uniformly through the tube matrix by keeping the local steam velocity in the range of 30 to 60 m/s. This should serve to keep the steam-side pressure drop through the condenser in the range of 170 to 1700 Pa (0.05 to 0.5 in. Hg).

Water-Pressure Loss

Although the pressure drop through the tubes on the water side can be estimated from basic fluid-flow relations, it is simpler to employ Fig. 16.6, which includes allowances for losses at the tube entrances and exits and in the plenum chambers at the ends. Since the Reynolds number is well above the transition region for

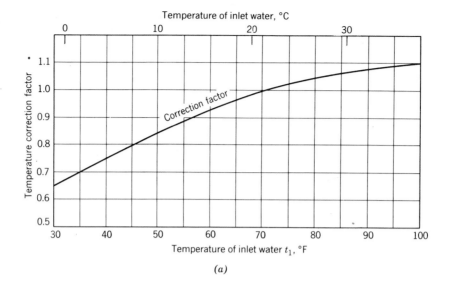

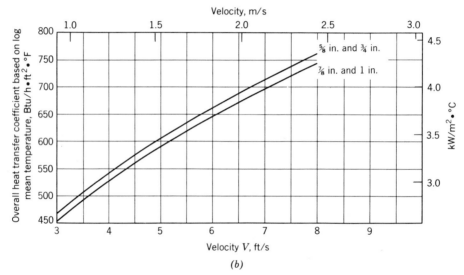

Figure 16.5 Effects of cooling-water velocity on the overall heat transfer coefficient for water-cooled surface condensers with new, clean, bright, oxide-free 18-gauge admiralty tubes, for a cooling-water temperature of 21°C (70°F). A correction factor for the cooling-water temperature is given by the curve at the top, while correction factors for the temperature drop through the tube wall are given in the table below. The heat transfer coefficient is based on the tube external surface area. (Courtesy Allis-Chalmers Manufacturing Co.)

	Tube-Wall Gauge				
Tube Materials	18 Bwg	17 Bwg	16 Bwg	15 Bwg	14 Bwg
Admiralty metal	1.0	0.97	0.93	0.88	0.82
Arsenical copper	1.0	0.97	0.93	0.88	0.82
Aluminum[a]	1.0	0.97	0.93	0.88	0.82
Aluminum brass	.96	0.93	0.89	0.84	0.78
Muntz metal	.96	0.93	0.89	0.84	0.78
Aluminum bronze	.90	0.87	0.84	0.79	0.72
90-10 copper nickel	.90	0.87	0.84	0.79	0.72
70-30 copper nickel	.83	0.80	0.76	0.71	0.64
Type 304 stainless steel	.58	0.56	0.54	0.51	0.47

[a]ASTM Spec. B-234, Alloys M-1A, M-1A clad, GS-11A.

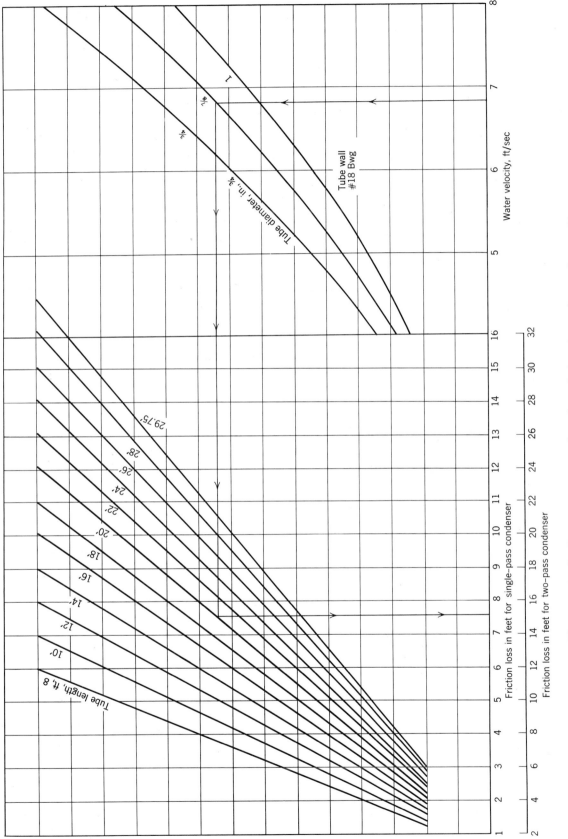

Figure 16.6 Cooling-water pressure drop through the tubes and water boxes of large surface condensers for steam power plants. (Courtesy Allis-Chalmers Manufacturing Co.)

the velocities covered in this chart, the effects of water temperature on the pressure drop are quite small.

Structural Design

The differential pressures across the shell and header sheets generate very large forces. The steam region runs approximately 97 kPa (14 psi) below atmospheric pressure; the water region normally runs from 35 to 170 kPa (5 to 25 psi) above atmospheric pressure, depending on the cooling-water supply system, the static head, and the characteristics of the installation. The condenser shell is often made round or oval to resist the external pressure load. In large units, installation space considerations favor a nearly rectangular cross section for which heavy ribbing is required to prevent the shell from buckling inward. The water boxes at the ends are normally provided with a number of transverse webs, partly to couple the header sheet to the outer end wall of the water box and thus carry the water-pressure loads, and partly to provide a box or beam type of structure to support the header sheet and maintain its shape. The difference between atmospheric pressure and the vacuum on the steam side of the header sheet is absorbed through compression stresses in the condenser tubes with differential expansion between the tubes and the shell ordinarily accommodated by a shell expansion joint of the double diaphragm type.

Tie bolts between the water box cover and the header sheet are provided at intervals across the face of the header sheet to assist in carrying the water-pressure load and the separating forces which it induces.

The thickness of the header sheets and the shell are dependent on the size of the unit, and are ordinarily chosen according to the schedule given in Table 16.3.

Example 16.1. Condenser for a Large Steam Turbine. It is instructive to consider a typical preliminary design problem involving an estimate of the size, weight, and cost of a condenser similar to that illustrated in Fig. 16.4 for a 225,000-kW steam-turbine generator unit. The details of the calculations are summarized in Table 16.4. The information derived from the specifications is tabulated first. The rest of the values represent design choices or calculations. In either case the procedure was consistent with the preceding discussion. For example, the steam-inlet temperature and the cooling-water exit temperature were chosen to be 12.2°C (22°F) and 8.3°C (15°F), respectively, higher than the water-

TABLE 16.3 Recommended Thickness for Condenser Shells and Header Sheets[a]

Total Area of Tube Sheet, in.²	Thickness of Tube Sheet, in.	Thickness of Shell, in.
Up to 1,964	0.875	
Up to 2,460		0.375
1,965 to 3,739	1.00	
2,461 to 5,670		0.500
3,740 to 8,495	1.125	
5,671 to 11,500		0.625
8,496 to 30,791	1.25	
11,501 to 29,000		0.75
90,792 and over	1.50	
29,001 and over		0.875

[a]Reprinted by permission of Allis-Chalmers Mfg. Co.

inlet temperature. The overall heat transfer coefficient as given by Fig. 16.5 for clean, bright, tubes was reduced by the fouling factor given by Table H5.4 for lake water.

The inlet-face area of the tube bank was made 150% of the inlet-face area required to give a steam velocity at the inlet face of 76 m/s (250 ft/s) (see line 25 of Table 16.4). This appeared to be sufficient to provide lanes between banks of tubes, as in Fig. 16.4. Note that in estimating the weight of the condenser components, the weight of the shell was doubled to allow for the extra weight of ribbing, support structure, inlet and outlet flanges, and access parts. Note also that the sum of the weights of the tubing, header sheets, and shell as given by line 35 of Table 16.4 is only about 60% of the value given by Table 11.2. Much of the difference stems from the more economical use of material in very large units.

In reviewing the results of the calculations in Table 16.4 it is apparent that the condenser is very large and unwieldy. The size could be greatly reduced by increasing the condenser pressure a little, thus increasing both the available temperature difference and the steam density. The allowance made for steam lanes is liberal, and offers a further means of reducing the size in arriving at a good detailed design.

CONDENSER-EVAPORATORS FOR DESALINATION PLANTS

The acute need for freshwater in arid areas and on islands has led to the construction of many plants for desalting seawater. Most of these utilize low-pressure steam from electric power plants as their source of heat so that the

TABLE 16.4 Design Calculations for a Steam-Turbine Condenser

Flow configuration: two-pass crossflow; design heat load: 1.2×10^9 Btu/h (352 MW)
Heat transfer matrix geometry: 18-Bwg admiralty tubes 7/8-in. OD (1.25-mm wall, 22.2-mm OD)

Line Number	Item	Source[a]	English Units Hot Fluid	English Units Cold Fluid	SI Units Hot Fluid	SI Units Cold Fluid
1.	Fluid		Steam	Water	Steam	Water
2.	Change in enthalpy, Btu/lb		1,042	15.0	2,420	35 kJ/kg
3.	Flow area per tube, ft^2			0.0329		0.00306 m^2
4.	Surface area per tube, ft^2/ft	Table H7.1	0.2291	0.2034	0.0751	0.0668 m^2/m
5.	Temperature in, °F		92	0.0	33.3	21.1°C
6.	Temperature out, °F		92	85.0	33.3	29.4°C
7.	Temperature rise (or drop), °F		0	15.0	0	8.3°C
8.	Inlet temperature differential, °F		22.0			12.2°C
9.	Greatest temperature differential, °F		22.0			12.2°C
10.	Least temperature differential, °F		7.0			3.9°C
11.	LMTD, °F	Fig. H4.1	13.0			7.2°C
12.	Density (mean), lb/ft^3		0.00226	62.4	0.0362	1,000 kg/m^3
13.	Flow velocity, ft/s	Assumed	250	6.0	76.2	1.83 m/s
14.	Flow rate, lb/ft$^2 \cdot$ s	⑫ × ⑬	0.565	374	2.76	1,825 kg/m$^2 \cdot$ s
15.	Total flow rate, lb/s	$1.2 \times 10^9/3{,}600 \times$ ②	320	22,200	145	10,000 kg/s
16.	Flow-passage area required, ft^2	⑮ / ⑭	565	59.3	52.5	5.51 m^2
17.	Number of tubes required per pass	⑯ / ③		18,000		
18.	Total number of tubes	2 × ⑰		36,000		
19.	U, Btu/h $\cdot$ ft$^2 \cdot$ °F/ft	Fig. 16.5 and fouling factor = 0.001 from Table H5.4	392		2,230 W/m$^2 \cdot$ °C	
20.	Surface area for condensing, ft^2	$1.2 \times 10^9/$ ⑪ × ⑲	236,000		21,900 m^2	
21.	Surface area for air cooling, ft^2	0.05 × ⑳	12,000		1,100 m^2	
22.	Total surface area, ft^2	⑳ + ㉑	248,000		23,000 m^2	
23.	Surface area per tube, ft^2	㉒ / ⑱	6.9		0.64 m^2	
24.	Tube length, ft	㉓ / ④	30.0		9.15 m	
25.	Tube matrix width, ft	1.5 × ⑯ / ㉔	27.7		8.44 m	
26.	Overall pressure drop, ft H$_2$O	Fig. 16.6		17		352 Pa
27.	Weight of tubing, lb	$0.432 \times 1.09 \times$ ⑱ × ㉔	510,000		232,000 kg	
28.	Header-sheet area, in.^{2}	$1.5 \times (0.875 \times 1.4)^2 \times 36{,}000$	81,300		52.5 m^2	
29.	Header-sheet thickness, in.	Table 16.3	1.50		38 mm	
30.	Shell thickness, in.	Table 16.3	0.875		22 mm	
31.	Header sheet weight, lb	2 × ㉘ × ㉙ 500/1,728	61,000		27,700 kg	
32.	Tube matrix height, ft	㉘ /144 × ㉕	20.4		6.22 m	
33.	Shell height, ft	1.5 × ㉜	30.0		9.15 m	
34.	Shell weight, lb	$(2 \times 0.875/12) \times 500 \times$ (surface A)	340,000		155,000 kg	
35.	Weight of tubes, headers, and shell, lb	㉗ + ㉛ + ㉞	911,000		414,000 kg	
36.	Total condenser weight, lb	Table 11.2	1,500,000		682,000 kg	

[a]Circled numbers are symbols to indicate the line from which to obtain the quantity to be used in the operation.

bulk of the fuel cost is chargeable to the production of electricity. The economics vary with the site. Obvious factors are the relative demands for electricity and desalted water, but less quantifiable factors such as corrosion, scale formation, and biofouling also enter in because they may degrade the heat transfer performance. The usual practice is to design the steam turbines for a discharge temperature of no more than 70°C (158°F) because higher temperatures may give trouble with scale deposits. The rate at which biofouling develops depends on the local species, but increases fairly rapidly with temperature up to around 60°C (140°F), above which it falls off very rapidly.

Multistage Evaporators

To obtain as much freshwater as possible from a given heat input it has been found best to employ a series of stages in the evaporator. Thus the heat in the vapor produced in the first stage can be utilized when it condenses to evaporate more water from the already heated brine in the next stage (which is operated at a lower temperature), and so on through a series of stages. It has been found possible to operate with a temperature difference in each stage of only 6 to 8°C (11 to 15°F) between the saltwater, or brine, and the condensing vapor. The stages are often mounted one above the other as in a fractionating tower so that the brine and product vapor streams cascade downward through as many as 10 stages. Vacuum pumps are used to remove dissolved and entrained noncondensables given off by the brine, so that each stage is kept at the proper pressure.

To minimize the temperature drop from the heated surface to the vapor, the units are designed so that much of the vaporization takes place from the surface of thin liquid films or small droplets, as in a flash evaporator.

Enhanced Heat Transfer Surfaces

The high premium associated with minimization of the temperature difference between the condensing vapor and the evaporating brine has led to extensive work on surfaces designed to enhance heat transfer.[9] One of the most widely used special surfaces for tubes that are to be installed vertically is a doubly fluted tube having a diameter of 50 to 75 mm (2 to 3 in.), a wall thickness of about 2 mm (0.05 in.), and a flute depth about equal to the thickness of the remaining wall so that in cross section the tube looks as if it were corrugated, though the tubes are usually formed by extrusion.[10] The heat transfer coefficient for a typical fluted tube together with data for a plain tube of the same size are shown in Fig.

16.7 for low-pressure steam condensing outside vertical tubes with hot brine sprayed down the inside of the tubes. In this case, fluting the tube increased the heat transfer coefficient by a factor of five even though it was based on the total area. As discussed at the end of Chapter 3, fluting the surface brings surface tension forces into play, with the result that capillary action spreads the liquid in a thin film over the ridges between grooves (see Fig. 3.25), thus minimizing the temperature drop through the liquid film. At the same time, the relatively large hydraulic radius of the fluid column in the grooves greatly increases the drainage rate which, for laminar flow, varies as the fourth power of the hydraulic radius. The short vertical length and the larger diameter gave better performance than obtained with the 15.875-mm (5/8-in.) diameter, 3.66-m (12-ft) long tubes of Fig. 16.3 because the liquid film thickness on the tube wall was greatly reduced by the short drainage path. Implicit in this is the fact that stripping the condensate off a long vertical tube by inserting tightly fitted baffle plates should improve the average heat transfer coefficient. This has been found to be the case, but it increases the mechanical

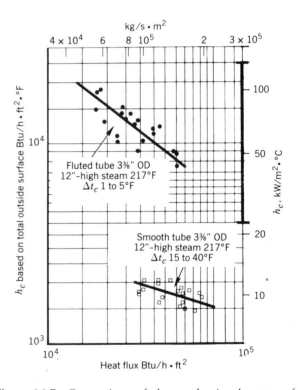

Figure 16.7 Comparison of the condensing heat transfer coefficients for 300-mm (12-in.) long, 85.8-mm (3.375-in.) diameter plain and fluted tubes for operation on low-pressure steam in vertical tube flash evaporators for desalination units. (Hammond, Ref. 14.)

complexity of the condenser. The alternative is to use shorter tubes; this is the reason that evaporators for desalination plants commonly use vertical tubes only 2 or 3 m (6.5 to 10 ft) long.

Tubes that are to be operated horizontally are often made with circumferential grooves or corrugations to facilitate condensate drainage if the vapor is condensed on the shell side of the tube. Tests show that condensate "rain" on the lower tubes in a horizontal tube bank leads to an increase in the liquid film thickness and thus to a progressively lower heat transfer coefficient as one moves down through the tube matrix for either plain or grooved tubes. The loss in the local heat transfer coefficient is rapid in the first few rows but largely levels off by the 14th row where its value drops to between 85 and 55% of the value for the top row, depending somewhat on test conditions.[11]

While corrosion is the dominant consideration in the choice of condenser tube material, it is also important to minimize the temperature drop through the tube wall. Because of the large surface areas required and the vital effects of capital costs, for desalination installations the tubes are often made of aluminum instead of one of the copper alloys normally employed in steam power plants. This requires closer control of the water chemistry to keep corrosion of the aluminun to an acceptable level and limits the peak temperature that can be used to about the same level as the upper temperature limit acceptable from the scaling standpoint.

If evaporation is to occur inside the tube, a large diameter is desirable to provide an open path for the vapor to minimize the pressure drop and the associated temperature drop in tubes at least 2 m long. The liquid is supplied to the evaporator side by spraying the brine into the tubes at the top of the evaporator with low-pressure nozzles designed to distribute the liquid uniformly around the circumference of the tubes. The resulting overall heat transfer coefficients are commonly around 8500 W/m^2 · °C (1500 Btu/h · ft^2 · °F) at 74°C (165°F). The coefficient drops off somewhat as the temperature is reduced because the viscosity of water increases.

Similar fluted or grooved surfaces have been used in condensers for conventional steam power plants.[12,13] The heat transfer performance has been improved as predicted, but the lower operating temperature makes such installations more subject to biofouling at some sites. Operational experience has demonstrated that deposits from biofouling usually can be removed even from grooved tubes while the plant is in operation by feeding properly sized sponge rubber balls into the water-inlet header from which they are carried through the tubes by the pressure drop across the condenser. However, up to the time of writing, concerns for fouling and cleanability are still factors, and these, coupled with the higher cost of the tubing, have limited the use of grooved tubes in steam power plants to a few test cases. Another factor is that the greatest benefits of grooved tubes are obtained with vertical tubes because of the better condensate drainage situation, and the vertical geometry is usually less favorable for use with steam turbines.

Vertical Tube Evaporators

By mounting doubly fluted tubes vertically, gravity serves to drive the film of evaporating brine down the tube in a uniformly thin layer and makes the condensate on the opposite side of the tube drain off rapidly. A fluted surface inhibits wave formation in the descending films, while surface tension acts as it does in heat pipes (see Chapt. 6) to keep the thickness of both water films over the ridges between grooves uniformly small, and hence their resistance to heat flow is minimized. This effect is

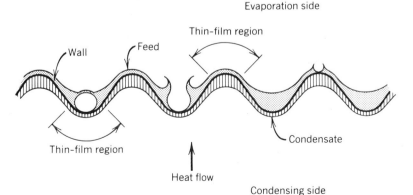

Figure 16.8 Section through a doubly-fluted tube wall showing the liquid surface configuration produced by surface tension and the resulting thin film of liquid over the ridges on the tube wall. (Courtesy Oak Ridge National Laboratory.)

indicated in Fig. 16.8. Observation of these tubes during operation discloses that on the evaporator side the release of vapor bubbles in the groove acts to slosh liquid up onto the crests of the flutes, keeping them wet so that their area contributes to the heat transfer.[14] Tube lengths are usually 1.5 to 3 m (5 to 10 ft), giving a convenient height for stacking a half-dozen stages one above the other. A good brine flow rate for 75-mm-diameter tubes has been found to be around 0.2 kg/s per tube (3 gpm), although in some cases values as low as half that have been used without serious trouble with dry spots and the resulting scale formation. If scale does form, it is usually $CaSO_4$, $CaCO_3$, or $Mg(OH)_2$ from the brine, although corrosion products from pipes or casings may also be a problem.

Design and Operating Data

A substantial amount of detailed design and operating data is available in the literature, much of it prepared under the auspices of the Office of Saline Water (OSW) of the U.S. Department of the Interior. Some of the OSW work on enhanced heat transfer has been referenced in Chapters 3, 5, and 14 because it has proved valuable in the areas treated in those chapters. A few of the other good sources of data particularly applicable to desalination system design may be found in References 14 to 17. Information on full-scale system design and operating

experience may be found in References 17 to 20. Among the more significant sets of information in the latter references are the sets of data from experience with corrosion and fouling; some plants have been operated successfully for over 20 years.

OTEC Plants

The data cited here are equally applicable to plants designed to obtain electric power from ocean thermal energy conversion (OTEC).[21] The inherently high capital cost of OTEC plants has limited work in that field to conceptual designs, some fine work on basic heat transfer for enhanced surfaces, and a few not very successful experiments with small OTEC power plants. In any event, the heat transfer problems in the design of boilers and condensers for OTEC plants are sufficiently similar to those of plants for desalination that further treatment here is not needed.

DESIGN OF CONDENSERS FOR REFRIGERATING AND AIR-CONDITIONING SYSTEMS

Condensers for refrigerating systems present some specialized and interesting design problems. The type of condenser to be used depends on the desired capacity

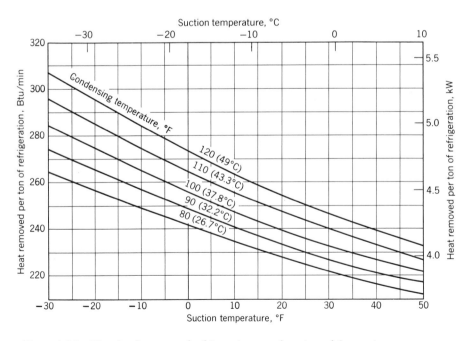

Figure 16.9 Heat load per ton of refrigeration as a function of the suction temperature for a Freon-12 condenser. (*ASHRAE Guide and Data Book*, 1961, p. 552.)

and the type of application. Water-cooled, shell-and-tube units are commonly used in large installations, whereas air-cooled tube-and-fin units are employed where small capacities are required.

In a refrigerating system the liquid refrigerant is expanded to the vapor state in the evaporator, and, after compression in the compressor, is converted to the liquid state in the condenser. Therefore the quantity of heat to be removed in the condenser is greater than the quantity of heat extracted in the evaporator because of the heat addition during the compression process. In the design of a condenser for a refrigerating system, the heat extraction, or cooling capacity, of the evaporator is given in "tons of refrigeration." (A ton of refrigeration equals 12,000 Btu/h, or 352 kW). The corresponding heat transfer rate through the condenser can be estimated if the evaporation and condensing temperatures are given. To facilitate such calculations, charts have been prepared which give the heat transfer rate through the condenser per ton of refrigeration in the evaporator as functions of the compressor inlet temperature and the condensing temperature. Figure 16.9 shows such a chart for Freon-12. Note that the values given are based on saturated vapor at the compressor inlet and adiabatic compression. Values for other refrigerants such as Freon-22 or ammonia differ only a little from those in Fig. 16.9 so that if a chart for the refrigerant specified is not readily available, Fig. 16.9 can serve to give a good approximation.

Once the heat transfer capacity of the condenser is determined, the procedure for sizing the condenser is essentially the same as that described for steam condensers, although the heat transfer coefficient for the condensing vapor is much lower.[22]

Example 16.2. *Condenser for a 10-ton Freon Refrigerating System.* Freon-12 is to be evaporated at $-8°F/$ 20 psia ($-22.2°C/138$ kPa) and condensed at $90°F/$ 114.5 psia ($32.2°C/789$ kPa) with the heat of condensation to be rejected to cooling water taken from a small cooling tower at $70°F$ ($21.1°C$). A shell-and-tube construction with two or more passes on the water side is to be employed. Brass tubes having a diameter of $5/8$ in. (15.9 mm) are specified because they are compatible with both Freon and water and are readily cleaned. The lower condensing heat transfer coefficient of Freon as compared to steam leads to a lower cooling-water velocity through the tubes for a good balance between water pumping power requirements and heat exchanger cost. A value of 5 ft/s (1.52 m/s) is usual with a temperature rise in the cooling water of $8°F$ ($4.4°C$). Determine the total heat transfer rate, the cooling-water flow rate, the number of tubes, the tube length, and the water-side pressure drop.

Solution. Heat removed in the condenser per ton of refrigeration from the evaporation, from $-8°F$ ($-22°C$) suction temperature and $90°F$ ($32.2°C$) condensing temperature, as obtained from Fig. 16.9, is 257 Btu/min or 15,400 Btu/h · ton (4.5 kW/h).

The heat transfer coefficient for vapors condensing on horizontal tubes is given by Eq. 3.32, that is,

$$h_v = 0.95 \left(\frac{L k^3 \rho^2 g}{W \mu} \right)^{1/3}$$

In this equation the only unknown quantity is W/L, the mass flow rate of condensate per foot of tube length from the lowest tube in the bank. A first approximation to W/L may be estimated from the total number of tubes, the total flow rate of condensate, and an estimated tube length of 8 ft assuming four shell-side passes having a tube length of 2 ft per pass, that is,

$$W/L = \frac{2680 \times 4}{52 \times 8} = 25.8 \text{ lb/h} \cdot \text{ft (38.5 kg/m)}$$

If we take the physical properties of condensing Freon-12 as $k = 0.04$ Btu/h · ft² · °F/ft, $\rho = 81.5$ lb/ft³, $\mu = 0.6$ lb/h · ft, we obtain

$$\left(\frac{k^3 \rho^2 g 3600^2}{\mu} \right)^{1/3} = \left(\frac{0.04^3 \times 81.5^2 \times 4.18 \times 10^8}{0.6} \right)^{1/3}$$
$$= 661$$

and the condensing heat transfer coefficient becomes $h_v = 0.95 \times 661/2.94 = 211$ Btu/h · ft² · °F (1197 W/m² · °C).

The full set of calculations is summarized in Table 16.5. If the tube length obtained with the initial design assumptions gives too long and slender a unit, or one too short with too many tube-to-header joints, new values have to be chosen for the number of passes and the calculations repeated until they give a well-proportioned unit from the shop fabrication and installation standpoints. The tube length of 7.65 ft (line 31, Table 16.5) is sufficiently close to the 8 ft assumed so that another iteration is not necessary.

The tube center-to-center spacing in the tube sheets should be at least 1.25 times the tube diameter to give an adequate thickness in the ligament between the holes in the tube sheet for good fabrication in the shop. A greater tube spacing may sometimes be required to provide an adequate vapor-flow-passage area to the

TABLE 16.5 Design Calculations for a Freon Condenser

Flow configuration: four-pass; design heat load: 154,000 Btu/hr (45.1 kW/h)
Heat transfer matrix geometry: 5/8-in. OD 16-Bwg (15.9-mm OD, 1.65-mm thick) brass tubes

Line Number	Item	Source[a]	Hot Fluid	Cold Fluid
1.	Fluid		Freon-12	City water
2.	Change in enthalpy, Btu/lb		57.46	8.0
3.	Flow area per passage, in.^{2}	Table H7.1		0.1924
4.	Surface area per passage, ft^2/ft	Table H7.1	0.1636	0.1296
5.	Temperature in, °F		90.0	70.0
6.	Temperature, out, °F		90.0	78.0
7.	Temperature rise, °F	⑤ and ⑥		8.0
8.	Inlet temperature difference, °F	⑤	20.0	
9.	Greatest temperature difference, °F	⑤ and ⑥	20.0	
10.	Least temperature difference, °F	⑤ and ⑥	12.0	
11.	LMTD, °F	Fig. H4.1	15.6	
12.	Pressure (mean), psia		14.5	50.0
13.	Density (mean), lb/ft^3		2.72	62.4
14.	Flow velocity, ft/s	Assumed	55.0	5.0
15.	Flow rate, lb/ft^2 · s	⑬ × ⑭	150	312
16.	Flow rate per passage, lb/s	③ × ⑮ /144		0.415
17.	Total flow rate, lb/h	154,000/ ②	2,680	19,260
18.	Number of tubes per pass	⑰ /3,600 × ⑯		13.0
19.	Equivalent diameter of flow passage, in.	Table H7.1		0.495
20.	Re	Fig. H3.1		20,900
21.	Friction factor	Fig. H3.4		0.029
22.	Heat transfer coefficient, Btu/h · ft^2 · °F	Eq. 3.32 and Fig. H5.3	211	1,065
23.	Dynamic head, psi	Fig. H3.2		0.155
24.	Fouling resistance	Table H5.4	0.0005	
25.	1/h	1/ ㉒	0.00473	0.00119
26.	Conductance of tube wall, Btu/h · ft^2 · °F	Table H2.1	12,000	
27.	1/U	㉔ + ㉕ + ㉕ + 1/ ㉖	0.00656	
28.	U, Btu/h · ft^2 · °F	1/ ㉗	152	
29.	Total surface area (external), ft^2	154,000/ ⑪ × ㉘	65.0	
30.	Surface area per tube, ft^2	㉙ /4 × ⑱	1.25	
31.	Tube length, ft	㉚ / ④	7.65	
32.	Water-pressure drop, psi	㉑ × ㉓ × 4 × ㉛ × 12/ ⑲		3.6

[a]Circled numbers are symbols to indicate the line from which to obtain the quantity to be used in the operation.

tubes, although this is not likely to present a problem for the small-diameter tube bundle of this example. It is worthwhile to check this point, however, assuming that the Mach number in the vapor may be as high as 0.1. Since the molecular weight of Freon is large, the velocity of sound is low (about 550 ft/s, or 168 m/s). From the vapor density and flow rate it is evident that a vapor-flow-passage area of only about 0.01 ft^2 (0.001 m^2) will suffice. Since the tube-bundle diameter will be approximately 8 in. (23 mm), it is apparent that the vapor-flow-passage area should be adequate, and the space normally provided between the tube bundle and the shell for ease

of assembly should be sufficient to distribute the vapor axially through the shell.

Multiple Tube Row Effects

Example 16.2 did not include an allowance for the additional liquid film thickness on the lower rows of tubes resulting from the effects of condensate dripping off the upper tube rows and "raining" down on the tubes beneath. As discussed in the section above on horizontal tubes for desalination units, this factor will increase the

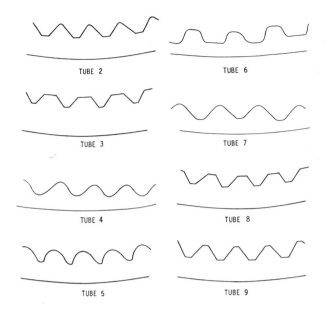

Figure 16.10 Fin profiles for the spiral "micro-fin" Freon condenser tubes of Table 16.6 and Fig. 16.11. (Courtesy Bergles, Ref. 23.)

surface area required by 10 to 30% depending on the number of tube rows involved.

Augmentation of Heat Transfer

The poorer heat transfer characteristics of the Freons relative to water have led to many efforts in the past 40 years to enhance heat transfer in air-conditioning equip-

ment by using turbulating devices such as the coiled springs of Fig. 3.18. As in the investigation for condensing steam summarized in Fig. 16.3, small, closely spaced fins (i.e., shallow grooves) such as those of Fig. 16.10 and Table 16.6 appear to be the most effective devices for improving the rate of condensation inside round tubes.[23] A representative set of data from a series of tests with these geometries is presented in Fig. 16.11. All of the grooved tubes gave better performance than the plain tube (Tube No. 1), and, as is the case for the mist-flow region in evaporators, the improvement from grooving becomes progressively greater with increasing vapor quality (see Fig. 5.27). An interesting result of this series of tests is that the best performance was obtained with helix angles around 20°; the performance with either 9° or 24° was not as good.[23]

FEEDWATER HEATERS

Large steam power plants employ regenerative feedwater heating systems both to increase the cycle efficiency and to reduce the size of the main condenser. The second consideration is more important than it appears to be at first glance, because the condensers for large steam turbines become so large that installation problems become awkward indeed. Thus having a third of the steam flow to the feed heaters is a real help in this respect.

A conventional U-tube, shell-and-tube construction is generally employed for the lower pressure feedwater heaters, the feedwater being on the tube side and the condensing steam on the shell side. The high-pressure

TABLE 16.6 Selected Geometrical Parameters of the Tubes of Figs. 16.10 and 16.11[a,b]

Tube Number:	1	2	3	4	5	6	7	8	9
Outside diameter, OD	9.525	9.525	9.525	9.525	9.525	9.525	9.525	9.525	9.525
Root diameter		8.738	8.712	8.890	8.890	8.915	8.890	8.941	8.839
Tip diameter		8.407	8.433	8.407	8.534	8.712	8.560	8.534	8.458
Fin height		0.19	0.16	0.16	0.17	0.10	0.15	0.18	0.18
No. of fins		65	60	65	70	60	65	65	65
Spiral angle		20	20	25	20	8–10	23–25	20	20
Tip geometry		***R	*F	R	**FR	F	FR	F	FR
Valley geometry		F	F	R	FR	F	F	F	F
Pitch		0.424	0.520	1.406	0.432	0.406	0.432	0.484	0.417
Peak width			0.16			0.10		0.12	
Valley width		0.10	0.11			0.18	0.092	0.11	0.085
Surface area, cm^2		390.19	362.32	362.32	399.48	306.58	380.90	408.77	390.19
A_{aug}/A_{smooth}	1.00	1.43	1.34	1.34	1.47	1.13	1.40	1.50	1.43

[a]All dimensions in mm or cm^2.
[b]Key: *F = flat tip geometry; **FR = flat fin curving toward round; and ***R = round fin geometry.

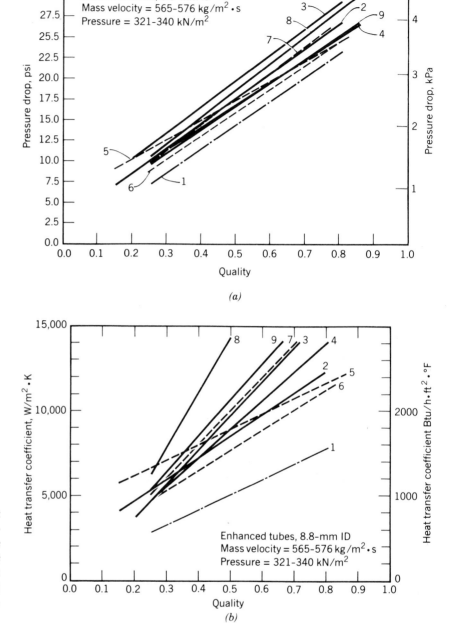

Figure 16.11 Comparison of the heat transfer and total pressure drop characteristics for Freon-113 condensing inside tubes having spiral "microfins." The geometries of the 10 different tubes, each 9.52-mm (0.375-in.) OD by 1 m (3.28 ft) long, are defined by Table 16.6 and Fig. 16.10. (Courtesy Bergles, Ref. 23.)

feedwater heaters used for upper stage feedwater heating are similar except that the high pressures (as much as 5000 psi on the tube side and over 1000 psi on the shell side) pose major problems in the structural design. A typical unit showing one approach to the solution of these problems is shown in Fig. 16.12. Note particularly the massive construction at the header end, where the tube sheet is made integral with the cylindrical header

and a removable cover is provided to permit access to the tube sheet. This closure employs an annular shear key rather than tension bolts to carry the load imposed by the high pressure. While not indicated specifically in this view, the unit is assembled by slipping the casing over the tube bundle and attaching it with a circumferential weld to a skirt integral with the tube sheet. A backing strip is provided behind the weld to prevent

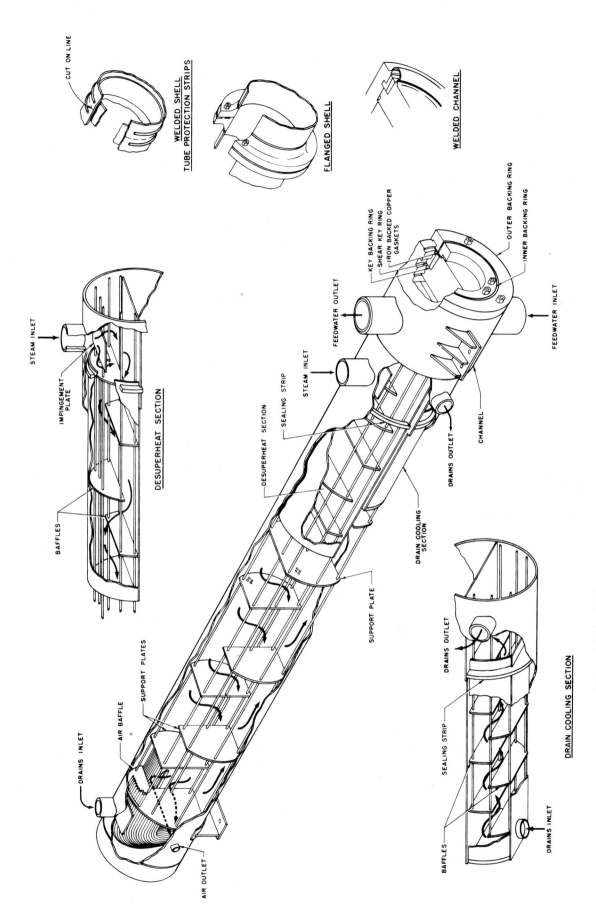

CUT ON LINE

WELDED SHELL
TUBE PROTECTION STRIPS

FLANGED SHELL

WELDED CHANNEL

STEAM INLET

IMPINGEMENT PLATE

DESUPERHEAT SECTION

BAFFLES

DRAINS INLET

AIR BAFFLE

SUPPORT PLATES

AIR OUTLET

DESUPERHEAT SECTION

SEALING STRIP

STEAM INLET

SUPPORT PLATE

DRAIN COOLING SECTION

DRAINS OUTLET

CHANNEL

FEEDWATER OUTLET

FEEDWATER INLET

KEY BACKING RING
SHEAR KEY RING
IRON BACKED COPPER GASKETS
OUTER BACKING RING
INNER BACKING RING

DRAINS OUTLET

SEALING STRIP

BAFFLES

DRAINS INLET

DRAIN COOLING SECTION

Figure 16.12 High-temperature horizontal feedwater heater for a large steam power plant. (Courtesy Yuba Heat Transfer Corp.)

335

damage to the tubes during both the welding operation and disassembly, when the weld is cut with a torch. The all-welded construction greatly eases the stress problems without seriously hampering maintenance, since control of feedwater impurities is so good that there is little difficulty with corrosion or deposits and, hence, little or no occasion to remove the shell.

The unit shown in Fig. 16.12 is designed for horizontal operation with a horizontal baffle plate extending through the length of the U-tube bundle. Superheated steam enters at the right end and is directed through the tube bundle in counterflow to take advantage of the superheat. This procedure is unusual, since most feedwater heaters are supplied with wet steam and employ baffle plates only to support the tubes and maintain their spacing. Except for the special cases such as this one, where the steam supplied is superheated, the heat transfer design of feedwater heaters is essentially similar to that of condensers.

Units similar in basic construction to the feedwater heater of Fig. 16.12 are employed in other high-pressure applications, including petroleum refineries and chemical processing plants.

REFERENCES

1. A. E. Bergles, "Techniques to Augment Heat Transfer," of *Handbook of Heat Transfer Applications*, 2nd ed., McGraw-Hill Book Co., New York, 1985, Chapt. 3.

2. P. J. Marto, "Recent Progress in Enhancing Film Condensation Heat Transfer on Horizontal Tubes," *Heat Transfer 1986*, vol. 1, *Proceedings of the 8th International Conference*, 1986, p. 161.

3. J. H. Royal and A. E. Bergles, "Augmentation of Horizontal In-Tube Condensation by Means of Twisted Tape Inserts and Internally-Finned Tubes, *Journal of Heat Transfer, Trans. ASME*, vol. 100, 1978, p. 17.

4. A. E. Bergles, "Recent Developments in Convective Heat Transfer Augmentation," *Applied Mechanics Reviews*, vol. 26:675–682, 1973, p. 233.

5. A. P. Colburn, "Problems in Design and Research on Condensers of Vapors and Vapor Mixtures," *Proceedings of General Discussion on Heat Transfer*, Institute of ME and ASME, 1951, p. 1.

6. J. D. Ristroph and E. B. Powell, "Contamination of Condensate by Heat-Exchanger-Tube Alloys," *Trans. ASME*, vol. 75, 1953, p. 729.

7. Corrosion-Related Failures in Power Plant Condensers, Report no. NP-1468, EPRI, Electric Power Research Institute, August 1980.

8. "Assessment of Condenser Leakage Problems," Report no. NP-1467, EPRI, Electric Power Research Institute, August 1980.

9. Ref. 4, p. 675.

10. L. G. Alexander and H. W. Hoffman, "Performance Characteristics of Advanced Evaporator Tubes for Long-Tube Vertical Evaporators," OSW R&D Report no. 644, 1971.

11. D. M. Eissenberg, "An Investigation of the Variables Affecting Steam Condensation on the Outside of a Horizontal Tube Bundle," Report no. ORNL-TM-4036, Oak Ridge National Laboratory, 1972.

12. L. W. Boyd et al., "Efficiency Improvement at Gallatin Unit 1 with Corrugated Condenser Tubing," Paper no. 83-JPGC-Pwr-4, ASME, 1983.

13. D. Katsman, "Spiral, High-Strength, Corrosion-Resistant Tube Application in PWR Power Plant Condensers," Paper no. 83-JPGC-Pwr-17, ASME, 1983.

14. R. P. Hammond, "Conceptual Design of a 250-MGD Vertical Tube Evaporator for Seawater Desalination, Phase I, Selection of Flowsheet, Layout, and Design Criteria," Report no. ORNL-TM-1901, Oak Ridge National Laboratory, September 1967.

15. G. Jansen and P. C. Owzarski, "Boiling Heat Transfer in Falling Film Evaporators with Corrugated Surfaces," OSW R&D Report no. 693, 1971.

16. D. M. Eisenberg, "Analysis of Heat Transfer Fouling in Seawater Evaporators," Report no. ORNL-TM-3173, Oak Ridge National Laboratory, 1970.

17. R. Van Winkle, "Preliminary Measurements of the Performance of a Large Horizontal Condenser Bundle in the MSF Module," Report no. ORNL-TM-3840, 1972.

18. P. M. Rapier et al., "Analysis and Summary Report of Operation—Guantanamo Naval Base Desalination Facility, OSW R&D Report no. 769, 1972.

19. C. D. Hornburg et al., "Commercial Desalting Plants—Data and Analysis, vol. 1: Summary and Comparisons," OSW R&D Report no. 906, 1972.

20. R. E. Bailie and O. J. Morin, "VTE Desalting Plant, St. Croix, U.S. Virgin Islands: Analysis of Operational Data," OSW R&D Report no. 778, 1972.

21. G. E. Ioup, ed., *Proceedings of the Fourth Annual Conference on Ocean Thermal Energy Conversion*, OTEC, March 22–24, 1977, Supported by The Division of Solar Energy, U.S. Energy Research and Development Administration, 1977.

22. F. L. Young and W. J. Wohlenberg, "Condensation of Saturated Freon-12 Vapor on a Bank of Horizontal Tubes," *Trans. ASME*, vol. 64, 1942, p. 787.

23. J. C. Khanpara, A. E. Bergles, and M. B. Pate, "Augmentation of R-113 In-Tube Condensation with Micro-Fin Tubes, Heat Transfer in Air Conditioning and Refrigeration Equipment," *ASME HTD*, vol. 65, 1985, 21.

17

Heat Exchangers for Liquid Metals and Molten Salts

Liquid metals and molten salts have been found to be excellent heat transport media for systems designed to operate at temperatures from 260 to 1100°C (500 to 2000°F).[1-3] The size of the piping and of major pieces of equipment together with the pumping power requirements can be kept much lower than if gases were employed. The wall thickness for both the piping and the casings for pumps, heat exchangers, and other items of equipment can be kept much lower than those required for high-pressure steam systems operating in this temperature range. There are no coking problems such as those that impose an upper temperature limit of about 290°C (550°F) for oil and about 370°C (700°F) for Dowtherm. On the other hand, the special corrosion characteristics of molten salts and liquid metals require that the structural materials be selected with care. Furthermore, the systems must be designed for a high degree of leaktightness to minimize contamination of the liquid by water vapor or oxygen if corrosion rates are to be kept small. With proper design, construction, and operation, systems of this sort have been operated at temperatures of 650°C (1200°F) and higher with corrosion rates of less than 2.5 μm/yr (0.0001 in./yr). The heat exchangers and systems must also be designed to provide for both preheating and good drainage to avoid difficulties with liquid freezing, particularly for molten salts.

The first commercial application of a liquid metal in heat transfer equipment was in mercury vapor power plants about 1923;[4] however, a molten salt was not used commercially as a heat transfer medium until 1937,[5] when a mixture of sodium nitrite, sodium nitrate and potassium nitrate was employed in a Houdry catalytic cracking unit, and proved quite successful. The same material has since been used for a number of other applications in the petroleum refinery and chemical-processing industries.

The advent of the fission reactor opened up a whole new set of applications for liquid metals and molten salts as heat transfer media for nuclear power plants.[6-8] Fluids that have been given particular attention include molten sodium, potassium, NaK (a sodium-potassium alloy), lithium, lead, bismuth, mercury,[9] the chlorides and fluorides of the alkali and alkaline earth metals,[10] and hydroxides. The sodium nitrite–sodium nitrate–potassium nitrate salt has not been given much attention for reactor applications, partly because a few explosions have occurred with this material as a result of decomposition when used in heat treating baths at temperatures above 510°C (950°F).

Liquid metals give exceptionally high heat transfer coefficients, thus making possible very compact heat exchangers. Some idea of the exceptionally high performance obtained from a few of the units that have been developed as compared to the performance of more conventional types is given by Table 17.1. These data are presented because it appears that light weight and

TABLE 17.1 Comparison of High-Temperature Liquid Heat Exchangers With More Conventional Units[a]

	PWR[b] Steam Boiler	ART[c] Fluoride-to-NaK Heat Exchanger	ART[c] NaK-to-Air Radiator	Calder Hall Steam Generator	Proposed Steam Generator for a Gas-Cooled Reactor
Configuration	Axial tubes in cylindrical U-shaped casing	(See Fig. 2.18) Helical spiral tube bundles in spherical shell	(Similar to Fig. 17.12) Round tubes with plate fin	(Similar to Fig. 1.5) Serpentine finned tubes in cylindrical shell	Axial bare tubes in cylindrical shell
Shell-side fluid flow	Crossflow	Axial counterflow	Special two-pass crossflow	Crossflow	Axial
Structural material	Carbon and stainless steel	Inconel	Inconel	Carbon steel	1% Cr–½% Mo
External dimensions, ft	4 × 10 × 28		1.5 × 2.5 × 6	17.3 × 73	9 × 125
Volume of heat transfer matrix, ft³	280	7.3	41.4	10,000	4,200
Tube OD, in.	0.75	0.23	0.187	1.5 and 2.0	0.50
Tube length (mean), ft	50.0	5.4	3.0	150	116
Number of tubes	921	3,120	5,760	450	6,538
Weight of heat transfer surface, lb	20,700	940	3,000	350,000	200,000
Weight of casing, lb	118,000			360,000	660,000
Total weight, lb	138,800	2,100	4,400	710,000	860,000
Energy transferred, MW (thermal)	77	55	55	46	1,320
LMTD, °F	35	136	865	113	238
Power density, W/cm³	9.7	266	47	0.16	11.1
W/cm³ · °F, LMTD	0.28	1.95	0.054	0.0014	0.047
Cost[d] per unit, $	235,585	324,000	320,000	761,000	2,200,000[e]
Total cost, $/lb heat transfer matrix	11.00[f]	344	107.00	2.17[f]	11.00[f]
Total cost, $/kW heat transferred	3.06	5.90	5.82	16.50	1.67
Total cost, $/kW for an LMTD = 100	1.07	8.02	50.30	18.70	3.97
Total cost, $/ft²	26.00	320.00	21.50	6.22	22.20
Surface area (external), ft²	9,050	1,017	14,900	122,000	99,300
Volume of shell, ft³				16,300	6,900

[a]Courtesy Oak Ridge National Laboratory.
[b]First utility pressurized water reactor (near Pittsburgh).
[c]Aircraft reactor test (Oak Ridge National Laboratory; components built and tested, but not reactor).
[d]Cost in 1960 dollars.
[e]Cost estimated by scaling from data for Calder Hall unit.
[f]Complete with headers, casing, etc.

small size can be important advantages, even in stationary power plants, for size and weight affect both the cost and the ease of replacement. For example, if a radioactive heat exchanger in a nuclear plant must be replaced and the activity level is such that shielding is required, a compact unit can be handled much more readily than a bulky unit. Note that two of the cases chosen for detailed treatment in this chapter, while small, represent two of the highest power density heat exchangers ever built, thus emphasizing the exceptional degree of compactness possible with liquid metals.

APPLICATIONS

Design and development work on heat exchangers for use with alkali metals began in the latter 1940s with the

Submarine Intermediate Reactor (SIR) program which used sodium and NaK to couple an intermediate neutron spectrum reactor to a steam power plant. (NaK is a term used for any of a series of low-melting-point sodium-potassium alloys.) This was soon followed by similar work on the Experimental Fast Breeder Reactor (EBR) and the Aircraft Nuclear Propulsion (ANP) programs. These programs pioneered the basic technology for the design, construction, and operation of alkali metal systems, including component development work on heat exchangers, pumps, valves, and instrumentation. The SIR and ANP programs were phased out in the latter 1950s, largely because of the success of pressurized water reactors in submarines and of ballistic missiles in replacing bomber aircraft. The work on the Liquid Metal Fast Breeder Reactor (LMFBR) has continued in the United States, but has been pursued much more aggressively in the USSR, England, France, Germany, and Japan, so that the latter countries dominate the field at the time of writing.

When the space program was initiated in the latter 1950s, one of the projects was the development of Space Nuclear Auxiliary Power (SNAP) units. Inasmuch as any thermodynamic cycle would have to reject its waste heat to space by thermal radiation, keeping the radiator size to an acceptable level became a major consideration and required the use of a high-temperature working fluid. A major effort was directed toward a sodium-cooled reactor that supplied heat to a mercury boiler, turbine, and condenser system. However, mercury corrosion problems limited the peak temperature in Fe-Cr-Ni alloys to about $500°C$ ($932°F$), a temperature that by 1970 proved to be too low to give an attractive power plant. In 1959 the author suggested the use of potassium as the thermodynamic cycle working fluid, and in 1960 design and development of hardware for the potassium vapor cycle was initiated. Boilers, turbines, condensers, and pumps were built and operated in reduced-scale systems with peak temperatures ranging from 810 to $1210°C$ (1490 to $2210°F$),[11] but funding for the programs became small and irregular after cutbacks in this phase of the space program began in 1966.

In the 1970s, efforts to conserve energy by increasing the efficiency of utility plants through an increase in the peak temperature of the thermodynamic cycle included a small effort on a potassium vapor topping cycle rejecting heat to a conventional steam cycle to make possible an overall thermal efficiency of over 50%. This system is adaptable to both fossil fuels and fission reactor plants[12,13] and has special advantages for use with fusion reactors if the plasma physics problems can be solved in a way that will provide an economically attractive plant for electric utility services.[14]

A tremendous amount of money and engineering effort has gone into the above programs. The resulting experience has shown that, from the heat exchanger design standpoint, the heat transfer and fluid-flow problems are little different from those for more conventional heat exchangers except for the exceptionally high heat transfer coefficients for the liquid metals. However, in these applications the penalties for leaks are far higher, and the thermal stress problems far more severe than in any previous experience. It is now clear that a much higher level of professional competence and a much broader technical background are required to design components for high-temperature systems than is the case for conventional heat exchangers. This applies to every phase of the design, development, fabrication, and operation if the units produced are to have both a reasonable capital cost and the requisite long life with the extremely high reliability that is essential for all of the applications mentioned in the above discussion. Thus the emphasis in this chapter is not on the heat transfer and fluid-flow problems themselves, but on factors affecting reliability. One of many indications that this emphasis is appropriate is given by Reference 15, which states that the heat exchangers were the principal cause of plant shutdowns in the first five years of operation of the Phenix (the first French LMFBR). As an engineer with 35 years of experience in this field put it, "Liquid metal systems are singularly unforgiving of any small errors in design, fabrication, or operation." Yet going to an unimaginative "conservative" design is likely to yield equipment that is so expensive that the system becomes unattractive, a key factor in the cancellation of the U.S. CRBR and the virtual demise of the U.S. LMFBR program.

Case histories are needed to illustrate not only the heat transfer and fluid-flow design of heat exchangers for liquid metals and molten salts but also the vital yet subtle and elusive problems that arise in their development. The author was able to find only a few cases in which a comprehensive picture of the whole complex of problems is readily available in the open literature.[16–18] Although these were small test units, their development program logged over 60,000 h of performance and endurance tests, including extensive, deliberately severe thermal cycling.[18] The exceptionally high-performance requirements these units were designed to meet brings into sharp focus many problems that might not be recognized in designs in which the conflict between the various requirements is less acute. These cases, together with several others much less completely documented, were chosen to provide instructive examples of design oversights that are so difficult to avoid. If one considers the amount of trouble that has been experienced with

PWR steam generators (see Chapts. 10 and 15), units whose design entailed far less extrapolation of previous engineering experience, one can appreciate the appropriateness of the emphasis in this chapter on design problems that were recognized only after failures were encountered in operation.

HEAT EXCHANGERS FOR TEST FACILITIES

The heat transfer equipment required for component development test stands presents some interesting problems in the design of high-temperature heat exchangers, since a high degree of dependability must be obtained with essentially no development effort. A typical case is the potassium boiler shown in Fig. 17.1, which was used in a test facility for potassium vapor turbines for space power plants. At the conditions for which this unit was designed, the heat of vaporization and the vapor pressure of potassium, together with its density and viscosity in both the liquid and vapor states, are much the same as the corresponding properties of water at about 120°C (250°F). Thus the general fluid-flow behavior should be much the same as it would be for

water, except for possible difficulties with bubble nucleation (see Table 5.1).

The boiler illustrated in Fig. 17.1 was designed to operate in a gas-fired furnace with the tube axes vertical and the large header drum at the top. Liquid potassium entered the 1-in. diameter tubes at the bottom, boiling took place in the tubes, and the vapor-liquid mixture flowed upward into the header drum at the top, where the liquid and vapor were separated. The four 5-in. diameter pipes served as downcomers to return the liquid to the lower header drum. To avoid local hot spots the gas flames were directed against the furnace walls so that they could not impinge on the tubes, and the bulk of the heat transfer to the tubes was by thermal radiation from the furnace walls. Differential thermal expansion between the tubes and the downcomers was accommodated by column buckling of the tubes in compression.

Figure 17.2 shows the potassium condenser for the same potassium vapor turbine test stand as the boiler of Fig. 17.1. The vapor entered the unit through the central vertical pipe at the top, flowed laterally through the transverse duct at the center into the top header drums at either side, and then downward into the tubes. These tubes were relatively large in diameter (about 2 in.) because of the low pressure and density of the vapor, that is, about that of the steam leaving the last turbine stage in a conventional steam plant. The condensate drained from the tubes into the large central header drum at the bottom. The unit was air-cooled, and the tubes had fins of the type shown in Fig. 2.7g. The tubes were bent to accommodate differential thermal expansion between them. The unit was supported by the large, water-cooled, central, horizontal pipe fitted with mounting feet at either end.

No difficulties were experienced with corrosion or mass transfer in some 12,000 h of operation, demonstrating that excellent quality control was exercised in the fabrication, erection, and operation of the system. There were no problems with the condenser, but the boiler presented a number of problems that were completely unforeseen. These stemmed from the fact that no nucleation sites (see Chapt. 5) were incorporated in the design because this new requirement peculiar to liquid metal boilers was not recognized during the design stage. As a consequence, when the boiler was started up, boiling was sometimes highly irregular because the liquid in some of the boiler tubes would superheat by large amounts before nucleation occurred; local explosive boiling followed. This overloaded the vapor separator and sent slugs of liquid into the turbine, causing some erosion of the turbine blades. In addition, the irregular and local character of the liquid superheating led to

Figure 17.1 Potassium boiler designed for use in a gas-fired furnace. The unit was of all-welded stainless steel construction and was designed for a heat input rate of about 3000 kWt (10,000,000 Btu/h) at about 815°C (1500°F). (Courtesy Struthers Wells Corp.)

Figure 17.2 Potassium condenser designed for use in the same potassium vapor turbine test stand as the boiler of Fig. 17.1. (Courtesy Struthers Wells Corp.)

temperature differences of as much as 150°C (302°F) between adjacent tubes; the resulting differential expansion was accommodated by column buckling of the hot tubes. This led to gross distortion of many of the tubes, severe thermal strain cycling, and eventually the failure of six tubes at one time or another in the course of the turbine test program. In some cases, leakage of potassium from a cracked tube led to a truly unpleasant mess.

It was most unfortunate that the boiler designers were not aware that a bit earlier the boiling nucleation problem had given months of trouble in a small boiling potassium system designed to investigate possible corrosion problems at the Oak Ridge National Laboratory. In that case it was finally diagnosed correctly, several different solutions were devised, and their efficacy demonstrated.[19] Unfortunately, as so often happens, the information didn't reach some who needed it; this illustrates one of the reasons for the selection of the information presented in this chapter. Interestingly, poor nucleation apparently had occurred in mercury boilers, but was not recognized. It had not caused serious trouble, partly because the poor wetting characteristics of mercury favored nucleation, and partly because in the mercury vapor topping cycle the combination of the reentry tube employed in the boiler design and the relatively low heat flux happened to be such that no serious trouble was encountered. Difficulties with explosive boiling sometimes caused slugging in the once-through mercury boiler designs used in the space program, apparently when they succeeded in getting the system so clean that the mercury was wetting the tube wall.

A COMPACT MOLTEN SALT-TO-NaK HEAT EXCHANGER

In the course of one phase of the aircraft reactor development program an effort was made to design, construct, and test the most compact, lightweight heat exchangers that could be built to give good reliability at a reasonable cost.[16,20,21] The power plant contemplated was based on a circulating sodium-zirconium-uranium fluoride fuel reactor designed to deliver 60 MWt with a peak temperature of 871°C (1600°F), a bright red heat. A primary heat exchanger within the reactor shield was to transmit the heat from the molten fluoride fuel to NaK in the secondary fluid circuit. The NaK was to be circulated through radiators, which were to reject the heat to an air-stream passing through a turbojet or a turboprop engine.

One of the overall objectives of the program was to

obtain the highest possible air temperature. At the same time, the pumping power and fluid system pressure drops in both the primary and the secondary fluid circuits had to be kept to reasonable levels. Promising system designs called for a liquid-to-liquid heat exchanger that would yield a temperature difference between the fuel salt and NaK circuits of 28 to 56°C (50 to 100°F) with fluid stream temperature rises of 167 to 222°C (300 to 400°F). These requirements made a counterflow type of heat exchanger essential.

Fabrication Techniques

The design of the heat exchanger was conditioned by the fabrication techniques available, particularly those for connecting the tubes to the header sheets. The two most promising processes appeared to be heliarc welding and furnace brazing with iron-chrome-nickel-silicon-boron alloys that melt around 1100°C (2000°F). As discussed in Chapter 2, a very good dry hydrogen atmosphere is necessary for the brazing process. Some units were fabricated using one process, some the other; and a few were first welded and then back-brazed. Extensive endurance tests were run to see which of these techniques was the best. It was found that imperfections were sufficiently common with either welding or brazing alone to cause occasional leaks.

One of the most important design problems is associated with the stress concentration at the root of a weld in the tube sheet. Figure 2.5 shows a photomicrograph through such a weld, and clearly indicates the severe stress concentration at the end of the crack terminating at the weld. While the effects of this stress concentration can be reduced by rolling the tube into the tube sheet, this is not readily accomplished with small-diameter tubes. The residual compression stress induced in the tube wall by such a rolling operation tends to relax and anneal out at high temperature, particularly under the thermal transients associated with abrupt changes in the temperature of the fluid passing through the tubes. Thus there is a strong incentive to braze the tube into the tube sheet. When this is done, a good metallurgical bond can be obtained between the tube and tube sheet all the way through. It has been found that if joints are first welded and then back-brazed, a very high degree of joint integrity can be obtained. In fact, approximately 7000 welded and back-brazed tube-to-header joints were endurance-tested without a single instance of a leak through such a joint.[18,21]

Development of a Design

The unit had to be shielded, and it was extremely important to minimize the shield weight. This meant that it

was essential to obtain the maximum possible output per cubic foot. Since the strength of the best iron-chrome-nickel alloys available for the heat exchanger structure falls off rapidly at temperatures above 650°C (1200°F), the fluid pressure drops were limited by stresses in the heat exchanger shell and in the walls separating the two fluid circuits. Although an effort can be made to balance the pressures, the pressure difference between the two fluid circuits is inherently high at one end or the other in a counterflow heat exchanger if the pressure drop in both fluid streams is high. This situation imposed limitations on the fluid velocities.

One way of obtaining an exceptionally compact liquid-to-liquid heat exchanger is to make use of as much heat-transfer surface area per cubic foot as possible. This implies the use of small-diameter, closely spaced tubes, hence Fig. 17.3 illustrates the effects of tube diameter on the power density obtainable with a given temperature difference. The advantages of small tube diameter and

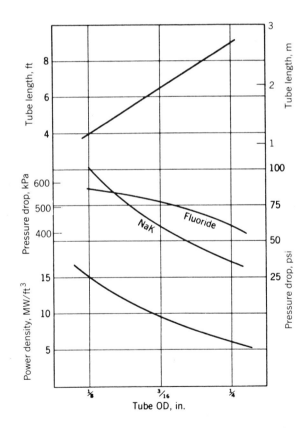

Figure 17.3 Effects of tube diameter on the power density obtainable with a counterflow tubular heat transfer matrix for exchanging heat between a molten fluoride salt and NaK. The shell-side flow is axial between Inconel tubes having a constant tube wall thickness of 0.50 mm, or 0.020 in. (Courtesy Oak Ridge National Laboratory.)

close tube spacing are accentuated if the heat exchanger is designed for use with liquid metals;[22] their high thermal conductivity leads to extremely high heat transfer coefficients, particularly for small-diameter passages. This raised a question regarding the minimum practicable tube diameter. In most of the conventional heat exchangers, experience shows that fouling of the tube walls and tube plugging are likely to make it impractical to use tubes having a diameter less than 12 mm (1/2 in.). However, liquid alkali metal systems can be kept so clean that fouling is not a problem.

Other factors of consequence in selecting a tube diameter include the effects on both cost and reliability of the increased number of tube-to-header welds required as the tube diameter is decreased, and the cost of the tubing. In tube diameters under about 19 mm (3/4 in.) the cost of tubing tends to be more a function of the total footage than of the weight.

Overall power plant performance considerations led to the conclusion that a tube diameter in the range of 6 to 12 mm (1/8 to 1/4 in.) was necessary to obtain the performance desired. With such a small diameter, and the very close tube spacing that it entailed (0.5 mm, or 0.020 in. between tubes), it was recognized that it would be difficult indeed to provide accurate tube spacing, adequate header-sheet ligament thickness, and acceptable pressure drops in the crossflow region on the shell side at the inlet and exit of the heat exchanger. One way of coping with these problems was to bend the tubes in the region near the header sheet in such a way as to spread them out so that the header-sheet area was about three times the cross-sectional area of the tube bundle. Tubes in alternate rows were offset to bring them in line with those in the intermediate rows, and thus open lanes for flow between the tubes in the crossflow region at the ends. In pursuing this approach the model shown in Fig. 17.4 was constructed and tested for pressure drop.

Analyses disclosed that the configuration of Fig. 17.4 made it possible to increase the power density in the heat exchanger by roughly a factor of four over that for a multipass, crossflow unit. This improvement accrued partially from the close tube spacing within the heat transfer matrix and partially from the elimination of the plenum regions required on either side of a multipass crossflow tube bundle.

Tube spacing in a unit of the type shown in Fig. 17.4 can be accomplished by any one of several devices. Helical wires can be wound around each tube and the tubes installed on an equilateral triangular pitch. This arrangement has several disadvantages, one of the greatest being the difficulties in attaching the wires to the tubes. Some thought was given to obtaining tubes with

Figure 17.4 Air flow test model of the header region for a compact sodium-to-NaK heat exchanger with flattened wire to maintain the proper spacing for axial flow between the tubes. The flattened wire spacers were trimmed to length after the picture was taken. (Courtesy Oak Ridge National Laboratory.)

extruded ridges or fins, but it appeared that the tube quality would suffer if this approach were employed. Flattened wire spacers can be placed between the tubes, as shown in Figs. 17.4, 17.5, and 2.18. Horizontal spacers were placed in one plane and the vertical spacers in another to reduce the obstruction to flow. Calculations of lateral tube deflections as well as shop experience indicated that the spacer "combs" should be installed at intervals of about 40 tube diameters along the length of the tube. The obstruction of the flow-passage area can be kept to a sufficiently low value so that the pressure drop is acceptable, that is, roughly 50% greater than the pressure drop associated with the flow for the ideal system with no spacers.[23]

The header layout posed a number of important questions. Experience with welding tubes into header sheets indicated that, for the small-diameter, thin-walled tubes under consideration, the minimum ligament thickness should not be less than 2.5 mm (0.10 in.). This thin ligament posed some difficult stress problems in the design of the header sheet to accommodate the pressure differentials that might prevail in the course of ordinary operation or under off-design conditions. It is often possible to design a power plant so the pressure differential across the header sheet at the high-temperature end of the heat exchanger will be normally small, allowing the stresses in the "hot" header sheet to be kept at acceptable values in spite of the low allowable stresses in that temperature range. This will make the pressure differential and the resulting stresses substantially higher in the "cold" header sheet, but the allowable stresses also will be higher because of the lower temperature. However, both header sheets must be designed to accommodate such off-design conditions as the abrupt

stoppage of a pump in either circuit or an inadvertent mishandling of controls which would impose pressure differentials much higher than normal. Stress analyses indicated that the header-sheet thickness would have to be from five to eight times greater for flat header sheets than for cylindrical header drums in which bending stress would be avoided. Furthermore, the gradual creeping and deflection of flat header sheets under the pressure differential across the sheet would lead to flexing of the tubes, which in turn would tend to give difficulty with tube cracking in the highly stressed zone close to the header sheet. There can be no such difficulty with cylindrical drum headers because any creep deformation of the cylindrical drums does not change the geometry of the drum and tube configuration.

On the basis of the preceding conclusions derived from the preliminary design studies, the first unit was designed, built, and tested with 3-mm (1/8-in.) OD tubes. This test showed no appreciable increase in fluid pressure drop during the course of a 3000-h test, and thus demonstrated the practicality of making use of tubes as little as 3 mm (1/8-in.) in diameter in sodium and NaK circuits.

Precepts for Detailed Design

After demonstrating the practicality of the basic tube matrix geometry indicated in Fig. 17.4, the next step was to evolve a detailed design for a unit to be installed in a reactor system. In the course of the work that led to the design of the unit shown in Fig. 2.18 it became apparent that the designer's problem was difficult indeed because a whole series of conditions had to be satisfied. Subsequent experience shows that these precepts have widespread application; hence a brief summary is in order. For the design requirements given:

1. The tube diameter should be no greater than 6.3 mm, or 0.25 in.

2. The tubes should be grouped into bundles, each having a header drum and outlet pipe at either end designed to fit within the pressure vessel. This facilitates assembly and inspection and minimizes the number of pressure vessel penetrations.

3. Normal differential thermal expansion between the tubes and the shell should be accommodated elastically.

4. Ligaments between tubes in the tube sheet should be at least 2.5 mm, or 0.1 in. thick for good welding.

5. Where bent tubes are employed to open up the tube spacing in the header sheet, the length of the

¼ in. tubes, 0.025 in. spacers

Section through
horizontal spacers

Section through
vertical spacers

Figure 17.5 Flattened wire spacers for a tube matrix similar to that of Fig. 17.4. (Courtesy Oak Ridge National Laboratory.)

moment arm between the point where the tube enters the tube sheet and the centerline of the main section of the tube should be limited to avoid excessive bending stresses from the fluid drag forces on the tubes.

6. The tube sheets should be curved in order to hold pressure stresses to reasonable values.

7. Adequate flow-passage area into and out of the tube matrix should be provided for the shell-side fluid.

8. Adequate flow-passage area should be provided in the header drums.

9. The shell walls should be curved to hold pressure stresses to reasonable values.

10. The shell walls should not be too thick if the thermal stresses are to be kept reasonably low during temperature transients.

11. Thermal sleeves should be provided where the tube-side fluid pipes penetrate the heat exchanger shell.

12. Flow bypass between the tube matrix and the shell should be less than 10%, and preferably less than 5%.

13. Adequate provision for spacing the tubes should be made to assure good velocity and temperature distributions.

14. Steep temperature gradients in the shell should be avoided, as, for example, at the parting plane between the cold and hot legs of a U-type of heat exchanger.

Selection of a Tube Configuration

Attempts to satisfy these conditions in layouts showed the three most difficult requirements to be met were provision for differential thermal expansion, provision for adequate ligaments between the tubes in the tube sheet, and provision for the loads imposed on the tubes by the fluid drag forces on the main runs of tubing. Six types of heat exchanger devised to accommodate these three basic requirements are shown in Fig. 17.6. The first type, the "hockey stick," was employed by the Knolls Atomic Power Laboratory in heat exchangers for the SIR, the Submarine Intermediate Reactor nuclear power plant

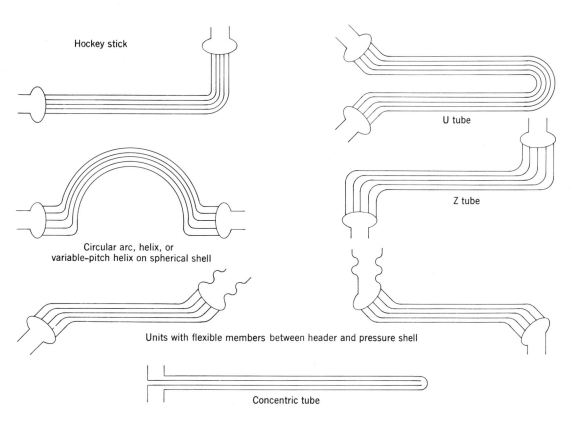

Figure 17.6   Tube and header configurations for heat exchangers designed for axial flow between closely spaced tubes. (Courtesy Oak Ridge National Laboratory.)

built for the Sea Wolf in the early 1950s. The U-tube type of heat exchanger has been used in boilers for pressurized-water reactor plants. Variations of the circular-arc heat exchanger include helical spirals. In any of these variations, differential thermal expansion between the tubes and the pressure shell can be absorbed by providing adequate clearance between the tube bundle and the shell so that the increased length of tube can be accommodated by radial expansion. A fourth type of unit, the Z-tube, was designed to absorb the differential expansion in bending of short lateral sections of tube at the headers. A fifth type incorporates a flexible member between one header drum and the pressure shell. Such a member could have a form similar to that of a sylphon bellows, which could be designed to deform either in compression or in bending. The sixth type, the concentric tube, has been used in mercury boilers for power plants. The fluid enters through a central tube and returns through an annulus between it and an outer thimble. This type of heat exchanger reduces both thermal and drag stresses to an absolute minimum, but a disadvantage is that the spacing both between tubes in the header sheet and in the tube matrix must be rather large, and hence the power density obtainable is lower than in the other types.

A review of many different layouts for the reactor-heat exchanger-shield assembly indicated that the hockey-stick configurations did not appear attractive. A variation of the circular-arc tube bundle gave the lowest system weight. Variations of the Z-tube arrangement of Fig. 17.6 gave a somewhat heavier system, but appeared to be easier to fabricate. However, it was also clear that differential thermal expansion between the tubes and the casing in the Z-tube bundle would make it particularly subject to thermal stress cracking in the short tube section projecting from the header sheet. To facilitate fabrication, the Z-type was chosen for use in the preliminary component tests because it gave one of the most severe sets of thermal strain cycling conditions arising from differential thermal expansion and thus provided an excellent means for evaluating the ability of a typical high-temperature structure to resist this sort of thermal strain cycling. It was believed that thermal strain cycling would prove to be the most likely cause of failure in high-temperature liquid systems, and this did indeed turn out to be the case. The sylphon bellows header outlet arrangements were found to be unsatisfactory in that it proved impossible to make adequate provision for both pressure stresses and differential thermal expansion.

Example 17.1. Design Calculations. Since a wealth of test data are available for small molten salt-to-NaK

heat exchanger test units similar to the Z-tube unit of Fig. 17.7a, detailed design calculations are presented in Table 17.2 for such a unit.[17] It must be remembered that the circular-arc configuration accommodates free differential expansion, whereas the Z-tube induces severe bending strains in the short tube legs going into the header sheets, and this may lead to tube failures from thermal strain cycling. This problem is discussed in some detail in Chapter 9. Figure 17.7b has been included to emphasize this problem and to show that tube failures of this sort actually do occur in the vicinity of the header. No difficulty of this sort was experienced with circular-arc tube bundles subjected to much more severe temperature cycling tests.

The first step in preparing a detailed design for a test unit was to estimate the heat transfer coefficient from the physical property data available. The characteristics of a whole series of heat exchangers having tube diameters of 4.5 to 7.5 mm (3/16- to 5/16-in.) OD were then calculated for a series of tube lengths and tube spacings.

This study showed that there was no way to maintain a Reynolds number well above 2000 for part-load conditions and yet avoid excessive pressure drops in either the fluoride or the NaK circuits at the full design power output. It was believed that the flattened wire spacers between the tubes would induce sufficient turbulence to avoid a serious loss in heat transfer coefficient for the molten fluoride at Reynolds numbers in the 1000 to 6000 range. Clearly, tests were essential. It was decided that a 6-mm (0.25-in.) diameter tube represented the best design compromise. The set of design calculations indicated in Table 17.2 was prepared for a test unit, and several units were built. These were Z-tube units similar to those in Figs. 17.6 and 17.7, with tube spacers similar to those in Figs. 17.4 and 17.5. To gain fabrication and endurance test experience, other units differing somewhat in size, tube spacer configuration, etc., were also built and tested.

Performance Test Results

Since data from a single test is always open to question, heat transfer coefficeints were calculated and plotted in Fig. 17.8 from data obtained during performance and endurance testing of twelve different heat exchangers of the design indicated in Fig. 17.7 and Table 17.2. These data were obtained from runs on six different test stands. Note that the test data indicate a heat transfer coefficient over 30% below that indicated by the Dittus-Boelter correlation. Some insight into the reasons for this is given by Fig. 17.9, which shows the heat transfer data obtained

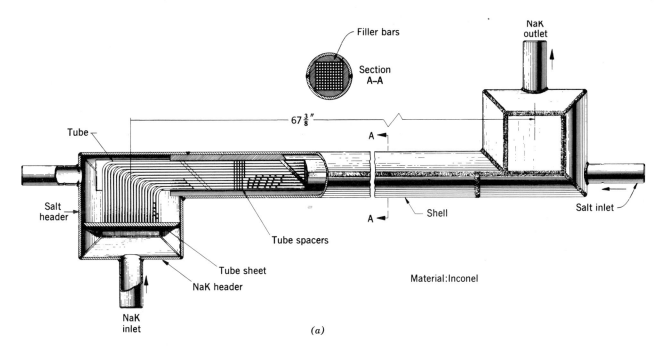

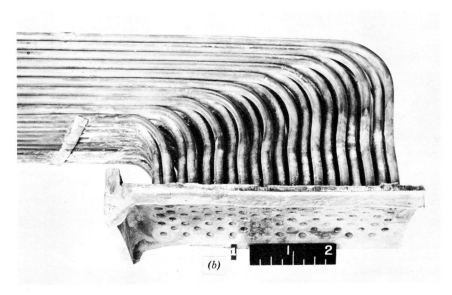

Figure 17.7 Salt-to-Nak heat exchanger test unit: (*a*) layout drawing; (*b*) photograph showing tube deformation in the hot header region after a severe thermal strain cycling test. (Courtesy Oak Ridge National Laboratory.)

with fluoride flowing through a single, small, electrically heated tube. In examining Fig. 17.9 it appears that, even at Reynolds numbers above 6000, under the ideal conditions in a single round tube the fluoride tends to give a heat transfer coefficient of about 12% less than would be expected from the Dittus-Boelter equation. No satisfactory explanation for this has been found. The rapid drop in heat transfer coefficient in the Reynolds number region from 6000 down to 1500 is consistent with the effects noted for other fluids in this transition region from laminar to fully developed turbulent flow, as is the essentially constant heat transfer coefficient indicated by the data for the laminar flow region (below Re = 1500). Both of these effects are discussed in Chapter 3.

TABLE 17.2 Summary of Performance Calculations for the Molten Salt-to-NaK Heat Exchanger Test Unit of Fig. 17.7a

Line Number	Item	Source[a]	Values	
Dimensional Data (Given):				
1.	Tube OD, in.	Given	0.230	
2.	Tube ID, in.	Given	0.180	
3.	Tube centerline spacing, in.	Assumed	0.25	
4.	Tube pitch	Given	Square	
5.	Shell-side equivalent diameter, in.	Table H6.1	0.069	
6.	Mean tube length, ft	Assumed	6.0	
7.	Number of tubes	Given	100	
8.	Heat load, Btu/h	Given	5.12×10^6	
Performance				
			Fuel no. 30	NaK
9.	Inlet temperature, °F	Fuel given, NaK from ㉒	1,600	1,496
10.	Outlet temperature, °F	Fuel given, NaK from ㉒	1,250	1,146
11.	Specific heat, Btu/lb · °F	Table H2.3	0.251	0.25
12.	Density lb/ft³	Table H2.3	201	45.0
13.	Weight flow, lb/h	⑧/350 × ⑪	58,400	58,500
14.	Weight flow, lb/s	⑬/3,600	16.2	16.2
15.	Flow-passage area, in.²	(③ × √⑦)² × factor from Fig. H6.2	2.1	2.55
16.	Flow rate, lb/s · ft²	⑭ × 144/⑮	1,110	915
17.	Re	Fig. H3.1	2,070	120,000
18.	Heat transfer coefficient, Btu/h · ft² · °F	Figs. H5.3, H5.15, and 17.8	2,230	13,000
19.	Surface area, ft²	⑥ × 100 × πD/12	36.0	29.2
20.	Film temperature drop, °F	⑧/⑱ × ⑲	63.8	14.0
21.	Wall temperature drop, °F	0.025 × ⑧/32.1 × k_w × 12	26.6	
22.	LMTD, °F	⑳ + ⑳ + ㉑	104	
23.	Dynamic head, psi	Fig. H3.2	0.66	2.0
24.	Friction factor	Fig. H3.4	0.040	0.017
25.	Passage length-diameter ratio	6 × 12/② or ⑤	1,040	400
26.	Pressure drop, psi	㉓ × ㉔ × ㉕	27.5	13.6

[a]Circled numbers are symbols to indicate the line from which to obtain the quantity to be used in the operation.

The most striking feature of Figs. 17.8 and 17.9 is that the data for flow outside the tubes define a line with the same slope as that for the ideal Dittus-Boelter correlation down to a Reynolds number of 700, yet falls below the fluoride data for round tubes at high Reynolds numbers and above it at low Reynolds numbers. In searching for reasons for this disparity it was decided that the fluoride flowing through the space between the tubes in a square pitch array probably tends to stratify; that is, there is poor mixing between the fluid flowing in the narrow ligaments between the tubes and that in the relatively open channel at the center of the roughly square channel bounded by the four surrounding tubes. This flow stratification, or "channeling," tends to reduce the effective temperature difference between the tube wall and the fluoride stream close to the wall. Thus a heat transfer coefficient calculated from test data using the temperature difference between the bulk free stream and the wall tends to be lower than for a circular channel at Reynolds numbers above 5000.

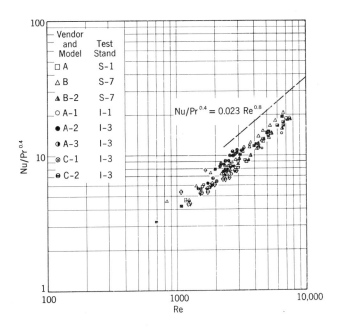

Figure 17.8 The heat transfer characteristics of a molten salt flowing on the shell side of 12 different Z-tube heat exchangers tested in six different systems. (Courtesy Oak Ridge National Laboratory.)

Another point of interest is that the data of Fig. 17.8 show clearly that the flattened wire spacers between tubes are effective in maintaining sufficient turbulence so that the slope of the curve for the heat transfer coefficient as a function of the Reynolds number is that characteristic of turbulent flow down to Reynolds numbers below 700. Tests with water showed a similar loss in heat transfer performance between flow inside round tubes and flow between tubes in the tube bundle at the higher Reynolds numbers, and the same slope for the heat transfer coefficient versus Reynolds number curve for flow between tubes down to a Reynolds number of 700.[17]

The foregoing deductions about the influence of the flattened wire spacers on flow conditions between the tubes are borne out by the data of Fig. 17.10 for the friction factor for the regions between spacers plotted as a function of Reynolds number for the range from about 300 to about 6000. The deviation of the experimental points from the ideal curves probably indicates stratification of the flow into both laminar and turbulent regions, with the laminar flow region in the thin ligaments between tubes gradually increasing in extent

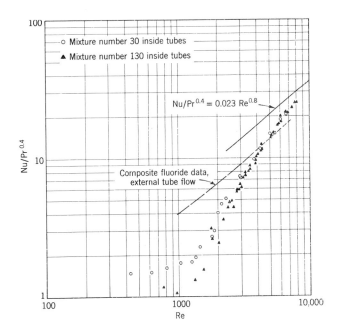

Figure 17.9 The heat transfer characteristics of the molten salt used in the tests of Fig. 17.8 when flowing inside of round tubes. (Courtesy Oak Ridge National Laboratory.)

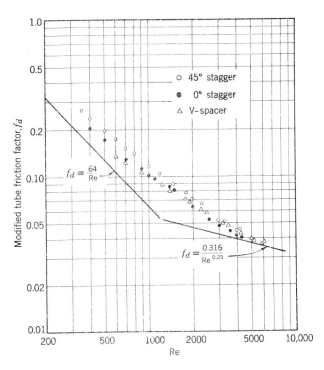

Figure 17.10 Friction factor determined for axial flow outside tubes in a 25-tube bundle with flattened wire spacers similar to those of Fig. 17.5. (Courtesy Oak Ridge National Laboratory.)

as the Reynolds number is reduced until the entire flow-passage cross section is occupied by laminar flow.

Quite a number of different flattened wire spacer configurations similar to those of Figs. 17.4 and 17.5 were considered and tested. The pressure losses arising from skin friction on the walls of the tubes were calculated and subtracted from the total pressure drop to give the pressure losses across the spacers alone, and pressure-loss coefficients for various spacer configurations were obtained and plotted in Fig. 17.11. The spacer configurations referred to in Fig. 17.11 were identified by their appearance as viewed from the ends of the spacers. In the "flat" array, the spacer wires all lay in a plane normal to the axes of the tubes. In the "staggered" array, the spacer wire axes were normal to the tubes, but the plane of the spacer array was inclined relative to the tubes. The "inclined" spacers were placed with their axes inclined relative to the tubes, but their ends lay in lines perpendicular to the tube axes. In the V-spacer array the wires were staggered from the center of the bundle so that the pattern at the ends of the wires formed a V. From the data in Fig. 17.11 it was decided that the most promising configuration was that given by alternating 45° staggered

spacers with 45° inclined spacers at intervals of 9 in. in the direction of flow.

NaK-TO-AIR-RADIATOR

The approach to the design of a NaK-to-air radiator for the aircraft reactor test system closely paralleled that for the salt-to-NaK heat exchanger presented in the preceding section. The problems peculiar to the NaK-to-air radiator stem partially from the much larger temperature difference between the two fluids, especially at the air-inlet face, and partially from the large difference in heat transfer coefficients which calls for an extended surface on the air side. A performance comparison of the many types of heat transfer matrix that could have been used was made, but is too complex to present here. A wide range of tube diameters, tube spacings, and fin spacings was considered and estimates of the performance in each case were prepared. The principal figures of merit were the weight, volume, number of tube-to-header joints, and the NaK and air-pressure drops required to give the desired heating effectiveness for a given set of NaK and airflow rate conditions. Suffice it to say that of the four principal heat transfer matrix geometries considered, the round tube-and-plate fin configuration of Fig. 17.12 was chosen, because it seemed to have the best overall characteristics, and, in any given area (i.e., heat transfer performance, fabrica-

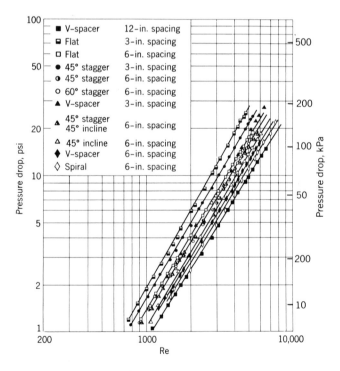

Figure 17.11 Effects of 11 different configurations of flattened wire spacers on the pressure drop for axial flow through a tube matrix similar to that of Fig. 17.5. (Courtesy Oak Ridge National Laboratory.)

Figure 17.12 Round tube-and-plate fin, NaK-to-air radiator used in development tests. The fin matrix was 8 × 8 × 16 in., the tube OD was ³⁄₁₆ in., the fin thickness 0.010 in., and the fin spacing 15/in. (Courtesy Oak Ridge National Laboratory.)

bility, weight, or ability to withstand thermal stresses), it appeared to be nearly as good as the configuration that seemed to be the best in that area.

A major factor contributing to the good performance of this geometry was the successful development of a new fin material. To obtain good thermal conductivity together with good oxidation resistance and strength at high temperatures, a type-310 stainless-steel-clad copper sheet stock was developed.[21] The overall thickness was 0.010 in. (0.25 mm), of which the copper core thickness was 0.006 in. (0.15 mm), while the thickness of each of the two layers of stainless steel cladding was 0.002 in. (0.05 mm). Another contributing factor was a clever design (proposed by a tool-maker in the shop) for the return bend region which gave a good approximation to pure counterflow (see Fig. 17.15).

Example 17.2. Design Calculations. The dimensional and performance data calculated for one of these test units is presented in Table 17.3. For the purposes of

TABLE 17.3 Summary of Performance Calculations for the NaK-to-Air Heat Exchanger Test Unit of Fig. 17.12

Line Number	Item	Source[a]	Values	
Dimensional Data (Given):				
1.	Tube OD, in.	Given	0.188	
2.	Tube ID, in.	Given	0.138	
3.	Tube centerline spacing, in.	Given	2/3 × 2/3	
4.	Tube pitch	Given	Square	
5.	Fin thickness, in.	Given	0.006 Cu + 0.004 SS	
6.	Fin spacing, fins per in.	Given	15.0	
7.	Inlet face size, in	Given	8 × 16	
8.	Air-passage length, in	Given	8	
9.	Number of tubes	6 × 12	72.0	
10.	Tube length, in.	2 × 16 + 10	42.0	
11.	Air-side surface area, ft²	⑥ × 16 × 2 × 64/144	200	
12.	Heat load, Btu/h	Given	2,250,000	
Performance			NaK	Air
13.	Inlet temperature, °F	㉓ given	1,475	100
14.	Outlet temperature, °F	Given	1,200	1,350
15.	Weight flow, lb/s	⑫ /(⑭ − ⑬) × 0.25 × 3,600; given	9.1	2.0
16.	Flow-passage area, ft²	② and ⑨; ①, ③, ⑤, ⑥, and ⑦	0.0075	0.52
17.	G′, lb/s · ft²	⑮ / ⑯	1,208	3.85
18.	d_e, in.	②, ⑤, and ⑥	0.128	0.114
19.	Re	Fig. H3.1	70,000	2,170
20.	h, Btu/h · ft² · °F	Figs. H5.5; H5.15; H5.9	16,000	29.0
21.	Fin efficiency, %	Figs. 3.1 and H7.2		86.0
22.	LMTD, °F	⑫ / ⑳ × ㉑ × ⑪	450	
23.	LTD, °F	Fig. H4.1	125	
24.	f_d	Figs. H3.4 and H5.11	0.020 0.044 (fins) 0.20 (tubes)	
25.	Dynamic head, psi	Figs. H3.2 and H3.3	3.2	0.043
26.	Pressure drop, psi	㉔ × ㉕ × l/d_e	19.5	0.24

[a]Circled numbers are symbols to indicate the line from which to obtain the quantity to be used in the operation.

this calculation the air inlet and outlet temperatures, the air flow, and the NaK outlet temperature were assumed and the NaK inlet temperature and flow rate were calculated. Since the controlling resistance is on the air side, as a first approximation the temperature drop through the wall and the NaK film can be neglected. Thus the first step was to calculate the air mass flow rate and the heat transfer coefficient on the air side using the air velocity through the constricted throat between the adjacent tubes. The physical properties at the mean metal temperature in the tube matrix were used rather than those at the mean air temperature (see Eq. 3.24). This approach overestimates the heat transfer coefficient since the average velocity over the plates is somewhat less than that at the restricted cross section between the tubes. On the other hand, the presence of the tubes does induce some additional turbulence which tends to increase the heat transfer coefficient. Because the fins were interrupted at 2-in. intervals in the direction of flow, a correction factor from Fig. H5.9 was applied to allow for the entrance effect. The fin efficiency was estimated from Fig. H7.2 for an equivalent circular fin having the same area as the square-fin element of the unit being designed. The LMTD and LTD were then calculated, and from these values the NaK inlet temperature and flow rate were determined, assuming pure counterflow conditions.

The air pressure drop was calculated by summing the pressure loss for flow over banks of bare tubes and for flow between parallel plates. The NaK pressure loss was estimated using the friction factor for flow through cold-drawn round tubes. The header losses were estimated by assuming the loss of one dynamic head at the tube outlets.

Performance Tests

The performance of several test units is shown in Figs. 17.13 and 17.14 for comparison with the performance estimates. Perhaps fortuitously, the measured heat transfer coefficient was within 10% of the estimate in Table 17.3. It is interesting to note that the slope of the curve for the heat transfer coefficient on the air side approximated the ideal value for turbulent flow even at Reynolds numbers down to 400. Note that the slope of the curve for the product of the local heat transfer coefficient and the fin efficiency is flatter than that for the primary surfaces, a feature characteristic of extended surfaces having a fin efficiency substantially less than 100%. The turbulating effect of the crossflow over the round tubes is also indicated by the pressure-drop data of Fig. 17.14, which shows a slope between that for turbulent and that for laminar flow.[15]

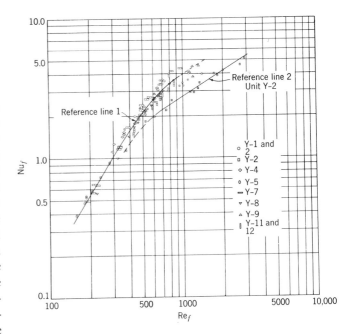

Figure 17.13 Heat transfer performance of 12 NaK-to-air radiators similar to that of Fig. 17.12. (Courtesy Oak Ridge National Laboratory.)

The measured air-pressure drop ran about 50% higher than the calculated value, probably indicating that it was a mistake to choose a value equivalent to that for smooth drawn brass tubing. Similarly, during the initial endur-

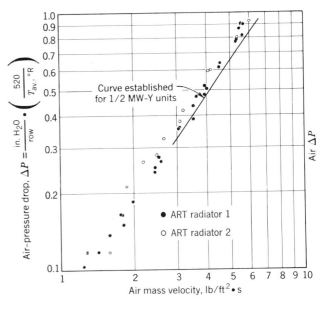

Figure 17.14 Overall air pressure drop through 12 NaK-to-air radiators similar to Fig. 17.12. (Courtesy Oak Ridge National Laboratory.)

ance test operations, the pressure drop on the NaK side was about double the estimated value. Furthermore it tended to increase, and in some instances even doubled after several months of operation. This effect apparently stemmed from a mass transfer phenomenon which entailed solution of the structural metal in the hot zone and its deposition in the form of tiny needlelike crystals on the tube walls in the radiator (which constituted the cold zone in the NaK system). Microscopic examination disclosed that these crystals were spaced at relatively wide intervals, but apparently they had a pronounced roughening effect on the walls, and hence led to much larger pressure drops even though there was no change in the flow-passage area for the bulk of the channel. Subsequent work has confirmed the inadvisability of using nickel base alloys such as Inconel at high temperatures in alkali metal systems. The stainless steels are much less susceptible to mass transfer.

Tube Failures in Endurance Tests

A series of radiators similar to the one illustrated in Fig. 17.12 were built and tested. As the test program progressed, tube failures occurred which could not be accounted for by any of the thermal strains envisioned in the preliminary stress analyses. Since this experience was immensely instructive, the whole matter of differential thermal expansion and thermal stress in the radiators deserves review. To begin with, the temperature distribution in a crossflow heat transfer matrix leads to an odd three-dimensional distortion and a complex stress distribution. Figure 17.15 shows the temperature distribution for a typical case. It must be remembered that the radiator matrix is three-dimensional, so that thermal expansion occurs not only in the air-flow direction, but also transverse to the air flow—both along the tubes and along the fins. As a result, it is necessary to allow for differential thermal expansion between strip or plate fins and the header sheet, between various tubes in the heat transfer matrix, and between such tubes and any support structure or baffles that may be required. Furthermore, provision must be made for changes in temperature distribution associated with temperature transients in the liquid circuit. Slugs of cold or hot fluid may pass through the system and these will be inclined to move through the heat transfer matrix with different transit times for different regions. Another factor that may lead to cracked tubes is that of plugging of passages on the high-temperature liquid side by foreign matter. The temperature in the vicinity of a plugged tube will be reduced drastically, and such a tube will tend to contract relative to the rest of the matrix. This actually occurred, and caused a

number of radiator failures. Another region that may be subjected to severe stresses lies in the vicinity of the tube-fin joint. The radial temperature gradient through the tube wall and along the fin tends to induce high stresses similar to those in a gun barrel. In view of the irregularities introduced by the fins, and most particularly of irregularities introduced by the nonuniformity of the braze fillet at the fin root, there will also be axial temperature gradients and hence local bending and shear stresses introduced along the length of the tube.

In laying out the test program for the later units, it was decided that the most important source of trouble lay in differential expansion between the fins and the header, which occurs because the fins may run cooler than the header drum by as much as 167°C (300°F). This differential expansion causes bending of the tubes in the same manner as indicated in Fig. 9.10. The difference between the average fin temperature and the average metal temperature in the header drum depends on the heat load on the radiator, that is, on the air mass flow through the radiator. This temperature difference is greatest at the air-inlet face by a factor which may be as much as 3 to 1 depending on the depth of the radiator in the air-flow direction (see Fig. 17.15). The thermal strains associated with this differential thermal expansion are cycled each time the air flow is changed. Thus a radiator can be subjected to a severe thermal cycle simply by turning on and off the air flow, or, for that matter, by changing the air flow. The large thermal cycles (i.e., large changes in the temperature differences within the matrix structure) are, of course, much more serious from the stress standpoint than the small thermal cycles. The adverse effects of this factor can be reduced by slitting, or interrupting, the fins at intervals to reduce the structural redundancy. This was done in the later radiators, and was found to be sufficiently effective so that the last unit tested withstood six times the number of severe thermal cycles anticipated for the normal design life of a full-scale unit.[16]

Summary of Design Stress Criteria

The stress analysis philosophy which emerged from the design and test work can be summarized as follows:

1. The structure should be designed so that the most highly stressed regions (i.e., those regions in which yielding is likely to occur) should be of fairly simple geometry so any plastic flow will be fairly well distributed throughout a large region instead of concentrated in a localized zone.

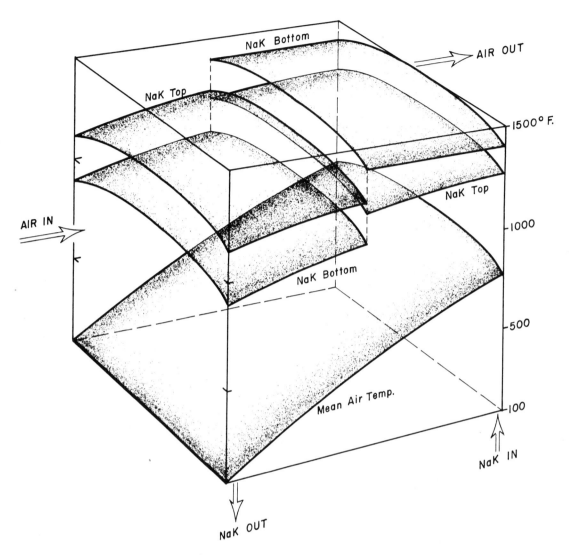

Figure 17.15 Temperature distribution through the NaK-to-air radiator of Fig. 17.12 showing the effects of the two-pass crossflow arrangement. (Courtesy Oak Ridge National Laboratory.)

2. The zone in which plastic strain will occur should not be subjected to a total amount of strain cycling greater than that which will consume 10 to 20% of the life of the metal for the amount of strain per cycle anticipated.

3. Where steady-state stresses are determining, the total integrated strain should not exceed 0.2%.

4. Where rupture life is determining, the steady-state stresses should not exceed 50% of the rupture stress for the design life of the element.

DESIGN OF A HEAT EXCHANGER FOR A MOLTEN-SALT REACTOR POWER PLANT

In stationary power plants a long life and good maintainability are much more important considerations than small size and light weight. To illustrate the somewhat different problems involved, it seems desirable to include an example for such an application. The author has proposed a plant that is particularly interesting because it shows promise of yielding the exceptionally high overall thermal efficiency of 55%. This is a binary vapor cycle power plant that employs a 1000 MW thermal output molten-salt reactor with a fuel outlet temperature of 982°C (1800°F), delivering heat through an intermediate heat exchanger to an inert salt, which leaves the heat exchanger at 871°C (1600°F). The inert salt heats a potassium boiler that delivers potassium vapor at 838°C (1540°F) to a turbine. The potassium vapor discharges from the turbine at 593°C (1100°F) to a condenser in which it gives up its heat of vaporization to the boiler, superheater, and reheater of a steam plant designed to operate with a 566°C (1050°F) peak steam temperature.[12]

In designing the fuel-to-inert salt heat exchanger, a temperature drop of 56°C (100°F) was chosen for both the fuel and the inert salt circuits. The heat exchanger layout chosen was a counterflow, shell-and-tube configuration with the tubes arranged on an equilateral triangular pitch in 20 individual tube bundles. A U-tube arrangement was chosen to provide for thermal expansion. The inert salt would flow inside the tubes and the fuel from the reactor on the shell side. Niobium tubes 3/8-in. OD and 0.035 in. thick were chosen to give the desired corrosion resistance and strength at the high operating temperature. A tube spacing of 1.25 times the tube OD, or larger, seemed appropriate.

A variety of power plant design requirements limited the pressure drop for the inert salt flow through the heat exchanger tubes to 69 kPa (10 psi), and that for the fuel flow on the shell side to 103 kPa (15 psi).

Example 17.3. The basic design specifications including the physical properties of the two fluids are summarized in Table 17.4. The design problem was to determine the flow rates of the fuel and the inert salt, the number of tubes, and the tube length to provide a basis for estimating the size and cost of the heat exchanger.

Solution. The flow rates of both the fuel and the salt can be calculated by equating the power output to the heat transported by these fluids, that is, $Q = Wc_p \delta t$. The power output Q to each tube bundle is

$$Q = (50,000 \text{ kW}) \left(0.946 \frac{\text{Btu/sec}}{\text{kw}} \right)$$

$$= 47,300 \text{ Btu/s}$$

Hence the flow rate of fuel W'_f is

$$W'_f = \frac{47300}{0.33 \times 100} = 1,435 \text{ lb/s}$$

TABLE 17.4 Design Specification for a Molten Salt Heat Exchanger

Tube Dimensions
Tube OD = 0.375 in.
Tube ID = 0.305 in.
Wall thickness = 0.035 in.
Internal flow area per tube = 0.0731 in.^{2}
External surface = 0.0982 ft^2/ft
Tube spacing = 1.25 × tube OD = 0.470 in.
Pitch = equilateral triangle pitch
Shell-side flow area per tube = 0.0796 in.^{2}
Shell-side passage equivalent diameter = 0.27 in.
Tube metal conductivity = 34 Btu/h · ft^2 · °F/ft

	Shell Side	Tube Side
Fluid	Fuel	Inert salt
Inlet temperature, °F	1800	1600
Outlet temperature, °F	1700	1700
Density ρ, lb/ft^3	218	129
Specific heat c_p, Btu/lb · °F	0.33	0.488
Viscosity μ, lb/h · ft	16.4	8.8
Thermal conductivity k, Btu/h · ft^2 · °F/ft	1.75	2.44
Pr = $c_p \mu/k$	3.09	1.76
Allowable Δp, psi	15	10

Similarly, the flow rate of the inert salt W'_s is

$$W'_s = \frac{47,300}{0.488 \times 100} = 970 \text{ lb/s}$$

where the subscript s refers to the inert salt that flows inside the tubes. The subscript f is used for the fuel.

The tube length depends on the number of tubes chosen, but this in turn must also satisfy both the pressure-drop and the heat transfer requirements. Therefore, the next step in the calculations was to obtain a good approximation for the number of tubes, and then modify it later if necessary. A first estimate of the number of tubes can be obtained from the pressure-drop considerations for the flow of fluid W'_s in pounds per second, inside the tubes, using the relation

$$n = \frac{W'_s}{6.31} \left(\frac{f_d L}{\rho_s D^5 \Delta P_s} \right)^{1/2}$$

A suitable tube length can be chosen on the basis of

practical considerations, and f_d can be estimated. By taking a tube length of 3048 mm (10 ft), and a friction factor of 0.025, a first estimate for the number of tubes for the allowable pressure drop of 69 kPa (10 psi) is n = 1,730 tubes. On this basis, the flow rates of fuel and inert salt per tube are

$$W'_s, \text{tube} = \frac{970}{1,730} = 0.560 \text{ lb/s} \cdot \text{tube}$$

$$W'_f, \text{tube} = \frac{1,435}{1,730}$$

$$= 0.830 \text{ lb/s} \cdot \text{tube (outside the tube)}$$

The mass flow rates of fuel and inert salt are

$$G'_s = \frac{W'_s/\text{tube}}{A_s} = \frac{0.560 \times 144}{0.0731} = 1,100 \text{ lb/s} \cdot \text{ft}^2$$

$$G'_f = \frac{W'_f/\text{tube}}{A_f} = \frac{0.830 \times 144}{0.0796} = 1,500 \text{ lb/s} \cdot \text{ft}^2$$

The Reynolds numbers for the fuel and inert salt are

$$\text{Re}_s = \frac{1,100 \times 3,600 \times 0.305}{8.8 \times 12} = 11,450$$

$$\text{Re}_f = \frac{1,500 \times 3,600 \times 0.27}{16.4 \times 12} = 7,400$$

The heat transfer coefficients now can be calculated for the fuel and inert salt. Figure 17.9 indicates that the heat transfer coefficient for molten fluoride salts is about 12% less than the values calculated from the Dittus-Boelter equation for Reynolds number above 6000. Hence we can approximate the heat transfer coefficients by the relation $Nu = 0.02 \text{ Pr}^{0.4}\text{Re}^{0.8}$ or $h = 0.02(k/D)\text{Pr}^{0.4}\text{Re}^{0.8}$. Substituting the numerical values gives

$$h_s = 0.02\frac{2.44 \times 12}{0.305}(1.76)^{0.4}(11,450)^{0.8}$$

$$= 4,230 \text{ Btu/h} \cdot \text{ft}^2 \cdot °\text{F}$$

$$h_f = 0.02\frac{1.75 \times 12}{0.27}(3.09)^{0.4}(7,400)^{0.8}$$

$$= 3070 \text{ Btu/h} \cdot \text{ft}^2 \cdot °\text{F}$$

The conductance of the tube wall is

$$\text{Wall conductance} = \frac{k_w \times 12}{b_w} = \frac{34 \times 12}{0.035}$$

$$= 11,650 \text{ Btu/h} \cdot \text{ft}^2 \cdot °\text{F}$$

The overall coefficient of heat transfer based on the external surface of the tube is

$$\frac{1}{U} = \frac{L}{3,070} + \frac{1}{11,650} + \frac{1}{4230(0.305/0.375)}$$

$$\frac{1}{U} = 10^{-3}[0.326 + 0.086 + 0.291]$$

$$= 10^{-3} \times 0.703$$

$$U = 1,420 \text{ Btu/h} \cdot \text{ft}^2 \cdot °\text{F}$$

The total tube external surface required can be calculated by equating the power output to the heat transfer rate from fuel to the inert salt, that is,

$$Q = A \times U \times \Delta t_m$$

$$47,300 \times 3,600 = A \times 1420 \times 100$$

$$A = 1,200 \text{ ft}^2$$

The external surface per tube is

$$A_{\text{tube}} = \frac{1,200}{1,730} = 0.695 \text{ ft}^2/\text{tube}$$

The tube length L is

$$L = \frac{0.695}{0.0982} = 7.08 \text{ ft}$$

The exact values of the pressure drops can now be calculated from the relation

$$\Delta P = f_d \frac{G'^2}{2g\rho}\frac{L}{D}$$

The pressure drop for the inert salt, taking $f_d = 0.033$, becomes

$$\Delta p_s = 0.033\frac{1,100^2}{2 \times 32.2 \times 129}\frac{7.08}{0.305}\frac{12}{144} = 9.3 \text{ psi}$$

and for the flow of fuel on the shell side, taking $f_d = 0.035$,

$$\Delta p_f = 0.035\frac{1,500^2}{2 \times 32.2 \times 218}\frac{7.08}{0.27}\frac{12}{144} = 12.3 \text{ psi}$$

The pressure drops thus calculated for the flow of fuel and inert salt are close to and do not exceed the allowable pressure drops. Therefore, the figure of 1,730 tubes chosen as a first approximation is a reasonable value. If the pressure drops were greatly different, the calculations could be repeated with a different number of tubes. If, with the number of tubes chosen, the tube-side pressure-drop requirement were satisfied but the shell-side pressure drop were greatly different, the tube spacing could be altered.

Since a U-tube arrangement was specified, the matrix length will be about 1,067 mm (3.5 ft), and the number of holes in the header sheet is 3,460.

STEAM GENERATORS

The dominant problems in the design of steam generators heated by a liquid metal or a molten salt are those characteristic of heat exchangers utilizing these high-temperature working fluids because they impose some unique boundary conditions. The high heat transfer coefficients of liquid metals are advantageous in that they make it possible to obtain a surface area less than 10% that for a fossil-fuel-fired plant. Further, a high heat flux can be obtained throughout the bulk of the steam generator with no danger of tube burn-out because (except in rare transients) the tube-wall temperature cannot exceed that of the liquid metal, usually about 600°C (1112°F). On the other hand, these high heat transfer coefficients pose some serious difficulties by causing large and rapid changes in surface temperatures in certain transients under emergency conditions, and thus severe thermal stresses. The more important of these effects are emphasized in this section.

High Heat Fluxes

Some of the key similarities and differences between steam generators heated by a gas and those heated by a liquid metal can be envisaged by looking at Fig. 15.13, which shows the temperature distribution in a once-through steam generator for a high-temperature gas-cooled reactor. The temperature difference between the two fluid streams is smallest at the sharp temperature discontinuity between the economizer and the boiler regions, that is, at the pinch point. This presents a similar limitation for both the gas-cooled and liquid metal-cooled reactor systems; in both cases the temperature difference at the pinch point must be at least 10°C to keep the tube length in this region from becoming excessive. On the other hand, the large temperature differences near the temperature discontinuity between the boiler and superheater regions—which is no problem in gas-cooled reactor systems—introduces truly formidable problems for high-temperature liquid systems. With a

heat transfer coefficient about 10 times that for a gas, the temperature difference in this region must be kept relatively low to avoid an excessive heat flux. If this is not done, excessive thermal stresses and/or a shift from nucleate to film boiling with severe flow instabilities are likely. As a consequence, the peak temperature of the hot fluid stream in high-temperature liquid systems is usually about 150°C (270°F) lower than for a high-temperature gas-cooled reactor. Even with this reduction, the designer must exercise great care and ingenuity to avoid excessive local heat fluxes.

Leaks

A second major concern stems from the consequences of a leak. Murphy's first law is that if something unpleasant can happen, it will—and this has certainly proved to be the case with leakage in every type of heat exchanger the author has encountered. Leakage in a steam generator for a PWR or a GCR, while troublesome, does not make a mess, but leakage in a steam generator for a sodium-cooled or molten-salt reactor may make a very nasty mess. Not only may it be difficult to clean up the crud from the reaction products, but there may be insidious effects of intergranular corrosion at unsuspected points remote from the leak, severely reducing the integrity of the system to a degree that may be difficult to determine or even suspect.

Sodium-Water Reactions

The possibility of leaks makes people worry that there will be an explosive reaction if water leaks into sodium in a steam generator. A major handicap faced by alkali metal systems is the impression made upon most of us as junior high school science students when our teachers introduced us to sodium by tossing a small chunk of the metal into a beaker of water. The vivid picture left by these demonstrations has made spontaneous fires and violent explosions seem almost synonymous with alkali metals. In point of fact, if there is no water around, a liquid alkali metal fire is much like a charcoal fire. Unlike hydrocarbons, the liquid metals have a very low vapor pressure so that oxidation can take place only at the metal surface; the liquid metal does not vaporize to yield a combustible or explosive mixture of air and vapor. As a matter of fact, an oxide blanket tends to form at the metal surface, and this interferes with the oxidation process so that, if the liquid metal is in a stagnant puddle in an air atmosphere, the reaction rate will be greatly reduced by the oxide. The oxide density is greater than that of the liquid metal in the puddle so that it tends to sink, thus reducing its inhibiting effects on the reaction.[24]

If no air is present, the reaction between sodium or NaK and water is rapid and violent, and both hydrogen and steam are evolved. If air is present, the hydrogen mixes with the air, an explosive mixture is formed, and, once a flame front starts in a hydrogen-air mixture, it may develop into a detonation wave that causes the explosion to be even more violent.

Some of the early experiments by many different groups[24-33] are particularly instructive. They showed that, if a leak occurs in a sodium- or NaK-to-water heat exchanger fitted with a blow-off valve or rupture diaphragm and a stack, the peak pressures developed are only a little higher than those for which the valve or diaphragm was designed. In small-scale experiments at the Mine Safety Appliances Company it was found that, if the liquid metal temperature were above 316°C (600°F), there were no violent pressure surges, and there was smooth relief of the gas evolved via the pressure relief valve.[25] Further, if the relief valve can be located in the water system, the peak local temperature in the reaction zone can be kept relatively low because the heat released will go into vaporizing the water.[26] However, if the relief valve is placed in the water system, the liquid metal system must be designed to take the same peak pressure as the water system—a step that is usually unattractive.

The tests at the Mine Safety Appliances Co. were carried out with sodium, NaK, and potassium. The shape of the pressure-time curves for typical simulated leaks differed somewhat between sodium and NaK, but there were no evident differences between those for NaK and those for potassium.[26]

A particularly fine series of tests has been run in the United Kingdom using a full-scale tube bank.[28] It was found that about half of the hydrogen produced is blown off as gaseous hydrogen, and the balance remains as NaOH or NaH. The temperature in the reaction zone may reach 1300°C (2372°F), and this may so weaken adjacent tubes as to lead to failures. Whether or not such failures occur seems to be very much a function of the character of the jet that emerges from the leak. For example, tests involving 15.9-MPa (2300-psi) water jets in full-scale tube bundles led to failures when the jets impinged at close range on adjacent tubes.[28] These failures occurred in a matter of a few seconds, providing a mechanism for progressive failures of the sort that occurred in the Fermi reactor plant. This phenomenon, commonly referred to as "tube wastage," is a familiar type of failure in fossil fuel boilers. The threat it poses places a strong premium on detecting such a leak quickly and relieving the pressure on the water side to bring the pressures into equilibrium.[31] When this is done, the

reaction rate drops sharply, and the leak may even plug with sodium oxide.[26]

If steam free of liquid water leaks into the liquid metal side of the steam generator, the reaction between the steam and the liquid metal takes place at a greatly reduced rate. This stems in part from the much lower density of the steam so that the reaction rate at the jet interface is reduced, and in part because the hydrogen produced recirculates in the reaction zone, diluting the system.[32] The reaction proceeds quietly without the sharp pressure fluctuations and noise commonly associated with liquid sodium-water reactions.

The type of leak to be expected is from a small crack induced by thermal strain cycling. Such a leak would be small initially and develop slowly, and can be discovered at an early stage by using a sensor that will detect hydrogen in the sodium system.[30] Early detection is important because corrosion will open the crack and cause the leak to grow until a jet forms. Once that happens, the failure can get worse rapidly, but if the leak is detected in the microscopic crack stage an expensive mess can be avoided and repair may not be as difficult.

As work on sodium-cooled fast breeder reactors progressed it became evident that the two key sets of questions were the rate at which tube wastage would progress as a function of the size of the leak, and how small a leak could be detected reliably. Typical of the tests run to investigate the first set of problems are those reported in References 33 and 34 which entailed tests with leak rates as small as 0.07 g/s—roughly the flow rate of high-temperature water through an orifice 25 μm (0.001 in.) in diameter under a pressure of about 70 bar (1000 psi). Both of these programs showed that the tube wastage rate increased rapidly with the water leakage flow rate and the sodium temperature.

Investigations of means for leak detection showed that small leaks can be detected before serious damage has occurred. One of the unusual approaches to leak detection that proved quite sensitive to leaks as small as 0.07 g/s was an acoustic device sensitive to the sound spectrum from the sodium-water reaction at the leak.[35]

Double-Walled Tubes

A number of special design features have been proposed to avoid serious trouble if a tube leaks. Double-walled tubes with separate header sheets have been used in chemical processes where it has been vitally important to ensure that leakage between the fluids cannot occur. The space between the double-walled tubes can be vented and monitored for leaks. Problems with this arrangement include a drastic reduction in the heat flux, large increases in the cost, and difficulty in accommodating differential thermal expansion between the inner and outer tubes if the heat flux is high, as it is with liquid metals. The reduction in the overall heat transfer coefficient might be minimized by using mercury to fill the gap between the tubes, but this would introduce the toxicity and corrosion problems of mercury.

Separate Tube Banks

Separate banks of tubes for the steam and the high-temperature liquid could be installed one above the other in the same shell, and the void between them filled with a fluidized bed of inert particles such as alumina. This would relieve the differential thermal expansion problem of the double-walled tubes and avoid the awkward difficulties with double header sheets, but the costs are still high and keeping the bed fluidized entails both extra complexity and a power loss. Replacing the fluidized bed with arrangements for boiling potassium or cesium (which has a lower boiling point) from the lower bank of heated tubes and allowing it to condense on the upper bank of steam generator tubes would provide a refluxing situation similar to that in a heat pipe. The heat transfer coefficient would be about 100 times that for a fluidized bed, and there would be no pumping power loss. The inventory of potassium or cesium required could be kept small by good design so that, if there were a steam leak, the quantity of alkali metal that could react would be strictly limited, and the shell can be designed so that any hydrogen formed will—as a noncondensable—be concentrated in a small region where it would be far easier to detect than in a big sodium system, where it would be extremely diluted. This approach has the further advantage that the interior of the vessel can be compartmented into a number of regions. The baffle clearances between regions would be chosen so that they would function as orifices to reduce the potassium or cesium vapor pressure and temperature along the length of the tube bundles, thus preventing excessively high local heat fluxes. However, the surface area and the number of tube-to-header joints would still be more than double the values for a conventional heat exchanger, which would roughly double the cost.

Reentry Tube Steam Generator

The problems with a high heat flux are particularly acute in a potassium condenser-steam generator for a potassium vapor topping cycle because the hot fluid condenses

at a uniformly high temperature. An attractive means for coping with this is the use of a reentry, or bayonet, tube such as that shown at the bottom of Fig. 17.6 and employed in the conceptual design of References 13, 14, and 24. The condensing coefficient for the potassium vapor is about three times that for nucleate boiling of water, so that it is possible to operate with a small temperature difference between the condensing potassium and the steam leaving the steam generator. In the reentry tube concept a layer of superheated steam acts as a buffer between the relatively low-temperature feedwater and the high-temperature wall heated by the potassium vapor. This reduces the heat flux to a reasonable level, yet not so much that the heat transfer surface area required becomes too large. Probably the best premise for the design of the reentry tube is to carry out the requisite preheating and all of the annular flow boiling in the central tube and use a fluted wall to keep liquid on the wall up to a vapor quality of about 90%. This would provide a high vapor quality at the point where the flow direction would reverse for the return

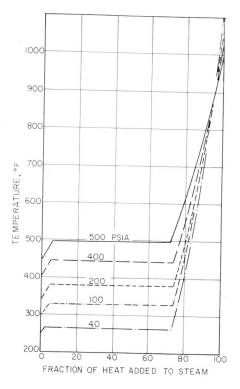

Figure 17.16 Local water and steam temperatures in a once-through steam generator as a function of the fraction of the heat added for a series of different steam pressures with a burner regulated to keep the fuel flow proportional to the steam flow.

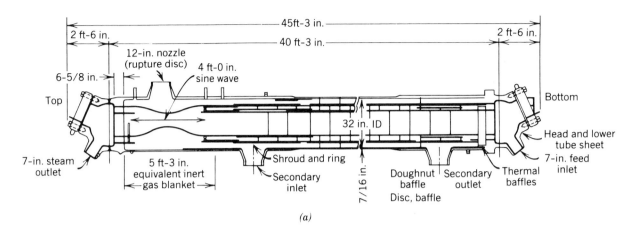

(a)

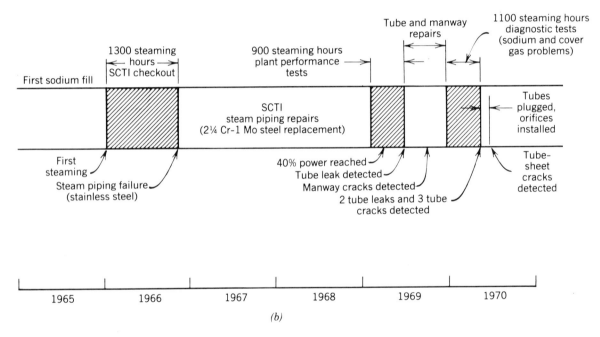

(b)

Figure 17.17 Longitudinal section through the Alco/BLH sodium-heated steam generator together with a chart showing the chronology of the test program. Range of operating conditions: steam temperature and pressure, 371 to 566°C (700 to 1,050°F)/15.1 MPa (2200 psia); feedwater-inlet temperature, 316°C (600°F); feedwater flow rate, 2.65 to 15.6 kg/s (21,000 to 124,000 lb/h); sodium-inlet temperature, 404 to 638°C (760 to 1,180°F); sodium temperature drop (inlet-to-outlet), 39 to 250°C (70 to 450°F); sodium-flow rate, 19.5 to 156 kg/s (154,000 to 1,233,000 lb/h); thermal power, 6 to 22 MW. (Courtesy Oak Ridge National Laboratory.)

through the outer annulus. At part load, boiling would be completed in the center tube well below the top. A single electrically heated tube of this type was built and tested at the Oak Ridge National Laboratory and functioned well with no trace of instability over the full range from zero to full load.[36]

Another application for which the reentry tube appears especially attractive is for molten fluoride salt systems in which the freezing point of the salt is around 450°C (842°F) so that it is difficult to avoid freezing of the salt in a conventional heat exchanger configuration under low-load and off-design conditions. The drastic change in the temperature distribution on the steam side with changes in load and steam pressure is shown in Fig. 17.16 for a typical system. Thus even though the heat transfer coefficient for the salt is lower than for condensing potassium so that the heat flux is lower, the drop in the boiling point of water at reduced loads and pressures makes freezing a progressively greater problem as the load is dropped. (Note also that the tube length required is greater for the molten salt than for condensing potassium vapor or liquid sodium.) An excellent, comprehensive set of tests was run with a tube heated by a molten salt at the University of Delft. As in the reentry tube boiler tests at Oak Ridge, there was no sign of flow instability over the full load range, and excellent correlation was obtained between analytical estimates and the measured performance.[37,38] The tube length in the Delft tests was about 10 m, but could have been shortened substantially if the inner tube had been fluted as in Table 15.8.

It should be mentioned that calculations for the reentry tube configuration are difficult if made by hand because of the interdependency of the temperatures in the central tube and the steam in the annulus. This interdependency requires both many small steps and multiple iterations. A clever computer program for accomplishing this is presented in References 37 and 38. It should also be noted that the cooling effect of the entering feedwater on the steam leaving the outer annulus is so pronounced that one cannot get as high a steam exit temperature as is desirable unless some form of thermal insulation is introduced between the central tube and the steam annulus in the vicinity of the steam outlet. In the Oak Ridge tests this insulation was accomplished by inserting a smaller tube in the inlet region of the reentry tube to introduce a thin steam blanket. The same effect was obtained in the Delft tests by coating the outer surface of the lower portion of the central tube with a layer of zirconia (which has a low thermal conductivity).

Conventional Design Practice

An early design for a sodium-heated, once-through steam generator is shown in Fig. 17.17.[39] Feedwater entered the boiler tubes from the header at the right and steam left through the header at the left end. Sodium (in a system simulating the secondary sodium circuit of a sodium-cooled reactor) entered the shell side at the left, flowed in a multipass counterflow configuration through a set of disc-and-doughnut baffles, and left the shell at the right end. The tubes were bent to a sine-wave form with a generous radius to provide for differential thermal expansion between the tubes and the shell with elastic

bending deflection in the tubes. The design for this unit was initiated in 1958 as a key element in the component development program of the USAEC, but construction and installation in the new test facility at Atomics International were not completed until 1966. A total of 3300 h of steaming operation were obtained over the next five years with long interruptions for repairs to elements of the test facility. As indicated in Fig. 17.17, the unit was removed for metallurgical examination because leaks developed in the tube header regions. The metallurgical examination revealed numerous cracks such as those in Fig. 17.18 in both header sheets. The metallurgists concluded from microscopic and chemical examination that these had been caused by caustic corrosion, that is, both sodium and water had been present in incipient cracks and caused crack propagation by intergranular corrosion.[39] The cause of the initial crack that permitted sodium and water to percolate into the initial crack and mix to produce the caustic corrosion could not be determined metallurgically.

The author examined the unit and went over the test data in an effort to appraise likely causes for initiation of the cracks. The tube-to-header joints were made with

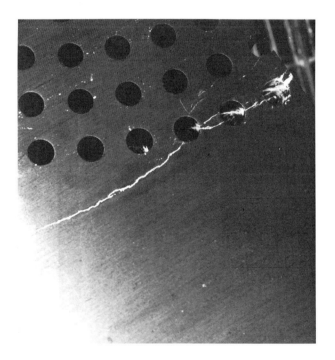

Figure 17.18 Close-up photo of a typical set of cracks in the inlet header sheet as viewed in a plane cut through the header sheet 3 cm from the inlet face. Fluorescent dye penetrant was used to enhance the image of the cracks. (Courtesy Oak Ridge National Laboratory.)

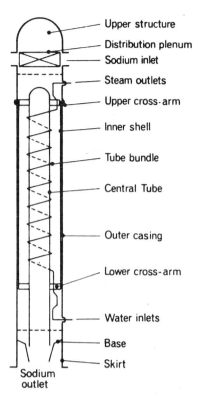

Figure 17.19 Schematic longitudinal section through the Creys-Malville steam generator. (Courtesy *Nuclear Engineering International*.)

TABLE 17.5 Summary of Design Data for the Creys-Malville Steam Generator Together With That for a PWR Steam Generator

Reactor type	LMFBR	PWR
Plant	Creys-Malville	Oconee
		Unit 1
Designer	French	B&W
Type	Once-through	Once-through
Tube shape	Helical	Straight
Hot fluid	Secondary sodium	Primary water
Fluid in tube	Steam	Steam
Type of tube	Plain	Plain
Tube material	Incoloy 800	Incoloy 800
Capacity, MWt	750	1,295
Steam flow, kg/s	340	706
Temperatures, °C		
Hot Fluid in	525	343
Hot fluid out	345	289
Feedwater	235	235
Steam out	487	299
Superheat	130	20
Pinch Δ T	45	10
Tube OD, mm	25	16
Tube ID, mm	20	14
Tube length, m	91	15
No. of tubes	1,430	15,530
Total tube length, km	130	233
Surface, m²	10,220	11,600
Surface, m²/kWt	0.0136	0.0089

fillet welds between the outer face of the header sheet and the ends of the tubes which projected about four tube-wall thicknesses out of the header sheet before welding. The tubes had an OD of 13.3 mm (0.523 in.) and a wall thickness of 2.67 mm (0.105 in.). Sodium filled the long crevices between the tubes and the 178-mm (7-in.) thick header sheets. The header sheets were washed by sodium on the shell side and by feedwater or steam on the opposite side, sometimes with widely different temperatures. In the author's opinion, the initial cracks stemmed from severe thermal strain cycling that occurred when there were abrupt drops in the pressure in the steam system that produced temperature distributions such as those indicated by Fig. 17.16.

Creys-Malville LMFBR Steam Generators

All of the specific designs for heat exchangers discussed in this chapter up to this point have been subjected to enough operation to show up weaknesses in their design. The steam generator described in this section has not yet seen much operation at the time of writing, but it represents the culmination of about 20 years of development experience in the French LMFBR program,[15,40,41] and has undoubtedly profited from the extensive experience in other countries as well. The key design features appear to avoid all of the sources of failure in previous designs, thus making this unit of particular interest.

A schematic section through the steam generator is shown in Fig. 17.19; Table 17.5 summarizes the design data and compares it with data for a PWR unit. The 25-mm (1.0-in.) OD Incoloy 800 tubes are about 91 m (300 ft) long. They are wrapped in spiral form around a central cylinder in 17 concentric layers, with 13 tubes in the inner layer and 29 in the outer. The coils are supported and positioned relative to each other by stays hung from a spider at the top and arranged to form radial ribs 45° apart around the central cylinder. Screws are used to clamp the tubes to the stays as each layer of tubes is wound into place. The tubes pass radially through thermal sleeves in the outer shell to enter the feedwater inlet and steam-outlet headers, thus placing all of these joints out where they are accessible.

REFERENCES

1. R. W. Ohse, ed., *Handbook of Thermodynamic and Transport Properties of Alkali Metals*, International Union of Pure and Applied Chemistry, Chemical Data Series no. 30, Blackwell Scientific Publications, Oxford, UK, 1985.

2. R. N. Lyon et al., *Liquid Metals Handbook*, ONR, AEC, and Bureau of Ships, Supt. of Documents, Washington, D.C., June 1952.

3. H. G. MacPherson et al., "Molten-Salt Reactor Program Quarterly Progress Report for Period Ending July 31, 1960," Report no. ORNL-3014, Oak Ridge National Laboratory, October, 1960.

4. W. L. R. Emmet, "The Emmet Mercury Vapor Process," *Trans. ASME*, vol. 46, 1924, p. 253.

5. W. E. Kirst, W. M. Nagle, and J. B. Cartner, "A New Heat Transfer Medium for High Temperatures," *Trans. AIChE*, vol. 36, 1940, p. 371.

6. P. S. Lykoudis and Y. S. Touloukian, "Heat Transfer in Liquid Metals," *Trans. ASME*, vol. 80, 1958, p. 653.

7. H. E. Brown et al., "Temperature and Velocity Distribution and Transfer of Heat in a Liquid Metal," *Trans. ASME,* vol. 79, 1957, p. 279.

8. W. B. Harrison and J. R. Menke, "Heat Transfer to Liquid Metals Flowing in Asymmetrically Heated Channels," *Trans. ASME*, vol. 71, 1949, p. 797.

9. H. A. Johnson et al., "Heat Transfer to Lead-Bismuth and Mercury in Laminar and Transition Pipe Flow," *Trans. ASME,* vol. 76, 1954, p. 513.

10. T. B. Douglas, "Specific Heats of Liquid Metals and Liquid Salts," *Trans. ASME*, vol. 79, 1957, p. 23.

11. A. P. Fraas, "Reactors for Space," Recent Advances in Engineering Science, vol. 4, *Proceedings of the Fifth Technical Meeting of the Society of Engineering Science*, Huntsville, Ala., October 30–November 1, 1967, Gordon & Breach Science Publishers, New York, p. 1.

12. A. P. Fraas, "A Potassium-Steam Binary Vapor Cycle for a Molten-Salt Reactor Power Plant," *Journal of Engineering for Power, Trans. ASME*, vol. 88(4), 1966, p. 355.

13. A. P. Fraas, "A Potassium-Steam Binary Vapor Cycle for Better Fuel Economy and Reduced Thermal Pollution," *Journal of Engineering for Power, Trans. ASME*, vol. 95(1), 1973, p. 53.

14. A. P. Fraas, "Comparative Study of the More Promising Conversion Systems, and Tritium Recovery and Containment Systems for Fusion Reactors," Report no. ORNL/TM-4999, Oak Ridge National Laboratory, November 1975.

15. E. Cambillard et al., "45,000 Hours of Operating Experience with the Phenix Steam Generators and Intermediate Heat Exchangers," ASME Paper no. 82-NE-3, ASME Nuclear Engineering Conference, July 25–28, 1982.

16. A. P. Fraas, "Design Precepts for High Temperature Heat Exchangers," *Nuclear Science and Engineering*, vol. 8, 1960, p. 22.

17. M. M. Yarosh, "Evaluation of the Performance of Liquid Metal and Molten Salt Heat Exchangers," *Nuclear Science and Engineering*, vol. 8, 1960, p. 32.

18. R. E. MacPherson et al., "Development Testing of Liquid Metal and Molten Salt Heat Exchangers," *Nuclear Science and Engineering*, vol. 8, 1960, p. 14.

19. R. E. MacPherson, "Techniques for Stabilizing Liquid Metal Pool Boiling, II-B/11," *Proceedings of the Conference Internationale sur La Surete des Reacteurs a Neutrons Rapids*, Aix-en-Provence, France, September 19–22, 1967, Commisariat a l'Energie Atomique, France.

20. J. H. DeVan, "Compatibility of Structural Metals with Boiling Potassium," CONF-760503-P1, Paper presented at the International Conference on Liquid Metal Technology in Energy Production, Champion, Pa., May 1976, p. 418.

21. P. Patriarca et al., "Fabrication of Heat Exchangers and Radiators for High Temperature Reactor Applications," Report no. ORNL-1955, Oak Ridge National Laboratory, June 1956.

22. A. J. Friedland and C. F. Bonilla, "Analytical Study of Heat Transfer Rates for Parallel Flow of Liquid Metals through Tube Bundles," *Journal of AIChE*, vol. 7, 1961, p. 107.

23. B. W. Le Tourneau et al., "Pressure Drop for Parallel Flow through Rod Bundles," *Trans. ASME*, vol. 79, 1957, p. 1751.

24. E. C. King and C. A. Wedge, "Reaction of NaK and H_2O," NP-1423, Mine Safety Appliances Co., 1950.

25. E. C. King and C. A. Wedge, "The Reaction of NaK and H_2O," NP-3646, Mine Safety Appliances Co., 1952.

26. E. C. King, "The Reaction of NaK and H_2O," NP-3334, Mine Safety Appliances Co., 1951.

27. J. A. Bray, "A Review of Some Sodium/Water Reaction Experiments," *Journal of the British Nuclear Energy Society*, vol. 10(2), April 1971, p. 107.

28. J. Graham, *Fast Reactor Safety*, Academic Press, New York, 1971.

29. A. Lacroix et al., Safety Investigations on Sodium-Water-Steam Generators—Calculation Methods and Experimental Results," *Proceedings of the Conference Internationale sur La Surete des Reacteurs a Neutrons Rapids*, Aix-en-Provence, France, September 19–22, 1967, Commisauat a l'Energie Atomique.

30. D. J. Hayes and G. Horn, "Leak Detection in Sodium-Heated Boilers," *Journal of the British Nuclear Energy Society*, vol. 10(1), January 1971, p. 41.

31. C. C. Addison and J. A. Manning, "The Reaction of Water Vapor with Liquid Sodium, Sodium Peroxide,

Sodium Monoxide, and Sodium Hydride; Vapor Pressures in the Sodium Hydroxide-Water System," *Journal of the Chemical Society*, December 1964.

32. C. A. Greene, "Preliminary Results from Small Leak Tube Wastage Heat Exchanger Tests," *Proceedings of the ANS Meeting on Fast Reactor Safety*, Beverly Hills, Calif., April 2–4, 1974, p. 151.

33. H. Nei and O. K. Ohshima, "Wastage of Steam Generator Tubes During Small Leak of Steam into Sodium," *Proceedings of the ANS Meeting on Fast Reactor Safety*, Beverly Hills, Calif., April 2–4, 1974, jp. 136.

34. H. Nei and O. K. Oshima, "Acoustic Detection for Small-Leak Sodium-Water Reaction," *Proceedings of the ANS Meeting on Fast Reactor Safety*, Beverly Hills, Calif., April 2–4, 1974, p. 173.

35. A. P. Fraas, "The Safety and Environmental Problems Posed by Operation of a Potassium Vapor Cycle," ASME Paper no. 73-WA/Ener-6, Presented at the Winter Annual Meeting, November 11–15, 1973.

36. R. S. Holcomb and M. E. Lackey, "Performance Charac-teristics of a Short Reentry Tube Steam Generator at Low Steam Output," Report no. ORNL-TM-3236, Oak Ridge National Laboratory, June 1971.

37. N. W. S. Bruens et al., "Modeling of Nuclear Steam Generator Dynamics," Paper presented at the International Conference on Materials for Nuclear Steam Generators, Gatlinburg, Tenn., September 9–12, 1975.

38. B. Viresema, "Aspects of Molten Fluorides as Heat Transfer Agents for Power Generation," Doctoral thesis, Technische Hogeschool Delft, February 1979.

39. G. M. Slaughter and J. H. DeVan, "Interim Information Report on Posttest Examination of Alco/BLH Sodium-Heated Steam Generator," Report no. ORNL-TM-3636, Oak Ridge National Laboratory, May 1972.

40. "Construction of the World's First Full-Scale Fast Breeder Reactor," *Nuclear Engineering International*, vol. 23(272), June 1978, p. 43.

41. "Making the Creys-Malville FBR Steam Generators," *Nuclear Engineering International*, vol. 24(290), September 1979.

18

Heat Exchangers Operating on Radiant Energy

Heat exchangers that depend on heat transfer by radiant energy transmission have become sufficiently important to merit a separate chapter to treat the special problems involved. Since 1960, large sums of government funds have been spent on radiant energy transfer in the solar energy programs for both space and terrestrial applications, in the temperature control of spacecraft crew compartments, radiators to dissipate the waste heat from space power plants, and for environmental test chambers designed to simulate the temperatures near absolute zero encountered in space. While thermal radiation plays an important role in other applications such as steam boilers, it completely dominates the design for heat exchangers in the space and solar energy programs, and hence these are the prime areas of concern here. This chapter is divided into three sections, each treating an application involving a representative set of problems: radiators for space power plants, solar thermal energy collectors, and environmental chambers for testing spacecraft.

RADIATORS FOR SPACE POWER PLANTS

A particularly challenging field is concerned with the design of radiators for space power plants intended to produce electricity for the propulsion, guidance, and communications equipment of space vehicles. Although specialized, this application presents some intriguing problems that provide an excellent exercise in the application of fundamental heat exchanger design considerations and relations.

The only way that heat can be dissipated from a spacecraft is by thermal radiation. The large areas required to dissipate the waste heat from the power generation process and the heat losses from the energy used in the electronic and life-support systems make the radiators for these systems among the largest and heaviest components of space vehicles. The relative proportions of the principal components are indicated by Fig. 18.1, which shows a large manned spacecraft. In this instance, most of the electric energy generated would go into ion propulsion, but a portion would go into the "hotel load" for electronic and life-support equipment. Although the waste heat from the latter would be small relative to the amount of waste heat from the thermodynamic cycle, the temperature at which it would be radiated is so much lower that the two radiators are roughly the same size in this case.

The size and weight of the main radiator for the power plant can be kept to a tolerable level only by going to a thermodynamic cycle that will yield a reasonably good efficiency with a heat rejection temperature of at least 538°C (1000°F), a red heat.[1] This is because the rate of heat radiation varies as the fourth power of the absolute

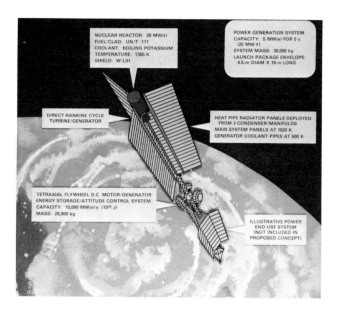

Figure 18.1 Radiator configuration for a nuclear electric space power plant installation in a spacecraft. (Courtesy Oak Ridge National Laboratory.)

temperature, and the penalties in radiator weight and size are so great for temperatures lower than this that the system becomes unattractive as well as too expensive. Even by utilizing a heat engine that runs so hot that it can yield a good thermal efficiency while rejecting its heat from a red-hot radiator, it is difficult to keep the radiator within a reasonable envelope for the launch vehicle. An appreciation for this situation is given by Fig. 18.2,

which compares the size of the radiators for several power plant outputs with the size of a typical launch vehicle to show the magnitude of the launch problem. Although folding radiator structures have been considered, the difficulties involved in making the joints leaktight at high temperatures appear to be insuperable. The type of power plant assumed in preparing Fig. 18.2 is one in which potassium is boiled using heat from a fission reactor. The potassium vapor is passed through a turbine, and—if condensed at the same pressure as in a conventional steam plant, that is, about 1.0 psia—the condenser temperature is about 538°C (1000°F). The condenser for such a power plant affords an interesting application of the basic condensate flow relationships outlined in Chapter 16 together with those for thermal radiation given in Chapter 3.

Liquid Flow Under 0-g Conditions

Control of free liquid surfaces presents problems. Unless a space vehicle is spun to induce an artificial gravitational field, the principal forces acting on droplets of condensate are surface tension and fluid dynamic forces. Surface tension forces are ordinarily much smaller than fluid dynamic forces, except at very low loads.

One way of assuring condensate flow through a condenser under 0-g conditions in space is to employ a jet condenser in which a subcooled jet of liquid is injected coaxially with the vapor stream at a high velocity into a converging channel. The momentum of the liquid and vapor suffices to carry the stream through a converging region where condensation takes place, and a liquid

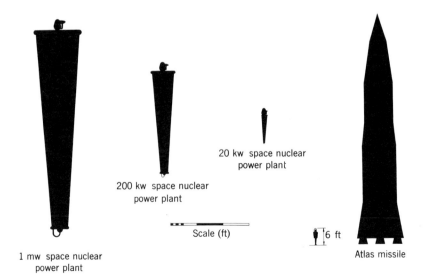

Figure 18.2 Comparison of radiator sizes for three typical space power plants with an Atlas missile. The reactor and shield assembly is the egg-shaped unit at the top; the radiator is a long truncated cone. (Courtesy Oak Ridge National Laboratory.)

stream emerges sufficiently subcooled to assure freedom from vapor bubbles. This stream of liquid can then be cooled in a radiator. Unfortunately, this approach requires that the radiator operate at an average temperature much below the saturation temperature at the inlet to the condenser.

If a direct surface condenser rather than a jet condenser is employed to maximize the radiator temperature, it is advantageous to use uniformly tapered tubes so that the velocity is high throughout all but the very last portion of the tube length so that fluid-friction forces will serve to drive the condensate toward the outlet. The condensing flow in such a tapered tube is difficult to analyze. Two-phase annular flow prevails throughout most of the length of the tube so that the friction factor for the vapor flow depends on the local Reynolds number and liquid flow rate, as indicated in Fig. 5.15. Moderately complicated expressions are required to establish the Reynolds number to determine the local friction factor for each of the two phases.[2] Limited space prevents a detailed discussion of this problem, but Fig. 18.3 shows the ratio of the local velocity to the inlet velocity as a function of axial position in the tube for a typical case. For analytical reasons the curves of Fig. 18.3 were plotted for a variety of flow rates expressed in terms of the parameter ψ, which is simply the ratio of the vapor-inlet mass flow rate to the design value; that is, ψ is the fraction of the full-power design flow rate into the tube. Note that the vapor velocity can be kept to a high value at the design flow rate up to the last 5% of the tube length, but that at flow rates below the design value the vapor velocity falls to zero well before the fluid reaches the tube outlet. Thus the condenser would tend to load up with liquid at part load unless the system were designed so that the condenser temperature, and hence the condensation rate, would drop with load sufficiently to avoid this difficulty. (One step toward accomplishing this end is to provide a condenser scavenging pump with sufficient excess capacity to maintain a "dry sump" condenser.[3])

A second set of curves, shown in Fig. 18.4, indicates the Reynolds numbers for both the vapor and liquid streams as functions of position in the tube. The upper set is for the vapor, and the lower set for the liquid. Two sets of curves are shown for the liquid, the solid curves being for an inlet vapor quality of 100% and the dashed curves for an inlet vapor quality of 80%. Note that for design conditions, the Reynolds number for the gaseous phase is in the turbulent region in all but the last portion of the tube, whereas the Reynolds number for the liquid phase would be in the laminar flow region except in the last 10% or so of the tube. Note also that, for the

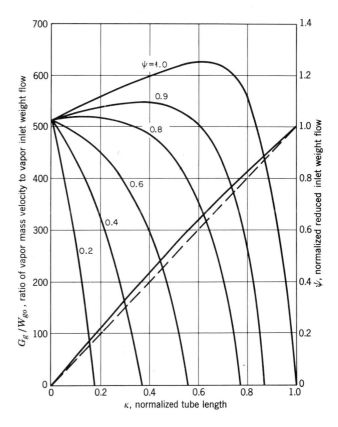

Figure 18.3 Ratio of local mass flow rate to inlet weight flow rate for condensing flow of potassium vapor in uniformly tapered tubes with a fixed heat removal rate per unit of length. Condenser inlet pressure = 1.5 psia, tube length = 4.1 ft, tube inlet ID = 0.60 in., tube exit ID = 0.15 in., two opposed fins with a tip-to-tip span of 1.8 in. (Courtesy Oak Ridge National Laboratory.)

constant heat flux condition assumed, at part load the Reynolds number for the liquid is the same as for full load up to the point where the liquid fills the tube, at which point the flow regime shifts from an annular film to a filled circular channel. Again it is evident that some means should be provided to reduce the heat flux at part load to keep the radiator from loading up with liquid.

A third set of curves showing the pressure gradient along the tube for the design point conditions ($\psi = 1.0$) is shown in Fig. 18.5. Curves are shown for two-phase flow both with no allowance for the momentum recovery and with such an allowance to give the net pressure drop.

Typical Tube Arrangements and View Factors

Perhaps the first special requirement of a space radiator is that the tubes lie on surfaces that give a good view

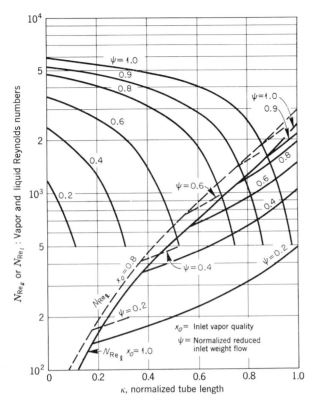

Figure 18.4 Local Reynolds numbers for the vapor and liquid streams (upper and lower sets of curves respectively) as a function of position in a tapered tube for the conditions of Fig. 18.2. (Courtesy Oak Ridge National Laboratory.)

the fins rather than in the tubes. Another helpful step is to arrange the tubes on the surface of a cylinder, as in configuration 5 of Fig. 18.6, and to provide a reflector arranged as shown for the Type-C tube at the top. This arrangement makes the rear face of the finned tube nearly as effective as the front face in dissipating thermal energy. If the surface of the reflector is clean and bright, about three-quarters of the incident radiation from the back side of the tube-and-fin array should be reflected into space by specular reflection from the involute. About 25% of the incident energy will be absorbed and reemitted or reflected by diffuse reflection. Of this amount, approximately half will be emitted to space, and about half will reimpinge on the surface of the tube. Thus the overall effectiveness of the portion of the tube and fins facing the reflector will be about 85% of that radiating directly to space. Several arrangements for the fins can be employed, but a tee-shaped arrangement similar to the Type-C tube shown in Fig. 18.6—but without the top fin, which is not very effective—appears to be close to that for minimum overall weight. Note that the front face of the tube should be thickened to provide meteoroid protection. This precaution is not required at the rear, since the reflector on the opposite

factor toward space. Figure 18.6 shows some typical configurations that have appeared attractive. Configurations 1–3 make use of Type-A tubes with two axial opposed fins. Note that the view factor characteristic of configuration 3 would be 0.866 if end effects are neglected, or somewhat higher it they are included. A similar arrangement with four banks of tubes arranged at 90° intervals around a central manifold would—if the length were infinite—have a view factor of 0.707 (i.e., the ratio of the perimeter of the outer envelope to the perimeter of the faces of the tube banks). Configurations 4–6 are designed to employ either Type-B tubes brazed to a shell that forms the surface or Type-C finned tubes backed by reflectors.

For surfaces having a projected area of 10 m^2 (100 ft^2) there is a probability of about 0.04 that a meteoroid of sufficient size to puncture a 0.10-in.-thick stainless steel tube will strike the surface during the course of a year.[4,5] One means of reducing the probability that such an impact will cause a condenser leak is to make use of finned tubes so that most of the surface area will be in

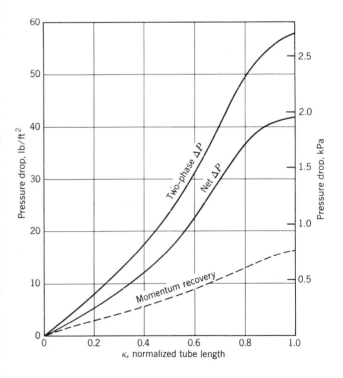

Figure 18.5 Pressure drop from the inlet of a tapered tube for the conditions of Fig. 18.2. (Courtesy Oak Ridge National Laboratory.)

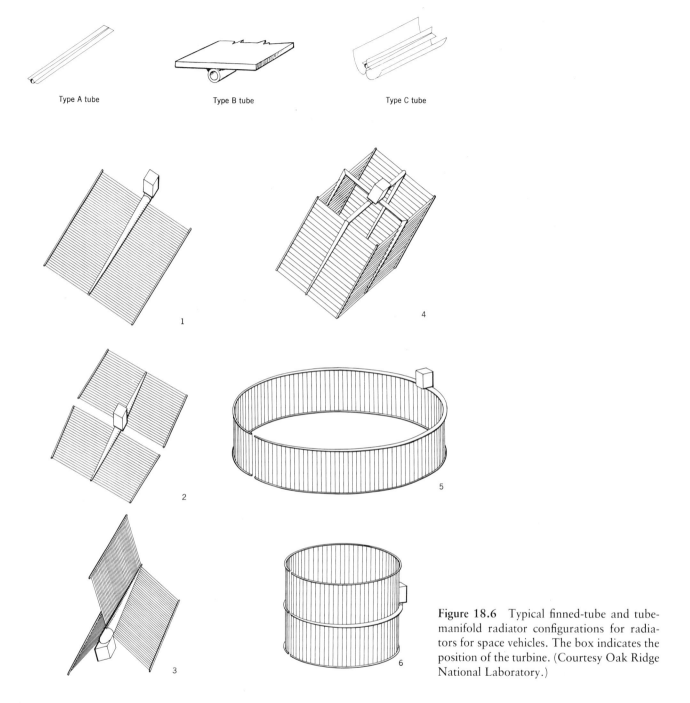

Type A tube   Type B tube   Type C tube

Figure 18.6 Typical finned-tube and tube-manifold radiator configurations for radiators for space vehicles. The box indicates the position of the turbine. (Courtesy Oak Ridge National Laboratory.)

side of the radiator will act as a bumper to shatter and disperse any meteoroid that penetrates it.[5]

Selection of a Fin Material for Minimum Weight

The suitability of a fin material depends on fabrication considerations, its density, and its thermal conductivity.

Ideally, the material's coefficient of thermal expansion should be close to that of the tube material; its strength should be fairly good at the operating temperature; it should be ductile to resist shock and vibration; and it should be readily brazed to the tube material. If all these conditions can be met, the suitability of the material is directly proportional to its thermal conductivity and inversely proportional to its density. Thus the ratio of the thermal conductivity to the density is a good figure of merit for comparing different fin materials. Values for this parameter are given in Table H2.2. It is interesting to note that copper is nearly as good as beryllium. Since copper is both readily available and easily brazed, whereas beryllium is neither, copper is likely to be preferable even though it gives a somewhat greater fin weight.

Example 18.1. *Radiator for a Space Vehicle.* A set of preliminary design calculations are summarized in Table 18.1 to show how the proportions of a condenser-radiator such as configuration 6 of Fig. 18.6 can be roughed out, and to point up the problems involved.[3]

Tapered tubes with a tee-fin and reflector configuration similar to the Type-C tube of Fig. 18.6 but with the top fin removed, form the basic heat-transfer elements. Note that this radiator (shown in Fig. 18.7) was built and tested with potassium vapor in air rather than a vacuum. After correcting for the thermal convection effects in air, the test results confirmed the design calculations.

Solution. The original design calculations are presented in Table 18.1. Starting with a given heat load and radiator operating temperature, values were assumed for the surface emissivity, fin efficiency, and reflector efficiency so that the average heat flux and the required surface area could be estimated. This value, together with an assumed vehicle diameter of 10 ft, gives the corre-

TABLE 18.1 Preliminary Design Calculations for a 10-Ft Diameter Cylindrical-Drum Radiator Similar to Configuration 6 of Fig. 13.12

Line Number	Item	Source[a]	Value
1.	Power to be dissipated, kW		860
2.	Power to be dissipated, Btu/h		2.93×10^6
3.	Ideal dissipation at 1500°R to 500°R sink, Btu/h · ft²	Fig. H5.16	8562
4.	Emissivity of treated surface	Assumed	0.92
5.	Fin efficiency	Assumed	0.90
6.	Reflector efficiency	Assumed	0.87
7.	Area required, ft²	②/③ × ④ × ⑤ × (0.5 + ⑥/2)	435
8.	Height of cylinder for 10-ft-diameter vehicle, ft	⑦/10π	14.0
9.	Tube length, ft	⑧/2	7.0
10.	Vapor temperature, °R	Given	1500
11.	Vapor density, lb/ft³	Ref. H11	0.0035
12.	Latent heat of vaporization, Btu/lb	Ref. H11	887
13.	Vapor quality, %	Given	91.7
14.	Potassium flow rate, lb/s	Given	1.0
15.	Vapor volume flow rate, ft³/s	⑬ × ⑭/11	262
16.	Vapor velocity at tube inlet, ft/s	Assumed	400
17.	Vapor flow-passage area at tube inlets, ft²	⑮/⑯	0.655
18.	Number of tubes	Assumed	96.0
19.	Tube inlet flow passage area per tube, ft²	⑰/⑱	0.0068
20.	Tube inlet ID, in.	√ ⑲/0.786	1.1
21.	Liquid density at tube outlet, lb/ft³	Table H2.3	44.0
22.	Tube outlet liquid flow rate, ft³/s	⑭/㉑	0.0228
23.	Tube outlet ID, in	Assumed	0.30
24.	Tube outlet area (per tube), ft²	0.786 × ㉓²/144	0.00049
25.	Tube outlet area (total), ft²	㉔ × 96	0.047
26.	Tube outlet velocity, ft/s	㉒/㉕	0.048
27.	Fin span, in.	10 × π × 12/2.41 × ⑱	3.2
28.	Fin cross-section shape		Triangular
29.	$0.408 \, w\sqrt{h/kb}$ for 90% fin efficiency	Fig. H7.3	0.5
30.	Equivalent heat transfer coefficient, Btu/h · ft² · °F	$[(Q/A)_{1500°R} - (Q/A)_{1400°R}]/100$	20.0
31.	Mean fin height, in.	㉗/2 − (0.6 + 0.2)/②	1.2
32.	Fin root thickness, in.	$(0.408 × ㉛/㉙)^2 × ㉚/200$	0.095

[a]Circled numbers are symbols to indicate the line from which to obtain the quantity to be used in the operation.

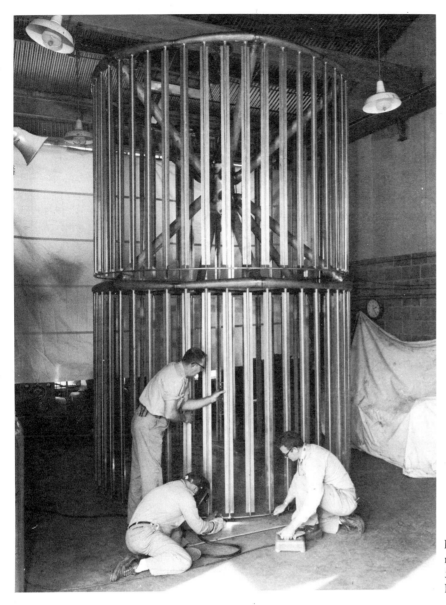

Figure 18.7 Potassium condenser-space radiator designed to reject 860 kWt at 560°C (1040°F). (Courtesy Oak Ridge National Laboratory.)

sponding height for the cylindrical envelope. Experience shows that, for the given vapor-inlet conditions, the tube length should be between 5 and 10 ft to give a near-minimum weight and a good balance between the weight of the tubes, the fins, the manifolding, and the armor for meteoroid protection. Thus a tube length of 7 ft (line 9 of Table 18.1) to give two banks of vertical tubes seems in order.

The large volume of vapor to be handled makes it desirable to use as high a vapor velocity as possible without a substantial pressure loss penalty. As in other types of equipment, this is commonly given by a Mach number of about 0.25 (see Fig. 3.12); hence a velocity of around 122 m/s (400 ft/s) appears to be a good

preliminary choice. Dividing the vapor volume flow rate by this assumed velocity defines the inlet area of the tubes. The choice of tube diameter is arbitrary. The larger the tube diameter, the stiffer the structure and the fewer the header joints, but these advantages are offset by the rapidly increasing weight for meteoroid armor or for fins. For the same ratio of total area to vulnerable area, the fin height must increase linearly with tube diameter, and the fin weight increases as the square of the fin height. A reasonable compromise appears to be given by 96 tubes arranged in two banks of 48 each. The inlet diameter of the tubes then follows directly from the vapor volume flow rate and the inlet velocity (lines 15 to 20). Manufacturing and structural considerations lead to a minimum tube ID at the outlet of about 0.30 in. This gives a low liquid exit velocity (line 26), and should make for a low pressure drop in the condensate manifold as well as ease the problems of scavenging the radiator under 0-g conditions.

The effective area for dissipating heat is the envelope of the fin configuration. This is essentially a 45° right triangle with the hypotenuse extending across the top of the tee formed by the fins, and the two sides extending between the fin tips. The perimeter of this envelope is 2.414 times the fin span at the top of the tee. Thus the fin span can be determined from the surface area required, the tube length, and the number of tubes around the perimeter (line 18).

Since the resulting fin height is large and the heat load high, a substantial weight savings can be effected by tapering the fin from root to tip. Assuming a triangular cross section for the fin, the value of the parameter $0.408w\sqrt{h/kb}$ can be determined from Fig. H7.3 to give the assumed 90% fin efficiency. To evaluate this parameter, an equivalent heat transfer coefficient (line 30) is obtained by dividing the difference in radiant heat flux between 1400 and 1500°R by 100°F. A better, but much more laborious approximation, can be obtained by assuming a fin thickness, imposing the given heat load on the tube, and estimating the heat dissipation first from the tube and then from each of a number of increments of fin height, calculating the temperature drop along the fin for each increment, and factoring in the lower local surface heat flux. This operation can be continued until sufficient fin height is provided to dissipate the imposed heat load.

The fin root thickness found in line 32 is rather large, and implies that, to save weight, it might be better to use a lower fin efficiency, a larger number of tubes, or a fin thickness that varies as the square of the fin height. The procedure roughed out in Table 18.1 can be refined and used as the basis for a parametric study to obtain an improved set of proportions.

Because of the high thermal conductivity of liquid potassium, the condensing film coefficient is so high (probably about 10,000 Btu/h · ft^2 · °F) that the film temperature drop would be small on the vapor side. Furthermore, the bulk of the heat flowing into the fins would come from condensation close to the fin root; there would be relatively little circumferential heat flow through the tube wall into the fin.

Example 18.2. *Design of a Radiator for a 5-MWe Plant.* The above example was for the highest capacity, high-temperature space radiator built and ground tested

up to the time of writing. The configuration was chosen to fit on a Titan launch vehicle and provide 140 kWe of electrical power for a space vehicle. However, for a 5-MWe electrical output the radiator becomes so large that a different configuration must be considered. Further, recent data on the accumulation of orbital debris from spacecraft (mostly Russian) indicates that this debris represents a greater hazard than meteoroids,[5] so that the probability of a radiator tube puncture for Earth orbits up to 1200 km becomes substantial for design power outputs above around 100 kWe. Thus the effects of the different boundary conditions on the design of radiator having 25 times the heat rejection capacity is instructive, for these factors have led to a quite different structure with very different elements. This radiator was a major component in a conceptual design for a 5-MWe nuclear electric space power plant employing a boiling potassium reactor coupled to a potassium vapor cycle for a system built of refractory alloy so that the radiator could operate at 746°C (1375°F). The principal design requirements were that the radiator should be launched in a package that would fit in the cargo bay of the NASA shuttle, all elements of the launch package should withstand 5 g during the launch, and assembly in orbit should be as simple as possible. In addition, to minimize its vulnerability to hypervelocity particles such as meteoroids,[5] it was decided that the bulk of the radiating surface should be mounted on heat pipes so that a leak in any given heat pipe would result in a reduction of much less than 1% in the capacity of the power plant.

Solution. The basic layout of the spacecraft is shown in Fig. 18.1. Three configurations were considered for the radiator, the largest component in the system. These were two banks of tubes in a flat array as in Fig. 18.6-2, three banks in a triflute array as in Fig. 18.6-3, and four banks in a cruciform arrangement. The view factors cited in the discussion accompanying Fig. 18.6 would be unity for the flat plate, 0.866 for the triflute arrangement, and 0.707 for the cruciform arrangement. (See also Refs. 11 and 13 of Chapt. 3.) Compromising between envelope size and view factor led to the choice of the triflute arrangement shown in Fig. 18.1. Three turbines operating in parallel would discharge vapor into three manifolds extending the length of the power plant package parallel to its centerline. The vapor flowing in each manifold would give up its heat by condensing on the heat input ends of two banks of potassium heat pipes. The pipes in the bank extending inward to the centerline of the system would be welded into the manifold. To stay within the launch envelope, the tubes in the outer bank would have long tapered inner ends that would fit into sockets welded into the manifold. For launching, the outer tube banks would be stowed in the gaps between the inner sets of tube banks; once in orbit these outer tube banks would be rotated into their radial operating positions and inserted into the sockets provided in the manifolds. To obtain a good thermal bond in the tapered joint, they might be driven in with small explosive charges with a thin layer of a very ductile metal foil in the joint to give a nearly perfect thermal bond. Indium has been used for this purpose very successfully in joining copper bus bars at low temperatures; silver should perform well in the temperature range considered here.

Design conditions:
  Heat load = 21.5 MWt
  Potassium condensing at 746°C (1,375°F or 1,835°R)
  Launch package envelope, 4.5 m dia. × 18.2 m long (14.76 × 59.7 ft)
  Employ heat pipes to reduce vulnerability
  Vapor velocity in manifold = 0.24 × sonic velocity
  Configuration of Fig. 18.1

Vapor flow rate:
  v = 2.18 m³/kg (35 ft³/lb), H_{fg} = 1,937 kJ/kg (831.7 Btu/lb)
  Sonic velocity = 542 m/s (1,780 ft/s)
  Total flow of vapor = 21.5/1937 = 11.1 kg/s (24.5 lb/s)
                      = 24.2 m³/s (860 ft³/s)
  Velocity = 0.24 × 542 = 131 m/s (430 ft/s)
  Flow area = 0.186 m² (2 ft²)
  Flow area per manifold = 0.0622 m² (0.67 ft²)
  Manifold inlet ID = 0.282 m (0.925 ft)

Assume:
  25.4-mm (1-in.) ID heat pipes with 38.1-mm (1.5-in.) high Cu fins
  Tubes on 101.6-mm (4-in.) centerlines
  44.5°C (80°F) drop from vapor in manifold to heat pipe fin surface

Heat dissipation:
  Ideal black body (from Fig. H5.16) = 49 kW/m² (15,500 Btu/h · ft²)
  Configuration view factor = 0.866
  Fin efficiency = 0.865 (see Fig. H7.3)
  Emissivity = 0.9
  Effective Q/A = 49 × 0.866 × 0.85 × 0.9 = 33.1 kW/m² (10,470 Btu/ft²)

Radiator size (to fit in shuttle):
  Area required = 21,500/33.1 = 650 m² (7,000 ft²)
  L = 17.8 m (58.5 ft)

Total width of a radial panel = 650/(3 × 2 × 17.8) = 6.1 m (20 ft)
Radius at in-board end of finned portion of outboard heat pipe = 2.24 m (7.35 ft)
Length of finned portion of outer heat pipe = 6.1 − 2.24 = 3.86 m (12.65 ft)
Heat load per outer heat pipe = Q/A × A = 33.1 × 3.86 × 0.2 = 26.2 kW (88,800 Btu/h)

Figure 6.8 indicates that a 1-in.-diameter potassium heat pipe has a capacity of 31 kW at 730°C, hence the choice of a 1.0-in. heat pipe is satisfactory.

Heat transfer from manifold vapor to the heat pipe in its socket:
  Total wall thickness = 1 + 1 = 2 mm (0.080 in.)
  k = 21.61 W/m² · °C (150 Btu/h · ft² · °F/in.)
  For a wall ΔT = 33.4°C (60°F), Q/A = $\Delta T k/L$ = 33.4 × 21.6/0.002 = 360 kW/m² (123,000 Btu/h · ft²)
  Area required = 26.2/360 = 0.0727 m² (0.795 ft²)
  Mean perimeter of tube socket = 48.5 × 3.14 = 152 mm (0.50 ft)
  Length of socket = 608 mm (2 ft) (part would be inside manifold)

Effects of Wavelength on Emissivity and Reflectivity

The values used for the emissivity in Examples 18.1 and 18.2 fall in the range of the values usually given in handbooks for lightly sandblasted and oxidized stainless steel. However, for many spacecraft applications including radiators, there is a strong incentive to take a more critical look at this parameter because the radiant energy input from the sun is likely to be an important factor, and both the emissivity and the reflectivity vary widely with both the choice of material and the wavelength. Crew compartments in low orbit present especially difficult temperature-control problems because almost half their orbit lies in the shadow of the Earth, so that drastic changes in temperature could occur unless the design is well chosen. The data given in handbooks for the reflectivity and emissivity of typical surfaces are ordinarily just average values for the usual range of interest. (See Table H2.5 in the back of this text, for example.) However, these average values do not provide sufficient information if one is to evolve the best design for a particular application because the reflectivity and emissivity not only vary with wavelength for any given material, but they may do so in surprisingly different ways. As a consequence, it is possible in a space radiator, for example, to choose a coating that will give both a good emissivity for the wavelengths of thermal radiation from the radiator yet have a good reflectivity for the shorter wavelengths of solar radiation. Similarly, some coatings for the solar energy collection panels treated later in this chapter will give a high absorptivity for solar radiation, yet low emissivity for the longer wavelength thermal radiation at the panel operating temperature.

Emission Spectra for the Sun and Black Bodies

The emission spectra for the sun for operation both at the Earth's surface and in Earth orbit are shown in Fig. 18.8[6] along with a curve for radiation from an ideal black body. The difference between the spectrum in space and that for the Earth's surface is cross-hatched and the various molecular species responsible for absorption are noted in the figure. Curves for the spectrum of the radiation from ideal black bodies radiating at various temperatures are shown in Fig. 18.9.[7] A second scale at the bottom of Fig. 18.8 shows the fraction of the energy in the solar spectrum in space that falls below any given wavelength; for example, only about 10% of the total energy flux lies in the wavelength range below 0.5 μm, while over 90% lies below a wavelength of 2 μm, hence over 80% of the energy lies in the 0.5- to 2.0-μm range. On the other hand, Fig. 18.9 shows that the bulk of the

energy emitted from an ideal black body at 500 K (227°C, or 441°F) is at longer wavelengths, that is, in the wavelength range from 2 to 30 μm.

Reflectivity and Emissivity as Functions of Wavelength

One ordinarily thinks of reflectivity in terms of specular reflection such as that from a mirror, and the data given for reflectivity in handbooks are commonly for specular reflectivity, that is, for light reflection where the angle of reflection is equal to the angle of incidence. However, the diffuse reflectivity (or hemispherical reflectivity) of a white surface may be as high as the specular reflectivity of a polished metal surface. In fact, the upper surface of the fuselages of passenger aircraft is commonly painted white to reduce the heat load on the air-conditioning system when the aircraft is on the ground on a hot sunny summer day because the diffuse reflectivity of the white paint is significantly higher than that for the bare surface of the aluminum, which cannot be kept highly polished.

A substantial effort has been devoted to crew compartment temperature control in Earth-orbiting spacecraft as the vehicle passes alternately into the Earth's shadow and then into bright sunlight. It happens that most of the white oxides are characterized by a high diffuse reflectance for solar spectrum radiation, yet have a high emissivity in the infrared. A considerable amount of testing was carried out in the 1960s in the course of detailed investigations of these properties for use in

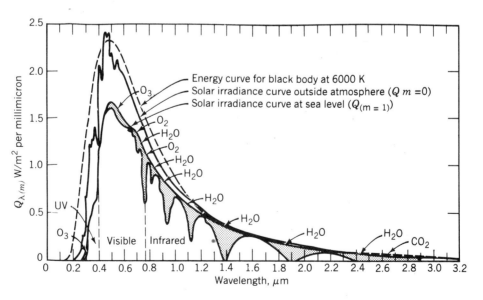

Figure 18.8 Energy distribution in the solar spectrum in space, at the Earth's surface, and for an ideal black body at 6000 K. (Morrison, Ref. 6.)

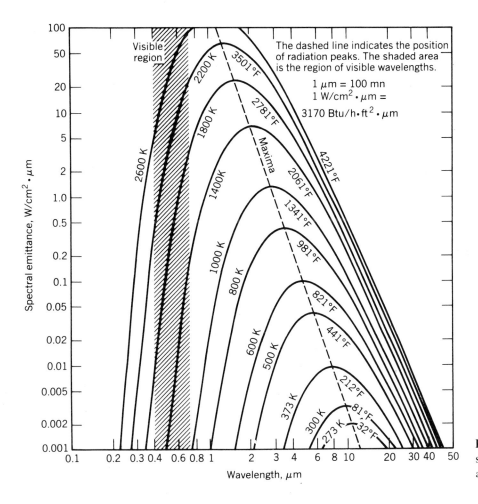

Figure 18.9 Black-body spectral intensities for source temperatures between 273 and 2600 K. (Bolz and Tuve, Ref. 7.)

spacecraft design, but the matter is highly specialized so that the data are hard to find. The best collection found by the author was that in Reference 8, but other useful data may be found in sources such as References 9 to 11.

The most significant data are of the sort shown in Fig. 18.10, which presents values for both the emissivity ϵ and the reflectance R as functions of wavelength for alumina and zirconia.[12] Silica, magnesia, magnesium aluminate, titania, thoria, calcium zirconate, zinc oxide, and many other white oxides have similar characteristics, although their characteristic curves shift somewhat to the right or left and up or down; hence, for any given application, one or another may appear preferable. In addition, these materials differ in the degree to which their reflectance and/or emissivity may deteriorate under ultraviolet light, solar proton bombardment, or neutron and gamma radiation from a fission reactor, particularly if they are applied as pigments in a paint with an organic vehicle.[13,14] The effects of gamma dose on the solar absorption α of some typical coatings are shown in Fig.

18.11, whereas Fig. 18.12 shows the effects of both ultraviolet and gamma radiation on titania bonded with an epoxy resin. Fairly extensive testing indicates that neutron and gamma or fast proton radiation at the levels to be expected in spacecraft have little effect on the reflectance of most coatings, but most painted coatings are damaged by ultraviolet radiation. The least sensitive coating has been found to be ZnO, data for which are shown in Fig. 18.13.[15] NASA has chosen this as the pigment in paints for use on the relatively low-temperature surfaces of space vehicles. ZnO has also been applied with sodium or potassium silicate as the binder for use on surfaces operating at temperatures up to 300°C; as can be seen in Fig. 18.14, this combination has shown little deterioration after extensive exposure to ultraviolet light.

It is difficult to judge by looking at Figs. 18.8–18.14 just how the properties of the various coatings are related on the one hand to the solar spectrum and on the other to the emission spectra of black bodies operating at the

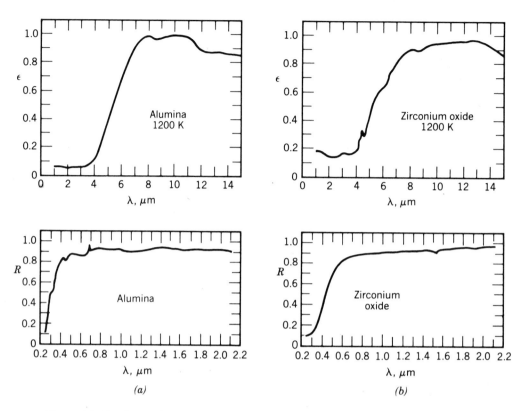

Figure 18.10 Spectral emittance and reflectance curves for alumina and zirconia. (Clark and Moore, Ref. 12.)

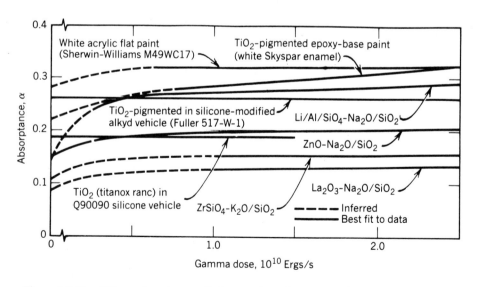

Figure 18.11 Effects of nuclear radiation on the solar energy absorptance of typical reflective coatings tested in vacuum. (Breuch and Pollard, Ref. 8, p. 365.)

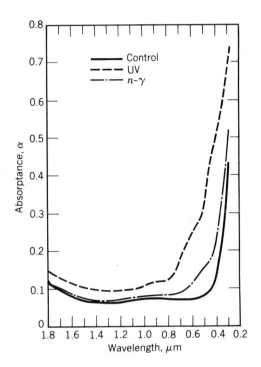

Figure 18.12 Effects of ultraviolet and neutron-gamma radiation on the solar spectrum absorptance of zirconium-potassium silicate coatings. (Gilligan and Caren, Ref. 13.)

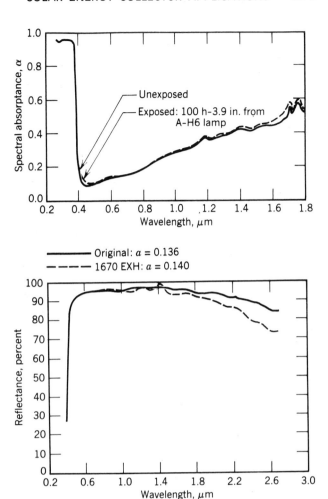

Figure 18.13 Spectral absorptance and reflectance curves for zinc oxide. (Olson et al., Ref. 14.)

temperatures contemplated for nuclear electric space power plants. Figure 18.15 was prepared to aid in visualizing these effects. In comparing the reflectance of ZnO and Al_2O_3 with the solar spectrum it appears at first glance that the alumina is more attractive than the ZnO. However, if one looks at the emissivity curve for alumina and compares it with the black-body spectral curves for 500 and 1000 K it is apparent that the emissivity would be poor in that range of wavelengths. On closer inspection, if one looks at the extra scale at the bottom showing the fraction of the energy in the solar spectrum below any given wavelength, one can see that less than 10% of the energy in the solar spectrum lies below the 0.4-μm wavelength at which the reflectance of the ZnO falls off sharply; hence ZnO would reflect over 80% of the incident solar radiation even though the reflectance is not nearly as good as that of the alumina at the shorter wavelengths. Further, the total emittance of alumina or zirconia at 1200 K runs only 0.3 to 0.5 (see p. 251 of Ref. 8), whereas that for ZnO runs about 0.9 in the temperature range up to about 800 K. Thus the ZnO should be the better coating for the radiator of a Brayton cycle.

As a point of interest, the emission spectrum for the

Earth is also shown in Fig. 18.15 to give an idea of what its effect might be on a radiator constructed to take advantage of a special coating such as ZnO. The total energy in the thermal radiation from the Earth is so small that it barely appears in the lowest decade in the chart.

SOLAR ENERGY COLLECTOR APPLICATIONS

From the cost-effectiveness standpoint, the most favorable application for solar energy is for residential heating and domestic hot water, but even in this low-temperature area it is difficult to keep the costs for the energy collection and storage system to a level competitive with other sources of heat. Some idea of the allowable cost

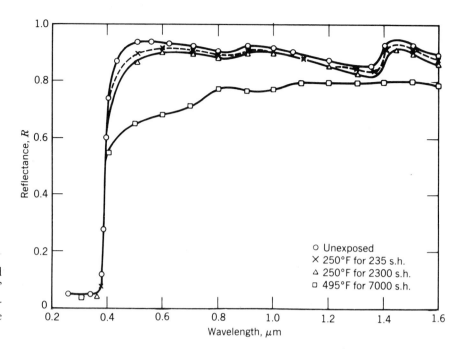

Figure 18.14 Effects of simulated ultraviolet exposure in "sun hours" (s.h.) in space on the spectral reflectance of zinc oxide-potassium silicate coatings. (Zerlaut et al., Ref. 15.)

for a solar energy collection panel can be obtained by considering an installation in southern Arizona where the annual solar energy input is about 2000 kWh/m². Various losses in the collection process make it difficult to collect as much as half of this. Thus, for 1987 costs, if the cost of energy from natural gas is $6/10⁶ Btu, the value of 1000 kWh/m² per year would be about $20/ year. The cost of a plate coil (see Table 11.2) would be about $10/ft² of total area, or about $20/ft² of face area, or about $200/m² just for the plate coil, hardly an attractive investment. This indicates the reason that materials much less expensive than metal, for example, polyolefins, are commonly used for solar energy collectors of this type. Further, the cost of the collector is only about half that for the complete solar hot water system, including a large storage tank with capacity sufficient for at least two days' supply, a pump, and controls. Under any circumstances, it is evident that every step to cut costs and increase the effectiveness of the energy collection process must be taken if a solar hot water heater is to have favorable costs relative to other sources of energy. Special coatings provide one option.

Extensive tests of different coatings for solar energy collectors have been carried out in connection with the large solar energy program initiated in the early 1970s.[16,17] One of the most promising of the plated coatings tested in the study of Reference 17 was a black chrome coating deposited electrolytically on nickel-plated steel; the reflectivity of a set of these coatings laid down

at different current densities is shown in Fig. 18.16. The absorptivity for sunlight was about 95% and the emissivity about 11% with no noticeable deterioration after a high-humidity test, MIL-STD-810B.

To keep costs low, many of the coatings investigated have been in the form of pigmented paints. For metal panels, the first cost of a painted coating appears to be about a third that of a plated coating, but the polymer vehicle of the paint makes it much more subject to deterioration with time because of moisture effects and/or damage from ultraviolet light. When taking the probable life of the coating into account, the better coatings were found to be mixed oxides of chrome, iron, manganese, and copper in silicone vehicles. These have given absorptivities of about 0.90 and emissivities of around 0.3 for the region of interest. The importance of this difference in absorptivity and emissivity is shown in the following example.

Example 18.3. *Flat-Plate Solar Energy Collector.* A solar energy collector array is to provide 60°C (140°F) hot water for a house in Florida with a water supply temperature of 20°C (68°F) and an average heat load per day of 20 kWh. Plate coils similar to those of Fig. 2.1 having a coating similar to that of Fig. 18.16 are to be employed. They would be mounted on the roof facing south at an angle of 45° so that they would be approximately normal to the sun's rays at solar noon in January. As a base case, both the solar energy absorbed on a clear

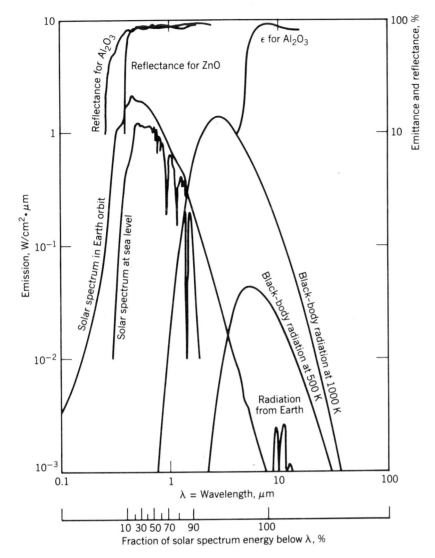

Figure 18.15 Reflectance and emittance curves superimposed on curves for both the solar spectrum and ideal black-body radiation.

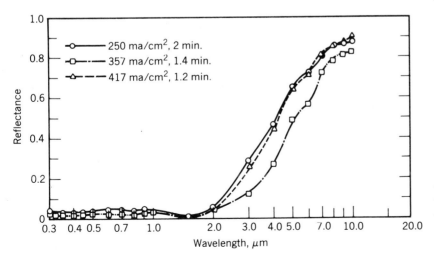

Figure 18.16 Reflectance of a black chrome coating on nickel-plated steel for electrical deposition with a time-current density product of 500 ma · min/cm^2. (Courtesy Honeywell, Inc.)

day and the energy re-emitted per square meter should be estimated for both noon and 3:00 p.m. on a clear day when the solar energy flux would run about 900 W/m². Neglect for the moment the heat losses to conduction and thermal convection, and losses in the incident radiation that would be present if there were cover glass plates designed to reduce heat losses from the top surface caused by air convection.

The energy absorbed at noon:

$$Q/A = 900 \times 0.95 = 855 \text{ W/m}^2$$

From Fig. H5.16, the heat flux from an ideal black body at 333 K is 695 W/m² (220 Btu/h · ft²), hence the heat reradiated for an emissivity of 0.11 would be

$$Q/A = 695 \times 0.11 = 76 \text{ W/m}^2$$

At 3:00 p.m. the panel would be at an angle of 45° to the incident rays of the sun. Neglecting the probably greater reflectivity of the surface at that angle and the reduced solar flux because of increased absorption in the atmosphere, the energy absorbed by the panel would be

$$Q/A = 0.707 \times 855 = 604 \text{ W/m}^2$$

Thus the net heat input to the water in the plate coil would be 769 W/m² at noon and 528 W/m² at 3:00 p.m. However, if the coating had an emittance equal to its absorptance, the heat reradiated would have been 660 W/m², the net energy to the water would have been only 195 W/m² at noon, and there would have been a net energy loss of 56 W/m² at 3:00 p.m. Of course, glass cover plates would reduce the loss to reradiation, but these would add substantially to the cost of the installation.

Solar Energy Collectors with Parabolic Trough Reflectors

The heat loss from a simple flat-plate collector caused by reradiation rises rapidly with an increase in the temperature at which the energy is collected so that the collection efficiency falls off rapidly as shown in Fig. 18.17. To reduce the reradiation loss, the best procedure is to employ a concentrator in the form of a parabolic reflector or a Fresnel lens. This has the disadvantage that in most areas about half of the incident solar energy is diffuse rather than direct because of high thin clouds, haze, or scattered passing clouds, so that reflectors are used mostly in desert areas where there is relatively little

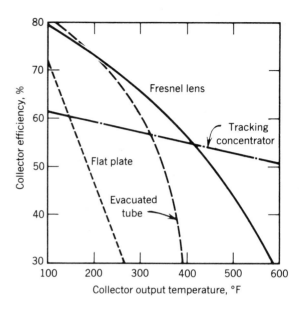

Figure 18.17 Estimated performance of several types of solar collectors. Ambient temperature = 21°C (70°F). Solar intensity = 948 W/m² (300 Btu/h · ft²), all available as direct radiation. (Courtesy Barber-Nichols Engineering Co.)

cloudiness. Further, the concentrator must be moved continually so that it tracks the sun to keep the rays focused on the receiver. As a consequence, the bulk of the problems are outside the scope of this text.

CRYOGENICALLY COOLED WALLS FOR SPACE CHAMBERS

Quite a number of large space chambers have been built to make it possible to conduct rigorously thorough tests both of components for space missions and complete spacecraft under conditions closely simulating those in space. One of the important requirements has been to simulate the 4 K temperature characteristic of space. This has called for lining the chamber with liquid helium-cooled panels. Layers of reflective insulation fill the gap between the panels and the walls of the tank. Panels cooled with liquid nitrogen are placed at an intermediate position in the layers of reflective thermal insulation to reduce further the load on the liquid helium refrigeration system. The hard vacuum in the space chamber greatly eases the thermal insulation problem, but it also imposes a stringent leaktightness specification to minimize the pumping load on the vacuum system. Thus the principal heat load on the helium system usually is from the bank of lamps used to simulate the sun; the beams from these lamps are focused on the equipment under test. The following example gives some instructive illustrations of the basic heat transfer relations involved in this specialized application.

Example 18.4. Helium-Cooled Liner for a Space Chamber. A 10-m diameter × 15 m high cylindrical vacuum chamber for testing spacecraft is to be lined with liquid helium-cooled walls that are to operate at 4 K. The maximum total heat load expected in the test program projected is from a spacecraft having a 30-kWt isotope power source. If the resulting heat flux were uniformly distributed over the 600 m^2 of cooled lining, the local heat flux would be 50 W/m^2, but the form factor for the spacecraft installation will give local heat fluxes that will run double the average. In addition, peak local heat loads at the wall will be imposed by beams from lamps simulating solar radiation; the energy intensity in these beams will run 1300 W/m^2. The helium-cooled walls are to be fabricated of copper sheet in a form similar to that of the plate coils of Fig. 2.1 except that, to minimize the inventory of liquid helium, the passages will be 5 mm square with the diagonal of the square cross section in the plane of the plate.

Solution: The critical heat flux in regions exposed to the maximum heat flux appears to be the limiting factor in the design of the copper plates. Choose the spacing for the helium passages on this basis. (Neglect the heat leakage into the outer face of the panels through the reflective insulation between them and the liquid nitrogen-cooled panels that would line the interior of the vessel.)

The critical heat flux for liquid helium boiling near 4 K as given by Fig. 5.20 is about 1.0 W/cm^2 (3171 Btu/h · ft^2). Doubling the average heat flux from the spacecraft gives local values of 100 W/m^2 which could add to the 1300 W/m^2 from the simulated sun to give a peak local heat flux of 1400 W/m^2, or 0.14 W/cm^2 (444 Btu/h · ft^2). This is only 14% of the critical heat flux, but a factor of safety of about 3 seems essential. On this basis, the surface area of the helium flow passages should be almost half the total face area of the panels. With the internal surface area of each 5-mm square passage running 2 cm^2/cm of length, a passage centerline spacing of 30 mm appears to be a good choice if thermal conduction in the copper sheet is good enough to distribute the heat from the fin fairly uniformly between the front and rear faces of the helium passage.

Figure 5.20 indicates that the film temperature drop for a heat flux of 0.3 W/cm^2 would be about 0.3 K, hence thermal conduction in the copper should be sufficiently good to distribute the heat flux over the perimeter of the helium channel with a temperature variation of less than, say, 0.1 K. The thickness of the copper walls has been tentatively chosen as 0.025 mm (0.010 in.) to give a reasonable degree of structural stability. Checking the temperature drop from the fin juncture with the helium passage wall to its midpoint, the heat flux from the fin is

$$Q = 0.14 \times 0.3 = 0.042 \text{ W/cm}$$

The heat flux entering the helium passage wall is

$$Q/A = 0.042/(0.025 + 0.025) = 0.84 \text{ W/cm}^2$$

The thermal conductivity of copper at 4 K is very high (see Table H2.7), about 113 W/cm · K) (6520 Btu/h · ft · F), hence the resulting temperature drop in the passage wall is

$$T = 0.84 \times 0.25/113 = 0.0018 \text{ K}$$

This is less than 2% of the 0.1 K that would probably be acceptable.

REFERENCES

1. A. P. Fraas, "Reactors for Space," *Recent Advances in Engineering Science*, vol. IV, Gordon & Breach Scientific Publishers, New York, 1969, p. 1.

2. R. B. Korsmeyer, "Condensing Flow in Finned, Tapered Tubes," Report no. ORNL-TM-534, Oak Ridge National Laboratory, 1963.

3. A. P. Fraas, "Design and Development Tests of Direct-Condensing Potassium Radiators," *AIAA Specialists Conference on Rankine Space Power Systems*, vol. 1, USAEC Report CONF-651026, October 1965.

4. I. J. Loeffler et al., "Recent Developments in Space Power System Meteoroid Protection," AIAA Paper no. 64-759, Third Biannual Aerospace Power Systems Conference, September 1964.

5. A. P. Fraas, "Protection of Spacecraft from Meteoroids and Orbital Debris," Report no. ORNL/TM-9904, Oak Ridge National Laboratory, March 1986.

6. R. B. Morrison, *Design Data for Aeronautics and Astronautics*, John Wiley and Sons, New York, 1962.

7. R. E. Bolz and G. L. Tuve, eds., *Handbook of Tables for Applied Engineering Science*, 2nd ed., CRC Press, Boca Raton, Fla., 1973.

8. *Symposium on Thermal Radiation of Solids*, March 4–6, 1964, NASA SP-55, (Air Force ML-TDR-64-159).

9. "Energy Conversion for Space Power," *Progress in Aeronautics and Rocketry*, vol. 3, Based on a Symposium of the American Rocket Society held in Santa Monica, Calif., September 27–30, 1960, Academic Press, New York, 1961.

10. "Thermophysics and Temperature Control," *Progress in Aeronautics and Astronautics*, vol. 18, Based on the American Institute of Aeronautics and Astronautics Thermophysics Specialist Conference held at Monterey, Calif., September 13–15, 1965, Academic Press, 1966.

11. H. M. Shafey et al., "Experimental Study on Spectral Reflective Properties of a Painted Layer," *AIAA Journal*, vol. 20(12), December 1982, p. 1747.

12. H. E. Clark and D. G. Moore, "Method and Equipment for Measuring Thermal Emittance of Ceramic Oxides from 1200 to 1800 K," *Symposium on Thermal Radiation of Solids*, NASA SP-55, 1964, p. 241.

13. J. E. Gilligan and R. P. Caren, "Some Fundamental Aspects of Nuclear Radiation Effects in Spacecraft Thermal Control Materials," *Symposium on Thermal Radiation of Solids*, NASA SP-55, 1964, p. 351.

14. R. L. Olson et al., "The Effects of Ultraviolet Irradiation on Low α/ϵ Surfaces," *Symposium on Thermal Radiation of Solids*, NASA SP-55, 1964, p. 421.

15. G. A. Zerlaut et al., "Ultraviolet Irradiation of White Spacecraft Coatings in Vacuum," *Symposium on Thermal Radiation of Solids*, NASA SP-55, 1964, p. 391.

16. J. F. Kreider and F. Kreith, *Solar Energy Handbook*, McGraw-Hill Book Co., New York, 1981.

17. H. Y. B. Mar et al., "Optical Coatings for Flat Plate Solar Collectors," Final Report of Honeywell, Inc. to Energy Research and Development Administration, Contract no. NSF-C-957 (AER-74-09104), 1975.

19

Cooling Towers

Cooling towers have been widely used to dispose of waste heat from industrial processes and from refrigeration or air-conditioning systems where it has been cheaper or more convenient to reject heat to the atmosphere rather than to water in a nearby river, lake, or ocean. In many instances the choice has been largely a matter of comparative costs, but since World War II our rapidly growing electrical power system has begun to outstrip the heat sink capacity of some of our rivers so that, even with the broad latitude in site selection open to central stations, in many cases there has been no alternative but to build large cooling tower installations. This has been particularly true in the Southwest.

The design of cooling towers is much more empirical than that of any other type of heat exchanger. There are so many imponderables, in fact, that some engineers hint at witchcraft. Much of the difficulty stems from idiosyncrasies in the behavior of the atmosphere in the vicinity of a cooling tower. Under certain conditions with little or no wind the combined effects of the geometry of the terrain and buildings in the vicinity may lead to air recirculation through the tower and, hence, a loss in performance. These effects are so much a function of the local topography, the prevailing wind direction, the weather, and the like that they are difficult to predict, and the performance of a cooling tower of a given design may vary widely from one location to another. Because of these and other complexities, this chapter is intended to

do little more than present the basic relationships involved and show something of their use in the specification, selection, and acceptance testing of cooling towers.

TYPES OF COOLING TOWERS

Cooling towers were developed from spray ponds in an effort to obtain a system that would take up less space.[1] It should be noted at this point that the heat dissipation capacity per unit of area of a small pond can be increased about 20 times by installing a simple spray system, and about 1000 times by building a cooling tower. Cooling towers have a further advantage over spray ponds in that they reduce the water consumption for a given heat load by a factor of about five, because they can be designed to eliminate the loss of water carried off by the wind in droplet form.

Natural Convection Cooling Towers

The simplest type of cooling tower is just a small water spray pond surrounded with walls having inwardly sloping louvers, as in Fig. 19.1. In this type of cooling tower the air changes direction in passing through the louvers as it leaves, and the suspended droplets of water

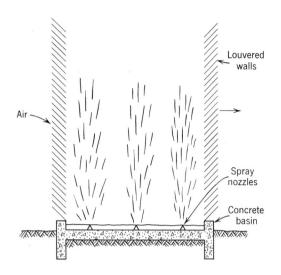

Figure 19.1 Section through a simple cooling tower formed by enclosing a spray pond with louvered walls.

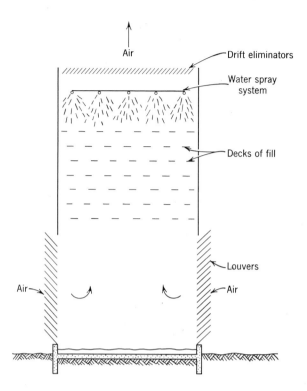

Figure 19.2 Section through a natural convection cooling tower with "fill" to increase the effective water droplet surface area by multiple splashing.

impinge on the louver slats, deposit out, and drain back into the basin at the base of the tower.

Both the water pumping requirements and the capacity of the cooling tower per unit of base area can be improved by providing horizontal surfaces within the cooling tower to reduce the average velocity of the falling droplets and to increase the time that the droplets are exposed to the cooling air stream in falling through the tower. Another advantage of this approach is that it makes possible counterflow performance and hence a lower temperature in the exit water stream. The latter advantage can be realized by using low-pressure water sprays at the top, placing horizontal decks of latticework in the tower, and arranging the walls as in Fig. 19.2 so that the air enters horizontally and is discharged vertically. A desirable feature of this arrangement is that the vertical upward movement of the air also slows the fall of the water droplets and thus increases the effective surface area for any given water flow rate. In this type of cooling tower the surfaces within the tower are called *packing* or *fill*, and are usually staggered so that water droplets can fall only a few feet before striking a surface. Some typical grids or *decks* made of redwood strips nailed to 1×2 in. redwood stringers are shown in Fig. 19.3. Grids of plastic are replacing redwood.

The increasing temperature of the air moving upward through the tower induces thermal convection circulation, which is especially helpful if the wind velocity is very low. For large installations, many cooling towers of the sort shown in Fig. 19.2 may be placed side by side in a long row at right angles to the direction of the

prevailing wind. This is a particularly effective arrangement in coastal regions where the winds are predominantly onshore or offshore.

The British initiated the construction of large thermal convection towers, many of them over 300 ft high. A typical set is shown in Fig. 19.4. The air enters through the annular gap provided by a ring of supporting columns at the base, is heated as it passes through fill near the base (see Fig. 19.5), and then rises under the action of thermal convection forces to the top of the stack. This type of cooling tower is well suited for sites in which the ambient air temperature rarely exceeds 27°C (80°F), and a substantial steady wind is available throughout the year. Other conditions favoring these large natural draft towers are a large temperature range (and hence a substantial air temperature rise), no need for a close approach temperature, a large winter heat load, and low capital charges.[2] The stacks themselves are made of reinforced concrete having a wall thickness of only two to six inches. The geometric figure is a hypoid so that two layers of straight steel reinforcing rods can be arranged so that they are inclined in opposite directions from the vertical to form a network. This arrangement facilitates fabrication, yet gives a strong structure.

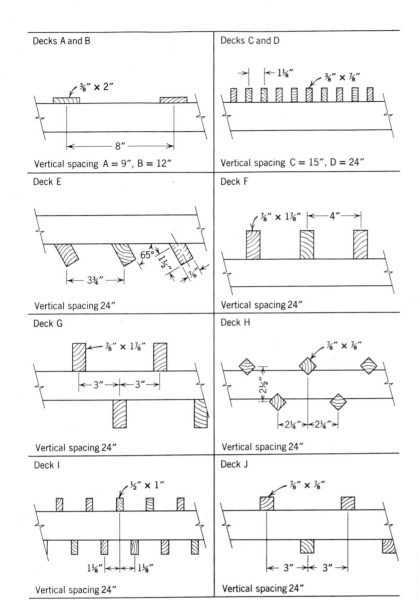

Figure 19.3 Typical geometries used in constructing redwood "fill" for cooling towers. (Kelly and Swenson, Ref. 6.)

Forced Convection Cooling Towers

The capacity of a given size of cooling tower may be increased by the installation of a fan as in Figs. 1.19 and 19.6. The added cost of the fan and the electrical power that it consumes is usually more than offset by reductions in the capital charges per unit of heat rejection capacity. Fans may be mounted just outside the base of the tower so that the air flow is directed horizontally inward to give a *forced draft* installation; or *induced draft* fans may be mounted at the top so that they draw air up through the tower and discharge it vertically upward from the top, as in Figs. 1.19 and 19.6. It is easier to provide a sturdy fan-mounting structure for the first arrangement, but the second is generally preferred, because it not only is less sensitive to wind direction, but by discharging the air upward in a strong vertical jet it reduces the recirculation of the warm moist exit air back into the tower under unfavorable wind and atmospheric conditions. (The recirculation problem is discussed further in a later section.) Low-speed fans as much as 12 m (40 ft) in diameter are employed to minimize the power consumption. For induced draft installations the motor is often mounted to one side of the fan and the power transmitted through a horizontal shaft to a gear box on the vertical shaft under the fan. This drive arrangement is used for the towers in both Figs. 1.19 and 19.6.

The towers of Figs. 1.19 and 19.6 differ in that the

Figure 19.4 Large natural convection cooling towers with hypoid concrete shells constructed for a large British steam power plant. Each tower handles 3,500,000 gal/h. (Courtesy Film Cooling Towers, Ltd.)

air flows vertically upward through the fill in Fig. 1.19, whereas it flows horizontally through the fill in the tower of Fig. 19.6. The horizontal flow configuration permits a reduction in tower height, and hence in cost, for locations in which a taller tower is not required by the nature of the terrain, nearby buildings, etc.

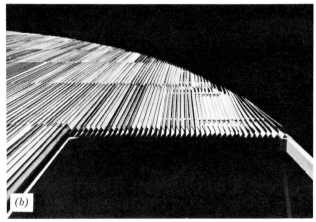

Figure 19.5 View of (a) water distributing troughs and corrugated asbestos cement sheet packing and (b) asbestos cement eliminators for cooling towers similar to those of Fig. 19.4. (Courtesy Film Cooling Towers, Ltd.)

BASIC RELATIONS

The prime function of a cooling tower is to reduce the temperature of a stream of hot water to as low a value as practicable. This temperature drop in the stream of water flowing through the tower is known as the *temperature range*. The cooling is accomplished partially by raising the temperature of the surrounding air and partially by evaporating a portion of the stream of hot water. The relative amounts of heat going into increasing the temperature of the air and into evaporating the water depend on the humidity of the air entering the cooling tower. Another factor, probably the most important single figure of merit for a cooling tower, is the extent to which the water exit temperature approaches the wet bulb temperature of the entering air—the latter being the minimum temperature to which the water could be cooled in an ideal installation. In any given tower this temperature difference, known as the *approach temperature*, varies with the entering air wet bulb temperature, the water-flow rate, and the heat load.

Heat Balance

Since the amount of water carried off by the air in the form of suspended droplets is normally negligible and the specific heat of water is unity, the product of the drop in the water temperature Δt_w and the liquid water-flow

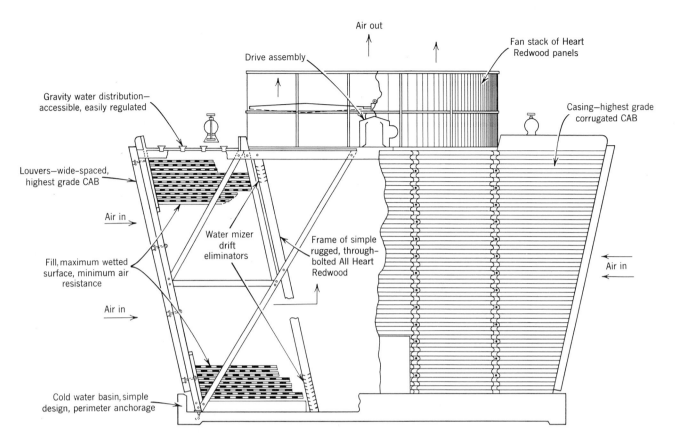

Figure 19.6 Induced draft vertical cooling tower constructed of redwood and corrugated asbestos board. (Courtesy Foster Wheeler Corp.)

rate L_w is equal to the product of the enthalpy rise in the air ΔH_a and the air-flow rate G, that is,

$$\Delta t_w L_w = \Delta H_a G \qquad (19.1)$$

where the water- and air-flow rates L_w and G are in pounds per hour per square foot of horizontal cross-sectional area in a vertical tower.

The enthalpy of moist air is almost solely a function of the wet bulb temperature. This is so nearly true that only the wet bulb temperature lines are ordinarily drawn on psychometric charts; enthalpy scales for the air-water vapor mixture are often drawn close to the ends of the constant wet bulb temperature lines. These wet bulb temperature lines can then be used with the enthalpy scales for obtaining approximate values of the enthalpy, or a straight edge can be used for an accurate determination. Thus the wet bulb temperatures for the inlet and outlet air give a good measure of the enthalpy rise. The dry bulb temperatures are mainly significant from the standpoint of the water consumption.

Heat Transfer

The heat balance of Eq. 19.1 includes no terms that define the size of the cooling tower. If the cooling tower is considered analogous to a heat transfer matrix with a water film surface area that depends on the water and air flow rates and on the characteristics of the packing or fill, heat will be given up to the air by two mechanisms: conventional convection heat transfer and vaporization of the water. It has been found that the rate of heat loss from the water by vaporization is analogous to the heat transfer coefficient for thermal convection, since both depend on the rate at which mixing takes place between the thin gas film immediately adjacent to the heat transfer surface and the bulk air stream passing over the surface. Test data indicate that the coefficient for heat loss by vaporization from the water film to an air stream is approximately equal to the conventional convection heat transfer coefficient h divided by the specific heat of the air.[3] Thus the enthalpy transfer coefficient K, for the rate of heat loss by vaporization, is given approximately by $K = h/c_p$.

It has been shown[4] that in any given element of tower volumn dV having the surface area a per unit of volume, the heat given up per pound of water can be related to the heat transferred to the air by convection and to the heat loss by vaporization by the following:

$$dt_w L_w = [h(t_w - t_a) + K \Delta H_v(x_s - x)] a \, dV$$

(19.2)

where ΔH_v is the enthalpy of vaporization, x_s the water vapor content of saturated air, and x the water vapor content of the air stream.

Equation 19.2 may be manipulated to put it into a somewhat different and more convenient form[4] by substituting $c_p K$ for h, taking $H_s = c_p t_a + \Delta H_v x_s$ as the enthalpy of saturated air at the local water temperature, and $H_a = c_p t_a + \Delta H_v x$ as the average enthalpy of the local air stream (i.e., neglecting the enthalpy of superheat in the moist air), and neglecting the difference in the water-inlet and -outlet flow rates arising from water evaporation, as follows:

$$dt_w L_w = K(c_p t_w - c_p t_a + \Delta H_v x_s - \Delta H_v x) a \, dV$$

$$= K(H_s - H_a) a \, dV$$

(19.3)

where H_s is the enthalpy of saturated air at the local water temperature and H_a the average enthalpy of the air stream.

Effective Temperature Difference

The relationship of Eq. 19.3 applies to the local conditions within the cooling tower. To determine the overall performance requires an integration similar to that carried out for the logarithmic mean temperature difference of a conventional heat exchanger. In examining Eq. 19.3 it can be seen that H_s and H_a depend on the water temperature, but that the other quantities do not. Rearranging terms accordingly and integrating between the inlet and outlet gives

$$\int_{t_{w1}}^{t_{w2}} \frac{dt_w}{H_s - H_a} = \int_0^V \frac{Ka \, dV}{L_w} = \frac{KaV}{L_w} \quad (19.4)$$

where V is the active volume in cubic feet per square foot of horizontal cross-sectional area. The enthalpy of saturated air does not vary linearly with temperature, and this prevents the simplifications usually possible in heat exchanger performance analysis. The nature of the problem can be visualized by examining Fig. 19.7. This

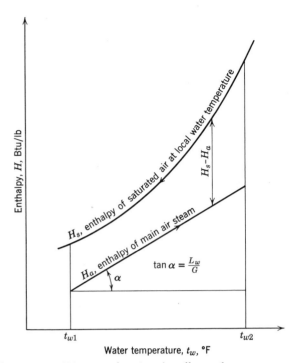

Figure 19.7 Diagram showing the effects of water temperature on the difference in enthalpy between the air flowing through a counterflow cooling tower and the enthalpy of saturated air at the local water temperature.

figure shows the enthalpy of saturated air as a function of the local water temperature, together with the actual enthalpy of the moist air in a tower. Since the specific heat of the water is essentially constant, and since the amount of heat given up by the water equals that absorbed by the air for any given change in water temperature, the local enthalpy of the moist air H_a varies linearly with the local water temperature. Because the derivation for the log mean temperature difference depends on a linear relationship between the enthalpy and the temperature for both fluids, a log mean enthalpy difference cannot be employed. The obvious course is to integrate Eq. 19.4 using graphical or numerical techniques.

Tower Characteristics

In examining Fig. 19.7 and Eq. 19.4 it can be seen that the vertical distance between the two curves represents the enthalpy difference $H_s - H_a$ in the integral of Eq. 19.4. Thus a second curve, such as that of Fig. 19.8, can be plotted for $1/(H_s - H_a)$ as a function of the local water temperature, and the value of the integral can be determined by obtaining the area under the curve. The resulting quantity KaV/L_w, known as the *tower charac-*

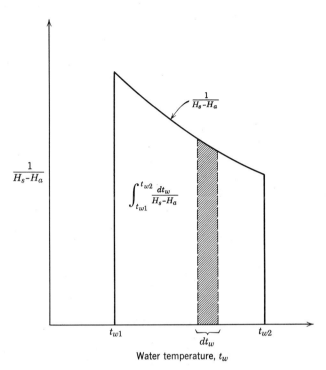

Figure 19.8 The parameter $1/(H_s - H_a)$ plotted as a function of the local water temperature in a counterflow cooling tower.

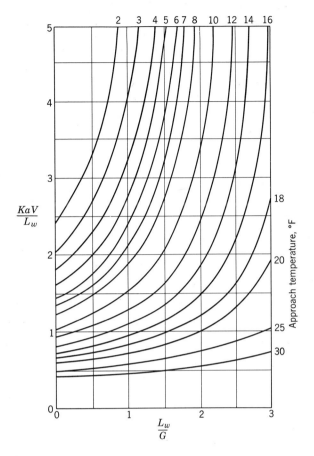

Figure 19.9 One method of presenting the performance of a cooling tower. (Lichtenstein, Ref. 4.)

teristic, is thus a function of the inlet and exit air wet bulb temperatures and the inlet and exit water temperatures. These can be expressed in terms of the approach temperature, the temperature range of the water, and the ratio of the water-flow rate to the air-flow rate. If many calculations for a variety of conditions are to be made, much time can be saved by using charts such as that shown in Fig. 19.9. A separate chart similar to Fig. 19.9 must be prepared for each representative combination of inlet air wet bulb temperature and water temperature range in the region of interest so that a set of 50 or 100 charts are needed for fast, accurate calculations.

The amount of heat that can be removed per unit of cooling tower volume depends on the geometry of the tower fill and the water distribution system. The resulting effective heat transfer surface area is a particularly difficult factor to evaluate because there is no good way to determine the average surface area of the droplets formed by splash or spray. Values for a commonly range from 2 to 6 ft^2/ft^3 (6.5 to 20 m^2/m^3).

When a set of performance charts similar to Fig. 19.9 is not available, the graphical or numerical integration of Eq. 19.4 becomes very time-consuming if a variety of cases are to be investigated. While the log-mean-enthalpy method based on the inlet and outlet enthalpy differences would underestimate the value of the tower charac-

istic KaV/L, the curve for H_s in Fig. 19.7 could be replaced by a straight line drawn in the manner shown in Fig. 19.10 to give the same area under the curve.[5] The position of this line may be defined by introducing an enthalpy correction δh where

$$\delta H = \frac{H_{s1} + H_{s2} - 2H_{sm}}{4}$$

H_{s1} and H_{s2} are the values of H_s at the outlet and inlets respectively, and H_{sm} is the value of H_s evaluated at the mean water temperature $(t_{w1} + t_{w2})/2$.

If ΔH_1 and ΔH_2 are the inlet and outlet enthalpy differences between the H_s and H_a curves shown in Fig. 19.10, an approximate log-mean-enthalpy difference, ΔH_m can now be defined as

$$\Delta H_m = \frac{\Delta H_2 - \Delta H_1}{2.3 \log \left[(\Delta H_2 - \delta h)/(\Delta H_1 - \delta h) \right]} \quad (19.5)$$

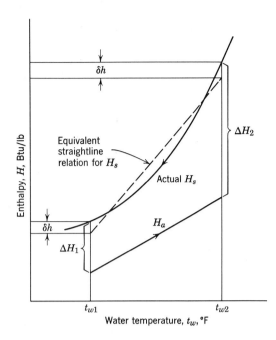

Figure 19.10 Method of approximating the curve for H_s with a straight line to simplify the calculations.

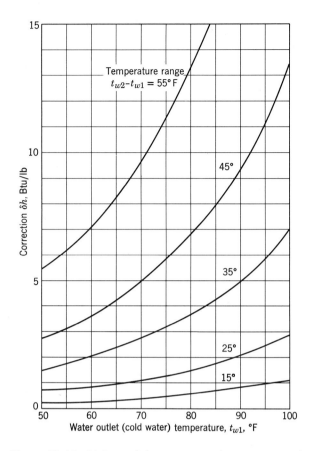

Figure 19.11 Values of the correction factor δ_h. (Based on data given by Berman, Ref. 5.)

The chart in Fig. 19.11 gives the values of the enthalpy correction δh as a function of the water outlet temperature for each of a series of values of the temperature range.

The tower characteristic can be calculated from

$$\frac{KaV}{L_w} = \frac{t_{w2} - t_{w1}}{\Delta H_m} \qquad (19.6)$$

The error involved in estimating the tower characteristic using the corrected log-mean-enthalpy method is small and normally acceptable.

Effects of Fill Geometry

An interesting set of data has been published on the performance of the various cooling tower fill matrix geometries shown in Fig. 19.3.[6] One of the most significant results is that, for a given matrix geometry, the tower characteristic KaV/L_w is almost directly proportional to the height of the packed section. This effect is shown in Fig. 19.12, in which the abscissa gives the height of the packed section in terms of the number of decks, or grids, of the fill. Cross-plots of the data of Fig. 19.12 on logarithmic coordinates showed that the tower characteristic varies as some power of the flow ratio L_w/G, so that the performance of the fill matrices tested could be approximated well by

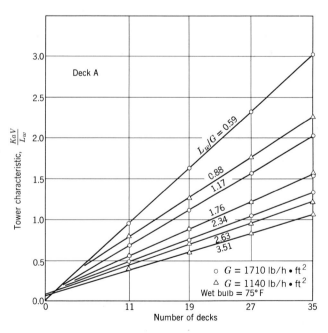

Figure 19.12 Effects of the number of decks (or the tower height) on the tower characteristic for fill matrix A of Fig. 19.3. (Kelly and Swenson, Ref. 6.)

TABLE 19.1   Values of *A* and *n* in Eq. 19.7

Deck	*A*	*n*
A	0.060	0.62
B	0.070	0.62
C	0.092	0.60
D	0.119	0.58
E	0.110	0.46
F	0.100	0.51
G	0.104	0.57
H	0.127	0.47
I	0.135	0.57
J	0.103	0.54

$$\frac{KaV}{L_w} = 0.07 + AN \left(\frac{L_w}{G} \right)^{-n} \qquad (19.7)$$

where N is the number of decks and A and n are constants for any given matrix. Table 19.1 summarizes the values found for these coefficients to relate the data for the geometries of Fig. 19.3 for a hot-water-inlet temperature of 120°F (49°C).

It happens that secondary effects cause some changes in tower performance with water-inlet temperature that are not included in Eq. 19.7. These effects vary substan-

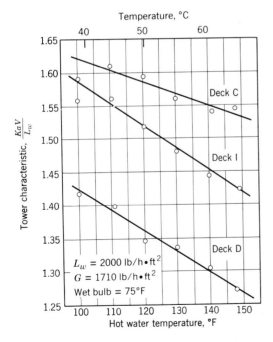

Figure 19.13 Effects of water inlet temperature on the tower characteristic for fill matrices C, D, and I of Fig. 19.3. (Kelly and Swenson, Ref. 6.)

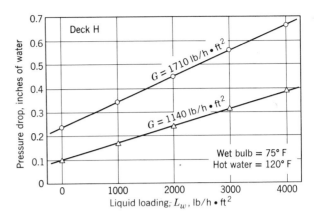

Figure 19.14 Effects of liquid loading on the air pressure drop for fill matrix H of Fig. 19.3. (Kelly and Swenson, Ref. 6.)

tially from one fill matrix geometry to another, but Fig. 19.13 shows data for three typical cases. Note that these effects may change the tower characteristic by as much as 15%.

Air-Pressure Drop

The air-pressure drop through a forced draft cooling tower partially depends on the geometry of the fill and partially on the water-flow rate, since the falling droplets have a total surface area which may be substantially greater than that of the fill. This implies that the expression for the pressure drop should have two terms: the first a function of the fill geometry and air-flow rate and the second a function of the fill geometry and both the air- and water-flow rates. Data for towers with the fill matrices of Fig. 19.3 were plotted to give curves such as those in Fig. 19.14. These were analyzed and it was found that the test data could be correlated well by

$$\frac{\Delta P}{N} = BG^2 \frac{0.0675}{\rho_a}$$
$$+ CL_w G_{eq}^2 \frac{0.0675}{\rho_a} \sqrt{(\text{mean free fall})} \quad (19.8)$$

where ΔP is the air-pressure drop in pounds per square foot, ρ_a is the density of the dry air in pounds per cubic foot of air-vapor mixture, and B and C are constants as given in Table 19.2. The quantity "mean free fall" is the mean vertical distance in feet that the water droplets fall between slats in the fill. This factor is a function of both the free flow area of the fill and the vertical spacing of the decks. Values for this quantity are given in Table

TABLE 19.2   Values of B and C in Eq. 19.8

Deck	Vertical Deck Spacing S, ft	Plan Solidity Fraction	Vertical Mean FreeFall, ft	$B \times 10^8$	$C \times 10^{12}$
A	0.75	0.250	3.00	0.34	0.11
B	1.00	0.250	4.00	0.34	0.11
C	1.25	0.333	3.75	0.40	0.14
D	2.00	0.333	6.00	0.40	0.14
· E	2.00	0.404	4.95	0.60	0.15
F	2.00	0.219	9.13	0.26	0.07
G	2.00	0.292	6.85	0.40	0.10
H	2.00	0.550	3.64	0.75	0.26
I	2.00	0.444	4.50	0.52	0.16
J	2.00	0.292	6.85	0.40	0.10

19.2 for the matrices of Fig. 19.3. The quantity G_{eq} is an equivalent air mass flow rate corresponding to the velocity of the air relative to the falling water droplets, and hence depends on both the air-flow rate and the average distance that a water droplet falls unimpeded. The relation between the G_a, G_{eq}, and the mean free fall is shown in Fig. 19.15.

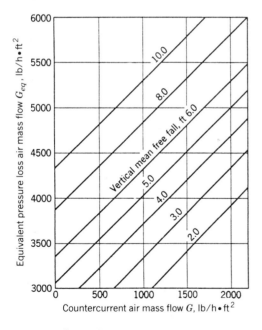

Figure 19.15   Effects of the countercurrent air mass flow rate and the vertical mean free fall of the water droplets on the equivalent pressure loss air mass flow rate G_{eq}. (Kelly and Swenson, Ref. 6.)

Effects of Design Conditions on Tower Size

Temperature range and the wet bulb and approach temperatures all affect the size of the cooling tower; these effects are included implicitly in Eq. 19.4. To show the general trend of these effects, calculations have been made for the cooling tower size relative to that for a reference unit evaluated for 75°F wet bulb, 25°F range, and a 10°F approach temperature.[4] For a given inlet air wet bulb temperature and a given water temperature drop, Fig. 19.16 indicates that the size of the tower varies widely with the design approach temperature. This curve is yet another example showing how difficult it is to bring the exit temperature of the hot fluid down close to the inlet temperature of the cold fluid. Figure 19.17 shows the effect of the inlet air wet bulb temperature on the size of tower required for a typical approach temperature and a typical water temperature range. Similarly, Fig. 19.18 illustrates the effect of temperature range in the water on the relative size of the tower.

Practical Limitations on Air- and Water-Flow Rates

The amount of cooling obtainable with a tower under the conditions of prime interest is insensitive to the water- or air-flow rates through the tower if the ratio of the water flow to the air flow is kept constant. As may be deduced from Fig. 19.7, for a given air-inlet condition the enthalpy of the exit air depends on the ratio of L_w/G, which is consistent with the characteristics of conventional heat exchangers. Thus there is a strong incentive to increase the fluid-flow rates through the

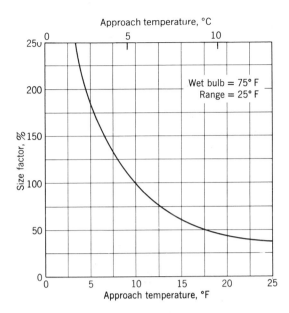

Figure 19.16 Effects of approach temperature on the relative size of a series of cooling towers. (Lichtenstein, Ref. 4.)

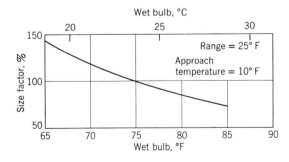

Figure 19.17 Effects of ambient air wet bulb temperature on the relative size of a series of cooling towers. (Lichtenstein, Ref. 4.)

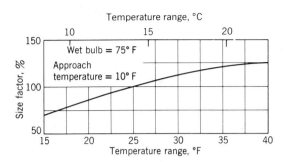

Figure 19.18 Effects of temperature range on the relative size of a series of cooling towers. (Lichtenstine, Ref. 4.)

tower until difficulty is experienced with poor water-flow distribution at high water-flow rates or with excessive fan power requirements. With packed towers, the balance between capital charges and fan power costs usually limits the fan power, and hence the maximum air-flow rate to about 1800 lb/h · ft². A poor water-flow distribution over the packing is commonly experienced at water-flow rates in excess of around 3000 lb/h · ft² floor area, because if the water-flow rate is increased too far, the water cascades in thick streams instead of falling as a spray, so that the effective surface area is reduced. This condition is called *flooding*. On the other hand, if the water flow drops to about 600 lb/h · ft², or less, surface tension causes the water flow to channel. This gives a poor water-flow distribution, and, hence, a marked drop in performance. The water-flow rate limits imposed by these conditions of course vary with the internal geometry of the tower, but the values given represent the usual limits for good designs.

Water Consumption

The water consumption of a cooling tower depends not only on the heat load but also on the ratio of the amounts of heat carried off by increasing the temperature of the air and by evaporation of the water. In a well-designed tower the amount of water lost in the form of suspended droplets in the exit air stream is usually less than 1% of the water consumption. Thus the water consumption ΔW_w of a cooling tower can be related to the air-flow rate and the air-inlet and -outlet water vapor contents x_1 and x_2 as follows:

$$\Delta W_w = W_a(x_2 - x_1) \qquad (19.9)$$

If the air entering the tower is saturated, as much as a third of the heat removed from the water may go into heating the air, while the balance will go into evaporating the water. Thus the water consumption will be only about two-thirds of what would be required if the entire heat load went into evaporation of the water. On the other hand, under unusual conditions at light loads, with a low temperature range and very dry air, evaporation of the water may actually reduce the air dry bulb temperature so that heat is removed from rather than added to the air, and the amount of heat going to evaporate the water actually exceeds the heat load on the tower.

Dissolved Solids in Make-Up Water

If, as is usually the case, there is an appreciable concentration of dissolved solids in the make-up water, this concentration increases as the water is evaporated. To avoid scaling of the surfaces within the tower, it is best to drain off a portion of the water to reduce the concentration of the solids. This operation is analogous to the blowdown of boilers. In this way the concentration of solids can be kept to a value such that objectionable deposits do not form. In some instances it may be worthwhile to employ a water-softening treatment for the make-up water.

Depending on the type of application, the cool water may be drawn from the basin at the base of the tower and pumped to heat exchangers at other points in the plant, or the hot fluid in the plant may be piped to the base of the cooling tower and banks of tubes arranged so that they are cooled directly by the falling water as it reaches the base of the tower.

Recirculation

As mentioned earlier in this chapter, under some wind conditions a portion of the warm moist air leaving the tower may recirculate back through the tower inlet and thus degrade the performance.[7] If a large number of units is arranged in a long row a similar difficulty is likely when the wind direction is roughly parallel to the length of the row. In an effort to understand the effects of the many parameters a detailed test procedure was worked out based on portable instrumentation, and this equipment was used to test 30 different cooling towers.[8] Every effort was made to make the tests as consistent as possible considering the wide variety of units tested: that is, forced and induced draft towers, crossflow and counterflow towers, tower lengths from 36 to 361 ft, tower widths from 13 to 68 ft, tower frame heights from 19 to 55 ft, and fan stack heights from 7 to 17 ft.

The test results[9] showed that the forced draft towers gave recirculation rates about double those for the induced draft towers. Since the bulk of the towers tested were of the induced-draft type, the test results for these were plotted as functions of the many variables. The only two parameters to show any well-defined effects were the water-flow rate and the tower length. (Where a multiplicity of units is arranged in a long row, the tower length is taken as the length of the entire set.) Analysis of the data showed that good correlation for the effects of these two factors could be obtained by applying a correction to the wet bulb temperature. This correction for a 20°F cooling range and a 10°F approach is shown in Fig. 19.19. The table beneath the chart gives the factors to be applied to the curves in Fig. 19.19 to obtain the proper correction for other range and approach values. Note that two curves are given in Fig. 19.19: one, the recommended recirculation allowance, and the other the maximum allowance likely to be required.

Design Compromises

The design of a cooling tower ordinarily entails not only compromises in the detailed design of the cooling tower itself but also in the selection of the design conditions. The cost of the cooling tower, of the water consumed, and of the fan power required for a given approach temperature must be carefully balanced against the value of a reduction in the approach temperature.[10] This is particularly true for air-conditioning applications in which the heaviest loads on the tower, and the most difficult conditions to meet, occur during a relatively small number of days per year so that if these desired conditions can be compromised somewhat, substantial savings in the cost of the tower can be effected.

Procedure for Estimating the Size of a Cooling Tower

A variety of techniques for the design of cooling towers are in use—many of them involving the use of charts.

If a person who is not well versed in the design of cooling towers wishes to make a preliminary estimate of the size and cost of a forced draft tower for a special application and has little information at hand other than that included here, the following procedure can be employed:

1. Choose the design conditions, that is, the water-inlet and -outlet temperature and the inlet air wet bulb temperature.

2. Prepare an enthalpy-temperature diagram similar to Fig. 19.7. Choose the exit air enthalpy such that the slope of the line for the air enthalpy is equal to the slope of the curve for the enthalpy of saturated air at the water outlet temperature. (This is an arbitrary step, but one that usually yields a reasonably good set of proportions because it is generally desirable to have the driving force for heat transfer a minimum at the bottom of the tower. At the same time the air enthalpy rise should be as large as possible to minimize the air pumping power requirement.)

3. Plot a curve similar to that in Fig. 19.8 for $1/(H_s - H_a)$ as a function of water temperature.

4. Using Simpson's rule, determine the area under the curve obtained from step 3, and from it the required value for the tower characteristic KaV/L_w.

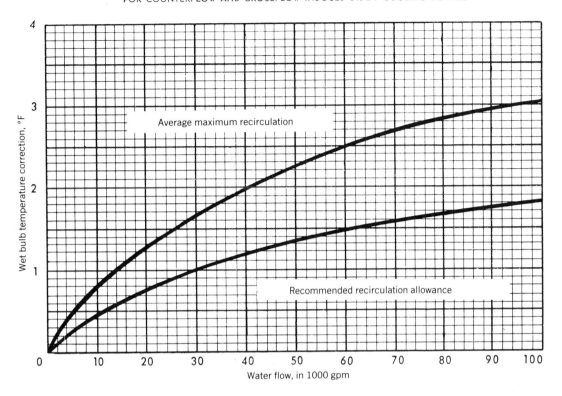

RECOMMENDED RECIRCULATION ALLOWANCES

FOR COUNTERFLOW AND CROSSFLOW INDUCED DRAFT COOLING TOWERS

Correction factors

Approach to ambient WBT, °F	Range, °F									
	5	10	15	20	25	30	35	40	45	50
5	0.29	0.47	0.64	0.80	0.97	1.14	1.30	1.47	1.63	1.80
6	0.31	0.49	0.68	0.85	1.03	1.20	1.37	1.56	1.73	1.91
7	0.33	0.51	0.71	0.89	1.08	1.25	1.44	1.63	1.83	2.01
8	0.35	0.53	0.74	0.93	1.12	1.30	1.50	1.70	1.91	2.10
9	0.37	0.55	0.76	0.97	1.16	1.35	1.56	1.77	1.97	2.18
10	0.39	0.57	0.78	1.00	1.20	1.40	1.62	1.83	2.04	2.25
11	0.41	0.59	0.81	1.04	1.24	1.45	1.66	1.88	2.09	2.31
12	0.43	0.61	0.84	1.07	1.27	1.49	1.70	1.92	2.13	2.36
13	0.45	0.63	0.86	1.10	1.30	1.52	1.74	1.96	2.17	2.40
14	0.46	0.65	0.88	1.13	1.33	1.55	1.77	1.99	2.21	2.44
15	0.47	0.67	0.90	1.15	1.36	1.57	1.80	2.02	2.25	2.47
16	0.49	0.69	0.93	1.18	1.39	1.61	1.83	2.06	2.29	2.52
17	0.51	0.70	0.95	1.20	1.42	1.64	1.86	2.10	2.33	2.57
18	0.52	0.72	0.97	1.22	1.44	1.66	1.89	2.13	2.37	2.61
19	0.53	0.74	0.99	1.24	1.46	1.68	1.92	2.16	2.40	2.64
20	0.54	0.75	1.00	1.26	1.48	1.70	1.95	2.19	2.43	2.67
21	0.55	0.77	1.02	1.28	1.50	1.73	1.98	2.22	2.46	2.70
22	0.56	0.79	1.04	1.30	1.52	1.76	2.00	2.25	2.49	2.73
23	0.57	0.80	1.05	1.31	1.54	1.78	2.02	2.27	2.52	2.76
24	0.58	0.81	1.06	1.32	1.56	1.80	2.04	2.29	2.54	2.79
25	0.58	0.82	1.07	1.33	1.57	1.82	2.06	2.31	2.56	2.81

Figure 19.19 Recommended allowances for recirculation losses. *Note:* Recirculation allowances shown in the curve are based on a 20°F cooling range and a 10°F approach to any wet bulb temperature. Recirculation allowances for other performance conditions can be obtained by means of the correction factors shown in the table. (Courtesy Cooling Tower Institute, Ref. 9.)

5. Choose a fill matrix geometry from Fig. 19.3, and obtain the data for its principal characteristics from Table 19.1.

6. From Fig. 19.13 estimate the effect on the tower characteristic KaV/L_w of the deviation of the water-inlet temperature from the 120°F for which the data of Table 19.1 were prepared.

7. Using the air enthalpy curve of step 2, determine L_w/G from heat balance considerations. (L_w/G is equal to the slope of the air enthalpy line.)

8. Substitute the values obtained from step 7 for KaV/L_w, A, (L_w/G), and n in Eq. 19.7 to obtain the number of decks N and, hence, the height of the packed portion of the tower.

9. For the L_w/G of step 7, determine the air-flow rate for a water flow of 2500 lb/h · ft². If the resulting air flow exceeds 1600 lb/h · ft², determine the water flow that corresponds to an air flow of 1600 lb/h · ft².

10. Determine the cross-sectional area of the tower using the water-flow rate per unit of area given by step 9 and the total water-flow rate established in step 1.

11. Select the horizontal dimensions of the tower to give the area defined by step 10, keeping the dimension in the direction of the prevailing wind to no less than the packed height and no more than 40 ft. The overall height should be the packed height plus about one and one-half times the depth in the direction of the prevailing wind.

Evaluation of Acceptance Tests

More often than not a cooling tower is completed and is ready for acceptance tests at a time when weather conditions give much lower wet bulb temperatures than the critical temperatures for which the tower was designed. Thus it is necessary to carry out the acceptance tests and apply corrections to determine whether the cooling tower meets the design conditions. The relation presented in Eq. 19.4 is appropriate for such corrections and forms the basis for the standard correction procedure that has been adopted by the Cooling Tower Manufacturers Institute.[11]

Cost

It is evident from the preceding material that it is difficult to estimate the cost of a cooling tower for a given set of requirements because there are so many variables.[12] To provide some basis for rough preliminary estimates,

Table 11.2 indicates the cost of small cooling towers for air-conditioning systems. Note that the water-flow rate is probably the best single parameter to use in estimating the size and cost of larger cooling towers.

Example 19.1. *Estimate of the Size of a Cooling Tower.* A cooling tower is to cool 120°F water to yield an approach temperature of 10°F when the entering air wet bulb temperature is 75°F. The water-flow rate is to be 1,000,000 lb/h, and the ratio L_w/G can be taken as 1.25.

1. Determine the tower characteristic using the numerical integration method.

2. Determine the tower characteristic using the log-mean-enthalpy method with and without the enthalpy correction applied.

3. Choose a fill matrix geometry from Fig. 19.3, and, using the data for its principal characteristics given in Table 19.1, determine the number of decks N and, hence, the height of the packed portion of the tower.

4. Assume an allowable water-flow rate through the cooling tower of 2000 lb/h · ft², and determine the air-flow rate and the cross-sectional area of the matrix.

Solution. For an approach temperature of 10°F and a wet bulb temperature of 75°F, the temperature of the water at the cooling tower outlet is 85°F. Hence the temperature range is $120 - 85 = 35$°F.

Values for H_s, H_a, and $1/(H_s - H_a)$ are plotted in Fig. 19.20 as functions of the local water temperature as it falls through the tower. In constructing this chart the enthalpy of saturated air H_s at the local water temperature was obtained from Table H2.6. The enthalpy of the air stream H_a increases linearly with the water temperature, and the total increase of enthalpy ΔH was evaluated from $\Delta H = (L_w/G) \Delta t_w = 1.25 \times 35 = 43.75$ Btu/lb. Since the air enters the tower at a wet bulb temperature of 75°F, the air enthalpy at the air inlet is 38.61 Btu/lb, while that at the air outlet is $38.61 + 43.75 = 82.36$ Btu/lb. From these values ($H_s - H_a$) and $1/(H_s - H_a)$ were evaluated.

The area under the curve $1/(H_s - H_a)$ in Fig. 19.20 gives the value of the tower characteristic KaV/L_w. Table 19.3 summarizes a numerical solution for the tower characteristic, using the numerical integration method, and this gives $(KaV/L_w) = 2.184$.

As a check, the tower characteristic may be calculated by the log-mean-enthalpy method. The inlet and outlet enthalpy differences between the H_s and H_a curves are

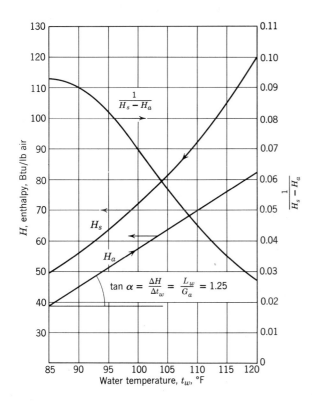

Figure 19.20 Parameters for Example 19.1 plotted as a function of the local water temperature.

$\Delta H_2 = 119.54 - 82.36 = 37.18$; $\Delta H_1 = 49.43 - 38.61 = 10.82$. From Fig. 19.11 the enthalpy correction δh for an 85°F water-outlet temperature and a 35°F temperature range is 4.3 Btu/lb. The corrected log mean-enthalpy difference from Eq. 19.5 is

$$\Delta H_m = \frac{37.18 - 10.82}{2.3 \log (37.18 - 4.3/10.82 - 4.3)}$$

$$= \frac{26.36}{2.3 \log 5.05} = \frac{26.36}{1.615} = 16.3$$

and the corrected tower characteristic from Eq. 19.6 becomes

$$\frac{KaV}{L_w} = \frac{35}{16.3} = 2.15$$

which is within 2% of the value obtained by numerical integration.

If the correction term δh were not included, the uncorrected log-mean-enthalpy difference would be

$$\Delta H_m \text{(uncorrected)} = \frac{37.18 - 10.82}{2.3 \log (37.18/10.82)}$$

$$= \frac{26.36}{2.3 \log 3.43} = 2.13$$

and

$$\frac{KaV}{L_w} = \frac{35}{21.3} = 1.645$$

Note that the uncorrected log-mean-enthalpy difference underestimates the required tower characteristic by about 24% for this particular problem.

The next step in the calculation is to choose a fill matrix geometry from Fig. 19.3 and obtain values for A and n from Table 19.1. Choosing Type-C fill, Eq. 19.7 becomes

TABLE 19.3   Numerical Integration for the Tower Characteristic of Example 19.1

Water Temperature t_w, °F	Enthalpy at t_w, H_s	Enthalpy of Air, H_a	$H_s - H_a$	$1/(H_s - H_a)$	Average $1/(H_s - H_a)$
85	49.43	38.61	10.82	0.0924	
90	55.93	44.86	11.07	0.0902	0.0913
95	63.32	51.11	12.11	0.0824	0.0863
100	71.73	57.36	14.43	0.0692	0.0758
105	81.34	63.61	17.73	0.0564	0.0628
110	92.34	69.86	22.48	0.0444	0.0504
115	104.98	76.11	28.87	0.0346	0.0395
120	119.54	82.36	37.18	0.0270	0.0308
					0.4369

$$\frac{KaV}{L} = \int \frac{1}{H_s - H_a} dt_w = 5 \times 0.4369 = 2.184.$$

$$\frac{KaV}{L_w} = 0.07 + 0.092\, N \left(\frac{L_w}{G}\right)^{-0.6}$$

Substituting the above values of KaV/L_w and L_w/G in this equation, we obtain

$$2.184 = 0.07 + 0.097\, N(1.25)^{-0.6}$$

Since the inlet temperature of the water is the same as that for which Table 19.1 was calculated, no temperature correction from Fig. 19.13 is required. Solving the preceding equation for the number of decks gives $N = 25$. Since the vertical spacing for the C-type deck is 15 in., the height of the packed portion of the tower becomes,

$$\text{Packed height} = \frac{25 \times 15}{12} = 31.3 \text{ ft}$$

The next step is to calculate the air-flow rate and the size of the tower. Since L_w/G was 1.25, the air-flow rate for $L_w = 2{,}000$ lb/h · ft^2 floor area is

$$G = \frac{2000}{1.25} = 1{,}600 \text{ lb/h} \cdot \text{ft}^2$$

The floor area required for a total water-flow rate of 1,000,000 lb/h

$$\text{Floor area} = \frac{1{,}000{,}000}{2{,}000} = 500 \text{ ft}^2$$

Thus the required tower cross-sectional area can be obtained by making it 22 × 24 ft. The overall height may be taken as the packed height, plus one and one-half of the depth in the direction of the prevailing wind, that is, $31.3 + 1.5 \times 22 = 64$ ft.

REASONS FOR LARGE UNCERTAINTIES

Large disparities between the predicted and measured performance of cooling towers have been notorious for over 50 years. A wealth of data has been accumulated by the manufacturers, but much of this is jealously guarded as proprietary information. The increasing need for large cooling towers to handle the heat rejection from power plants has aroused new interest in the matter, so that the long-standing programs of the Cooling Tower Institute have been supplemented by new programs sponsored by the Electric Power Research Institute (EPRI) and the American Society of Mechanical Engineers.[13] Much of this newer effort has stemmed from the hope that the vastly increased complexity of calculations made possible by the computer would permit far more elegant mathematical models that would make it possible to reduce the uncertainties from perhaps 30% to not more than 10%. A number of such models have been evolved and developed into computer programs.[14] However, even performance predictions obtained with these new programs do not always check well with test results.[15]

In view of the fact that many very competent engineers have tackled the problems of cooling tower performance prediction with limited success, one is led to suspect that random, unpredictable variations in flow distribution may be responsible for the disparities between the predicted performance and the test results. The deviations from ideality appear to be of the same character as those discussed in Chapter 8. This view is reinforced by the fact that the deviations from predicted performance are greater in natural draft than in forced-draft cooling towers, a trend one should expect because the dynamic head of the wind represents a substantially greater fraction of the pressure drop through the natural draft towers than for the forced draft towers. Not only can changes in the wind velocity and direction change both the draft and the amount of recirculation, but they can also change the flow pattern inside the cooling tower, so that large segments of the fill may lie in eddies which make that region ineffective because of low air velocity or even internal recirculation. The possibilities for such eddies can be envisioned readily for any given cooling tower by trying to sketch flow patterns that might develop in the inlet region for some of the less favorable of the full range of possibilities of wind direction and velocity. One can't predict with certainty whether such eddies will form or what their magnitude might be, but the possibility for large deviations from an ideal flow distribution is undoubtedly there, and a large-scale turbulent flow pattern is as unpredictable as the weather, whose vagaries also depend on random combinations of truly large-scale turbulence and thermal convection effects.

Effects of Water-Flow Distribution

It happens that an irregularity in the water-flow distribution is a deviation from ideality the effects of which can be examined quantitatively by using characteristic curves included in this chapter. Using Fig. 19.9, consider the effects of a local region having a water flow 25%

greater than the average when the tower is supposed to be operating with $L_w/G = 2$ and an approach temperature of $10°F$ which makes the value of the parameter $KaV/L_w = 3.4$. In the higher-than-normal water-flow region the value of L_w/G would be 2.5, while KaV/L_w would equal 2.72. This yields an approach temperature of about $15°F$ if no allowance is made for the reduction in the local air flow resulting from the extra resistance to air flow caused by the increased water flow. From Fig. 19.14 it appears that the higher water flow would increase the air-pressure drop by about 15%, which would lead to a reduction in the air flow by about 7%. This would cause a further increase in the value of L_w/G to about 2.7, giving an approach temperature of about $16.5°F$. Examination of Fig. 19.16 indicates that this local 25% oversupply of water reduces the performance of the affected section of the tower to that of a tower half the size! This example provides a glimpse of the degree to which cooling towers may be surprisingly sensitive to seemingly small deviations from ideality.

ENVIRONMENTAL PROBLEMS

The plume of moist vapor emitted from the stacks of cooling towers may cause difficulties such as hazardous patches of fog on roads, and fine droplets entrained from the water sprays may be troublesome immediately downwind of the tower. As a consequence, part of the elaborate process of satisfying EPA licensing requirements is estimating the characteristics of the plume from the cooling tower under the full range of wind and possible temperature inversion conditions to be expected at the proposed site. To help satisfy these requirements, extensive wind-tunnel tests (such as those of Ref. 16) have been run to investigate the effects of terrain, and elaborate computer programs (such as that of Ref. 17) have been developed.

WIND LOADS

The structural problems posed by wind loads are handled readily by conventional building design practice for the smaller and medium-sized cooling towers, but the huge natural draft cooling towers such as those of Fig. 19.4 are in a quite different class; they are thin shells something like aircraft structures in which large-scale buckling may occur as a consequence of aerodynamic loads in a high wind.[18] "Thin as an eggshell" is an idiom that gives a vivid impression of fragility, yet the thickness of an eggshell is about 1% of its diameter, whereas the

127-mm (5-in.) thickness of the reinforced concrete shells of the Ferrybridge cooling towers was only about 0.14% of their 91-m (300-ft) diameter. The possibility of large-scale buckling of these thin concrete shells was appreciated and wind-tunnel tests were run in England to provide data on the aerodynamic forces that would be generated in hurricane-force winds. This work even included tests of thin shells electroformed of nickel to investigate buckling. In spite of this work, three of a set of eight large cooling towers at the Ferrybridge plant of the Central Electricity Generating Board buckled and collapsed in a severe storm on November 1, 1965. These towers were 114 m (375 ft) tall and 91 m (300 ft) in diameter at the base, and the concrete shells were 127 mm thick. The large investment not only in the Ferrybridge towers but also in the many other similar units in the United Kingdom led to a thorough investigation.[18] No one of the many factors that were found to have contributed to the failures was considered responsible except that the investigating committee felt that the complexity of the problems and the possibility of an unfavorable accumulation of comparatively small uncertainties made it advisable to use a much larger factor of safety than the factor of 1.25 that had been used.

DRY COOLING TOWERS

Construction of new power plants is restricted in some areas because the water supply is not even sufficient for a wet cooling tower which consumes about 10,000 gpm (631 L/s)/1000 MWe, only about 2% of the water flow to conventional steam turbine condensers.[19] It is not surprising to find this to be the case in the U.S. Southwest, but it is also true in the English Midlands. There one of the first large dry cooling tower installations was made in the 1960s[20] because the small rivers that drain that island nation were barely adequate to meet domestic and industrial needs in the industrialized areas. Although the cost of dry cooling towers is roughly four times that of wet cooling towers, so that their use leads to an increase in the capital cost of a steam plant by about 30%, there is no alternative in some areas.

Typical Installations

A surprisingly large number of plants using dry cooling towers have been built since 1970 or are under construction at the time of writing. In the arid western plains and valleys of the United States there are several installations.[21] One of the earliest of the large installations is a 300-MWe plant in the Dakotas that uses forced-draft cooling. Another is the 330-MWe Wyodak plant in the coal fields of Wyoming; it condenses the steam directly in air-cooled coils. This approach has the disadvantage that large pipes are required to handle the large valume of low-pressure steam, but it avoids the temperature losses involved if an intermediate cooling water circuit is employed. Another example is at the San Juan III power plant near Farmington, New Mexico; this uses an intermediate water circuit. It also employs a supplemental wet section in the tower for use under heavy loads on particularly hot days; the relatively small water consumption for these peak load conditions is acceptable, and the reduced dry cooling capacity permits a marked savings in capital costs. A third approach has been tested at the Kern Station in Bakersfield, California; it condenses the steam from the turbines by evaporating ammonia and then condenses the ammonia in a dry cooling tower. This system avoids the temperature loss associated with the temperature drop in a water circuit because, as in a heat pipe, there is little temperature difference between the boiling and condensing ammonia. It happens that the density of steam drops so rapidly below about 38°C (100°F) that it becomes uneconomical to build turbines large enough to handle it. This is not the case for ammonia, because its vapor density is about 200 times higher than that of steam at 38°C. Thus the ammonia system has the additional advantage in a cold climate that the ammonia can be boiled at around 40°C and expanded through a turbine in a bottoming cycle to near the ambient air temperature to produce additional power and increase the plant thermal efficiency. A 30-MWe plant of this type has been put into service by Electricite de France. Use of ammonia also avoids the freezing problem in the winter in cold climates; this was another reason that an ammonia vapor cycle with a dry cooling tower was chosen to recover energy from the exhaust of a gas turbine in a pipeline pumping station in Alberta, Canada. The Kalina steam-ammonia vapor cycle also is well suited to use with dry cooling towers.[22] The excellent heat transfer properties of ammonia make it especially well suited to these applications. Some quite large dry cooling tower installations have been made in other countries. Several have been made in South Africa, one of which employs six natural draft dry cooling towers 150 m (500 ft) tall, each rejecting the heat from a 670-MWe steam turbine unit. Another plant in the arid province of Shanxi, China uses dry cooling towers for two 200-MWe units.

Design Problems

Quite a number of studies have been directed at the principal problem of dry cooling towers, namely, their high cost.[23-26] Ideally, a very finely divided heat transfer matrix with plate spacings perhaps half the smallest of those in Fig. 14.3 would halve the surface area, the metal weight requirements, and hence the cost of the heat exchangers together with the size and cost of the tower structure.[23] However, the fine passages would be sensitive to fouling by dirt and windblown plant debris, and would be hard to clean; hence this approach does not appear to be practicable for most sites. The only metals having a high thermal conductivity that would be suitable are aluminum and copper. Copper provides excellent resistance to corrosion, but is expensive. Aluminum may be suitable in some installations, but in the first big dry cooling tower installation in England[20] corrosion of the aluminum terminated the operation in two years. The subtle problems of dirt and corrosion are site-dependent, and the cost-benefit ratios for the various possibilities can be evaluated only after extensive field test experience.

REFERENCES

1. G. R. Nance, "Fundamental Relationships in the Design of Cooling Towers," *Trans. ASME*, vol. 61, 1939, p. 721.

2. J. W. Hubenthal, "A Comparison between European and United States Cooling Towers," Cooling Tower Institute Bulletin TPR-123, Houston, Texas, 1962.

3. A. L. London, W. E. Mason, and L. M. K. Boelter, "Performance Characteristics of a Mechanically Induced Draft Counterflow Packed Cooling Tower," *Trans. ASME*, vol. 62, 1940, p. 41.

4. J. Lichtenstein, "Performance and Selection of Mechanical Draft Cooling Towers," *Trans. ASME*, vol. 65, 1943, p. 779.

5. L. D. Berman, *Evaporative Cooling of Circulating Water*, Pergamon Press, Elmsford, N.Y., 1961.

6. N. W. Kelly and L. K. Swenson, "Comparative Performance of Cooling Tower Packing Arrangements," *Chemical Engineering Progress*, vol. 52, 1956, p. 263.

7. J. Lichtenstein, "Recirculation in Cooling Towers," *Trans. ASME*, vol. 73, 1951, p. 1037.

8. J. L. Willa et al., "Instrumenting a Field Study of Industrial Water-Cooling Tower Performance," *Trans. ASME*, vol. 79, 1957, p. 1679.

9. "Recirculation," Cooling Tower Institute Bulletin PFM-116, Houston, Texas, 1958.

10. M. W. Larinoff, "Cooling Towers for Steam-Electric Stations—Selection and Performance Experience," *Trans. ASME*, vol. 79, 1957, p. 1685.

11. "Acceptance Test Procedure for Industrial Water-Cooling Towers, Mechanical Draft Type," Cooling Tower Institute Bulletin ATP-105, Houston, Texas, 1959.

12. S. Katell and J. H. Faber, "An Economic Evaluation of Cooling Water Costs," *Cost Engineering*, vol. 2, 1957, p. 70.

13. M. R. Lefevre and D. R. Moran, "Evaporative Water Cooling Towers, *Mechanical Engineering*," vol. 108(8), August 1986, p. 29.

14. "VERA2D: Program for 2-D Analysis of Flow, Heat, and Mass Transfer in Evaporative Cooling Towers," Report no. EPRI-CS-2923, Electric Power Research Institute, March 1983.

15. "Heat Transfer Characteristics of a Dry and Wet/Dry Advanced Condenser for Cooling Towers," Report no. EPRI CS-2476, Electric Power Research Institute, June 1982.

16. J. Andreopoulos, "Wind Tunnel Experiments on Cooling Tower Plumes," Parts I and II, ASME Paper nos. 86-WA/HT-31, 1986.

17. "Studies of Mathematical Models for Characterizing Plume and Drift Behavior from Cooling Towers," Report no. EPRI CS-1683, Electric Power Research Institute, vols. 1 to 5, January 1981.

18. "Report of the Committee of Inquiry into Collapse of Cooling Towers at Ferrybridge," Monday, 1 November, 1965, Central Electricity Generating Board.

19. J. A. Bartz and J. S. Maulbetsch, "Are Dry-Cooled Power Plants a Feasible Alternative?" *Mechanical Engineering*, vol. 103(10), October 1981, p. 34.

20. P. J. Christopher and V. T. Forster, "Rugeley Dry Cooling Tower System," *Proceedings of the Institute of Mechanical Engineers*, Part 1, vol. 184(11), 1969–1970, p. 197.

21. J. A. Bartz and J. S. Maulbetsch, "A Substitute for Water: Dry Cooling of Power Plants," *Mechanical Engineering*, vol. 108(4), April 1986, p. 55.

22. A. L. Kalina and H. M. Leibowitz, "Applying Kalina Technology to a Bottoming Cycle for Utility Combined Cycles," ASME Paper no. 87-GT-35, May 31, 1987.

23. F. K. Moore and T. Hsieh, "Concurrent Reduction of Draft Height and Heat-Exchange Area for Large Dry Cooling Towers," *Journal of Heat Transfer, Trans. ASME*, vol. 96(2), 1974, p. 279.

24. "Test Report: Wet/Dry Cooling Tower Test Module," Report no. EPRI CS-1565, Electric Power Research Institute, October 1980.

25. "Comparative Economics of Indirect and Direct Dry/Wet-Peaking Cooling Tower Systems," Report no. EPRI CS-2925, Electric Power Research Institute, March 1983.

26. "Power Plant Waste Heat Rejection Using Dry Cooling Towers," Report no. EPRI CS-1324-SY, Electric Power Research Institute, February 1980.

20

Heat Exchanger Tests

A variety of tests are employed in the development of heat exchangers. These range from evaluations of heat transfer, pressure drop, and velocity and temperature distributions using small models to acceptance tests of large full-scale units. The selection and delineation of the tests to be run, the design of the test setups, the conduct of the tests, and the analysis and interpretation of the results present a host of subtle problems. A nice balance must be achieved between the cost of the test on the one hand, and the value of the information to be obtained on the other. This chapter outlines the problems involved, some of the approaches that have proved effective, and a useful set of techniques.

PERFORMANCE TESTS ON MODELS

When after a series of design studies a preliminary design for a new type of heat exchanger is selected as the basis for a unit to be built, the engineer is confronted with a difficult decision. He knows that there are uncertainties in the heat transfer coefficients and pressure-loss coefficients that he has employed in his analysis. On the one hand, the incremental cost of overdesign of the heat exchanger, to allow for the most adverse possible accumulation of uncertainties, may in some cases increase the cost of the heat exchanger by 50%; on the

other hand, a unit of inadequate size would not meet the performance guarantees. Not only might such an inadequate unit entail a loss to the manufacturer, but it might very well detract so seriously from the performance of the plant in which it is installed as to entail a loss by the customer that is many times the cost of the heat exchanger. For large units, the cost of building a test rig capable of conducting a thorough test program might be enormous; just the cost of the heat source required can be many times the cost of the heat exchanger. Fortunately, extensive experience shows that significant tests can be run on properly scaled models.[1-4] In fact, it is often possible to carry out a better set of tests with a properly designed model than is possible with the full-scale heat exchanger, and the tests can be carried out at a tiny fraction of the cost of testing a full-scale unit. It is possible to build the models more quickly, and, if necessary, they can be modified more readily than a full-scale heat exchanger, thus saving much precious time.

Similarity of Fluid-Flow Patterns and Heat Transfer Coefficients

The proper design of a heat exchanger model test entails a good grasp of the basic relations involved—particularly the principles of similitude. Basically, the pressure drop and heat transfer depend on the Nusselt, Prandtl, Reynolds, and Mach numbers. If the full-scale heat exchanger is to employ a toxic or dangerous fluid such as mercury, hydrogen, or sulfuric acid—or if it is desired to make a simple and inexpensive model with no special attempts to make it leaktight—it may be desirable to conduct the test with a fluid different from that to be employed in the full-scale unit. This may also be the case if it is important to achieve a high Reynolds and/or Mach number that could not otherwise be obtained with an existing set of equipment.[5] In examining the basic heat transfer equation, $(hD/k) = 0.023\text{Re}^{0.8}\text{Pr}^{0.4}$, it is apparent that if the Prandtl number and Reynolds number can be the same in the model as in the full-scale unit, the heat transfer coefficient can be easily corrected for changes in the thermal conductivity of the fluid. Tables H2.3 and H2.4 list the Prandtl number for a variety of gases and liquids. Note that the Prandtl number for all the gases over the temperature range considered falls within a range of plus or minus 20% of the mean. Thus any convenient gas may be used in model tests to simulate the gas that is to be employed in the full-scale unit (e.g., air can be used in place of helium in model tests). It is also apparent from Table H2.3 that the Prandtl number for most of the more common aqueous solutions and lighter hydrocarbon liquids varies almost as much with temperature as it does from one liquid to another. Thus one liquid often can be simulated by running another at the appropriate temperature. From Table H2.3 it can also be seen that the liquid metals fall in a class by themselves and cannot be simulated in their heat transfer performance by water or hydrocarbons.

Similarity for Boiling and Condensing

The heat transfer coefficients for boiling liquids and condensing vapors depend on factors such as the heat of vaporization, wetting, surface tension, and the vapor-liquid density ratio. Because of this dependency, for boilers and condensers great care should be exercised in simulating one liquid with another; as a minimum requirement, the ratio of the specific volumes of the vapor and liquid, their wetting characteristics, and their heats of vaporization should be roughly the same.

Choice of Model Size

The majority of large heat exchangers make use of forced convection flow in both fluid streams and employ large numbers of tubes in an orderly array. A first and obvious step in reducing the size of the unit to be tested is to take a representative core element from the large heat transfer matrix. There should be little difference in the heat transfer coefficient for a cluster of 100 tubes as compared

to that for a cluster of 10,000 tubes as long as the Reynolds numbers and flow distributions are the same and the flow passages are geometrically similar. This similarity suggests that the size of the flow passages can also be decreased, that is, 1.0-in. tubes could be simulated with 0.25-in. tubes. If this were done, and if the same fluids were employed in the model as in the full-scale unit, to maintain the same Reynolds number the mass flow rate in the model should be inversely proportional to the size of the tubes, that is,

$$\frac{G_{(\text{model})}}{G_{(\text{fsu})}} = \frac{D_{(\text{fsu})}}{D_{(\text{model})}} \qquad (20.1)$$

where the subscript fsu denotes the "full-scale unit." For a given temperature rise and Reynolds number, the quantity of heat transferred from one fluid to another per tube for a given tube L/D can be related by

$$\frac{Q_{(\text{model})}}{Q_{(\text{fsu})}} = \frac{GD^2_{(\text{model})}}{GD^2_{(\text{fsu})}} \qquad (20.2)$$

Substituting for G from Eq. 20, gives

$$\frac{Q_{(\text{model})}}{Q_{(\text{fsu})}} = \frac{D_{(\text{model})}}{D_{(\text{fsu})}} \qquad (20.3)$$

Thus the amount of heat required of the heat source for the test rig is directly proportional to the flow-passage equivalent diameter, and the mass flow rate should be made inversely proportional to the fluid-passage diameter. With gases this is often accomplished by increasing the pressure, because increasing the velocity may lead to pumping problems or to difficulties with compressibility effects.

The design of a heat exchanger scale model test is usually conditioned by the test equipment available. To achieve the desired Reynolds number in the model, it may be easier to make use of scaled-up fluid passages if, for example, large low-pressure fans are available, whereas a scaled-down model would require new high-head blowers to achieve the desired Reynolds number. The capacity of the heat source and/or sink is also likely to present problems; in fact, the cost of these items may far exceed the cost of the rest of the test equipment. It is for this reason that plant steam is so often used as the heat source.

It is sometimes desirable to go to much larger diameter tubes in the model than in the full-scale matrix if a detailed temperature distribution survey is to be made in a complex geometry such as a finned tube. Increasing the scale of the flow passages in the model, while using only a small number, may make it possible to conduct detailed temperature and flow distribution surveys that would be quite impractical in the full-scale unit.

Geometric Similarity Considerations

The extent to which geometric similarity between the test model and the full-scale unit should be preserved is an important consideration. The flow passages through the heat transfer matrix certainly should be geometrically similar. For a tube bundle the ratios of the tube diameter to the tube spacing should be the same both transverse and parallel to the flow. If fins are used, the fin efficiency should be much the same, and hence the fin material and thickness should be selected so that the parameter $w\sqrt{h/kb}$ (see Fig. H7.2) is nearly the same. If complex fin matrices such as those of Fig. 2.8 are to be compared, great care should be employed to ensure that the test models have the same finish, the same degree of regularity, and the same detailed geometric proportions as the full-scale units they are to represent.

It is important that the passage length-diameter ratio be sufficient so that corrections for entrance effects can be made with little error. A full-scale unit that is to employ a length-diameter ratio of 300 could be simulated readily with a model having a passage length-diameter ratio of only 100, and a small correction could be applied to give a good prediction of the performance of the full-scale unit. However, it is unwise to make use of a passage length-diameter ratio in the region of 10 to 20 in the model if a passage length-diameter ratio in excess of 100 is required in the full-scale unit, or vice versa, because the entrance effects would be quite large. While a correction could be applied, the associated error would approach that of an analytical estimate, and hence there would be little point in running the test. See Fig. H5.9 for L/D corrections.

Effects of Large Differences in Heat Transfer Coefficient

A brief review of the nature of the overall heat transfer coefficient will suffice to show that the design of a test should be conditioned by the fluid side giving the poorer thermal conductance. For example, a test section for an automotive water-to-air radiator should be chosen to give exact geometric simulation on the air side, since even with an extended surface the product of the air heat transfer coefficient, the fin efficiency, and the surface area usually is far less than the product of the water-side heat transfer coefficient and the water-side surface area. In a case such as this it has often proved both expedient and proper to make use of steam as a heat source rather than hot water, especially in cases for which the estimated water-side conductance is five or more times the air-side conductance (based on the water-side surface area).[3] This is particularly true where, as in this case, the flow passages on the high heat transfer coefficient side are long smooth tubes for which the heat transfer coefficients can be predicted with confidence and a suitable correction applied. Thus the performance of the special fin geometry on the low heat transfer coefficient side can be evaluated to determine the effects of turbulence promoters or interrupted surfaces which are not amenable to analysis.

Effects of Heat Losses on Model Size

Heat losses represent an important consideration in establishing the size of a small model. The conduct of the test and the analysis of the results can be carried out most readily if the unit is sufficiently large so that, if covered with a few inches of thermal insulation, the heat losses will not be more than a few percent of the heat transferred. If special considerations make it essential to employ a smaller unit, it is possible to install guard heaters between the inner and outer layers of thermal insulation so that heat losses from the test section can be kept to quite small values. However, such an arrangement requires a rheostat on each guard heater and a relatively large number of thermocouples to ensure that the guard heaters do not distort the temperature distribution and put heat into the system rather than minimize the heat loss from the test section.

Typical Cases

Some of the compromises commonly made in designing heat exchanger model tests are indicated by the typical cases outlined in Table 20.1. Note that in every case the capacity of the test unit was less than 10% of the capacity of the full-scale unit. Of course, in most instances the test was run to obtain data for a whole family of full-scale units. The relation between the test unit and the full-scale heat exchanger contemplated is indicated in Table 20.1. This reduction in capacity in all instances was achieved, at least partially, through a reduction in the size of the heat transfer matrix. In many instances a further reduction in capacity was achieved by employing a smaller temperature difference, and in one instance a marked reduction in capacity was obtained by employing air at atmospheric pressure in place of helium at high pressure. In the last case, it was possible to reduce the heat flux by a factor of about 20 while maintaining the

TABLE 20.1 Comparison of Heat Transfer Matrix Elements Used in Typical Evaluation Tests With the Corresponding Full-Scale Exchangers

Application		Recuperator for Mobile Gas Turbines[a]	Aftercooler for Merlin Aircraft Engine[b]	Shell-and-Tube Heat Exchanger[c]	Fuel Element for a Gas-Cooled Reactor[d]	Salt-to-NaK Heat Exchanger for a Molten-Salt Reactor[e]
Capacity, Btu/h	Core element	2×10^5	42,000	34,000	30,000	8×10^6
	Full-scale	6×10^8	2×10^6	40×10^6	500,000	1.9×10^8
Matrix size, in.	Core element	$8.4 \times 9.1 \times 3$	$3 \times 3 \times 8.7$	$3 \times 6 \times 5$	3.8(dia.) $\times$ 40	$2.5 \times 2.5 \times 65$
	Full-scale	$100 \times 100 \times 12$	$8.5 \times 12 \times 8.7$		3.8(dia.) $\times$ 40	$2.5 \times 50 \times 65$
Matrix geometry	Core element	Plate-fin	Flattened tube-and-fin	Tube bundle	7-rod cluster	Tube bundle
	Full-scale	Plate-fin	Flattened tube-and-fin	Tube bundle	7-rod cluster	Tube bundle
Cold fluid	Core element	Air	H_2O-glycol	Water	Air	NaK
	Full-scale	Air	H_2O-glycol	Water	Helium	NaK
Passage equivalent diameter, in.	Core element	0.10	0.08	0.28	1.1	0.18
	Full-scale	0.10	0.08	0.2 to 1.0	1.1	0.18
Passage L/D (per pass)	Core element	30	37	20	35	360
	Full-scale	120	105	50 to 400	35	360
Number of passes	Core element	1	2	1	1	1
	Full-scale	2	4	1 to 4	1	1
Maximum temperature, °F	Core element	225	100	100	400	1,500
	Full-scale	600	150	300	1,100	1,500
Maximum pressure, psia	Core element	17	20		17	70
	Full-scale	100	50		300	0
Mass flow, lb/h · ft^2	Core element	42,000	500,000	800,000	36,000	4×10^6
	Full-scale	40,000	750,000	2×10^6	48,000	4×10^6
Reynolds number	Core element	8,400	3,000	500	75,000	120,000
	Full-scale	8,000	4,000	500	75,000	120,000
Hot fluid	Core element	Steam	Air	Turbine oil	Not applicable	Molten salt
	Full-scale	Combustion products	Air and gasoline vapor	Hydrocarbons		Molten salt
Passage equivalent diameter, in.	Core element	Not applicable	0.17	0.27	Not applicable	0.20
	Full-scale		0.17	0.2 to 1.0		0.20
Passage L/D	Core element	Not applicable	50	20	Not applicable	325
	Full-scale		50	10 to 100		325

TABLE 20.1 (*Continued*)

Application		Recuperator for Mobile Gas Turbines[a]	Aftercooler for Merlin Aircraft Engine[b]	Shell-and-Tube Heat Exchanger[c]	Fuel Element for a Gas-Cooled Reactor[d]	Salt-to-NaK Heat Exchanger for a Molten-Salt Reactor[e]
Number of passes	Core element	Not applicable	1	1	Not applicable	1
	Full-scale		1	1 to 20		1
Maximum temperature, °F	Core element	Not applicable	275	150	Not applicable	1,600
	Full-scale		650	450		1,600
Maximum pressure, psia,	Core element	Not applicable	15		Not applicable	70
	Full-scale		45			70
Mass flow, lb/h · ft^2	Core element	Not applicable	25,000	1×10^6	Not applicable	5×10^6
	Full-scale		28,000	1×10^6		5×10^6
Reynolds number	Core element	Not applicable	4,000	500	Not applicable	4,000
	Full-scale		4,500	500		4,000

[a]Ref. 3 (See also Fig. 14.3).
[b]See Fig. 8.24.
[c]Ref. 2.
[d]Ref. 6 (See also Fig. 8.29.
[e]Ref. 4 (See also Figs. 2.18 and 17.6.)

temperature rise per unit of passage length-diameter ratio at the same level as in the full-scale heat transfer matrix. It is interesting to note that in all but one instance the flow regime for one or both of the fluids involved fell in the transition region (Reynolds numbers of from 500 to 5000). It was this factor that made tests particularly important as there was no other good way to establish the effects of the geometric irregularities peculiar to the particular units of interest in this crucial flow regime.

These examples show that, in principal, heat exchanger model tests ordinarily entail building a small element of the heat transfer matrix with fluid-flow passages of the same size as in the full-scale unit, and testing it in such a way that the two fluid streams pass through it in the same relation as they would if the matrix element were not isolated from the rest of the heat exchanger. In a sense, it is like using a core drill to obtain a typical specimen of rock when exploring geological structures.

INSTRUMENTATION

Temperature Measurements

Heat exchanger tests ordinarily entail the measurement of a large number of temperatures. Although as few as four temperatures might suffice (i.e., the inlet and outlet temperatures for the hot and cold fluids), it is usually desirable, if not essential, to include additional temperature measurements, partly as a check on the primary instruments and partly to evaluate the temperature distribution. For example, a single-pass crossflow heat exchanger inherently has substantial variations in the temperature of the fluid stream across the outlet. While a mixing plenum or a substantial channel mixing length may be employed downstream of the outlet, it is still necessary to obtain a temperature traverse to make sure that mixing has been sufficient to give a uniform temperature distribution in the plane of the temperature measurement.

Thermocouples

Thermocouples are by far the most widely used temperature indicators. If properly installed, they give a relatively inexpensive, yet accurate, indication of temperature that lends itself to easy reading at a central location. Their thermal inertia is low; hence their response to temperature changes is much more rapid than that of most of the other temperature-sensing devices.[7]

While thermocouples are better suited to the measurement of metal surface temperatures than most other forms of instrumentation, it is difficult to make an installation that yields a true temperature of the metal surface. The thermocouple lead wires normally project into the gas stream so that they act as fins and may lead to local deviations in the surface temperature of as much as several hundred degrees Fahrenheit if the base surface is red hot. Even if laid flat and carried downstream along the surface for some distance, they may induce sufficient turbulence locally to cause an appreciable error. The best way to get a good metal surface temperature measurement is to employ relatively thick metal walls and drill holes in them, as indicated in Fig. 20.1, so that the installation in no way disturbs the fluid flow over the heat transfer surface, and so conduction along the thermocouple wire has a negligible effect on the temperature of the point being measured.[8] Unfortunately, the walls of most heat exchangers are too thin to permit this type of installation. As a result, it is usually not practicable to separate the heat transfer coefficients for the two fluid streams; only an overall coefficient can be obtained directly.

An indirect measure of the heat transfer coefficient on one side can be obtained by varying the fluid flow on that side while leaving the fluid-flow rate fixed on the other side. This is a particularly effective technique if applied to the low-coefficient side of a heat exchanger in which the thermal conductance is substantially lower on one side than the other, that is, the air side of a water-to-air heat transfer matrix.

Temperature measurements in regions where the heat flux is nearly zero, for example, in the inlet and outlet fluid streams, can be made by installing a thermocouple or some type of thermometer in a well, or thimble, that extends into the fluid stream. If a substantial amount of thermal inertia is not objectionable, thermocouples may be attached directly to the pipe wall by clamping or peening the thermocouple in place. The wires should be bound tightly to the pipe for a distance of an inch or two, and wrapped with thermal insulation to avoid fin effects. A good method for making such an installation is shown in the lower, left portion of Fig. 20.2.

Thermopiles

It is sometimes convenient to connect the thermocouples so that the temperature difference between the inlet and outlet can be read directly by taking a single reading

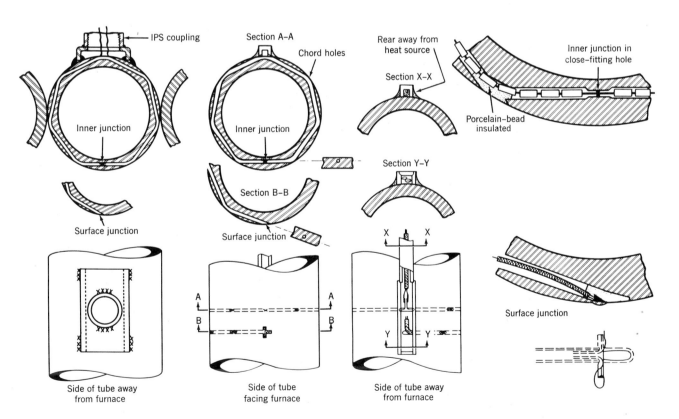

Figure 20.1 Recommended method for installing thermocouples in boiler tubes to obtain true metal temperatures in a furnace. (Courtesy The Babcock and Wilcox Co.)

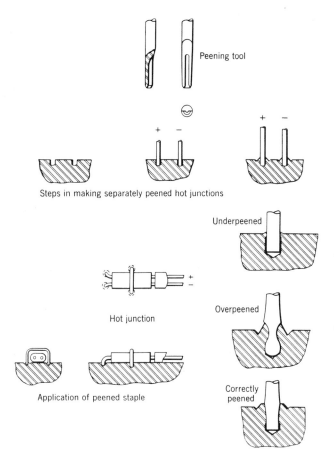

Peening tool

Steps in making separately peened hot junctions

Hot junction

Underpeened

Overpeened

Application of peened staple

Correctly peened

Figure 20.2 Recommended method for installing thermocouples in pipes in isothermal or nearly isothermal zones. *Note:* Use drill one size larger than wire diameter. Drill to depth of 1.5 wire diameters. Space holes at minimum practical distance. (Courtesy The Babcock and Wilcox Co.)

instead of two. In this way the error inherent in measuring the temperature difference can be cut in half. If desired, a thermopile of from 5 to 10 thermocouples may be employed. These may be clustered in a bundle or arranged in a grid to yield an average temperature. In either case, the signal to the potentiometer is increased in direct proportion to the number of thermocouples connected in series, and the error in reading is inversely proportional to the number of thermocouples in the pile. Such an arrangement often must be used in conducting tests on a unit such as an automotive air radiator, for which the water temperature drop may be only 6°C. With a 10-couple pile it is possible to get a voltage output equivalent to a 60°C temperature difference which can be used to give a temperature difference accurate within 1% (i.e., 0.06°C).

Errors Caused by Thermal Radiation

If gas temperatures above 250°C are to be measured, thermal radiation to or from the thermocouples may be important.[9-11] For temperatures of the order of 600°C such errors may be as much as 20°C. This difficulty may be reduced by placing the thermocouple inside a thermal radiation shield consisting of from one to four concentric steel tubes with their axes parallel to the direction of flow. The tubes should have a length-diameter ratio sufficiently large so that the angle subtended at the thermocouple by the open end of the inner tube is trivial compared to the total solid angle. Since the emissivity of silver is only about 0.03, the error caused by thermal radiation can be reduced very effectively by encasing the thermocouple in a single silver tube. An excellent arrangement is that obtained by flattening a 1/8-in.-diameter silver tube over a junction formed by making an oversize weld bead and then grinding it down to form a flat disc slightly thicker than the wires.[10] The error associated with this type of thermocouple using silver, gold, and platinum shields was determined. Some of the data are shown in Fig. 20.3 along with corresponding values for both bare thermocouples and with the multiple stainless steel cylindrical shields just mentioned.

Infrared Temperature Measuring Equipment

One of the most valuable applications of infrared temperature measuring equipment is to tests in which the temperature distribution is an important consideration. If the surface temperature is above 200°C and the surfaces involved can be observed visually,[12-14] a detailed picture of the temperature distribution can be obtained without spending the time and money to install a large number of thermocouples. Some types of instrument are designed so they can be sighted at any point in the field of view and the temperature determined. A few of these devices can discriminate temperature differences of as little as 1°C between areas as small as 1 mm in diameter over a range from room temperature to 800°C. As indicated in Fig. 20.4, it is also possible to take photographs of large areas and obtain something approaching this degree of accuracy in the measurements. In either case, thermocouples should be used at some points in the field of view to provide reference points. A high level of technical competence is required in applying this type of equipment, since many subtle effects must be considered. For example, the emissivity of the surfaces may change with the viewing angle or the accumulation of small amounts of oxide or deposits.[15,16]

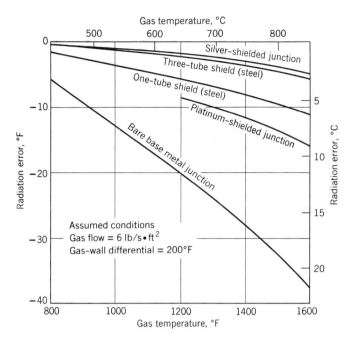

Figure 20.3 Effects of temperature on the radiation error associated with the use of typical thermocouples. (Dahl and Fiock, Ref. 10.)

An application of infrared photography to temperature measurement is presented in Fig. 20.4. This delineates the temperature distribution in a ground test prototype of a radiator designed for a space power plant. (One panel of the prototype is shown in Fig. 1.18.) The effects of temperature on the intensity of the image can be inferred from the 16°C temperature difference between the roots and the tips of the fins. Note the remarkably uniform temperature distribution. Over

1000 thermocouples would have been needed to have given the minimum information required, but even this number would not have given as detailed a picture, and they would not have been as convenient to use.

Pressure Measurement

The pressure drop across the heat transfer matrix is usually just as important as the heat transfer perfor-

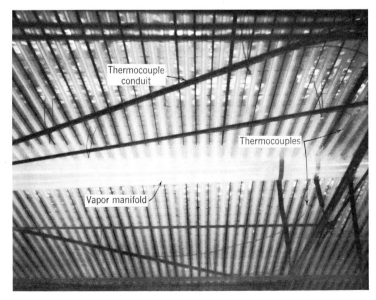

Figure 20.4 Photograph made with polaroid infrared sensitive film using the light emitted from a double bank of axially finned tubes in which potassium vapor is condensing at 800°F. (The eye can barely distinguish a dull red at 1000°F). The photo was made by looking up at nearly horizontal tube banks in the configuration shown in Fig. 18.6-1. The highlights on the fins are caused by reflected radiation from the center manifold. (Courtesy of the Oak Ridge National Laboratory.)

mance. The test rig may be designed so that the duct cross section is the same as that of the inlet face of the heat transfer matrix under test, in which case simple static pressures in the duct may be satisfactory. If not, allowances should be made for differences in the velocity pressure with changes in flow-passage size. It is, of course, important that at least 10 diameters of straight duct preceed the heat transfer matrix to give a good velocity distribution across the face of the duct. If particularly good pressure-drop data are to be obtained, it may be desirable to make use of a piezometer ring, that is, an interconnected set of static pressure holes around the perimeter of the duct in a plane normal to the direction of flow. The pressure drop across the heat exchanger can be measured directly with a manometer or a differential pressure gauge.

Flow Measurement

A variety of flow meters are employed. Venturii are used for both liquid- and gas-flow measurement, particularly for the larger passage sizes. If air is used as one of the fluids, the most simple and accurate means of flow measurement is a flow nozzle mounted at the air inlet from the room. This avoids the errors stemming from turbulence and the poor velocity distribution downstream of bends and other irregularities in piping and ducts.[17] If the temperature isn't too high, the blower should be on the outlet side of the heat transfer matrix to avoid effects it might have on the turbulence and velocity distribution in the fluid stream entering the heat transfer matrix. Pitot tubes or electronic sensors may be employed[18] if it is difficult to assure a truly uniform velocity distribution across the heat transfer matrix. A condition of this sort is to be avoided, of course; it makes the reduction of data a much more tedious task and introduces uncertainties associated with the effects of flow and temperature distribution.

Several types of flow meter employing an element buoyed up by fluid dynamic forces have attractive features. Rotometers are widely used for both gas and liquid flow. They are especially well suited to liquid flow, particularly for flow rates of from 1 to 100 gpm. Ball flow meters employing a sphere in a curved tube of glass or plastic give a lower pressure drop than rotometers and can be improvised quickly to give good accuracy over a moderate flow range.[19] They can be calibrated easily with water by taking weight or volumetric flow measurements. For that matter, if water is the fluid to be metered, it is sometimes most convenient to take weight flow measurements, as, for example, in tests of small condensers.

HEAT TRANSFER PERFORMANCE TESTS

Temperature Stabilization

In preparing for heat transfer performance tests the test procedure should be worked out concurrently with the design of the test rig to facilitate the conduct of the test and the reduction of the test data. Depending on the size of the unit, the heat capacity of various components in the test setup, and heat losses, it is usually necessary to stabilize from 15 min to several hours for each point in order to obtain good equilibrium conditions and consistent, repeatable data. One means of assuring that equilibrium has in fact been achieved is to take readings at from 5- to 15-min intervals for a fixed set of flow and temperature conditions. This procedure should be continued until three successive readings show negligible changes in the various temperatures in the system. It is important that the rig be designed to facilitate this sort of stabilization. For example, if water from a main is to be employed, it is important that it not be subject to pressure changes caused by variations in the water requirements of other equipment. If room air is used, care should be taken to avoid temperature irregularities arising from the opening and closing of doors or windows. Fluctuations in steam pressure or the voltage in electric power lines for electric heaters may also give difficulty.

Transient Techniques

Where suitable recording instrumentation is available, it is sometimes possible to avoid long periods for rig stabilization by employing transient techniques.[20] This procedure requires a careful analysis of the heat capacity of the various elements of the system so that rates of temperature change can be used as indices of heat transfer.[21]

ANALYSIS OF TEST RESULTS

The test results may be organized and presented in a variety of fashions, depending on the application. In some instances it may be desirable to present the results in very general terms, for example, in the form of curves for the Colburn modulus versus Reynolds number, as in Fig. 14.5. At the other extreme it may be better simply to present the total amount of heat rejected as a function of fluid-flow rate. One of the most convenient methods of presentation is in the form of a chart (such as Fig.

14.10) that permits the direct selection of the appropriate size of heat transfer matrix with no need for complex calculations and the possibility for error that they introduce.

Heat Balances

Probably the most potent single tool in the analysis of heat transfer test data is the heat balance obtained by comparing the heat given up by the hot fluid with the heat absorbed by the cold fluid. The difference between these two quantities can be compared with the estimated heat losses. If, as is often the case, the heat losses do not account for the difference between the two, an error must have occurred in the measurement of either the flow rates or the temperature differences in the fluid streams. It is at this point that the more liberally instrumented test rig is advantageous. Various temperatures and temperature changes can be compared for consistency. The effects of changes in temperature level or flow rate on the disparity in the heat balance can be examined for clues as to the probable source of the errors. The effects of the manner in which conditions are changed and a point is approached may be important. That is, in approaching a given point, it may make a difference whether the flow is being increased or decreased, the temperature raised or lowered, etc. No general rules can be given; only experience and a sharpened sense for detecting inconsistencies can minimize the time lost in the tedious calibration of instrumentation and repeated check runs with the equipment.

Correlation of Data

It is always important to correlate the experimentally evaluated performance of the heat exchanger with that estimated analytically. This is desirable partly to check both the design and testing techniques and partly to provide the engineer with a better background for future work of a similar nature. It may also be important to compare results of the experiment at hand with other data for somewhat similar heat transfer matrices (e.g., those of a competitor).

In reducing the test data, care must be exercised in making allowances for factors such as large differences in the physical properties of the fluid between the bulk free stream temperature and the surface temperature. Probably the first step in deciding how to organize and reduce the data is to determine the extent to which the physical properties of the fluids vary over the temperature range covered in the test.

FLOW TESTS

In the course of the development of a full-scale heat exchanger it is sometimes desirable to carry out flow tests of header regions or other complex geometries to determine the gross flow distribution or the pressure drop. Tests of this sort may be carried out with simple models, since no provisions for heat addition or extraction need be employed. In these tests the only requirement is that the model be geometrically similar and that the Reynolds number be in the range of interest. Thus the tests may be carried out with water or air rather than with fluids that would be difficult to handle. Air is especially suited to tests of this sort, because the models can be inexpensive in construction and small leaks will not make a mess. If there is no severe flow separation, Pitot traverses can be used to establish the direction of the flow as well as the velocity distribution. If there is severe flow separation, flow visualization work can be carried out by using adhesive tape to attach tufts of thread or yarn to the passage walls, or by attaching a tuft to a wire probe that can be moved about in the flow field. Smoke can also be employed, but it is tricky to use and usually not very satisfactory. The smoke filaments tend to be dispersed so rapidly by turbulence that the technique is applicable only for relatively low Reynolds numbers and simple geometries. Any of these tests can be run more conveniently if the models are made of a transparent plastic such as Lucite.

Water is a better working fluid than air for some types of flow visualization work. Where three-dimensional effects may be important, it is easy to include small air bubbles or particles in suspension to indicate the flow path and the nature of the turbulence. For two-dimensional flow, an excellent insight into the flow behavior can be obtained by sprinkling aluminum powder on the surface of water flowing through a model of the channel under study. The simplest sort of setup can serve if a photographic record is not desired. However, to obtain pictures of the quality shown in Figs. 3.10 and 14.2 takes much care in designing and building the test equipment and much patience and skill in carrying out the tests. The models must be large enough so that surface tension forces do not distort the flow pattern. This requires that the flow channel width be not less than about 12 mm (0.5 in.) at any point, and it is usually best to add a small amount of detergent to the water to obtain good dispersion of the aluminum powder. The flow pattern also will be badly distorted by surface waves if the flow velocity exceeds about 0.3 m/s (1 ft/s); thus it is difficult to get Reynolds numbers much above 5000. Other points to watch include the use of a good stilling pool ahead of the test section; a long cloth sack over the inlet pipe will help in this respect. For good photographs the bottom of the channel should be covered with black cloth or black paper which can be held in place by a glass plate. The exposure time should be neither too long nor too short; a good rough rule of thumb is to use $\theta = D/8V$, where θ is the exposure time in seconds, D the transverse dimension of the flow obstruction, and V the water velocity.

It is sometimes necessary to investigate the details of the flow in the boundary layer. The technique just described, which uses aluminum powder on water, proved effective as an aid to understanding the behavior of a fluid flowing transversely across fins, as in Fig. 3.21. Hot wire anemometers have proved to be potent instruments for investigating the fine structure of turbulent flow, but they are difficult devices to use and are better suited to the needs of sophisticated research specialists than to those of design personnel. Injection of a dye may be a good technique for some types of problem. Just a trace of iodine solution can be allowed to bleed into a starch solution to give a sharply defined plume extending downstream from the point of injection. The movement and rate of dispersion of the dye plume give a good indication of the character and intensity of the turbulence in the vicinity. By adding a small amount of sodium thiosulfate to the starch solution to react with the iodine, the discoloration can be confined to the dye plume, and many injections can be made without affecting the transparency of the bulk liquid.

Another flow visualization technique that has proved useful for the investigation of the fine structure of the flow over turbulating devices and tube spacers entails coating the surface with a fluorescent dye dissolved in a medium weight motor oil.[22] The dye collects in stagnation zones, and shows up clearly when viewed in ultraviolet light. Figure 20.5 is a typical photograph that was obtained in this way.

An interesting technique that makes use of the doubly refracting properties of certain dye solutions has been applied to the study of laminar flow. Through the use of polarized light it is possible to obtain interference fringe patterns similar to those obtained in photoelastic stress analysis.[23] The flow patterns can be determined from these by mathematical analysis.

STRUCTURAL TESTS

Structural tests on heat exchanger components such as pressure vessels or heads can be carried out using any of a variety of techniques. Probably the most common

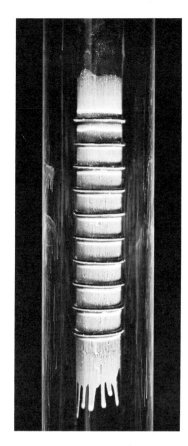

Figure 20.5 Stagnation regions indicated by bright bands of fluorescent powder. Air flowed axially through the annulus between a 2-in. glass pipe and a 1-in.-diameter steel rod fitted with turbulators in the form of 0.085-in.-diameter O-rings on a 0.5-in. pitch. The eddies caused the fluorescent powder to accumulate in the stagnation zone immediately downstream of each turbulator. (Courtesy Oak Ridge National Laboratory).

approach is to build a fractional- or full-scale model of the structural element in question and determine the stress distribution with strain gauges. These can be applied in gauge lengths of as little as 6 mm. In one instance some 1300 strain gauges were installed on a $\frac{1}{5}$ scale model of a complex pressure vessel; the cost of the test amounted to approximately 3% of the cost of the completed vessel, but it increased enormously the degree of confidence that could be placed in the design.

Both heat exchanger pressure vessels and complete heat exchangers are often tested hydraulically as part of the acceptance tests to ensure that they can withstand the design pressures. These tests are usually carried to a 40% overpressure. During the test, strain gauges may be used to check design calculations, particularly in highly stressed zones or in complex regions which are difficult to analyze. Precautions should be taken to ensure the safety of personnel in the event of a burst type of failure.

Pressure tests are usually accompanied by leak tests to assure structural soundness of the parts and that the detail design of the flanged joints is such that, even though the joint may be sufficiently strong to withstand the overpressure, deflections under pressure do not induce leaks.

ENDURANCE TESTS

Endurance testing may be required for lightweight heat exchanger units designed for automotive or aircraft applications. Vibration, pressure stresses, or thermal stresses may induce failures of a nature not disclosed by preliminary design analyses. Such endurance tests should be carefully designed to induce essentially the same pressure and thermal stress cycles anticipated for the full-scale unit. Where relaxation is a factor in high-temperature units, the time interval between cycles may be made much shorter than that anticipated for the full-scale unit if this can be justified by relaxation data. For example, when high thermal stresses are a factor, it is usually possible to induce most of the plastic flow associated with a thermal cycle in 15 or 20 min so that 1-h cycles would be sufficiently long in a model test to simulate 24-h or longer cycles in a full-scale unit.

Endurance tests to investigate the effects of tube fouling by deposits also may be much in order. Care must be taken in such tests to assure that the operating conditions in the full-scale unit are well simulated.

LEAK TESTS

One of the most exasperating problems in heat exchanger fabrication and maintenance is the detection, location, and correction of small leaks. Large leaks can be located easily, but small leaks are usually hard to detect and still harder to locate precisely so that they can be repaired. Yet with some types of fluid such as acids, molten metals, and radioactive liquids or gases, a very high degree of leaktightness must be maintained. For such systems, specifications may require that the system be tested with a gas under pressure and that the total leakage from the entire system be 10^{-5} cm^3/s, that is, about 1 cm^3/day, or even lower. Special testing techniques must be used to determine whether a unit meets such a difficult requirement.[24]

Soap Bubble Tests

A simple, yet reasonably sensitive, method of detecting small leaks and locating their source entails pressurizing the component with 0.5 atm of compressed air and painting it with a soap solution. As the soap solution is applied, the surface should be observed carefully for the formation of tiny bubbles. Solutions especially prepared for this purpose are available commercially, at least one being in the form of an aerosol spray.

A more sensitive indication of a leak can be obtained by pressurizing the unit with air and immersing it in hot water containing a small amount of detergent to reduce the surface tension. This technique is sufficiently sensitive to detect individual leaks as small as 10^{-5} cm^3/s, that is, leaks giving one bubble about a millimeter in diameter per minute.

Rate of Pressure Rise

If a higher degree of leaktightness is required, the unit can be evacuated and the rate of pressure rise determined. Specifications may call for the pressure rise to be less than 1 μm/h at a pressure of about 10 μm Hg. For a given system volume the rate of system pressure rise for a given in-leakage rate is independent of pressure in the range of interest for vacuum test work, that is, for system pressures from 10 to 1000 μm Hg. If the in-leakage rate exceeds the specified value for a vacuum test, it is necessary to locate the principal leak (or leaks) and make repairs.

Helium Leak Tests

The most sensitive technique available uses helium escaping from a small probe moved about over the surfaces that are thought likely to be responsible for the leak. When the probe is in the vicinity of the leak, the helium enters the vacuum system with the air that is leaking in, and even a small fraction can be detected easily with a mass spectrometer located in the exhaust line to the vacuum pump. The signal increases sharply when the probe is moved close to the outer surface of the leak. Commercial helium leak detectors are sufficiently sensitive to detect helium leaks as small as 10^{-10} cm^3/s. The rate of in-leakage of other gases would be inversely proportional to the square root of the molecular weight. The volumetric leakage rate for liquids is smaller than that for gases by a factor of at least 1000.

Halide Leak Tests

A somewhat similar technique is often used for equipment to contain gaseous halides or halide compounds.

The unit can be filled with the halide or halide compound and pressurized. A probe or "sniffer" connected to the suction side of a small pump can be moved over the suspected surface to collect any gas or vapor that may be leaking out. The air drawn from the "sniffer" by the pump is fed to a Primus torch which normally burns with a colorless flame. A small amount of halide colors the flame green, thus indicating a leak. This technique is sufficiently sensitive to indicate a leak of about 10^{-4} cm^3/s.

In carrying out leak tests such as the preceding it is usually necessary to make the initial tests with the unit partially disassembled and the openings blocked off with special caps made for the purpose. The heat transfer matrix and the shell may be leak tested separately. After each component has met its leaktightness requirements, the unit can be assembled, the joints made, and the final leak test conducted.

Heat Exchanger Acceptance Tests

Acceptance tests including both performance evaluation and endurance testing have been standard requirements for aircraft components as well as complete aircraft since the 1920s. Too much was at stake to risk flying a plane unless vital structural elements and mechanical equipment such as the engine had been thoroughly tested; hence the Army Air Corps set up an impressive set of facilities at Wright Field for giving components a thorough workout before trying to fly them. Similarly, both the Air Force and NASA have set up elaborate facilities for exhaustive testing of components for spacecraft. Components for nuclear power plants for utility service have been so large that no central facility for testing them was set up until the 1960s, when the liquid metal fast breeder reactor program was initiated in the United States and the USAEC made the large investment that was necessary. Similar facilities were subsequently established by the French government for their LMFBR program. Tests in these facilities have yielded invaluable experience and insights.[25,26] (See also Reference 54 of Chapter 9 and Reference 19 of Chapter 17.) Even with these large facilities, it is generally necessary to reduce the size of the units tested by employing modules having the full-scale tube length but with the number of tubes reduced to fit the capacity of the available heat source.

REFERENCES

1. R. A. Lane and E. L. Morrison, "Southwark Station Boiler Air-Flow Model Tests and Operation Results," *Trans. ASME*, vol. 71, 1949, p. 941.

2. G. A. Omohundro, O. P. Bergelin, and A. P. Colburn, "Heat Transfer and Fluid Friction During Viscous Flow Across Banks of Tubes," *Trans. ASME*, vol. 71, 1949, p. 27.

3. W. M. Kays, and A. L. London, "Heat Transfer and Flow-Friction Characteristics of Some Compact Heat Exchanger Surfaces," *Trans. ASME*, vol. 72, 1950, p. 1075.

4. R. E. McPherson et al., "Development Testing of Liquid Metal and Molten Salt Heat Exchangers," *Nuclear Science and Engineering*, vol. 8, 1960, p. 14.

5. D. R. Chapman, "Some Possibilities of Using Gas Mixtures Other than Air in Aerodynamic Research," NACA Technical Report no. 1259, 1956.

6. H. W. Hoffman, J. L. Wantland, and W. J. Stelzman, "Heat Transfer with Axial Flow in Rod Clusters," *International Developments in Heat Transfer*, Part three, ASME, 1961, p. 553.

7. A. J. Hornfeck, "Response Characteristics of Thermometer Elements," *Trans. ASME*, vol. 71, 1949, p. 121.

8. *Steam*, 37th ed., The Babcock and Wilcox Co., 1955.

9. W. M. Rohsenow and J. P. Hunsaker, "Determination of the Thermal Correction for a Single Shielded Thermocouple," *Trans. ASME*, vol. 69, 1947, p. 699.

10. A. I. Dahl and E. F. Fiock, "Shielded Thermocouples for Gas Turbines," *Trans. ASME*, vol. 71, 1949, p. 153.

11. J. G. Bartas and E. Mayer, "Estimation of Temperature Patterns in Multiply Shielded Systems," *Trans. ASME*, vol. 79, 1957, p. 1722.

12. *Bibliography of Temperature Measurement*, USNBS monograph no. 27, 1960.

13. G. W. McDaniel and A. P. DiMattia, "Inspecting Infrared Optical Materials and Systems by Means of the Evaporograph," *Applied Optics*, vol. 1, 1962, p. 483.

14. M. Camac and R. M. Feinberg, "High-Speed Infrared Bolometer," *Review of Scientific Instruments*, vol. 33, 1962, p. 964.

15. A. S. Tenney III, "Industrial Radiation Thermometry," *Mechanical Engineering*, vol. 109(10), October 1986, p. 36.

16. T. G. Conway, "Using Thermal Images to Measure Temperature," *Mechanical Engineering*, vol. 109(6), June 1987, p. 32.

17. A. Linford, "Flow Measurement and Meters," 2nd ed., E. and F.N. Spon, London, 1961.

18. T. Ramey, "Measuring Air Flow Electronically," *Mechanical Engineering*, vol. 107(5), May 1985, p. 29.

19. H. L. Shulman and K. A. Van Wormer, Jr., "Flow Measurement with Ball Flow Meters," *Journal of AIChE*, vol. 4, 1958, p. 380.

20. W. M. Kays, A. L. London, and R. K. Lo, "Heat Transfer and Friction Characteristics for Gas Flow Normal to the Tube Banks—Use of a Transient-Test Technique," *Trans. ASME*, vol. 76, 1954, p. 387.

21. J. H. Stang and J. E. Bush, "The Periodic Method for Testing Compact Heat Exchanger Surfaces," *Journal of Engineering for Power, Trans. ASME*, vol. 96(1), 1974, p. 87.

22. D. L. Loving and S. Katzoff, "The Fluorescent Oil Film Method and Other Techniques for Boundary Layer Flow Visualization," NASA Memo 3-17-59L, March 1959.

23. F. N. Peebles and J. W. Prados, "Two-Dimensional Laminar-Flow Analysis Using a Doubly Refracting Liquid," *Journal of AIChE*, vol. 5, 1959, p. 225.

24. W. Steckelmacher, "Leak Detection," *Nuclear Engineering*, vol. 4, 1959, p. 450.

25. D. van Essen, "Experimental Results of the 85 MWTH SNR-300 Intermediate Heat Exchanger, Thermal Hydaulics and Effects of Nuclear Steam Generators and Heat Exchangers," ASME Book no. G00310, HTD-vol. 51, 1986.

26. "Testing Fast Reactor Components in the Tripot Facility," *Nuclear Engineering International*, November 1982, p.2 6.

HANDBOOK

This handbook has been compiled to provide the designer with a convenient set of tables and charts to facilitate heat exchanger design work. Consistent sets of SI and English units are used throughout the book and this handbook in an effort to eliminate those annoying errors that so commonly creep in when diverse units are converted to a common basis.

At the beginning of each of the following nine sections there is a brief explanation of the tables and charts in that section together with some suggestions on ways in which they can be used.

Nomenclature, Constants, and Conversion Factors

NOMENCLATURE, CONSTANTS, AND CONVERSION FACTORS

Although the nomenclature and symbols employed in the heat transfer, fluid-flow, stress analysis, and heat exchanger design fields is standarized to a substantial degree, many variations are found in the literature. The symbols and units used here are chosen to give a consistent set by taking the most widely used symbols for the most important quantities, and then by selecting each of the balance, partially on the basis of the extent to which it is used and partially for its conformity with the rest of the symbols in the set.

In using the conversion factors, a physical quantity in one set of units can be converted to another by multiplying by the appropriate entry in the table. Many conversion factors that do not appear directly in the tables may be obtained by combining the appropriate factors.

TABLE H1.1   Nomenclature

A	Area, m^2 (ft^2)
a	Surface area per unit of volume, m^2/m^3 (ft^2/ft^3)

TABLE H1.1   (*Continued*)

b	Thickness, m (ft)
C	Coefficient
c_p	Specific heat, $kJ/kg \cdot °C$ ($Btu/lb \cdot °F$)
D	Diameter, m (ft)
D_e	Equivalent diameter of noncircular cross sections, m (ft)
d	Diameter, in.
d_e	Equivalent diameter of noncircular cross sections, in.
E	Modulus of elasticity, Pa (psi)
F	Form factor (for radiant heat transmission)
$\mathfrak{F}$	Radiant heat transmission factor (includes ϵ and F)
f_d	Flow friction factor (based on equivalent passage diameter)
f_r	Flow friction factor (based on hydraulic radius)
G	Mass flow rate, $kg/m^2 \cdot s$ ($lb/ft^2 \cdot h$)
G'	Mass flow rate, $lb/ft^2 \cdot s$
G'_{tr}	Mass flow rate at transition from laminar to turbulent flow (at a Reynolds number of 2000) in 25-mm (1.0-in.) ID round tubes, $lb/ft^2 \cdot s$
Gr	Grashof number
GTD	Greatest temperature difference, °C (°F)
g	Acceleration of gravity, m/s^2 (ft/s^2)
H	Enthalpy, kJ/kg (Btu/lb)
ΔH_p	Change in enthalpy to preheat liquid to the boiling point, kJ/kg (Btu/lb)
H_v	Enthalpy of vaporization, kJ/kg (Btu/lb)
ΔH_s	Enthalpy of superheat, kJ/kg (Btu/lb)
h	Heat transfer coefficient, $W/m^2 \cdot °C$ ($Btu/h \cdot ft^2 \cdot °F$)
ITD	Inlet temperature difference, °C (°F)
j	Colburn modulus, $\left(\dfrac{h}{c_p G}\right)\left(\dfrac{c_p \mu}{k}\right)^{2/3}$

TABLE H1.1 (*Continued*)

j'	Modified Colburn modulus,

$$j' = \left(\frac{h}{Gc_p}\right)\left(\frac{c_p\mu}{k}\right)^{2/3}\left(\frac{\mu_w}{\mu_b}\right)^{0.14}$$

J	Mechanical equivalent of heat, 4.184 J/g · cal (ft · lb/Btu)
K	Enthalpy transfer coefficient h/c_p, kg/m² · s (lb/ft² · h)
k	Thermal conductivity, W/m · °C (Btu/h · ft² · °F/ft)
L	Length, m (ft)
L_w	Liquid water-flow rate in cooling towers, kg/s · m² (lb/h · ft²)
l	Length, (in.)
LMTD	Log mean temperature difference, °C (°F)
LTD	Least temperature difference, °C (°F)
M	Molecular weight
Nu	Nusselt number, $\dfrac{hD}{k}$
N	Number of units in an array
n	Number in a series
P	Pressure, Pa (lb/ft²)
Pr	Prandtl number
p	Pressure, (psi)
ΔP	Pressure drop, Pa (lb/ft²)
Δp	Pressure drop, (psi)
Pe	Peclet number
Q	Heat-flow rate, W (Btu/h)
q	Dynamic head, Pa (lb/ft²)
R	Radius, m (ft)
r	Radius, in.
R_h	Hydraulic radius, m (ft)
r_h	Hydraulic radius, in.
Re	Reynolds number, $\rho VD/\mu$
R_g	Gas constant, 8314 J/kg · mol · K (1544 ft · lb/lb · mol · °R)
S	Stress, Pa (psi)
s	Spacing, m (in.)
T	Temperature, K (°R)
t	Temperature, °C (°F)
δt	Temperature rise (or drop), °C (°F)
Δt	Temperature difference, °C (°F)
Δt_m	Log mean temperature difference, °C (°F)
U	Overall heat conductance between two fluid streams, W/m² · °C (Btu/h · ft² · °F)
u	Heat conductance of one of several layers between two fluid streams, W/m² · °C (Btu/h · ft² · °F)
V	Fluid velocity, m/s (ft/s)
v	Specific volume, m³/kg (ft³/lb)
W	Weight flow rate, kg/s (lb/h)
w	Fin width (or height), m (ft)
w'	Fin width (or height), in.

TABLE H1.1 (*Continued*)

X	Vapor quality (ratio of vapor weight to mixture weight)
x	Humidity, kg water/kg dry air (lb water/lb dry air)
x_s	Humidity of saturated air, lb water/lb dry air
α	Coefficient of thermal expansion, m/m · °C (in./in. · °F)
δ	Deflection, m (in.)
ϵ	Emissivity
η	Heating (or cooling) effectiveness
ϕ	Temperature correction factor for h
μ	Viscosity, Pa · s (lb/h · ft)
ν	Kinematic viscosity, m³/s (ft²/h)
ρ	Density, kg/m³ (lb/ft³)
σ	Ratio of gas density to that of air at standard conditions

Subscripts:

a	Air
b	Bulk fluid conditions
c	Cold fluid
e	Equivalent
f	Film (refers to arithmetical mean temperature between the wall and the bulk free stream)
h	Hot fluid
w	Wall
g	Gas
i	Internal
id	Ideal
l	Liquid
m	Mean
o	Outer
p	Flow passage
s	Saturated
sh	Superheated
0	Initial, or base condition
1	Inlet
2	Outlet

TABLE H1.2 Abbreviation for Units

Atmosphere	atm
Barrel	bbl
British thermal unit	Btu
Calorie	cal
Centimeter	cm
Centimeters of mercury	cm Hg
Centimeters per second	cm/s
Cubic centimeter	cm³
Cubic foot	ft³
Cubic feet per minute	cfm
Cubic feet per second	cfs

TABLE H1.2 (*Continued*)

Cubic inch	in.^{3}
Cubic meter	m^3
Cubic millimeter	mm^3
Cubic yard	yd^3
Degrees centigrade	°C
Degrees Fahrenheit	°F
Degrees Rankine	°R
Degrees Kelvin	K
Feet per minute	fpm
Feet per second	fps
Foot	ft
Foot-pound	ft · lb
Gallon	gal
Gallons per hour	gph
Gallons per minute	gpm
Gallons per second	gps
Gram	g
Gram-calorie	g-cal
Horsepower	hp
Horsepower-hour	hp · h
Hour	h
Inch	in.
Inches of mercury	in. Hg
Joule	J
Kilogram-calorie	kg-cal
Kilogram	kg
Kilogram-meter	kg · m
Kilojoule	kJ
Kilowatt	kW
Kilowatt-hour	kWh
Liter	L
Megawatt	MW
Meter	m
Meters per second	m/s
Microinch	μin.
Micron (micro meter)	μm
Millimeters of mercury	mm Hg
Millimeter	mm
Minute	min
Ounce	oz
Pascal	Pa
Pound	lb
Pounds per cubic foot	lb/ft^3
Pounds per square foot	psf
Pounds per square inch	psi
Pounds per square inch, absolute	psia
Pounds per square inch, gage	psig
Pound-foot	lb · ft
Quart	qt
Revolutions per minute	rpm
Second	s
Square centimeter	cm^2
Square foot	ft^2
Square inch	in.^{2}
Square meter	m^2

TABLE H1.2 (*Continued*)

Watt	W
Watt-hour	Wh
Yard	yd

General Abbreviations of Words:

Absolute	abs
Average	av
Centimeter-gram-second (system)	cgs
Diameter	dia.
Exponential	exp
Inside diameter	ID
Logarithm (common)	log
Logarithm (natural)	ln
Outside diameter	OD
Specific gravity	sp gr
Standard International System	SI
Square	sq

TABLE H1.3 **Constants**

$e = 2.71828$
$\pi = 3.14159$
$\ln x = 2.30 \log x$

Acceleration of gravity	$g = 980.665$ cm/s^2
	$= 32.174$ ft/s^2
	$= 4.17 \times 10^8$ ft/h^2
Mechanical equivalent of heat	$J = 4.1840 \times 10^7$ erg/cal
	$= 778.16$ ft · lb/Btu
Gas constant (universal)	$R_u = 1544$ ft · lb/lb-mol · °R
	$= 4.968 \times 10^4$ lb · ft^2/s · lb-mol · °R
	$= 8.314 \times 10^7$ g · cm^2/s · g-mol · K
	$= 8.314 \times 10^3$ kg · m^2/s^2 · kg-mol · K
Stefan-Boltzmann constant	$\sigma = 1.355 \times 10^{-12}$ cal/s · cm^2 · K^4
	$= 0.1713 \times 10^{-8}$ Btu/h · ft^2 · °R^4

TABLE H1.4 **Conversion Factors**

Multiply	By	To Obtain
Length:		
in.	2.54	cm
in.	0.0254	m
m	39.37	in.
m	3.28084	ft
ft	0.3048	m

TABLE H1.4 (*Continued*)

Multiply	By	To Obtain
Area:		
in.^{2}	6.452	cm^2
circular mil	5.0671×10^4	mm^2
cm^2	0.155	in.^{2}
ft^2	0.0929	m^2
m^2	10.76	ft^2
acre	4,046.9	m^2
hectare	10,000	m^2
Volume:		
in.^{3}	16.387	cm^3
m^3	6.2898	bbl
m^3	35.314	ft^3
m^3	1.308	yd^3
yd^3	4.8089	bbl
cm^3	0.06102	in.^{3}
ft^3	0.028317	m^3
ft^3	0.1781	bbl
ft^3	7.4805	gal
ft^3	28.317	L
fluid oz (U.S.)	0.02957	L
gal (U.S.)	3.7853	L
gal (U.K.)	4.546	L
gal (U.S.)	231.0	in.^{3}
gal (U.S.)	0.13368	ft^3
Area per Unit Volume:		
ft^2/ft^3	3.28	m^2/m^3
Mass:		
g	15.432	grains
g	0.03527	oz
grain	0.0648	g
lb	453.59	g
kg	2.2046	lb
lb	0.4536	kg
ton (long)	2,240.0	lb
ton (metric)	2,205.0	lb
ton (short or net)	2,000.0	lb
ton (metric)	1,000.0	kg
Force:		
lbf	4.4482	N
kgf	9.80665	N
Density:		
lb/in.^{3}	27.68	g/cm^3
lb/ft^3	16.018	kg/m^3

TABLE H1.4 (*Continued*)

Multiply	By	To Obtain
lb/ft^3	0.016018	kg/L
lb/gal (U.S.)	0.11983	kg/L
g/cm^3	62.43	lb/ft^3
g/cm^3	1.0	sp gr @ 4°C
specific gravity	62.4	lb/ft^3 @ 60°F
specific gravity	8.338	lb H_2O/gal @ 60°F
Pressure:		
atm	101,325	Pa
bar	100,000	Pa
atm	14.696	psi
bar	14.5045	psi
kg/m^2	9.8067	Pa
kg/cm^2	14.223	psi
psi	6,894.8	Pa
psf	47.88	Pa
psi	2.309	ft of water at 60°F
psi	2.0360	in. Hg at 32°F
psi	27.81	in. H_2O at 60°F
psi	51.713	mm Hg at 32°F
psi	0.0703	kg/cm^2
g/cm^2	0.01422	lb/in.^{2}
in. H_2O	249.09	Pa
in. Hg	3,386.6	Pa
torr (mm Hg)	133.32	Pa
Energy:		
Btu	778.2	ft · lb
Btu	107.6	kg · m
Btu	1,055	J
Btu	0.252	kg-cal
Btu	3.93×10^{-4}	hp · h
Btu	2.93×10^{-4}	kwh
J	1×10^7	erg
J	1.0	W/s
J	2.39×10^{-4}	kg-cal
J	2.778×10^{-4}	Wh
J	0.7376	ft/lb
kgf/m	9.8067	J
kwh	3,413	Btu
ft · lb	3.239×10^{-4}	kg-cal
g-cal	4.186	J
g · cm	9.294×10^{-8}	Btu
g · cm	980.6	erg
g · cm	7.233×10^{-5}	ft · lb
therm	10^5	Btu
quad	10^{15}	Btu
quad	1.0551×10^{12}	MJ
Specific Heat:		
Btu/lb · °F	1.0	g-cal/g · °C

TABLE H1.4 (*Continued*)

Multiply	By	To Obtain
Btu/lb · °F	4.1868[a]	kJ/kg · °C
kg-cal/kg · °C	4.1868[a]	kJ/kg · °C
J/g · °C	0.23885	Btu/lb · °F

Power:

W	1.0	J/s
Btu/s	1.055	kW
Btu/s	1.4147	hp
kW	3,413	Btu/h
kgf · m/s	9.8067	W
hp	2,545	Btu/h
hp	3.3×10^4	ft · lb/min
hp	550	ft · lb/s
hp	0.7457	kW
hp (British)	0.7457	kW
hp (metric)	0.73548	kW
kg-cal/s	4.1868	kW
ton refrigeration	3.5169	kW

Heat Flux:

Btu/h · ft^2	3.1546	W/m^2
W/m^2	0.31709	Btu/h · ft^2
kg-cal/s · m^2	4.1868	kW/m^2
cal/s · cm^2	4.1868	W/cm^2

Mass Flux:

lb/h · ft^2	1.356×10^{-3}	kg/s · m^2
lb/s · ft^2	4.881	kg/s · m^2
kg/s · m^2	737.61	lb/h · ft^2

Flow Rate:

lb/h	1.256×10^{-4}	kg/s
gpm	0.063088	L/s
gpm	2.228×10^{-3}	ft^3/s

TABLE H1.4 (*Continued*)

Multiply	By	To Obtain
Heat Transfer Coefficient:		
Btu/h · ft^2 · °F	5.6784×10^{-4}	W/cm^2 · °C
Btu/h · ft^2 · °F	5.6784	W/m^2 · °C
Btu/h · ft^2 · °F	4.882	kg-cal/h · m^2 · °C
Btu/h · ft^2 · °F	1.356×10^{-4}	g-cal/cm^2 · s · °C
Btu/s · ft^2 · °F	20.442	kW/m^2 · °C

Thermal Conductivity:

Btu/h · ft^2 · °F/in.	12.4	kg-cal/h · m^2 · °C/cm
Btu/h · ft · °F	1.7308	W/m · °C
Btu/h · ft^2 · °F/in.	0.14423	W/m · °C
kg-cal/cm · s · °C	0.242	Btu/h · ft · °F
kg-cal/h · m · °C	1.163	W/m · °C

Viscosity:

centipoises	0.001	Pa · s
centipoises	10^{-2}	poises (g/s · cm)
centipoises	3.6	kg/h · m
centipoises	2.419	lb/h · ft
centipoises	2.09×10^{-5}	lb · s/ft^2
lb/h · ft	4.134×10^{-4}	Pa · s
lb/s · ft	1.4882	Pa · s

Kinematic Viscosity:

centistokes	density, g/cm^3	centipoises

Enthalpy:

Btu/lb	0.556	g-cal/g
Btu/lb	2.326	kJ/kg
kg-cal/kg	4.186	kJ/kg
J/g · °C	0.23885	Btu/lb · °F

[a]The international steam table calorie is defined as 4.1868 J, while the thermochemical calorie is defined as 4.184 J. The former value is used here.

Physical Properties Affecting Heat Transfer

PHYSICAL PROPERTIES

The designer of heat exchangers for unusual applications is often handicapped by lack of data on physical properties for one or both of the fluids to be used. Quite complete data for a broad temperature range are presented in this section for a sufficiently diverse set of gases and liquids so that one of them should be similar to almost any fluid likely to be of interest. Where limited data can be obtained for a fluid not included in these tables, but not for the temperature range desired, the points available can be plotted on suitable coordinates along with data from these tables for similar fluids to provide a guide for extrapolation to the desired temperatures.

The correction factor ϕ_l in Fig. H5.4 indicates the relation between the turbulent heat transfer coefficient for the liquid at the temperature in question and that for water at 200°F for the same mass flow rate and passage diameter, that is,

$$\phi_l = \frac{\text{Heat transfer coefficient for liquid at given temperature}}{\text{Heat transfer coefficient for water at 200°F}}$$

The correction factor ϕ_g in Fig. 5.6 is similar to ϕ_l except that it relates the turbulent heat transfer coefficient for a gas to that of air at 200°F for the same mass flow rate and passage diameter. This helps to show the relative heat transfer performance of different fluids, indicates the effects of temperature, and facilitates estimation of the heat transfer coefficient. A further explanation of their application is given at the beginning of Section H5.

Of the properties affecting heat transfer, only the viscosity and vapor pressure are strongly temperature-dependent. Figures H2.2 and H2.3 show these effects. Pressure has little effect except in the region close to saturation. Consequently, all the data given are for atmospheric pressure, except for Figs. H2.4–H2.6. As can be seen from these three figures, the viscosity, specific heat, and thermal conductivity of a vapor (as well as its density) vary widely with pressure in the region close to the saturation temperature and pressure. Since the applications of saturated vapors are specialized, and since about 30 charts would be required to give a good set for the fluids of Tables H2.3 and H2.4, only these three charts for steam are presented here; the designer working with other vapors near the saturation region is simply cautioned to obtain similar data for the fluid of interest to him.

Note that special alloys such as high nickel and refractory alloys are expensive and hence are normally used only where their superior strength at high temperatures justifies the extra cost. Thus the physical properties given for these alloys in Table H2.1 are for temperatures near the upper limit of the range in which their use is likely.

TABLE H2.1 Physical Properties of Representative Structural Materials at 20°C (70°F) Except Where Temperature is Given[a-c]

Material	Thermal Conductivity		Coefficient of Thermal Expansion		Modulus of Elasticity		Density		Yield Point		Elongation, % in 5 cm
	Btu/h · ft · °F	W/m · °C	in./in. · °F	m/m · °C	psi × 10⁻⁶	GPa	lb/ft³	g/cm³	psi	MPa	(2 in.)
Aluminum:											
Commercially pure	134	232.0	12.7	22.9	10	69	170	2.7	13,000	90	120.0
Wrought alloy (6061-T6)	90	155.8	12.5	22.5	10.6	73	173	2.8	38,000	262	18.0
Sand cast (108°F)	73	126.4	12	21.6		0	175	2.8	14,000	97	2.5
Die cast (360°F)	65	112.5	11.4	20.5	10.3	71	166	2.7	25,000	172	2.0
Beryllium:											
Commercially pure	92	159.3	6.9	12.4	37	255	114	1.8	20,000	138	0.5
2Be-0.4Ni-97Cu	54	93.5	9.2	16.6	18	124	515	8.3	90,000	621	4.0
Ceramics (commercial, dense):											
Alumina	2.3	4.0	4.9	8.8	50	345	235	3.8	200,000	1,379	0
Beryllia	44	76.2	5	9.0	39	269	180	2.9	100,000	690	0
Fused quartz	0.62	1.1	0.3	0.5	10.4	72	137	2.2	160,000	1,103	0
Graphite (ATJ)	60	103.9	1.5	2.7	1.3	9	108	1.7	1,420	10	0
Pyrex	0.66	1.1	1.9	3.4	9.8	68	138	2.2	100,000	690	0
Tungsten carbide	53	91.7	3	5.4	80	552	880	14.1	325,000	2,241	0
Copper:											
Electrolytic, hard-drawn	225	389.5	9.5	17.1	17	117	556	8.9	45,000	310	6.0
Admiralty metal (28Zn-1Sn)	70	121.2	11.2	20.2	15	103	530	8.5	20,000	138	65.0
Brass (65Cu-35Zn)	60	103.9	10	18.0	15	103	530	8.5	60,000	414	8.0
Cupro-nickel (70Cu-30Ni)	18	31.2	9	16.2	22	152	560	9.0	70,000	483	15.0
Aluminum bronze (10Al)	45	77.9	9.2	16.6	18	124	475	7.6	25,000	172	30.0
Glidcop AL-20 (425°C)	196	339.3	10.9	19.6	17	117	530	8.5	26,000	179	
Iron:											
Grey cast iron	31	53.7	5.8	10.4	13.2	91	480	7.7	15,000	103	0.2
Low carbon steel	28	48.5	6.4	11.5	29.5	203	490	7.9	39,000	269	66.0
Croloy (1Cr-1/2Mo)	27	46.7	6.4	11.5	29.5	203	490	7.9	45,000	310	70.0
Stainless steel, (type 304)	10	17.3	9.5	17.1	28	193	500	8.0	34,000	234	63.0
(type 304, 600°C)	10	17.3	9.5	17.1	21	145	500	8.0	10,000	69	
Lead (commercially pure)	20	34.6	16.3	29.3	2.6	18	700	11.2	800	6	30.0
Magnesium wrought alloy (AZ 61A-P)	35	60.6	14.4	25.9	6.5	45	113	1.8	31,000	214	16.0
Molybdenum, TZM (0.5Ti-0.1Zr)	85	147.1	2.7	4.9	50	345	637	10.2	130,000	897	4.0

TABLE H2.1 (*Continued*)

Material	Thermal Conductivity		Coefficient of Thermal Expansion		Modulus of Elasticity		Density		Yield Point		Elongation, % in 5 cm
	Btu/h · ft · °F	W/m · °C	in./in. · °F	m/m · °C	psi × 10^{-6}	GPa	lb/ft^3	g/cm^3	psi	MPa	(2 in.)
Nickel:											
Commercially pure	35	60.6	7.2	13.0	30	207	555	8.9	21,000	145	35.0
Monel (67Ni-30Cu)	14.5	25.1	7.8	14.0	26	179	550	8.8	35,000	241	40.0
Inconel (13Cr-6.5Fe)	9	15.6	6.4	11.5	31	214	530	8.5	35,000	241	50.0
Incoloy 600 (15Cr-8Fe)	8.9	15.4	5.8	10.4	30	207	524	8.4	35,000	241	
Incoloy 625 (21Cr-9Mo-4Nb-4Fe)	5.7	9.9	7	12.6	30	207	525	8.4	50,000	345	
Incoloy 690 (29Cr-9Fe)	7	12.1	7	12.6	30	207	511	8.2	55,000	379	
Incoloy 718 (20Cr-20Fe-5Nb-3Mo)	6.5	11.3	6	10.8	30	207	511	8.2	150,000	1,034	
Incoloy 800 (40Fe-21Cr-1.5Mn)	6.7	11.6	7	12.6	28.5	197	496	7.9	27,000	186	
Hastelloy B	6.5	11.3	5.6	10.1	31	214	576	9.2	60,000	414	45.0
Haynes 188 (500°C)	11.5	19.9	7.7	13.9	27.6	190	560	9.0	43,800	302	
Niobium (Nb-1Zr)	40	69.2	3.9	7.0	12	83	535	8.6	30,000	207	
Tantalum T-111	31.5	54.5	3.5	6.3	27	186	1040	16.7	40,000	276	25.0
(8W-2Hf)(900°C)	30	51.9	6	10.8	24	166	1040	16.7	40,000	276	
Titanium	12	20.8	4.8	8.6	15.5	107	273	4.4	80,000	552	20.0
Ti-6Al-4V (500°C)	5.6	9.7	5.3	9.5	11	76	256	4.1	45,000	310	
Tungsten	116	200.8	2.4	4.3	50	345	1200	19.2	40,000	276	
Vanadium (800°C)	24	41.5	6	10.8	18	124	382	6.1	38,000	262	
Zircalloy 2 (1.5Sn)	10	17.3	3	5.4	11	76	406	6.5	20,000	138	

[a]Where available, the properties given are for as-drawn tubing.
[b]Values for yield point and elongation are typical, but vary widely with amount of cold work.
[c]Data from Refs. 15–19.

TABLE H2.2 **Ratio of Thermal Conductivity to Density for Typical Fin Materials**[a]

Material	100°F	500°F	1000°F
Aluminum (1060-H-12)	0.79	0.77	
Beryllium	0.81	0.63	0.51
Cast iron	0.065	0.057	0.050
Niobium	0.056	0.062	0.067
Copper	0.40	0.40	0.40
Graphite (ATJ)	0.55	0.40	0.3
Nickel	0.104	0.078	0.071
Magnesium	0.31		
Molybdenum	0.13	0.12	0.11
Steel, low-carbon	0.057	0.051	0.043
Steel, stainless	0.02	0.022	0.024

[a]Data from Refs. 1, 16, 18, and 19.

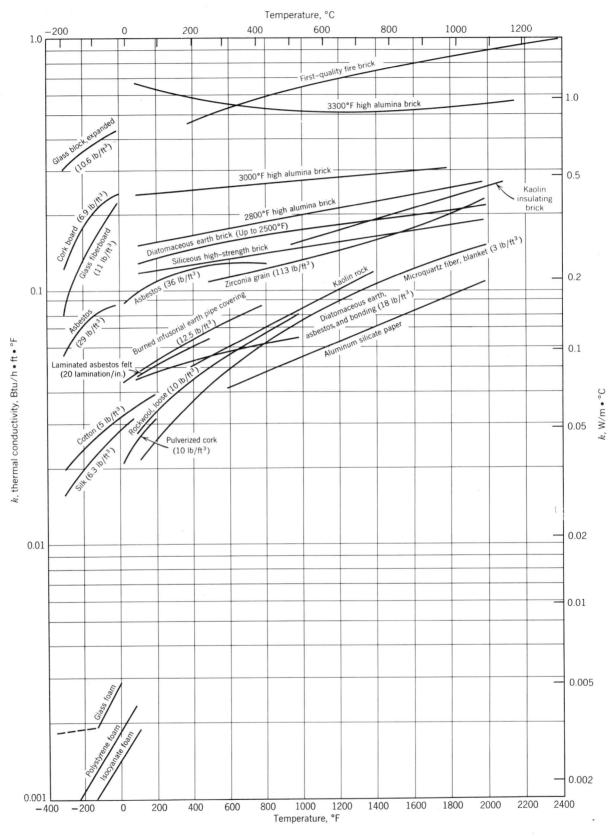

Figure H2.1 Effects of temperature on the thermal conductivities of thermal insulating materials. (Data from a number of sources—including Ref. 8, in Chapter 20 and Refs. 4, 14, and 19 of this handbook—were selected, plotted, and averaged to yield these curves.)

TABLE H2.3 Heat Transfer Properties of Liquids

Temperature		c_p		k		μ		ρ		$\dfrac{c_p\mu}{k}$
°F	°C	Btu/lb · °F	J/kg · °C	Btu/h · ft · °F	W/m · °C	lb/h · ft	Pa · s × 10⁻⁴	lb/ft³	kg/m³	
Water										
32		1.0293	4.3095	0.337	0.5833	4.32	17.9	62.54	1002	13.20
200	93	1.0039	4.2031	0.393	0.6802	0.738	3.05	60.20	964	1.88
400	204	1.0750	4.5008	0.382	0.6612	0.320	1.32	53.62	859	0.91
600	316	1.5250	6.3849	0.293	0.5071	0.215	0.889	42.37	679	1.08
Aqueous solution 30% ethylene glycol										
60	16	0.8820	3.6928	0.276	0.4777	6.04	25.0	64.90	1040	19.60
100	38	0.9000	3.7681	0.285	0.4933	3.27	13.5	64.30	1030	10.30
200	93	0.9340	3.9105	0.292	0.5054	1.23	5.08	62.10	995	3.93
300	149	0.9700	4.0612	0.285	0.4933	0.692	2.86	59.20	948	2.36
Ethylene glycol										
60	16	0.5560	2.3279	0.169	0.2925	62.1	257	69.40	1112	204.00
100	38	0.5810	2.4325	0.160	0.2761	25.1	104	68.70	1100	91.40
200	93	0.6440	2.6963	0.135	0.2337	5.67	23.4	66.20	1060	27.05
300	149	0.7060	2.9559	0.111	0.1921	2.30	9.49	63.30	1014	14.60
H_2										
−430	−257	1.9100	7.9968	0.064	0.1101	0.045	0.185	4.67	75	1.35
−410	−246	4.4400	18.5894	0.080	0.1378	0.020	0.084	3.69	59	1.14
N_2										
−210	−134	0.5000	2.0934	0.041	0.0710	0.162	0.670	34.50	553	1.98
−110	−79	0.4740	1.9845	0.095	0.1644	0.756	3.13	54.00	865	3.75
NH_3										
0	−18	1.0800	4.5217	0.290	0.5019	0.567	2.34	42.00	673	2.08
100	38	1.1700	4.8986	0.290	0.5019	0.172	0.711	35.60	570	0.69
Dowtherm A										
200	93	0.4320	1.8087	0.086	0.1494	2.71	11.2	62.60	1003	13.56
400	204	0.6000	2.5121	0.105	0.1817	1.14	4.71	56.80	910	6.51
600	316	0.7000	2.9308	0.104	0.1795	0.727	3.01	50.50	809	4.90
Methyl alcohol										
0	−18	0.5700	2.3865	0.124	0.2146	2.80	11.6	51.30	822	12.87
100	38	0.6150	2.5749	0.121	0.2086	1.15	4.75	48.10	770	5.87
200	93	0.6500	2.7214	0.117	0.2025	0.666	2.75	43.10	690	3.70
Freon-11										
0	−18	0.1980	0.8290	0.060	0.1038	1.64	6.78	98.27	1574	5.40
100	38	0.2120	0.8876	0.530	0.9173	0.920	3.80	90.19	1445	3.68
200	93	0.2250	0.9420	0.460	0.7962	0.637	2.63	80.94	1296	3.12
Freon-114										
0	−18	0.2300	0.9630	0.044	0.0762	1.45	6.00	98.62	1580	7.58
100	38	0.2412	1.0099	0.035	0.0611	0.809	3.34	88.37	1416	5.53
200	93	0.2627	1.0999	0.027	0.0467	0.600	2.48	79.00	1265	5.84
Gasoline										
0	−18	0.4470	1.8715	0.110	0.1904	2.60	10.7	49.70	796	10.58
200	93	0.5650	2.3655	0.103	0.1783	0.745	3.08	42.70	684	4.08
400	204	0.6830	2.8596	0.097	0.1674	0.336	1.39	36.80	589	2.37
Kerosene										
0	−18	0.430	1.800	0.101	0.1748	17.1	70.7	52.50	841	72.80
200	93	0.545	2.281	0.095	0.1644	1.59	6.57	47.40	759	9.12

TABLE H2.3 (*Continued*)

Temperature		c_p		k		μ		ρ		$\dfrac{c_p\mu}{k}$
°F	°C	Btu/lb · °F	J/kg · °C	Btu/h · ft · °F	W/m · °C	lb/h · ft	Pa · s × 10⁻⁴	lb/ft³	kg/m³	
Kerosene										
400	204	0.655	2.742	0.089	0.1544	0.625	2.58	42.40	679	4.58
600	316	0.745	3.119	0.083	0.1435	0.310	1.28	38.10	610	2.78
SAE 10 (petroleum lubricating oil)										
0	−18	0.4110	1.720	0.09	0.1618	47.30	19.554	55.60	891	207.50
200	93	0.5200	2.177	0.09	0.1530	11.9	49.1	52.25	837	69.70
300	149	0.5750	2.407	0.09	0.1475	4.50	18.6	48.75	781	22.50
HTS (NaNO₃, KNO₃, KNO₂)										
400	204	0.36	1.51	0.34	0.588	18.2	75.0	121	1932	103
500	260	0.36	1.51	0.34	0.588	12.0	49.6	118	1893	66.70
600	316	0.36	1.51	0.34	0.605	7.02	29.0	116	1855	37.10
ORNL molten-salt no. 14 (10% NaF, 43.5% KF, 44.5% LiF, 1.1% UF₄)										
1100	593	0.4880	2.043	2.30	3.980	12.6	52.1	132	2116	2.67
1200	649	0.4880	2.043	2.44	4.223	8.80	36.4	129	2066	1.76
ORNL molten-salt no. 30 (50% NaF, 46% Zr F₄, 4% UF₄)										
1200	649	0.2600	1.088	1.38	2.388	17.0	70.3	208	3329	3.20
1300	704	0.2555	1.069	1.41	2.431	12.8	52.9	205	3277	2.33
1400	760	0.2508	1.050	1.50	2.596	10.6	43.8	201	3224	1.77
Sodium										
200	93	0.3305	1.383	49.1	84.98	1.73	7.13	57.90	927	0.0116
400	204	0.3199	1.339	46.7	80.82	1.10	4.53	56.40	903	0.0075
600	316	0.3115	1.304	43.8	75.80	0.797	3.29	54.60	875	0.0057
800	427	0.3049	1.276	40.1	69.40	0.610	2.52	53.00	849	0.0046
1000	538	0.3020	1.264	37.2	64.38	0.560	2.32	51.20	820	0.0046
1200	649	0.3011	1.260	35.0	60.57	0.475	1.96	49.10	786	0.0041
1400	760	0.3033	1.269	32.7	56.59	0.415	1.72	47.70	764	0.0039
NaK (56% Na, 44% K)										
200	93	0.270	1.130	14.90	25.78	1.36	5.62	55.30	886	0.0245
400	204	0.260	1.088	15.30	26.48	0.920	3.80	53.80	862	0.0155
600	316	0.255	1.067	15.70	27.17	0.710	2.94	52.10	835	0.0115
800	427	0.251	1.050	16.00	27.69	0.500	2.07	50.60	811	0.0086
1000	538	0.250	1.046	16.00	27.69	0.490	2.03	49.00	785	0.0083
1200	649	0.251	1.050	16.00	27.69	0.410	1.69	47.20	756	0.0080
Potassium										
800	427	0.183	0.766	22.8	39.46	0.510	2.11	46.10	738	0.0041
1000	538	0.182	0.762	21.1	36.52	0.414	1.71	44.40	711	0.0036
1200	649	0.183	0.766	19.5	33.75	0.354	1.46	42.90	687	0.0033
1400	760	0.187	0.782	18.0	31.15	0.322	1.33	41.50	665	0.0033
Lithium										
400	204	1.0425	4.364	26.8	46.3	1.31	5.42	31.65	507	0.0510
600	316	1.0200	4.270	26.0	45.0	1.08	4.46	31.00	497	0.0443
800	427	1.0057	4.210	26.0	45.0	0.950	3.93	30.40	487	0.0432
1000	538	0.9962	4.170	26.0	45.0	0.840	3.47	29.60	474	0.0476
Mercury										
0	−18	0.0338	0.141	5.64	9.76	4.44	18.3	851.9	13646	0.0266
200	93	0.0326	0.136	6.00	10.38	2.96	12.2	833.4	13349	0.0161
400	204	0.0324	0.135	7.30	12.63	2.43	10.0	818.4	13109	0.0108
600	316	0.0342	0.143	7.90	13.67	2.27	9.38	802.6	12856	0.0098

TABLE H2.4 Heat Transfer Properties of Gases

Temperature		c_p		k		μ		ρ		$\dfrac{c_p\mu}{k}$
°F	°C	Btu/lb · °F	kJ/kg · °C	Btu/h · ft · °F	W/m · °C	lb/hr · ft	Pa · s × 10⁻⁴	lb/ft³	kg/m³	
Air										
−200	−129	0.2392	1.001	0.0079	0.0137	0.0252	0.1042	0.153	2.4508	0.78
0	−18	0.2401	1.005	0.0139	0.0241	0.0415	0.1716	0.0864	1.3840	0.711
200	93	0.2414	1.010	0.0184	0.0318	0.0519	0.2146	0.0602	0.9643	0.685
400	204	0.2451	1.026	0.0224	0.0388	0.0624	0.2580	0.0462	0.7400	0.683
600	316	0.2505	1.048	0.0263	0.0455	0.0721	0.2981	0.0375	0.6007	0.686
800	427	0.2567	1.074	0.0299	0.0518	0.0805	0.3328	0.0316	0.5062	0.688
1000	538	0.2631	1.101	0.0332	0.0575	0.0884	0.3654	0.0272	0.4357	0.7
1200	649	0.2692	1.127	0.0363	0.0628	0.0961	0.3973	0.0239	0.3828	0.712
1400	760	0.2755	1.153	0.0391	0.0677	0.1035	0.4279	0.0214	0.3428	0.728
N_2										
−200	−129	0.252	1.055	0.0079	0.0137	0.0237	0.0980	0.148	2.3707	0.755
0	−18	0.2484	1.040	0.0132	0.0228	0.0389	0.1608	0.0835	1.3375	0.734
200	93	0.249	1.042	0.0173	0.0299	0.0498	0.2059	0.0582	0.9322	0.716
400	204	0.2515	1.053	0.021	0.0363	0.0601	0.2485	0.0448	0.7176	0.719
600	316	0.2562	1.072	0.0248	0.0429	0.0696	0.2877	0.0362	0.5799	0.719
800	427	0.262	1.096	0.0283	0.0490	0.0775	0.3204	0.0305	0.4885	0.717
1000	538	0.2687	1.125	0.0317	0.0549	0.0849	0.3510	0.0263	0.4213	0.72
1200	649	0.2755	1.153	0.0345	0.0597	0.0925	0.3824	0.0221	0.3540	0.733
1400	760	0.282	1.180	0.0372	0.0644	0.0982	0.4060	0.0207	0.3316	0.745
O_2										
−200	−129	0.2175	0.910	0.0079	0.0137	0.0272	0.1124	0.169	2.7070	0.749
0	−18	0.2182	0.913	0.0135	0.0234	0.044	0.1819	0.096	1.5377	0.711
200	93	0.2223	0.930	0.018	0.0312	0.0583	0.2410	0.0665	1.0652	0.718
400	204	0.2305	0.965	0.0233	0.0403	0.0712	0.2943	0.0512	0.8201	0.705
600	316	0.2385	0.998	0.0278	0.0481	0.0825	0.3411	0.0415	0.6647	0.707
800	427	0.2463	1.031	0.0317	0.0549	0.0925	0.3824	0.035	0.5606	0.718
1000	538	0.2525	1.057	0.0352	0.0609	0.1018	0.4208	0.0302	0.4837	0.73
1200	649	0.257	1.076	0.0385	0.0666	0.1108	0.4580	0.0265	0.4245	0.739
1400	760	0.2615	1.094	0.0416	0.0720	0.119	0.4919	0.0236	0.3780	0.749
CO_2										
0	−18	0.248	1.038	0.0122	0.0211	0.038	0.1571	0.0835	1.3375	0.772
200	93	0.2495	1.044	0.0174	0.0301	0.0507	0.2096	0.0582	0.9322	0.726
400	204	0.2528	1.058	0.0215	0.0372	0.0613	0.2534	0.0448	0.7176	0.721
600	316	0.2587	1.083	0.0254	0.0440	0.0702	0.2902	0.0362	0.5799	0.715
800	427	0.2655	1.111	0.0288	0.0498	0.0785	0.3245	0.0305	0.4885	0.724
1000	538	0.272	1.138	0.0317	0.0549	0.086	0.3555	0.0263	0.4213	0.738
1200	649	0.2782	1.164	0.0347	0.0601	0.093	0.3845	0.0221	0.3540	0.746
1400	760	0.2834	1.186	0.0377	0.0653	0.0998	0.4126	0.0207	0.3316	0.75
H_2										
−400	−240	2.46	10.29	0.014	0.0242	0.0043	0.0178	0.045	0.7208	0.756
−200	−129	2.975	12.45	0.055	0.0952	0.0131	0.0542	0.0105	0.1682	0.709
0	−18	3.385	14.17	0.092	0.1592	0.0204	0.0843	0.0059	0.0945	0.738
200	93	3.45	14.44	0.122	0.2112	0.0248	0.1025	0.00415	0.0665	0.700
400	204	3.46	14.48	0.152	0.2631	0.0297	0.1228	0.0032	0.0513	0.676
600	316	3.47	14.52	0.18	0.3115	0.0342	0.1414	0.0026	0.0416	0.659
800	427	3.48	14.57	0.207	0.3583	0.0394	0.1629	0.0021	0.0336	0.662
1000	538	3.48	14.57	0.223	0.3860	0.0421	0.1740	0.00186	0.0298	0.656
1200	649	3.49	14.61	0.241	0.4171	0.0461	0.1906	0.00165	0.0264	0.667
1400	760	3.5	14.65	0.257	0.4448	0.0497	0.2055	0.00147	0.0235	0.676
He										
−200	−129	1.25	5.2335	0.052	0.0900	0.0395	0.1633	0.021	0.3364	0.949
0	−18	1.25	5.2335	0.08	0.1385	0.0434	0.1794	0.012	0.1922	0.678

TABLE H2.4 (*Continued*)

Temperature		c_p		k		μ		ρ		$\dfrac{c_p\mu}{k}$
°F	°C	Btu/lb · °F	kJ/kg · °C	Btu/h · ft · °F	W/m · °C	lb/hr · ft	Pa · s × 10⁻⁴	lb/ft³	kg/m³	
He										
200	93	1.25	5.2335	0.0985	0.1705	0.0545	0.2253	0.083	1.3295	0.691
400	204	1.25	5.2335	0.118	0.2042	0.066	0.2728	0.064	1.0252	0.699
600	316	1.25	5.2335	0.137	0.2371	0.077	0.3183	0.0051	0.0817	0.702
800	427	1.25	5.2335	0.156	0.2700	0.088	0.3638	0.0044	0.0705	0.705
1000	538	1.25	5.2335	0.176	0.3046	0.099	0.4093	0.0037	0.0593	0.703
1200	649	1.25	5.2335	0.194	0.3358	0.109	0.4506	0.0033	0.0529	0.702
1400	760	1.25	5.2335	0.212	0.3669	0.119	0.4919	0.0029	0.0465	0.701
Argon										
0	−18	0.124	0.5192	0.009	0.0156	0.049	0.2026	0.1135	1.8180	0.674
200	93	0.124	0.5192	0.012	0.0208	0.064	0.2646	0.0792	1.2686	0.661
400	204	0.124	0.5192	0.0147	0.0254	0.078	0.3225	0.0607	0.9723	0.658
600	316	0.124	0.5192	0.0172	0.0298	0.0905	0.3741	0.0492	0.7881	0.652
800	427	0.124	0.5192	0.0194	0.0336	0.102	0.4217	0.0415	0.6647	0.652
1000	538	0.124	0.5192	0.0218	0.0377	0.1125	0.4651	0.0358	0.5734	0.64
1200	649	0.124	0.5192	0.0234	0.0405	0.1225	0.5064	0.0314	0.5030	0.649
1400	760	0.124	0.5192	0.0252	0.0436	0.1315	0.5436	0.0281	0.4501	0.648
Neon										
0	−18	0.246	1.0300					0.06	0.9611	
200	93	0.246	1.0300	0.0324	0.0561	0.0884	0.3654	0.042	0.6728	0.67
400	204	0.246	1.0300	0.0384	0.0655	0.104	0.4299	0.032	0.5126	0.668
600	316	0.246	1.0300	0.0438	0.0758	0.119	0.4919	0.026	0.4165	0.668
800	427	0.246	1.0300	0.0488	0.0845	0.1325	0.5478	0.022	0.3524	0.668
1000	538	0.246	1.0300	0.0535	0.0926	0.145	0.5994	0.019	0.3043	0.6659
1200	649	0.246	1.0300	0.0585	0.1013	0.1585	0.6552	0.0166	0.2659	0.6659
1400	760	0.246	1.0300	0.0625	0.1082	0.17	0.7028	0.0149	0.2387	0.6659
CO_2										
0	−18	0.19	0.7955	0.0077	0.0133	0.031	0.1282	0.1315	2.1064	0.765
200	93	0.218	0.9127	0.0127	0.0220	0.0433	0.1790	0.0915	1.4656	0.743
400	204	0.238	0.9965	0.0177	0.0306	0.0548	0.2265	0.0702	1.1245	0.737
600	316	0.2554	1.0693	0.0226	0.0391	0.0652	0.2695	0.057	0.9130	0.736
800	427	0.2684	1.1237	0.0273	0.0473	0.074	0.3059	0.048	0.7689	0.727
1000	538	0.2793	1.1694	0.0317	0.0549	0.0827	0.3419	0.0415	0.6647	0.728
1200	649	0.2898	1.2133	0.0358	0.0620	0.091	0.3762	0.0364	0.5831	0.735
1400	760	0.2975	1.2456	0.0396	0.0685	0.0988	0.4084	0.0325	0.5206	0.742
NH_3										
0	−18	0.522	2.1855	0.0117	0.0203	0.0213	0.0881	0.0441	0.7064	0.633
200	93	0.532	2.2274	0.0192	0.0332	0.0303	0.1253	0.0307	0.4918	0.84
400	204	0.574	2.4032	0.028	0.0485	0.0394	0.1629	0.0236	0.3780	0.807
600	316	0.625	2.6168	0.0397	0.0687	0.0479	0.1980	0.0192	0.3075	0.755
800	427	0.675	2.8261	0.0537	0.0929	0.0557	0.2303	0.0161	0.2579	0.7
CH_4										
0	−18	0.507	2.1227	0.0157	0.0272	0.0237	0.0980	0.0455	0.7288	0.765
200	93	0.579	2.4242	0.0255	0.0441	0.0317	0.1310	0.0317	0.5078	0.72
400	204	0.674	2.8219	0.0358	0.0620	0.038	0.1571	0.0243	0.3892	0.715
600	316	0.772	3.2322	0.0505	0.0874	0.044	0.1819	0.0197	0.3156	0.672
Freon-11										
0	−18	0.124	0.5192	0.00412	0.0071	0.0232	0.0959	0.0398	0.6375	0.698
100	38	0.134	0.5610	0.00519	0.0090	0.0274	0.1133	0.0322	0.5158	0.707
200	93	0.145	0.6071	0.00627	0.0109	0.0312	0.1290	0.0278	0.4453	0.722

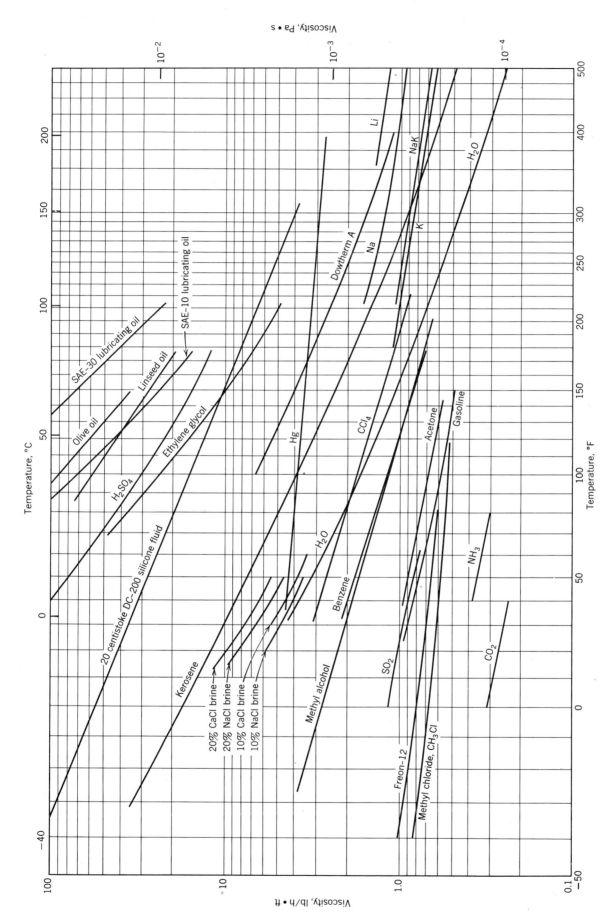

Figure H2.2 Effects of temperature on the viscosities of typical liquids. (Data from a number of sources—including Refs. 1–5 and 11–14 of this handbook—were selected, plotted, and averaged to yield these curves.)

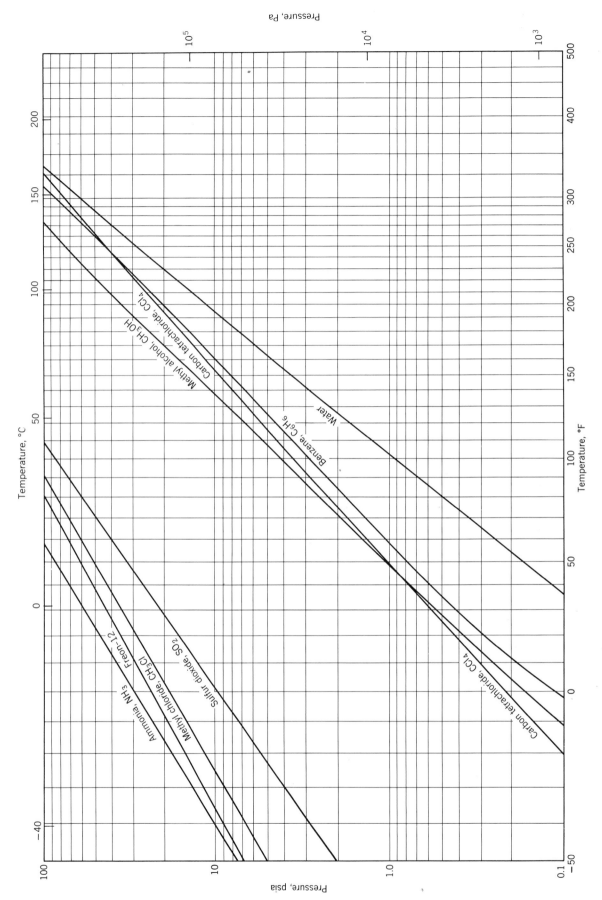

Figure H2.3 Effects of temperature on the vapor pressure of typical liquids. (Data from Refs. 1–4 and 11–14 of this handbook were selected, plotted, and averaged to yield these curves.)

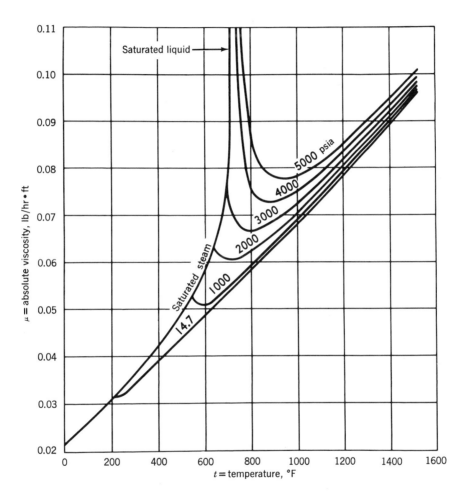

Figure H2.4 Effects of temperature and pressure on the viscosity of steam (Courtesy The Babcock and Wilcox Co.)

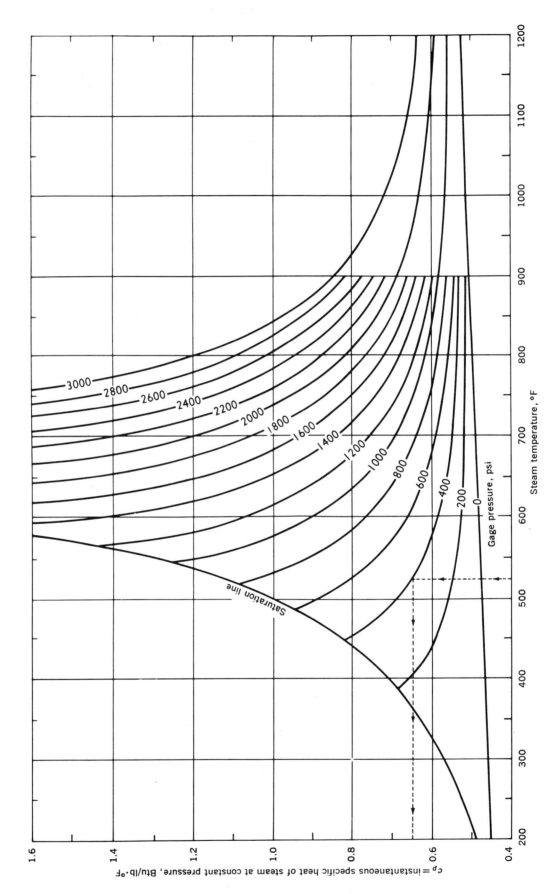

Figure H2.5 Effects of temperature and pressure on the specific heat of steam. (Courtesy The Babcock and Wilcox Co.)

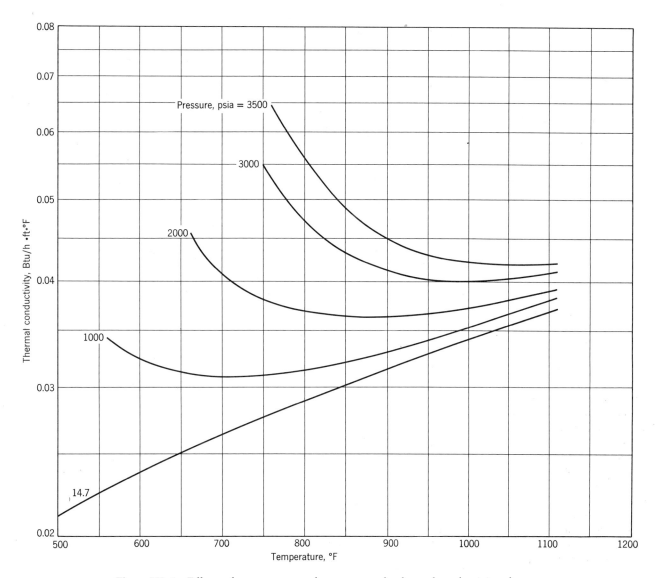

Figure H2.6 Effects of temperature and pressure on the thermal conductivity of steam. (Replotted from W. H. McAdams et al., "Heat Transfer to Superheated Steam at High Pressures," *Trans. ASME*, vol. 72, 1950, p. 428.)

TABLE H2.5 Emissivities for Typical Surfaces[a]

	Metals							
	Temperature, °F							
Surface	100	200	300	400	500	1000	2000	5000
					Emissivity, %			
Aluminum, polished, 98% pure	4	4	4	5	5	8	17	26
Aluminum, oxidized	20	20	21	22	23	33		
Copper, polished					2	4	6	
Copper, black oxide	92	91	90	89	83	77		
Chromium, polished sheet	8	11	14	15	17	27	37	43
Gold, electrolytic, polished	2	2	2	2	2	3	62	

TABLE H2.5 (*Continued*)

Metals

Surface	Temperature, °F							
	100	200	300	400	500	1000	2000	5000
	Emissivity, %							
Iron and steel, pure polished iron	6	6	6	7	8	12	22	35
Iron and steel, cast-iron, polished	21	21	21	21	21			
Iron and steel, polished steel	7	7	8	9	10	14	23	37
Iron and steel, rough steel plate	94	95	95	96	97	98		
Iron and steel, cast-iron, oxidized	58	61	62	64	66	75		
Iron and steel, matt wrought-iron, oxidized	95	95	95	95	95			
Iron and steel, oxidized steel, after long heating at dull red	85	88	90	92	93	96		
Lead, pure, polished	5	5	6	7	8			
Magnesium, polished	7	7	10	12	13	18	23	26
Molybdenum, polished	6	6	6	7	8	11	18	43
Nickel, electrolytic	4	4	5	5	6	10	16	28
Platinum, pure polished	4	5	5	6	6	10	19	27
Platinum, black	93	93	94	95	96	97	97	97
Rhodium, polished	5	6	7	7	7	8	9	16
Silicon, polished	72	72	72	72	72	72	72	72
Silver, polished or deposited	1	1	2	2	2	3	3	4
Tantalum, polished	6	6	7	7	7	7	9	25
Tellurium, polished	22	28	33	37	39	45	48	51
Tungsten, polished	2	2	2.5	3	3.5	7.5	15	35
Vanadium, polished	8	10	12	13	17	23	31	39
Zinc, pure polished	2	2	2	2	3	4	6	50
Zinc, matt zinc	21	21	21	21	21			
Alloys, brass, polished	10	10	10	10	10			
Alloys, brass, oxidized	46	48	50	53	56	75		
Alloys, nichrome wire, bright	65	65	66	66	67	71	79	
Alloys, nichrome wire, oxidized	95	95	96	96	97	98		
Alloys, stellite (Cr, Mo, Co)	12	12	13	13	14	18	24	28

Pigments

Surface	Temperature, °F						
	−250	125	750	1500	2000	2500	5000
	Emissivity, %						
Acetylene soot	97	99	99				99
Blue (Co_2O_3)	94	87	86				97
Red (Fe_2O_3)	91	96	70				59
Green (Cr_2O_3)	92	95	67				55
White (Al_2O_3)	94	98	79				12
White (Al_2O_3)				38	46	46	
White (ZrO_2)	95	95	77				16

TABLE H2.5 (*Continued*)

Paints

Surface	Temperature, °F				
	100	200	500	1000	5000
	Emissivity, %				
Lacs and oils, dark glossy varnish	89				
Clear lac on bright copper, bright copper with two thin coats	65	62	58		
White lac on bright copper, heavy coat	92	92			
White lac on bright copper	97	96	91	84	
White lac on bright copper, lampblack paint	96	96			
Aluminium paints, Al lac two coats polished on bright copper	26	26	26	25	

Miscellaneous Substances

Surface	Temperature, °F						
	100	500	1000	1500	2000	2500	5000
	Emissivity, %						
White paper	95						
Asbestos paper	93	94	94				
Black velvet	97						
Ordinary refractory brick					59		
White refractory brick		89	63		29		
Forty different refractories			63–84		77–91		
Limestone	95	83	75				
White marble	95	94	93				
Glazed procelain	92	99					
Pyrex glass		94	75				
Ice	92–96 (at 32°F)						
Polished graphite	42	97	97				
Pressed graphite		49	44		64		73

[a]Data excerpted from *International Journal of Heat and Mass Transfer*, vol. 5, 1962, pp. 67–76.

TABLE H2.6 Enthalpy in Btu per Pound of Dry Air for Air Saturated With Water Vapor at a Pressure of 29.92 in. Hg

Temperature, °F	0	1	2	3	4	5	6	7	8	9
90	55.93	57.33	58.78	60.25	61.77	63.32	64.92	66.55	68.23	69.96
100	71.73	73.55	75.42	77.34	79.32	81.34	83.42	85.56	87.76	90.03
110	92.34	94.72	97.18	99.71	102.31	104.98	107.73	110.55	113.46	116.46
120	119.54	122.72	125.98	129.35	132.80	136.4	140.1	143.9	147.8	151.8
130	155.9	160.3	164.7	169.3	174.0	178.9	183.9	189.0	194.4	199.9
140	205.7	211.6	217.7	224.1	230.6	237.4	244.4	251.7	259.3	267.1

TABLE H2.7 Thermal Conductivity of Metals at Cryogenic Temperatures in W/cm·K[a]

Temperature °K	°R	Aluminum	Cadmium	Chromium	Copper	Gold	Iron	Lead	Magnesium	Molybdenum
1	1.8	7.8	48.7	0.401	28.7	4.4	0.75	27.7	1.30	0.146
2	3.6	15.5	89.3	0.802	57.3	8.9	1.49	42.4	2.59	0.292
3	5.4	23.2	104	1.20	85.5	13.1	2.24	34.0	3.88	0.438
4	7.2	30.8	92.0	1.60	113	17.1	2.97	22.4	5.15	0.584
5	9	38.1	69.0	1.99	138	20.7	3.71	13.8	6.39	0.730
6	10.8	45.1	44.2	2.38	159	23.7	4.42	8.2	7.60	0.876
7	12.6	51.5	28.0	2.77	177	26.0	5.13	4.9	8.75	1.02
8	14.4	57.3	18.0	3.14	189	27.5	5.80	3.2	9.83	1.17
9	16.2	62.2	12.2	3.50	195	28.2	6.45	2.3	10.8	1.31
10	18	66.1	8.87	3.85	196	28.2	7.05	1.78	11.7	1.45
11	19.8	69.0	6.91	4.18	193	27.7	7.62	1.46	12.5	1.60
12	21.6	70.8	5.56	4.49	185	26.7	8.13	1.23	13.1	1.74
13	23.4	71.5	4.67	4.78	176	25.5	8.58	1.07	13.6	1.88
14	25.2	71.3	4.01	5.04	166	24.1	8.97	0.94	14.0	2.01
15	27	70.2	3.55	5.27	156	22.6	9.30	0.84	14.3	2.15
16	28.8	68.4	3.16	5.48	145	20.9	9.56	0.77	14.4	2.28
18	32.4	63.5	2.62	5.81	124	17.7	9.88	0.66	14.3	2.53
20	36	56.5	2.26	6.01	105	15.0	9.97	0.59	13.9	2.77
25	45	40.0	1.79	6.07	68	10.2	9.36	0.507	12.0	3.25
30	54	28.5	1.56	5.58	43	7.6	8.14	0.477	9.5	3.55
35	63	21.0	1.41	5.03	29	6.1	6.81	0.462	7.4	3.62
40	72	16.0	1.32	4.30	20.5	5.2	5.55	0.451	5.7	3.51
45	81	12.5	1.25	3.67	15.3	4.6	4.50	0.442	4.57	3.26
50	90	10.0	1.20	3.17	12.2	4.2	3.72	0.435	3.75	3.00
60	108	6.7	1.13	2.48	8.5	3.8	2.65	0.424	2.74	2.60
70	126	5.0	1.08	2.08	6.7	3.58	2.04	0.415	2.23	2.30
80	144	4.0	1.06	1.82	5.7	3.52	1.68	0.407	1.95	2.09
90	162	3.4	1.04	1.68	5.14	3.48	1.46	0.401	1.78	1.92
100	180	3.0	1.03	1.58	4.83	3.45	1.32	0.396	1.69	1.79

Temperature °K	°R	Nickel	Niobium	Platinum	Silver	Tantalum	Tin	Titanium	Tungsten	Zinc	Zirconium
1	1.8	0.64	0.251	2.31	39.4	0.115		0.0144	14.4	19.0	0.111
2	3.6	1.27	0.501	4.60	78.3	0.230		0.0288	28.7	37.9	0.233
3	5.4	1.91	0.749	6.79	115	0.345	297	0.0432	42.6	55.5	0.333
4	7.2	2.54	0.993	8.8	147	0.459	181	0.0576	55.6	69.7	0.442
5	9	3.16	1.23	10.5	172	0.571	117	0.0719	67.1	77.8	0.549
6	10.8	3.77	1.46	11.8	187	0.681	76	0.0863	76.2	78.0	0.652
7	12.6	4.36	1.67	12.6	193	0.788	52	0.101	82.4	71.7	0.748
8	14.4	4.94	1.86	12.9	190	0.891	36	0.115	85.3	61.8	0.837
9	16.2	5.49	2.04	12.8	181	0.989	26	0.129	85.1	51.9	0.916
10	18	6.00	2.18	12.3	168	1.08	19.3	0.144	82.4	43.2	0.984
11	19.8	6.48	2.30	11.7	154	1.16	14.8	0.158	77.9	36.4	1.04
12	21.6	6.91	2.39	10.9	139	1.24	11.6	0.172	72.4	30.8	1.08
13	23.4	7.30	2.46	10.1	124	1.30	9.3	0.186	66.4	26.1	1.11
14	25.2	7.64	2.49	9.3	109	1.36	7.6	0.200	60.3	22.4	1.13
15	27	7.92	2.50	8.4	96	1.40	6.3	0.214	54.8	19.4	1.13
16	28.8	8.15	2.49	7.6	85	1.44	5.3	0.227	49.3	16.9	1.12
18	32.4	8.45	2.42	6.1	66	1.47	4.0	0.254	40.0	13.3	1.08

TABLE H2.7   (*Continued*)

| Temperature | | | | | | | | | | | |
°K	°R	Nickel	Niobium	Plat-inum	Silver	Tan-talum	Tin	Tita-nium	Tung-sten	Zinc	Zicro-nium
20	36	8.56	2.29	4.9	51	1.47	3.2	0.279	32.6	10.7	1.01
25	45	8.15	1.87	3.15	29.5	1.36	2.22	0.337	20.4	6.9	0.85
30	54	6.95	1.45	2.28	19.3	1.16	1.76	0.382	13.1	4.9	0.74
35	63	5.62	1.16	1.80	13.7	0.99	1.50	0.411	8.9	3.72	0.65
40	72	4.63	0.97	1.51	10.5	0.87	1.35	0.422	6.5	2.97	0.58
45	81	3.91	0.84	1.32	8.4	0.78	1.23	0.416	5.07	2.48	0.535
50	90	3.36	0.76	1.18	7.0	0.72	1.15	0.401	4.17	2.13	0.497
60	108	2.63	0.66	1.01	5.5	0.651	1.04	0.377	3.18	1.71	0.442
70	126	2.21	0.61	0.90	4.97	0.616	0.96	0.356	2.76	1.48	0.403
80	144	1.93	0.58	0.84	4.71	0.603	0.91	0.339	2.56	1.38	0.373
90	162	1.72	0.563	0.81	4.60	0.596	0.88	0.324	2.44	1.34	0.350
100	180	1.58	0.552	0.79	4.50	0.592	0.85	0.312	2.35	1.32	0.332

[a]From: NSRDS-NBS-8 and NSRDS-NBS-16, "Thermal Conductivity of Selected Materials", C. Y. Ho, R. W. Powell, P. E. Liley, National Standard Reference Data System-National Bureau of Standards, Part 1, 1966, Part 2, 1968.

[b]Values in this table are in watts/cm °K. To convert to W/m·K multiply by 100; to convert to Btu/hr·ft·°R, multiply the tabular values by 57.818. These data apply only to metals of purity of at least 99.9%. In the table the third significant figure is for smoothness and is not indicative of the degree of accuracy.

Fluid Flow and Pressure Drop

FLUID FLOW AND PRESSURE DROP

In preparing charts involving fluid flow rate, it is difficult to decide whether to use the velocity in feet per second, the mass flow rate G' in pounds per second-square foot, or the mass flow rate G in pounds per hour-square foot. All three are widely used in the literature. G' is used in this handbook, since the conversion factor to put it in pounds per hour-square foot is obviously 3600. Similarly, dividing it by a single obvious factor, the density in pounds per cubic foot, serves to convert it to the velocity in feet per second. The same basic rationale holds for SI units. Use of a mass flow rate rather than a velocity has the advantage of making corrections for pressure effects unnecessary. For example, the chart of Fig. H3.1 would have been much more complex if the velocity had been used in place of the mass flow rate.

In using Fig. H3.1 to find the Reynolds number, simply find Re/d in the chart for the mass flow rate, fluid, and temperature desired, and multiply the value found for Re/d by the passage equivalent diameter in inches.

Dynamic heads for liquids and gases are given in pounds per square inch in Figs. H3.2 and H3.3 as a function of fluid-flow rate and density.

The friction factor f_d, for use in the equation $\Delta P = f_d(\rho V^2/2g)(L/D_e)$, is given in Fig. H3.4 as a function of the Reynolds number for passages having various degrees of roughness.

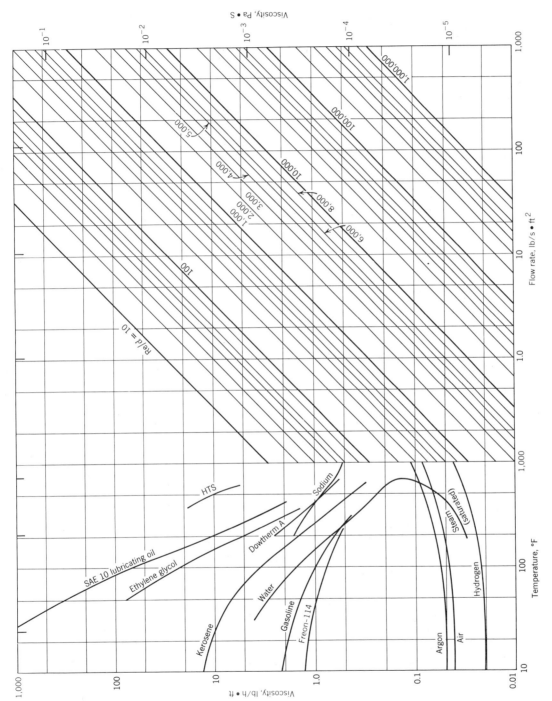

Figure H3.1 Reynolds number as a function of flow rate and viscosity for d in inches. (Viscosity data from Refs. 1–11 of this handbook were selected, plotted, and averaged to obtain these curves.)

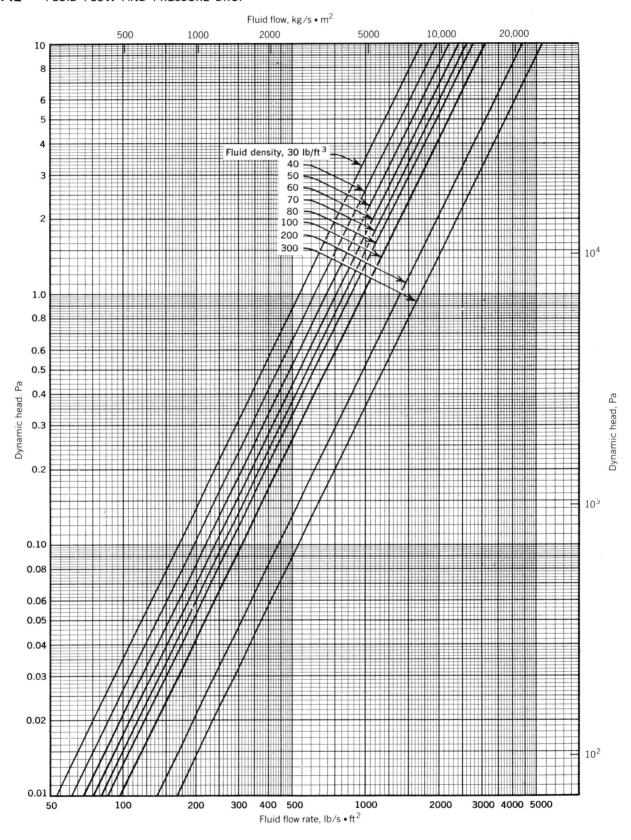

Figure H3.2 Dynamic head as a function of liquid density and flow rate.

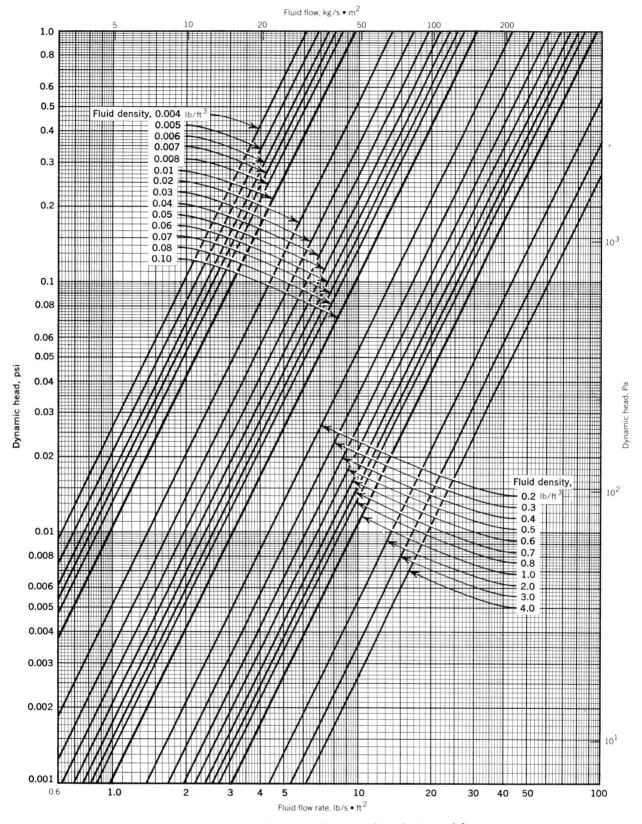

Figure H3.3   Dynamic head as a function of gas density and flow rate.

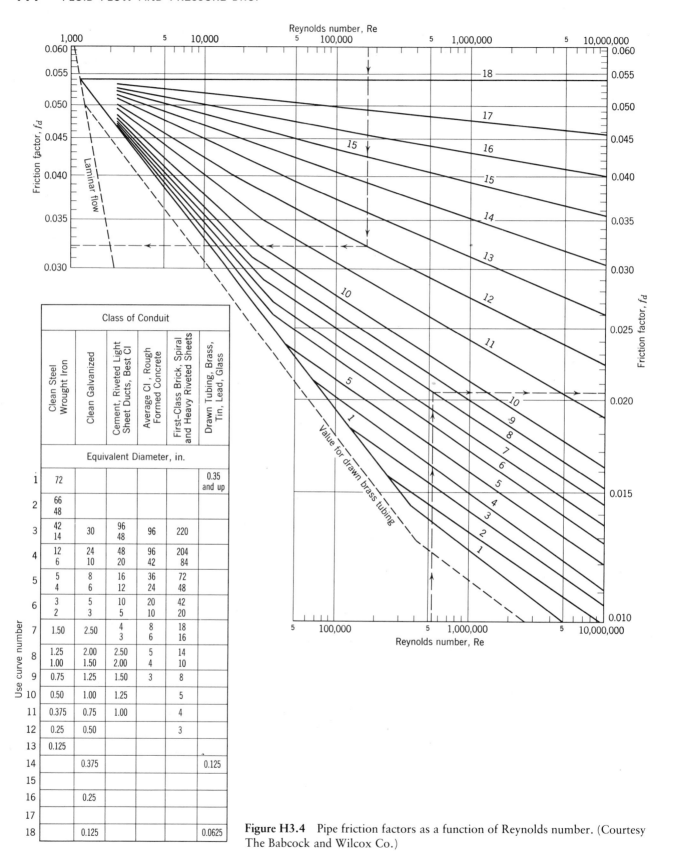

Figure H3.4 Pipe friction factors as a function of Reynolds number. (Courtesy The Babcock and Wilcox Co.)

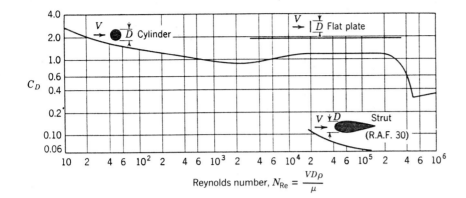

Figure H3.5 Drag coefficients for long flat plates, cylinders, and streamlined struts, (S. K. Vennard, *Elementary Fluid Mechanics,* 4th ed., John Wiley & Sons, New York, 1961.)

TABLE H3.1 Drag Coefficients for Typical Bodies[a]

Geometric Form	C_D
Flat plates normal to the flow:	
Rectangular, $a/b = 1$	1.10
$= 4$	1.19
$= 18$	1.40
Circular disc	1.11
Circular disc with central hole	1.22
Flat plates parallel to the flow (drag coefficient based on wetted area)	$0.45/(\log_{10} \text{Re})^{2.58}$
Hemispherical shells, axes parallel to flow:	
Open face to the rear	0.34
Open face forward	1.33
Conical shells, axis parallel to flow:	
Apex facing flow, 30° included angle	0.34
Apex facing flow, 60° included angle	0.51
I beams	
Web normal to flow	2.04
Web parallel to flow	0.86
Spheres, $\text{Re} < 2 \times 10^5$	0.5
$\text{Re} > 5 \times 10^5$	0.2
Ellipsoids, Long axis parallel to flow	0.6
$\text{Re} < 10^5$	0.4
$\text{Re} > 10^5$	0.09
Streamlined body, $l/d = 3$	0.10
$l/d = 5$	0.06
$l/d = 10$	0.083

[a]Data from Ref. 16 of Chapter 3, and K. Delany and N. E. Sorenson, NACA TN 3038, 1953.

TABLE H3.1 (*Continued*)

Bars with their axes normal to the flow:

Flow direction	Fineness ratio, (c_0/b_0)	Corner radius ratio, (r/b_0)	C_D at Re $= 10^5$	Flow direction	Fineness ratio, (c_0/b_0)	Corner radius ratio, (r/b_0)	C_D at Re $= 10^5$
(circle)	1:1	0.50	1.00	(diamond with r)	1:2	0.021	1.8
					1:2	0.083	1.7
(vertical ellipse)	1:2	. . .	1.6		1:2	0.167	1.7
(horizontal ellipse)	2:1	. . .	0.6	(square rotated, diamond)	1:1	0.015	1.5
(rounded rectangle tall)	1:2	0.021	2.2		1:1	0.118	1.5^a
	1:2	0.083	1.9		1:1	0.235	1.5
	1:2	0.250	1.6	(flat diamond)	2:1	0.042	1.1
(rounded square)	1:1	0.021	2.0		2:1	0.167	1.1^a
	1:1	0.167	1.2^a		2:1	0.333	1.1
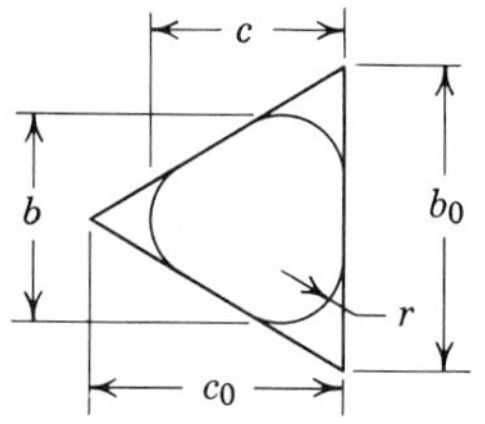	1:1	0.333	1.0	(triangle apex right)	1:1	0.021	1.2
(rounded rectangle wide)	2:1	0.042	1.4		1:1	0.083	1.3^a
	2:1	0.167	0.7^a		1:1	0.250	1.1
	2:1	0.500	0.4	(triangle apex left)	1:1	0.021	2.0
					1:1	0.083	1.9^a
					1:1	0.250	1.3

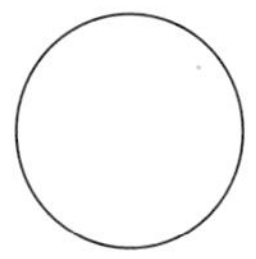

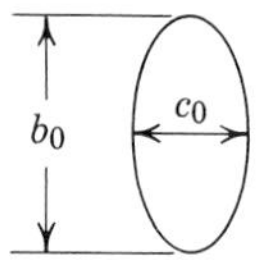

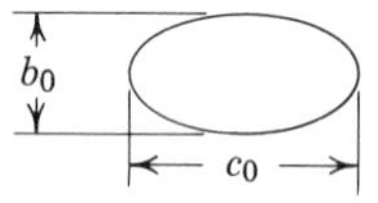

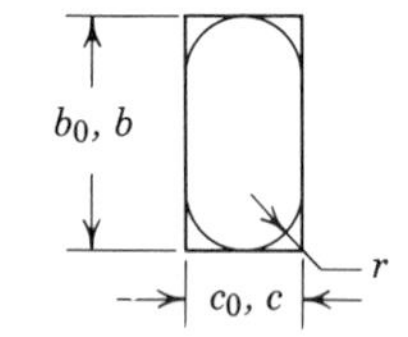

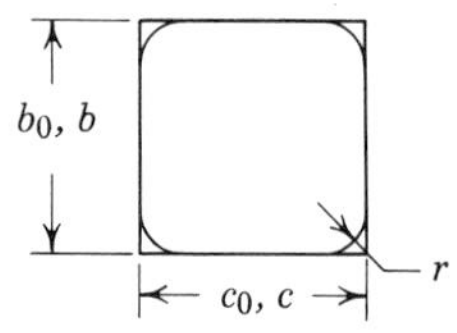

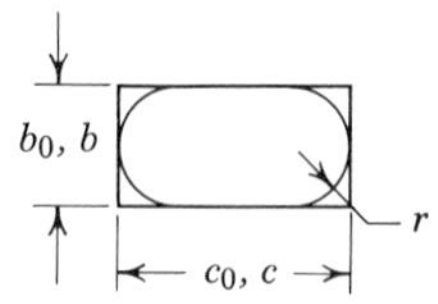

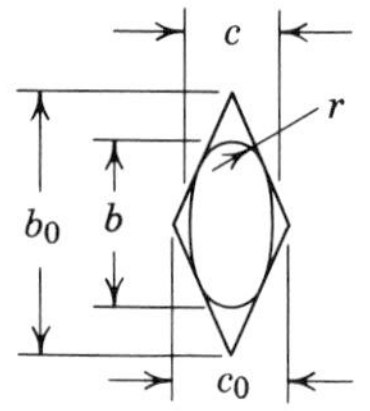

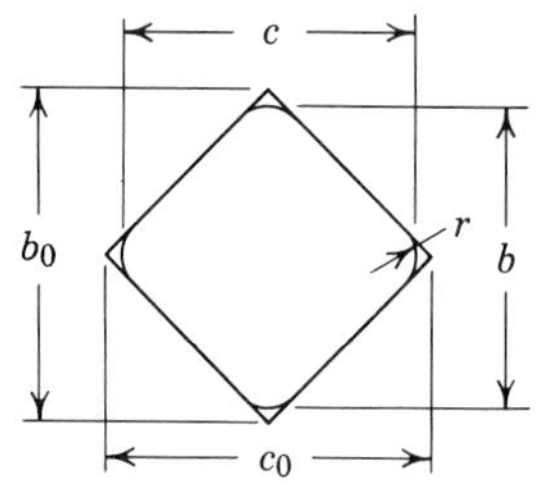

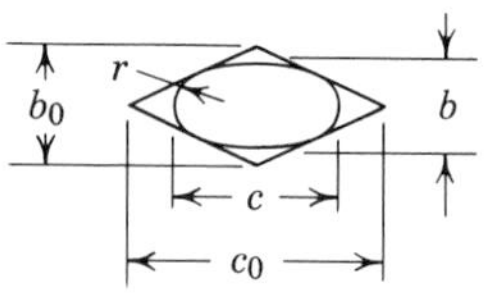

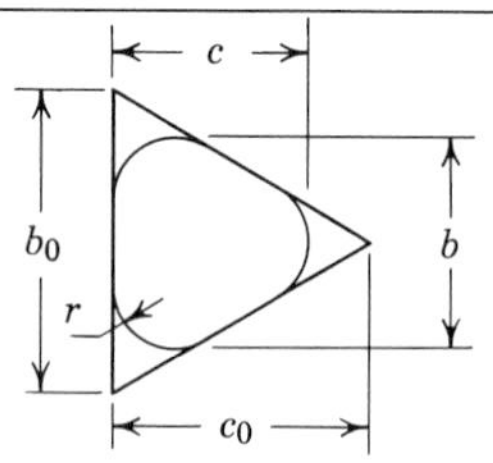

^{a}Re $= 2 \times 10^5$

TABLE H3.2   Equivalent Length in Pipe Diameters (*L/D*) of Various Valves and Fittings

		Description of Product		Equivalent Length in Pipe Diameters, *L/D*
Globe valves	Conventional	With no obstruction in flat-, bevel-, or plug-type seat	Fully open	340
		With wing or pin-guided disc	Fully open	450
	Y-pattern	With no obstruction in flat-, bevel-, or plug-type seat		
		Stem 60° from run of pipeline	Fully open	175
		Stem 45° from run of pipeline	Fully open	145
Angle valves	Conventional	With no obstruction in flat-, bevel-, or plug-type seat	Fully open	145
		With wing or pin-guided disc	Fully open	200
Gate valves	Conventional wedge disc, double disc, or plug disc		Fully open	13
			Three-quarters open	35
			One-half open	160
			One-quarter open	900
	Pulp stock		Fully open	17
			Three-quarters open	50
			One-half open	260
			One-quarter open	1200
	Conduit pipeline		Fully open	3
Check valves	Conventional swing		0.5[a]...Fully open	135
	Clearway swing		0.5[a]...Fully open	50
	Globe, lift or stop		2.0[a]...Fully open	Same as globe
	Angle, lift or stop		2.0[a]...Fully open	Same as angle
	In-line ball	2.5 vertical and 0.25 horizontal[a]....Fully open		150
Foot valves with strainer		With poppet lift-type disc	0.3[a]...Fully open	420
		With leather-hinged disc	0.4[a]...Fully open	75
Butterfly valves (6 in. and larger)			Fully open	20
Cocks	Straight-through	Rectangular plug port area equal to 100% of pipe area	Fully open	18
			Flow straight through	44
	Three-way	Rectangular plug port area equal to 80% of pipe area (fully open)	Flow through branch	140
Fittings	90° standard elbow			30
	45° standard elbow			16
	90° long radius elbow			20
	90° street elbow			50
	45° street elbow			26
	Square corner elbow			57
	Standard tee	With flow through run		20
		With flow through branch		60
	Close pattern return bend			50

[a]Minimum calculated pressure drop (psi) across valve to provide sufficient flow to lift disc fully. (Courtesy of The Crane Co.)

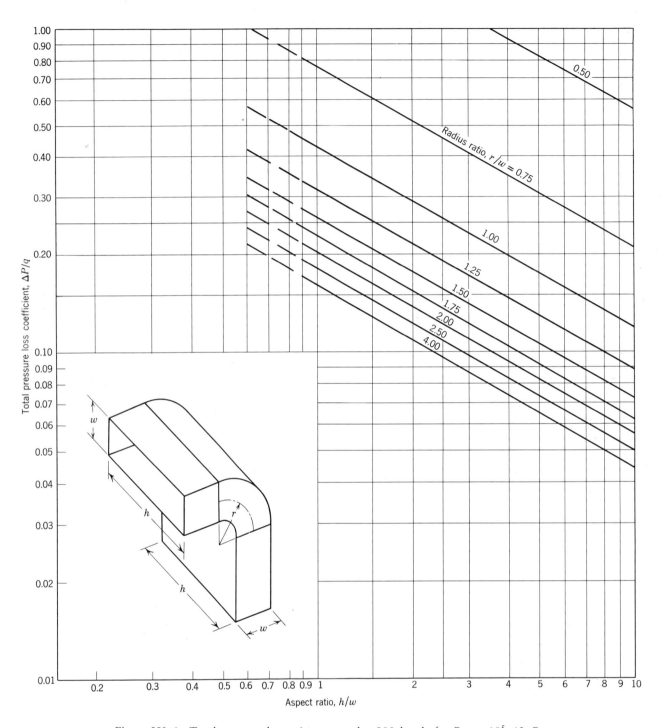

Figure H3.6 Total pressure losses in rectangular 90° bends for Re = 10^5. (J. R. Henry, "Design of Power Plant Installations—Pressure Loss Characteristics of Duct Components," NACA ARR no. L4F26, 1944.)

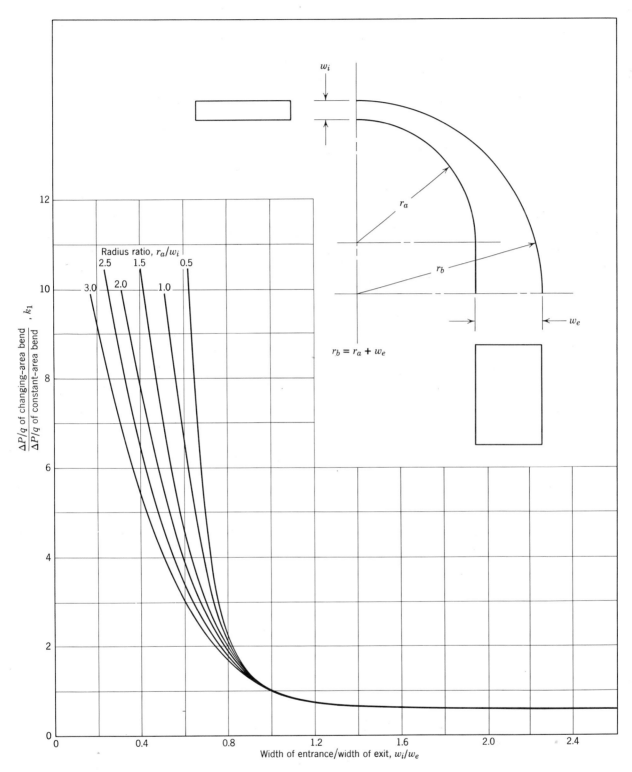

Figure H3.6 (*Continued*)

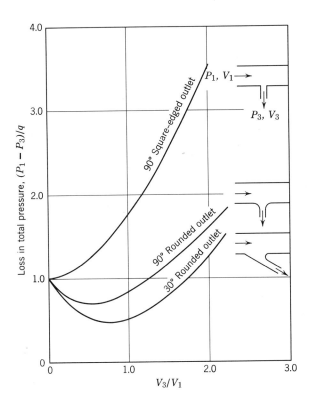

Figure H3.7 Head losses in manifold branches in terms of q, the dynamic head in the inlet duct.

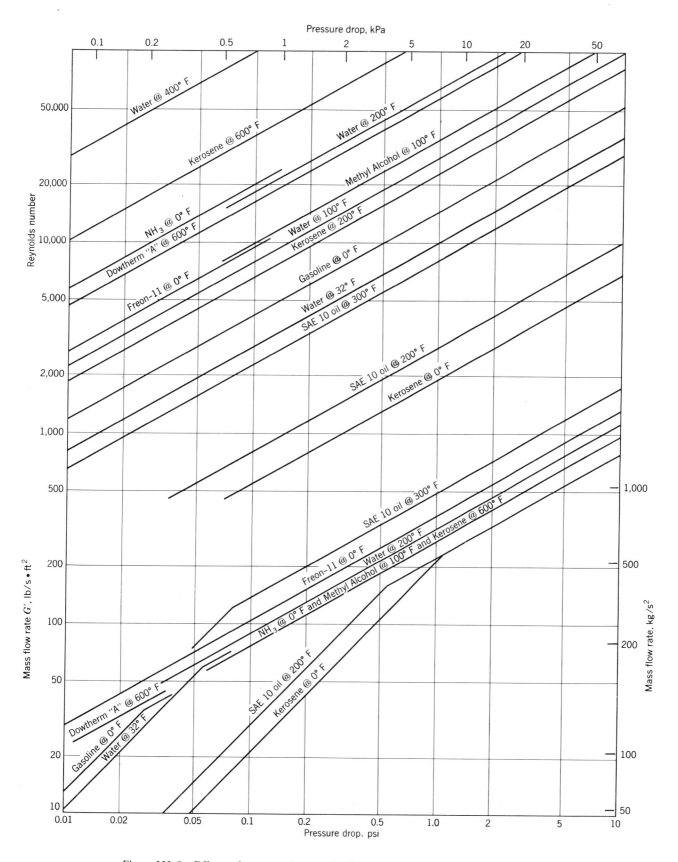

Figure H3.8 Effects of pressure drop on the flow rate and Reynolds number for typical liquids flowing through a 0.5-in. ID tube having an L/D of 200.

LMTD and Thermal Effectiveness

LMTD AND THERMAL EFFECTIVENESS

The LMTD is often tedious to compute because it involves the ratio of small differences in large quantities; hence a chart is included here to facilitate its evaluation. Furthermore, the effects of configurations other than pure counterflow or parallel flow entail correction factors that are very difficult to evaluate. Charts for these correction factors to be applied to the LMTD are presented for the principal cases of interest in shell-and-tube heat exchanger design. Since for many heat exchanger applications the parameter *thermal effectiveness* is more convenient to use than the LMTD, somewhat similar correction factors are presented for a variety of crossflow configurations, as is a chart designed to relate the LMTD to the thermal effectiveness.

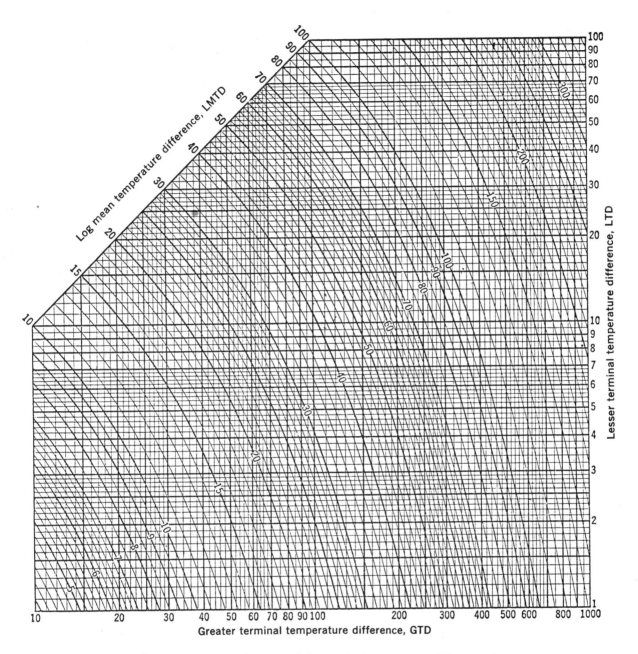

Figure H4.1 LMTD as a function of the terminal temperature differences for pure parallel or counterflow heat exchangers. If a point falls off the chart, multiply the scales by a factor such as 10 and the value then found on the chart by the same factor. (Courtesy Industrial Equipment Division, Baldwin-Lima-Hamilton Corp.)

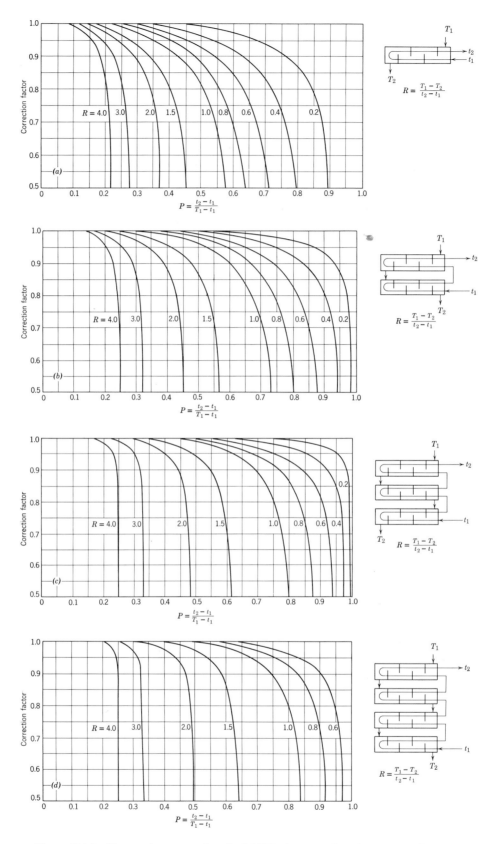

Figure H4.2 Factors for computing the LMTD for typical multipass crossflow heat exchanger geometries. The correction factor is the ratio of the LMTD for the crossflow geometry shown to the LMTD for pure counterflow. (Bowman, Mueller, and Nagle, Ref. 2 of Chapt. 4.)

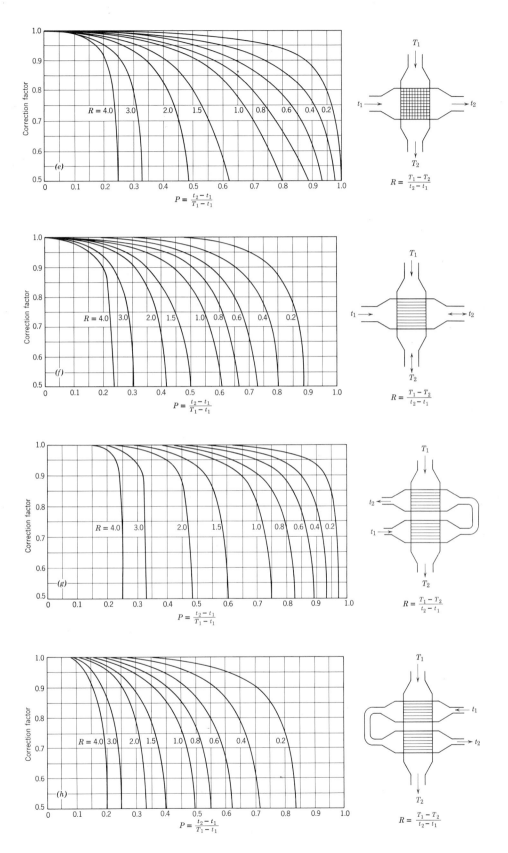

Figure H4.2 (*Continued*)

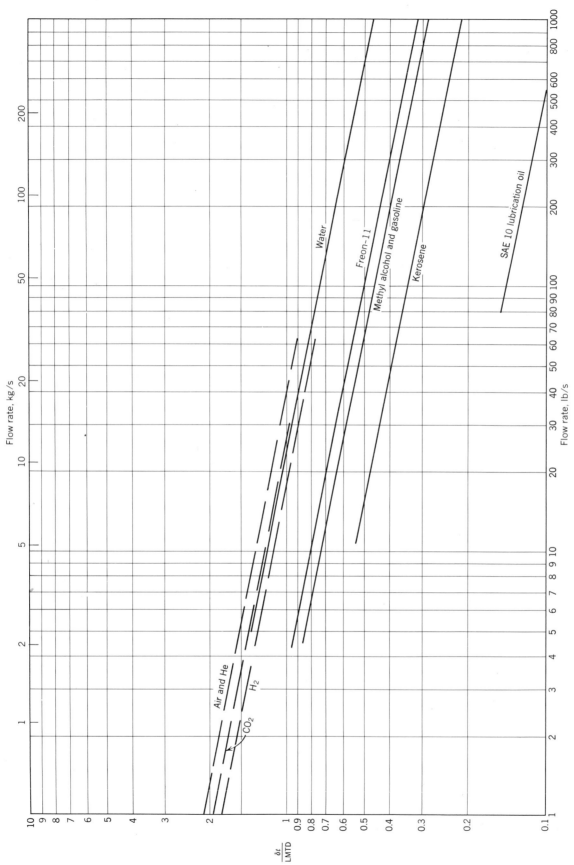

Figure H4.3 Ratio of the fluid temperature change to the LMTD as a function of the fluid-flow rate for counterflow heat exchangers operating at 200°F with 1.0-in. ID passages having a length-diameter ratio of 100. *Note:* Use the diameter correction factor from Fig. H4.4. At temperatures other than 200°F, use ϕ from Figs. H5.4 or H5.6 as follows:

$$\frac{\delta t}{\text{LMTD}} = \left(\frac{\delta t}{\text{LMTD}}\right)_{200°F} \phi \frac{(c_p)_{200°F}}{c_p}$$

(The curves for liquids were terminated at the left at Re = 2000.)

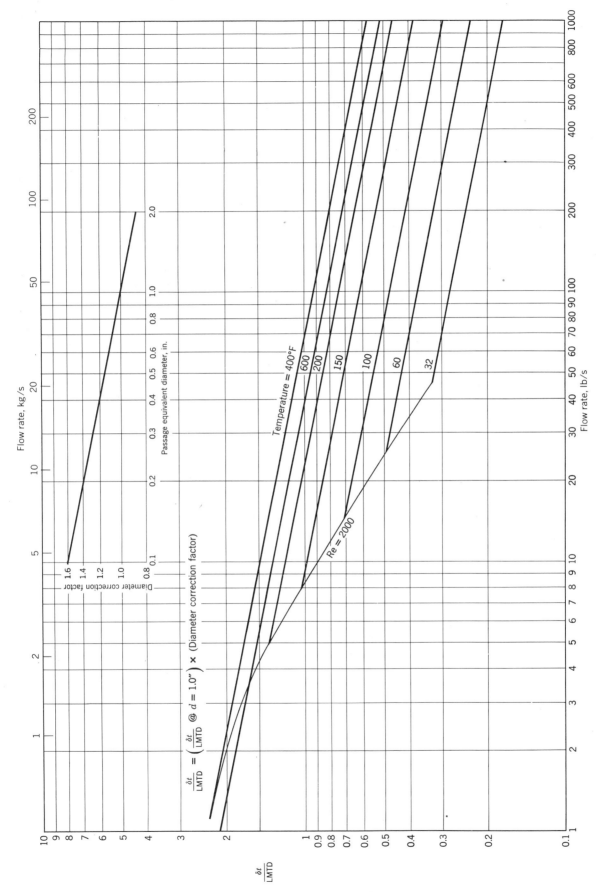

Figure H4.4 Effects of temperature and passage diameter on the ratio of the fluid temperature change to the LMTD for water flowing through 1.0-in. ID passages having a length–diameter ratio of 100.

457

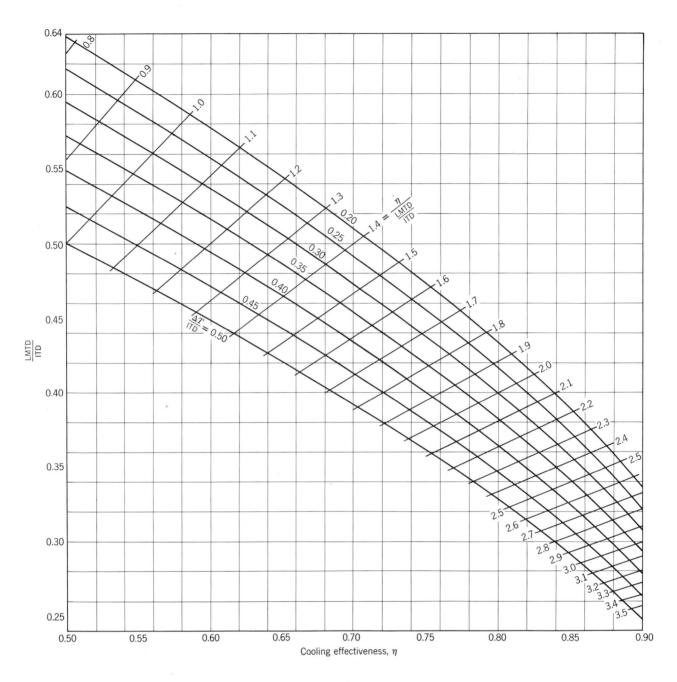

For counterflow heat exchangers

LMTD = Log mean temperature difference

ITD = Inlet-temperature difference

ΔT = Temperature change in primary fluid

Δt = Temperature change in secondary fluid

$$\eta = \frac{\Delta t}{ITD}$$

$$\frac{LMTD}{ITD} = \left(\eta - \frac{\Delta T}{ITD}\right) \Big/ \left(\ln \frac{1 - \Delta T/ITD}{1 - \eta}\right)$$

Figure H4.5 Relation between LMTD and thermal effectiveness. (Courtesy Oak Ridge National Laboratory.)

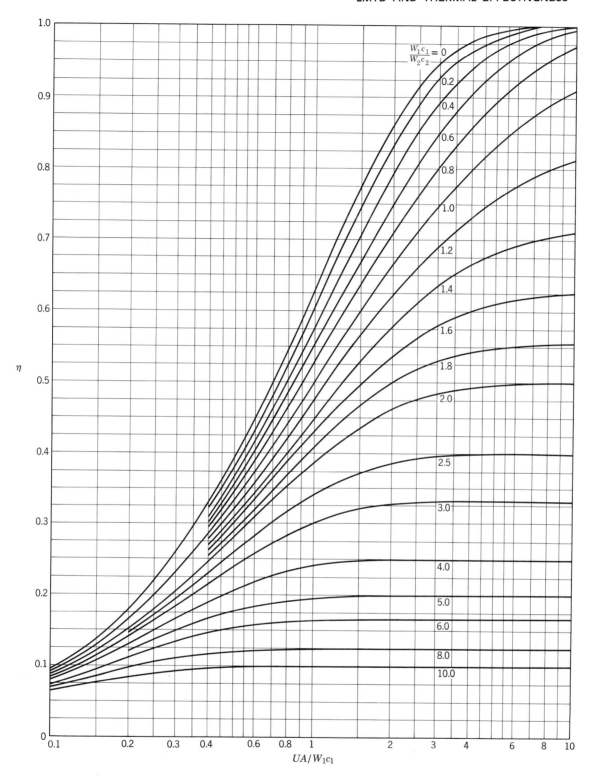

Figure H4.6 Thermal effectiveness in counterflow heat exchangers. For applications, see p. 77 (THEMA, Ref. 1.)

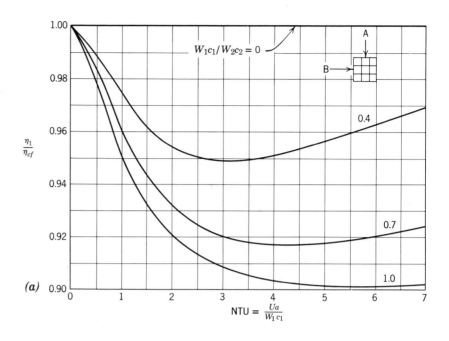

(a)

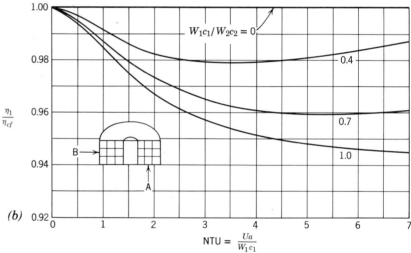

(b)

Figure H4.7 Thermal effectiveness correction factors for crossflow heat exchangers. The symbol η_{CF} is the thermal effectiveness for pure counterflow. See Chapter 4 for background data. Note that similar correction factors for three or more passes are closer to unity. (*a*) Correction factor for single-pass crossflow; both fluids unmixed. (*b*) Correction factor for two-pass counter crossflow; both fluids mixed between passes; unmixed in each pass. (*c*) Correction factor for two-pass counter crossflow; fluid A mixed between passes, unmixed in each pass; fluid B unmixed throughout; inverted order. (*d*) Correction factor for two-pass counter crossflow; both fluids unmixed throughout; both inverted order. (*e*) Correction factor for two-pass counter crossflow; both fluids unmixed throughout; both identical order. (*f*) Correction factor for two-pass counter crossflow; both fluids unmixed throughout; fluid A inverted order; fluid B identical order. (Stevens et al., Ref. 4 of Chapt. 4.)

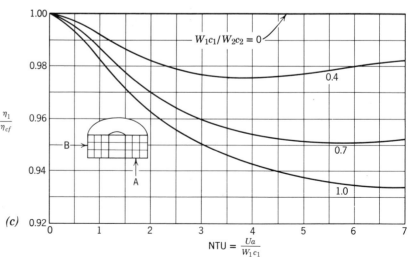

(c)

460

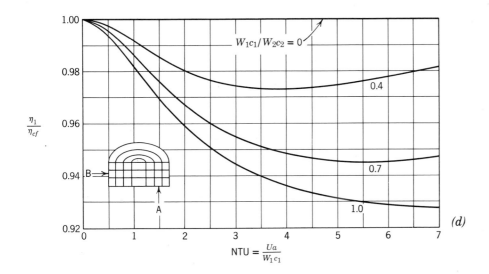

(d)

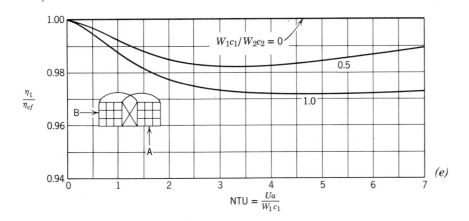

(e)

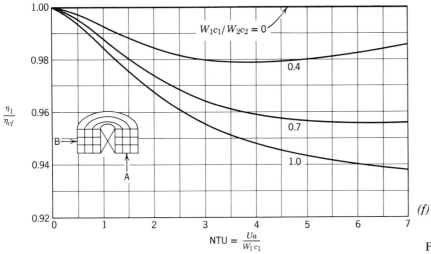

(f)

Figure H4.7   *(Continued)*

SECTION FIVE

Heat Transfer Coefficients

HEAT TRANSFER

Before determining the heat transfer coefficient for a fluid other than a liquid metal, the Reynolds number should be determined to establish the flow regime, that is, whether it is laminar or turbulent. The appropriate chart then can be used to obtain the heat transfer coefficient for long, straight, smooth passages.

For laminar flow, Fig. H5.1 gives the heat transfer coefficient for flow through circular passages with a constant heat flux and a uniform viscosity. Table H5.1 gives limiting Nusselt numbers for a variety of passage geometries for both constant heat flux and constant wall temperature conditions. Figure H5.2 gives a correction factor to be applied to values from Fig. H5.1 to allow for the change in velocity distribution resulting from differences between the viscosity of the bulk free stream and that at the wall.

For turbulent flow, the uncorrected heat transfer coefficient for a liquid or a gas can be obtained from Figs. H5.3 or H5.5. The correction factor for the temperature and the fluid used can be obtained from Figs.

H5.4 or H5.6. An additional correction factor for passage length-diameter ratio can be obtained from Fig. H5.9. Thus the corrected heat transfer coefficient for clean, smooth, surfaces becomes $h_{(corr.)} = (h$ from Fig. H5.3)(ϕ from Fig. H5.4 or H5.6)(L/D factor from Fig. H5.9). The correction factor ϕ is simply the ratio of the heat transfer coefficient for the fluid at the temperature in question to that for water or air at 200°F and the same mass flow rate G'. Explicitly, for the liquid heat transfer coefficients,

$$\phi_l = \frac{[0.023(c_p^{0.4}k^{0.6}/\mu^{0.4})G^{0.8}]_{\text{liquid}}}{[0.023(c_p^{0.4}k^{0.6}/\mu^{0.4})G^{0.8}]_{\text{water@200°F}}}$$

$$= \frac{(c_p^{0.4}k^{0.6}/\mu^{0.4})}{0.646}$$

Similarly, for the gas heat transfer coefficients,

$$\phi_g = \frac{(c_p^{0.4}k^{0.6}/\mu^{0.4})_{\text{gas}}}{(c_p^{0.4}k^{0.6}/\mu^{0.4})_{\text{air@200°F}}}$$

$$= \frac{(c_p^{0.4}k^{0.6}/\mu^{0.4})}{0.169}$$

Unlike the gases for which temperature correction factors are given in Fig. H5.6, the heat transfer coefficient for steam varies widely with the pressure because it is generally used with relatively little superheat. Because steam is so frequently used, Figs. H5.7 and H5.8 have been included to facilitate the estimation of heat transfer coefficients for steam.

For crossflow of any fluid over bare tube banks Fig. H5.11 or H5.12 can be employed, and Fig. H5.13 gives data for flow over banks of finned tubes. Heat transfer relations for thermal convection are listed in Tables H5.2 and H5.3, the latter giving a simplified form that represents a rougher approximation. The effects of surface fouling can be included by treating the surface film in the same way as a wall resistance. Typical fouling factors are presented in Table H5.4.

Heat transfer coefficients for liquid metals flowing through long, straight, smooth, circular passages can be obtained from Fig. H5.15 by first establishing the thermal conductivity of the fluid, the flow rate, and the passage diameter. For passage geometries other than circular, the use of equivalent passage diameter gives a fairly good approximation, especially since the principal barrier to heat flow is likely to be either the tube wall or the fluid film on the other side.

Heat fluxes for thermal radiation are given in Fig.

H5.16 for a surface emissivity of 1.0. The actual heat flux can be obtained by multiplying the value from Fig. H5.16 by the surface emissivity (see Table H2.5).

Chapter 3 presents the background for all of the charts and tables mentioned, except for Fig. H5.14, which is treated in the last part of Chapter 13.

TABLE H5.1 Ideal Nusselt Numbers for Laminar Flow Regions in Which the Viscosity is Constant, and the Transverse Temperature Distribution Does Not Vary in the Direction of Flow, For Example, Where $\dfrac{X}{L}\left(\dfrac{1}{\text{Re Pr}}\right) > 1.0$

Passage Shape	Constant Wall Temperature	Constant Heat Flux
Circle	3.66	4.36
Equilateral triangle	2.35	3.0
Square	2.9	3.6
Rectangle, $a/b = 1.5$	3.1	3.8
$= 2.0$	3.4	4.1
$=.4.0$	4.7	5.3
$= 10.0$	6.1	6.8
Parallel flat plates	7.6	8.2

Source: Jakob, Ref. 3 of Chapt. 3.

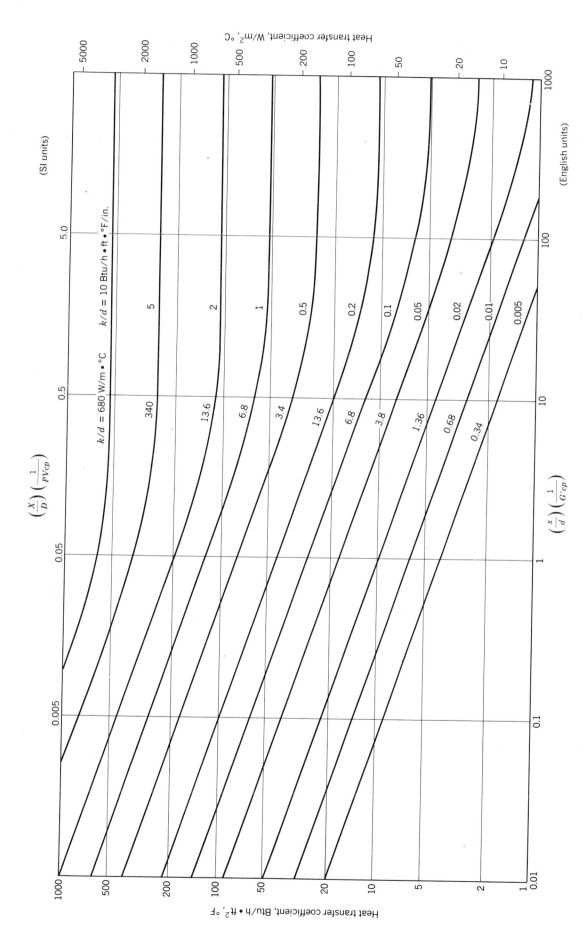

Figure H5.1 Local heat transfer coefficient for laminar flow through circular passages under constant heat flux conditions with the fluid viscosity independent of temperature. See Example 3.1 on p. 44. (Based on data in Ref. 20 of Chapt. 3.)

465

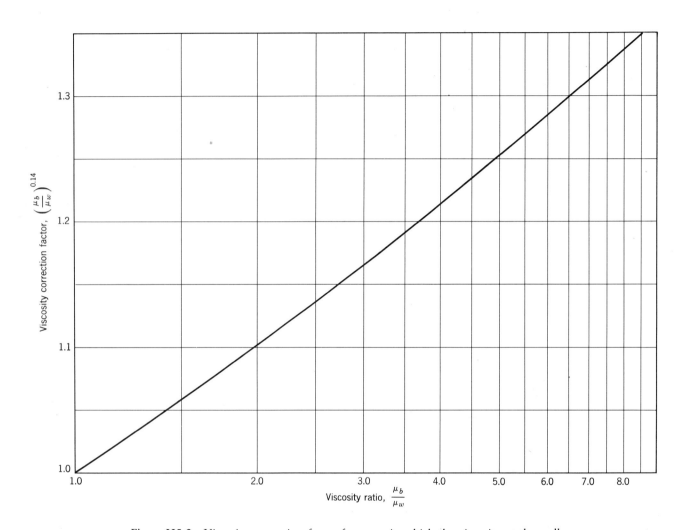

Figure H5.2 Viscosity correction factor for cases in which the viscosity at the wall differs substantially from that for the bulk free stream.

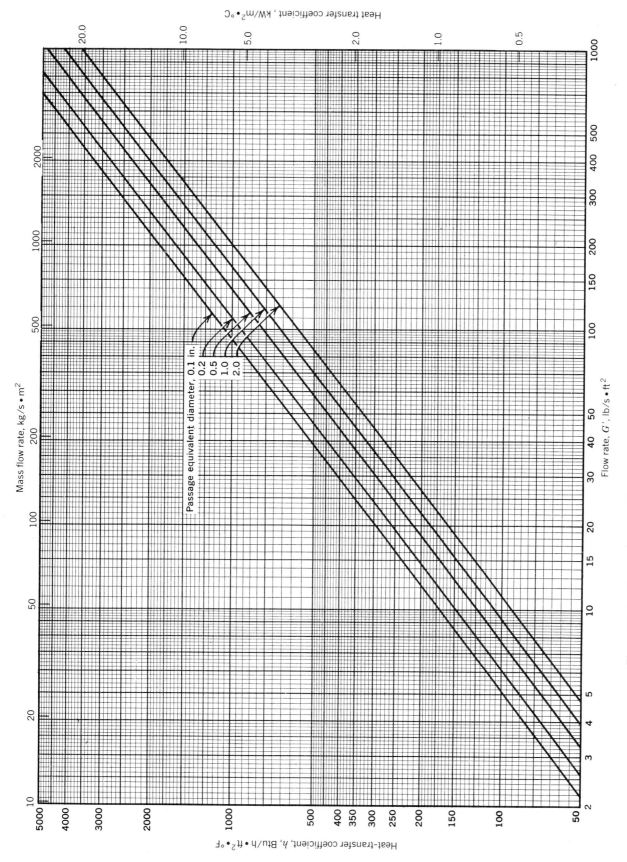

Figure H5.3 Heat transfer coefficients for water at 200°F under turbulent flow conditions.

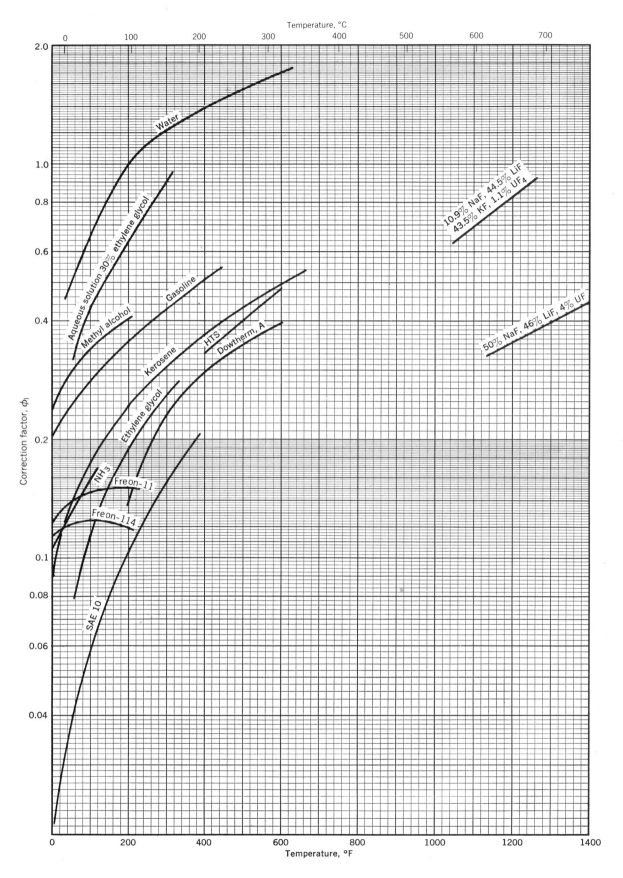

Figure H5.4 Correction factors for Fig. H5.3 to obtain turbulent heat transfer coefficients for typical liquids.

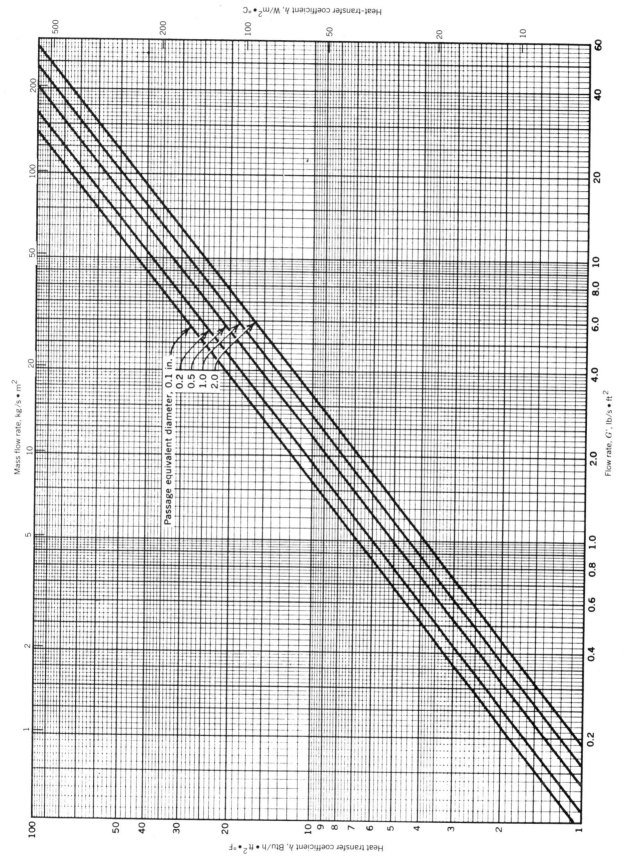

Figure H5.5 Heat transfer coefficient for air at 200°F under turbulent flow conditions.

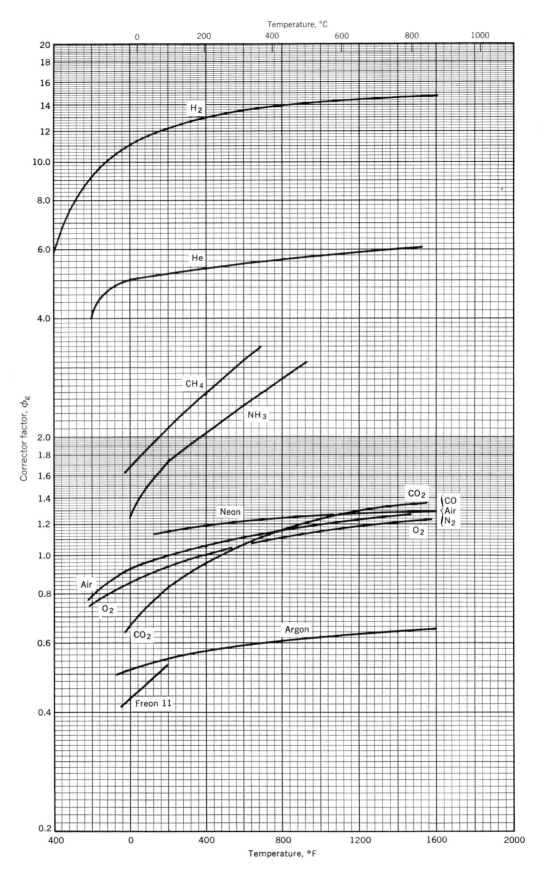

Figure H5.6 Correction factors for Fig. H5.5 to obtain turbulent heat transfer coefficients for typical gases.

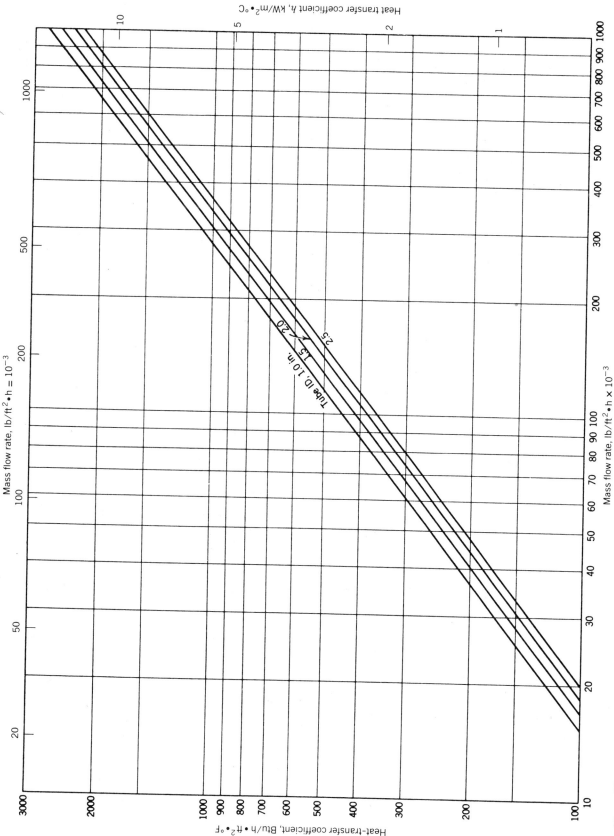

Figure H5.7   Heat transfer coefficient for superheated steam for conditions under which $c_p \mu^{0.2} = 1.0$. Since the heat transfer coefficient is directly proportional to $c_p \mu^{0.2}$, for other conditions the value for this parameter can be obtained from Fig. H5.8 and applied to the heat transfer coefficient given above.

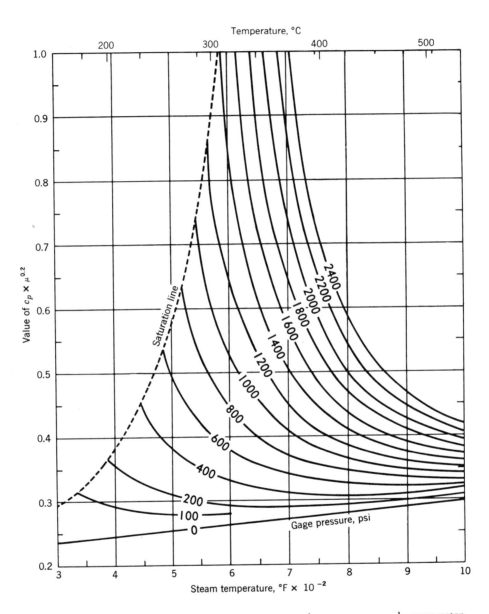

Figure H5.8 Effects of superheated steam pressure and temperature on the parameter $c_p \mu^{0.2}$. (Courtesy The Babcock and Wilcox Co.)

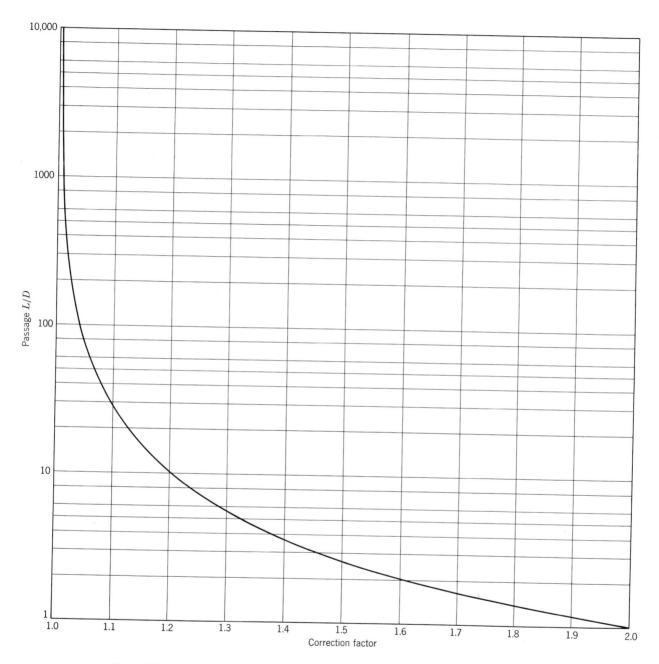

Figure H5.9 Correction factor to apply to Figs. H5.3, H5.5, and H5.7 to obtain the average heat transfer coefficient in the entrance region under turbulent flow conditions. (Based on data given in Ref. 30 of Chapt. 3.)

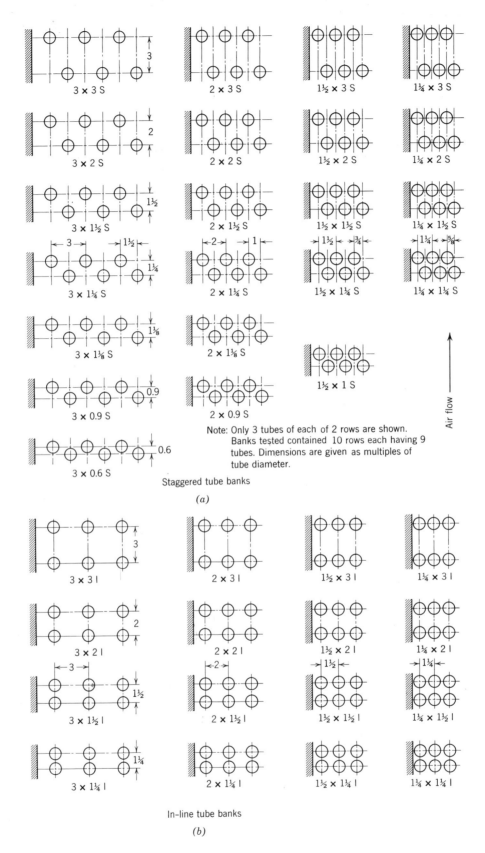

Staggered tube banks

(a)

In-line tube banks

(b)

Figure H5.10 Tube bank arrangements used to obtain the data for Figs. H5.11 and H5.12. (Pierson, Ref. 33 of Chapt. 3.)

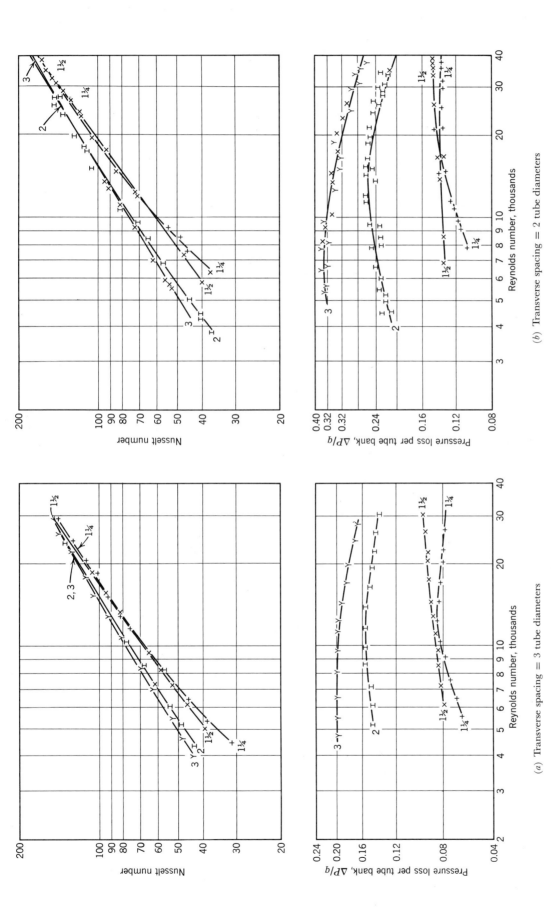

(a) Transverse spacing = 3 tube diameters

(b) Transverse spacing = 2 tube diameters

Figure H5.11 Heat transfer and pressure-loss coefficients for crossflow over in-line banks of bare tubes. Re and $\Delta P/q$ are based on the tube diameter and the mass flow rate through the minimum gap between the tubes; $\Delta P/q$ is the ratio of the pressure drop per bank to the dynamic head. The figures on the curves indicate the tube pitch parallel to the flow in tube diameters. (a) Transverse spacing = 3 tube diameters; (b) transverse spacing = 2 tube diameters; (c) transverse spacing = 1 1/2 tube diameters; (d) transverse spacing = 1 1/4 tube diameters. (Pierson, Ref. 3 of Chapt. 3.)

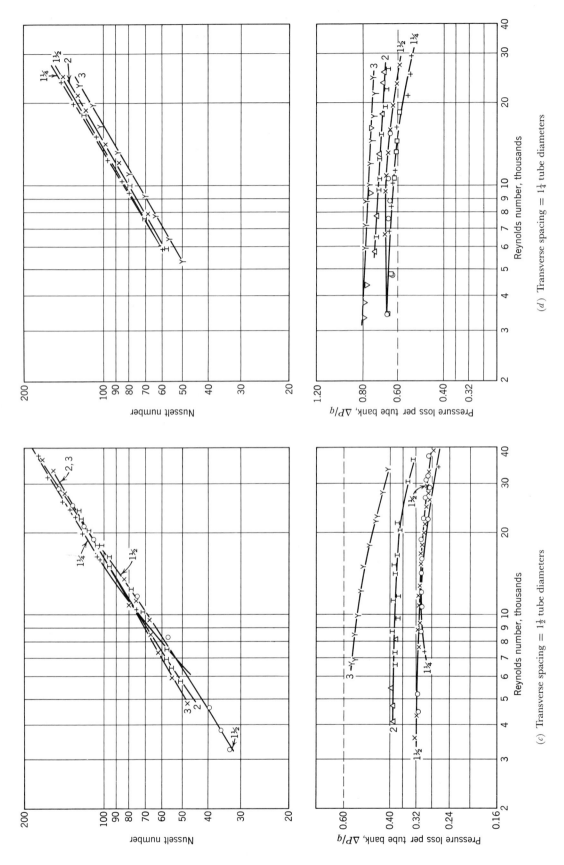

(d) Transverse spacing = $1\frac{1}{4}$ tube diameters

(c) Transverse spacing = $1\frac{1}{2}$ tube diameters

Figure H5.11 (*Continued*)

476

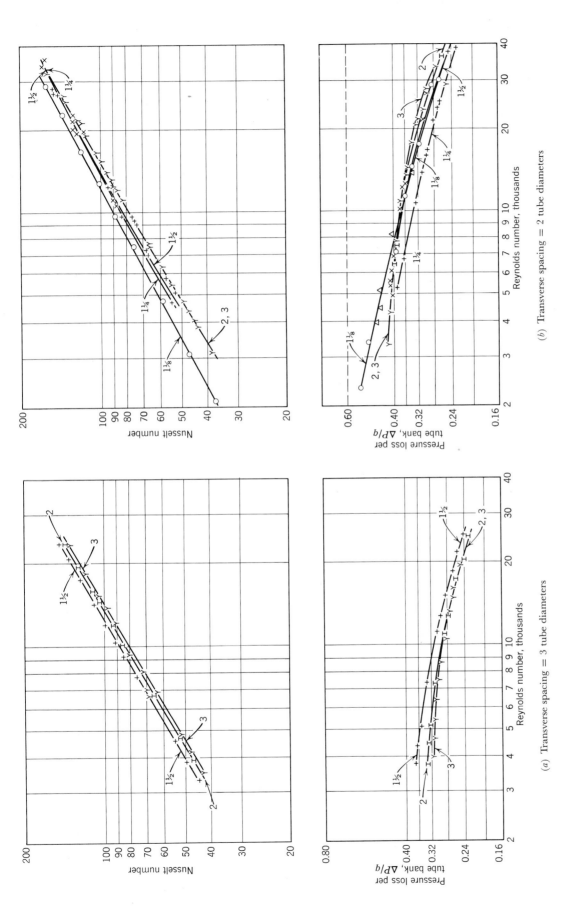

Figure H5.12 Heat transfer and pressure-loss coefficients for crossflow over staggered banks of bare tubes. Re and $\Delta P/q$ are based on the tube diameter and the mass flow rate through the minimum gap between the tubes; $\Delta P/q$ is the ratio of the pressure drop per bank to the dynamic head. The figures on the curves indicate the tube pitch parallel to the flow in tube diameters. (*a*) Transverse spacing = 3 tube diameters; (*b*) transverse spacing = 2 tube diameters; (*c*) transverse spacing = 1 1/2 tube diameters; (*d*) transverse spacing = 1 1/4 tube diameters. (Pierson, Ref. 33 of Chapt. 3.)

(a) Transverse spacing = 3 tube diameters

(b) Transverse spacing = 2 tube diameters

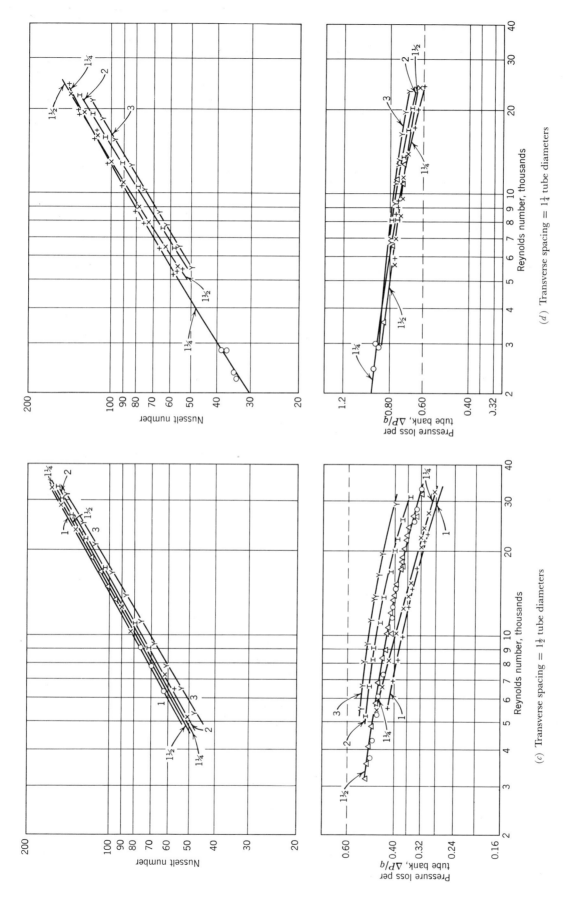

(c) Transverse spacing = $1\frac{1}{2}$ tube diameters

(d) Transverse spacing = $1\frac{1}{4}$ tube diameters

Figure H5.12 (*Continued*)

478

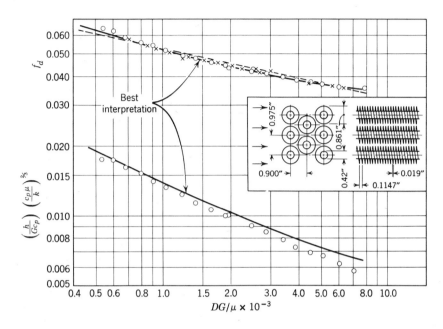

(a)
Surface CF-8.72(c)

Tube OD	0.42 in.
Fin pitch	8.72/in.
Fin thickness	0.019 in.*
Fin area/total area	0.876
Air-passage equivalent diameter	0.1452 ft
Free flow area/frontal area	0.494
Heat transfer area/total volume	136 ft^2/ft^3

*Fins slightly tapered, average value given.

Figure H5.13   Heat transfer coefficients and friction factors for crossflow over banks of finned tubes. The friction factor is based on a mean equivalent flow-passage diameter obtained by dividing the flow-passage volume by the total surface area to obtain a mean hydraulic radius and by multiplying the mean hydraulic radius by 4. The pressure drop is the product of the ratio of the flow-passage length to this equivalent diameter, the friction factor, and the dynamic head for the minimum free flow area. G is also based on the minimum free flow area. The fins were made of copper. (Kays and London, Ref. 11 of Chapt. 13.)

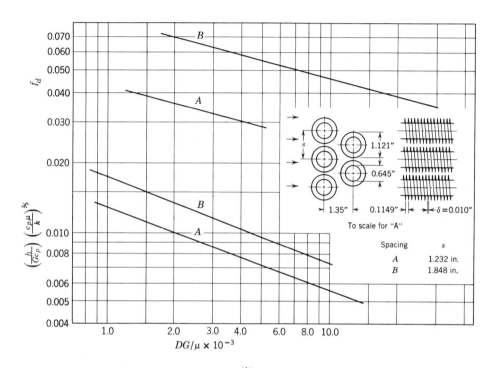

(b)
Surfaces CF-8.7-5/8J

	A	B
Tube OD		0.645 in.
Fin pitch		8.7/in.
Fin thickness		0.010 in.
Fin area/total area		0.862
Air-passage equivalent diameter	0.01797	0.0383 ft
Free flow area/frontal area	0.443	0.628
Heat transfer area/total volume	98.7	65.7 ft²/ft³

Figure H5.13 (*Continued*)

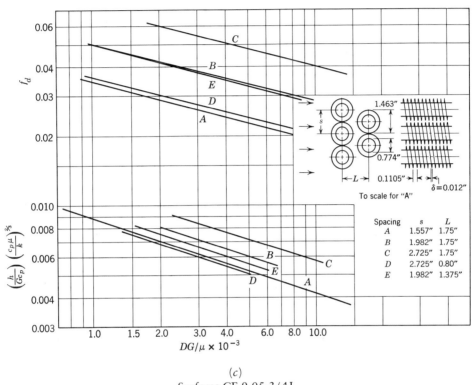

(c)
Surfaces CF-9.05-3/4J

		A	B	C	D	E
Tube OD	0.774 in.					
Fin pitch	9.05/in.					
Fin thickness	0.012 in.					
Fin area/total area	0.835					
		A	B	C	D	E
Air-passage equivalent diameter		0.0168	0.0269	0.0445	0.0159	0.0211 ft
Free flow area/frontal area		0.455	0.572	0.688	0.537	0.572
Heat-transfer area/total volume	108		85.1	61.9	135	108 ft²/ft³

Figure H5.13 *(Continued)*

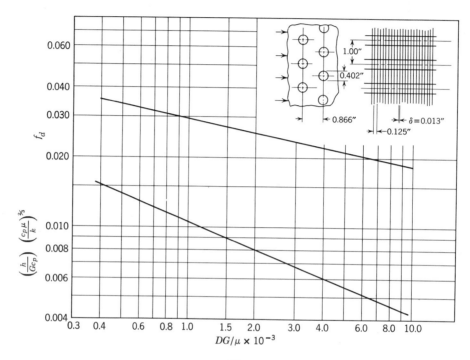

(d)
Surface 8.0-3/8T

Tube OD	0.402 in.
Fin pitch	8.0/in.
Fin thickness	0.013 in.
Fin area/total area	0.839
Air-passage equivalent diameter	0.01192 ft
Free flow area/frontal area	0.534
Heat transfer area/total volume	179 ft^2/ft^3

(Kays and London, Ref. 11 of Chapt. 13.)

Figure H5.13 (*Continued*)

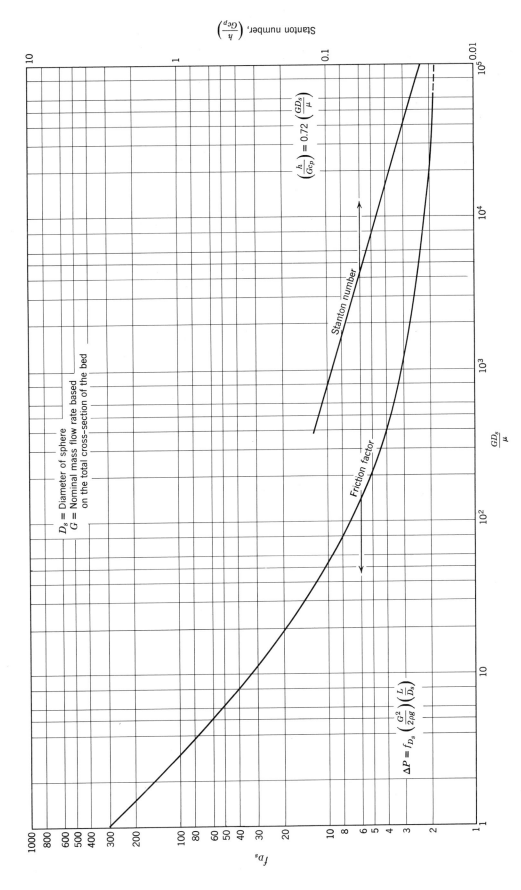

Figure H5.14 Heat transfer and friction factor for flow through randomly packed beds of spheres. (Denton, Ref. 22 of Chapter 13.)

483

TABLE H5.2   Equations for Heat Transfer Coefficients under Natural Convection[a]

Data can be correlated by equations of the form

$$Nu = c[Gr\ Pr]^n$$

$$\frac{hL}{k} = c\left[\left(\frac{L^3 \rho^2 \beta \Delta t g\ 3600^2}{\mu^2}\right)\left(\frac{c_p \mu}{k}\right)\right]^n$$

where Δt = temperature difference between the surface and the ambient fluid, °F

L = characteristic dimension, ft

and c_p, k, ρ, μ, and β are evaluated at a mean film temperature between the temperature of the surface and the temperature of ambient fluid. (β is the volumetric coefficient of expansion of the fluid.) The equation includes the quantity 3600^2 for English units; this quantity should be deleted for SI units.

	Gr Pr	c	n
Vertical plates or cylinders of large diameter, L = height, ft	Laminar: 10^4 to 10^9	0.59	1/4
	Turbulent: 10^9 to 10^{12}	0.13	1/3
Horizontal cylinders, L = outside diameter, ft	Laminar: 10^3 to 10^9	0.53	1/4
Horizontal square plates: hot plate facing up or cold plate facing down, L = side length, ft	Laminar: 10^5 to 2×10^7	0.54	1/4
	Turbulent: 2×10^7 to 3×10^{10}	0.14	1/3
Horizontal square plates: hot plate facing down or cold plate facing up, L = side length, ft	Laminar: 3×10^5 to 3×10^{10}	0.27	1/4

Natural convection to liquid metals [Na, NaK, Pb, Bi, Pb-Bi, and Hg] from horizontal cylinders in the laminar range can be correlated by

$$Nu = 0.53\left[\frac{Pr}{0.952 + Pr}\right]^{1/4} (Gr\ Pr)^{1/4}$$

[a]See Eq. 3.30 of Chapt. 3.

TABLE H5.3   Simplified Equations for Heat Transfer Coefficients for Natural Convection of Atmospheric Air

Note: The equations apply reasonably well to air at 100 to 1500°F, and to CO, O_2, N_2, and flue gases. For air at standard atmospheric pressure, using English units, $\rho^2 \beta g\ 3600^2/\mu^2 = 4.2 \times 10^6$ at 0°F = 1.8×10^6 at 100°F = 0.16×10^6 at 500°F.

	Gr Pr	Heat Transfer Coefficient, Btu/h · ft² · °F
Vertical plates or cylinders of large diameter, L = height, ft	10^4 to 10^9	$h = 0.29\ (\Delta t/L)^{1/4}$
	10^9 to 10^{12}	$h = 0.19\ \Delta t^{1/3}$
Horizontal cylinders, D = outside diameter, ft	10^3 to 10^9	$h = 0.27\ (\Delta t/D)^{1/4}$
	10^9 to 10^{12}	$h = 0.18\ \Delta t^{1/3}$
Horizontal plates: hot plate facing up or cold plate facing down, L = side length, ft	10^5 to 2×10^7	$h = 0.27\ (\Delta t/L)^{1/4}$
	2×10^7 to 3×10^{10}	$h = 0.22\ \Delta t^{1/3}$
Horizontal plates: hot plate facing down or cold plate facing up, L = side length, ft	3×10^5 to 3×10^{10}	$h = 0.12\ (\Delta t/L)^{1/4}$

TABLE H5.4 Normal Fouling Factors[a] for Heat Transfer Equipment[b]

Temperature of Heating Medium:	Up to 240°F		240 to 400°F[c]	
Temperature of Water:	125°F or Less		Over 125°F	
	Water Velocity, ft/s		Water Velocity, ft/s	
	3 ft/s	Over	3 ft/s	Over
Types of Water	and Less	3 ft/s	and Less	3 ft/s
Seawater	0.0005	0.0005	0.001	0.001
Distilled	0.0005	0.0005	0.0005	0.0005
Treated boiler feedwater	0.001	0.0005	0.001	0.001
Engine jacket	0.001	0.001	0.001	0.001
City or well water (such as Great Lakes)	0.001	0.001	0.002	0.002
Great Lakes	0.001	0.001	0.002	0.002
Cooling tower and artificial spray pond:				
Treated make-up	0.001	0.001	0.002	0.002
Untreated	0.003	0.003	0.005	0.004
Boiler blowdown	0.002	0.002	0.002	0.002
Brackish water	0.002	0.001	0.003	0.002
River water:				
Minimum	0.002	0.001	0.003	0.002
Mississippi	0.003	0.002	0.004	0.003
Delaware, Schuylkill	0.003	0.002	0.004	0.003
East River and New York Bay	0.003	0.002	0.004	0.003
Chicago sanitary canal	0.008	0.006	0.010	0.008
Muddy or silty	0.003	0.002	0.004	0.003
Hard (over 15 grains/gal)	0.003	0.003	0.005	0.005

Fouling factors—industrial oils:	
Clean recirculating oil	0.001
Machinery and transformer oils	0.001
Vegetable oils	0.003
Quenching oil	0.004
Fuel oil	0.005
Fouling factors—industrial gases and vapors:	
Organic vapors	0.0005
Steam (non-oil bearing)	0.0005
Alcohol vapors	0.0005
Steam, exhaust (oil-bearing from reciprocating engines)	0.001
Refrigerating vapors (condensing from reciprocating compressors)	0.002
Air	0.002
Coke oven gas and other manufactured gas	0.01
Diesel engine exhaust gas	0.01
Fouling factors—industrial liquids:	
Organic	0.001
Refrigerating liquids (heating, cooling or evaporating)	0.001
Brine (cooling)	0.001

[a]Fouling factor $= \dfrac{1}{\mu}, \dfrac{h \cdot ft^2 \cdot °F}{Btu}$.

[b]Reprinted by permission of Tubular Exchanger Manufacturers Association.

[c]Ratings in columns 3 and 4 are based on a temperature of the heating medium of 240 to 400°F. If the heating medium temperature is over 400°F and the cooling medium is known to scale, these ratings should be modified accordingly.

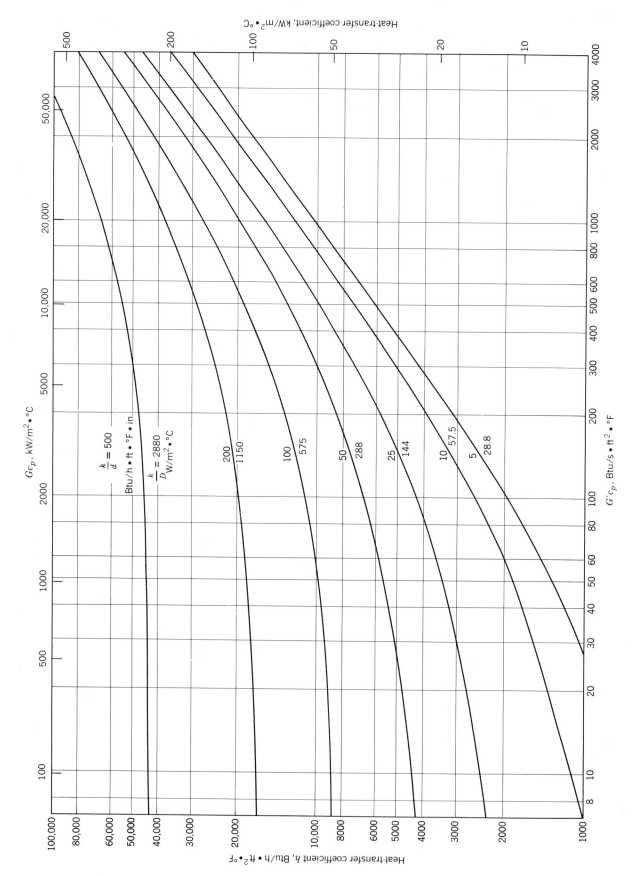

Figure H5.15 Heat transfer coefficient for liquid metals flowing in round tubes. (Based on Eq. 3.28.)

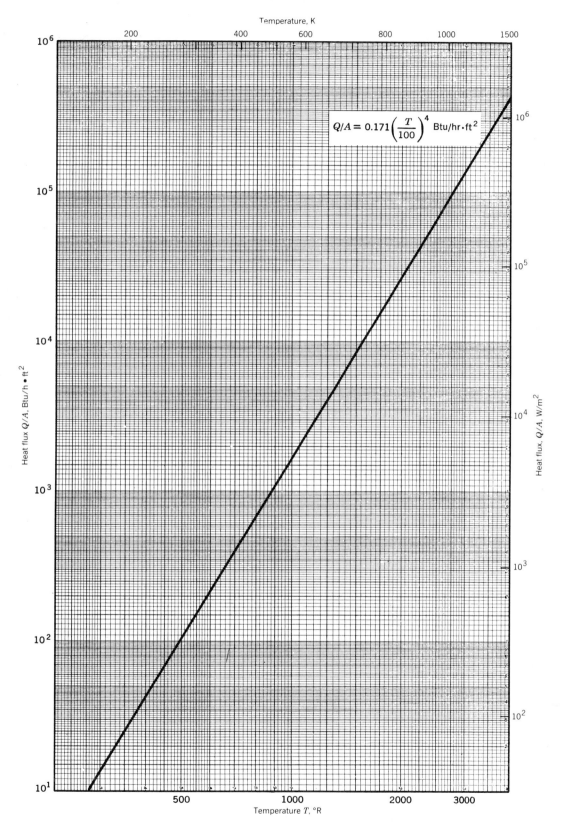

Figure H5.16 Radiant heat flux from a surface having an emissivity of 1.0. (Based on Eq. 3.5.)

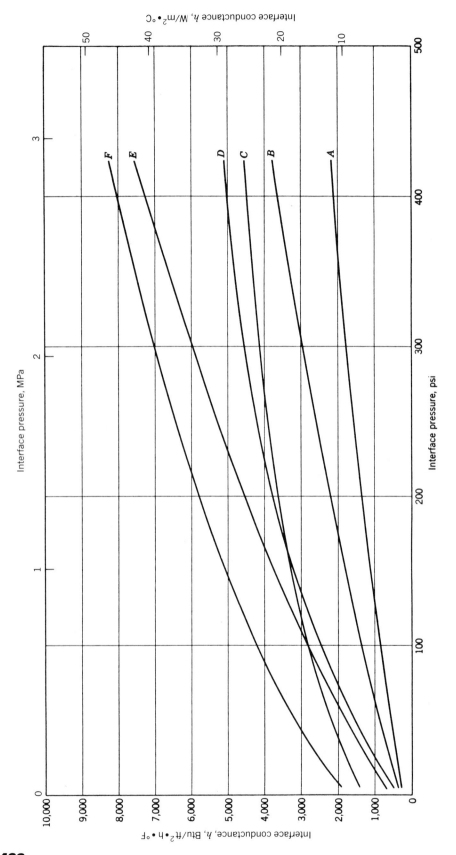

Figure H5.17 Effects of surface roughness, temperature, and interface pressure on the interface heat conductance of 7075T6 aluminum joints.

Curve	Surface Roughness, μin. rms	t_m, °F	Q/A, Btu/ft² · h	Test	Specimen
A	120 vs 120		9,050 to 14,200	14	1 and 2
B	10 vs 120	200	12,000 to 16,200	15	15 and 1
C	10 vs 10		12,500 to 14,700	8	15 and 16
D	120 vs 120		30,000 to 43,300	14	1 and 2
E	10 vs 120	400	44,100 to 72,800	15	15 and 1
F	10 vs 10		54,900 to 63,100	8	15 and 16

(Barzelay, Tong, and Holloway, Ref. 10 of Chapt. 3.)

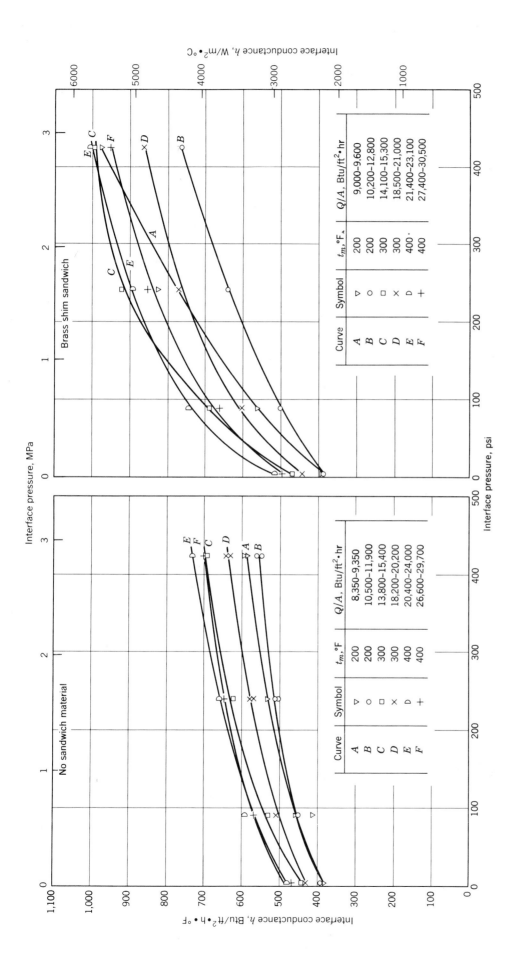

Figure H5.18 Effects of temperature, heat flux, and interface pressure on the interface heat conductance of stainless steel surfaces having a surface roughness of 100 μin., with and without a brass shim sandwiched between the surfaces. (Barzelay, Tong, and Holloway, Ref. 10 of Chapt. 3.)

Geometric Data for Tube Bundles and Header Sheets

TUBE AND HEADER CONFIGURATION DATA

The chart for equivalent flow-passage diameter given in Fig. H6.1 applies to axial flow outside of tubes in bundles, in which the wetted area of the shell is small as compared to that of the tube bundle. For example, the walls surrounding a bank of 100 tubes have an area equal to about 5% of the tube external surface area; hence some correction may be advisable for banks of less than 100 tubes.

Figure H6.3 is designed to serve as an aid in visualizing tube patterns for layout work. The pattern of the row of tubes around the outer perimeter of the largest bundle that fits within each of a series of circular envelopes is indicated by using alternately light and heavy lines for the tubes in successively larger rings. All dimensions and areas are given in terms of a unit tube centerline spacing.

Figure H6.4 goes beyond Fig. H6.3 in that it gives analytical expressions for the coordinates of hole centerlines and the radius of circles circumscribed around hole patterns for tubes on an equilateral triangular pitch. There are four different header sheet centerline positions in the tube array that give patterns that may be useful. In configuration 1 of Fig. H6.4, the center of the circumscribed circle is identical with the center of the central tube. In configuration 2, the center of the circumscribed circle lies midway between the two central tubes. In configuration 3, the center of the circumscribed circle lies midway between the centerlines of the three central tubes. Configuration 4 is similar to configuration 3 except that the centerline is displaced to take advantage of asymmetries in the hole pattern.

Configuration 1 is by far the most widely used, partially because it gives the most symmetrical patterns and partially because fabrication and inspection are more straightforward if one hole is concentric with the header sheet perimeter. Configuration 1 gives one of two types of hole pattern, depending on whether the diameter of the circumscribed circle is an odd or an even multiple of the tube spacing. These two types of hole pattern are presented in Fig. H6.3, together with tables for the geometric properties of the configurations shown.

Since it is sometimes desirable to employ a tube matrix in which the number of tubes differs from any obtainable with configuration 1, Table H6.1 has been included to present the minimum header sheet diameter in terms of the tube spacing for symmetrical tube matrices ranging from 1 to 564 tubes. The number designating the tube configuration of Fig. H6.4, that corresponds to each case, is tabulated together with the number of tubes enclosed by the circle having the diameter D/S.

A convenient chart for the surface area per cubic foot of tube bundle is presented in Fig. H6.5, while Table H6.2 has been included to aid in the design of baffles for shell-and-tube heat exchangers.

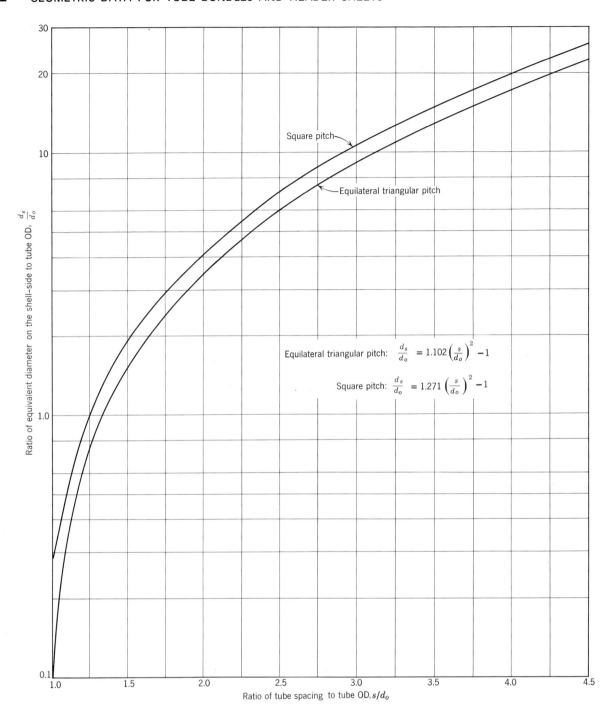

Equilateral triangular pitch: $\dfrac{d_s}{d_o} = 1.102 \left(\dfrac{s}{d_o}\right)^2 - 1$

Square pitch: $\dfrac{d_s}{d_o} = 1.271 \left(\dfrac{s}{d_o}\right)^2 - 1$

Figure H6.1   Equivalent passage diameter for axial flow between parallel tubes in tube bundles.

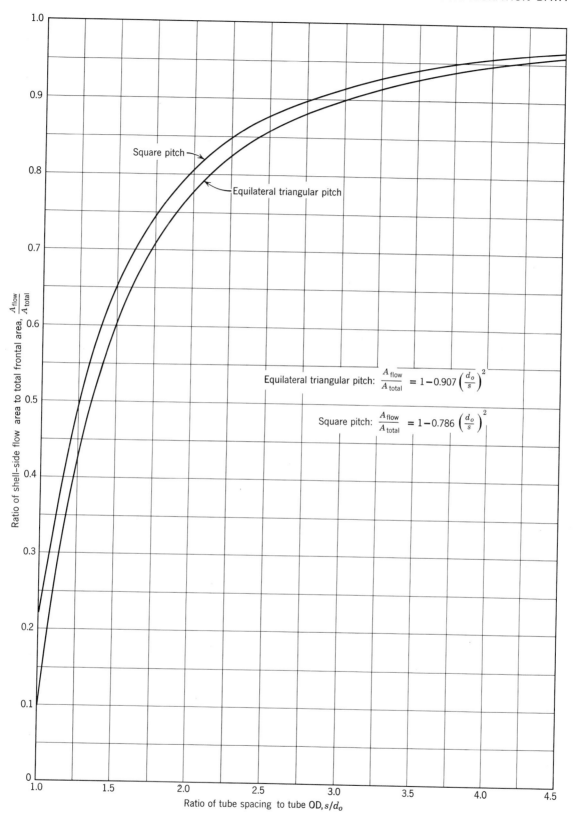

Equilateral triangular pitch: $\dfrac{A_{flow}}{A_{total}} = 1 - 0.907\left(\dfrac{d_o}{s}\right)^2$

Square pitch: $\dfrac{A_{flow}}{A_{total}} = 1 - 0.786\left(\dfrac{d_o}{s}\right)^2$

Figure H6.2   Ratio of shell-side, axial-flow passage area to total tube bundle cross-sectional area for bundles of parallel tubes.

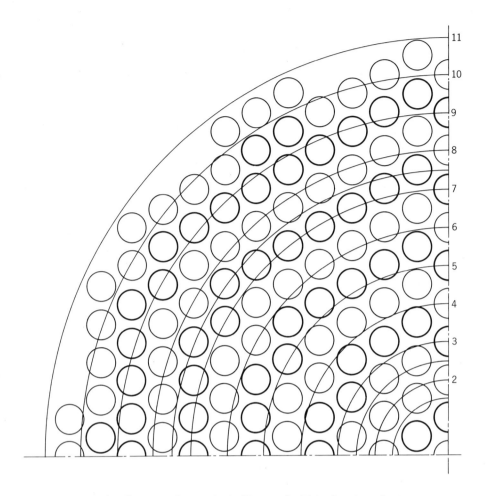

Configuration Properties in Terms of a Tube Spacing of 1.0

Radius	Area of Circle	Number of Tubes Enclosed	Idea Number of Tubes	Waste Area %
4	50.2	55	62.7	12.3
5	78.6	85	90.7	4.5
6	113.1	121	130.8	7.5
7	154.0	161	177.8	9.4
8	201.0	211	232.3	9.2
9	254.0	265	294.3	12.0
10	314.0	337	363.0	7.2
11	381.0	415	440.0	5.7

Note: Waste area $= \dfrac{\text{column 4} - \text{column 3}}{\text{column 4}}$.

Figure H6.3 Hole patterns for circular header sheets for equilateral triangular pitch.

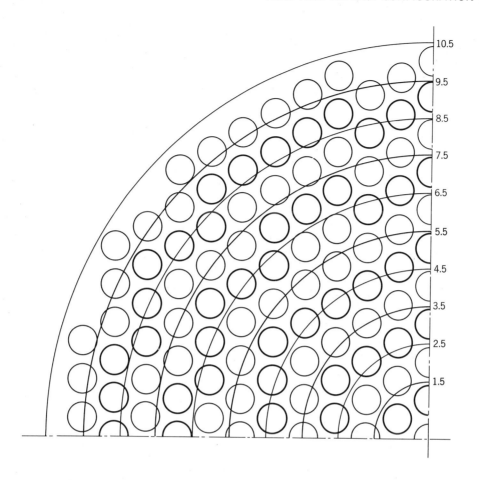

Configuration Properties in Terms of a Tube Spacing of 1.0

Radius	Area of Circle	Number of Tubes Enclosed	Ideal Number of Tubes	Waste Area %
1.5	7.07	7	8.18	14.4
2.5	19.6	19	22.7	16.3
3.5	38.5	37	44.5	16.9
4.5	63.7	61	73.7	17.2
5.5	95.2	91	110.0	17.3
6.5	132.6	139	153.2	9.3
7.5	177.0	187	204.3	8.5
8.5	227.0	241	262.0	8.0
9.5	284.0	301	323.6	7.0
10.5	346.0	367	400.0	8.3

Note: Waste area $= \dfrac{\text{column 4} - \text{column 3}}{\text{column 4}}$.

Figure H6.3   (*Continued*)

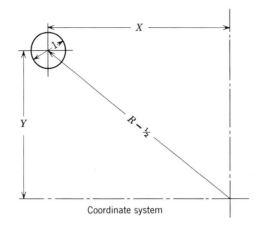

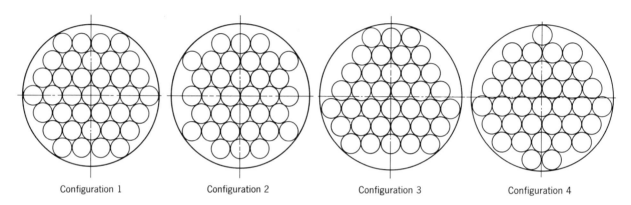

| Configuration 1 | Configuration 2 | Configuration 3 | Configuration 4 |

Figure H6.4 Four types of equilateral triangular hole pattern in circular header sheets based on four different locations of the center of the envelope circle. Coordinates of the tube holes in the outer perimeter of the bundle are also given

$$(R - 1/2)^2 = X^2 + Y^2$$

where X = function of P
Y = function of Q
P = tubes per row
Q = row number, where the longest row near the horizontal diameter is row 1.

Coordinates of tube holes in outer perimeter of hole pattern:

$$Y_U = Y \text{ in upper semicircle}$$
$$Y_L = Y \text{ in lower semicircle}$$

X	$\frac{1}{2}(P-1)$	$\frac{1}{2}(P-1)$	$\frac{1}{2}(P-1)$	$\frac{1}{2}(P-1)$
Y_U	$\frac{\sqrt{3}}{2}(Q-1)$	$\frac{\sqrt{3}}{2}(Q-1)$	$\frac{\sqrt{3}}{6}(3Q-2)$	$\frac{\sqrt{3}}{4}(2Q-1)$
Y_L	$\frac{\sqrt{3}}{2}(Q-1)$	$\frac{\sqrt{3}}{2}(Q-1)$	$\frac{\sqrt{3}}{6}(3Q-1)$	$\frac{\sqrt{3}}{4}(2Q-1)$

(Courtesy Oak Ridge National Laboratory)

TABLE H6.1 Diameter of the Envelope of the Tube Hole Pattern and Number of Tubes for the Basic Configurations of Fig. H6.4[a]

D = diameter of the envelope of the header-sheet hole pattern
S = tube center-to-center spacing
N = number of tubes

N	Configuration	Minimum D/S
1	1	1.000
2	2	2.000
3	3	2.154
4	2	2.732
7	1	3.000
8	2	3.646
10	2	4.000
12	3	4.056
13	1	4.464
14	2	4.606
19	1	5.000
22	2	5.582
23	4	5.770
24	2	6.000
27	3	6.034
31	1	6.292
37	1	7.000
38	2	7.244
42	3	7.430
44	4	7.764
48	2	8.000
55	1	8.212
56	2	8.810
57	4	8.858
60	2	8.938
61	1	9.000
63	3	8.082
64	2	9.186
69	3	9.326
70	2	9.660
73	1	9.718
74	2	9.888
76	2	10.000
85	1	10.166
88	2	10.644
92	2	10.848
96	4	11.038
102	3	11.264
104	2	11.536
109	1	11.584
110	2	12.000
114	3	12.016
121	1	12.136
126	2	12.532
129	3	12.718

TABLE H6.1 *(Continued)*

N	Configuration	Minimum D/S
130	2	12.790
131	4	12.906
133	4	12.948
135	4	13.032
136	2	13.124
139	1	13.166
141	3	13.220
151	1	13.490
154	2	14.000
156	3	14.012
158	2	14.076
163	1	14.114
168	3	14.316
170	2	14.528
174	3	14.614
176	4	14.812
178	2	14.892
187	1	15.000
188	4	15.344
190	4	15.414
199	1	15.422
202	2	15.798
206	2	15.934
208	2	16.000
211	1	16.100
212	2	16.132
213	3	16.144
217	4	16.256
219	3	16.275
220	2	16.524
225	3	16.534
230	2	16.716
235	1	16.874
241	1	17.000
246	3	17.290
253	1	17.370
254	2	17.644
258	2	17.704
262	2	17.822
264	2	18.000
270	3	18.010
274	4	18.198
276	3	18.244
283	1	18.436
284	2	18.578
288	2	18.692
295	1	18.776
301	1	19.000
306	3	19.148
313	1	19.330
316	2	19.520
321	3	19.584
324	2	19.736

TABLE H6.1 *(Continued)*

N	Configuration	Minimum D/S
325	4	19.862
327	3	19.904
329	4	19.994
330	2	20.000
333	3	20.008
337	1	20.078
339	3	20.218
349	1	20.288
351	4	20.640
352	2	20.672
361	1	20.698
362	4	20.944
364	2	20.974
367	1	21.000
372	2	21.074
376	2	21.224
378	3	21.232
379	1	21.298
380	4	21.366
382	4	21.390
384	3	21.428
390	3	21.526
392	2	21.664
394	4	21.802
396	2	21.808
397	1	21.880
400	2	21.952
406	2	22.00
409	1	22.072
421	1	22.166
426	3	22.572
433	1	22.634
434	2	22.794
437	4	22.858
442	2	22.932

TABLE H6.1 *(Continued)*

N	Configuration	Minimum D/S
447	3	23.030
450	2	23.114
453	3	23.120
455	4	23.288
459	3	23.300
461	4	23.422
463	4	23.466
465	3	23.480
468	2	23.606
472	2	23.650
475	1	23.716
476	2	23.870
480	2	23.914
482	2	24.000
483	4	24.060
499	1	24.066
504	3	24.438
506	2	24.516
511	1	25.580
514	2	25.644
518	2	24.812
520	4	24.848
522	3	24.860
524	4	24.974
526	4	25.016
528	3	25.028
530	2	25.062
534	3	25.110
540	3	25.194
547	1	25.332
550	2	25.556
559	1	25.576
562	2	25.880
564	4	25.934

[a]Excerpted from A. P. Fraas and M. E. LaVerne, "Heat Exchanger Design Charts," ORNL-1330, Oak Ridge National Laboratory, USAEC, 1952.

Figure H6.5 Tube external surface area per cubic foot of heat-transfer matrix as a function of tube diameter and spacing for bare tubes on an equilateral triangular pitch.

Equilateral triangular pitch:

$$\frac{\text{Tube pitch}}{\text{Tube OD}} = \frac{s}{d_0} \qquad \frac{\text{Tube external surface, ft}^2}{\text{Matrix volume, ft}^3} = \frac{43.6}{(s/d_0)^2} \frac{1}{d_0}$$

Note: For square pitch, multiply chart value by 0.864.

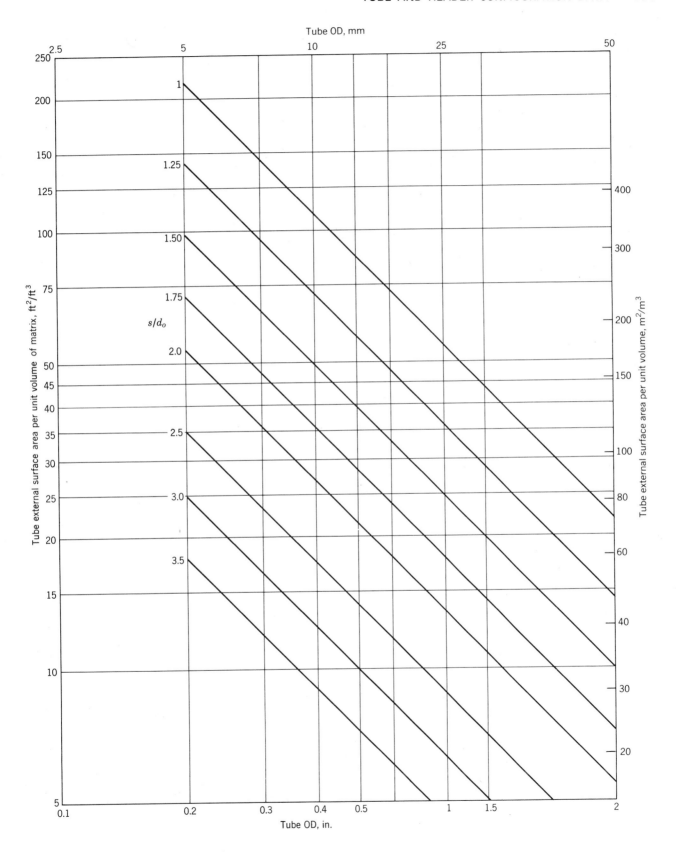

TABLE H6.2 Circular Segments and Chords[a]

h/D	A	h/D	A	h/D	A	h/D	A	h/D	A	h/D	A	h/D	A	h/D	A	h/D	A	h/D	A
0.001	0.00004	0.050	0.01468	0.100	0.04087	0.150	0.07387	0.200	0.11182	0.250	0.15355	0.300	0.19817	0.350	0.24498	0.400	0.29337	0.450	0.34278
0.002	0.00012	0.051	0.01512	0.101	0.04148	0.151	0.07459	0.201	0.11262	0.251	0.15441	0.301	0.19908	0.351	0.24593	0.401	0.29435	0.451	0.34378
0.003	0.00022	0.052	0.01556	0.102	0.04208	0.152	0.07531	0.202	0.11343	0.252	0.15528	0.302	0.20000	0.352	0.24689	0.402	0.29533	0.452	0.34477
0.004	0.00034	0.053	0.01601	0.103	0.04269	0.153	0.07603	0.203	0.11423	0.253	0.15615	0.303	0.20092	0.353	0.24784	0.403	0.29631	0.453	0.34577
0.005	0.00047	0.054	0.01646	0.104	0.04330	0.154	0.07675	0.204	0.11504	0.254	0.15702	0.304	0.20184	0.354	0.24880	0.404	0.29729	0.454	0.34676
0.006	0.00062	0.055	0.01691	0.105	0.04391	0.155	0.07747	0.205	0.11584	0.255	0.15789	0.305	0.20276	0.355	0.24976	0.405	0.29827	0.455	0.34776
0.007	0.00078	0.056	0.01737	0.106	0.04452	0.156	0.07819	0.206	0.11665	0.256	0.15876	0.306	0.20368	0.356	0.25071	0.406	0.29926	0.456	0.34876
0.008	0.00095	0.057	0.01783	0.107	0.04514	0.157	0.07892	0.207	0.11746	0.257	0.15964	0.307	0.20460	0.357	0.25167	0.407	0.30024	0.457	0.34975
0.009	0.00113	0.058	0.01830	0.108	0.04576	0.158	0.07965	0.208	0.11827	0.258	0.16051	0.308	0.20553	0.358	0.25263	0.408	0.30122	0.458	0.35075
0.010	0.00133	0.059	0.01877	0.109	0.04638	0.159	0.08038	0.209	0.11908	0.259	0.16139	0.309	0.20645	0.359	0.25359	0.409	0.30220	0.459	0.35175
0.011	0.00153	0.060	0.01924	0.110	0.04701	0.160	0.08111	0.210	0.11990	0.260	0.16226	0.310	0.20738	0.360	0.25455	0.410	0.30319	0.460	0.35274
0.012	0.00175	0.061	0.01972	0.111	0.04763	0.161	0.08185	0.211	0.12071	0.261	0.16314	0.311	0.20830	0.361	0.25551	0.411	0.30417	0.461	0.35374
0.013	0.00197	0.062	0.02020	0.112	0.04826	0.162	0.08258	0.212	0.12153	0.262	0.16402	0.312	0.20923	0.362	0.25647	0.412	0.30516	0.462	0.35474
0.014	0.00220	0.063	0.02068	0.113	0.04889	0.163	0.08332	0.213	0.12235	0.263	0.16490	0.313	0.21015	0.363	0.25743	0.413	0.30614	0.463	0.35573
0.015	0.00244	0.064	0.02117	0.114	0.04953	0.164	0.08406	0.214	0.12317	0.264	0.16578	0.314	0.21108	0.364	0.25839	0.414	0.30712	0.464	0.35673
0.016	0.00268	0.065	0.02166	0.115	0.05016	0.165	0.08480	0.215	0.12399	0.265	0.16666	0.315	0.21201	0.365	0.25936	0.415	0.30811	0.465	0.35773
0.017	0.00294	0.066	0.02215	0.116	0.05080	0.166	0.08554	0.216	0.12481	0.266	0.16755	0.316	0.21294	0.366	0.26032	0.416	0.30910	0.466	0.35873
0.018	0.00320	0.067	0.02265	0.117	0.05145	0.167	0.08629	0.217	0.12563	0.267	0.16843	0.317	0.21387	0.367	0.26128	0.417	0.31008	0.467	0.35972
0.019	0.00347	0.068	0.02315	0.118	0.05209	0.168	0.08704	0.218	0.12646	0.268	0.16932	0.318	0.21480	0.368	0.26225	0.418	0.31107	0.468	0.36072
0.020	0.00375	0.069	0.02366	0.119	0.05274	0.169	0.08779	0.219	0.12729	0.269	0.17020	0.319	0.21573	0.369	0.26321	0.419	0.31205	0.469	0.36172
0.021	0.00403	0.070	0.02417	0.120	0.05338	0.170	0.08854	0.220	0.12811	0.270	0.17109	0.320	0.21667	0.370	0.26418	0.420	0.31304	0.470	0.36272
0.022	0.00432	0.071	0.02468	0.121	0.05404	0.171	0.08929	0.221	0.12894	0.271	0.17198	0.321	0.21760	0.371	0.26514	0.421	0.31403	0.471	0.36372
0.023	0.00462	0.072	0.02520	0.122	0.05469	0.172	0.09004	0.222	0.12977	0.272	0.17287	0.322	0.21853	0.372	0.26611	0.422	0.31502	0.472	0.36471
0.024	0.00492	0.073	0.02571	0.123	0.05535	0.173	0.09080	0.223	0.13060	0.273	0.17376	0.323	0.21947	0.373	0.26708	0.423	0.31600	0.473	0.36571
		0.074	0.02624	0.124	0.05600	0.174	0.09155	0.224	0.13144	0.274	0.17465	0.324	0.22040	0.374	0.26805	0.424	0.31699	0.474	0.36671

h/D	Area	h/D	Area	h/D	Area	h/D	Area	h/D	Area	h/D	Area	h/D	Area	h/D	Area	h/D	Area	h/D	Area
0.025	0.00523	0.075	0.02676	0.125	0.05666	0.175	0.09231	0.225	0.13227	0.275	0.17554	0.325	0.22134	0.375	0.26901	0.425	0.31798	0.475	0.36771
0.026	0.00555	0.076	0.02729	0.126	0.05733	0.176	0.09307	0.226	0.13311	0.276	0.17644	0.326	0.22228	0.376	0.26998	0.426	0.31897	0.476	0.36874
0.027	0.00587	0.077	0.02782	0.127	0.05799	0.177	0.09384	0.227	0.13395	0.277	0.17733	0.327	0.22322	0.377	0.27095	0.427	0.31996	0.477	0.36971
0.028	0.00619	0.078	0.02836	0.128	0.05866	0.178	0.09460	0.228	0.13478	0.278	0.17823	0.328	0.22415	0.378	0.27192	0.428	0.32095	0.478	0.37071
0.029	0.00653	0.079	0.02889	0.129	0.05933	0.179	0.09537	0.229	0.13562	0.279	0.17912	0.329	0.22509	0.379	0.27289	0.429	0.32194	0.479	0.37171
0.030	0.00687	0.080	0.02943	0.130	0.06000	0.180	0.09613	0.230	0.13646	0.280	0.18002	0.330	0.22603	0.380	0.27386	0.430	0.32293	0.480	0.37270
0.031	0.00721	0.081	0.02998	0.131	0.06067	0.181	0.09690	0.231	0.13731	0.281	0.18092	0.331	0.22697	0.381	0.27483	0.431	0.32392	0.481	0.37370
0.032	0.00756	0.082	0.03053	0.132	0.06135	0.182	0.09767	0.232	0.13815	0.282	0.18182	0.332	0.22792	0.382	0.27580	0.432	0.32491	0.482	0.37470
0.033	0.00791	0.083	0.03108	0.133	0.06203	0.183	0.09845	0.233	0.13900	0.283	0.18272	0.333	0.22886	0.383	0.27678	0.433	0.32590	0.483	0.37570
0.034	0.00827	0.084	0.03163	0.134	0.06271	0.184	0.09922	0.234	0.13984	0.284	0.18362	0.334	0.22980	0.384	0.27775	0.434	0.32689	0.484	0.37670
0.035	0.00864	0.085	0.03219	0.135	0.06339	0.185	0.10000	0.235	0.14069	0.285	0.18452	0.335	0.23074	0.385	0.27872	0.435	0.32788	0.485	0.37770
0.036	0.00901	0.086	0.03275	0.136	0.06407	0.186	0.10077	0.236	0.14154	0.286	0.18542	0.336	0.23169	0.386	0.27969	0.436	0.32887	0.486	0.37870
0.037	0.00938	0.087	0.03331	0.137	0.06476	0.187	0.10155	0.237	0.14239	0.287	0.18633	0.337	0.23263	0.387	0.28067	0.437	0.32987	0.487	0.37970
0.038	0.00976	0.088	0.03387	0.138	0.06545	0.188	0.10233	0.238	0.14324	0.288	0.18723	0.338	0.23358	0.388	0.28164	0.438	0.33086	0.488	0.38070
0.039	0.01015	0.089	0.03444	0.139	0.06614	0.189	0.10312	0.239	0.14409	0.289	0.18814	0.339	0.23453	0.389	0.28262	0.439	0.33185	0.489	0.38170
0.040	0.01054	0.090	0.03501	0.140	0.06683	0.190	0.10390	0.240	0.14494	0.290	0.18905	0.340	0.23547	0.390	0.28359	0.440	0.33284	0.490	0.38270
0.041	0.01093	0.091	0.03559	0.141	0.06753	0.191	0.10469	0.241	0.14580	0.291	0.18996	0.341	0.23642	0.391	0.28457	0.441	0.33384	0.491	0.38370
0.042	0.01133	0.092	0.03616	0.142	0.06822	0.192	0.10547	0.242	0.14666	0.292	0.19086	0.342	0.23737	0.392	0.28554	0.442	0.33483	0.492	0.38470
0.043	0.01173	0.093	0.03674	0.143	0.06892	0.193	0.10626	0.243	0.14751	0.293	0.19177	0.343	0.23832	0.393	0.28652	0.443	0.33582	0.493	0.38570
0.044	0.01214	0.094	0.03732	0.144	0.06963	0.194	0.10705	0.244	0.14837	0.294	0.19268	0.344	0.23927	0.394	0.28750	0.444	0.33682	0.494	0.38670
0.045	0.01255	0.095	0.03791	0.145	0.07033	0.195	0.10784	0.245	0.14923	0.295	0.19360	0.345	0.24022	0.395	0.28848	0.445	0.33781	0.495	0.38770
0.046	0.01297	0.096	0.03850	0.146	0.07103	0.196	0.10864	0.246	0.15009	0.296	0.19451	0.346	0.24117	0.396	0.28945	0.446	0.33880	0.496	0.38870
0.047	0.01339	0.097	0.03909	0.147	0.07174	0.197	0.10943	0.247	0.15095	0.297	0.19542	0.347	0.24212	0.397	0.29043	0.447	0.33980	0.497	0.38970
0.048	0.01382	0.098	0.03968	0.148	0.07245	0.198	0.11023	0.248	0.15182	0.298	0.19634	0.348	0.24307	0.398	0.29141	0.448	0.34079	0.498	0.39070
0.049	0.01425	0.099	0.04028	0.149	0.07316	0.199	0.11102	0.249	0.15268	0.299	0.19725	0.349	0.24403	0.399	0.29239	0.449	0.34179	0.499	0.39170
																		0.500	0.39270

[a]Rules for using table: (1) Divide height of segment by the diameter; multiply the area in the table corresponding to the quotient, height/diameter, by the diameter squared. When segment exceeds a semicircle, its area is: Area of circle minus the area of a segment whose height is the circle diameter minus the height of the given segment. (2) To find the diameter when given the chord and the segment height; the diameter = [(1/2 chord)²/height] + height. (Reprinted by permission from J. H. Perry, *Chemical Engineers Handbook*, 3rd ed., McGraw-Hill Book Co., New York, 1950.)

Dimensional and Related Data for Pipes, Tubes, and Fins

TUBE DATA

This section includes dimensional, weight, and materials compatibility data for commercial tubing and pipe, together with recommended values for minimum tube bend radii and header-sheet hole sizes and tolerances.

The fin surface area per foot of length of finned tubes having 10 circular fins per inch is given as a function of fin OD and tube OD in Fig. H7.1. The surface area for any other fin spacing can be obtained by simply multiplying the value given in Fig. H7.1 by the desired number of fins per inch divided by 10. The fin efficiency can be obtained from Fig. H7.2 and applied to the fin area from Fig. H7.1. The tube base surface area is simply the surface area for the bare tube multiplied by one minus the product of the number of fins per inch and the fin thickness.

To help select materials, the galvanic series for metals in seawater has been included in Table H7.4, and the compatibility of 18 structural metals with over 100 fluids is indicated in Table H7.5.

TABLE H7.1 Dimensional Data for Commercial Tubing[a]

OD of Tubing, in.	Bwg Gauge	Thickness, in.	Internal Flow Area, in.^{2}	Sq Ft External Surface per Foot Length	Sq Ft Internal Surface per Foot Length	Weight per Ft Length, Steel, lb[b]	ID Tubing, in.	Moment of Inertia, in.^{4}	Section modulus, in.^{3}	Radius of Gyration, in.	Constant, C[c]	OD/ID	Transverse Metal Area, in.^{2}
1/4	22	0.028	0.0295	0.0655	0.0508	0.066	0.194	0.00012	0.00098	0.0792	46	1.289	0.0195
1/4	24	0.022	0.0333	0.0655	0.0539	0.054	0.206	0.00011	0.00083	0.0810	52	1.214	0.0159
1/4	26	0.018	0.0360	0.0655	0.0560	0.045	0.214	0.00009	0.00071	0.0824	56	1.168	0.0131
3/8	18	0.049	0.0603	0.0982	0.0725	0.171	0.277	0.00068	0.0036	0.1164	94	1.354	0.0502
3/8	20	0.035	0.0731	0.0982	0.0798	0.127	0.305	0.00055	0.0029	0.1213	114	1.233	0.0374
3/8	22	0.028	0.0799	0.0982	0.0835	0.104	0.319	0.00046	0.0025	0.1227	125	1.176	0.0305
3/8	24	0.022	0.0860	0.0982	0.0867	0.083	0.331	0.00038	0.0020	0.1248	134	1.133	0.0244
1/2	16	0.065	0.1075	0.1309	0.0969	0.302	0.370	0.0022	0.0086	0.1556	168	1.351	0.0888
1/2	18	0.049	0.1269	0.1309	0.1052	0.236	0.402	0.0018	0.0072	0.1606	198	1.244	0.0694
1/2	20	0.035	0.1452	0.1309	0.1126	0.174	0.430	0.0014	0.0056	0.1649	227	1.163	0.0511
1/2	22	0.028	0.1548	0.1309	0.1162	0.141	0.444	0.0012	0.0046	0.1671	241	1.126	0.0415
5/8	12	0.109	0.1301	0.1636	0.1066	0.602	0.407	0.0061	0.0197	0.1864	203	1.536	0.177
5/8	13	0.095	0.1486	0.1636	0.1139	0.537	0.435	0.0057	0.0183	0.1903	232	1.437	0.158
5/8	14	0.083	0.1655	0.1636	0.1202	0.479	0.459	0.0053	0.0170	0.1938	258	1.362	0.141
5/8	15	0.072	0.1817	0.1636	0.1259	0.425	0.481	0.0049	0.0156	0.1971	283	1.299	0.125
5/8	16	0.065	0.1924	0.1636	0.1296	0.388	0.495	0.0045	0.0145	0.1993	300	1.263	0.114
5/8	17	0.058	0.2035	0.1636	0.1333	0.350	0.509	0.0042	0.0134	0.2016	317	1.228	0.103
5/8	18	0.049	0.2181	0.1636	0.1380	0.303	0.527	0.0037	0.0118	0.2043	340	1.186	0.089
5/8	19	0.042	0.2298	0.1636	0.1416	0.262	0.541	0.0033	0.0105	0.2068	358	1.155	0.077
5/8	20	0.035	0.2419	0.1636	0.1453	0.221	0.555	0.0028	0.0091	0.2089	377	1.126	0.065
3/4	10	0.134	0.1825	0.1963	0.1262	0.884	0.482	0.0129	0.0344	0.2229	285	1.556	0.260
3/4	11	0.120	0.2043	0.1963	0.1335	0.809	0.510	0.0122	0.0326	0.2267	319	1.471	0.238
3/4	12	0.109	0.2223	0.1963	0.1393	0.748	0.532	0.0116	0.0309	0.2299	347	1.410	0.220
3/4	13	0.095	0.2463	0.1963	0.1466	0.666	0.560	0.0107	0.0285	0.2340	384	1.339	0.196
3/4	14	0.083	0.2679	0.1963	0.1529	0.592	0.584	0.0098	0.0262	0.2376	418	1.284	0.174
3/4	15	0.072	0.2884	0.1963	0.1587	0.520	0.606	0.0089	0.0238	0.2410	450	1.238	0.153
3/4	16	0.065	0.3019	0.1963	0.1623	0.476	0.620	0.0083	0.0221	0.2433	471	1.210	0.140
3/4	17	0.058	0.3157	0.1963	0.1660	0.428	0.634	0.0076	0.0203	0.2455	492	1.183	0.126
3/4	18	0.049	0.3339	0.1963	0.1707	0.367	0.652	0.0067	0.0178	0.2484	521	1.150	0.108
3/4	20	0.035	0.3632	0.1963	0.1780	0.269	0.680	0.0050	0.0134	0.2532	567	1.103	0.079
7/8	10	0.134	0.2892	0.2291	0.1589	1.061	0.607	0.0221	0.0505	0.2662	451	1.441	0.312
7/8	11	0.120	0.3166	0.2291	0.1662	0.969	0.635	0.0208	0.0475	0.2703	494	1.378	0.285
7/8	12	0.109	0.3390	0.2291	0.1720	0.891	0.657	0.0196	0.0449	0.2736	529	1.332	0.262
7/8	13	0.095	0.3685	0.2291	0.1793	0.792	0.685	0.0180	0.0411	0.2778	575	1.277	0.233
7/8	14	0.083	0.3948	0.2291	0.1856	0.704	0.709	0.0164	0.0374	0.2815	616	1.234	0.207
7/8	16	0.065	0.4359	0.2291	0.1950	0.561	0.745	0.0137	0.0312	0.2873	680	1.174	0.165
7/8	18	0.049	0.4742	0.2291	0.2034	0.432	0.777	0.0109	0.0249	0.2925	740	1.126	0.127
7/8	20	0.035	0.5090	0.2291	0.2107	0.313	0.805	0.0082	0.0187	0.2972	794	1.087	0.092
1	8	0.165	0.3526	0.2618	0.1754	1.462	0.670	0.0392	0.0784	0.3009	550	1.493	0.430
1	10	0.134	0.4208	0.2618	0.1916	1.237	0.732	0.0350	0.0700	0.3098	656	1.366	0.364
1	11	0.120	0.4536	0.2618	0.1990	1.129	0.760	0.0327	0.0654	0.3140	708	1.316	0.332
1	12	0.109	0.4803	0.2618	0.2047	1.037	0.782	0.0307	0.0615	0.3174	749	1.279	0.305
1	13	0.095	0.5153	0.2618	0.2121	0.918	0.810	0.0280	0.0559	0.3217	804	1.235	0.270
1	14	0.083	0.5463	0.2618	0.2183	0.813	0.834	0.0253	0.0507	0.3255	852	1.199	0.239
1	15	0.072	0.5755	0.2618	0.2241	0.714	0.856	0.0227	0.0455	0.3291	898	1.167	0.210
1	16	0.065	0.5945	0.2618	0.2278	0.649	0.870	0.0210	0.0419	0.3314	927	1.149	0.191
1	18	0.049	0.6390	0.2618	0.2361	0.496	0.902	0.0166	0.0332	0.3366	997	1.109	0.146
1	20	0.035	0.6793	0.2618	0.2435	0.360	0.930	0.0124	0.0247	0.3414	1060	1.075	0.106
1-1/4	7	0.180	0.6221	0.3272	0.2330	2.057	0.890	0.0890	0.1425	0.3836	970	1.404	0.605
1-1/4	8	0.165	0.6648	0.3272	0.2409	1.921	0.920	0.0847	0.1355	0.3880	1037	1.359	0.565
1-1/4	10	0.134	0.7574	0.3272	0.2571	1.598	0.982	0.0741	0.1186	0.3974	1182	1.273	0.470
1-1/4	11	0.120	0.8012	0.3272	0.2644	1.448	1.010	0.0688	0.1100	0.4018	1250	1.238	0.426
1-1/4	12	0.109	0.8365	0.3272	0.2702	1.329	1.032	0.0642	0.1027	0.4052	1305	1.211	0.391
1-1/4	13	0.095	0.8825	0.3272	0.2775	1.173	1.060	0.0579	0.0926	0.4097	1377	1.179	0.345

TABLE H7.1 (*Continued*)

OD of Tubing, in.	Bwg Gauge	Thickness, in.	Internal Flow Area, in.^{2}	Sq Ft External Surface per Foot Length	Sq Ft Internal Surface per Foot Length	Weight per Ft Length, Steel, lb^b	ID Tubing, in.	Moment of Inertia, in.^{4}	Section modulus, in.^{3}	Radius of Gyration, in.	Constant, C^c	$\dfrac{OD}{ID}$	Transverse Metal Area, in.^{2}
1-1/4	14	0.083	0.9229	0.3272	0.2838	1.033	1.084	0.0521	0.0833	0.4136	1440	1.153	0.304
1-1/4	16	0.065	0.9852	0.3272	0.2932	0.823	1.120	0.0426	0.0682	0.4196	1537	1.116	0.242
1-1/4	18	0.049	1.042	0.3272	0.3016	0.629	1.152	0.0334	0.0534	0.4250	1626	1.085	0.185
1-1/4	20	0.035	1.094	0.3272	0.3089	0.456	1.180	0.0247	0.0395	0.4297	1707	1.059	0.134
1-1/2	10	0.134	1.192	0.3927	0.3225	1.955	1.232	0.1354	0.1806	0.4853	1860	1.218	0.575
1-1/2	12	0.109	1.291	0.3927	0.3356	1.618	1.282	0.1159	0.1546	0.4933	2014	1.170	0.476
1-1/2	14	0.083	1.398	0.3927	0.3492	1.258	1.334	0.0931	0.1241	0.5018	2181	1.124	0.370
1-1/2	16	0.065	1.474	0.3927	0.3587	0.996	1.370	0.0756	0.1008	0.5079	2299	1.095	0.293
2	11	0.120	2.433	0.5236	0.4608	2.410	1.760	0.3144	0.3144	0.6660	3795	1.136	0.709
2	13	0.095	2.573	0.5236	0.4739	1.934	1.810	0.2586	0.2586	0.6744	4014	1.105	0.569
2-1/2	9	0.148	3.815	0.6540	0.5770	3.719	2.204	0.7592	0.6074	0.8332	5951	1.134	1.094

^{a}Reprinted with permission of Tubular Exchanger Manufacturers Association.

^{b}Weights are based on low-carbon steel with a density of 0.2833 lb/in.^{3}. For other metals multiply by the following factors:

Aluminum	0.35	Nickel-chrome-iron	1.07
AISI 400 series stainless steels	0.99	Admiralty	1.09
AISI 300 series stainless steels	1.02	Nickel and nickel-copper	1.13
Aluminum bronze	1.04	Copper and cupro-nickels	1.14

^{c}Liquid velocity $= \dfrac{\text{lb per tube per hour}}{C \times \text{sp gr of liquid}}$ ft/s (sp gr of water at 60°F $= 1.0$).

TABLE H7.2 Dimensional Data for Pipe

Nominal Pipe Size, in.	Outside Diameter, in.	Schedule Number or Weight	Wall Thickness, in.	Inside Diameter, in.	Surface Area Outside, ft²/ft	Surface Area Inside, ft²/ft	Metal Area, in.²	Flow Area, in.²	Pipe, lb/ft	Water, lb/ft
3/4	1.05	40	0.113	0.824	0.275	0.216	0.333	0.533	1.131	0.231
		80	0.154	0.742	0.275	0.194	0.434	0.432	1.474	0.187
1	1.315	40	0.133	1.049	0.344	0.275	0.494	0.864	1.679	0.374
		80	0.179	0.957	0.344	0.250	0.639	0.719	2.172	0.311
1-1/4	1.660	40	0.140	1.38	0.434	0.361	0.668	1.496	2.273	0.648
		80	0.191	1.278	0.434	0.334	0.881	1.283	2.997	0.555
1-1/2	1.900	40	0.145	1.61	0.497	0.421	0.799	2.036	2.718	0.882
		80	0.200	1.50	0.497	0.393	1.068	1.767	3.632	0.765
2	2.375	40	0.154	2.067	0.622	0.541	1.074	3.356	3.653	1.453
		80	0.218	1.939	0.622	0.508	1.477	2.953	5.022	1.278
2-1/2	2.875	40	0.203	2.469	0.753	0.646	1.704	4.79	5.794	2.073
		80	0.276	2.323	0.753	0.608	2.254	4.24	7.662	1.835
3	3.5	40	0.216	3.068	0.916	0.803	2.228	7.30	7.58	3.20
		80	0.300	2.900	0.916	0.759	3.106	6.60	10.25	2.86
3-1/2	4.0	40	0.226	3.548	1.047	0.929	2.680	9.89	9.11	4.28
		80	0.318	3.364	1.047	0.881	3.678	8.89	12.51	3.85
4	4.5	40	0.237	4.026	1.178	1.054	3.17	12.73	10.79	5.51
		80	0.337	3.826	1.178	1.002	4.41	11.50	14.99	4.98
5	5.563	10 S	0.134	5.295	1.456	1.386	2.29	22.02	7.77	9.53
		40	0.258	5.047	1.456	1.321	4.30	20.01	14.62	8.66
		80	0.375	4.813	1.456	1.260	6.11	18.19	20.78	7.88
6	6.625	10 S	0.134	6.357	1.734	1.664	2.73	31.7	9.29	13.74
		40	0.280	6.065	1.734	1.588	5.58	28.9	18.98	12.51
		80	0.432	5.761	1.734	1.508	8.40	26.1	28.58	11.29
8	8.625	10 S	0.148	8.329	2.258	2.180	3.94	54.5	13.40	23.59
		30	0.277	8.071	2.258	2.113	7.26	51.2	24.7	22.15
		80	0.500	7.625	2.258	1.996	12.76	45.7	43.4	19.8
10	10.75	10 S	0.165	10.420	2.81	2.73	5.49	85.3	18.7	36.9
		30	0.279	10.192	2.81	2.67	9.18	81.6	31.2	35.3
		Extra heavy	0.500	9.750	2.81	2.55	16.10	74.7	54.7	32.3
	12.75	10 S	0.180	12.390	3.34	3.24	7.11	120.6	24.2	52.2
		30	0.330	12.09	3.34	3.17	12.88	114.8	43.8	49.7
		Extra heavy	0.500	11.75	3.34	3.08	19.24	108.4	65.4	47.0
14	14.0	10	0.250	13.5	3.67	3.53	10.80	143.1	36.7	62.0
		Standard	0.375	13.25	3.67	3.47	16.05	137.9	54.6	59.7
		extra heavy	0.500	13.00	3.67	3.40	21.21	132.7	72.1	57.5
16	16.0	10	0.250	15.50	4.19	4.06	12.37	188.7	42.1	81.7
		Standard	0.375	15.25	4.19	3.99	18.41	182.7	62.6	79.1
		extra heavy	0.500	15.00	4.19	3.93	24.35	176.7	82.8	76.5
18	18.0	10 S	0.188	17.624	4.71	4.61	10.52	243.9	35.8	105.6
		Standard	0.375	17.25	4.71	4.52	20.76	233.7	70.6	101.2
		extra heavy	0.500	17.00	4.71	4.45	27.49	227.0	93.5	98.3

TABLE H7.2 (*Continued*)

Nominal Pipe Size, in.	Outside Diameter, in.	Schedule Number or Weight	Wall Thickness, in.	Inside Diameter, in.	Surface Area		Areas and Weights Cross-sectional		Weight	
					Outside, ft²/ft	Inside, ft²/ft	Metal Area, in.²	Flow Area, in.²	Pipe, lb/ft	Water, lb/ft
20	20.0	10 S	0.218	19.564	5.24	5.12	13.55	300.6	46.1	130.2
		Standard	0.375	19.25	5.24	5.04	23.12	291	78.6	126.0
		extra heavy	0.500	19.00	5.24	4.97	30.6	283.5	104.1	122.8
22	22.0	10	0.250	21.50	5.76	5.63	17.1	363	58.1	157.2
		Standard	0.375	21.25	5.76	5.56	25.5	355	86.6	153.6
		extra heavy	0.500	21.00	5.76	5.50	33.8	346	114.8	150.0
24	24.0	10	0.250	23.50	6.28	6.15	18.7	434	63.4	187.8
		Standard	0.375	23.25	6.28	6.09	27.8	425	94.6	183.8
		extra heavy	0.500	23.00	6.28	6.02	36.9	415	125.5	179.9
26	26.0	Standard	0.375	25.25	6.81	6.61	30.2	501	102.6	216.8
		extra heavy	0.500	25.00	6.81	6.54	40.1	491	136.2	212.5
30	30.0	10	0.312	29.376	7.85	7.69	29.1	678	98.9	293.5
		Standard	0.375	29.250	7.85	7.66	34.9	672	118.7	291.0
		extra heavy	0.500	29.00	7.85	7.59	46.3	661	157.6	286.0
34	34.0	Standard	0.375	33.250	8.90	8.70	39.6	868	134.7	376
		extra heavy	0.500	33.00	8.90	8.64	52.6	855	178.9	370.3
36	36.0	Standard	0.375	32.25	9.42	9.23	42.0	976	142.7	422.6
		extra heavy	0.500	35.00	9.42	9.16	55.8	962	189.6	416.6
42	42.0	Standard	0.375	41.25	11.0	10.8	49.0	1336	166.7	578.7
		extra heavy	0.500	41.00	11.0	10.73	65.2	1320	221.6	571.7

[a]Reprinted with permission, from "Design Properties of Pipe," © 1958, Chemetron Corp.

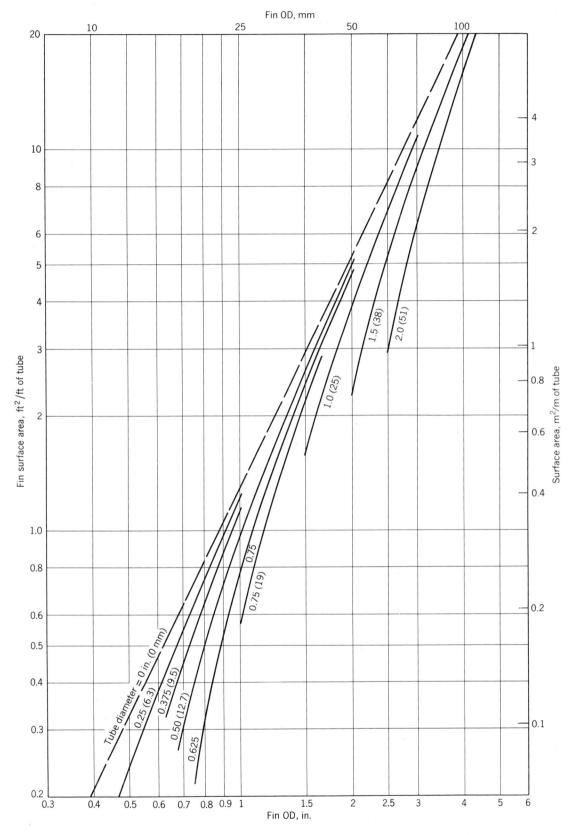

Figure H7.1 Fin surface area for tubes having 10 circular disc fins per inch of length. (The tube surface area between fins is not included, because it varies with the fin thickness.)

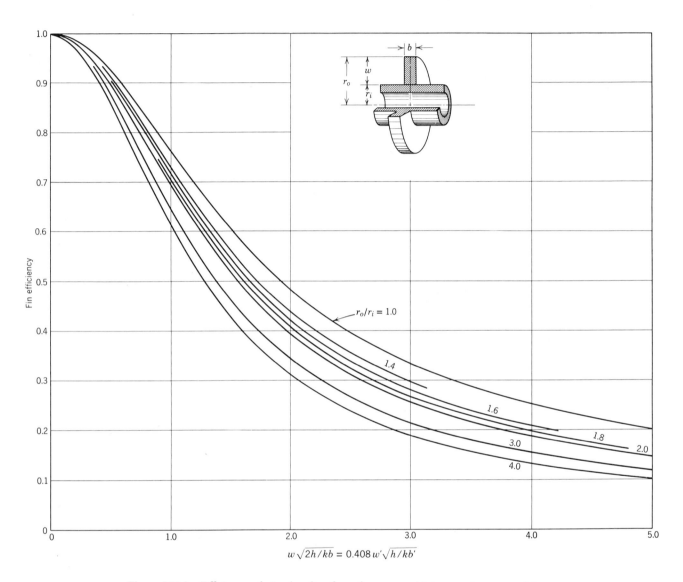

Figure H7.2 Efficiency of circular disc fins of constant thickness where w and b are in feet or meters. For English units one can also use the alternative expression, where w' and b' are in inches. See p. 41 for detailed discussion of fin efficiency. (Gardner, Ref. 7 of Chapt. 3.)

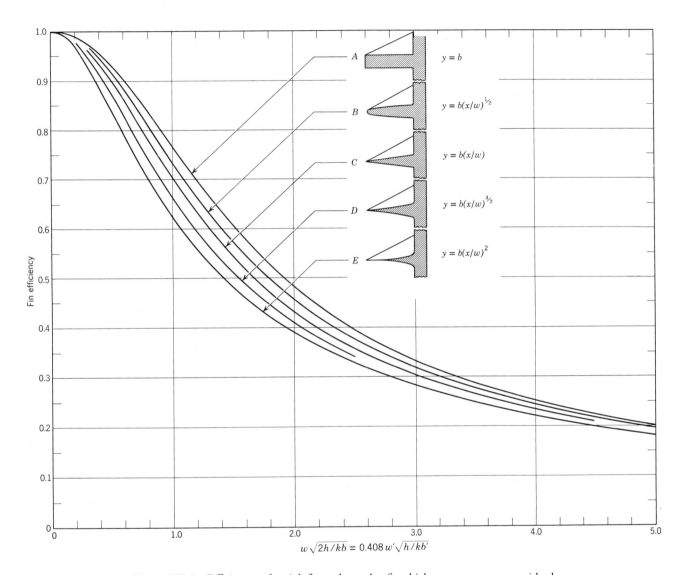

Figure H7.3 Efficiency of axial fins where the fin thickness y may vary with the distance x from the root of the fin where $y = b$. The fin height w and thickness b are in feet or meters. For English units one can also use the alternative expression, where w' and b' are in inches. (Gardner, Ref. 7 of Chapt. 3.)

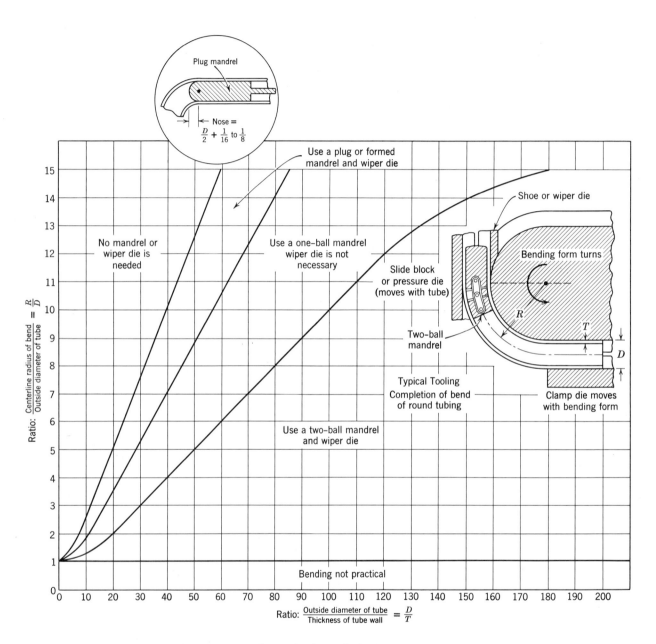

Figure H7.4 Effects of tube-wall thickness and tooling on minimum bend radii. (Courtesy International Nickel Co.)

TABLE H7.3   Recommended Hole Sizes for Tube Header Sheets

| | Tube Fit Allowances | | | Tube Sheet Hole Diameters | | | |
| | | | | Nominal Tube Hole Diameter | | | |
Nominal Tube OD, in.	Standard Tube OD Tolerance, in.	Nominal Tube Sheet Hole Clearances, in.	Special Close Fit, in.	Nominal Tube OD in.	Standard Fit, in.	Close Fit, in.	Tube Hole Tolerance, in.
0.250	±0.003	0.008	0.004	0.250	0.258	0.254	±0.002
0.375	±0.003	0.008	0.004	0.375	0.383	0.379	±0.002
0.500	±0.004	0.008	0.004	0.500	0.508	0.504	±0.002
0.625	±0.004	0.008	0.004	0.675	0.683	0.629	±0.002
0.750	±0.004	0.010	0.005	0.750	0.760	0.755	±0.002
1.000	±0.005	0.010	0.006	1.000	1.010	1.006	±0.002
1.250	±0.005	0.015	0.008	1.250	1.265	1.258	±0.002
1.500	±0.005	0.018	0.009	1.500	1.518	1.509	±0.003

TABLE H7.4

galvanic series of metals and alloys in sea water

ANODE

Magnesium

Zinc

Aluminum, Alclad 3003

Aluminum 3003

Aluminum 6061

Aluminum 6063

Aluminum 5052

Low steel

Alloy steel

Cast iron

Stainless Steel, type 410 (active)

Stainless Steel, type 430 (active)

Stainless Steel, type 304 (active)

Stainless Steel, type 316 (active)

Ni-resist

Muntz metal

Yellow brass

Admiralty

Aluminum brass

Red brass

Copper

Aluminum bronze

Composition G bronze

Cupro-Nickel, 10%

Cupro-Nickel, 30%—low iron

Cupro-Nickel—high iron

Nickel

Inconel

Tantalum

Silver

Stainless Steel, type 410 (passive)

Stainless Steel, type 430 (passive)

Stainless Steel, type 304 (passive)

Stainless Steel, type 316 (passive)

Monel

Hastelloy c

Zirconium

Titanium

CATHODE

TABLE H7.5   Compatibility of Structural Materials With Typical Fluids[a]

LEGEND

A = EXCELLENT
B = GOOD
C = FAIR
D = NOT SUITABLE
E = EXPLOSIVE
I = IGNITES
• = INFORMATION NOT AVAILABLE

No.	Chemicals	% Concentration	°F Temperature	Carbon Steel	Copper	Red Brass	Muntz	Admiralty	Copper Silicon	90-10 Cupro-Nickel	70-30 Cupro-Nickel	Aluminum	304 Stainless Steel	316 Stainless Steel	Nickel	Monel	Inconel	Hastelloy	Titanium	Zirconium	Tantalum	No.
1.	Acetaldehyde	100	70	A	E	E	E	E	E	E	E	A	A	A	A	A	A	A	B	•	A	1.
2.	Acetic Acid (Aerated)	100	70	D	D	D	D	D	D	C	C	B	A	A	D	A	B	A	A	A	A	2.
3.	Acetic Anhydride	100	70	D	B	C	D	C	B	B	B	A	B	B	B	B	B	A	A	A	B	3.
4.	Acetone	100	70	A	A	A	A	A	A	A	A	A	A	A	B	A	A	B	A	•	A	4.
5.	Acetylene	100	70	A	E	E	E	E	E	E	E	A	A	A	A	A	A	A	A	•	A	5.
6.	Aluminum Chloride	10	70	D	D	D	D	D	D	D	D	D	D	D	C	B	D	A	A	A	A	6.
7.	Aluminum Hydroxide	10	70	B	B	B	B	B	B	B	B	B	B	B	B	B	B	B	•	•	B	7.
8.	Ammonia (Anhydrous)	100	70	A	A	A	A	A	A	A	A	A	A	A	B	A	B	B	A	•	A	8.
9.	Ammonium Chloride	10	70	D	D	D	D	D	D	D	D	D	C	B	B	B	B	B	A	A	A	9.
10.	Ammonium Sulfate	10	70	C	C	C	C	C	C	C	C	C	D	C	C	B	A	B	B	A	A	10.
11.	Ammonium Sulfite	10	70	D	B	B	B	D	B	B	B	B	C	C	D	D	D	•	A	•	A	11.
12.	Amyl Acetate	100	70	B	A	A	C	A	B	A	A	A	A	A	A	A	A	B	A	•	A	12.
13.	Aniline	100	70	A	D	D	D	D	D	D	D	D	D	A	A	B	B	B	B	A	•	A	13.
14.	Aroclor	100	70	B	A	A	A	A	A	A	A	A	B	B	A	A	A	A	A	•	A	14.
15.	Barium Chloride	30	70	B	B	B	D	C	B	B	B	B	B	B	B	B	B	B	A	A	A	15.
16.	Benzaldehyde	100	70	B	B	B	B	B	B	B	B	B	B	B	B	B	B	B	A	•	A	16.
17.	Benzene	100	70	A	A	A	A	A	A	A	A	B	B	B	B	B	B	B	A	•	A	17.
18.	Benzoic Acid	10	70	D	B	B	B	B	B	B	B	B	B	B	B	B	B	B	A	•	A	18.
19.	Boric Acid	10	70	D	B	B	B	B	B	B	B	C	A	A	B	B	B	A	A	•	A	19.
20.	Butadiene	100	70	A	A	A	A	A	A	A	A	A	A	A	A	A	A	A	A	•	A	20.
21.	Butane	100	70	A	A	A	A	A	A	A	A	A	A	A	A	A	A	A	A	•	A	21.
22.	Butanol	100	70	A	A	A	A	A	A	A	A	A	A	A	A	A	A	A	A	•	A	22.
23.	Butyl Acetate	100	70	A	B	B	B	B	B	B	B	A	B	B	A	B	A	B	A	•	A	23.
24.	Butyl Chloride	100	70	A	A	A	A	A	A	A	A	A	A	A	A	A	A	A	A	•	A	24.
25.	Calcium Chloride	20	70	B	B	B	D	C	B	B	B	B	C	B	A	A	A	B	A	A	A	25.
26.	Calcium Hydroxide	10	70	B	B	B	B	B	B	B	B	B	D	B	B	B	B	B	A	•	A	26.
27.	Carbon Dioxide (Wet)	100	70	C	C	C	C	C	C	C	C	C	B	A	A	A	A	A	A	•	A	27.
28.	Carbon Tetrachloride (Dry)	100	70	B	B	B	B	B	B	B	B	B	B	B	A	A	A	B	A	A	A	28.
29.	Carbonic Acid	100	70	C	C	C	C	C	C	C	C	C	B	B	B	C	A	A	A	•	A	29.
30.	Chlorine Gas (Dry)	100	70	B	B	B	B	B	B	B	B	C	B	B	B	B	A	B	I	A	A	30.
31.	Chloroform (Dry)	100	70	B	B	B	B	B	B	B	B	B	B	B	A	A	B	A	A	•	A	31.
32.	Chromic Acid	20	70	D	D	D	D	D	D	D	D	D	C	B	D	D	B	B	B	A	A	32.
33.	Citric Acid	20	70	D	C	C	D	C	C	C	C	A	C	B	B	B	A	C	A	A	A	33.
34.	Creosote	100	70	B	B	B	B	B	B	B	B	B	B	B	B	B	B	B	A	•	A	34.
35.	Dibutylphthalate	100	70	A	A	A	A	A	A	A	A	B	B	B	B	B	B	B	A	•	A	35.
36.	Dichlorobenzene	100	70	B	B	B	B	B	B	B	B	B	B	B	B	B	B	B	B	•	A	36.
37.	Dichlodifluoromethane (F-12)	100	70	A	A	A	A	A	A	A	A	A	A	B	B	B	B	A	A	•	A	37.
38.	Diethanolamine	100	85	A	B	B	B	B	B	B	B	A	A	A	A	A	A	A	A	•	A	38.
39.	Diethyl Ether	100	70	B	B	B	B	B	B	B	B	B	B	B	B	B	B	B	A	•	A	39.
40.	Diethylene Glycol	100	70	A	B	B	B	B	B	B	B	B	A	A	B	B	B	B	A	•	A	40.
41.	Diphenyl	100	160	B	B	B	B	B	B	B	B	A	B	B	B	B	B	B	A	•	A	41.
42.	Diphenyl Oxide	100	85	B	B	B	B	B	B	B	B	B	B	B	B	B	B	B	A	•	A	42.

TABLE H7.5 (*Continued*)

LEGEND

A = EXCELLENT
B = GOOD
C = FAIR
D = NOT SUITABLE
E = EXPLOSIVE
I = IGNITES
• = INFORMATION NOT AVAILABLE

No.	Chemicals	% Concentration	°F Temperature	Carbon Steel	Copper	Red Brass	Muntz	Admiralty	Copper Silicon	90-10 Cupro-Nickel	70-30 Cupro-Nickel	Aluminum	304 Stainless Steel	316 Stainless Steel	Nickel	Monel	Inconel	Hastelloy	Titanium	Zirconium	Tantalum	No.
43.	Ethane	100	70	A	A	A	A	A	A	A	A	A	A	A	A	A	A	A	A	•	A	43.
44.	Ethanolamine	100	70	B	B	B	B	B	B	B	B	B	A	B	B	B	B	B	B	•	A	44.
45.	Ether	100	70	B	B	B	B	B	B	B	B	B	B	B	B	B	B	B	A	•	A	45.
46.	Ethyl Acetate (Dry)	100	70	B	B	B	B	B	B	B	B	B	B	B	B	B	B	B	A	•	A	46.
47.	Ethyl Alcohol	100	70	B	B	B	B	B	B	B	B	B	B	B	B	B	B	A	A	A	A	47.
48.	Ethyl Ether	100	70	B	B	B	B	B	B	B	B	B	B	B	B	B	B	B	A	•	A	48.
49.	Ethylene	100	70	A	A	A	A	A	A	A	A	A	A	A	A	A	A	A	A	•	A	49.
50.	Ethylene Glycol	100	70	B	B	B	B	B	B	B	B	B	B	B	B	B	B	B	A	•	A	50.
51.	Fatty Acids	100	400	D	D	D	D	D	D	D	D	D	A	D	A	B	C	B	A	•	A	51.
52.	Ferric Chloride	20	70	D	D	D	D	D	D	D	D	D	D	D	D	D	D	B	A	D	A	52.
53.	Ferric Sulfate	10	70	D	D	D	D	D	D	D	D	D	B	B	B	D	D	A	A	•	A	53.
54.	Ferrous Sulfate	10	70	D	B	B	D	B	B	B	B	B	B	B	D	D	D	B	A	•	A	54.
55.	Formaldehyde	50	200	B	B	B	D	B	B	B	B	C	B	B	B	B	B	B	A	•	A	55.
56.	Furfural	100	70	B	B	B	D	B	B	B	B	B	B	B	B	B	B	B	A	•	A	56.
57.	Glycerine	100	70	A	A	A	A	A	A	A	A	A	A	A	A	A	A	A	A	•	A	57.
58.	Hexane	100	70	A	A	A	A	A	A	A	A	A	A	A	A	A	A	A	A	•	A	58.
59.	Hydrochloric Acid (Aerated)	38	70	D	D	D	D	D	D	D	D	D	D	D	D	D	D	B	D	D	A	59.
60.	Hydrofluoric Acid (Aerated)	40	70	D	C	D	D	D	D	D	D	C	D	D	D	C	D	A	D	D	D	60.
61.	Iodine	20	70	D	D	D	D	D	D	D	D	D	D	D	D	D	D	B	D	•	A	61.
62.	Isopropanol	100	70	A	B	B	B	B	B	B	B	B	B	B	B	B	B	B	A	•	A	62.
63.	Lactic Acid	50	70	D	B	B	D	C	B	B	B	D	B	A	B	C	A	A	A	A	A	63.
64.	Linseed Oil	100	70	A	B	B	B	B	B	B	B	B	A	A	B	B	B	B	A	•	A	64.
65.	Lithium Chloride	30	200	B	B	B	D	B	B	B	B	D	B	A	A	A	A	A	•	•	A	65.
66.	Lithium Hydroxide	10	200	B	B	B	D	B	B	B	B	D	B	B	B	B	B	B	•	•	A	66.
67.	Magnesium Chloride	30	70	B	B	B	D	C	B	B	B	C	B	B	A	B	A	A	A	A	A	67.
68.	Magnesium Hydroxide	10	70	B	B	B	B	B	B	B	B	D	B	B	B	B	B	B	A	•	B	68.
69.	Magnesium Sulfate	30	200	B	B	B	B	B	B	B	B	C	A	A	B	B	B	A	A	A	A	69.
70.	Methane	100	70	A	A	A	A	A	A	A	A	A	A	A	A	A	A	A	A	A	A	70.
71.	Methallyamine	100	70	C	B	B	B	B	B	B	B	B	B	B	B	B	C	B	B	•	A	71.
72.	Methyl Alcohol	100	70	B	B	B	B	B	B	B	B	B	B	B	B	A	B	A	A	A	A	72.
73.	Methyl Chloride (Dry)	100	70	A	A	A	A	A	A	A	A	E	A	A	B	B	B	B	A	•	A	73.
74.	Methylene Chloride (Dry)	100	70	B	B	B	B	B	B	B	B	B	B	B	B	B	B	B	B	•	A	74.
75.	Monochlorobenzene (Dry)	100	70	B	B	B	B	B	B	B	B	A	B	B	A	A	A	B	B	•	A	75.
76.	Monochlorodifluoro Methane (F-22)	100	70	A	A	A	A	A	A	A	A	A	A	A	A	A	A	A	A	•	A	76.
77.	Monoethanolamine	100	200	B	B	B	B	B	B	B	B	B	B	B	B	B	B	•	•	•	A	77.

TABLE H7.5 (*Continued*)

Table H7.5 (*Cont.*)

LEGEND

A = EXCELLENT
B = GOOD
C = FAIR
D = NOT SUITABLE
E = EXPLOSIVE
I = IGNITES
• = INFORMATION NOT AVAILABLE

No.	Chemicals	% Concentration	°F Temperature	Carbon Steel	Copper	Red Brass	Muntz	Admiralty	Copper Silicon	90-10 Cupro-Nickel	70-30 Cupro-Nickel	Aluminum	304 Stainless Steel	316 Stainless Steel	Nickel	Monel	Inconel	Hastelloy	Titanium	Zirconium	Tantalum	No.
78.	Naphtha	100	70	A	B	B	B	B	B	B	B	A	B	B	B	B	B	B	B	•	A	78.
79.	Naphthalene	100	70	A	B	B	B	B	B	B	B	B	A	A	A	A	A	B	B	•	A	79.
80.	Nickel Chloride	20	70	D	B	B	D	B	B	B	B	D	B	B	D	B	D	A	A	A	A	80.
81.	Nickel Sulfate	10	200	D	B	B	D	B	B	B	B	D	B	B	B	B	B	B	B	A	A	81.
82.	Nitric Acid	50	200	D	D	D	D	D	D	D	D	D	B	B	D	D	D	D	A	B	A	82.
83.	Nitrous Acid	10	70	D	D	D	D	D	D	D	D	D	B	B	D	D	D	•	•	•	A	83.
84.	Oleic Acid	100	70	B	B	B	C	B	B	B	B	B	B	B	A	A	A	B	B	B	B	84.
85.	Oxalic Acid	10	70	D	B	B	C	B	B	B	B	C	B	B	C	B	B	B	D	B	A	85.
86.	Perchloric Acid (Dry)	100	70	D	D	D	D	D	D	D	D	B	B	B	D	D	D	•	•	•	A	86.
87.	Perchloroethylene	100	70	A	B	B	C	B	B	B	B	B	B	B	A	A	A	•	A	•	A	87.
88.	Phenol	10	120	B	B	B	B	B	B	B	B	A	B	B	B	A	B	A	A	•	A	88.
89.	Phosphoric Acid (Aerated)	50	200	D	D	D	D	D	D	D	D	D	B	B	D	D	B	A	C	D	B	89.
90.	Phthalic Anhydride	100	300	B	B	B	B	B	B	B	B	B	B	B	B	B	B	•	•	•	A	90.
91.	Potassium Bicarbonate	30	200	B	B	B	C	B	B	B	A	D	B	B	B	B	B	A	•	•	A	91.
92.	Potassium Carbonate	40	200	B	B	B	B	B	B	B	B	D	B	B	B	B	B	A	•	•	A	92.
93.	Propylene Glycol	100	70	B	B	B	B	B	B	B	B	B	B	B	B	B	B	A	•	•	A	93.
94.	Pyridine	100	70	A	B	B	B	B	B	B	B	B	B	B	B	B	B	B	B	•	A	94.
95.	Silver Chloride	10	70	D	D	D	D	D	D	D	D	D	D	D	D	D	C	B	B	•	A	95.
96.	Silver Nitrate	10	70	D	D	D	D	D	D	D	D	D	B	B	D	D	B	B	A	A	A	96.
97.	Sodium Acetate	10	70	D	B	B	B	B	B	B	B	C	B	B	B	B	B	B	B	•	A	97.
98.	Sodium Hydroxide	50	300	D	D	D	D	D	D	D	D	D	D	D	A	B	B	B	B	B	D	98.
99.	Sodium Nitrate	40	70	B	B	B	C	B	B	B	B	A	A	B	B	A	B	A	B	•	A	99.
100.	Sodium Sulfate	10	200	B	B	B	B	B	B	B	A	B	A	B	B	B	B	A	B	•	A	100.
101.	Sulfur Dioxide (Dry)	100	300	B	B	B	C	B	B	B	B	B	B	B	B	B	B	A	B	•	A	101.
102.	Sulfuric Acid (Aerated)	60	200	D	D	D	D	D	D	D	D	D	D	D	D	D	D	B	D	A	A	102.
103.	Toluene	100	200	A	A	A	A	A	A	A	A	A	A	A	A	A	A	A	A	A	A	103.
104.	Trichloroethylene (Dry)	100	150	B	B	B	C	B	B	B	B	B	B	B	A	A	B	A	A	A	A	104.
105.	Turpentine	100	70	B	B	B	B	B	B	B	B	B	B	B	B	B	B	B	B	•	A	105.
106.	Vinyl Chloride (Dry)	100	70	A	B	B	D	C	B	B	B	A	B	A	A	A	A	A	A	•	A	106.
107.	Water (Fresh)	100	70	C	A	A	A	A	A	A	A	B	A	A	A	A	A	A	A	A	A	107.
108.	Water (Sea)	100	70	C	B	B	C	A	B	A	A	B	A	A	B	A	B	B	A	A	A	108.
109.	Xylene	100	200	B	A	A	A	A	A	A	A	B	A	A	A	A	A	A	A	A	•	109
110.	Zinc Chloride	10	70	D	D	D	D	D	D	D	D	C	B	B	B	A	D	B	A	A	A	110.
111.	Zinc Sulfate	20	70	D	B	B	D	B	B	B	B	D	B	A	B	B	A	B	A	•	A	111.

(Courtesy The Patterson Kelley Co., Inc.)

TABLE H7.6 Standard and Actual Values of Japanese PWR Secondary Water Chemistry Indicators[a]

	Standard	Actual (Sample)
Condensate		
Cation conductivity, μS	<0.2	0.1
Dissolved oxygen, ppb	<50	1–4
Feedwater		
pH	8.8–9.3	9.1
Conductivity, μS	<5	3.8
Dissolved oxygen, ppb	<5	0
Total iron, ppb	<20	8–12
Total copper, ppb	<5	0–1
Total nickel, ppb	<5	0–1
Hydrazine, ppb	>2	160–170
Steam generator blowdown		
pH	8.5–9.1	9.0
Conductivity	<5	
Cation conductivity, μS	0.0142.0	0.1–0.2
Silica, ppb	<500	6–13
Chloride, ppb	<100	3–7
Sodium, ppb	<100	0.5–1.2
Free alkali, ppb	<150	
Turbidity, ppm	<1	0.2–0.3
Steam		
Silica, ppb	<20	10–13

[a]Ref. 27 of Chapt. 10. (Courtesy *Nuclear Engineering International.*)

TABLE H7.7 Material Ability to Withstand Adverse Operating Conditions

Condition	Material Resistance[a]			
	Admiralty	Type 304 SS	90-10 Cu-Ni	Titanium
Corrosion	2	5	4	6
Erosion-corrosion	2	6	4	6
Pitting, operating	2	4	6	6
Pitting, stagnant	2	1	5	6
High water velocity	3	6	4	6
Bubble impingement (inside tubes)	2-1	6	3	6
Steam impingement (outside tubes)	2-1	6	3	6
Stress corrosion	1	5	6	6
Chloride attack	3	1	6	6
Ammonia attack	2	6	4	6
Biological fouling	5	2	4	3
Total	27	48	49	63
Average	2.5	4.4	4.5	5.7

[a]Scale of 1 (lowest) to 6 (highest).

Stress Analysis

STRESS ANALYSIS

This section includes data and charts that have proved helpful in heat exchanger stress analysis. Figure H8.2 presents a header-sheet stress parameter which is based on the relationships presented in Chapter 9. The chart of Fig. H8.3, for tubes loaded as simple cantilever beams, also can be used—with appropriate allowances—to estimate stresses and deflections for other loading configurations, for example, lateral displacement of one end of a tube when the axes at the ends are kept parallel.

Deflections and stresses in circular shells under lateral loads are heavily dependent on end conditions and other details of the geometry. Some indication of their magnitude can be deduced for many cases from Fig. H8.4, which gives data for one idealized condition, that is, an open-ended cylinder loaded uniformly along its length by forces acting along a diameter.

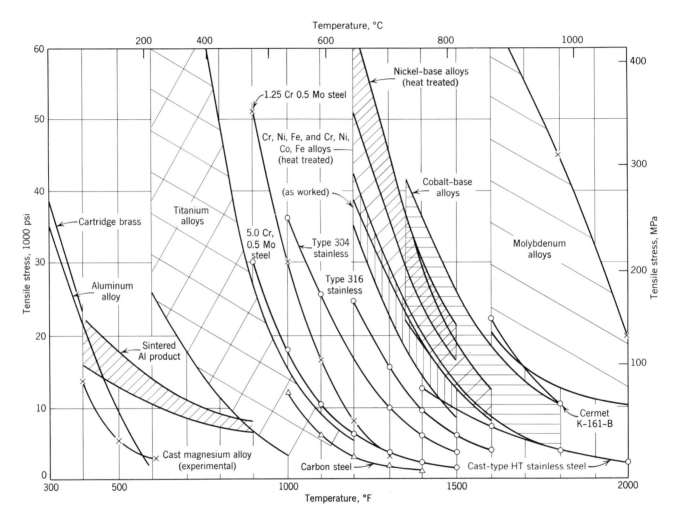

Figure H8.1 Stress for rupture in 1000 h as a function of temperature for typical materials. (*Metals Handbook,* Supplement, ASM, 1954.)

TABLE H8.1 Maximum Working Pressures Recommended for Steel Tubing[a,b]

Tube OD, in.	Tube-Wall Thickness, in.							
	0.025	0.035	0.049	0.065	0.083	0.095	0.109	0.125
1/8	6,000	8,430	11,750					
3/16	4,000	5,610	7,850					
1/4	2,980	4,200	5,890	7,850				
5/16	2,410	3,370	4,710	6,290	8,010			
3/8	1,990	2,800	3,920	5,240	6,660	7,640	8,720	9,990
13/32	1,845	2,585	3,620	4,800	6,130	7,015	8,050	9,235
7/16	1,715	2,400	3,360	4,500	5,720	6,550	7,490	8,580
1/2	1,490	2,100	2,940	3,925	5,000	5,720	6,540	7,500
9/16	1,330	1,870	2,620	3,490	4,450	5,090	5,810	6,660
5/8	1,200	1,680	2,350	3,140	4,000	4,580	5,230	6,000
11/16	1,090	1,530	2,135	2,835	3,620	4,145	4,755	5,455
3/4	1,000	1,400	1,960	2,600	3,320	3,800	4,360	5,000
7/8	860	1,200	1,680	2,225	2,845	3,255	3,740	4,285
1	750	1,050	1,470	1,950	2,490	2,850	3,270	3,750
1 1/8	665	935	1,310	1,735	2,215	2,535	2,910	3,330

[a]Reprinted by permission of the Superior Tube Co.

[b]*Note:* The above values are in pounds per square inch and are based on an allowable fiber stress of 15,000 psi. For approximate bursting pressures, multiply these by the factor 3.3. The formula used is Barlow's, as follows:

$$P = \frac{2St}{OD}$$

where P is bursting pressure, S fiber stress, psi, t wall thickness, in., and OD outside diameter, in.

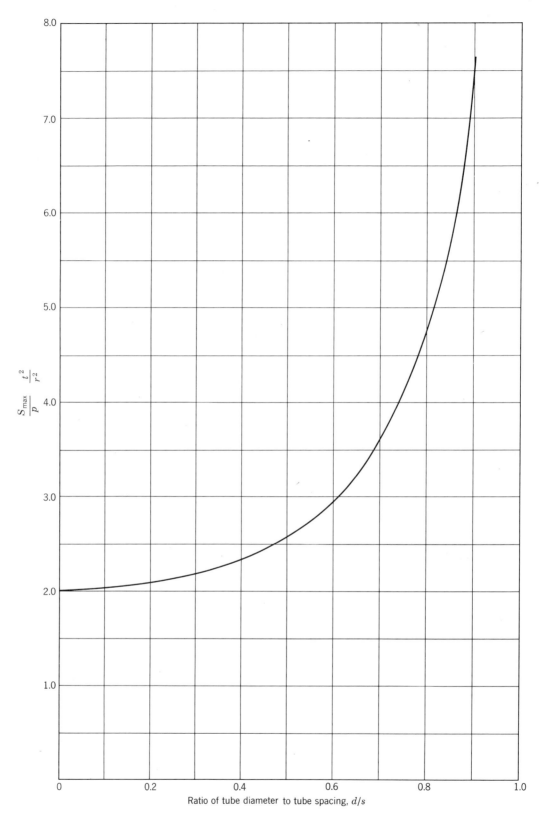

Figure H8.2 Effects of the ratio of tube diameter to tube spacing on the maximum stress in tube header sheets in which the tubes are on an equilateral triangular pitch. (Calculated from relations given in the ASME Pressure Vessel Code.)

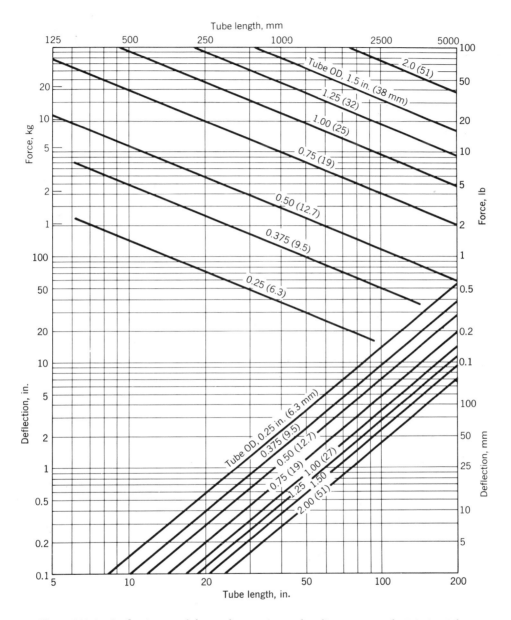

Figure H8.3 Deflections and forces for maximum bending stresses of 15,000 psi for steel tubes loaded at the free end as simple cantilever beams. These data also can be used for more complex loading conditions; for example, the stresses will be the same and the deflection twice as great in a tube four times as long if it is rigidly mounted at the ends and loaded at the center. The wall thickness is 10% of the OD. (Courtesy Oak Ridge National Laboratory.)

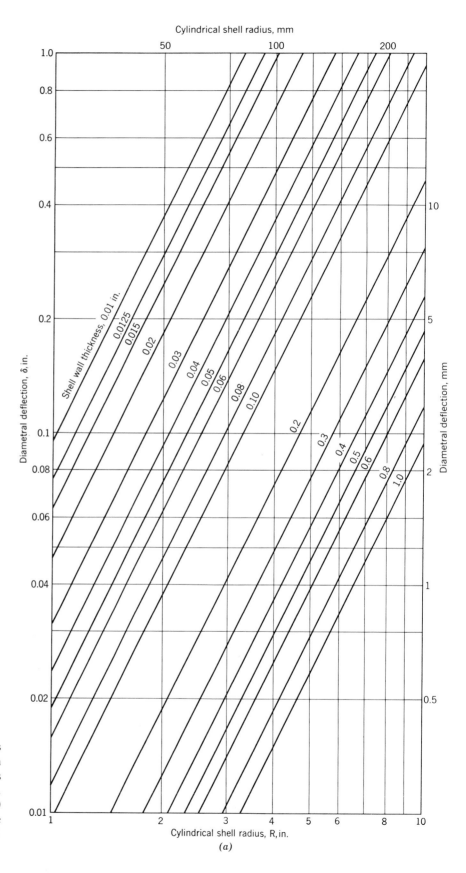

Figure H8.4 Diametral deflections and maximum bending stresses in open-ended cylindrical steel shells under loads applied across a diameter. For tubes having radii larger than 10 in., the diametral deflection can be found by using 1/10 the radius and 1/100 the thickness.

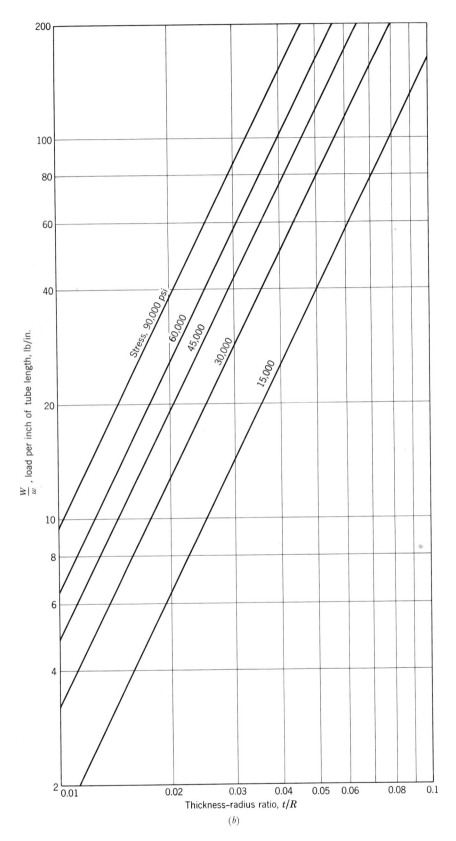

(b)

Figure H8.4 (*Continued*)

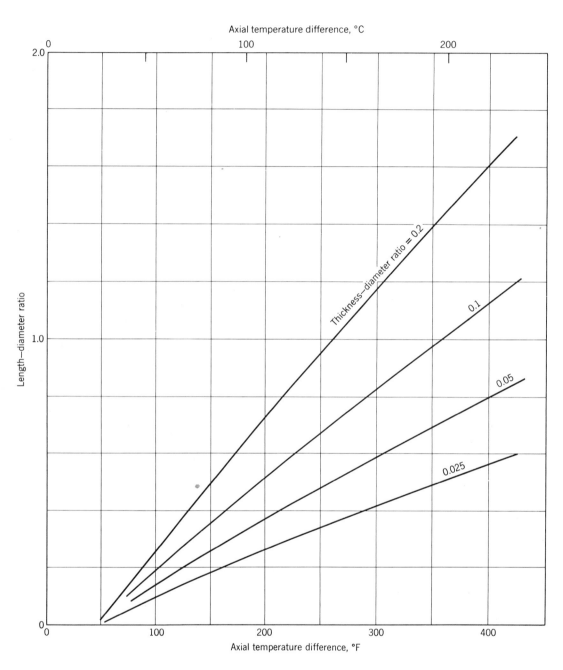

Figure H8.5 Thermal sleeve length-diameter ratio for plain carbon steel to give an allowable shear stress of 10,000 psi. The assumed axial temperature distribution is exponential. (Calculated from Ref. 28 of Chapt. 9.)

TABLE H8.2 Equations for Temperature Gradient and Thermal Stress for Various Shapes

Equations are given for various simple shapes in common use. These equations can also be used to determine the approximate stress in more complex structures. The stress equations give the maximum thermal stresses induced based on elastic deformation.

1. For flat plate, heat generated at center:

$$\Delta T = \frac{pw^2}{4k}$$

$$S = \frac{\Delta T}{2}\frac{E\alpha}{1-v} = \left(\frac{pw^2}{8k}\right)\left(\frac{E\alpha}{1-v}\right)$$

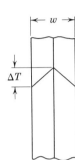

where k = thermal conductivity, Btu/h · ft · °F
p = the equivalent heat generated per unit volume, Btu/h · ft^3
S = stress, psi
E = modulus of elasticity, psi
w = thickness, ft
ΔT = temperature difference, °F
v = Poissons ratio
α = coefficient of thermal expansion, in./in.

2. For flat plate, uniform heat generation:

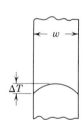

$$\Delta T = \frac{pw^2}{8k}$$

$$S = \frac{2}{3}\Delta T\frac{E\alpha}{1-v}$$

$$= \frac{2}{3}\left(\frac{pw^2}{8k}\right)\left(\frac{E\alpha}{1-v}\right)$$

3. For square bar, uniform heat generation:

$$\Delta T = \frac{pw^2}{12k}$$

$$S = \frac{5}{8}\cdot\frac{\Delta T E\alpha}{1-v}$$

$$= \frac{5}{12}\cdot\left(\frac{pw^2}{8k}\right)\left(\frac{E\alpha}{1-v}\right)$$

4. For rod, uniform heat generation:

$$\Delta T = \frac{pR^2}{4k}$$

$$S = \frac{\Delta T}{2}\cdot\frac{E\alpha}{1-v} = \frac{pR^2}{8k}\cdot\frac{E\alpha}{1-v}$$

5. For cylinder with heat generated at one surface:

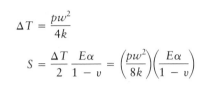

$$\Delta T = \frac{q}{2.74\,k}\log\frac{D_2}{D_1}$$

$$= \frac{0.287\,p(D_2{}^2 - D_1{}^2)}{k}\log\frac{D_2}{D_1}$$

$$S = \Delta T\left(\frac{E\alpha}{1-v}\right)f_1$$

$$= \left(\frac{q}{k}\frac{E\alpha}{1-v}\right)f_2$$

where q = heat generated per unit length, Btu/ft

D_1/D_2	For Heat Flow in		For Heat Flow Out	
	f_1	f_2	f_1	f_2
0.2	0.730	0.0933	0.270	0.0345
0.3	0.686	0.067	0.317	0.0303
0.4	0.645	0.047	0.358	0.0261
0.5	0.61	0.0337	0.390	0.0216
0.6	0.575	0.0234	0.420	0.0171
0.7	0.555	0.0158	0.442	0.0125
0.8	0.535	0.0095	0.465	0.0082
0.9	0.515	0.0045	0.485	0.0042
1.0	0.5		0.5	

These formulas are approximate, not exact.

Source: S. Timoshenko, *Theory of Plates and Shells*, McGraw-Hill Book Co., New York, 1940, p. 54.

TABLE H8.3a Natural Frequencies for Typical Modes of Vibration in Terms of the Frequency Constant $C = fL^2/r$ for Continuous Steel Beam of k Equal Spans, Extreme Ends Clamped[a]

f = natural frequency, cps
L = span length, in.
r = radius of gyration = $\sqrt{I/A}$, in.

Uniform Beam, Extreme Ends Clamped	Number of Spans = k	$C/10^4 = (fL^2/r)/10^4$ Mode Numbers				
		1	2	3	4	5
	1	72.36	198.34	388.75	642.63	959.98
	2	49.59	72.36	160.66	198.34	335.20
	3	40.52	59.56	72.36	143.98	178.25
	4	37.02	49.59	63.99	72.36	137.30
	5	34.99	44.19	55.29	66.72	72.36
	6	34.32	40.52	49.59	59.56	67.65
	7	33.67	38.40	45.70	53.63	62.20
	8	33.02	37.02	42.70	49.59	56.98
	9	33.02	35.66	40.52	46.46	52.81
	10	33.02	34.99	39.10	44.19	49.59
	11	32.37	34.32	37.70	41.97	47.23
	12	32.37	34.32	37.02	40.52	44.94

Source: MacDuff and Felgar, Ref. 35 of Chapt. 7.
[a]See p. 128.

TABLE H8.3b Frequency Correction Factor for Typical Materials Other Than Steel for Use With Table H8.3a

E = Young's modulus for material, lb/in.^{2}
ρ = mass density for material, lb/ft^3
E_s = Young's modulus for steel = 30×10^6 psi
ρ_s = mass density for steel = 490 lb/ft^3

Material			$K_m = \left(\dfrac{E}{\rho}\dfrac{\rho_s}{E_s}\right)^{1/2}$
Steel			1.000
Aluminum alloys 2S, 3S, 4S, 17S, 24S, 25S, 51S, 52S			0.985
Brass, bronze, copper			0.673
Nickel			0.940
Monel metal			0.872
Magnesium			0.965
Titanium	Temperature, °F	TI-50 A	TI-75 A
	80	0.985	0.975
	200	0.966	0.945
	400	0.932	0.910
	600	0.896	0.873
	800	0.866	0.835
	1000	0.828	0.784

Source: MacDuff and Felgar, Ref. 35 of Chapt. 9.

Example: Determine the natural frequency in first mode bending of a 100-in. long, 5/8-in. OD, no. 18 Bwg brass tube installed between two header sheets with four equally spaced baffles acting as intermediate supports to give a continuous beam having five 20-in. spans.

$$f = \frac{C \times r \times K_m}{L^2}$$

$$= \frac{34.99 \times 10^4 \times 0.2043 \times 0.673}{20 \times 20}$$

$$= 120$$

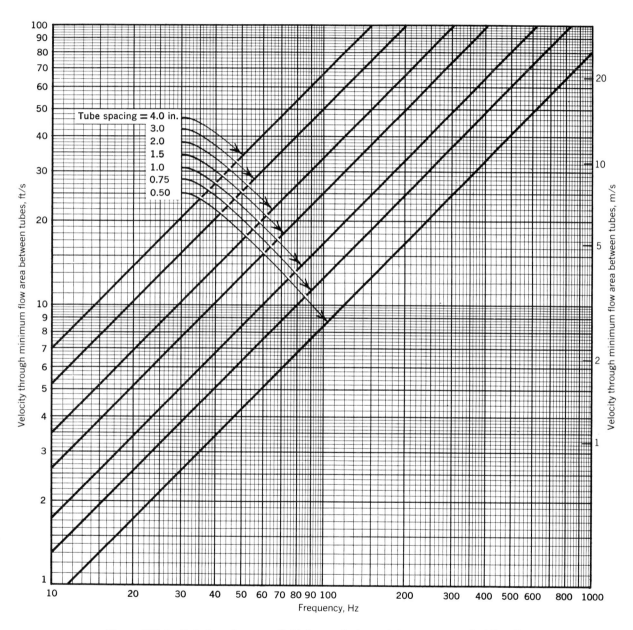

Figure H8.6 Relations between fluid-flow velocity, minimum vortex-shedding (i.e., flow oscillation) frequency, and tube spacing for a Strouhal number of 0.5 for crossflow through tube banks. (See p. 126 and Example 14.3, p. 187.)

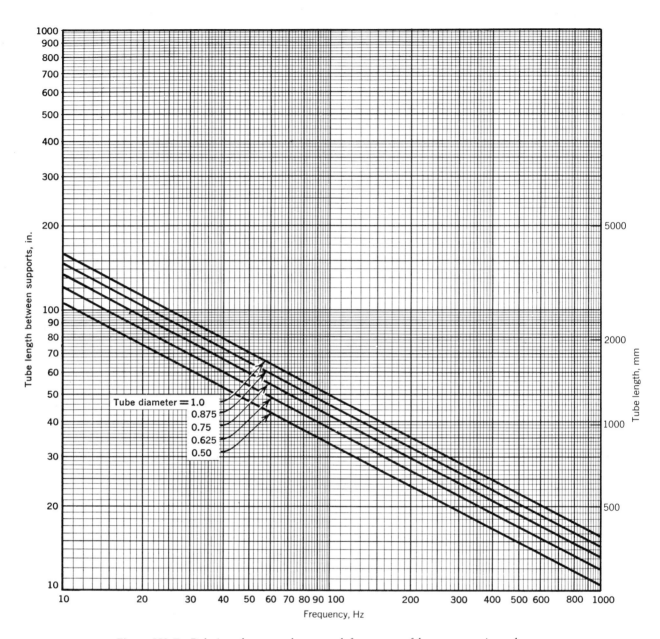

Figure H8.7 Relations between the natural frequency of bare, empty iron-chrome nickel alloy tubes, the tube diameter, and the length between fixed-end supports for first-mode bending vibrations. For continuous beams having a multiplicity of spans with no rotational restraint at the intermediate supports, the frequency is about half as large (see Table H8.3.)

Cost Estimation

COST ESTIMATION

This section presents a set of tables and charts designed to help prepare rough cost estimates. Allowances should be made for increases in the cost index, special materials or requirements, etc. The problems and inherent limitations of cost estimating are discussed in general terms in Chapter 11, together with methods for use of the following charts.

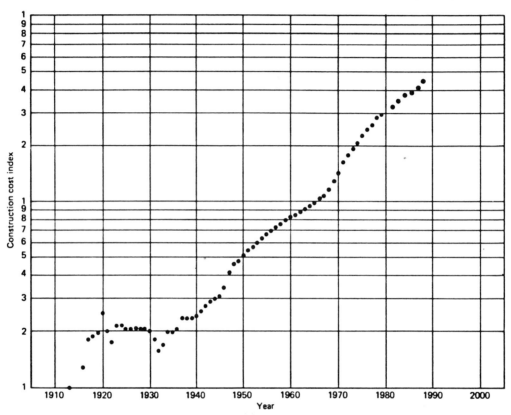

Figure H9.1   U.S. construction cost index as a function of year. (Data plotted from *Engineering News Record*, July 23, 1987.)

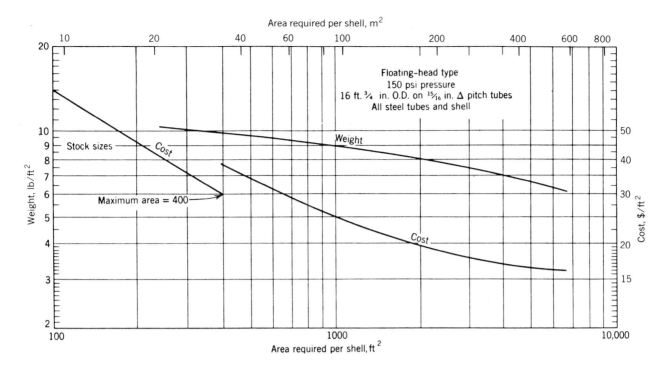

Figure H9.2 Chart for estimating the cost and weight of shell-and-tube heat exchangers based on Ref. 20 with the costs revised to 1983 levels on the basis of Ref. 21, and with the material cost factors of Ref. 21 as presented in Table H9.1 from Ref. 21. See Ref. 21 for much more detailed data on cost factors including wider ranges. By far the largest factors are those of Table H9.1 for the effects of the choice of materials.

To use this chart, multiply cost from the graph by the product of all the factors required to convert the standard exchanger to the case at hand:

Tube Diameter and Pitch Factors

3/4" OD	15/16"	△	Pitch	1.0
3/4" OD	1"	□	Pitch	1.08
1" OD	1-1/4"	△	Pitch	1.08
1" OD	1-1/4"	□	Pitch	1.12

Construction Factors

U-tube removable	0.85
Fixed tube sheet	0.80
Straight tube kettle reboiler	1.35

Pressure Factors

Psia	400 ft^2	1000 ft^2	4000 ft^2
150	1.0	1.0	1.0
300	1.02	1.06	1.2
450	1.1	1.2	1.26
600	1.15	1.3	1.47

Tube Length Factors

8 ft	1.3
12 ft	1.1
16 ft	1.0
20 ft	0.95

Source: Chilton, Ref. 20 of Handbook. (Reprinted by permission of McGraw-Hill Book Co.)

TABLE H9.1 Material Cost Factors to be Used with Fig. H9.2[a]

Material	M_1, Tubing-Price Ratio Relative to Welded Carbon-Steel Tubes Welded	M_1, Tubing-Price Ratio Relative to Welded Carbon-Steel Tubes Seamless	M_2, Price Ratio for Shell, Channel and Tube Sheet Relative to Carbon Steel	Material	M_1, Tubing-Price Ratio Relative to Welded Carbon-Steel Tubes Welded	M_1, Tubing-Price Ratio Relative to Welded Carbon-Steel Tubes Seamless	M_2, Price Ratio for Shell, Channel and Tube Sheet Relative to Carbon Steel
Carbon steel	1.0 (Base)	2.50	1.0 (Base)	Carpenter 20 Mo-6	18.90		
Carbon steel, low alloys:				AL-6-X	12.20		
1/2 Mo	1.04	2.60	1.04	AL-29-4	12.00		
1 Mo	1.05	2.70	1.05	AL-29-4-2	11.80		
2 1/2 Ni	1.15	2.90	1.15	AL-29-4-C	5.0		
3 1/2 Ni	1.20	3.10	1.20	Nickel 200		20.90	18.40
2 Ni-1 Cu		3.30	1.30	Monel 400 (Alloy 400)		15.50	14.50
Carbon steel, chromium-molybdenum alloys:				Inconel 600 (Alloy 600)	19.40		15.30
1 Cr-1/2 Mo	Not standard	2.60	2.10	Inconel 625 (Alloy 625)		32.70	27.40
1 1/4 Cr-1/2 Mo	Not standard	2.70	2.10	Incoloy 800 (Alloy 800)	11.00	21.80	9.00
2 1/4 Cr-1 Mo	Not standard	3.00	2.40	Incoloy 800H (Alloy 800H)		18.00	
3 Cr-1 Mo	Not standard	.3.20	2.50	Incoloy 825		23.50	
5 Cr-1/2 Mo	Not standard	4.40	3.50	Hastelloy B-2	34.90	48.60	38.40
7 Cr-1/2 Mo	Not standard	5.50	Not standard	Hastelloy C-4	28.70	40.00	31.30
9 Cr-1 Mo	Not standard	6.10	Not standard	Hastelloy C-276	29.10	38.10	31.00
Stainless steels:				Hastelloy G	15.30	24.70	18.10
304	2.80	6.50	3.70	Hastelloy X	16.70	27.10	21.30
304L	3.00	7.50	4.70	Titanium (Grade 2)	11.00	22.00	11.00
309	5.80	14.50	7.70	Titanium (Grade 7)	21.00	42.00	
310	7.40	12.00	9.80	Titanium (Grade 12)	14.00	28.00	
310L	7.60	12.40	10.10	Zirconium 702	35.00	43.70	36.80
316	4.70	10.10	6.20	Zirconium 705	39.00	48.70	40.00
316L	4.80	11.00	6.40	Aluminum	Not standard	1.60	1.60
317	8.10	13.30	8.10	Naval rolled brass	Not standard	3.50	3.50
317L	8.30	13.60	8.30	Admiralty	Not standard	3.60	3.60
321	4.20	9.50	5.60	Aluminum brass	Not standard	3.70	3.70
329 (Carpenter 7 Mo)	10.50	17.20	10.50	Aluminum bronze (5%)	Not standard	4.10	4.10
330	7.90	12.90	9.50	Copper (arsenical or deoxidized)	Not standard	4.20	4.20
347	5.50	13.70	7.30	90-10 cupro-nickel	3.50	4.60	4.60
405	6.00	15.00	6.90	70-30 cupro-nickel	4.20	5.50	5.50
410	6.90	17.20	7.90	Union Carbide high-flux tubing:			
430	5.40	10.60	6.20	SA-214, welded	4.40		
439	5.00	11.20	5.80	SA-334-1, welded	4.70		
444 (Alloy 18-2)	7.80	8.80	9.00	SA-334-3, seamless		9.00	
446	4.70	10.00	5.40	SA-214, fluted, welded	7.00		
904L (Sandvik 2RK-65)	15.30	19.20	17.00	SA-334-1, fluted, welded	7.40		
Sandvik 2RE-69		14.50		SA-210, fluted, seamless		8.40	
Sandvik 3RE-60		10.10					
Sandvik 253 MA		12.70					
Sandvik SAF 2205		11.80					
Sanicro 28	16.10	20.20	18.20				
E-brite-26-1 (XM-27)	9.00		10.00				
Ferralium (Alloy 255)	12.00	23.90	14.00				
Carpenter 20 Cb-3	15.10		16.00				

[a]Ref. 21. (Courtesy *Chemical Engineering*.)

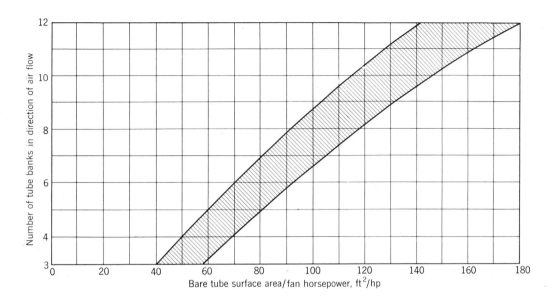

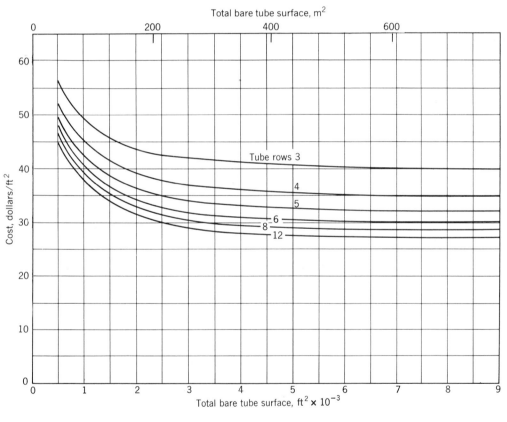

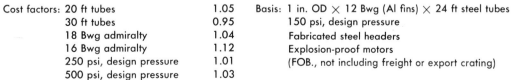

Cost factors: 20 ft tubes   1.05   Basis: 1 in. OD $\times$ 12 Bwg (Al fins) $\times$ 24 ft steel tubes
           30 ft tubes   0.95   150 psi, design pressure
           18 Bwg admiralty   1.04   Fabricated steel headers
           16 Bwg admiralty   1.12   Explosion-proof motors
           250 psi, design pressure 1.01   (FOB., not including freight or export crating)
           500 psi, design pressure 1.03

Figure H9.3 Chart for estimating the costs and fan power requirement of air-cooled banks of finned tubes. *Note:* 8 fins/in., 16.9 ft^2 total surface/ft^2 bare tube external surface. (Based on Ref. 20 but with costs increased to 1983 levels on the basis of Ref. 21 and Refs. 13–17 of Chapt. 11.)

Figure H9.4 Cost in 1987 dollars of double-pipe heat exchangers for single- and multi-tube units for outer pipe sizes from 3 to 16 in. Tables H9.2 and H9.3 provide data on the surface areas in standard tube matrix configurations for both bare- and axially finned tubes. (Note that the base for the material cost factors provided in the inset on this figure are for a base of 1.0 for seamless carbon-steel tubing, but even so are generally much lower than the cost factors of Table H9.1 given with Fig. H9.1 for shell-and-tube units. The differences probably stem in part from differences in quality-control requirements.) (Courtesy Chemical Engineering Co.)

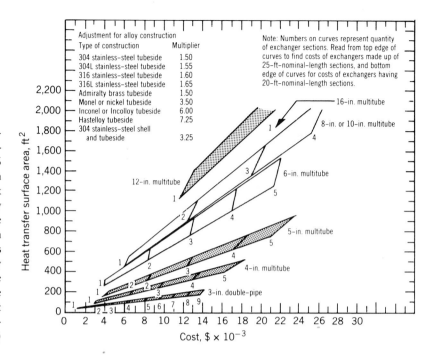

TABLE H9.2 Surface Areas for Standard 20-ft Nominal Length Sections of Double-Pipe Heat Exchangers Similar to That of Fig. 1.27 for Use With Fig. H9.4[a]

	Surface area, ft^2	
Standard size, dia.	Bare tube	Finned tube
Double-pipe		
2-in. shell, 1-in. pipe	10.9	78
3-in. shell, 1-in. pipe	10.9	144
3-in. shell, 1 1/2-in. pipe	20.8	141
4-in. shell, 1 1/2-in. pipe	20.8	261
4-in. shell, 2-in. pipe	26.1	224
4-in. shell, 2 1/2-in. pipe	31.6	191
Multitube:		
4-in. shell	80	263
6-in. shell	253	424
8-in. shell	450	840
12-in. shell	1137	
16-in. shell	1900	

[a]Ref. 22. (Courtesy *Chemical Engineering*.)

TABLE H9.3 External Surface Areas for Longitudinally Finned Tubes in Square Feet per Foot of Tube Length for Use with Fig. H9.4[a]

	No. of Fins	Fin Height, in.		
		1/2	3/4	1
Tube size, outside diameter, in.				
3/4	12	1.196	1.696	2.196
	16	1.529	2.196	2.863
7/8	12	1.229	1.729	2.229
	16	1.562	2.229	2.896
	20	1.895	2.729	3.562
1	12	1.262	1.762	2.262
	16	1.595	2.262	2.928
	20	1.928	2.762	3.595
Nominal iron-pipe size, in.				
1-1/2	24	2.497	3.497	4.497
	28	2.831	3.997	5.164
	36	3.497	4.997	6.497
2	24	2.622	3.622	4.622
	36	3.622	5.122	6.622
	40	3.955	5.622	7.288
2-1/2	24	2.753	3.753	4.753
	36	3.753	5.253	6.753
	48	4.753	6.753	8.753

[a]Ref. 22. (Courtesy *Chemical Engineering*.)

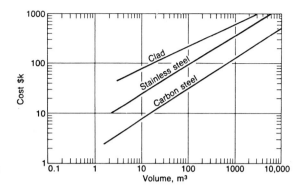

Figure H9.5 The cost of tanks in 1984 dollars. (Klumpar and Soltz, Ref. 23. Courtesy Chemical Engineering.)

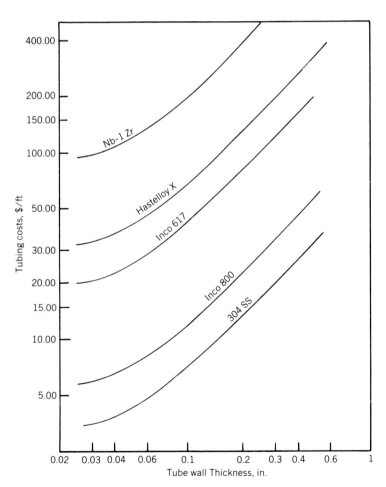

Figure H9.6 The cost of seamless tubing. (Based on data gathered at ORNL in 1975 and escalated with 1987 data for a few tube sizes to provide a basis for estimating costs in 1987 dollars.)

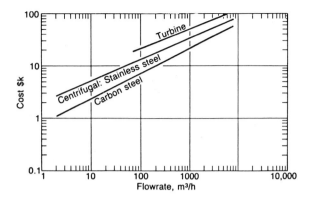

Figure H9.7   The cost of pumps. (Klumpar and Soltz, Ref. 23. Courtesy Chemical Engineering.)

REFERENCES

1. "Standards of the Tubular Heat Exchanger Manufacturers Association," New York.

2. *Heat Exchangers,* The Patterson-Kelley Co., East Stroudsburg, Pa., 1960.

3. "Flow of Fluids through Valves, Fittings, and Pipe," Technical paper no. 410, Crane Co., Chicago, 1957.

4. R. B. Scott, *Cryogenic Engineering,* D. Van Nostrand Co., New York, 1959.

5. J. A. Lane et al., *Fluid Fuel Reactors,* Addison-Wesley Co., Reading, Ma., 1958.

6. J. Hilsenrath and Y. S. Touloukian, "The Viscosity, Thermal Conductivity, and Prandtl Number for Air, O_2, N_2, NO, H_2, CO, CO_2, H_2O, He, and A," *Trans. ASME,* vol. 76, 1954, p. 967.

7. R. C. Reid and T. K. Sherwood, *The Properties of Gases and Liquids,* McGraw-Hill Book Co., New York, 1958.

8. R. G. Vines, "Measurement of the Thermal Conductivities of Gases at High Temperatures," *Journal of Heat Transfer, Trans. ASME,* vol. 82–2, 1960, p. 48.

9. F. G. Keyes and D. J. Sandell, Jr., "New Measurements of the Heat Conductivity of Steam and Nitrogen," *Trans. ASME,* vol. 72, 1950, p. 767.

10. J. H. Keenan and J. Kaye, *Gas Tables,* John Wiley & Sons, New York, 1948.

11. W. D. Weatherford, Jr. et al., "Properties of Inorganic Energy-Conversion and Heat Transfer Fluids for Space Applications," U.S. Air Force, WADD Technical report, 61–96, 1961.

12. *Handbook of Chemistry and Physics,* 44th ed., The Chemical Rubber Publishing Co., Cleveland, Ohio, 1962.

13. N. A. Lange, *Handbook of Chemistry,* 10th ed., McGraw-Hill Book Co., New York, 1961.

14. J. H. Perry et al., *Chemical Engineers Handbook,* 4th ed., McGraw-Hill Book Co., New York, 1963.

15. U.S. Atomic Energy Commission, *Reactor Handbook, Materials,* vol. 1, Interscience Publishers, Inc., New York, 1960.

16. *Metals Handbook,* 8th ed., American Society for Metals, 1961.

17. *Nickel and Nickel-Base Alloys,* Technical bulletin T-13, The International Nickel Co., New York, 1948.

18. *Alcoa Aluminum Handbook,* Aluminum Co. of America, Pittsburgh, Pa., 1962.

19. L. S. Marks, *Mechanical Engineers Handbook,* 6th ed., McGraw-Hill Book Co., New York, 1958.

20. C. H. Chilton, *Cost Engineering in the Process Industries,* McGraw-Hill Book Co., New York, 1960.

21. G. P. Purohit, "Estimating Costs of Shell-and-Tube Heat Exchangers," *Chemical Engineering,* August 22, 1983, p. 56.

22. M. J. McDonough, "Hairpin Exchangers: Double-Pipe and Monotube," *Chemical Engineering,* July 20, 1987, p. 87.

23. I. V. Klumpar and K. M. Soltz, "Updating Fixed Capital Estimating Factors," *Cost Engineering,* vol. 28(1), January 1986.

INDEX